Introduction to Geography

Introduction to Geography

Edward F. Bergman

Lehman College of the City University of New York

Tom L. McKnight

University of California, Los Angeles

Prentice Hall, Englewood Cliffs, NJ 07632

Library of Congress Cataloging-in-Publication Data

Bergman, Edward F.

Introduction to geography / Edward F. Bergman, Tom L. McKnight.
p. cm.
Includes bibliographical references and index.
1. Geography. I. McKnight, Tom L. (Tom Lee), (date)
II. Title.
G128.B37 1993 910—dc20 92-26294 CIP
ISBN 0-13-472416-X

Acquisition Editor: Ray Henderson
Editor in Chief: Timothy Bozik
Development Editor: Robert J. Weiss
Production Editor: Debra A. Wechsler
Marketing Manager: Kris Kleinsmith
Copy Editor: Barbara Ligouri
Interior Designer: Judith A. Matz-Coniglio
Cover Designer: Amy Rosen
Design Director: Florence Dara Silverman
Prepress Buyer: Paula Massenaro
Manufacturing Buyer: Lori Bulwin
Page Layout: Claudia Durrell
Photo Research: Teri Stratford
Photo Editor: Lorinda Morris-Nantz
Illustrators: Sanderson Associates, Patricia Caldwell Lindgren, and Network Graphics
Editorial Assistant: Joan Dello Stritto
Cover Photos: Background (forest) by Bill Ross/Westlight; Japanese woman and children by Koos Rusting; market scene from Morocco National Tourist Office; ship by Albert Normandin/The Image Bank; women with baskets from Agency for International Development.

A Simon & Schuster Company
Englewood Cliffs, New Jersey 07632

Printed in the United States of America
10 9 8 7 6 5 4 3 2

ISBN 0-13-472416-X

Prentice-Hall International (UK) Limited, *London*
Prentice-Hall of Australia Pty. Limited, *Sydney*
Prentice-Hall Canada Inc., *Toronto*
Prentice-Hall Hispanoamericana, S.A., *Mexico*
Prentice-Hall of India Private Limited, *New Delhi*
Prentice-Hall of Japan, Inc., *Tokyo*
Simon & Schuster Asia Pte. Ltd., *Singapore*
Editora Prentice-Hall do Brasil, Ltda., *Rio de Janeiro*

Dedicated to the memories of our fathers,
Donald K. Bergman, Alva F. McKnight,
and George T. Weiss.

Contents

CHAPTER 3

Biota and Soils *68*

CHAPTER 4

Topography *91*

CHAPTER 5

The Principles of Cultural Geography *131*

CHAPTER 6

Population, Population Increase, and Migration *167*

CHAPTER 7

Food and Mineral Resources *206*

CHAPTER 8

The Geography of Languages and Religions *238*

CHAPTER 9

Cities and Urbanization *275*

CHAPTER 10

A World of Nation-States *311*

CHAPTER 11

National Paths to Economic Growth *348*

CHAPTER 12

Regional International Organizations: Knitting or Unraveling? *384*

CHAPTER 13

The Globalization of Economics and Politics 416

Preface

The illustrations on the cover of this book tell a story of ties among peoples and world regions today. Vancouver dockworkers load sulfur for Moroccan farmers, who grow vegetables to feed Frenchmen. Moroccans also combine sulfur with local natural resources to produce fertilizers and industrial raw materials that Morocco exports throughout Asia. Global transportation and communication facilitate such exchanges, which take advantage of the different physical endowments of places and the different cultures of peoples.

THREE THEMES IN THIS TEXTBOOK

Geography explores not only *where* people and activities are located, but *why* they are located where they are. Physical geography sets the stage, and cultural, political, and economic forces distribute human activities across that stage. This textbook emphasizes three themes throughout the study of geography.

One theme is that as communication and transportation ties have multiplied among peoples and regions, what happens *at* places depends more and more on what happens *among* places. This is the dynamism of modern geography. It is important to know the current distributions of people, languages, religions, cities, and economic activities, but to understand geography we must understand the forces that cause these distributions and may redistribute them in the future.

Every day the news reports events in which what happens is directly related to where it happens, and these events trigger changes in geography. An earthquake forces thousands to flee a Central American city. A bountiful harvest in Australia reduces food prices and thus improves the diet available to Indonesians. Canadian scientists synthesize a substitute for a mineral previously imported. New U.S. immigration laws welcome professionals whose skills are needed in their own poor homelands. U.S. movies and music diffuse U.S. languages and culture around the world, while Americans themselves adopt new foods such as sushi. Japanese multinational corporations create jobs and wealth in Thailand and Mexico. Protestant Christianity wins converts throughout Latin America, accepting women as priests and introducing family planning. Meanwhile, in some African and Asian countries, Islamic fundamentalists win political power and curb women's rights. Caribbean tourist destinations thrive. Refugees flee countries wracked by civil or religious wars. These events reorganize world cultural, political, and economic landscapes. Today's dynamic geography doesn't just *exist*; it *happens*.

The second theme woven throughout this book is that geographers exercise the scientific method of gathering and analyzing information. Geographers seek explanations for distributions, seek and compare patterns, and investigate cause-and-effect relationships. In the study of physical geography, for example, we ask why certain climatic types appear at certain latitudes and relative positions on the continents, and why these patterns are repeated in the patterns of types of vegetation and soil. What are the causes?

The discovery of patterns and comparisons often challenges geographers with paradoxes. The map of the world's dense rainforest vegetation, for example, does not correlate with the map of the richest soils. The countries with the highest percentages of their labor forces working in agriculture do not have the greatest food supplies. The world distribution of wealth is not the same as the world distribution of raw material resources. As we investigate these paradoxes, we develop critical thinking skills while learning important information.

Thus, when we map any phenomenon—whether a soil type, a religion, or a form of urbanization—we ask what processes or forces caused this distribution and how these forces might redraw the map in the future. If the map corresponds with the maps of any other phenomena, we investigate possible cause-and-effect relationships.

The third theme emphasized in this book is that geography is not restricted to the study of exotic lands and peoples. Its principles can be studied in our hometowns and even on campus. How do local temperatures and precipitation vary through the year? Where have new arrivals in the community come from, and why have they moved? Where do local food crops and manufactured goods find markets? Does the town recycle waste? Do local political districts reveal gerrymandering?

People in many different professions use geography, even though the professionals may not call themselves geographers. Soil analysts investigating the success of new crops in a local field, planners designing new suburbs, transportation consultants routing new highways, direct mailers targeting zip codes of people with certain incomes or ethnic backgrounds, and diplomats negotiating international fishing treaties are all studying or "doing" geography. Much fine journalism exemplifies superb geography.

FEATURES OF THE BOOK

This book introduces the principal content of both physical and cultural/human geography, as well as the major tools and techniques of the field. The first four chapters focus on physical geography, and the next nine deal with cultural/human geography. Each chapter offers several aids:

- Key terms are boldfaced and listed at the end of each chapter, and each term is again defined in the Glossary at the end of the book.
- Each chapter finishes up with a summary.
- "Questions for Investigation and Discussion" encourage exploration beyond the text into the library or out into the community.
- Distinctive "Focus" boxes highlight individual problems or case studies.
- "Critical Thinking" boxes discuss problems or pose questions for which there are no simple or clear answers. These require readers to develop their own viewpoints.
- A short list of up-to-date readings provides sources for further information.
- A rich variety of visual devices illustrates the text: maps of patterns and of flows, cartograms, photographs, drawings, reproductions of works of art, bar graphs and pie charts, histograms, and tables. The captions guide the reader's eye through or into the figures, and many captions include thought-provoking questions.

Three Interchapter Essays

Three interchapter essays also highlight the text. The first introduces the book's focus on "The Scientific Methods" by discussing the history and procedures of the method and then demonstrating specific examples from geographic studies. This essay follows the discussion of climate in Chapter 2 because Chapter 3 notes how the maps of world vegetation and soils resemble that of climatic types and investigates cause-and-effect relationships among these phenomena. Throughout the text, maps are compared, and relationships among facts are investigated.

A second essay, titled "The Influence of the Environment on Human Life," is placed just after the first four chapters, which describe and analyze the natural environment. This essay explains why human adaptation to the environment is more cultural than biological, and it discusses several theories of human-environmental interaction: human perception of the environment, environmental determinism and possibilism, cultural ecology, and behavioral geography. This discussion bridges the physical geography of the first four chapters and the human and cultural geography of the following chapters.

The third essay, "Protecting the Global Environment," is placed at the end of the book to bring the discussion of human-environmental relationships full circle. Today environmental protection challenges all humankind and requires international collaboration.

SUPPLEMENTS

This text is supplemented by a package of 175 figures selected from the text and made available in formats convenient for classroom use: 50 color transparencies, 50 black-and-white transparencies, and 75 color slides.

Two additional media supplements demonstrate how geography lends understanding to the daily news.

The New York Times Contemporary View Program

A "custom edition" of *The New York Times* has been prepared especially for users of this book. It contains articles chosen to exemplify topics in the text. An article reporting "Spy Data Now Open for Studies of Climate," for example, explains how the end of the Cold War allows the release for study by Earth scientists of secret information that has been gathered by satellites. This information may provide measures of global warming and of changes in land use. The article "In Rwanda, Births Increase and the Problems Do, Too," examines population pressures in a small African country, and "Uzbeks, Free of Soviets, Dethrone Czar Cotton" reports why the newly won independence of this Central Asian republic led to changes in agriculture and ecology. Other selections report on topics as diverse as the impact of international cultural diffusion ("Iran is Unable to Stem West's Cultural Invasion") and how multinational corporations redistribute manufacturing around the globe ("BMW Details Plan to Build Cars in South Carolina").

ABC/Prentice Hall Video Library

This textbook is accompanied by a videocassette from ABC News. A variety of clips demonstrate again how many of the topics in the news reveal changing geography. Special focus boxes in the text introduce each video clip and pose questions to consider while watching it. These clips include segments on

- The Impact of Hurricane Hugo
- Environmental Pollution in Eastern Europe
- Economics and Politics of Water in the West
- America's Infant Mortality Crisis
- Atmospheric Ozone Depletion
- Japanese Culture
- The Debate Over Free Trade With Mexico.

The text is also accompanied by an *Instructor's Manual and Resource Guide*. This contains chapter-by-chapter suggestions for amplification of material in the text and for supplementary lecture topics, additional questions for investigation and discussion, additional bibliography, and a list of the relevant supplementary materials supplied.

ACKNOWLEDGMENTS

Countless colleagues, librarians, and generous individuals both in government and in the private sector helped with information for this text. Any errors in the text are the fault of the authors, but we would like to thank the following people for their special help: Steve Shivers, U.S. Department of the Interior; Christaud Geary, National Museum of African Art; Laveta Emory, Smithsonian Institution; Jennifer Locke, Larry Jones, and Pete Daniel, National Museum of American History; Priscilla Strain, National Air and Space Museum; Douglas Van den Bosch; Dan Beard, House Committee on the Interior and Insular Affairs; Ruud Weggemans; Pat O'Connell, Twin Cities Metropolitan Council; Clyde McNair and Mary Felder, Agency for International Development; William Rathbun, Seattle Art Museum; Tab Lewis, National Archives; Linda Carrico and David Morse, U.S. Bureau of Mines; Barbara Mathe and Pat Brunauer, American Museum of Natural History; Kerstin Erickson and Anne Stanley, European Community Information Service; William Usnik, World Education Services; Leo Dillon, Office of the Geographer, U.S. Department of State; Steve Janauschek, Planecon, Inc.; Jeffrey Hoover, Institute for East West Studies; Judith Zilczer, Hirschhorn Museum; John Rutter, U.S. Department of Commerce; Dr. Joseph E. Ryan and Mark Wolkenfeld, Freedom House; Major General Du Kuanyi, Chinese Mission to the United Nations; Alex de Sherbinin, Population Reference Bureau; Gary Cohen, *U.S. News & World Report*; Jerry Hagstrom, *The National Journal*; Dan Critchett, Morgan, Stanley and Co.; Robert Nead, American Express International Bank; Dan McCloskey, American Express Travel Related Services; Kevin Mischka; and Robert Gaiser, Nassau County Planning Commission. Plus the many friends and colleagues whose photographs are used and credited in the text.

We are grateful to our reviewers; they provided many thoughtful and valuable suggestions for improving the manuscript. We are especially thankful for those who made comments on the successive drafts of the manuscript. Our thanks to:

Robert Morrill, Virginia Polytechnic Institute & State University
Anthony Grande, Hunter College-City University of New York
Ron Marionneaux, Eastern Kentucky University
Kavita Pandit, University of Georgia-Athens
Christopher Exline, University of Nevada-Reno
Roland Williams, West Liberty State College
W.C. Jameson, University of Central Arkansas
Richard Allen Earl, Southwest Texas State University
Robert Cullison, Essex Community College
David Icenogle, Auburn University
Nancy Bain, Ohio University
Albert J. Larson, University of Nebraska-Lincoln
Thomas D. Anderson, Bowling Green State University
John Harmon, Central Connecticut State University
Lisle Mitchell, University of South Carolina-Columbia
Edward Babin, University of South Carolina-Spartanburg
Charles Gritzner, South Dakota State University
Charles Heatwole, Hunter College

The authors also wish to thank Patricia Caldwell Lindgren and John Sanderson for their careful and imaginative cartography. At Prentice Hall there are many persons who contributed to this project. Thanks to Dan Joranstaad for proposing our collaboration. Ray Henderson managed our project from the initial draft forward, and saw it to completion. Bob Weiss, our Developmental Editor, consistently focused our ideas and coaxed us to clarify our expression throughout the book. Debra Wechsler coolly and competently coordinated all aspects of book production. We gave these people our ideas, and they returned them as a book. We have enjoyed working with you very much. Thank you.

CHAPTER 1

Introduction to the Earth

Antarctica, Africa, and the Arabian peninsula can be seen beneath the swirling atmosphere in this picture taken by an astronaut from the Apollo 17 spacecraft. (Courtesy of NASA.)

hen Russian astronaut Sergei Krikalev landed on Earth on March 25, 1992, after 313 days in space, he might have wondered whether he had returned to the same planet. The physical landscape—the plains of Kazakhstan—had not changed. The political and cultural landscape, in contrast, had changed dramatically. Kazakhstan was no longer the Kazakh Republic of the Union of Soviet Socialist Republics. Instead, it had become an independent republic and was busily developing its own economic and cultural life. The Soviet state, whose seal decorated Krikalev's spacesuit, was gone, and his home city of Leningrad had reclaimed its historic name of St. Petersburg. Since Krikalev had rocketed into space, the Communist Party had fallen from power in the USSR, the constituent republics had proclaimed their sovereignty, and the USSR had dissolved (see Figure 1–1).

This episode demonstrates how rapidly world geography can change. The USSR had exercised centralized political and economic power over one-sixth of the Earth's land surface, and its collapse is remapping cultural, economic, and political activities throughout that territory and around the world. The disintegration of the USSR is one of the most spectacular events of recent years, but world geography is never static. Distributions of people and of activities are continuously changing. This book will help you understand current distributions and also the forces that drive changes.

Figure 1–1. Russian cosmonaut Sergei Krikalev returned from a 10-month space voyage on March 25, 1992, to discover that the USSR had dissolved. What other major changes occurred while he was in space? (Courtesy of ITAR-TASS/Sovfoto.)

Geography is one of the oldest fields of study. We may consider the first geographer to have been the first person who climbed a mountain or crossed a river to see what was on the other side, or who noticed that the environment or the people were different away from home and who tried to understand why. He or she must have posed the basic questions still asked by geographers: Where is it, and what is it like there? These two descriptive questions must be answered before geographers can go on to analyze questions of cause and meaning: Why is it there? How does that relate to other things? Location matters, and the events at different places relate to one another. Therefore we can define **geography** as the study of the interaction of all physical and human phenomena at individual places and of how interactions among places form patterns and organize space.

THE DEVELOPMENT OF GEOGRAPHY

Curiosity about various lands and peoples was one of the forces that drove Alexander the Great (356–323 B.C.), whose career of military conquest took him from Greece all the way to India. Alexander's boyhood tutor was the philosopher Aristotle (384–322 B.C.), who emphasized that whenever possible a person should personally investigate to learn the truth. Alexander took a group of scholars along with him, and he is often credited with treating all the peoples he conquered as equals.

Every language has a word for the study that we call "geography," but the Greek word that we use, which means "earth description," was the title of a book by Eratosthenes (c. 275–195 B.C.). He was the director of the library in Alexandria, a city in Egypt founded by Alexander. The library was the greatest center of learning in the Mediterranean world for hundreds of years. No copies of Eratosthenes's book have survived, but we know from references by other authors that Eratosthenes accepted the idea that the Earth is round and that he even calculated its circumference with amazing precision. Eratosthenes also drew fine maps of the world as it was known to him. Hipparchus (180–127 B.C.), a later director of the library, was the first person to draw a grid of imaginary lines on the Earth's surface to locate places precisely.

After the great age of Greek civilization gave way to the Roman era, the Roman Empire produced a number of geographical scholars and compilations of geographical learning. These culminated in the geography of Ptolemy, who also worked at the library in Alexandria between A.D. 127 and 150. Subsequently, in the long period between the fall of the Roman Empire and the European

Renaissance of the fifteenth century, Western Civilization accumulated little additional geographical knowledge, and when Ptolemy's book reappeared in Latin in 1406, it was still taken as the most authoritative source of geographical knowledge.

Geography in the Non-European World

Outside Europe, however, geography made considerable advances. The expansion of the religion of Islam (discussed in Chapter 8) allowed travel and research across the wide region from Spain to India and beyond. Islam requires each believer to pilgrimage to Mecca, in Arabia, at least once in his or her life, if possible. This obligation encouraged travel. Muslim scholars produced impressive texts describing and analyzing both physical environments and also customs and lifestyles of different peoples.

China also developed an extensive geographical literature. The oldest known work dates from the fifth century B.C. *The Tribute of Yu* describes the physical geography and the natural resources of the various provinces of the Chinese Empire. It interprets world geography as a nest of concentric squares of territory, ranging from an innermost zone of the imperial domain to an outermost zone inhabited by "barbarians." In A.D. 267 Phei Hsiu made an elaborate map of the empire, and he is often called the father of Chinese *cartography* (map making). In the following centuries Chinese Buddhists wrote occasional accounts of travels to India to visit places sacred to the history of Buddhism, which had spread into China. Chinese maritime trade extended throughout Southeast Asia and the Indian Ocean, and it even reached East Africa, but the Chinese eventually withdrew from more extensive exploration, and few Chinese descriptive geography texts of this activity survive.

The Japanese and the Koreans also engaged in East Asian trade. In fact the greatest world map of which we know that was produced before the European voyages of exploration was made in Korea in 1402. This map, the "Kangnido," was made by an unknown cartographer, but he or she drew on the combined resources of the libraries of Korea, Japan, and China, including Islamic sources that were known to the Chinese. As a result, this map includes not only East Asia but also India, the countries of the Islamic world, the African continent, and even Europe itself. It shows a far more extensive world knowledge than the Europeans had at the time.

The peoples of Africa, of the Western Hemisphere, and of other areas may have had geographical accounts of the parts of the world known to them, but these works either have not survived or else modern scholars have not yet fully inventoried and studied them.

The Revival of European Geography

Beginning in the fifteenth century, Europe's world exploration and conquest brought so much new geographical knowledge that Europeans were challenged to devise new methods of cataloging or organizing all of it. Sir Walter Raleigh's book *The History of the World* (1614) is a milestone in European intellectual life because Raleigh for the first time assumed the uniformity of the human mind. The leading role in his book is taken by the collective person "mankind," with Raleigh as "his" biographer.

Regional and Topical Geography Another key work of western intellectual history was the book *General Geography* (1650) by Bernhard Varen ("Varenius" in Latin), a German who taught at the University of Leiden, in Holland. Varen differentiated what he called "special geography" from "general geography." Special geography was the description and analysis of any place in each of ten categories: (1) "the stature of the natives;" (2) employment; (3) virtues, vices, learning, and wit; (4) customs; (5) speech and language; (6) political government; (7) religion and church government; (8) cities and renowned places; (9) history; (10) famous people and inventors. Today we call such an inventory-analysis of any individual place a **regional geography.** All Varen's categories are attributes of local culture, but today geographers usually begin regional geography studies with a description of the local physical environment—the climate, topography, soils, and other physical attributes—and then proceed with an inventory much like Varen's, weaving together all these aspects as they interrelate. The study of geography bridges the physical sciences and the social sciences.

What Varen called "general geography" was the discussion of universal laws or principles that apply to all places. Today we call this type of analysis **topical** or **systematic geography,** and individual geographers concentrate on topics as diverse as the geography of soils (*pedology*), of life forms (*biogeography*), of politics (*political geography*), of economic activities (*economic geography*), and of cities (*urban geography*). The Association of American Geographers, a professional scholarly organization, today recognizes 51 topical specialties. Regional and topical geography are complementary. A regional study covers all topics in the region under review, whereas a topical study notes how a particular topic varies across regions. This book takes a topical approach to the study of geography.

Varen's book was a standard reference for over 100 years. Even Sir Isaac Newton edited two editions in Latin, one of which was studied by students at Harvard in the eighteenth century.

The German philosopher Immanual Kant (1724–1804) taught that the study of geography was the natural complement to the study of history. History, he wrote, was the study of phenomena in their relations of time, or *chronology,* whereas geography was the study of phenomena in their relations of place, which he called *chorology.* In the nineteenth century scholarly geography developed as an ever-more rigorous inquiry into the reasons for the distribution of phenomena. Geography courses were offered at leading American and European universities.

Between 1845 and 1862 virtually all knowledge of the earth sciences and anthropology was interwoven in the great multivolume book *Cosmos* by the German explorer and naturalist Alexander von Humboldt (1769–1859). Von Humboldt's brilliant interrelating of the phenomena of physical nature and of the world of humankind had an enormous intellectual impact in the United States. This was partly because European Americans had long thought of America as a "New World," a veritable "Garden of Eden" over which they had been given dominion.

George Perkins Marsh, an American scholar and diplomat, expanded on von Humboldt's theme of the interconnections between humankind and the physical environment. While Marsh was serving as U.S. ambassador to several Mediterranean countries, he was impressed by the destruction by humankind of an environment that ancient authors had described as lush and rich. Marsh's book *Man and Nature, or Physical Geography as Modified by Human Action* (1864) was one of the earliest key works in what would become today's environmental movement.

THE ORGANIZATION OF THIS TEXT

This introduction to geography text first examines the physical geography of the Earth and then maps and analyzes the distributions of people and of their cultures and activities. The physical environment sets the stage for the activities of humankind. Geographers then investigate the interrelationships between humankind and the environment, and the distributions and interrelationships among human activities.

Our goals are always to know where things are and to understand both the physical environment and human activities at different places. Beyond that, we also want to understand why elements of the physical environment, people, and activities are distributed as they are. Understanding the causes of the distributions on the maps we study will help us understand how and why those maps change through time. We cannot always predict exactly how physical phenomena or human activities will redistribute themselves in the future, but we can safely predict that they will do so.

THE EARTH

The Earth is the ancestral, and thus far the only, home of humankind. People live on the surface of the Earth in a physical environment that is extraordinarily complex, extremely diverse, and yet ultimately fragile. This habitable environment exists over almost the entire face of the Earth, which means that its horizontal dimensions are exceedingly vast. Its vertical extent, however, is very limited; the vast majority of all earthly life inhabits a zone less than 3 miles (5 km) thick, and the total vertical extent of the life zone is less than 20 miles (32 km).

With a diameter of some 8000 miles (12,875 km) and a circumference of about 25,000 miles (40,000 km), the Earth is a very impressive mass from a human viewpoint. Its highest point is almost 30,000 feet (9000 m) above sea level, and the deepest spot in the oceans is about 36,000 feet (11,000 m) below sea level (see Figure

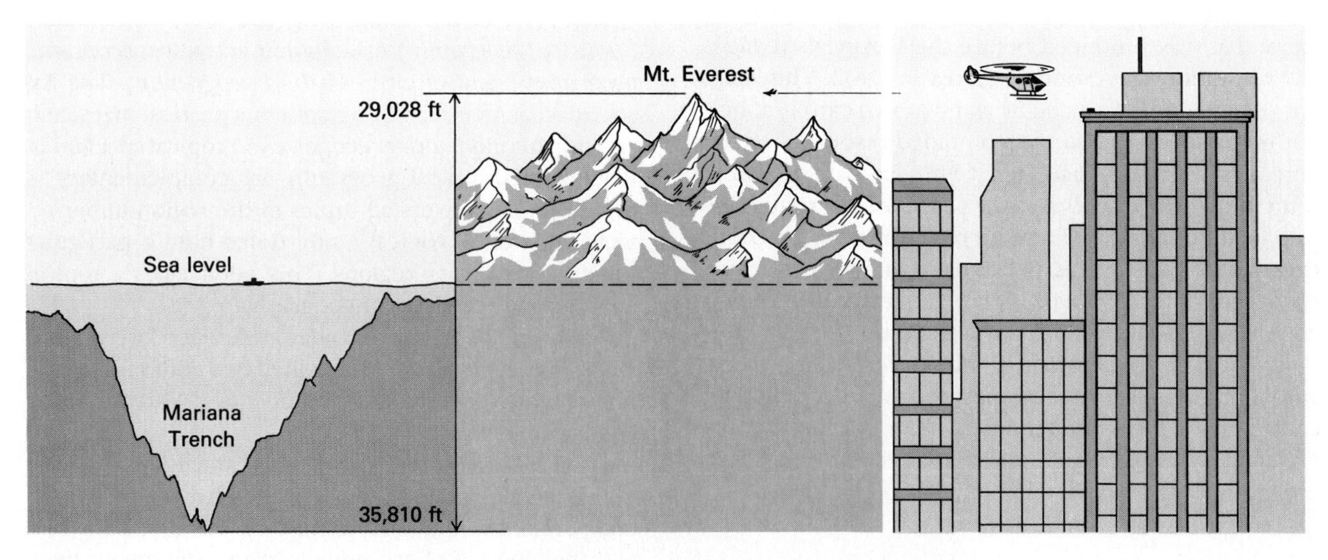

Figure 1–2. The Earth's maximum relief is about 65,000 feet (more than 20 km) from the top of Mt. Everest to the bottom of the Mariana Trench. In comparison with the diameter of the Earth, however, this would be equivalent in vertical dimension to a single brick on top of a 40-story building.

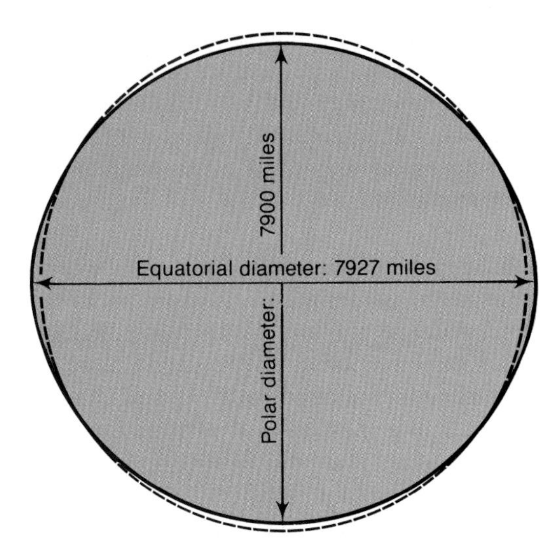

Figure 1–3. The Earth (surface represented by a solid line) is an oblate spheroid. The dashed line represents the shape of a perfect sphere. The variation from sphere to spheroid is so small that it has been exaggerated in this diagram to make it visible.

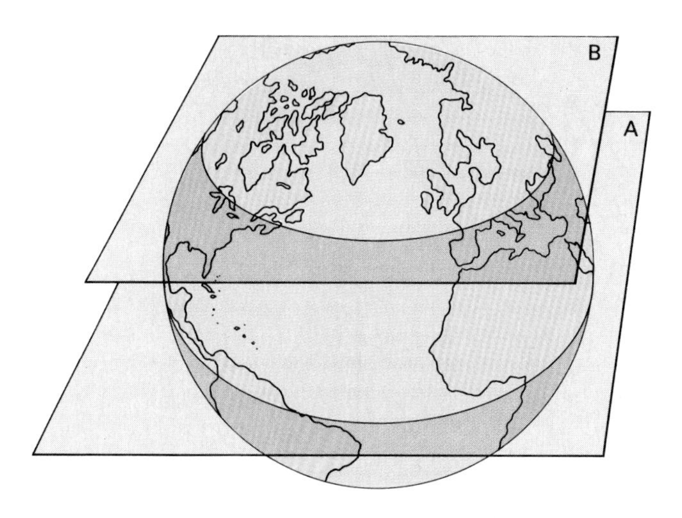

Figure 1–4. This drawing compares great and small circles. A great circle represents the intersection of the Earth's surface with a plane (such as A) that passes through the exact center of the Earth. A small circle results from the intersection of the Earth's surface with any plane (such as B) that does not pass through the center of the Earth.

1–2). Its shape is spherical, although it varies slightly from true sphericity due to the effect of centrifugal force. Any rotating body tends to bulge in the "middle" and flatten at the "ends"; thus the equatorial diameter of the Earth is slightly greater (27 miles or 43 km) than its polar diameter. This is a minuscule variation (0.3 percent) from perfect sphericity that can be ignored for most purposes, although it is useful in establishing a scientific locational system (see Figure 1–3).

The Geographic Grid

As implied in the preceding paragraph, scientists have developed a precise locational system. The basis of this system is the **geographic grid,** which consists of east-west lines and north-south lines that intersect at right angles.

If the Earth were an absolutely perfect sphere, the problem of describing precise surface locations would be much more formidable than it is. With no deviation from sphericity, no natural points of reference would exist from which measurements and surveys could be started. Our rotating Earth, however, with its slightly bulging middle and slightly flattened ends, provides the natural reference points for a systematic locational system.

The Earth rotates continuously about an **axis,** a diameter that connects the points of maximum flattening on the Earth's surface. These points are called the North Pole and the South Pole. If we visualize an imaginary plane perpendicular to the axis of rotation which passes through the Earth halfway between the poles, we will then have another valuable reference feature: the **plane of the equator.** Where this plane intersects the Earth's surface is the imaginary midline of the earth, called simply the **equator** (see Figure 1–4). We can now use the North Pole, South Pole, rotational axis, and equatorial plane as natural reference points for measuring and describing locations on the Earth's surface.

Latitude is angular distance measured north and south of the equator. As shown in Figure 1–5, we can project a line from a given point on the Earth's surface to the center of the Earth. The angle between this line and the equatorial plane is the measure of the latitude of the point. Like any other angular measurement, lati-

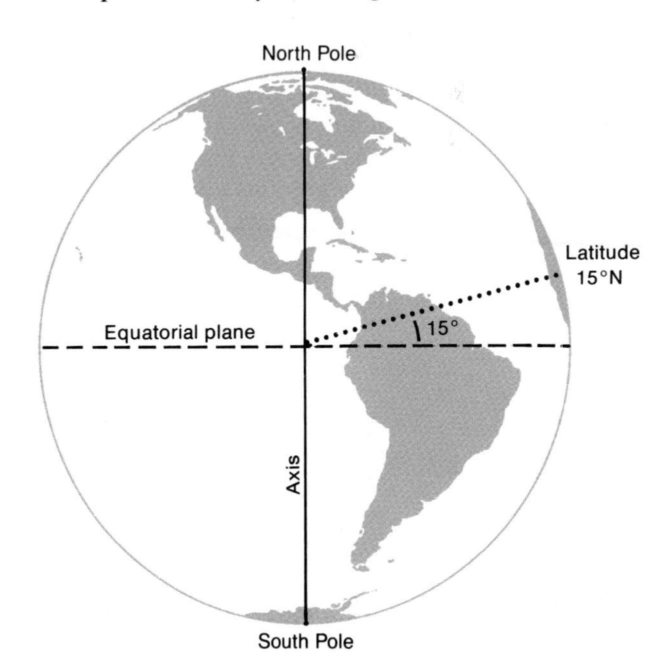

Figure 1–5. The latitude of any place on the Earth's surface is determined by measuring the angle between a radius from that place to the center of the Earth and the equatorial plane.

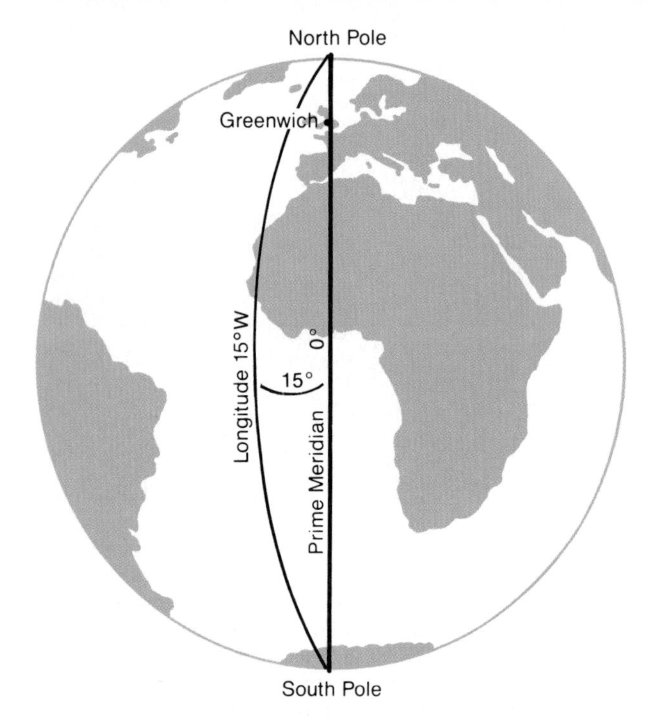

Figure 1–6. The longitude of any place on the Earth's surface is determined by measuring the angle between a plane connecting that point with both poles and the plane of the prime meridian.

tude is expressed in *degrees*, *minutes*, and *seconds*. There are 360 degrees (°) in a circle; 60 minutes (′) in a degree; and 60 seconds (″) in a minute. Latitude varies from 0° at the equator to 90° at the North Pole and South Pole.

A line connecting all points of the same latitude is called a **parallel** because it is parallel to all other lines of latitude. Because parallels are imaginary lines, there can be an infinite number of them. There can be a parallel for each degree of latitude, or for every minute, or for every second, or for any fraction of a second. However many latitude lines we visualize, their common property is that they are all parallel to one another.

Longitude is angular distance measured east and west on the Earth's surface. Again, this is angular distance, so longitude is also measured in degrees, minutes, and seconds. It is represented by imaginary lines extending from pole to pole and crossing all parallels at right angles. These lines, called *meridians*, are not parallel to one another except where they cross the equator. They are farthest apart at the equator, become increasingly close northward and southward, and finally converge completely at the poles.

The "baseline," or **prime meridian,** from which longitude is measured was arbitrarily chosen by an international conference as the meridian passing through the Royal Observatory in Greenwich (England), just east of London. Longitude is measured both east and west of the prime meridian to a maximum of 180° in each direction (see Figure 1–6). Thus, the location of any spot on the Earth's surface can be described with great precision by reference to detailed latitude and longitude data (see Figure 1–7).

Earth and Sun Relations

Life on Earth is dependent on solar energy, and the Earth's functional relationship to the sun is vitally important. The relationship is not static because of the perpetual motions of the Earth that continually change the distance between the Earth and the sun and the portion of the Earth that faces the sun.

The earth *rotates* toward the east on its axis, making one complete rotation in 24 hours. It also *revolves* around the sun in an elliptical orbit, completing the full orbit in slightly more than 365 days. If there were a plane that passed through the sun and through the Earth at every position in its orbit around the sun, this would be the **plane of the ecliptic.** If the Earth's axis were perpendicular to the plane of the ecliptic there would be no change of seasons, and day and night would be of nearly constant length. However, the axis is tilted at an angle of 23.5° from the perpendicular to the plane of the ecliptic, and it is always parallel to itself as it revolves around the sun, pointing toward Polaris, the North Star (see Figure 1–8). This constancy of orientation is called **po-**

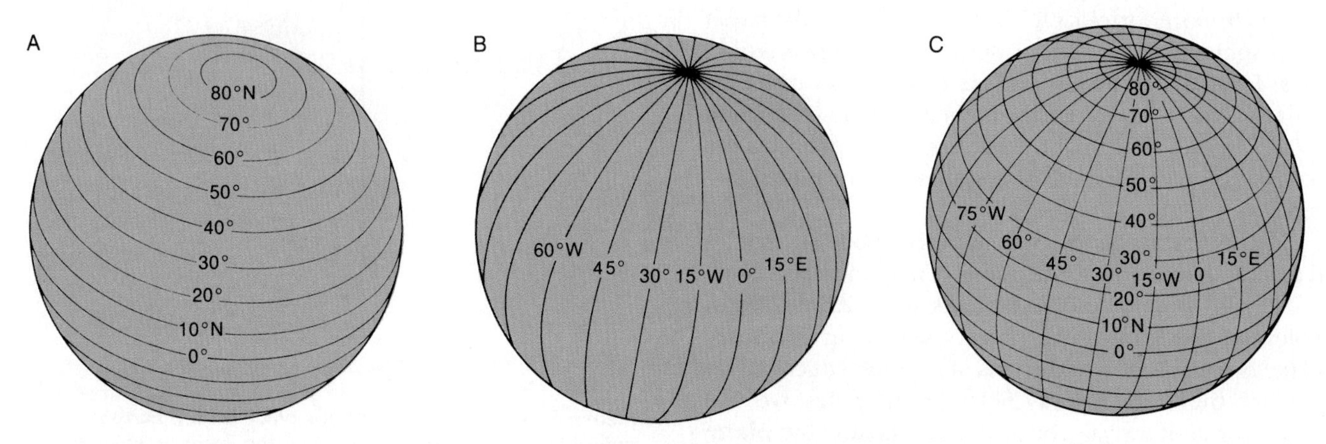

Figure 1–7. Development of the geographic grid: (A) parallels of latitude; (B) meridians of longitude; (C) the completed grid system.

larity or parallelism. The combined effect of rotation, revolution, inclination, and polarity is such that the angle at which the sun's rays strike the Earth changes throughout the year, creating the annual march of the seasons. The Tropic of Cancer at 23.5° north latitude and the Tropic of Capricorn at 23.5° south latitude define the region within which the noon sun is always at a high angle and which often is referred to simply as "the tropics" (see Figure 1–9). The sun appears to migrate from north to south and back again within this region, as its direct rays reach the Tropic of Cancer each year at the time of the June **solstice** (approximately June 21st), cross the equator on its journey south at the September **equinox** (about September 22nd), reach the Tropic of Capricorn at the time of the December solstice (approximately December 21st), and recross the equator on its northern journey at the March equinox (about March 20th) (see Figure 1–10).

Thus, daylight lasts longer in the Northern Hemisphere at the time of the June solstice (which is the "summer" season in the Northern Hemisphere but is "winter" in the Southern Hemisphere), and the reverse is true south of the equator. At the time of the equinoxes, however, all places on earth receive 12 hours of daylight and 12 hours of darkness (see Figure 1–11). Both the length of the period of daylight and the angle at which the sun's rays strike the Earth are major determinants of the amount of solar energy received at any particular latitude.

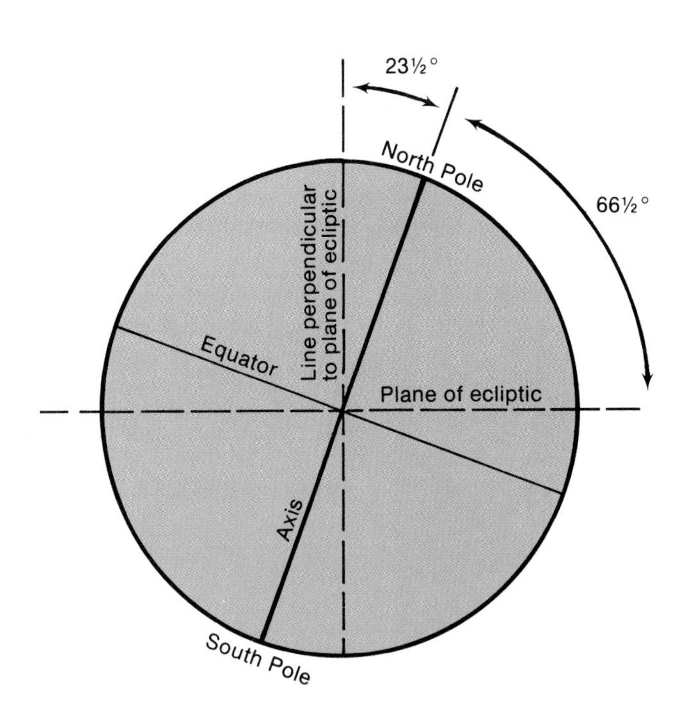

Figure 1–8. The Earth's axis is inclined at an angle of $66\frac{1}{2}°$ from the plane of the ecliptic, but this inclination is normally described as being $23\frac{1}{2}°$ from the vertical (a line perpendicular to the plane of the ecliptic).

Figure 1–9. The Tropic of Capricorn, like all parallels of latitude, is an imaginary line. However, as a significant parallel, its calculated location is often commemorated by an informational sign. This scene is near Alice Springs, in the center of Australia. (TLM photo)

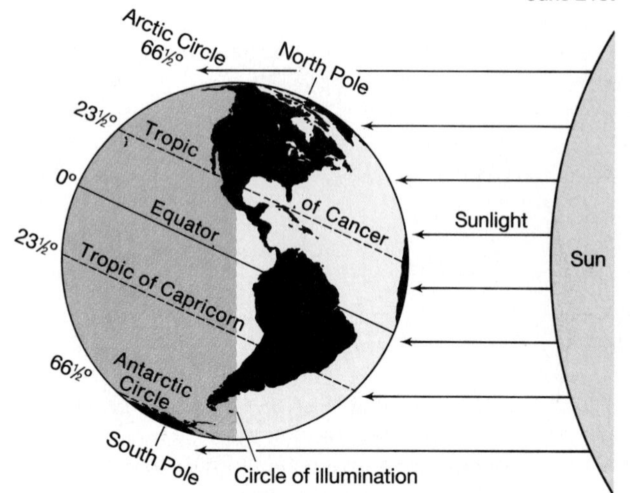

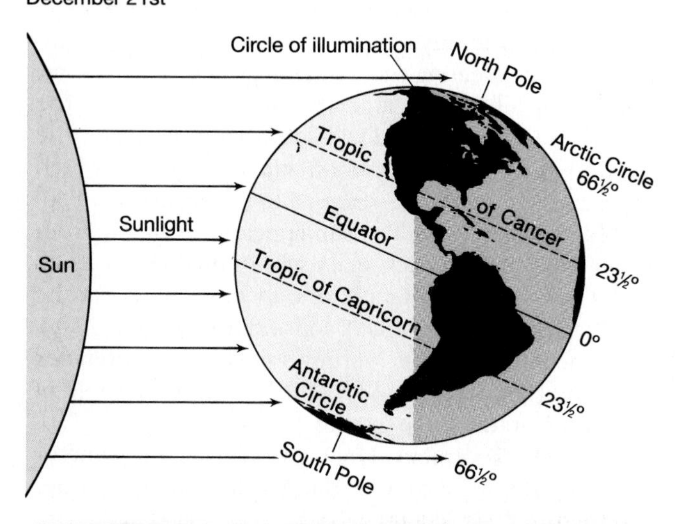

Figure 1–10. At the time of the solstices, the sun's noon rays strike directly at $23\frac{1}{2}°$ latitude. The June (Northern Hemisphere summer) solstice is a time when sunlight is concentrated in the Northern Hemisphere; the December (Northern Hemisphere winter) solstice is a time of Southern Hemisphere concentration. The arrows on the accompanying drawings give an indication of the sun's angle at various latitudes.

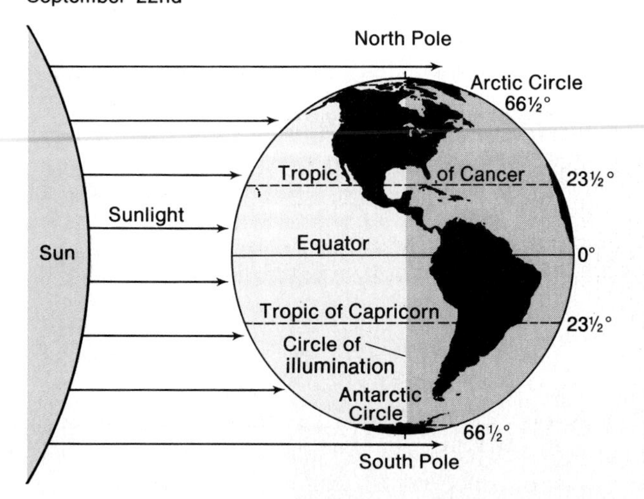

Figure 1–11. At the times of each equinox, the direct rays of the noon sun strike the equator. On those two days only, the length of daylight and darkness is equal all over the Earth. The arrows in this diagram indicate the solar angle at various latitudes.

ABC NEWS *The Sun and Today's Solar Eclipse*

The relationships between the Earth and the sun have been well documented by geographers, meteorologists, climatologists, and astronomers. One of the most interesting of these relationships is that of the solar eclipse. Solar eclipses are embedded in folklore, considered as omens both positive and negative, but also are very important from a scientific standpoint.

The sun plays an important role in the earth system, such as aiding in the process of photosynthesis and creating wind, rain, and everything else we call the weather. Although extremely rare, solar flares and eclipses have important impacts on the earth and on our day-to-day activities. Flares and eclipses have an unusual effect on the magnetic field that surrounds our planet. This has often led to peculiar distortions in global communications, radio and television reception, navigational equipment, and even electronic garage door openers!

The accompanying video segment demonstrates the fundamentals of solar eclipses, focusing on the solar eclipse that occurred on July 11, 1991. This eclipse lasted 6.5 minutes. Although solar eclipses occur approximately every 11 years, there will not be an eclipse lasting this amount of time for another 141 years.

While viewing the video, consider the following questions.

1. What are the most serious effects of solar phenomena on our planet? What, if anything, might we do to prevent these?
2. What would happen during a prolonged eclipse of 20 minutes? An hour? A month?

PORTRAYING THE EARTH

The surface of the Earth is the focus of the geographer's interest. Because the surface is so enormous and complex, it would be difficult to comprehend and analyze without tools and equipment to aid in systematizing and organizing the varied data. Many kinds of tools are useful in geographical studies, but probably the most important and most universal are maps. The mapping of the phenomena under study normally is essential to an understanding of spatial distributions and relationships.

A map is a depiction of an area in graphic form. It is a representation of an area at a reduced scale in which only selected data are shown. Maps display horizontal dimensions. Their basic attribute is the capability of showing distance, direction, size, and shape in their horizontal spatial relationships. In addition to these fundamental pieces of graphic data, most maps are designed to show other kinds of information as well. Thus, a map can show road patterns, or the distribution of population, or the ratio of sunshine to clouds, or the number of daisies per square meter of ground, or any of an infinite number of other facts or combinations of facts.

The Matter of Scale

A map is always smaller (usually extremely smaller) than the portion of the Earth's surface it represents, so any understanding of areal relationships (distance, size) depends on the proper use of scale. The **scale** of a map expresses the relationship between length measured on the map itself and the corresponding distance on the ground. Knowing the scale of a map makes it possible to measure distance, determine area, and make comparisons of size. Scale usually is shown on a map in one of two ways: graphic or fractional.

A **graphic scale** (Figure 1–12a) makes use of a line marked off in graduated distances. Its advantage is the simplicity of determining approximate distances on the ground by measuring them directly on the map.

A **fractional scale** (Figure 1–12b) compares map distance with ground distance by proportional numbers and is expressed as a fraction or ratio. A common fractional scale is 1/63,360, which is usually expressed as a ratio, 1:63,360. This means that one unit of distance on the map represents 63,360 of the same units on the ground. Thus, 1 inch on the map represents 63,360 inches on the ground, 1 millimeter on the map represents 63,360 millimeters on the ground, and so forth. With a fractional scale, the same units of measurement must be used in both numerator and denominator, or on both sides of the ratio, unless the required mathematical transition is performed. Accordingly, a scale of 1:63,360 could also be expressed as 1 inch: 1 mile, because 1 mile contains 63,360 inches. In this case, the understanding is that 1 inch on the map represents 1 mile on the ground.

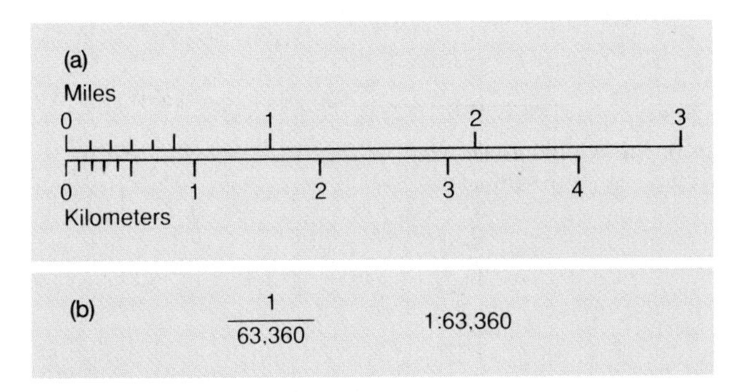

Figure 1–12. Examples of graphic (a) and fractional (b) scales.

The adjectives "large" and "small" often are used to describe a map scale, although the words are used in a comparative, rather than an absolute, sense. In other words, scales are large or small in comparison with other scales.

A *large-scale map* is one that has a relatively large fraction, which means that its denominator is small. Thus, 1/10,000 is a larger fraction than, say, 1/1,000,000, so a scale of 1:10,000 is large in comparison with one of 1:1,000,000 and would be called a large-scale map. A large-scale map portrays only a small portion of the Earth's surface but does so in considerable detail (see Figure 1–13). For example, if the page you are reading now were covered with a map at a scale of 1:10,000, it would be able to show just a small part of a single county, but it would be capable of doing so in great detail.

A *small-scale map*, in contrast, involves a small fraction, that is, one with a large denominator. Thus, a map at a scale of 1:10,000,000 would be classed as a small-scale map. If this page were covered with a map at a scale of 1:10,000,000, it would be able to portray about one-third of the United States, but in limited detail.

Equivalence Versus Conformality

Scale can never be represented with perfect accuracy on a map because of the impossibility of rendering the curve of a sphere on the flatness of a sheet of paper without distortion. A piece of paper cannot be closely fitted to a sphere unless it is wrinkled or torn, so data cannot be transferred directly from a globe to a map without deformation. In essence, this comes down to a choice between showing accurate shape or accurate size.

To portray shapes accurately involves **conformality.** This means that the proper angular relationships are maintained so that the shape of something on the map is the same as its actual shape on the earth. To accomplish this, however, usually means that the size of an area must be distorted.

The portrayal of proper size relationships involves **equivalence.** Equivalent maps maintain the same size ratio between any area on the map and the corresponding area on the ground at all points on the map. Equivalent maps are very desirable because areas are shown in correct proportion to one another, and misleading impres-

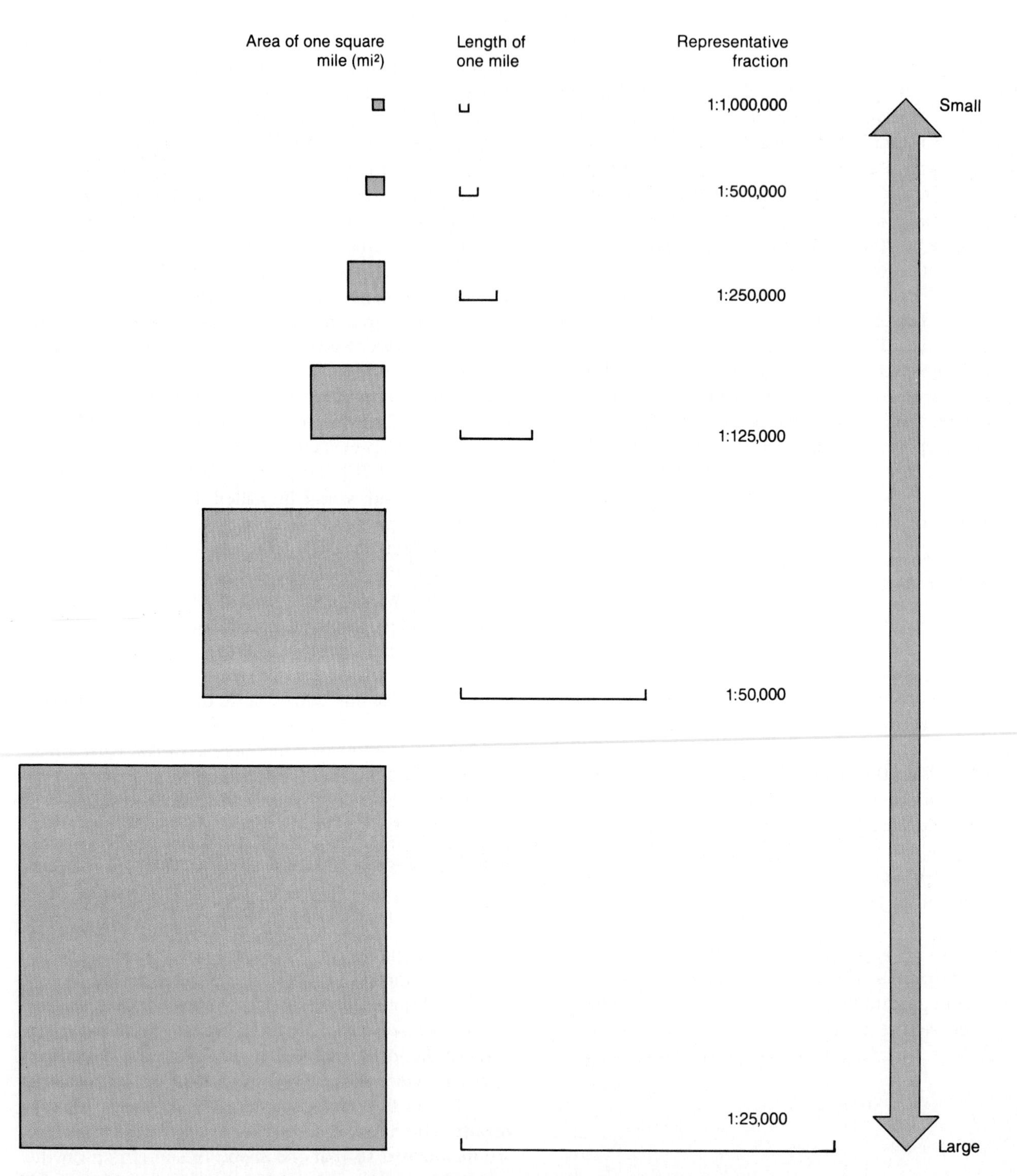

Figure 1–13. Comparisons of distance and area with different map scales. A small-scale map can portray a large part of the Earth's surface, whereas a large-scale map can show only a small part of the surface.

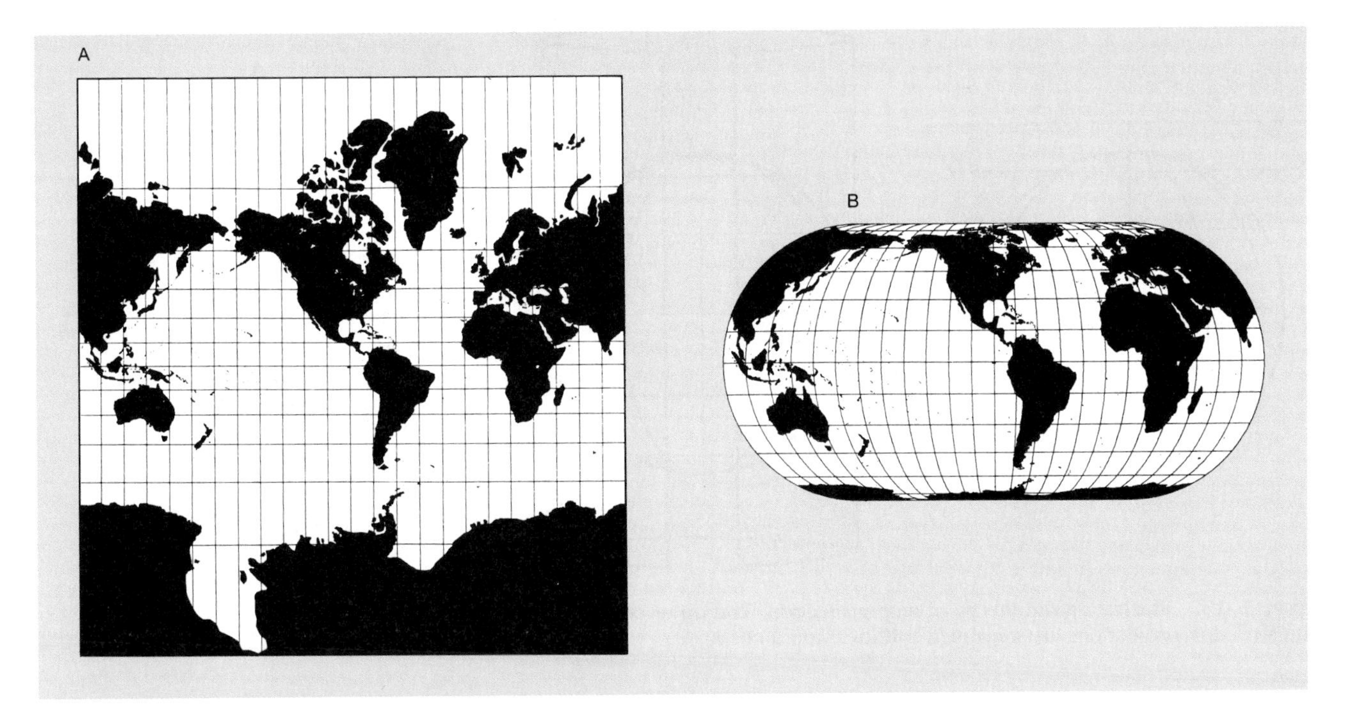

Figure 1–14. Portraying the whole world on a flat map requires considerable distortion. A *conformal* projection (A) displays correct shapes but enormously exaggerates sizes in high-latitude areas. An *equivalent* projection (B) can portray accurate sizes, but shapes are extremely distorted around the edges of the map.

sions of size are avoided. The problem with equivalent maps is that shapes must be sacrificed to maintain proper areal relationships; thus, equivalent maps often display disfigured shapes (see Figure 1–14).

Thus, the cartographer is faced with a difficult choice: size or shape? In fact, some maps are conformal, some are equivalent, none are both conformal and equivalent, and most represent some compromise between conformal and equivalent.

Map Projections

A **map projection** is a method whereby the rounded surface of the Earth is transformed for display on a flat surface. No matter how this is done, however, something will be wrong with the shapes and relative sizes of regions and the distances and directions among places.

The basic principle of projecting a map is direct and simple. Imagine a transparent globe on which are drawn meridians, parallels, and continental boundaries and that has a light bulb in its center. A piece of paper in the form of a geometric figure (such as a cylinder or a cone) is placed over the globe (see Figure 1–15). When the bulb is lighted, the lines drawn on the globe are "projected" outward onto the paper cylinder or cone. These lines can then be sketched on the paper, and the paper laid out flat, producing a map projection.

In actuality, very few map projections have ever been constructed by the direct projection of data from a globe to a piece of paper. Nearly all projections have been derived by mathematical computation. Their com-

ABC NEWS ***Dramatic Changes in Soviet Union Affect Map Makers***

Maps and globes are the fundamental tools for geographers. Many people assume that once a map is drawn, it ceases to change. But changes do occur. These changes require people who make maps to keep up with the shifting global picture. For example, the changes occurring in Eastern Europe and the former Soviet Union have rendered maps drawn up only a few years ago obsolete. (This issue is discussed in Chapter 10.)

In addition to political changes, physical changes are affecting maps. As the Kilauea volcano on the island of Hawaii continues to erupt, more land area for the island is created annually. In the state of Washington, Mount St. Helens radically altered the nearby landscape during and after the eruptions of the early 1980s.

To understand the importance of timeliness for maps and globes, investigate the following questions.

1. Find an atlas that includes a map of Africa in the 1950s. What has happened to the territories labeled French West Africa and French Equatorial Africa?
2. Find a map of Europe printed in 1980, and compare it with a current one from this text.

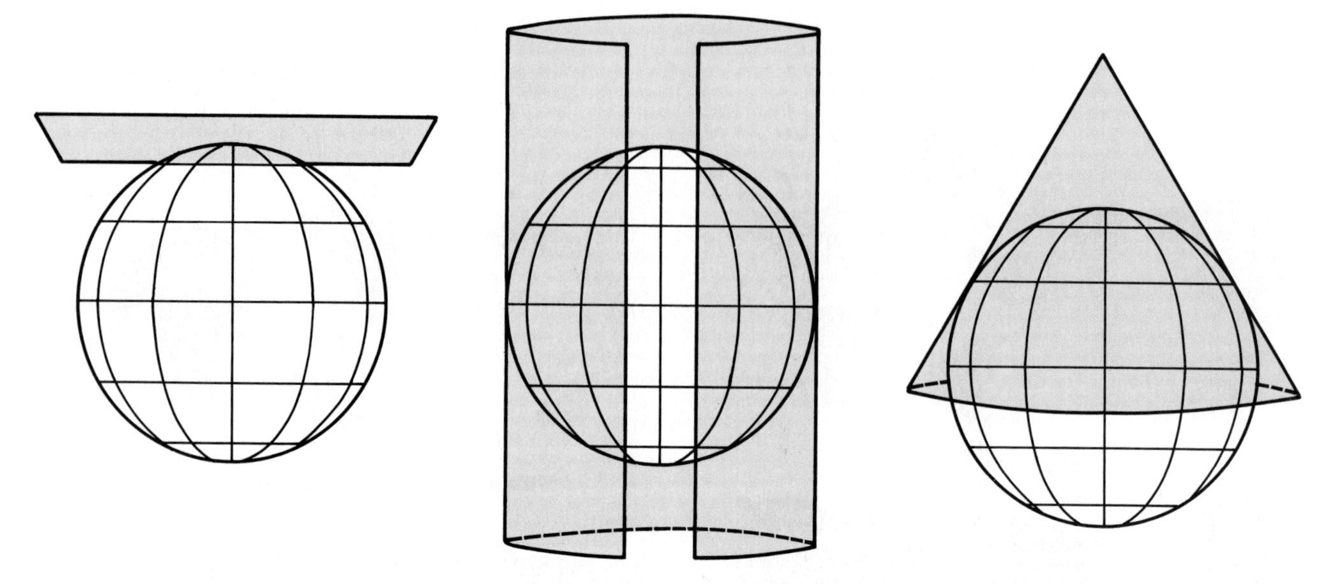

Figure 1–15. Illustrating the theory of map projection. The three common geometric figures used in projections are a plane, a cylinder, and a cone.

mon feature is that each projection shows the correct location of latitude and longitude on the Earth's surface. In other words, each projection consists of an orderly rearrangement of the geographic grid transposed from the globe to the map. The difference among projections is the difference among the grid arrangements.

There is no possible way to avoid distortion completely, so no map projection is perfect. Each of the many hundreds of different projections has been designed as a compromise from reality to achieve some purpose. Each projection, then, has some advantage over the others, but it also has its own particular limitations.

More than a thousand different map projections have been devised. Most of them can be grouped into just a few families, based on their derivation. Projections in the same family generally have similar properties and related distortion characteristics. Figure 1–16 provides examples of four major families: cylindrical, elliptical, azimuthal, and conic projections.

A **cylindrical projection** is derived from the concept of projection onto a paper cylinder that is tangential to, or intersecting with, a globe. Most cylindrical projections are designed so that the cylinder is tangent to the globe (that is, it just touches the globe) at the equator. This produces a right-angled grid network (which means that meridians and parallels meet at right angles) on a rectangular map. There is no distortion at the circle of tangency, but distortion increases progressively away from this circle. Thus, most cylindrical projections, display little distortion in low latitudes but enormous distortion in the polar regions. Cylindrical projections are generally used for maps of the whole world. The Mercator projection discussed in the box on page 14 is an example of cylindrical projection.

Elliptical projections are oval-shaped and display the whole world, although their central sections sometimes are used for maps of smaller areas. In most elliptical projections a central parallel (usually the equator) and a central meridian (generally the prime meridian) cross at right angles in the middle of the map, which is a point of no distortion. Distortion normally increases progressively toward the outer margins of the map. Parallels are usually arranged parallel to one another, but meridians (apart from the central meridian) are shown as curved lines.

An **azimuthal projection** is derived by the perspective extension of the geographic grid from a globe to a plane that is tangent to the globe at some point. There is no distortion at the point of tangency, but distortion increases progressively away from that point. No more than half the earth (a hemisphere) can be displayed with any success on an azimuthal projection. Thus, azimuthal projections have a logical "look" because they portray the Earth as you might view a globe, or an astronaut's perspective from space. They have the same drawback as globes in that only half the Earth can be seen at one time, although they can be useful for focusing attention on a specific region of the world.

A **conic projection** is conceived as having one or more cones set tangent to, or intersecting, a portion of the globe, and the geographic grid is projected onto the cone(s). Normally the apex of the cone is considered to be above a pole, which means that the circle of tangency coincides with a parallel, which becomes the standard parallel of the map. Distortion increases progressively away from this parallel. Consequently, these projections are best suited for regions of east-west orientation in the middle latitudes, which makes them particularly useful for maps of the United States, Europe, and China. It is impractical to use conic projections for more than one-fourth of the earth's surface (a semihemisphere). They are particularly well adapted for mapping smaller areas.

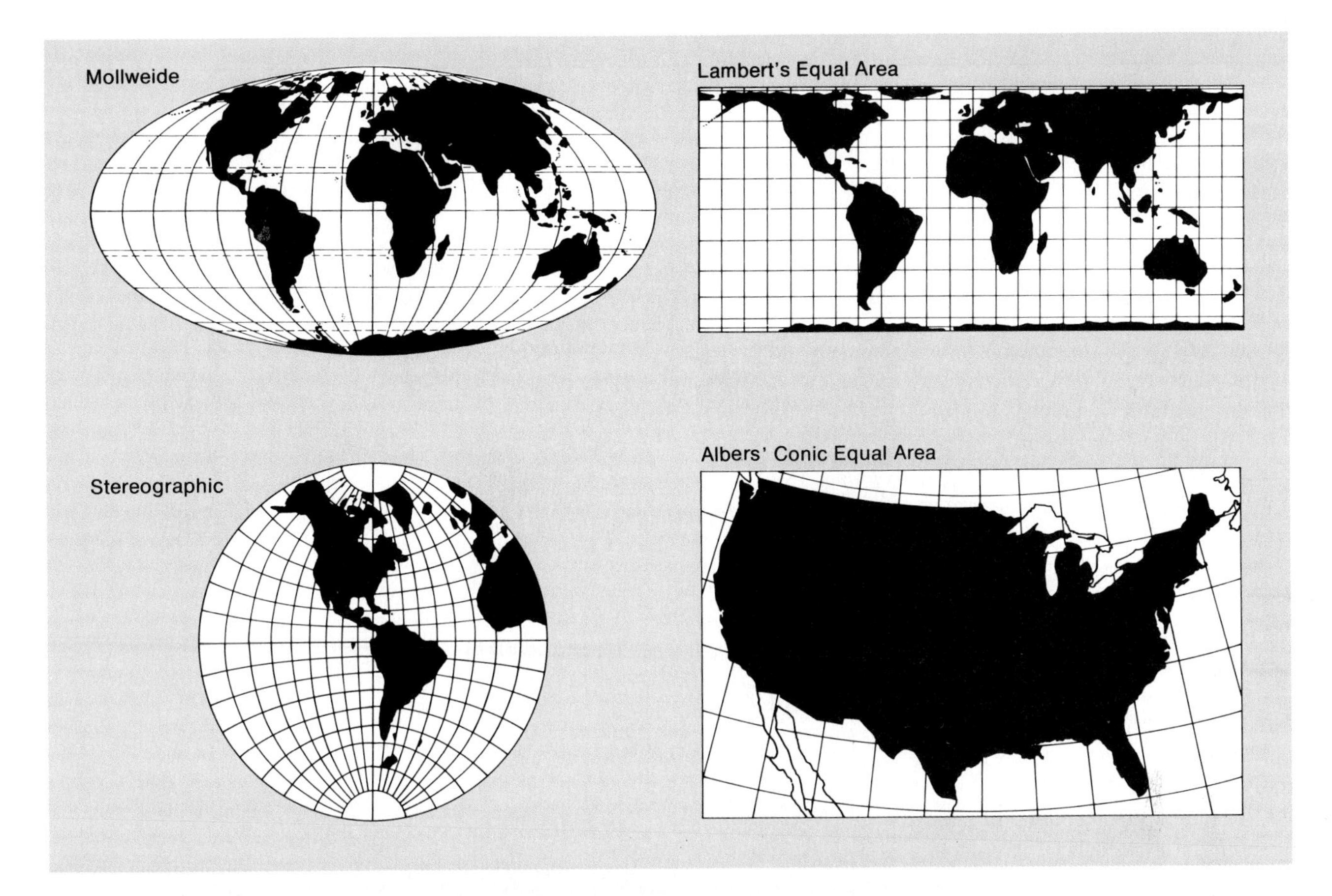

Figure 1–16. Some sample projections: (1) Lambert's Cylindrical Equal Area Projection is an example of the cylindrical family. (2) Mollweide's Projection is in the elliptical group. (3) The Stereographic Projection is a member of the azimuthal family. (4) Alber's Conic Equal Area is a conic projection.

There are dozens of other projections that do not fit into any of the four families discussed above. Each has its specific uses and specific limitations.

Geographic Information Systems

One major trend of the 1980s was a dramatic increase in the use of computer systems to store and manipulate the kinds of information we are used to seeing on maps. These systems, called **geographic information systems (GIS),** reflect both a philosophy and a technology. Although storing the information that might be printed on a map is a part of what we do with a GIS, it is only a small part.

Philosophy In a very real sense, a GIS is a way to think about data and information so that we can work with things and their locations. Knowing the spatial relationships among objects on the Earth's surface is important in many applications, ranging from environmental impact assessment to regional planning to civil engineering.

Consider the kinds of information that are presented on a map. The types of information contained on a map are called *themes*: roads, lakes, and so forth. For any theme, we need two kinds of information: Where is it, and what are its characteristics? The locations of things on the earth's surface are represented by their locations on the map. The characteristics of things—for example, indicating that a line on the map is a railroad rather than a stream—are portrayed in many ways, including using symbols and colors. Some of the themes exist at all possible places on the map. For example, we can estimate the elevation of any place on the map. A theme that can be found at any place on a map is called a *continuous theme.* Other themes are rather more like individual objects that are found at specific places: There is an airport here, and a swamp over there. Sometimes there is structure to a theme: Small streams feed into the large streams, and large streams feed into the river.

A GIS helps us capture the information in these various themes in a systematic manner, manipulate it to solve problems, and display the results in a way that communicates well to others. Some systems are particularly useful when working with continuous themes, like elevation, temperature, and rainfall. These systems may be valuable to a forest ecologist seeking to explore relationships between the regrowth of trees after a fire and the characteristics of various locations in the burned area. Other systems are particularly good at organizing large

CRITICAL THINKING: *Mercator, the Most Famous Projection*

Although some map projections were devised centuries ago, there has been a continuing development and refining of projections right up to the present. Thus, it is remarkable that the most famous of all projections was "invented" more than 400 years ago and is still widely used today without significant modification. This is the *Mercator projection*, originated in 1569 by the Flemish geographer and cartographer Gerardus Mercator.

Mercator produced some of the best maps and globes of his time. His place in history, however, is based largely on the fact that he developed a special-purpose projection that became inordinately popular for general-purpose use. The Mercator projection was essentially a navigational chart of the world, designed to facilitate oceanic navigation. His instructions accompanying the map stated clearly that its proper use would guide mariners by simple compass direction to their destination but that it would not necessarily be the quickest way to get there. Mercator used an illustration of the mythological figure Atlas holding the world on his shoulders to decorate a collection of maps. Books of maps have been called *atlases* ever since.

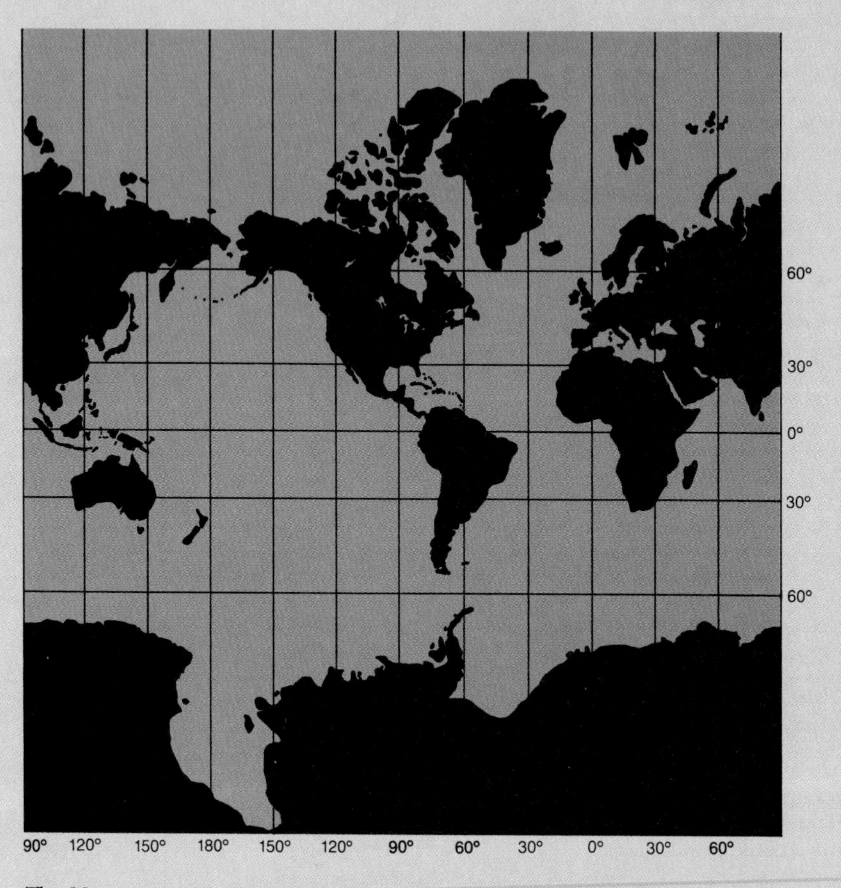

The Mercator projection, in all its simplicity and exaggeration.

The prime advantage of a Mercator map is that it shows loxodromes as straight lines. A *loxodrome* is a true compass heading, a line of constant compass direction. A navigator, whether on a ship or a plane, can plot the shortest distance between origin and destination (a *great circle*) on a projection in which great circles are shown as straight lines and can then transfer that route to a Mercator projection by marking spots on the meridians that the great circle crosses. These spots can then be connected by straight-line loxodromes, which approximate arcs of the great circle but consist of constant compass headings. This allows the navigator to chart an approximately shortest route between origin and destination by making periodic changes in the compass course of the airplane or ship as it generally follows a great circle.

The Mercator map is a cylindrical projection mathematically adjusted to attain conformality. Parallels and me-

volumes of data about objects in space. A common example of the latter is found in a city planning department, where information about every parcel of land in the city is maintained, including the name and address of the owner, the precise boundaries of the parcel, and the electric and water supplies that are delivered to the parcel.

To perform these tasks, a GIS must be able to support a number of phases of activity. First, there must be tools to get the data into the system. These tools may include various kinds of devices related to television cameras that can recognize materials on printed pages and record them in a way that the computer can understand. Next, there must be facilities in the system that permit us to store the data in an efficient way, so that we can retrieve the things we need as we need them. When such a **geographic database** has been built, the GIS permits us to analyze it in many ways: Are all the bridges that need maintenance before the engineers expected sited on the same kind of soil? Where are the homeowners who will be affected by the construction of the new shopping center? How might we rearrange

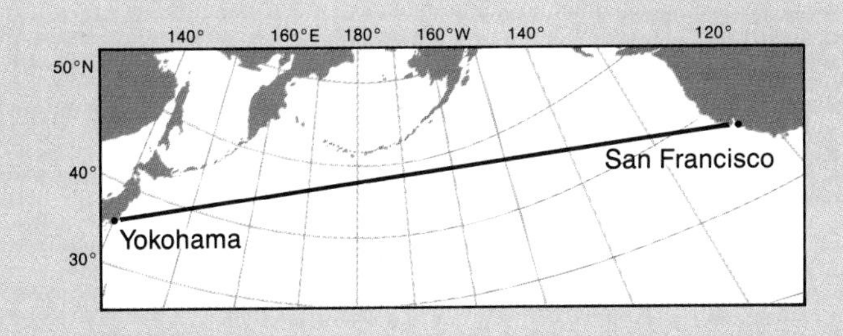

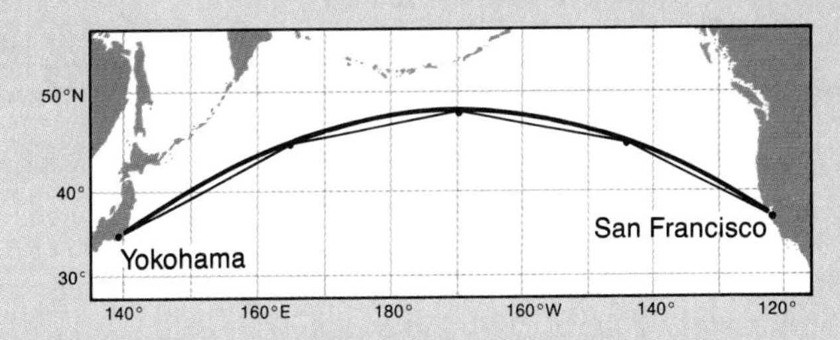

The prime virtue of the Mercator projection is for straight-line navigation. The shortest distance between San Francisco and Yokohama can be plotted on any of several projections in which great circles are shown as straight lines (left). This route can be transferred to a Mercator projection (right) with mathematical precision. Straight-line loxodromes can then be substituted for the curved great circle, allowing the navigator to maintain constant compass headings from point to point, approximating the curve of the great circle.

ridians form a perfectly rectangular grid on the map, with the equator as the circle of tangency. It is accurate at the equator and relatively undistorted in the low latitudes in general, but distortion of size increases rapidly in the middle and high latitudes. To visualize the construction of this projection, imagine that a light in the center of the globe projects the network of the geographic grid onto a cylinder tangent at the equator. This causes the meridians to appear as parallel lines rather than converging at the poles, resulting in extreme east-west distortion in the higher latitudes. In order to maintain conformality, Mercator compensated for the east-west stretching by spacing the parallels of latitude increasingly farther apart so that north-south stretching would occur at the same rate. This procedure allowed him to approximate shapes with reasonable accuracy, but at great expense to proper areal relationships. Area is distorted by 4 times at the 60th parallel of latitude and by 36 times at the 80th parallel. If the North Pole were shown on a Mercator projection, it would be as extensive as the equator. Rather than a single point, it would become a line 25,000 miles (40,250 km) long.

It is clear then that the Mercator projection is excellent for straight-line navigation, but it has serious limitations for most other uses. However, despite the obvious flaws associated with areal distortion in the high latitudes, Mercator projections have been widely used in U.S. classrooms and atlases. Indeed, several generations of U.S. students have obtained their principal view of the world from Mercator maps. This has created many misconceptions, not the least of which is confusion about the relative sizes of high-latitude landmasses. For example, on a Mercator projection the island of Greenland appears to be as large or larger than four of the recognized continents (Africa, Australia, Europe, South America). In actuality, however, Africa is 14 times larger than Greenland, South America is 9 times larger, even Australia is 3.5 times larger; it just doesn't appear that way on a Mercator map.

Questions

The Mercator map serves its purpose of navigation excellently, but what has been the result of using the Mercator map for reasons other than its original purpose? If a person studied only a Mercator map, how would that person interpret the importance of various countries? What would a Canadian think, a Zairian, a Dane? (Greenland is ruled by Denmark.)

Why has the Mercator map been so popular? Could it be simply that it prints in a convenient rectangle? Can you think of any other reason?

What would happen if an airline pilot drew a straight line from his or her home to his or her destination on a Mercator map and then tried to fly that route as the most direct. Plot a few such routes on a Mercator map, and then compare them with direct routes traced by a string stretched taut between those same two points on a globe. Try Tokyo-Los Angeles: Is Hawaii on the way? How about New York-London or Buenos Aires-Madrid.

the bus schedules and routes to serve the community better? Are the forests or lakes showing the effects of acid rain? After these analyses are complete, we then use the GIS to communicate our results by drawing maps on the computer screen or on paper, or printing statistical reports.

Technology The technologies that are used to create the computer's version of maps, keep them up to date, and use them to solve problems come from many disciplines. The tools of computer-aided design and computer graphics are a portion of the tool kit that is used to build a GIS, as are those of database management and statistics. In addition, the concepts of geographical analysis, where we focus on the relationships between objects and their locations, are a fundamental basis of a GIS.

Until recently, geographic information systems were rare and expensive. They required massive computer systems and large numbers of staff to keep them running. Today, although many systems are still running on huge computers, many are also running on personal computers. This transition has been due to several trends.

Focus on Using a GIS

Researchers at the University of California, Santa Barbara are using geographic information systems to develop models of crop growth and yield for an area in northeastern Italy. In this program, funded by the regional government, a number of data layers are used to identify crops and predict their yields. The various layers illustrated in Figure 1—precipitation, soils, temperature, and terrain—all affect the way that plants grow. Figure 2 provides an example of using a GIS to determine the interrelationships between two of these factors: soil and rainfall.

The changes that occur in plants during the growing season have been used in this analysis in several ways. Because different crops are planted, emerge from the soil, and become green at different times of the year, we can distinguish among the crops grown in northern Italy by observing them at intervals from space (see Figure 3). In addition, the GIS uses information from previous years to make decisions concerning planting and harvesting for the current year (see Figure 4). These changes in crop greenness through time can also provide estimates of eventual crop yield from a vantage point early in the growing season. Finally, these statistical analyses can be used with regional information on crop practices through time (see Figure 5) to enable farmers in northern Italy to compete more effectively in the global market, a topic discussed in more detail in Chapter 7.

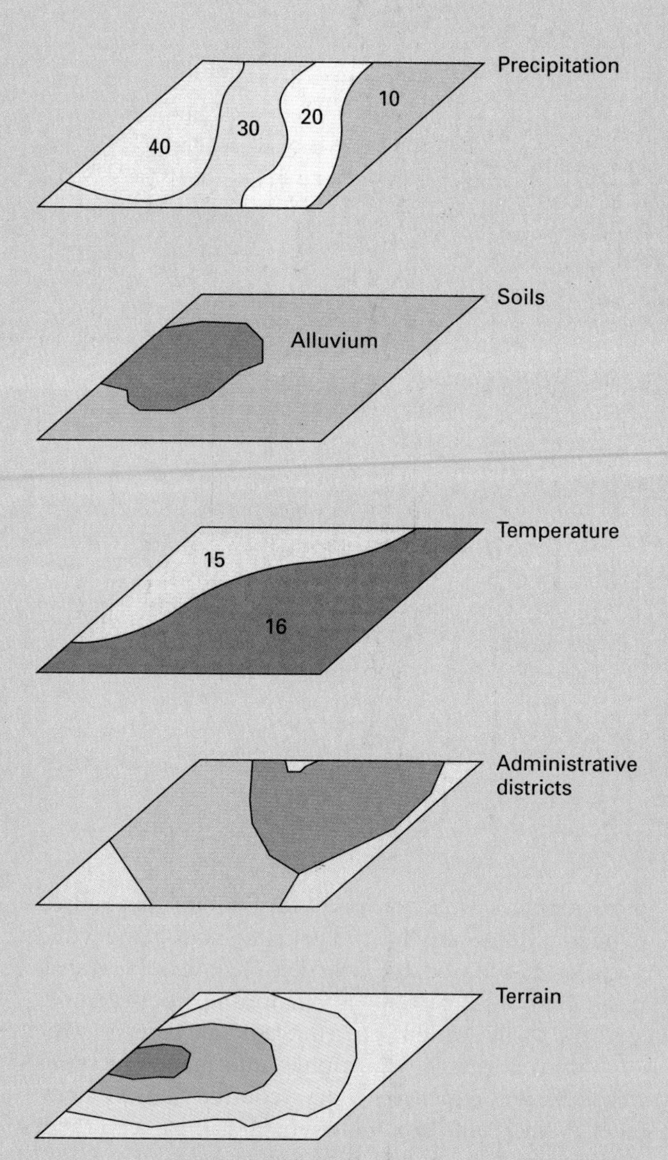

Figure 1. (Courtesy of J. Star and D. Ehrlich, University of California, Santa Barbara.)

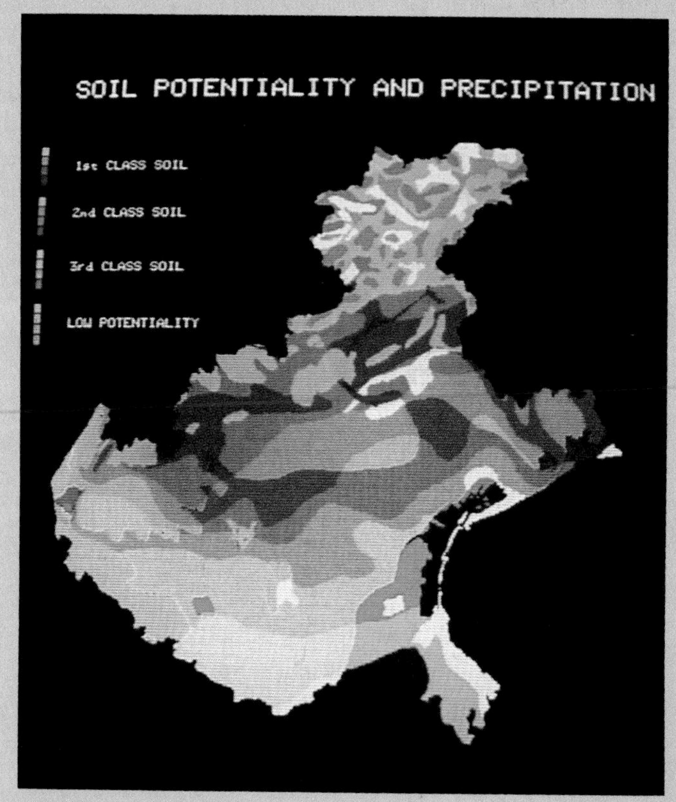

Figure 2. (Courtesy of D. Ehrlich, University of California, Santa Barbara.)

Figure 3. (Courtesy of Jeffrey Star, University of California, Santa Barbara.)

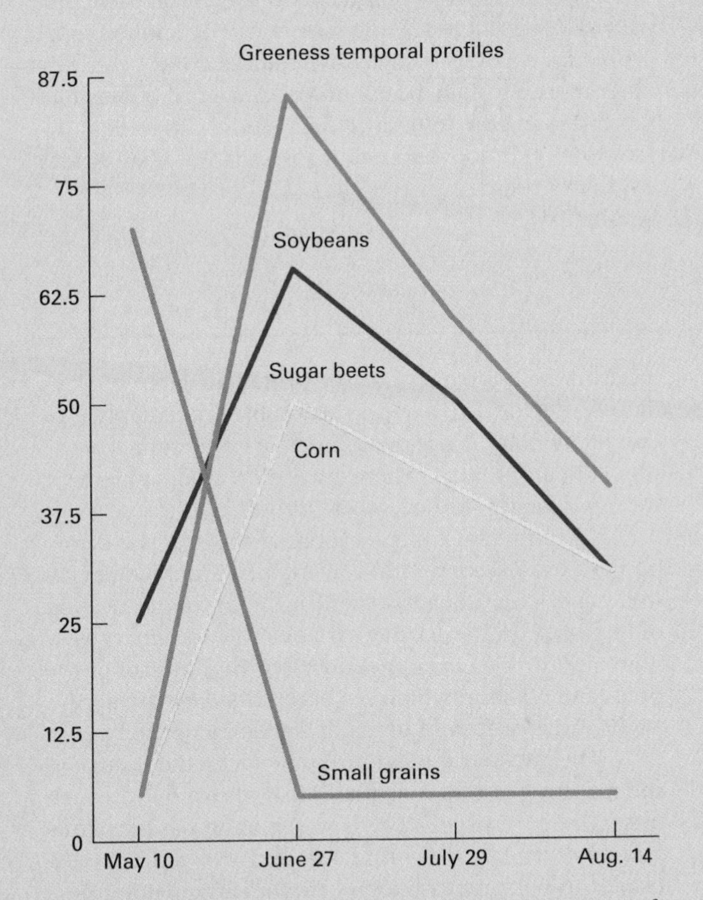

Figure 4. (Courtesy of J. Star and D. Ehrlich, University of California, Santa Barbara.)

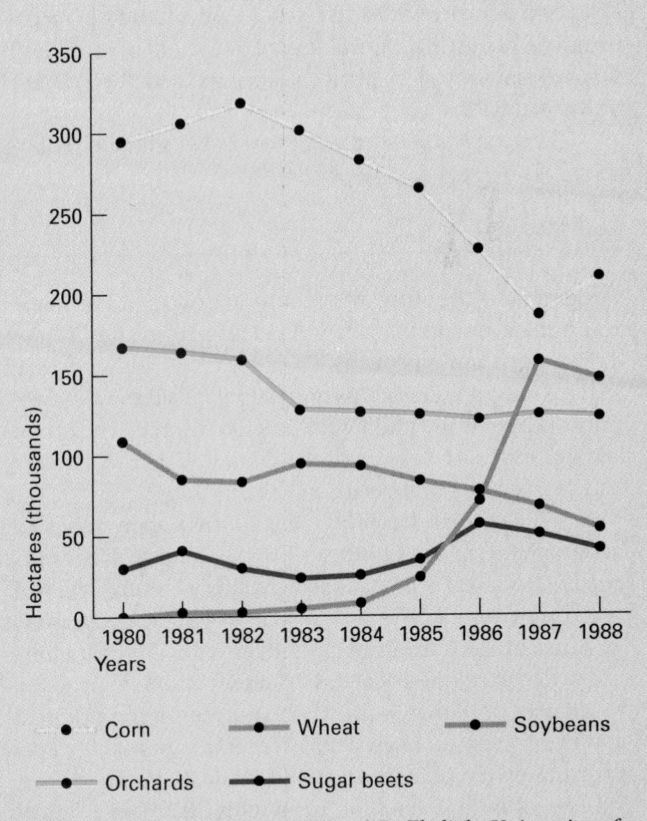

Figure 5. (Courtesy of J. Star and D. Ehrlich, University of California, Santa Barbara.)

First, the processing power and data-storage capacity of a personal computer today are equal to those of the massive systems of a previous decade. Whereas a group of people might have needed a special air conditioned room and a sizable staff simply to keep the computers running, today a few desktop boxes provide equal capabilities. Second, the quality of the GIS software today, in terms of the range of capabilities it provides and the required maintenance, is vastly improved (although some experts would argue that there is a long way to go). For some applications, computer programmers are no longer necessary to modify the GIS software to meet specific project needs. Third, we better understand how to get data and information into the systems. The government agencies and private firms that create some of the data for geographers are increasingly making it available in a computer-compatible form. And fourth, we better understand how to work cost-effectively with large volumes of spatial data and modern GISs.

Remote Sensing

Throughout most of history, maps have been the only tools available to depict anything more than a tiny portion of the earth's surface with any degree of accuracy. More recently, however, sophisticated technology has been developed that permits precision recording instruments to operate from high-altitude vantage points, providing a remarkable new set of tools for the study of the earth. The term **remote sensing** refers to any method of measuring or acquiring information by use of a recording device that is not in physical contact with the object under study.

The earliest form of remote sensing was aerial photography, which was first tried in 1858, although it was not widely used until three-quarters of a century later. The use of color in air photos dates from the 1920s, and color infrared film (which senses, in part, beyond the visible portion of the electromagnetic spectrum) was not developed until World War II. Sensing systems developed in more recent years use thermal infrared, microwave, radar, and sonar.

Although remote sensing systems frequently gather data from only one part of the electromagnetic spectrum at a time, modern, sophisticated sensors increasingly can gather data from several parts of the spectrum simultaneously. These systems usually are digital in format, which means that the data gathered are represented as a matrix of numbers that are transformed into colors or gray values for the final product.

The United States and other countries have launched many satellites that monitor the Earth and all activities across it using a variety of the most advanced electronic and photographic techniques. The data provided by these satellites are used to create daily weather forecasts, to monitor environmental changes in inaccessible areas or across great stretches of territory, to map deforestation, pollution, and new construction and land uses around the world, and even to spy on foreign nations.

Certain types of imagery are useful for particular purposes; no single sensing system can be applied to all problems. Analysis of the Earth's surface has been revolutionized through the use of aerial imagery, but remote sensing is still an adjunct to field study, geographic description, and conventional maps; it is not a substitute for any of these.

SUMMARY

Geography is the study of the interactions of all physical and human phenomena at individual places and of how interactions among places form patterns and organize space. Geographers first ask questions of simple description: "Where is it?" and "What is it like there?" They then ask questions of cause and meaning: "Why is it there?" and "How does that relate to other things?"

The scholar Varenius defined two approaches in geography. One was to describe and analyze places according to a series of standard categories. Today we call such inventory analyses of places regional geography. Regional studies traditionally begin with the characteristics of a place's physical environment and then proceed to aspects of human culture, thus bridging the physical and the social sciences. The alternative approach to organizing geographic information, topical geography, examines universal laws or principles that apply to all places. The topics of interest to geographers are also both physical-environmental and human-cultural, ranging from soil chemistry to the world geography of religions. Regional geography and topical geography are complementary. A regional study covers all topics in the region being studied, whereas a topical study notes how the topic being studied varies across regions.

Fundamental to geographic analysis is the development of a comprehensive and logical framework (latitude and longitude) to establish the accurate location of any spot on the Earth's surface. This imaginary geographic grid system is anchored by the position of the poles and equator, which are determined by the Earth's slight variance from a perfectly spherical shape.

The geometric relationship between the Earth and the sun varies continuously as both bodies move through interplanetary space. Moreover, the Earth has perpetual motions of its own—rotation and revolution—that change its position with respect to the sun and determine both the annual march of the seasons and the alternation of day and night.

Although only globes are capable of portraying the spherical Earth with complete accuracy, they are too cumbersome for most purposes. Thus, the most widely used tools for portraying spatial distributions and relationships on the Earth's surface have always been maps, which can show distance, direction, size, and shape in the horizontal dimension. A great many kinds of map projections have been devised in an effort to minimize the problem of distortion when transferring data from a spherical surface (Earth) to a flat surface (map). None has been completely satisfactory because of the insoluble problem of accommodating equivalence (equal areal relationships) with conformality (true shapes), but many types of projections are exceedingly useful.

In recent years there has been a phenomenal expansion of the technology of remote sensing, wherein precision instruments, operating from high-altitude vantage points, record information from the Earth's surface far below. Thus, maps and various kinds of remotely sensed images give us an increasingly graphic and accurate portrayal of the Earth.

KEY TERMS

geography (p. 2)
regional geography (p. 3)
topical or systematic geography (p. 3)
geographic grid (p. 5)
axis (p. 5)
plane of the equator (p. 5)
equator (p. 5)
latitude (p. 5)
parallel (p. 6)
longitude (p. 6)
prime meridian (p. 6)
plane of the ecliptic (p. 6)
polarity (p. 6)
solstice (p. 7)
equinox (p. 7)
scale (p. 9)
graphic scale (p. 9)
fractional scale (p. 9)
conformality (p. 9)
equivalence (p. 9)
map projection (p. 11)
cylindrical projection (p. 12)
elliptical projection (p. 12)
azimuthal projection (p. 12)
conic projection (p. 12)
geographic information systems (GIS) (p. 13)
geographic database (p. 14)
remote sensing (p. 18)

QUESTIONS FOR INVESTIGATION AND DISCUSSION

1. The sphericity of the earth is not perfect. Is this fact important in our study of physical geography?

2. What are the major differences between parallels and meridians?

3. What would be the effect on the annual march of the seasons if the earth's axis did not maintain parallelism during earthly revolution?

4. Why is the sun never directly overhead poleward of the tropic lines?

5. Explain the implications of the statement "No map is totally accurate."

6. A globe can portray the earth's surface more accurately than a map, but globes are rarely used. Why?

7. Compare and contrast equivalence and conformality.

8. Why are there so many different kinds of map projections?

9. What are the latitude and longitude of your town?

10. What is the highest point in your state?

ADDITIONAL READINGS

DENT, BORDEN D. *Principles of Thematic Map Design.* Reading, MA.: Addison-Wesley Publishing Co., Inc., 1984.

DRURY, S. A. *A Guide to Remote Sensing.* Melbourne: Oxford University Press, 1990.

GAILE, GARY L., and CORT J. WILLMOTT, eds. *Geography in America.* Columbus, Ohio: Merrill Publishing Co., 1989.

HAGGETT, PETER. *The Geographer's Art.* Oxford, England: Basil Blackwell, 1990.

JAMES, PRESTON, E., and GEOFFREY J. MARTIN. *All Possible Worlds: A History of Geographical Ideas.* New York: John Wiley & Sons, 1981.

KISH, GEORGE, ed. *A Source Book in Geography.* Cambridge, MA: Harvard University Press, 1978.

MONMONIER, M., and G. SCHNELL. *Map Appreciation.* Englewood Cliffs, N.J.: Prentice Hall, Inc., 1988.

MURPHEY, RHOADS. *The Scope of Geography.* New York: Methuen, 1982.

RIDLEY, B. K. *The Physical Environment.* Chichester, England: Ellis Horwood Limited, 1979.

ROBINSON, A. H., R. D. SALE, and J. L. MORRISON. *Elements of Cartography*, 4th ed. New York: John Wiley & Sons, 1978.

SABINS, F. J., JR. *Remote Sensing: Principles and Interpretation*, 2nd ed. San Francisco: W. H. Freeman & Company, Publishers, 1987.

CHAPTER 2

Weather and Climate

A lightning storm over Lava Beds National Monument in California provides a spectacular display of natural forces. (Courtesy of Barbara Filet/Tony Stone Worldwide.)

The Earth is different from other known planets in a variety of ways. One of the most notable differences is the presence around our planet of a substantial atmosphere with components and characteristics that are distinctive from those of other planetary atmospheres. Our atmosphere makes life possible on this planet. It supplies most of the oxygen that animals must have to survive, as well as the carbon dioxide needed by plants. It helps maintain a water supply, which is essential to all living things. It serves as an insulating blanket to ameliorate temperature extremes and thus provides a livable environment over most of the Earth. It also shields the Earth from much of the sun's ultraviolet radiation, which otherwise would be fatal to most life forms.

The atmosphere consists of a mixture of gases, along with small but varying quantities of solid and liquid particles that function as "impurities." It completely surrounds the planet and can be thought of as a gaseous envelope with the Earth tucked inside or as a vast ocean of air with the Earth at its bottom. It is held to the Earth by gravitational attraction and therefore accompanies our planet in all its celestial motions, such as rotation and revolution. The attachment of the atmosphere to the Earth, however, is a loose one, and the atmosphere has actions of its own.

Our vast and largely invisible atmospheric envelope is energized by solar radiation, stimulated by earthly motions, and affected by contact with the Earth's surface. It reacts by producing an infinite variety of conditions and phenomena known collectively as **weather.** Weather is in an almost constant state of change, sometimes in a seemingly erratic fashion. In the long-term view, however, it is possible to generalize the variations into a composite pattern, which is termed **climate.** Climate is the aggregate of day-to-day weather conditions over a long period of time.

Climate is probably the component of the environment that interests people most. It has direct and obvious influences on agriculture, transportation, and human life in general. Moreover, general climatic characteristics are enormously influential on the development of all major aspects of the physical landscape—soils, vegetation, animal life, *hydrography* (water features), and *topography* (landforms).

In our study of the atmosphere, the climatic elements provide the basic organizational framework. We first consider each of the elements—temperature, pressure, wind, moisture—separately, but it is important to keep in mind that they are all interrelated. Following individual treatment of the elements is an integrative section devoted to air masses, fronts, and storms, in which the interactions of climatic elements, climatic controls, and atmospheric processes are emphasized as a prelude to a consideration of the climatic pattern of the world.

INSOLATION AND TEMPERATURE

The sun is the only important source of energy for the Earth's atmosphere. Solar output that reaches the Earth or its atmosphere—known as **insolation**—varies significantly both in intensity and duration. Although the atmosphere as a whole maintains a radiation balance—as much energy leaves the atmosphere for space as enters it from the sun—this balance does not exist within different parts of the globe. For example, tropical areas achieve a large energy surplus, with more incoming than outgoing radiation. Conversely, the high latitudes experience an energy deficit, with more energy loss than gain.

Oceans respond very differently than do continents to the arrival of solar radiation. In general, land heats and cools faster and to a greater degree than does water. Therefore, both the hottest and coldest areas of the Earth are found in the interiors of continents, distant from the moderating influence of oceans. In the study of the atmosphere, probably no single geographical relationship is more important than the distinction between continental and maritime climates. A **continental climate** experiences greater seasonal extremes of temperature—hotter in summer, colder in winter—than does a **maritime climate** (see Figure 2–1).

The oceans act as great reservoirs of heat. In summer they absorb heat and store it. In winter they give off heat and warm up the air. Thus, they function as a sort of global thermostat, moderating temperature extremes. Oceanic temperatures do not vary a great deal from summer to winter, in contrast to the notable changes of continental temperatures.

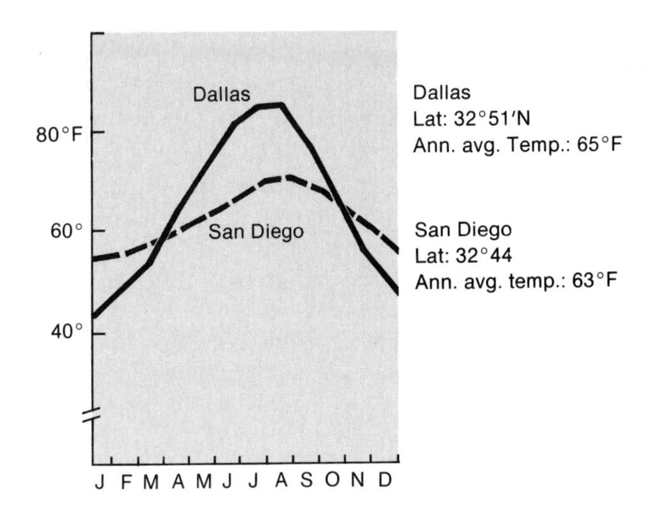

Figure 2–1. Temperature curves for San Diego and Dallas. San Diego, situated on the coast, experiences milder temperatures in both summer and winter than inland Dallas.

CRITICAL THINKING: *Global Warming: The Greenhouse Effect*

The term *greenhouse effect* refers to the fact that our atmosphere generally allows shortwave radiation from the sun to pass through to the Earth's surface but inhibits the escape of long-wave radiation from the Earth into space. This process increases the temperature of the atmosphere. In the last few years there has been increasing evidence that the atmosphere generally is warming up, and there are strong indications that an intensified greenhouse effect is at least partly responsible.

Is the warming a natural occurrence, or is it induced by human actions? Scientists disagree on the answer. It is well known that climate undergoes frequent natural fluctuations, becoming warmer or colder regardless of human activities. An increasing body of evidence, however, indicates that *anthropogenic* (human-induced) factors are responsible for some recent temperature increases. Many scientists attribute the rising temperatures in part to the increasing presence in the atmosphere of the so-called greenhouse gases: water vapor, carbon dioxide, methane, nitrous oxide, ozone, and chlorofluorocarbons. These gases have a low capacity for transmitting long-wave radiation. Thus, as their concentrations in the atmosphere increase, more terrestrial radiation is retained in the lower atmosphere, thereby raising its temperature.

Human activities are clearly responsible for the increasing release of most of these gases into the air. Chlorofluorocarbons (CFCs) are synthetic chemicals that were very popular for a variety of uses until recently. Nitrous oxide comes from chemical fertilizers and automobile emissions. Methane is produced by grazing livestock, rice paddies, burning wood, and the use of natural gas, coal, and oil. Most of these gases have been released at an accelerating rate in recent years.

Carbon dioxide, however, appears to be the principal culprit; some studies indicate that it accounts for about half the recent global temperature increase. Since the mid-1960s, concentrations of carbon dioxide in the atmosphere have increased by some 25 percent. Carbon dioxide is a principal byproduct of the combustion of fuels containing carbon, particularly coal and petroleum. The world consumption of carbon-containing, or fossil, fuels continues to increase. Indeed, some scientists estimate that to control the warming trend, the world's population will have to decrease its consumption of fossil fuels by at least 50 percent. Also contributing to global warming is the destruction of the world's forests, particularly the tropical rainforests. Trees absorb great amounts of carbon dioxide, and with fewer trees, more carbon dioxide floats into the atmosphere.

Although the long-term climatic result of the buildup of greenhouse gases cannot be predicted with certainty, some of the possible consequences are frightening. Certainly, temperature and precipitation patterns would change. Heat and drought would become more prevalent in much of the midlatitudes, and milder temperatures would prevail in higher latitudes. Some arid lands might receive more rainfall. Ice caps would surely melt, and global sea levels would rise. Current living patterns over much of the world would be affected.

What can be done to prevent this scenario? The key is to reduce emissions, particularly from smokestacks and internal combustion engines. Coal and petroleum are major offenders, and their use could be curtailed. In contrast, natural gas produces fewer emissions, and solar, wind, and nuclear energy sources are "clean," insofar as carbon dioxide is concerned. Greater use of these energy sources could be encouraged.

Questions

If the Earth's climate warms due to human activity, or if it changes in any ways that we cannot predict, is this necessarily bad? Is there anything inherently and necessarily wrong about changing the Earth's environment? Should we assume that anthropogenic change in itself is bad? Haven't human beings been changing their environment for tens of thousands of years? Some philosophers have argued that humans are very good at adapting to difficult circumstances, such as air or noise pollution. If this is true, should people focus more on reducing consumption of fossil fuels or adjusting to polluted air?

People in the poor countries often argue that the concern for the environment being voiced by people in the rich countries is only an excuse to keep the poor in a state of poverty by restricting their access to vital resources. The rich, having done so much damage, have no right to try to hold back those who remain poor. Are the rich willing to sacrifice to help the poor? If global tragedy such as flooding or heating occurs, who will probably suffer the most, people in the rich technologically advanced countries, or the world's poor? Explain your answer.

Heat Transfer

We have seen that areas close to the equator experience an energy surplus, whereas those in higher latitudes experience an energy deficit. We might expect, therefore, that the tropics would become progressively warmer and the high latitudes continually colder. Such temperature trends do not occur, however, because there is a persistent shifting of warmth toward the high latitudes and of coldness toward the low latitudes. This is accomplished by movements of air and water, that is, by circulation patterns in both the atmosphere and the oceans. The broad-scale, or planetary, circulation of air and water currents moderates the buildup of heat in equatorial regions and of cold in polar regions, thereby making both of those latitudinal zones more habitable.

Of the two mechanisms of global heat transfer, by far the more important is the general circulation of the atmosphere. Air movements take place in almost infinite diversity, but there is a broad planetary circulation pattern that serves as a general framework for moving warm air poleward and cool air equatorward. Our discussion of atmospheric circulation will be withheld for a few pages, following consideration of some fundamentals concerning pressure and wind.

Various kinds of oceanic water movements are categorized as **currents.** For our purposes in understanding heat transfer, we are concerned primarily with the broad-scale surface currents that make up the general circulation of the oceans. These major currents are usually set in motion by the frictional drag of wind moving horizontally over the ocean surface. In essence, ocean currents reflect average wind conditions over a period of several years, with the result that the major components of oceanic circulation are closely related to major components of atmospheric circulation.

In each of the five major ocean basins—North Pacific, South Pacific, North Atlantic, South Atlantic, and South Indian—there is a similar, simple pattern of major surface currents. It consists of a series of enormous elliptical loops elongated east-west and centered approximately at 30° of latitude (see Figure 2–2). These loops, called *gyres*, have a clockwise direction of flow in the Northern Hemisphere and counterclockwise movement in the Southern Hemisphere. This pattern produces a flow of warm tropical water toward the poles along the western edge of each ocean basin and a movement of cool high-latitude water toward the equator along the eastern margin of each basin.

Vertical Temperature Patterns

The lowest layer of the atmosphere is called the **troposphere,** and it is within this layer that most weather phenomena occur. The depth of the troposphere varies but usually ranges between 5 and 11 miles (8 and 18 km). Within the troposphere, temperatures usually decrease as the distance from the Earth's surface increases. Although the rate of temperature decline with height varies according to local circumstances, the normal expectable rate of decrease is about 3.6°F per 1000 feet (6.5°C per km). This is called the **average lapse rate.**

The most prominent exception to a normal lapse-rate condition is a **temperature inversion,** a situation in which temperature increases at higher altitudes. Inversions are relatively common in the troposphere, but they are usually of brief duration and restricted depth. They can occur near the surface or at higher levels.

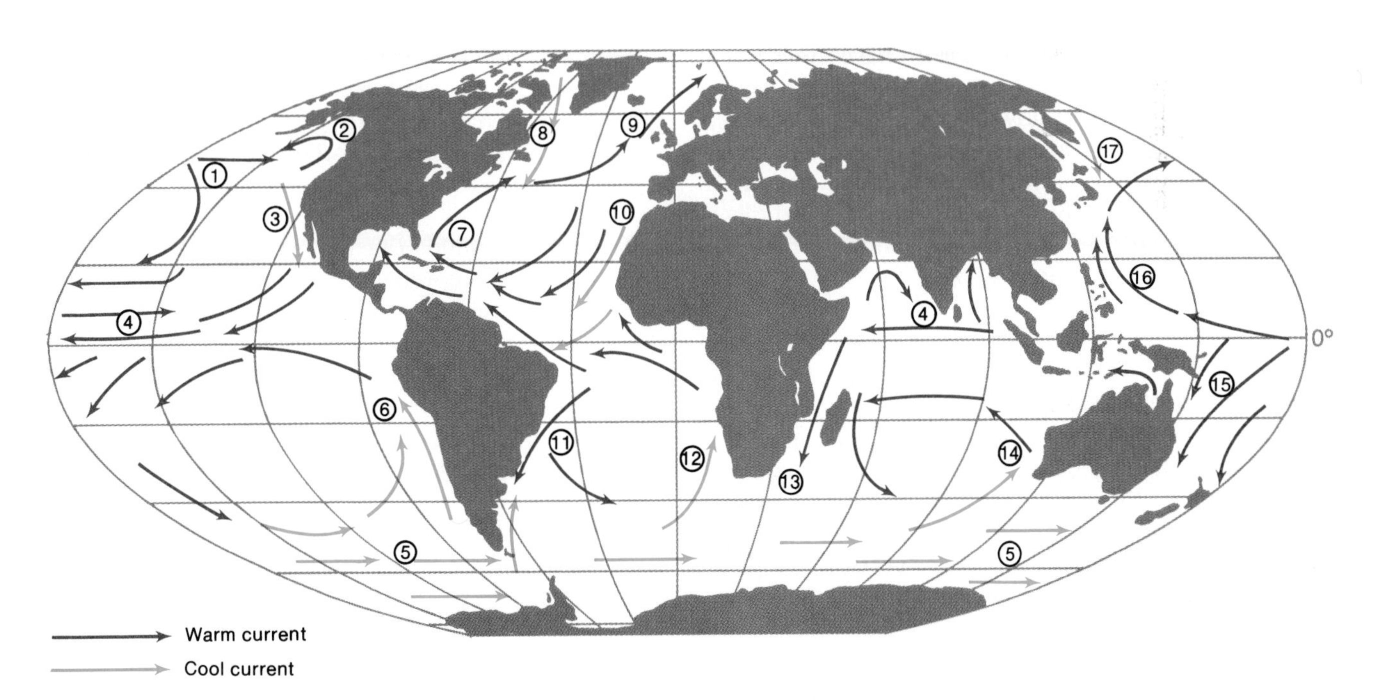

Figure 2–2. The major surface ocean currents. (1) North Pacific Drift, (2) Alaska Current, (3) California Current, (4) Equatorial Current, (5) West Wind Drift, (6) Peru Current, (7) Gulf Stream, (8) Labrador Current, (9) North Atlantic Drift, (10) Canaries Current, (11) Brazil Current, (12) Benguela Current, (13) Agulhas Current, (14) West Australian Current, (15) East Australian Current, (16) Kuroshio (Japan) Current, and (17) Oyashio (Kamchatka) Current.

Figure 2–3. A temperature inversion usually also functions as a "stability lid." In this easterly view across the Los Angeles lowland it is apparent that the murky air ("smog") is trapped beneath the lid, which virtually prohibits updrafts from penetrating it. (TLM photo)

In addition to the obvious reversal of the normal vertical temperature pattern, inversions have other important influences on weather. An inversion inhibits vertical air movements and greatly diminishes the possibility of precipitation. Inversions also contribute significantly to increased air pollution because they create stagnant air conditions that greatly limit the natural upward dispersal of pollutants from automobiles, industrial plants, and other sources (see Figure 2–3).

Global Temperature Patterns

Thus far we have examined a variety of facts, concepts, and processes associated with insolation and temperature. As students of geography, our goal is to gain an understanding of the world distribution pattern of the climatic elements—temperature, pressure, wind, and moisture. With the preceding pages as background, we now turn our attention to the worldwide distribution of temperature.

The basic maps of world temperatures display the seasonal extremes rather than the annual average. January and July are the months of lowest and highest temperatures for most places, so maps portraying the average temperatures of these two months provide a simple but meaningful expression of thermal conditions in winter and summer (see Figures 2–4 and 2–5). Temperature distribution is shown by means of **isotherms,** which are lines joining points of equal temperature.

Temperature responds sharply to altitudinal changes, as we have seen, so it would be misleading to plot *actual* temperatures on these maps, as stations at higher elevations would almost always be colder than low-altitude stations. The complexity introduced by hills and mountains would make the maps more complicated and difficult to comprehend. Consequently, the data for these maps are simplified by using only the temperatures that exist or would exist at sea level. This strategy produces artificial temperature values but eliminates the complication of terrain differences.

The broad patterns shown on these small-scale maps demonstrate the influence of many factors, but most conspicuously:

1. The dominating influence of latitude as shown by the general east-west trend of the isotherms;
2. The differential heating and cooling characteristics of land and water as shown by the occurrence of the highest and lowest temperatures over continents; and
3. The influence of warm and cool ocean currents is shown by the obvious bends in many isotherms in near-coastal areas where ocean currents are prominent.

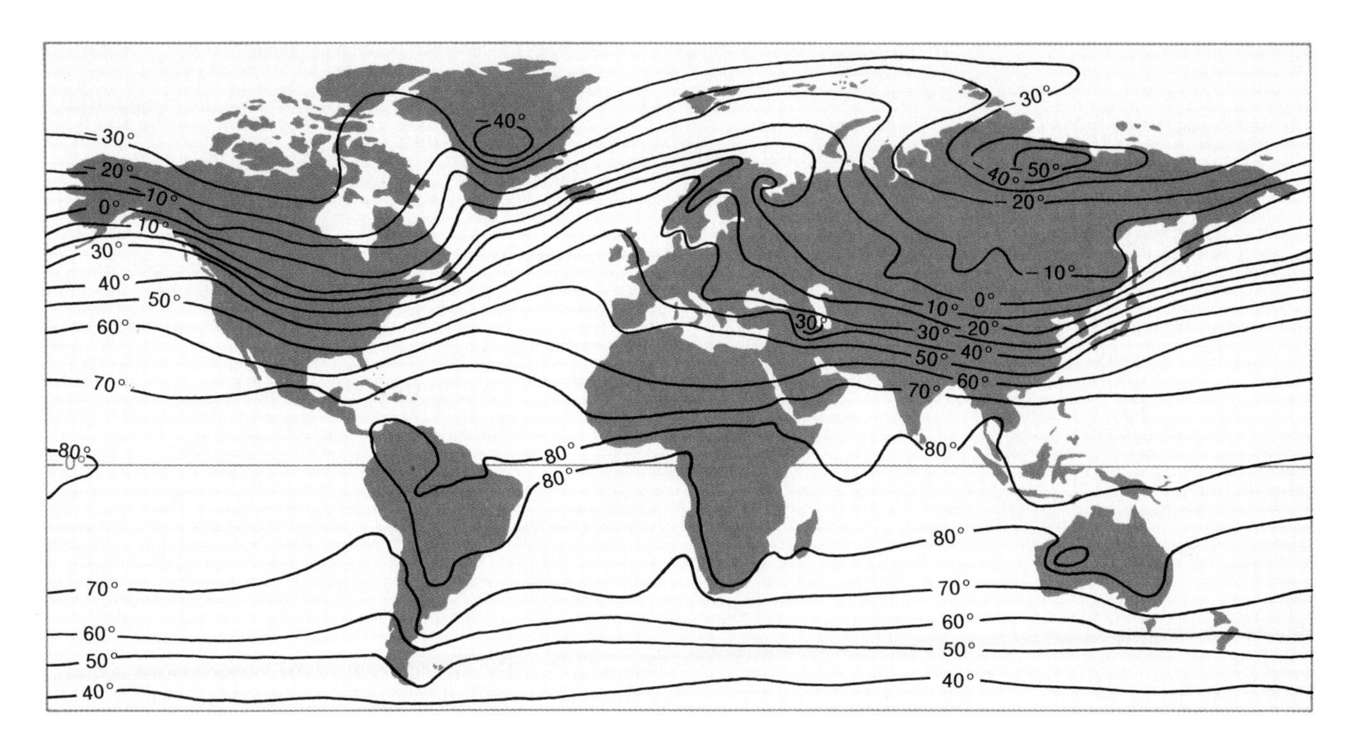

Figure 2–4. Average January sea-level temperatures.

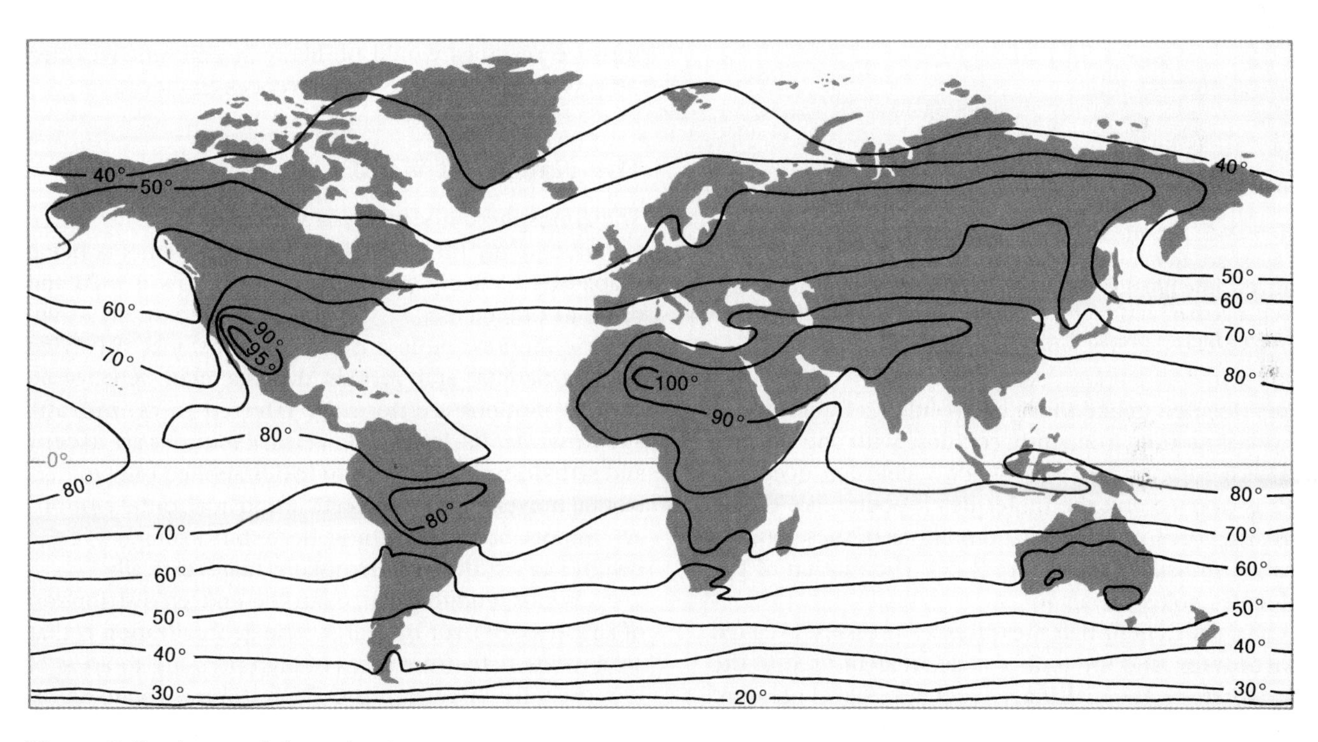

Figure 2–5. Average July sea-level temperatures.

PRESSURE AND WIND

To the layperson, **atmospheric pressure** is the most difficult of the four basic climatic elements to comprehend. The other three—temperature, wind, and moisture—are more readily understood because our bodies are much more sensitive to them. Nevertheless, pressure is a very important feature of the atmosphere. It is tied closely to the other weather elements, acting on them and responding to them. The interaction between pressure and wind is particularly important: Spatial variations in pressure are largely responsible for air movements.

The Nature of Atmospheric Pressure

Because the atmosphere is essentially a mixture of gases, it is composed of gas molecules and behaves like a gas.

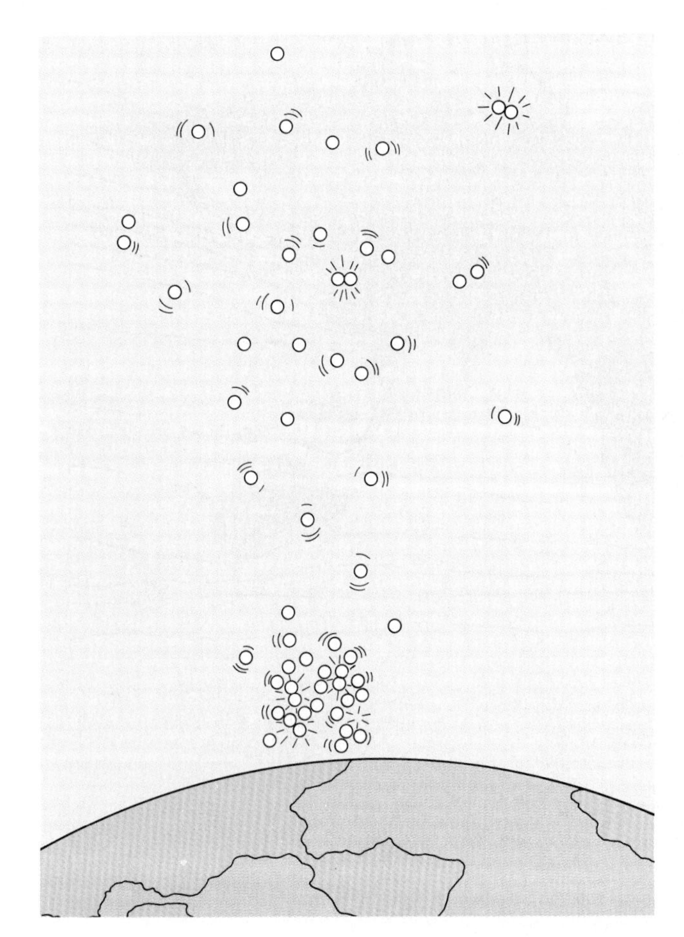

Figure 2–6. In the upper atmosphere, gaseous molecules are far apart and collide infrequently, which produces relatively low pressure. In lower layers, the molecules are closer together, and there are many more collisions, which produces higher pressure.

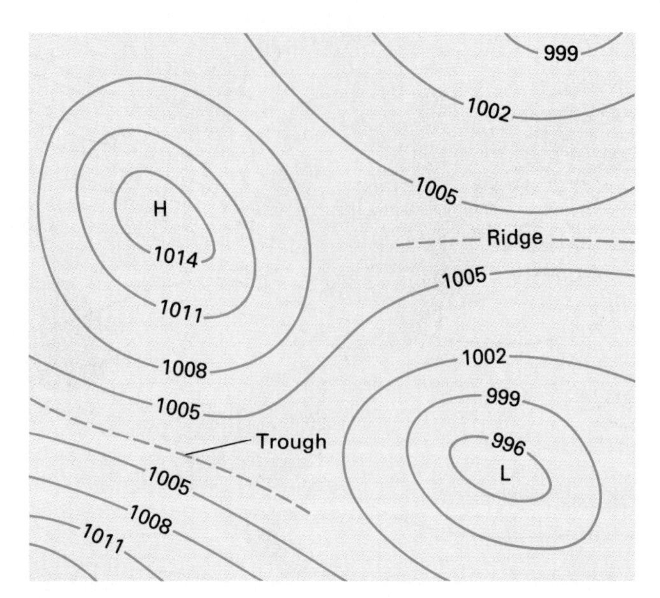

Figure 2–7. The arrangement of the isobars shows the position of pressure ridges and troughs as well as highs and lows.

Gas molecules, unlike those of a solid or a liquid, are in continual motion, frequently colliding with one another and with adjacent surfaces. When a collision occurs, a push is exerted and the molecules rebound like elastic balls. The force of millions and millions of these pushes is called *pressure* (see Figure 2–6). The motion of the molecules holds the air up.

An instrument that measures air pressure is called a **barometer,** and the unit of measurement used in the United States is the **millibar.** Average sea-level pressure is about 1013 millibars. When pressure data are recorded on weather maps, it is then possible to draw **isobars,** lines connecting points of equal pressure. The pattern of the isobars reveals the horizontal distribution of pressure in the region under consideration. Prominent on such maps are roughly circular or oval areas that are characterized by "high" and "low" pressure. These highs and lows represent relative conditions, that is, pressure that is higher or lower than the surrounding areas (see Figure 2–7).

On most maps of air pressure, actual pressure readings are revised to represent a common elevation, usually sea level. This is done because pressure decreases rapidly with altitude, and significant variations in simultaneous pressure readings would be likely at different weather stations simply because of differences in elevation.

The Nature of Wind

The atmosphere is virtually always in motion. Air is free to move in any direction, its specific movements being shaped by a variety of factors. Some airflow is mild and brief; in other cases, it is strong and persistent. Atmospheric motions often are three-dimensional, involving both horizontal and vertical displacement. Small-scale vertical motions are normally referred to as **updrafts** and **downdrafts**; large-scale vertical motions are **ascent** and **subsidence.** The term **wind** is applied only to horizontal movements. Both vertical and horizontal motions are important in the atmosphere, but a much greater volume of air moves horizontally than vertically.

Winds tend to balance out the uneven distribution of air pressure over the Earth. The generalization is that air flows initially from areas of high pressure to areas of low pressure. Indeed, if the Earth did not rotate and if there were no such thing as friction, that is precisely what would happen—a direct movement of air from high pressure to low pressure. However, rotation and friction both exist, and they have significant influence on wind flow. The direction of wind movement is determined principally by the interaction of three factors: pressure gradient, the Coriolis effect, and friction (see Figure 2–8).

Pressure Gradient The basic activating force for wind is called **pressure gradient**. If pressure is higher on one side of a parcel of air than on the other, the parcel will be "pushed" away from the higher toward the lower

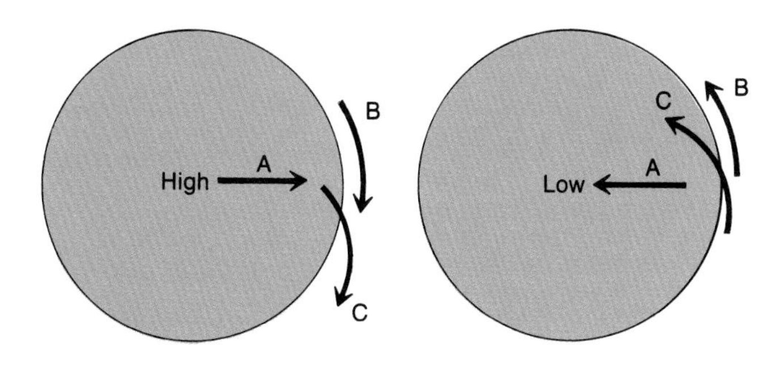

Figure 2–8. The direction of wind flow is determined by the balance among three factors: (A) The pressure gradient dictates that movement will be perpendicular to the isobars. (B) The Coriolis effect causes movement to parallel the isobars. (C) Friction determines that an intermediate course will be followed, moving across the isobars at an acute angle.

pressure. The pressure gradient force acts at right angles to the isobars in the direction of the lower pressure.

Coriolis Effect Any free-moving object on the Earth or in its atmosphere will drift sideways from the direction of movement (to the right in the Northern Hemisphere and to the left in the Southern Hemisphere) as a result of the Earth's rotation (see Figure 2–9). This phenomenon is called the **Coriolis effect**, and it has an important influence on the direction of wind flow. The practical result is that the wind would be shifted 90° from its pressure gradient path and would flow parallel to the isobars, providing no other factor impinged. In actuality, most winds in the troposphere follow this pattern; only near the ground is there another significant factor to further complicate the situation: friction.

Figure 2–9. The apparent deflection of the Coriolis effect is to the right in the Northern Hemisphere and to the left in the Southern Hemisphere. The dashed lines represent the gradient force; the solid lines represent actual movement.

Friction In the lower portion of the troposphere, **friction** appears. The frictional drag of the Earth's surface acts both to slow down wind movement and to modify its direction of flow. Instead of blowing directly across the isobars (in response to the pressure gradient force) or parallel to the isobars (due to the Coriolis effect), the wind takes an intermediate course between the two and crosses the isobars at a small angle.

Anticyclones and Cyclones

Distinct and predictable wind-flow patterns develop around all high-pressure and low-pressure centers. These patterns are determined by the combined influence of the factors discussed above: pressure gradient, Coriolis effect, and friction in the lower levels of the troposphere; and pressure gradient and Coriolis effect at higher elevations. Eight possible combinations result, four of which are known as cyclones, and the other four as anticyclones.

Anticyclones The combinations known as **anticyclones** represent high-pressure cells. They are described below and are illustrated in Figure 2–10.

1. In the friction layer (lower troposphere) of the Northern Hemisphere, there is a divergent clockwise flow, with the air spiraling out away from the center of the anticyclone.
2. Above the friction layer in the Northern Hemisphere, the winds move parallel to the isobars in clockwise fashion.
3. In the lower troposphere of the Southern Hemisphere, the pattern is a mirror image of example 1. The air diverges in a counterclockwise pattern.
4. Above the friction layer in the Southern Hemisphere, there is a counterclockwise movement parallel to the isobars.

Cyclones The other four combinations, known as **cyclones**, represent low-pressure cells. They are explained below and are also illustrated in Figure 2–10.

5. In the friction layer of the Northern Hemisphere, a converging counterclockwise flow exists.
6. Above the friction layer in the Northern Hemisphere, air movement parallels the isobars in a counterclockwise direction.
7. In the lower troposphere of the Southern Hemisphere, the winds move inward in a clockwise spiral.
8. In the upper air of the Southern Hemisphere, there is clockwise flow paralleling the isobars.

In addition to clockwise and counterclockwise motion, air flows follow certain vertical patterns within anticyclones and cyclones. Generally air descends in an-

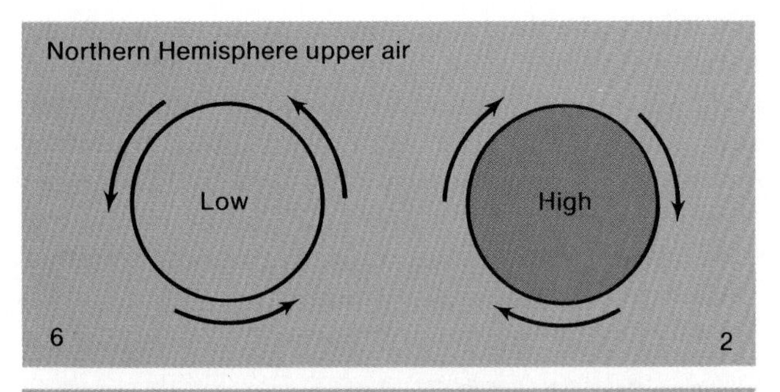

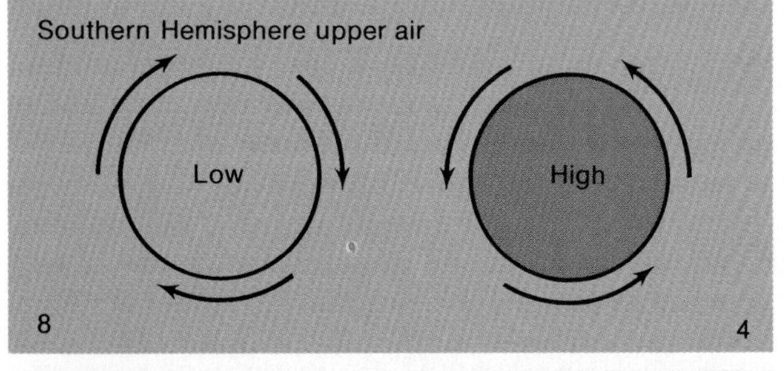

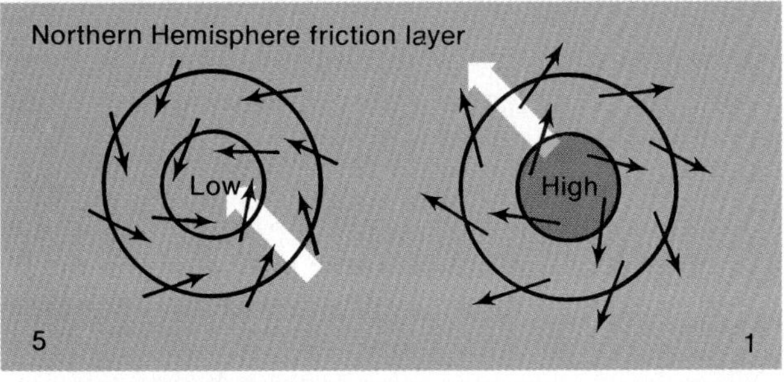

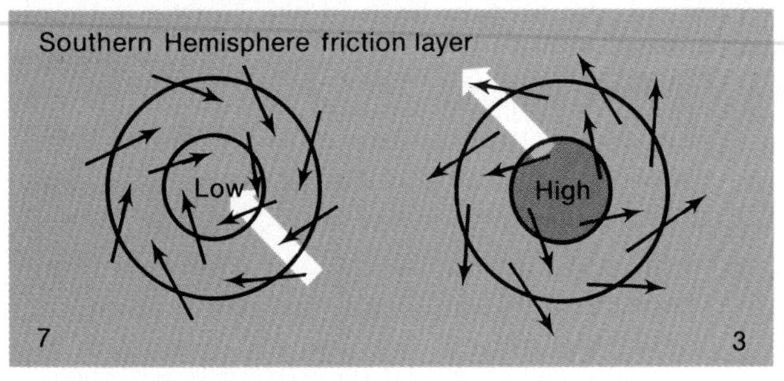

Figure 2–10. The eight basic patterns of air circulation around pressure cells.

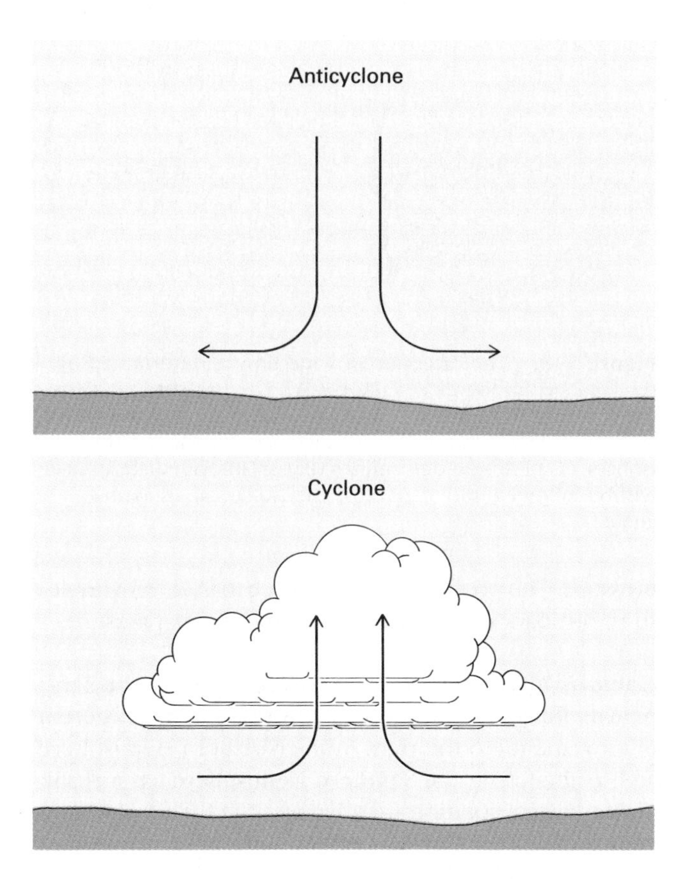

Figure 2–11. In an anticyclone (high-pressure cell), air subsides and diverges; in a cyclone (low-pressure cell), air converges and rises.

ticyclones and rises in cyclones, as pictured in Figure 2–11. The anticyclonic pattern can be visualized as upper air sinking down into the center of the high and then diverging near the surface. Opposite conditions prevail in a low-pressure cell, with the air converging horizontally into the cyclone and then rising.

The General Circulation of the Atmosphere

The atmosphere is an extraordinarily dynamic medium. It is constantly in motion, responding to the various forces described previously, as well as to a variety of more localized conditions. Some atmospheric motions are broad scale and sweeping; others are minute and momentary. Most important to an understanding of geography is the general pattern of circulation, which involves major semipermanent conditions of both wind and pressure. This circulation is the principal mechanism for both longitudinal and latitudinal heat transfer. Only the global pattern of insolation plays a greater role in determining world climates (see Figure 2–12).

The general pattern of atmospheric circulation has seven surface components, which are replicated on both sides of the equator. They are listed here in the north-to-south sequence of their occurrence in the Northern Hemisphere:

- Polar high pressure
- Polar easterlies
- Subpolar low pressure
- The westerlies
- Subtropical high pressure
- Trade winds
- Intertropical convergence zone
- Trade winds
- Subtropical high pressure
- The westerlies

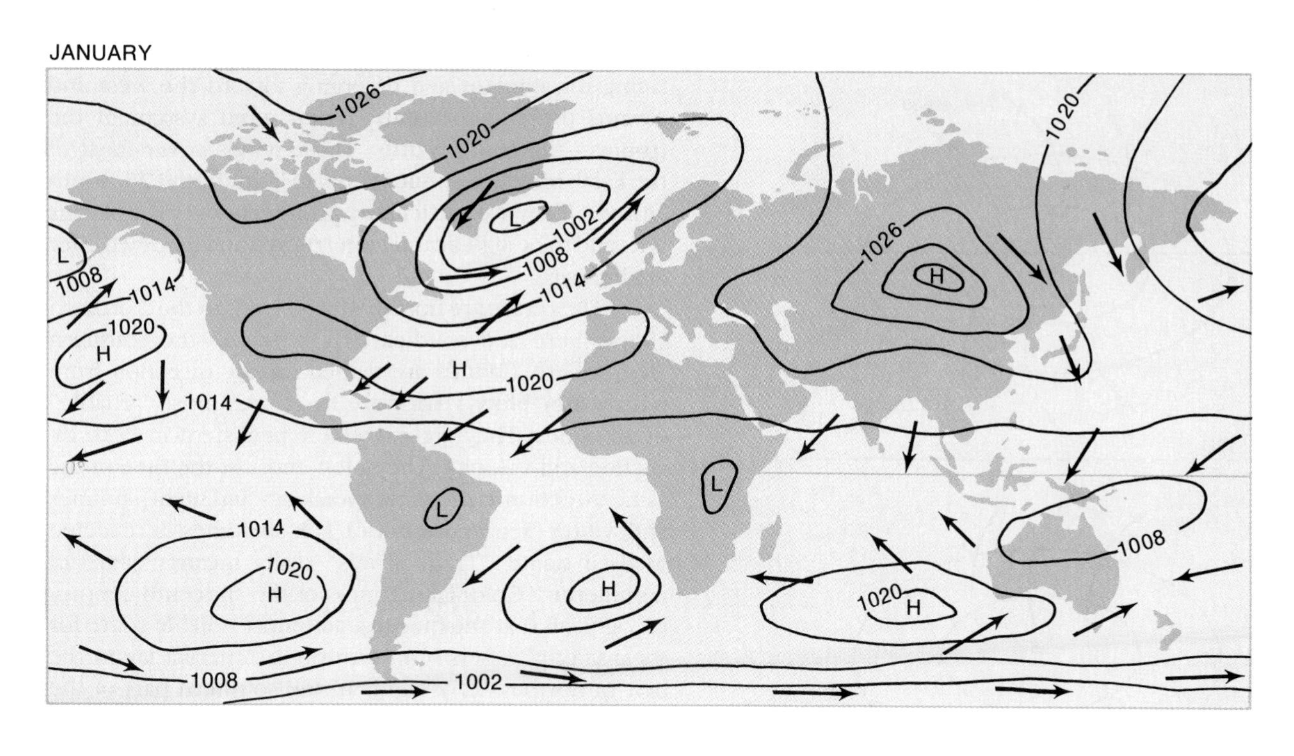

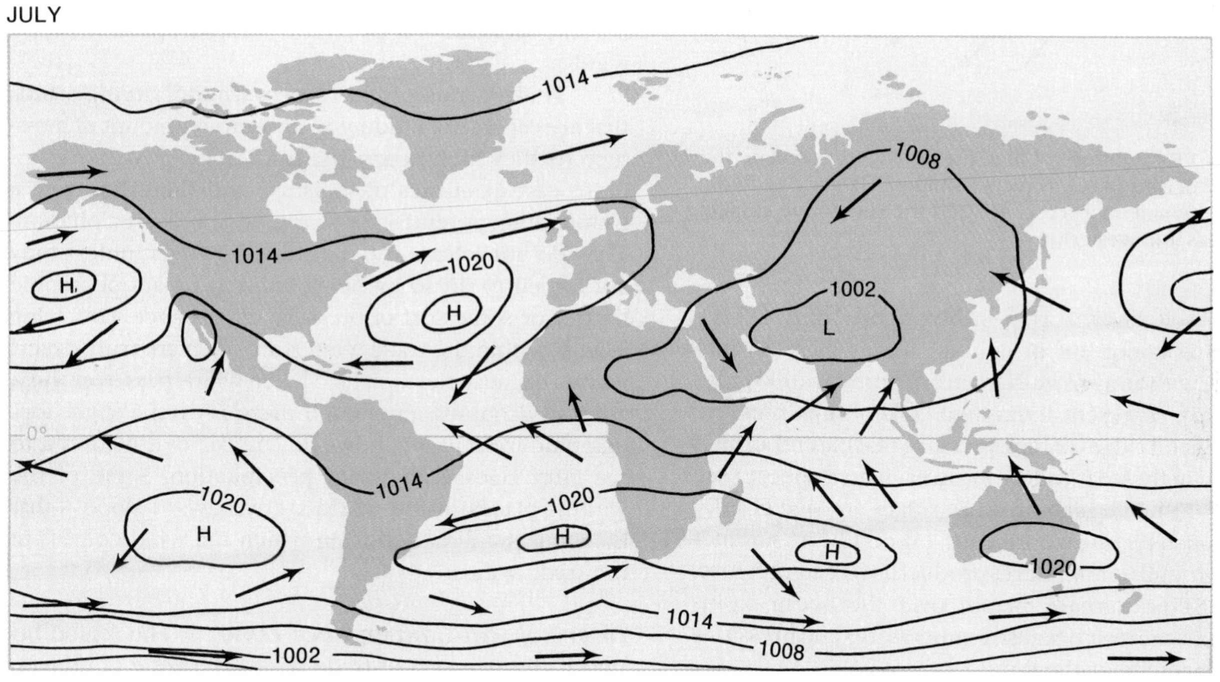

Figure 2–12. Average atmospheric pressure and wind conditions in January and July. Pressure is reduced to sea level and shown in millibars. Arrows indicate generalized surface wind movements.

- Subpolar low pressure
- Polar easterlies
- Polar high pressure

Tropospheric circulation can be thought of as a closed system, with neither a beginning nor an end, so its description can be initiated almost anywhere within the system. It seems logical to begin a discussion of atmospheric circulation as we did with oceanic circulation, in the subtropical latitudes of the five major ocean basins.

Subtropical High Pressure Each of the five basins (North Pacific, South Pacific, North Atlantic, South Atlantic, South Indian) has a large semipermanent high-

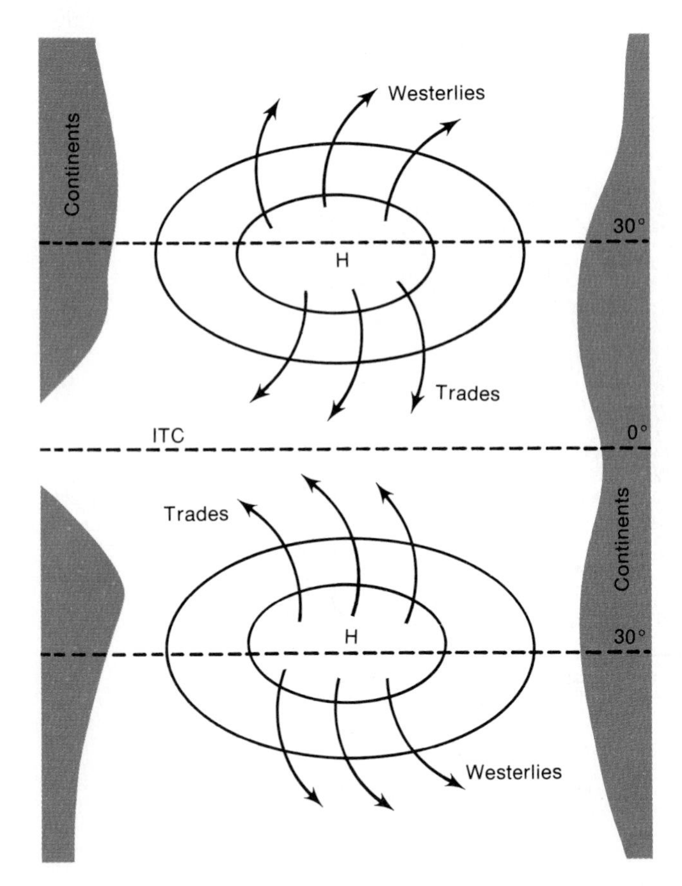

Figure 2–13. Subtropical high-pressure cells dominate the five major oceans in subtropical latitudes. They are centered at 25° to 30° and, in a sense, serve as the source for surface trade winds and westerlies.

pressure cell known as a **subtropical high (STH)**, centered at about 30° of latitude. These gigantic anticyclones, with an average diameter of perhaps 2000 miles (3200 km), represent intensified cells of high pressure in two general ridges of high pressure that extend around the world in these latitudes, one in each hemisphere (see Figure 2–13). The high-pressure ridges are significantly broken up over the continents, especially in summer when high land temperatures produce lower air pressure, but the STHs normally persist over the ocean basins throughout the year because temperatures and pressures there remain about the same.

Within the STHs, subsiding air produces weather that is nearly always clear, warm, and calm. The air circulation pattern around the STHs is typically anticyclonic: divergent clockwise in the Northern Hemisphere and divergent counterclockwise in the Southern Hemisphere. In essence the STHs can be thought of as gigantic wind-wheels whirling in the lower troposphere, fed with air sinking down from above and spinning off winds horizontally in all directions, but particularly on the northern and southern sides of the anticyclones. Indeed, the STHs are the source regions for two of the world's three major surface wind systems: the trade winds and the westerlies.

Trade Winds Issuing from the sides of the STHs facing the equator and diverging toward the west and toward the equator is the major wind system of the tropics—the **trade winds**. These winds cover most of the Earth between about 25° north latitude and 25° south latitude. They are particularly prominent over the oceans but tend to be significantly interrupted and modified over landmasses.

The trades are northeasterly winds in the Northern Hemisphere and southeasterly winds in the Southern Hemisphere. (Winds are named for the direction from which they blow.) They are by far the most "reliable" of all winds. They are extremely persistent in both direction and velocity. They blow most of the time in the same direction at the same speed, day and night, summer and winter (see Figure 2–14). This steadiness is reflected in their name: "Trade winds" really means "winds of commerce." Colonial mariners of the sixteenth century recognized that the quickest and most reliable route for their sailing vessels from Europe to America lay in the belt of northeasterly winds of the southern part of the North Atlantic Ocean. Similarly the trade winds were used by galleons in the Pacific Ocean, and the name became generally applied to these tropical easterly winds.

Trade winds originate as warming, drying winds that are capable of holding an enormous amount of moisture. As they blow across the tropical oceans, they evaporate vast quantities of moisture and therefore have a tremendous potential for storminess and precipitation. They do not release the moisture, however, unless they are forced to do so by being uplifted by a topographic barrier or some sort of pressure disturbance. Low-lying islands within the trade wind zone often are truly desert islands because the moisture-laden winds pass over them without dropping any rain. If there is even a slight topographic irregularity, however, the air that is forced to rise may release abundant precipitation. Some of the wettest places in the world are windward slopes (that is, facing the direction from which the wind comes) in the trade winds.

Intertropical Convergence Zone The region between the two sets of trade winds is a zone of convergence and weak horizontal airflow. The northeast trades and the southeast trades come together in the general vicinity of the equator, although the latitudinal position shifts northward and southward following the sun. This zone of meeting is usually called the **Intertropical Convergence Zone (ITC)**. The ITC is a low-pressure area that is associated with instability and rising air. Not surprisingly, then, it usually is characterized by towering clouds and frequent precipitation (see Figure 2–15).

The Westerlies The fourth component of the general atmospheric circulation is represented by the arrows

Figure 2–14. Many tropical coastal locations feel the incessant movements of the trades, as exemplified by these wind-blown palm trees at Cairns on the northeastern coast of Australia. (TLM photo)

that issue from the poleward sides of the STHs in Figure 2–13. This is the great wind system of the midlatitudes commonly called the **westerlies**. These winds flow basically from west to east around the world in the latitudinal zone between about 30° and 60° both north and south of the equator.

Discovery of these major wind belts (westerlies and trades) was enormously important in the days of sailing ships, and even today airlines take them into account. For example, Air France schedules almost an hour more to fly from Paris to New York (against the wind) than from New York to Paris (with the wind).

Near the surface, the westerlies are much less constant and persistent than the trades, which is to say that surface winds often do not actually flow from the west but may come from any point of the compass. At higher altitudes, however, the winds blow very prominently from the west. Moreover, a remarkable core of high-velocity winds, called the **jet stream**, usually occupies a position 30,000 to 40,000 feet (9 to 12 km) high in the westerlies. The belt of the westerlies can be thought of as a meandering "river" of air moving generally from west to east around the world in the midlatitudes, with the jet stream as its fast-moving core. The core, however, is not in the center of the westerlies; it is displaced poleward and is usually called the **polar front jet stream** because of its location near the polar front.

This jet stream core of the westerlies shifts its position with some frequency, which has considerable influence on the trajectory of general airflow in the mid-

Figure 2–15. The Intertropical Convergence Zone is characterized by instability, vertical air movement, towering cumuliform clouds, and considerable rainfall.

latitudes. Although the basic direction of movement is west to east, the path frequently shifts to the north and south due to the effects of undulations (wavy movements) referred to as **Rossby waves** (see Figure 2–16). Rossby waves can be thought of as separating cold polar air from warmer tropical air. Thus, the position and orientation of the waves play a major role in determining midlatitude weather, especially in winter.

All things considered, no other portion of the Earth experiences such short-run variability of weather as the midlatitudes. These variations are not caused by the westerlies themselves but by the Rossby waves and by the migratory pressure systems and storms that are associated with westerly flow and that will be discussed later in this chapter.

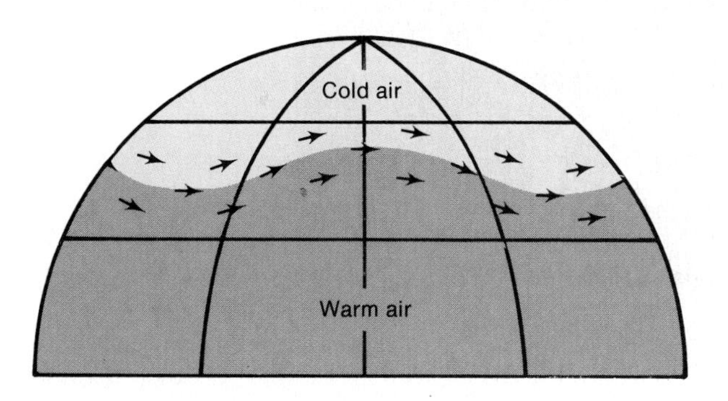

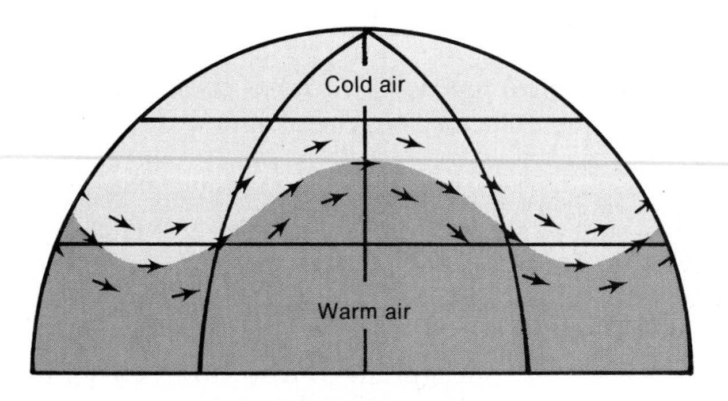

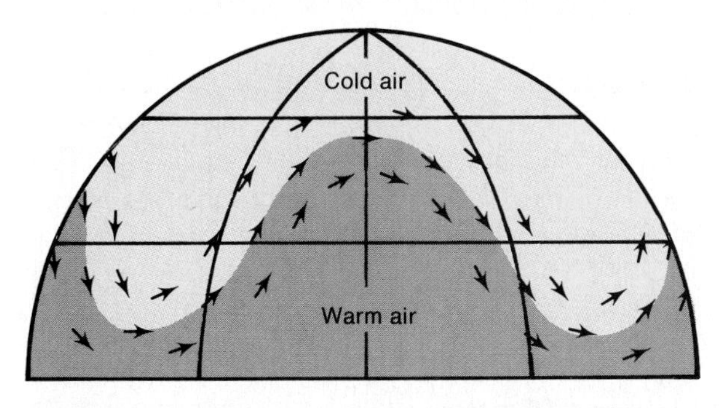

Figure 2–16. Rossby waves are major undulations in the general flow (particularly the upper-air flow) of the westerlies. When there are few waves, and their amplitude (north-south component of movement) is small, cold air usually remains poleward of warm air, as shown in diagram 1. When the waves have greater amplitude (as in diagram 3), cold air pushes equatorward, and warm air moves poleward.

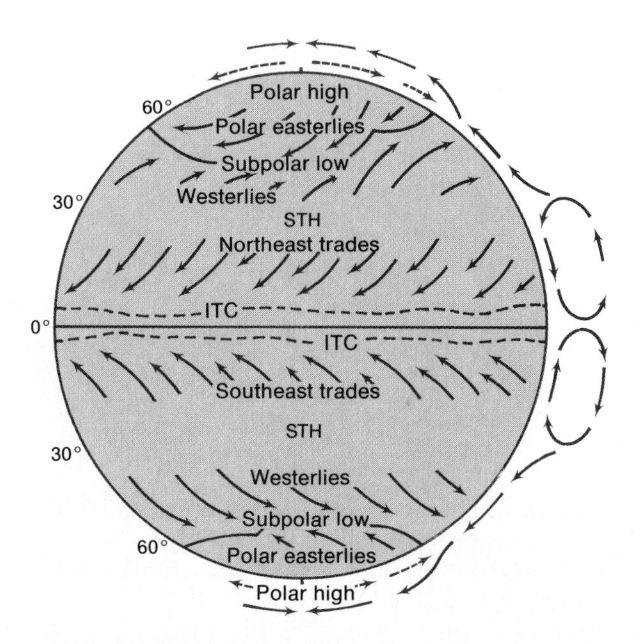

Figure 2–17. The general circulation of the atmosphere, disregarding the effect of major landmasses. (The vertical component is considerably exaggerated.)

Polar High Pressure Situated over both polar regions are high-pressure cells, called **polar highs**. Air movement associated with these cells is typically anticyclonic. Air from above sinks down into the high and diverges horizontally near the surface, clockwise in the Northern Hemisphere and counterclockwise in the Southern Hemisphere.

Polar Easterlies The third broad-scale global wind system occupies most of the area between the polar highs and about 60° of latitude. The winds move generally from east to west and are called the **polar easterlies**. They are typically cold and dry, although they vary in direction and strength.

Subpolar Low-Pressure Zone The final surface component of the general pattern of atmospheric circulation is a zone of low pressure at about 50° to 60° of latitude in both the Northern and Southern Hemispheres (see Figure 2–17). It is commonly called the **subpolar low** and often contains the **polar front**. This zone is a meeting ground and a region of conflict between the cold winds of the polar easterlies and the warmer westerlies. For this reason it is characterized by rising air, widespread cloudiness, precipitation, and generally unsettled or stormy weather conditions. Many of the migratory storms that travel with the westerlies originate in the polar front.

Monsoons

There are many variations to the generalized pattern of global wind and pressure systems discussed above, the most important of which are the monsoons. A **monsoon**

Figure 2–18. The principal monsoon areas of the world.

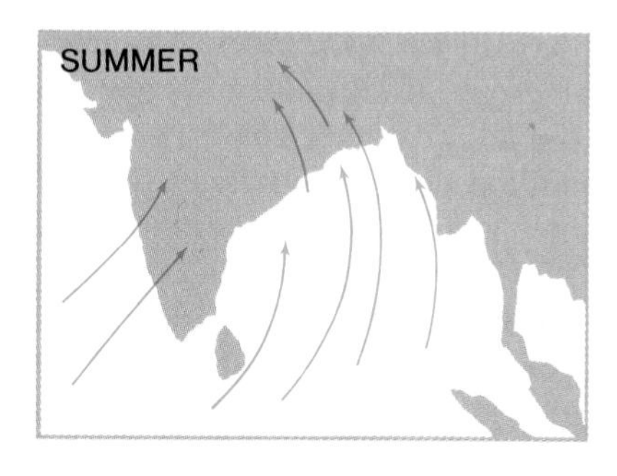

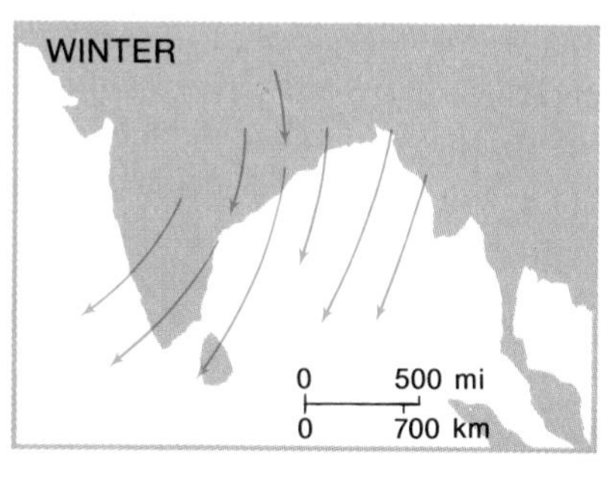

Figure 2–19. The South Asian monsoon is characterized by a strong onshore flow in summer and a somewhat less pronounced offshore flow in winter.

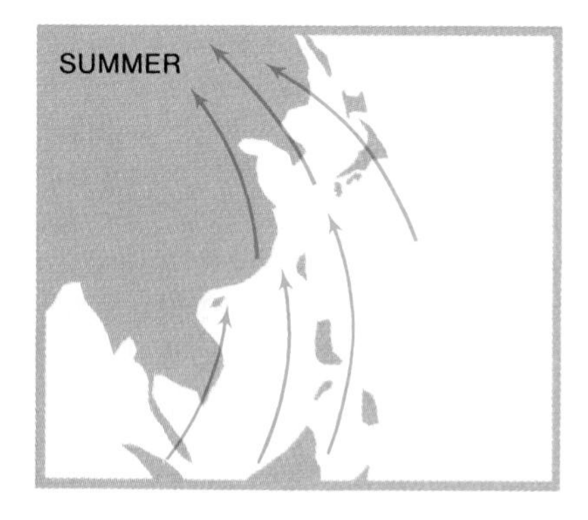

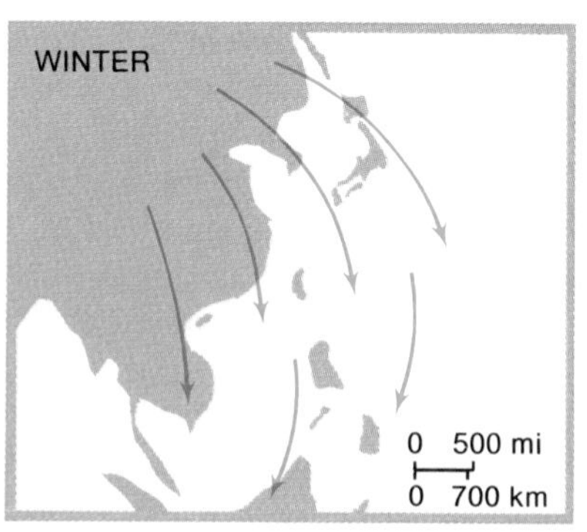

Figure 2–20. In East Asia, the outblowing winter monsoon is stronger than the onshore flow of the summer monsoon.

is characterized by a seasonal reversal of winds, accompanied by distinct precipitation patterns. During the summer, moist winds flow from the ocean toward the shore, bringing heavy rainfall. In contrast, during the winter the winds blow out to sea from the continents, which results in a pronounced dry season. The origins of the monsoons are not well understood, although evidence suggests that they are associated with occurrences in the upper atmosphere, particularly the jet stream.

It is difficult to overestimate the importance of monsoonal circulation to humankind. More than half of the world's population inhabits the regions in which climates are largely controlled by monsoons. Moreover, these are regions in which the majority of the populace depends on agriculture for its livelihood. The people's lives are intricately bound up with the monsoonal rains, which are essential for the production of both food and cash crops. The failure, or even late arrival, of monsoonal moisture inevitably causes widespread starvation and economic disaster.

The broad monsoon realm depicted in Figure 2–18 contains several circulatory systems. The two principal monsoons systems operate in South Asia and East Asia (see Figures 2–19 and 2–20). Although they overlap in Southeast Asia (the region that includes Vietnam, Thailand, and neighboring countries), they are essentially separate. Apparently the Himalayan mountain system serves as an effective barrier between them. Lesser systems are found in northern Australia and West Africa.

WATER VAPOR AND HUMIDITY

The fourth and final basic element of weather and climate is moisture. It occurs in the atmosphere in all three states in which matter can occur—as a solid (snow, sleet, hail, ice crystals in high clouds), liquid (rain, cloud droplets), and gas (water vapor).

Although we usually think of water as a liquid, it frequently exists in the gaseous state, where it is called **water vapor**. Water vapor is colorless, odorless, tasteless, invisible, and mixes with the other gases of the atmosphere. In terms of mass, it is a very minor constituent of the atmosphere, and its occurrence is quite variable. Most atmospheric water vapor is found within 1 mile (1.6 km) of the Earth's surface, and even there its distribution is scarce and uneven.

This erratic distribution is a reflection of the ease with which moisture can change from one state to an-

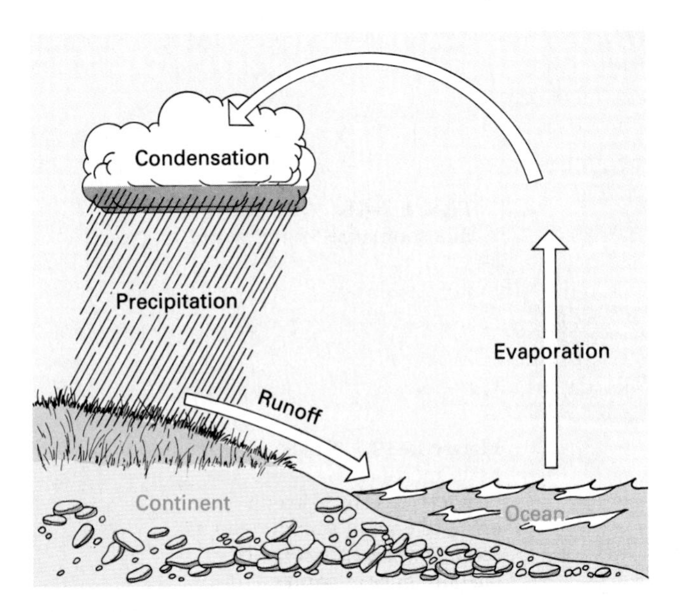

Figure 2–21. The hydrologic cycle is a continuous interchange of moisture between the atmosphere and the Earth.

other. Moisture can leave the Earth's surface as a gas (water vapor) and return to it as a liquid (water) or a solid (ice). Indeed, there is a continuous interchange of moisture between the Earth and the atmosphere, known as the **hydrologic cycle**, pictured in Figure 2–21. The essential feature of the hydrologic cycle is that liquid water (primarily from the oceans) evaporates into the gaseous air and subsequently is reconverted to a liquid or solid state and returns to the Earth as rain or some other form of precipitation.

The amount of water vapor in the air is referred to as **humidity**. It can be measured and expressed in several ways, each of which is useful for certain purposes. Two measures are especially useful: absolute humidity and relative humidity.

Absolute humidity is a quantitative measurement of the actual amount of water vapor in the air at a given time and place. It is expressed as the weight of water vapor in a given volume of air, normally in terms of grams of vapor per cubic meter of air.

Relative humidity is an expression of the amount of water vapor in the air in comparison with the total amount that could be there if the air were saturated. It is not a direct measure of quantity; rather it is a ratio that is expressed as a percentage. In essence, it is the percentage of saturation. Relative humidity is determined by the balance between vapor content and the air's capacity for holding moisture, which is in turn primarily dependent on its temperature. Warm air can hold more vapor than cool air; thus, an increase in temperature results in a decrease in relative humidity because the capacity for holding vapor is increased.

To understand the role of moisture in the atmosphere we must consider three processes that are basic to the hydrologic cycle: evaporation, condensation, and precipitation. The first two processes involve not only a change of state for moisture but also a transfer of heat. More specifically, evaporation requires and stores heat, and condensation releases it.

Evaporation

The conversion of moisture from liquid to gas—from water to water vapor—is called **evaporation**. This process involves molecular escape; molecules of water become detached from the liquid surface and escape into the overlying or surrounding air. This activity is speeded up by heat; that is, warm air produces more evaporation than cool air.

The energy absorbed by the escaping molecules is stored, normally to be released subsequently when the vapor changes back to a liquid. The stored energy is called *latent heat of vaporization*. The heat thus removed with the escaping molecules is subtracted from the remaining water, reducing its temperature in an activity often referred to as *evaporative cooling*. The effect of evaporative cooling is experienced when a swimmer leaves the water on a dry, warm day. The dripping wet body immediately loses moisture and heat through evaporation to the surrounding air, causing the swimmer to feel "cooled off." Thus, the amount and rate of evaporation from a water surface depends essentially on two factors: the temperature and the degree of saturation of the air.

Condensation

Condensation is the opposite of evaporation. It is the process whereby water vapor in the atmosphere is converted to water. It is a change of state from gas to liquid. For condensation to take place, the air must be saturated. This is nearly always accomplished by the cooling of the air to the point of saturation, known as the **dew point**. In addition, it is necessary to have a surface on which the condensation can take place. If no such surface is available, no condensation will occur.

Normally, plenty of surfaces are available for condensation. At ground level this obviously is no problem. In the air above the ground there is also usually an abundance of "surfaces," as represented by tiny particles of dust, smoke, bacteria, salt, pollen, and other compounds. These compounds are referred to as **condensation nuclei**, and they serve as collection centers for water molecules. These particles generally are too small to be seen with the naked eye, but they occur in countless billions in the atmosphere (Figure 2–22).

When the temperature cools to the dew point, water vapor molecules begin to condense around these nuclei. The droplets grow rapidly by adding increasing amounts of moisture, and as they become larger, they may bump into one another and coalesce. Continued

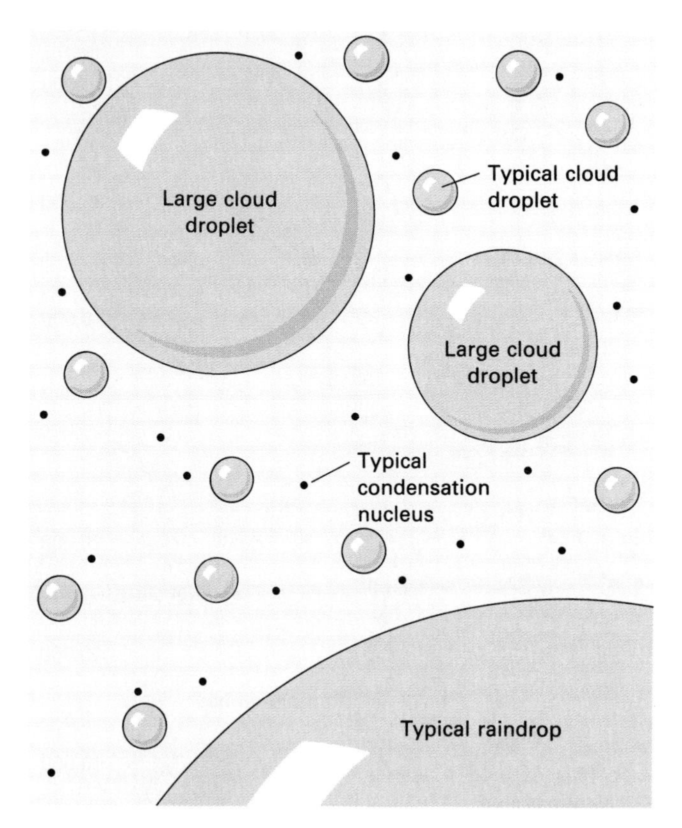

Figure 2–22. Comparative sizes of condensation and precipitation particles.

growth can make them large enough to be visible, forming haze or cloud particles.

Adiabatic Activities The only way in which large masses of air can be cooled to the dew point is by expansion in rising air, which is called **adiabatic cooling**. Thus, the only prominent mechanism for the development of clouds and the production of rain is through such adiabatic cooling. As noted previously, when air rises it reaches lower pressure, so it expands and cools. If it rises sufficiently, it cools to the dew point. This permits condensation to begin and clouds to form. As a parcel of air rises, it cools at the relatively steady rate of 5.5°F per 1000 feet (10°C per km). This is known as the *dry adiabatic lapse rate* (see Figure 2–23). As the rising air cools, its capacity for holding moisture decreases and its relative humidity increases. If it cools enough to reach saturation, condensation will begin.

As soon as condensation begins, latent heat is released. If the air continues to rise, cooling will continue, but release of the latent heat will reduce the rate of cooling. This diminished rate of cooling is at the *wet adiabatic lapse rate*, which averages about 2°F per 1000 feet (6.5°C per km).

Adiabatic warming occurs when air is caused to descend. Any descent immediately produces warming through compression as the air comes under higher pressure. This increases the capacity of the air for holding moisture and produces an unsaturated condition. Therefore, any descending air will warm at the dry adiabatic lapse rate.

Clouds Most atmospheric activities are invisible to the human eye. Clouds, on the other hand, are not only prominent but are sometimes spectacular. Clouds serve as the visible expression of condensation and often provide perceptible evidence of other things that are happening in the atmosphere. The basic importance of clouds is that they serve as the source of precipitation. Not all clouds precipitate, but all precipitation comes from clouds. Moreover, clouds serve as a radiation buffer between Earth and the sun.

A cloud consists of a visible collection of minute droplets of water or tiny crystals of ice. With a few exceptions, cloud formation is a result of adiabatic cooling in rising air. The base of a cloud often is sharp and clear-cut; it marks the lowest level at which condensation began. The cloud then grows upward and outward from this base.

Clouds are good indicators of the state of the atmosphere, providing at a glance some understanding of the current weather and often serving as harbingers of future conditions. On the basis of two factors—appearance and altitude—clouds can be classified into types, which is a useful tool in interpreting the atmosphere. Figure 2–24 shows the major cloud types and their locations in the atmosphere.

Cirriform clouds are thin and wispy, composed of ice crystals rather than water particles; they are found

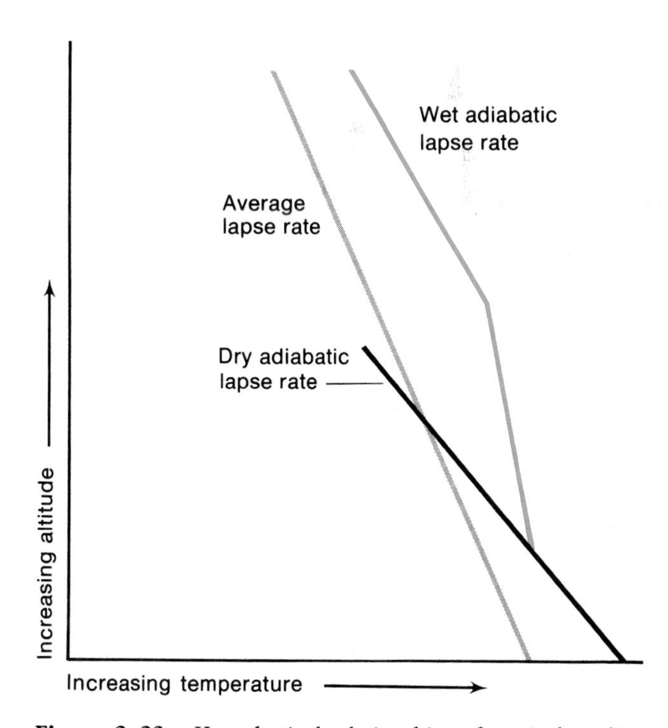

Figure 2–23. Hypothetical relationships of vertical cooling rates Unsaturated rising air cools at the dry adiabatic lapse rate. Saturated rising air cools at the wet adiabatic lapse rate. Nonrising air experiences vertical cooling at various rates, generalized as the average lapse rate.

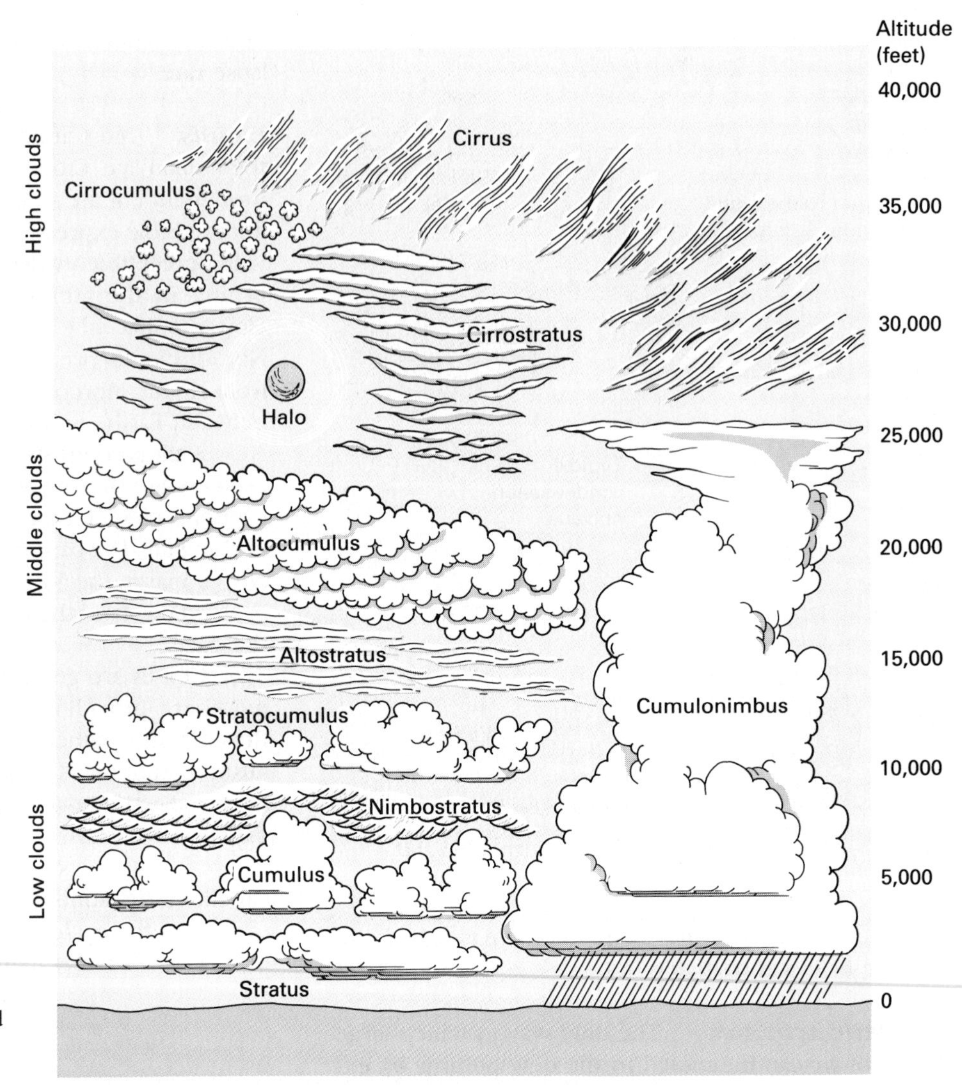

Figure 2–24. Idealized diagram of typical shapes and altitudes of the principal cloud types.

at high elevations. **Stratiform clouds** appear as grayish sheets or layers that cover most or all of the sky, rarely being broken up into individual cloud units. **Cumuliform clouds** are massive and rounded, usually with a flat base and limited horizontal extent, but often billowing upward to great heights.

Precipitation comes only from clouds that have the root *nimb* in their name, specifically *nimbostratus* or *cumulonimbus*. Normally these types develop from other types: that is, cumulonimbus develops from cumulus, and nimbostratus from stratus.

Fog is simply a cloud on the ground, that is, a cloud whose base is at or very near ground level. From a global standpoint, fogs represent a minor form of condensation. From a human standpoint, however, they can be very important because they can hinder visibility enough to make surface transportation hazardous or even impossible.

The Buoyancy of Air

We have seen that most condensation and almost all precipitation are the results of rising air. It is also clear that under certain circumstances air rises more freely and more extensively than at other times. This involves the **buoyancy**, or **stability**, of the parcel of air.

As with other gases, air tends to seek its own level. This means that a parcel of air will move vertically until it reaches a level at which the surrounding air is of equal density. If a parcel of air is warmer, and thus lighter, than the surrounding air, it will tend to rise. If the parcel is cooler than the surrounding air, it will tend to sink or at least to resist uplift.

Air can be forced to rise by various environmental conditions. If it resists vertical movement, it is said to be *stable*. If stable air is forced to rise, it will do so only as long as the force is applied. If the impelling force is removed, stable air will sink back to its former position. Air is said to be *unstable* if it rises without an impelling force or after the force has ceased to function.

In the atmosphere, stability is promoted when cold air is located beneath warm air, a condition most prominently displayed when a temperature inversion is in effect. When colder, denser air is situated below warmer, lighter air, upward movement is unlikely. The air resists vertical displacement. For this reason, a cold winter night

TABLE 2–1 ***The General Relationship of Buoyancy and Atmospheric Moisture***

Buoyancy	Clouds	Precipitation
Stable air	None or stratiform	None or drizzle
Unstable air	None or cumuliform	None or showers

is a typical stable air situation, although stability can also occur in the daytime. Stable air provides little opportunity for adiabatic cooling unless there is some sort of forced uplift. Thus, stable air is normally not associated with cloud formation and precipitation.

When a parcel of air is heated and becomes warmer than the surrounding air, it will become unstable. This is a typical condition on a warm summer afternoon. Unstable air will rise until it reaches an altitude where the surrounding air has the same temperature. While ascending, it will be cooled adiabatically. In this situation, clouds are likely to form.

The general relationship of air buoyancy to atmospheric moisture is summarized in Table 2–1.

PRECIPITATION

Condensation readily forms clouds, and all precipitation originates in clouds. However, most clouds do not yield precipitation. Even so, rain and other forms of precipitation are commonplace in the troposphere. What is it, then, that produces precipitation?

An average-sized raindrop contains several million times as much water as an average-sized cloud particle. Consequently, vast numbers of cloud particles must join together to form a drop large enough to overcome both turbulence (strong, irregular air currents) and evaporation and thus be able to fall to Earth. Two separate mechanisms are believed to be principally responsible for producing precipitation particles: *ice-crystal formation and collision/coalescence.*

Many clouds or portions of clouds extend high enough in the atmosphere to have temperatures well below the freezing point. These clouds therefore contain numerous ice crystals. The ice crystals serve as condensation nuclei and attract vapor to condense around them. If they become heavy enough to fall, they will pick up more moisture as they descend through the lower, warmer portions of the cloud (see Figure 2–25). They may then precipitate from the cloud as snowflakes, or they may be melted and fall as raindrops.

In many cases, particularly in the tropics, clouds have temperatures too high for the formation of ice crystals. In such clouds, rain is produced by the collision and merging of water droplets. The role of condensation is to enable droplets to grow large enough for the cloud to reach the coalescence stage. Different-sized droplets

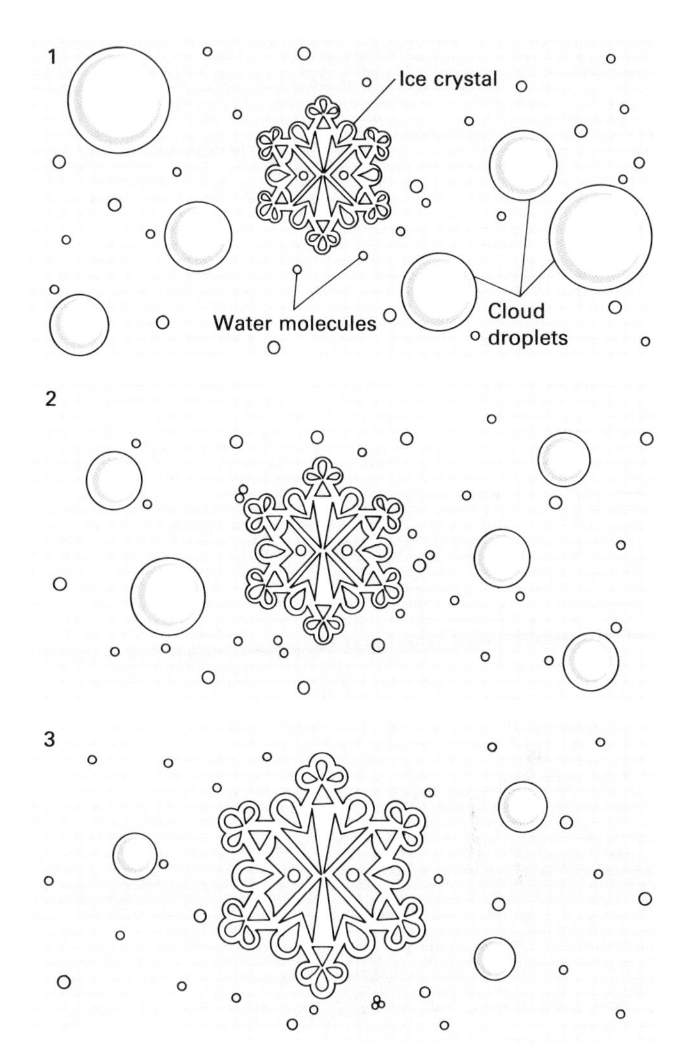

Figure 2–25. Precipitation by means of ice crystal formation in clouds. Ice crystals grow by attracting water vapor to them, causing cloud droplets to evaporate, thereby replenishing the water vapor supply. The process of growing ice crystals and shrinking cloud droplets may continue until the ice crystals are large and heavy enough to fall. Particle sizes are greatly exaggerated in this and the next figure.

fall at different velocities. The larger ones fall faster (due to greater gravitational attraction), overtaking and often coalescing with smaller ones, which are swept along in the descent (see Figure 2–26). This sequence of events favors the continued growth of the larger particles. The larger they grow, the faster they grow.

Forms of Precipitation

Several forms of precipitation can result from the processes just described. The form that is produced depends largely on the temperature of the air and on its degree of turbulence.

Rain is by far the most common and widespread form of precipitation. It consists of drops of liquid water. Most rain is the result of condensation and precipitation in ascending air that has temperatures above freezing,

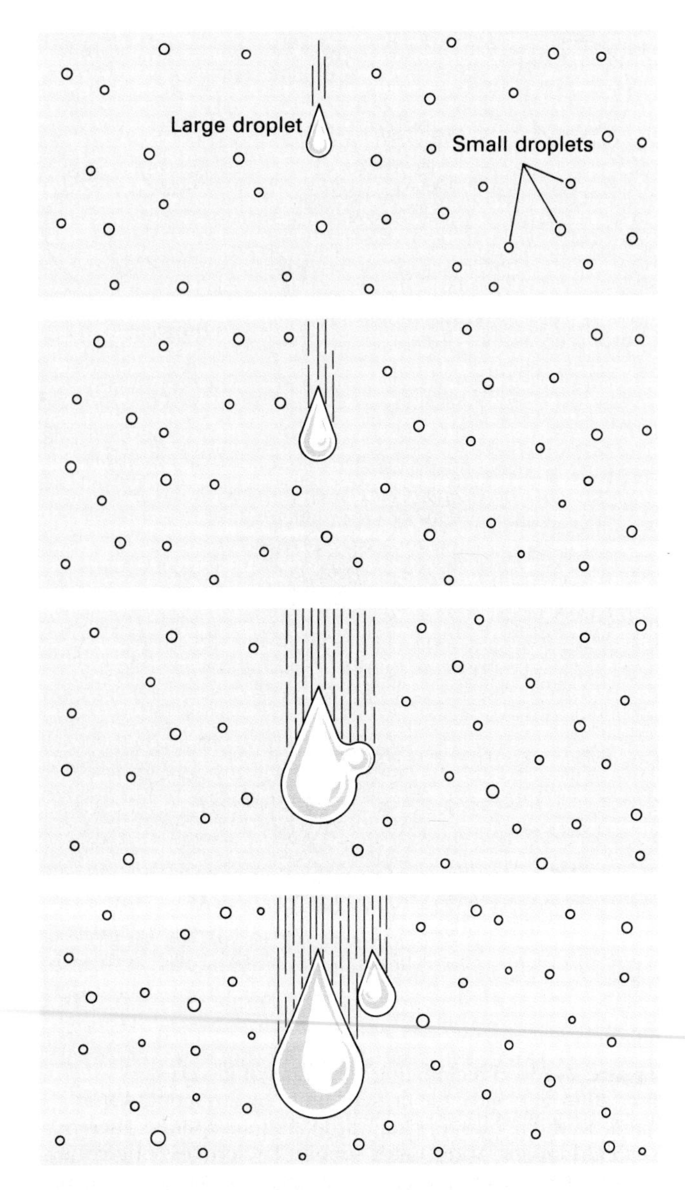

Figure 2–26. The process of collision/coalescence. Large droplets fall more rapidly than small ones, coalescing with some and sweeping others along in their descending path. As droplets become larger during descent, they sometimes break apart.

but some rain results from the thawing of ice crystals as they descend through warmer air.

Snow is the general name given to solid precipitation in the form of ice crystals, small pellets, or flakes. It is formed by direct conversion of water vapor into ice, without an intermediate water stage. **Sleet,** in contrast, is made up of small raindrops that freeze during descent.

Hail consists of rounded or irregular pellets or lumps of ice. It is produced in cumulonimbus clouds as a result of active turbulence and vertical air currents (see Figure 2–27). The lower part of the cloud provides liquid water, the upper part of the cloud supplies the refrigeration that freezes the water, and strong updrafts lift the droplets so that they can collide with the growing hailstones. The updrafts carry water droplets or small ice particles upward, where they grow by collecting moisture from supercooled cloud droplets. When the particles become too large to be supported in the air, they fall, adding more moisture on the way down. If they encounter another strong updraft they may be carried skyward again, only to fall another time. This sequence may be undergone repeatedly.

Types of Atmospheric Lifting and Precipitation

Air must rise for any significant amount of precipitation to originate. What are the natural forces that can cause this to occur? Four principal types of atmospheric lifting are recognized, although two or more of them often operate together: convective, orographic, frontal, and convergent lifting (see Figure 2–28).

Convective Lifting **Convective lifting** results from the unequal heating of different surface areas. Air above the warmest surface may then be heated by conduction more than the air around it. The heated air expands and rises vertically toward lower pressure. This spontaneous uplift is particularly notable if unstable air is involved, which is often the case on a warm summer day. If sufficient adiabatic cooling occurs, condensation will begin, and a cumulus cloud will form. With proper temperature, humidity, and stability conditions, the cloud is likely to grow into a towering cumulonimbus thunderhead, with a downpour of showery rain or hail, accompanied sometimes by lightning and thunder.

Orographic Lifting **Orographic lifting** is a forced ascent of air over a topographic barrier. Adiabatic cooling is an invariable result, and if the dew point is reached,

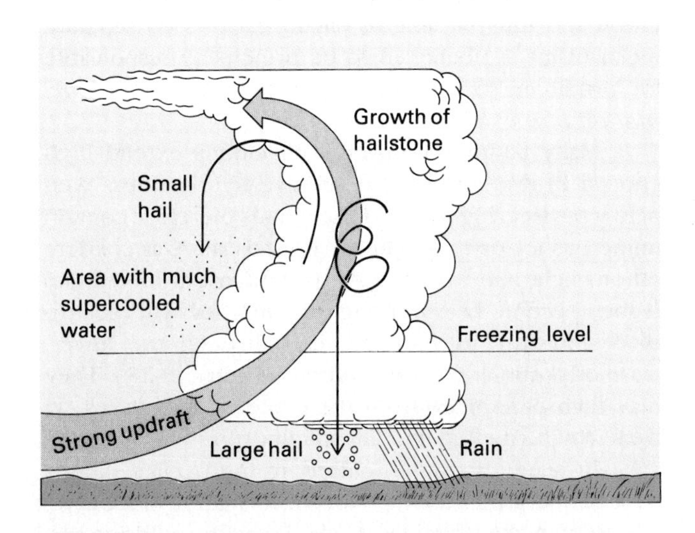

Figure 2–27. Small hailstones form in a relatively short time; large hailstones have spent a longer time in formation. (After John Oliver, *Climatology: Selected Applications.* New York: V. H. Winston and Sons, 1981, p. 140.)

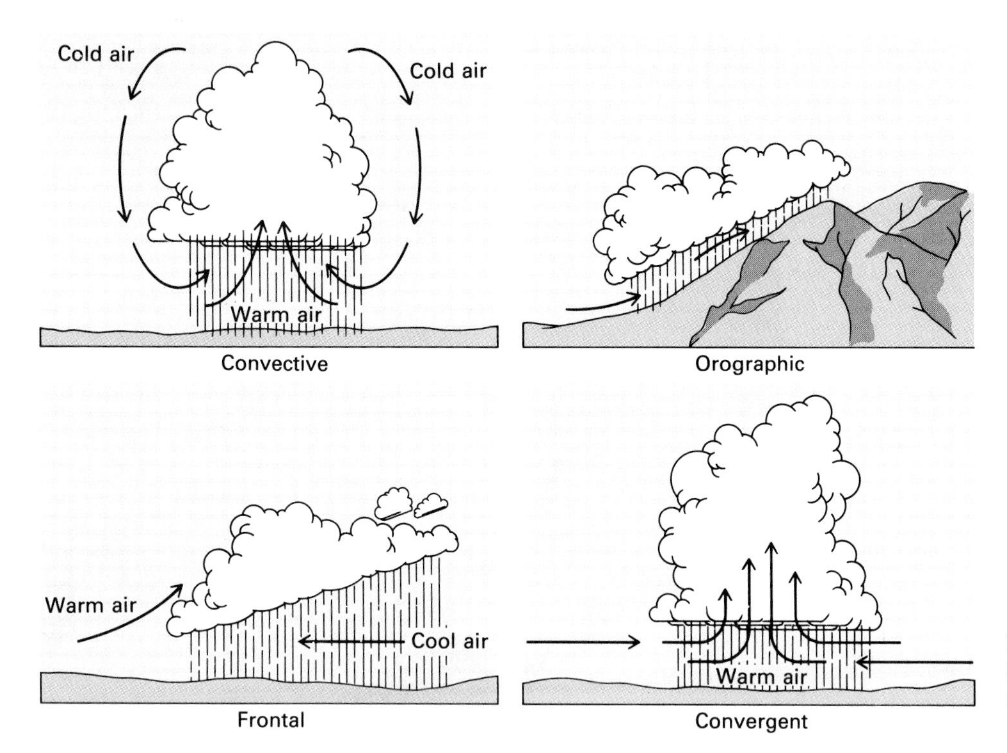

Figure 2–28. The four basic types of atmospheric lifting and precipitation.

precipitation will be produced. If significant instability has been triggered by the upslope movement, the air will keep rising when it reaches the top of the slope, and the precipitation will continue.

Frontal Lifting **Frontal lifting** is created when unlike air masses come together. They do not mix; instead, the warmer air always rises over the colder air. As the warmer air is forced to rise, it may be cooled to the dew point, with resulting clouds and possible precipitation.

Convergent Lifting Whenever air converges, it results in a general uplift because of the crowding. This enhances instability, and it is likely to produce showery precipitation. Although **convergent lifting** is less common than the other three types, it is significant in some situations. Convergent precipitation is particularly characteristic of the low latitudes, particularly in the Intertropical Convergence Zone, and in such tropical disturbances as hurricanes and easterly waves.

Global Distribution of Precipitation

As with other features of the environment, precipitation occurs unevenly over the Earth's surface. To a large extent, precipitation patterns are based on latitude, but many other factors are involved, and the overall pattern is very complex. To highlight these patterns, the accompanying world maps use **isohyets**, lines joining points of equal quantities of precipitation.

The most conspicuous feature of the worldwide precipitation pattern is that the tropical latitudes contain most of the wettest areas (see Figure 2–29). The warm trade winds carry enormous amounts of moisture, and where they are forced to rise, very heavy rainfall usually is produced. Equatorial regions, particularly, reflect these conditions, as warm, moist, unstable air is uplifted in the ITC, where warm ocean water is easier to vaporize. Also, where trade winds are forced to rise by topographic obstacles, considerable precipitation may result. Because the trades are easterly winds, this orographic effect is most pronounced on the east coasts of tropical landmasses, for example, the east coast of Central America, northeastern South America, and Madagascar. Where the normal trade-wind pattern is modified by monsoons, the onshore trade-wind flow may completely reverse its direction. Thus, the wet areas on the west coast of Southeast Asia, India, and the Guinea Coast of West Africa are caused by the onshore flow of southwesterly winds that are nothing more than trade winds being diverted from a "normal" pattern by the South Asian and West African monsoons.

The only other regions of conspicuously high precipitation are narrow zones along the west coasts of North and South America between 40° and 60° of latitude. These areas reflect a combination of frequent onshore westerly airflow, considerable storminess, and mountain barriers athwart the direction of the prevailing westerly winds. The presence of these north-south mountain ranges near the coast restricts the precipitation to a relatively small area and creates a pronounced rain-shadow effect (lack of precipitation on the downwind side of a mountain range) to the east of the ranges.

The principal regions of sparse precipitation on the world map are found in three types of locations.

1. In subtropical latitudes, dry lands are most prominent on the western sides of continents. High-pres-

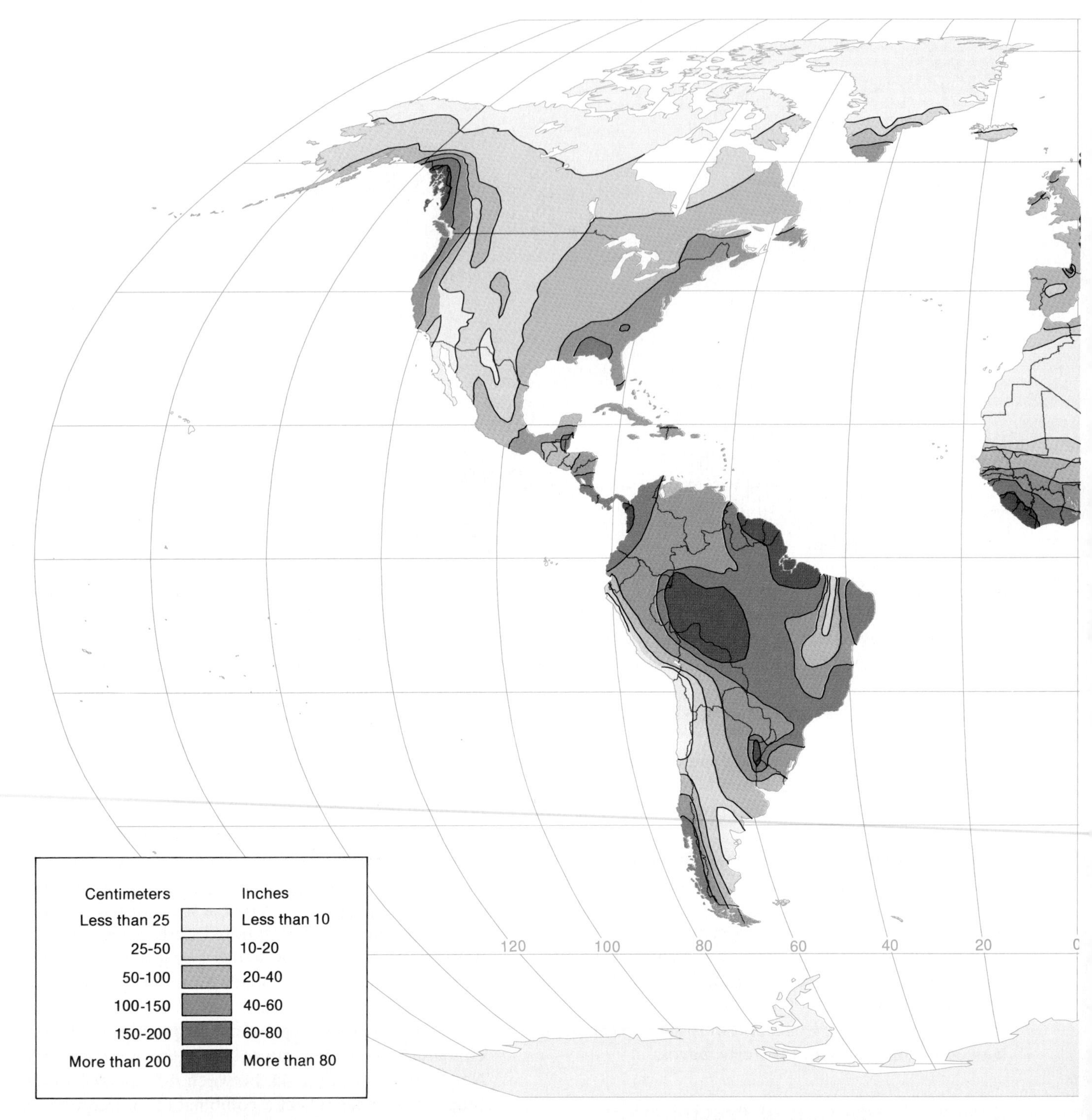

Figure 2–29. Average annual precipitation.

sure conditions dominate at these latitudes, which makes condensation and precipitation unlikely. In North Africa and Australia these dry zones are quite extensive, primarily because of the blocking effect of landmasses or highlands to the east, which inhibits any moisture being brought in from that direction.

2. Dry regions in the midlatitudes are most extensive in central and southwestern Asia, but they also occur in western North America and southeastern South America. In each case, the dryness is due to lack of access for moist air masses because of topographic barriers that block air flow from the ocean.
3. In the high latitudes there is not much precipitation anywhere. Water surfaces are scarce and cold, so little opportunity exists for evaporation of moisture into the air. Accordingly, polar air masses have low absolute humidities, and precipitation is slight. These regions can be referred to accurately as "cold deserts."

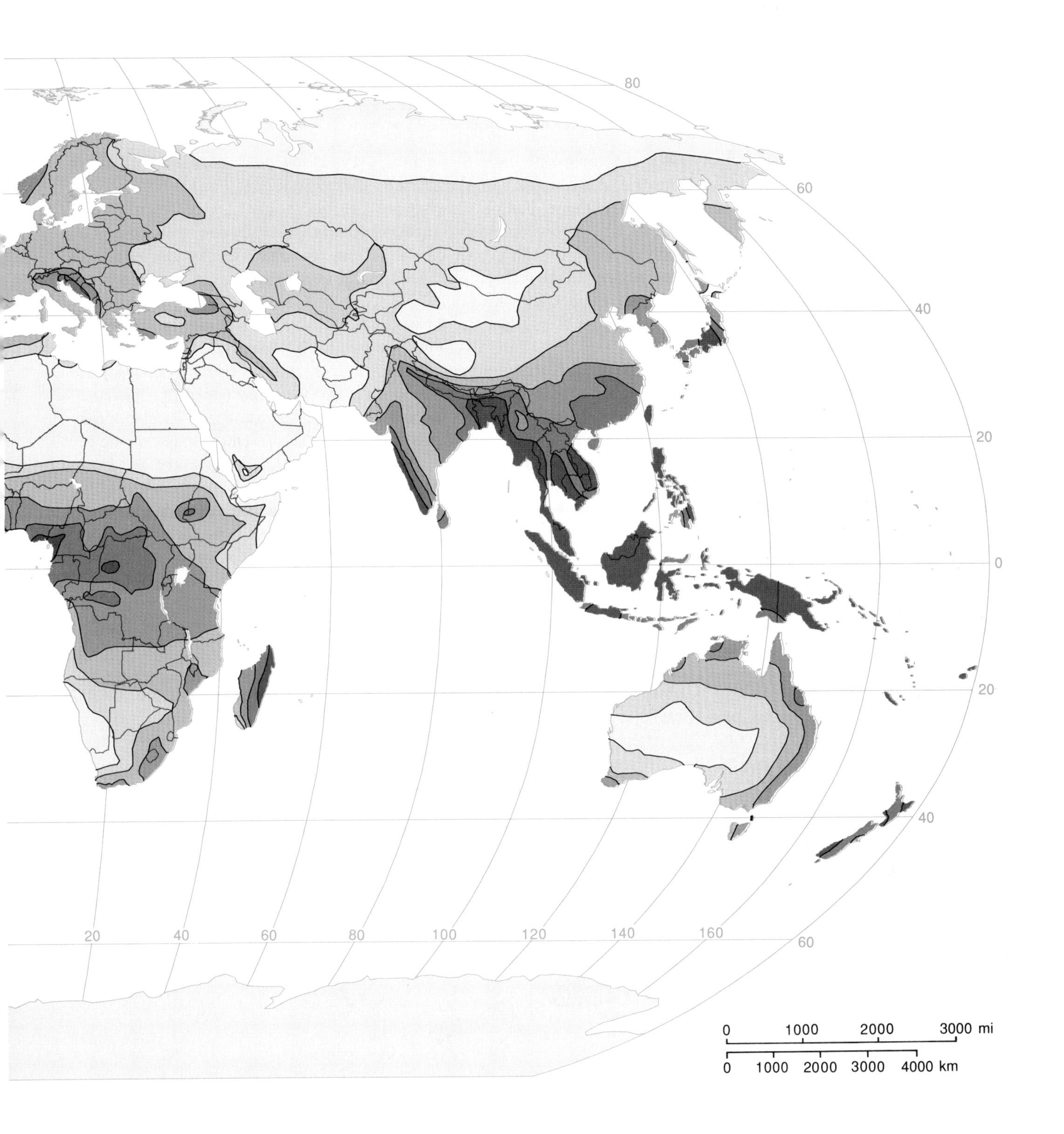

To understand climate properly, we must be aware of seasonal as well as annual precipitation patterns. Most parts of the world experience considerable variation from summer to winter in the amount of precipitation received. This variation is most pronounced over continental interiors, where strong summer heating at the surface induces greater instability and the potential for greater convective activity. Thus, most of the annual precipitation occurs during summer months, and winter is generally a time of anticyclonic conditions with diverging airflow. Coastal areas, however, often have a more balanced pattern of seasonal precipitation, which again reflects the fact that they are close to moisture sources.

Analysis of the maps of average January and July precipitation, which appear in Figure 2–30, produces the following generalizations:

1. The seasonal shifting of major pressure and wind systems, following the sun (northward in July and

Focus on Acid Rain

One of the most vexing and perplexing environmental problems of recent decades has been *acid rain*. This term refers to the deposition of either wet or dry acidic materials from the atmosphere on the Earth's surface. Although most conspicuously associated with rainfall, the pollutants may fall to Earth with snow, sleet, hail, or fog, or in the dry form of gases or particulate matter.

Although there is no universal agreement on the exact origins of acid rain, evidence indicates that the principal human-induced sources are sulfur dioxide emissions from smokestacks and nitrogen oxide exhaust from motor vehicles. These and other sulfur and nitrogen compounds are expelled into the air, where they may be carried hundreds or even thousands of miles by winds. During this time they may mix with atmospheric moisture to form sulfuric and nitric acids that sooner or later are precipitated.

Acidity is measured on the pH scale, which is based on the relative concentration of active hydrogen ions. The scale ranges from 0 to 14, where the lower end represents extreme acidity (battery acid has a value of 1) and the upper end, extreme alkalinity (lye has a value of 14). It is a logarithmic scale, which means that a difference of one whole number on the scale reflects a tenfold increase or decrease in absolute values. Rainfall in clean, dust-free air has a pH of about 5.6. Thus, it is slightly acidic due to the reaction of water with carbon dioxide to form a weak acid called carbonic acid. Today, however, precipitation with a pH of less than 4.5 (the level below which most fish perish) is being recorded, and an acid fog with a record low of 1.7 (8000 times more acidic than normal rainfall) was experienced in California in 1982.

Many parts of the Earth's surface have naturally alkaline conditions in soil or bedrock that buffer or neutralize acid precipitation. Soils developed from limestone, for example, contain calcium carbonate, which can neutralize acid. Granite soils, on the other hand, have no buffering component.

The most conspicuous damage by acid rain is done to aquatic ecosystems. Several hundred lakes in the eastern United States and Canada have become largely devoid of life in the last quarter century, primarily due to the deposition from acid rain. The precise effects of acid rain on forests are not clearly understood, but evidence increasingly indicates that it is primarily responsible for forest "diebacks" (extensive dying of trees) that are currently taking place on every continent except Antarctica. Crops and human health are also at risk. Even buildings and monuments are being destroyed. For example, acid deposition has caused more erosion to the Parthenon—an ancient marble temple in Athens, Greece—in the most recent quarter century than took place in the previous 24 centuries.

If this problem is so serious, why aren't people and governments doing more to resolve it? One reason is that much of the pollution is experienced at great distances from its source. Downwind locations receive unwanted acid deposition from upwind locations. Thus, Scandinavians and Germans complain about British pollution, Canadians

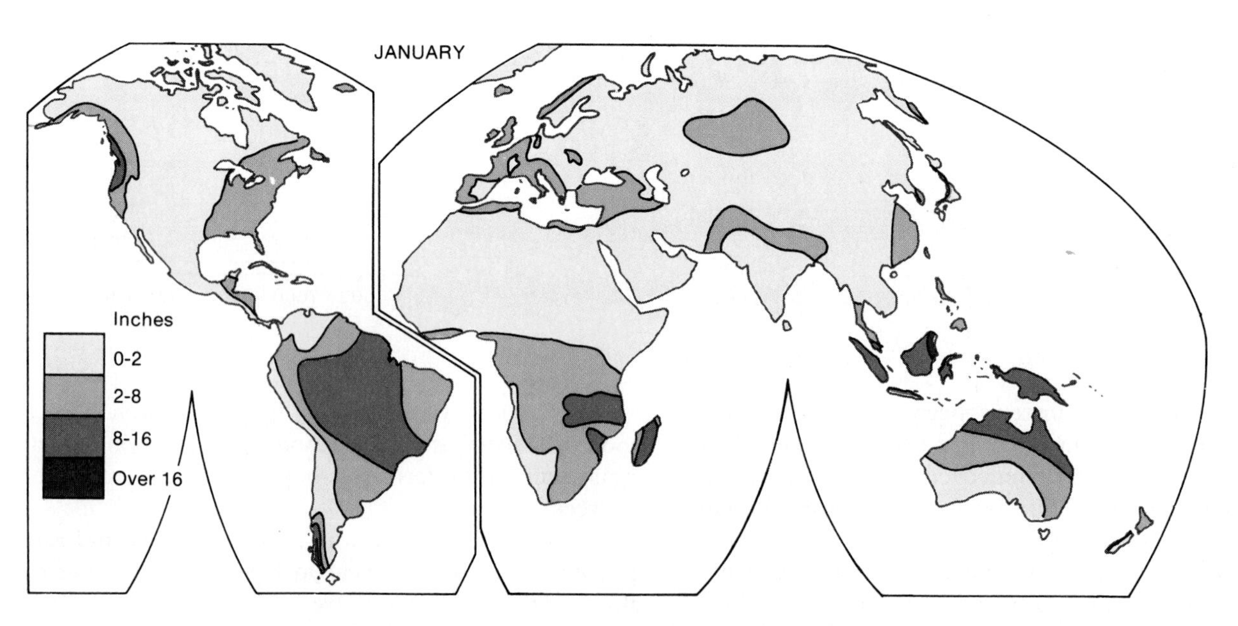

Figure 2–30. Average January and July precipitation over the land areas of the world.

blame U.S. sources, New Englanders accuse the Midwest, and so on. It is understandably difficult to persuade people in Ohio to finance expensive cleanup costs that will benefit forests in Maine.

One of the thorniest issues in relations among North American nations is Canadian dissatisfaction with the approach of the U.S. government toward mitigation of acid rain. In general, acid rain is viewed by Canadians as their gravest environmental concern, and it is believed that about half the acid rain that falls on Canada comes from U.S. sources, particularly older, coal-burning power plants in the Ohio and Tennessee river valleys. Since the U.S. Congress passed the Clean Air Act in 1970, coal-fired plants have invested more than $60 billion in mitigation efforts. As a result, sulfur emissions have declined by almost 35 percent, despite the fact that the use of coal has almost doubled. Still, the problem persists, and heroic efforts and staggering costs are obviously required to sanitize emissions further. Meanwhile, federal and provincial governments in Canada have passed laws requiring a 50 percent reduction in acid rain emissions by the year 1994.

Acid rain is now clearly in the public consciousness. Governments and citizen groups are mobilizing against this problem. Although the costs of reducing the acid-rain problem will be enormous, the costs of not doing so will be far greater.

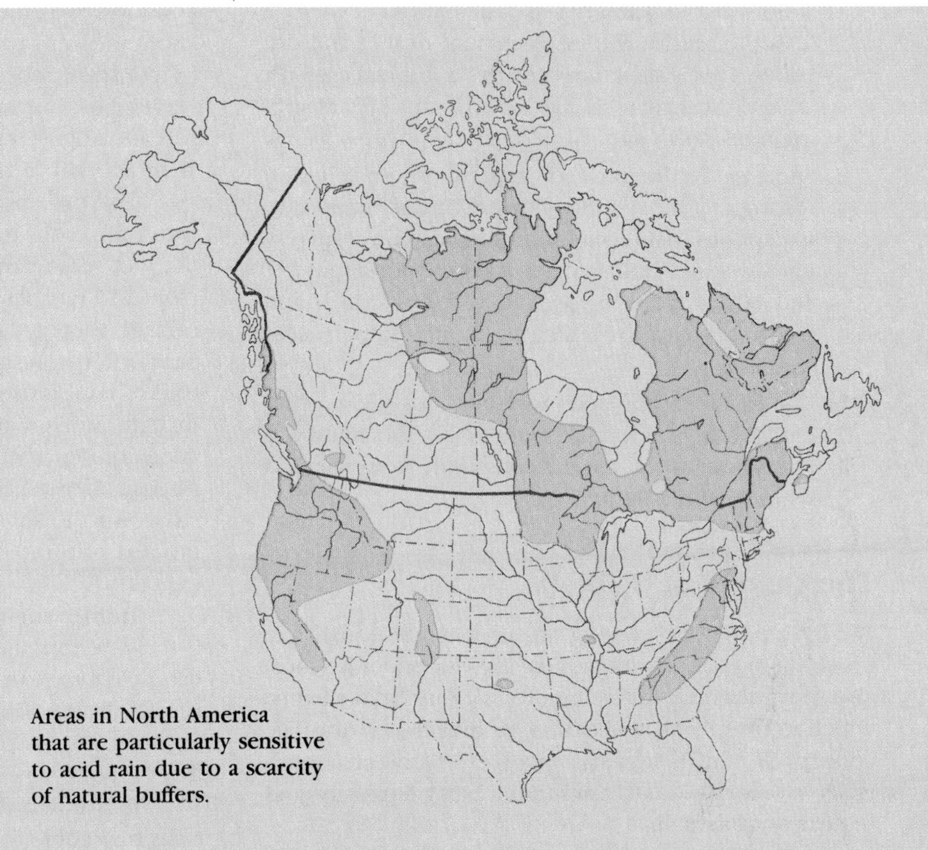

Areas in North America that are particularly sensitive to acid rain due to a scarcity of natural buffers.

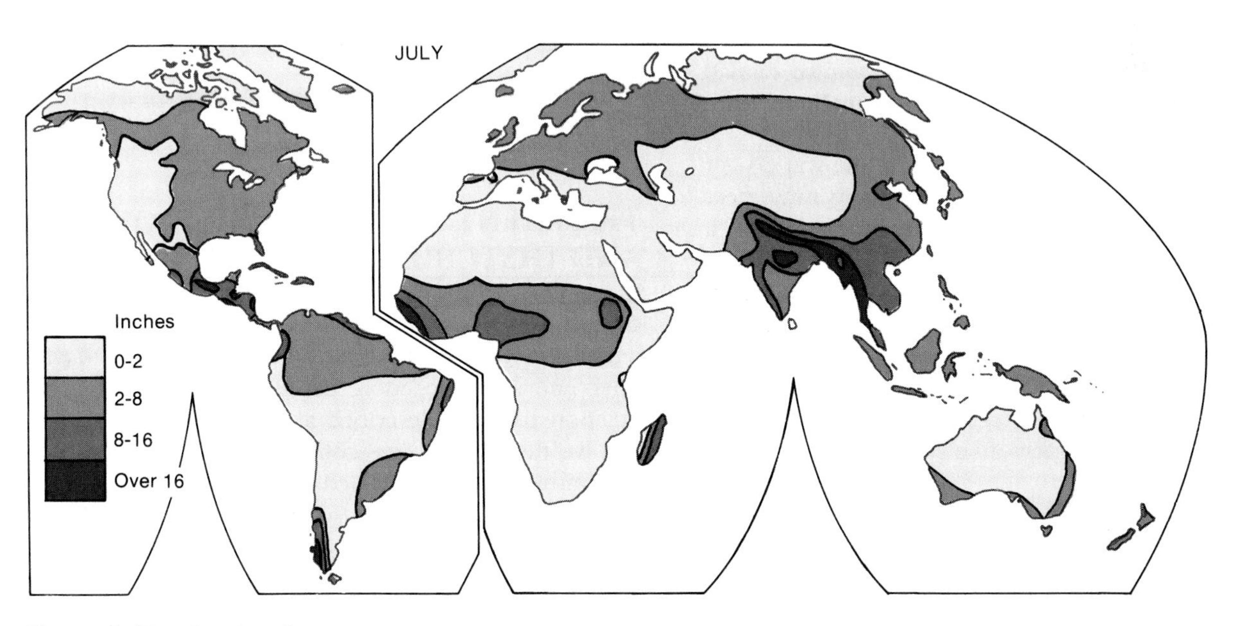

Figure 2–30. Continued

southward in January), is also experienced to a lesser extent in the displacement of wet and dry zones. This is seen most clearly in tropical regions, where the heavy rainfall belt of the ITC clearly migrates north and south in the opposite seasons.

2. Summer is the time of maximum precipitation throughout most of the world. The only important exceptions to this generalization occur in relatively narrow zones along west coasts between about 35° and 60° of latitude, as exemplified in North and South America, New Zealand, and southernmost Australia.
3. The most conspicuous variation in seasonal precipitation is found, predictably, in monsoon regions, where summer tends to be very wet and winters generally dry.

The Quality of Rainfall

We have noted the varying quantity and seasonality of precipitation in different parts of the world, but we must also recognize that precipitation may vary in its quality, which in turn has major effects on human activity. The concept of rainfall "quality" involves several characteristics, two of the most important being intensity and regularity of rainfall.

Intensity of Rainfall Regarding intensity, in temperate areas precipitation frequently takes the form of gentle rainfall, whereas in tropical regions it may come in torrential and destructive downpours. In northern Nigeria, for example, 90 percent of all rain falls in storms of more than 1 inch (25 mm) per hour. That is, by way of comparison, half the average monthly rainfall of London. In Ghana, cloudbursts regularly occur at a rate of 8 inches (200 mm) per hour, four times London's monthly total. In Java a quarter of the annual rainfall comes in showers of 2.4 inches (60 mm) per hour, more than Berlin gets in an average month.

Sudden storms are the worst possible way to get water. Much of the water gets lost, carrying away great quantities of precious topsoil. The first heavy drops in a downpour clog the pores of the soil with fine particles washed from the surface. After only a few minutes the soil cannot absorb more than a small fraction of the rain. More than two-thirds of the water may then run off in sheets and rivulets, which leads to tremendous erosion. Studies in the West African nation of Burkina Faso found that in 1 year nearly 90 percent of all erosion took place in just 6 hours.

Regularity of Rainfall Another component of the quality of rainfall is its regularity. In Western Europe and eastern North America, rain does not vary much from month to month. New York's rainiest month receives only one and a half times as much rain as its driest; London's two times; Berlin's two and a half. Delhi receives about the same total annual rainfall as London, but Delhi, by contrast, receives only 0.2 inch (10 mm) of rain in November, but more than 7 inches (175 mm) in both July and August. Zungeru, in central Nigeria, gets 54 times as much rain in the wettest month as in the driest.

Soil can absorb and hold only a limited amount of water for crops to use, like a bank account deposited in the wet season and drawn on in the dry. Though many tropical countries get what appears to be adequate annual rainfall, they get it in a lump sum, like a huge win at a casino. Most is squandered, and the little that is left in the soil bank is used up within a few months. For example, in Agra, India, where the Taj Mahal is located, rain exceeds the current needs of vegetation only in July and August. As the rains decrease in September, the water stored in the soil is used up in just 3 weeks, leaving 9 months of the year in which water supplies are inadequate.

Human societies can adjust to variations in rainfall if these variations are regular or predictable. Irrigation can help even out the supply of water over the growing season. In the Sahel, south of the Sahara, farmers grow fast-maturing crops such as sorghum and millet that shoot from seed to ripened ear in 3 or 4 months.

It is much more difficult, however, to cope with rains that come irregularly from year to year. In Europe and much of North America, the amount of annual rainfall varies by less than 15 percent per year on average. In many of the tropical and subtropical regions, by contrast, it fluctuates from 15 to 20 percent, and in the semiarid and arid lands, by up to 40 percent or more. In the lean years, crops fail. If the rains are late, the growing season is cut short, and yields are greatly diminished. Famine and disease usually follow. That is why, on the Indian subcontinent, the monsoon is awaited with such hope and trepidation, and a delay of a week or two brings panic.

TRANSIENT ATMOSPHERIC FLOWS AND DISTURBANCES

Over most of the Earth, particularly in the midlatitudes, day-to-day weather conditions are closely identified with phenomena that are more limited than the general circulation patterns described above. These phenomena involve the more-or-less cohesive flow of a body of air that moves as a unit, called an *air mass*, as well as a variety of atmospheric disturbances that are usually referred to as **storms**. Air masses and storms are transient and temporary, but in certain parts of the world they are so frequent and powerful that their interactions are major determinants of weather conditions and, to a lesser extent, of climatic characteristics.

Air Masses

Although the troposphere is a continuous body of mixed gases that surrounds the planet, it is by no means a uniform blanket of air. Instead, it is composed of many large, variable parcels of air, which are often distinct from one another in their characteristics. Such parcels are referred to as **air masses**.

An air mass is a large portion of the lower troposphere, usually with a diameter of about 1000 miles (1600 km), that has relatively uniform properties (primarily temperature, humidity, and stability) in the horizontal dimension, and that appears and travels as a recognizable entity.

An air mass develops its characteristics by remaining over a relatively uniform land or sea surface for enough time to acquire the temperature/humidty/stability characteristics of the underlying surface. For an air mass to form, air must stagnate over a large area that has a uniform surface with distinctive properties. The stagnation needs to last for only a few days. Air masses have their origin either in tropical/subtropical or in polar/subpolar regions, from which they often travel into the midlatitudes. Thus, air masses represent the means by which the tropics are cooled and polar areas are warmed.

Air masses are identified and classified on the basis of the source regions from which their original characteristics are derived (see Figure 2–31). The four most common types are *continental polar*, *maritime polar*, *continental tropical*, and *maritime tropical* (see Table 2–2).

When an air mass departs from its source region, its structure begins to change, due in part to thermal modification (heating or cooling from below), in part to dynamic modification (uplift, subsidence, convergence, turbulence), and perhaps also in part due to addition or subtraction of moisture. More importantly, the air mass modifies the weather of the region into which it moves; it introduces the characteristics of the source region into other regions.

Fronts

When unlike air masses meet, they do not mix readily; instead, a boundary zone develops between them. This

TABLE 2–2 ***Simplified Classification of Air Masses***

Type	Code	Source region properties
Continental polar	cP	Cold, dry, very stable
Maritime polar	mP	Cool, moist, relatively unstable
Continental tropical	cT	Hot, very dry, unstable
Maritime tropical	mT	Warm, moist, variable stability

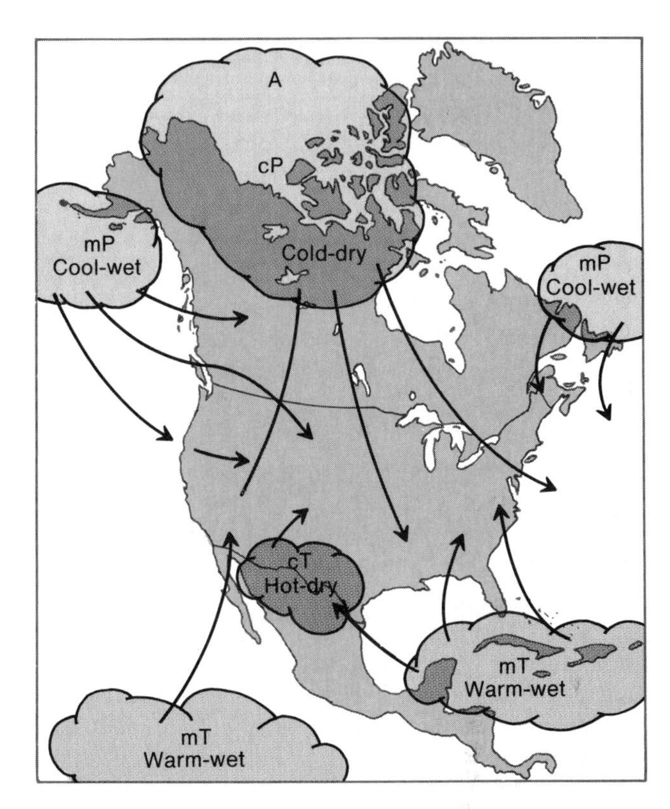

Figure 2–31. Major air masses that affect North America.

boundary is a **front**, a relatively narrow zone of discontinuity within which the properties of air masses change rapidly. The most conspicuous difference between air masses usually is temperature, so a frontal surface typically separates warm air from cool air. At any given level in the air there may be relatively uniform warm temperatures on one side of the front and relatively uniform cool temperatures on the other, with a fairly steep and abrupt temperature gradient through the frontal zone (see Figure 2–32). Air masses may also vary significantly in density, humidity, stability, and other characteristics, all of which show a steep gradient of change through the frontal zone.

An important attribute of frontal surfaces is that they are not vertical but rather lean to one side or the other with height. Indeed, they slope at such a low angle that they are much closer to horizontal than vertical in their orientation. The normal slope of a front averages about 1:150 (a rise of 1 vertical foot for every 150 horizontal feet). A front always slopes so that the warmer air overlies the cooler air.

A front usually is in constant motion, shifting the position of the boundary between the air masses but maintaining its function as a barrier between them. Usually one air mass is actively displacing the other; thus, the front advances in the direction dictated by the movement of the more active air mass.

The leading edge of an advancing warm air mass represents a **warm front**. The warm air ascends over the retreating cool air, its temperature decreasing adi-

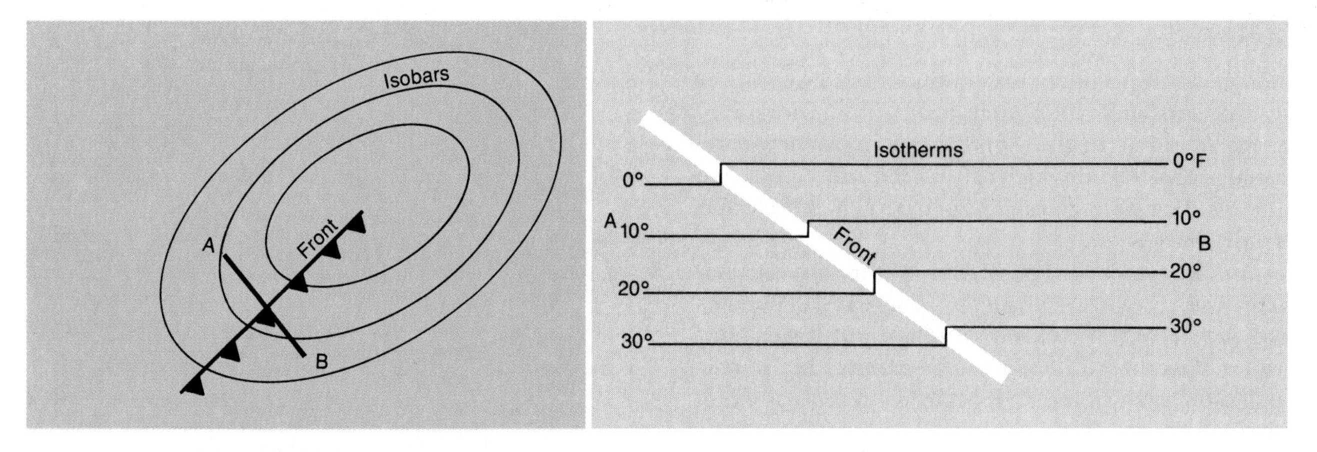

Figure 2–32. Map and vertical cross section through a hypothetical front, with the cooler air on the left and the warmer air on the right. The solid lines of the cross section represent isotherms.

abatically as it rises. The usual result is clouds and precipitation. The frontal uplift is very gradual, so cloud formation normally is slow, expansive, and not very turbulent. Precipitation usually occurs in a broad band; it is likely to be protracted and gentle, without much convective activity. As illustrated in Figure 2–33, most precipitation falls in advance of the surface position of the front. The surface position of a warm front is portrayed on a weather map either by a red line (if color is used) or (more typically) by a solid black line along which black semicircles are located at regular intervals, with the semicircles extending in the direction of the cool air (see Figure 2–34).

The leading edge of a cool air mass that is actively displacing warm air is a **cold front**. Cold fronts normally are steeper than warm fronts and move faster (see Figure 2–35). This combination of steepness and speed makes for more rapid lifting of the warm air, which increases instability and produces more blustery and violent weather. Both clouds and precipitation tend to be concentrated along and immediately behind the surface position of the front. On a weather map, the surface position of a cold front is shown either by a blue line or by a solid black line studded at intervals with solid triangles that extend in the direction of the warm air.

Figure 2–33. Along a warm front, warm air rises above cooler air. Widespread cloudiness and precipitation often develop along and in advance of the surface position of the front. Higher and less dense clouds often appear dozens or hundreds of miles ahead of the surface position of the front.

When neither air mass displaces the other, their common "boundary" is a **stationary front**. Weather is not readily predictable along such a front.

A more complex type of front is formed when a cold front overtakes a warm front, forming an **occluded front**. The development of occluded fronts will be discussed later.

Extratropical Cyclones

The middle latitudes represent the principal battleground of tropospheric phenomena, where air masses from polar and tropical regions meet and come into conflict, where most frontal conditions occur, and where weather is most dynamic and changeable from season to season and from day to day (see Figure 2–36). By far the most significant atmospheric disturbances of the mid-

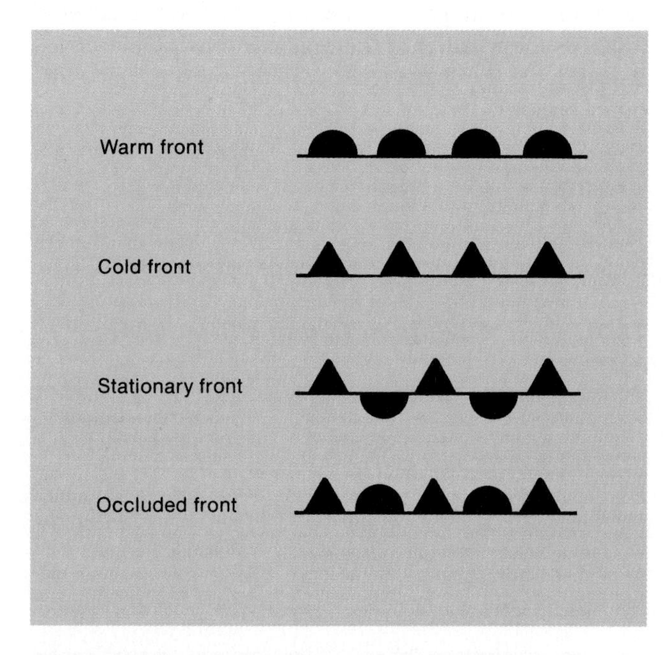

Figure 2–34. Weather map symbols for fronts.

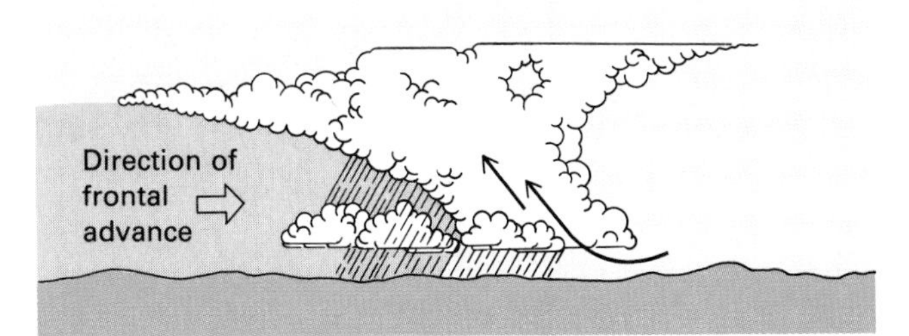

Figure 2–35. A cold front is usually steeper than a warm front. The warm air is forced upward by the advancing cold air behind the front. This often creates cloudiness and relatively heavy precipitation along and immediately behind the surface position of the front.

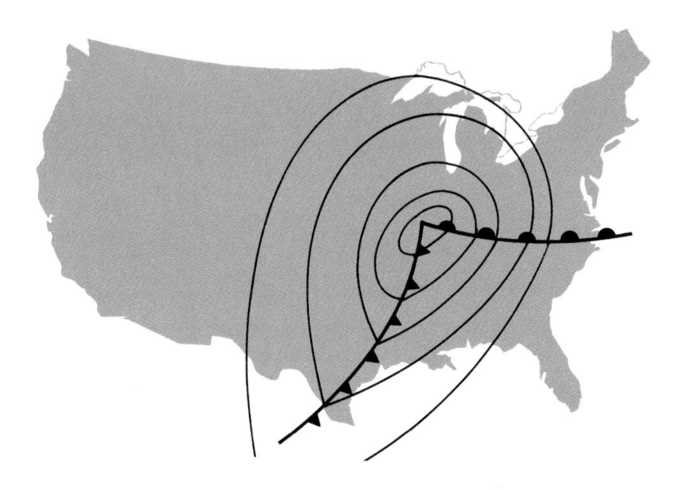

Figure 2–37. A typical mature extratropical cyclone in the Northern Hemisphere has a cold front trailing to the southwest and a warm front extending toward the east.

latitudes are **extratropical cyclones,** low-pressure systems located between 35° and 70° of latitude that are marked by conflict and a convergence of air masses. They are found almost entirely within the zone of westerly winds, and their general path of movement is toward the east. They bring changeable, unsettled weather that normally includes widespread, abundant, and often intense precipitation. Within their zones of latitude they are primarily responsible for most day-to-day weather changes.

Figure 2–37 depicts a mature extratropical cyclone in the middle of the United States, with its center approximately over St. Louis. The cyclonic wind-flow pattern (converging counterclockwise) attracts cool air from the Great Lakes area and southern Canada into the northern and western portions of the system, and warm air from the Gulf Coast area into the southern and eastern portions. The irregular convergence of these unlike air masses characteristically creates two fronts: a cold front extending southwesterly from the center of the low, and a warm front extending generally toward the east from the center of the low.

Along both frontal surfaces, warm air rises. It is uplifted by the advancing cold air along the cold front, and it actively overrides the cold air as it ascends the warm front. The typical result is two zones of cloudiness and precipitation that overlap around the center of the storm and extend outward in general association with the fronts.

Extratropical cyclones are transient phenomena that are on the move throughout their existence. This transiency is complicated by the fact that four different kinds of movement are involved (see Figure 2–38). These movements are related to one another but can be analyzed as discrete entities, as follows:

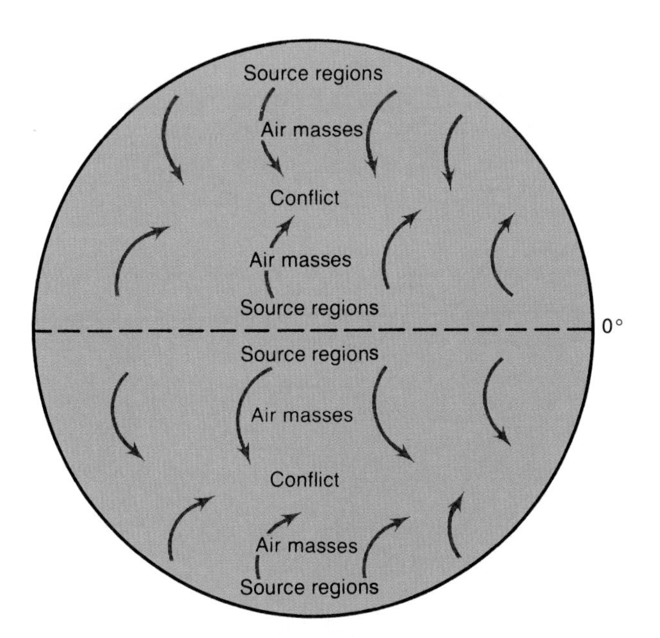

Figure 2–36. The midlatitudes are zones of air-mass conflict.

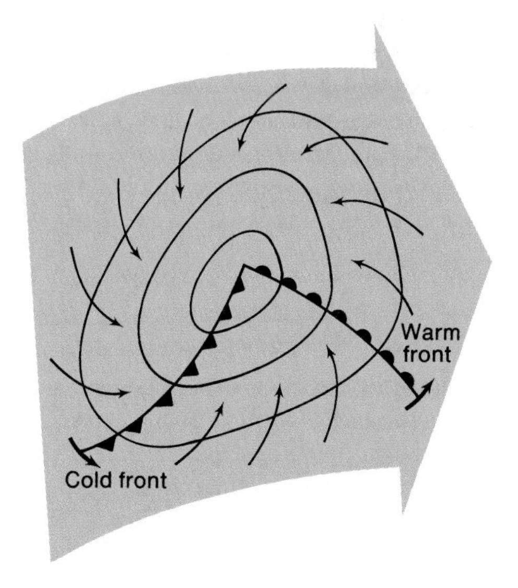

Figure 2–38. Four different varieties of motion occur in a typical extratropical cyclone. These are: (1) generally convergent airflow, (2) cold front advance, (3) warm front advance, and (4) movement of the entire system in the general flow of the westerlies.

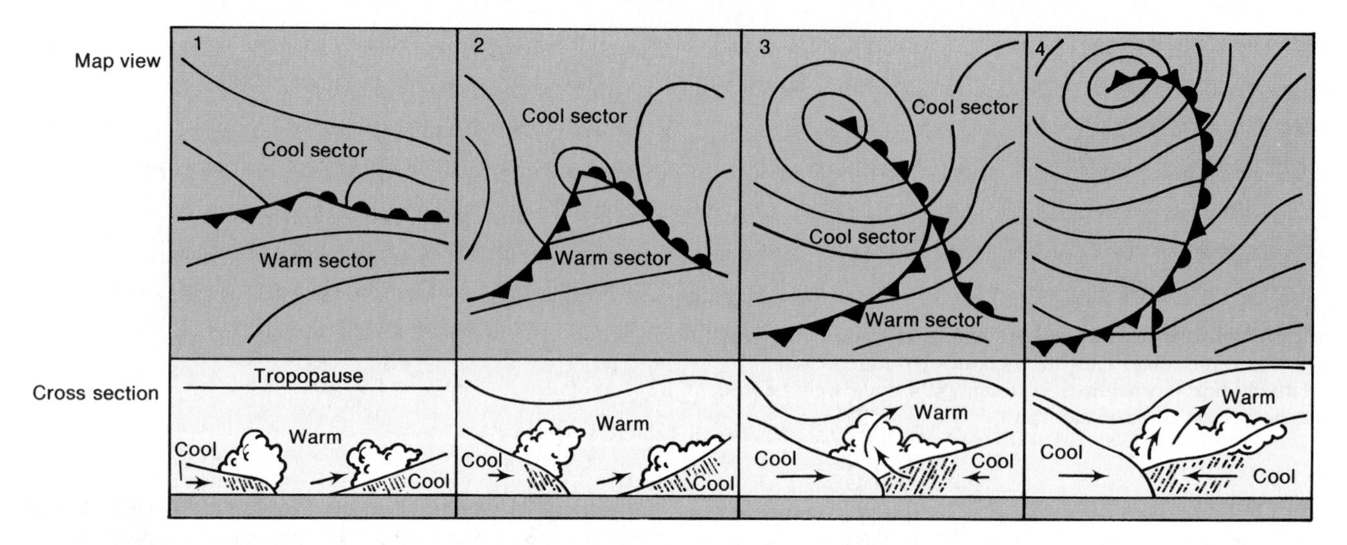

Figure 2–39. Stages in the life cycle of an extratropical cyclone: (1) early development, (2) maturity, (3) partial occlusion, (4) full occlusion.

1. The system has a cyclonic wind circulation, with air generally converging from all sides in a counterclockwise pattern in the Northern Hemisphere.
2. The entire storm system moves as a major disturbance in the westerlies, traversing the middle latitudes generally from west to east at an average speed of 20 to 30 miles (32 to 48 km) per hour.
3. The cold front normally advances faster than the storm is moving. Thus, it swings from its pivot in the center of the low, increasingly moving into and displacing the warm sector of the storm.
4. The warm front usually advances more slowly than the storm itself. This causes it to lag behind and has the effect of swinging the warm front backward in the direction of the advancing cold front.

These systems have a life cycle that averages between 10 days and 2 weeks. Eventually the storm dissipates because the cold front overtakes the warm front, forming an occluded front, as illustrated in Figure 2–39. This *occlusion* process usually results in a short period of intensified activity, followed by the dying out of the system.

Extratropical Anticyclones

Another major disturbance in the general flow of the westerlies is the **extratropical anticyclone.** This is an extensive, migratory high-pressure cell that has air subsiding into it from above and then diverging at the surface (clockwise in the Northern Hemisphere). No air mass convergence or conflict is involved, so anticyclones contain no fronts (see Figure 2–40). The weather is clear and dry with little or no opportunity for cloud formation. In winter, anticyclones are characterized by very cold temperatures. Anticyclones move in the same general direction and at about the same rate of speed as cyclones, and they have a similar longevity.

Extratropical cyclones and anticyclones function as migratory perturbations (irregularities) in the westerlies, alternating with one another in irregular sequence around the world in the midlatitudes (see Figure 2–41). Each can occur independently of the other, but they often occur together. This can be seen most clearly when an anticyclone closely follows a cyclone, as diagrammed in Figure 2–42. The winds diverging from the eastern margin of the high fit precisely into the flow of air converging into the western side of the low. It is easy to visualize the anticyclone as a polar air mass with its leading edge represented by the cold front of the cyclone.

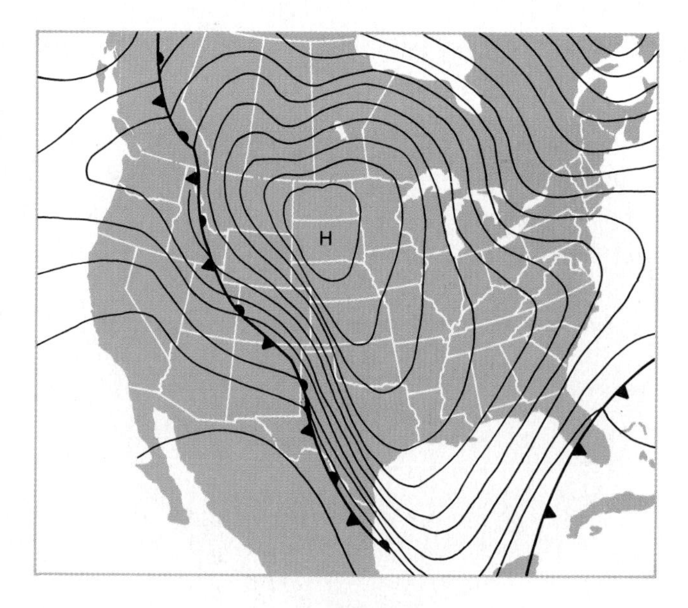

Figure 2–40. A typical well-developed extratropical anticyclone, centered over the Dakotas.

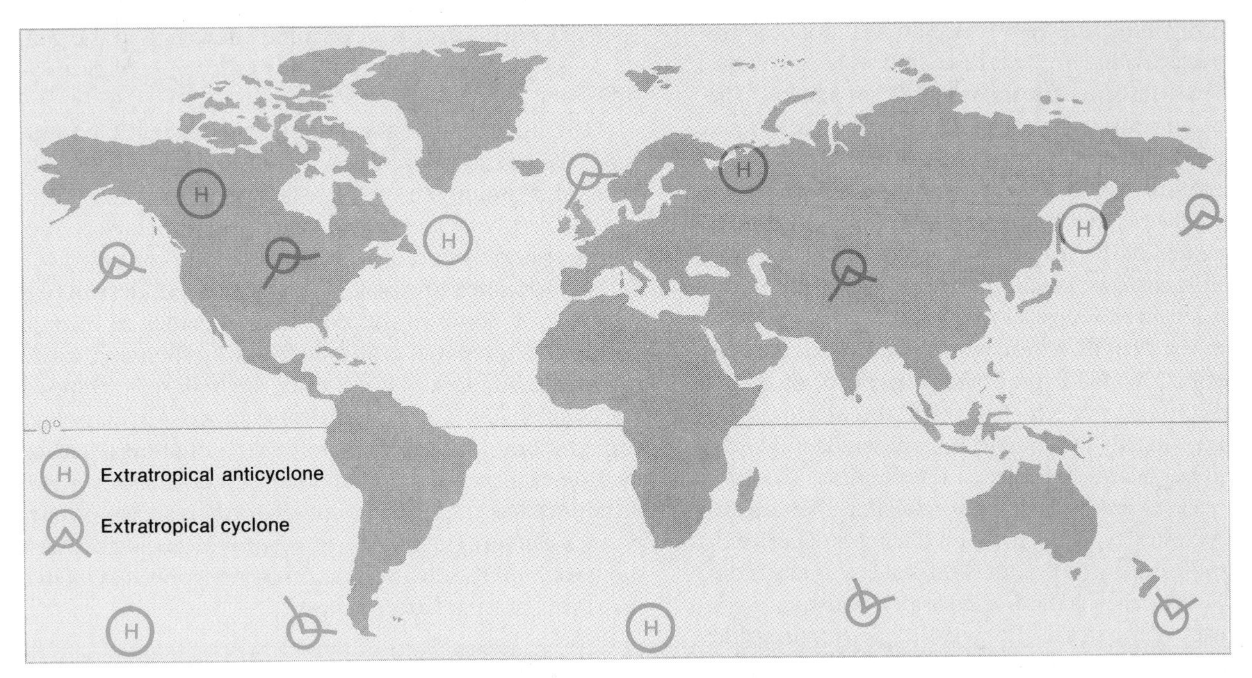

Figure 2–41. At any given time, the midlatitudes are dotted with extratropical cyclones and anticyclones. This map depicts a hypothetical situation in January.

The development of cyclones and the paths followed by both cyclones and anticyclones are determined largely by conditions aloft in the westerlies. The undulations of the westerly jet and the position of the Rossby waves are closely associated with the movements of these migratory pressure systems. Weather forecasting in the midlatitudes largely depends on predicting the characteristics and movements of these lows and highs.

Hurricanes

Whereas the midlatitudes constitute a zone of dynamic, changeable weather, the low latitudes are characterized by repetitive weather conditions day after day, month

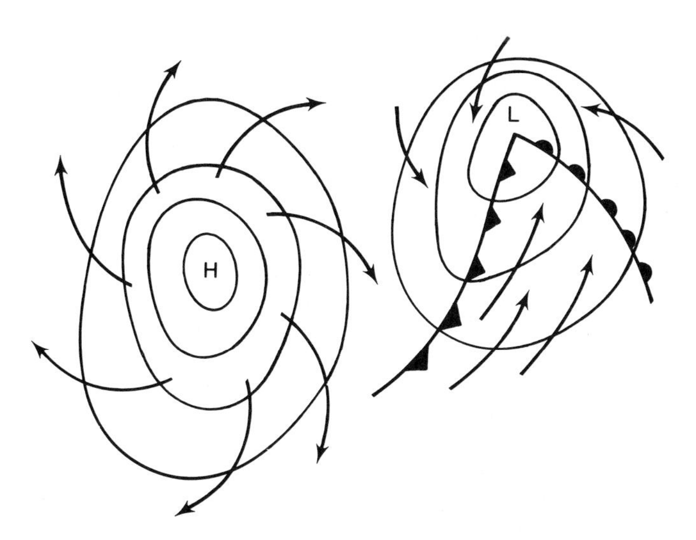

Figure 2–42. Extratropical cyclones and anticyclones often occur in juxtaposition in the middle latitudes.

ABC NEWS ***When Anarchy Reigned***

Hurricane Hugo was the most devastating storm experienced in the southeastern United States in over a decade. The intense damage caused by this storm cost Americans over $3 billion. The physical damage to structures, the loss of life, and the overall terror these storms cause rank them as one of humankind's most dreaded natural events. Because hurricanes begin over water bodies, there is generally more opportunity for meteorologists to observe the developing storm and warn those people who live in its path of its location and intensity. Thankfully, this reduces the human toll, but warnings do very little to minimize the cost in structures and property damage.

In the accompanying video segment, you will see the damage that occurred on the island of St. Croix in the U.S. Virgin Islands as a result of Hurricane Hugo. As you watch this tape, consider the following questions.

1. If hurricanes are so common in this region, why do so many people live here, and how do you think they perceive these storms? Can you draw any parallels to people who live in earthquake-prone areas, like southern California?
2. Explain the role of early warning systems in this instance. How could they be improved?

after month. Almost the only breaks in this monotonous regime are provided by transient atmospheric disturbances, of which by far the most significant affecting the tropics and subtropics are **tropical cyclones,** which are known as **hurricanes** in the United States.

Hurricanes are intense, revolving, rain-drenched, migratory, destructive storms that occur erratically in certain regions of the tropics and subtropics. They consist of prominent low-pressure centers that are essentially circular in shape and have a steep pressure gradient outward from the center. Strong winds spiral inward in cyclonic fashion. Winds must reach a speed of 65 knots (74 miles or 119 km per hour) for the storm to be considered officially as a hurricane, but winds in a well-developed hurricane often reach double that velocity.

Hurricanes are considerably smaller than extratropical cyclones, typically having a diameter of between 100 and 600 miles (160 and 1000 km). A remarkable feature of hurricanes is the presence of a nonstormy *eye* in the center of the storm. The weather pattern is rather symmetrical about the eye. The strong converging winds produce bands of dense cumuliform clouds that yield heavy rain. Within the eye, however, there is no rain or low cloud, and scattered high clouds may part to let in intermittent sunlight.

The origin of hurricanes is not yet understood in detail. They form only over warm oceans in the tropics but at least 5° away from the equator, usually in late summer and fall (see Figure 2–43). Once formed, they follow irregular tracks within the general flow of the trade winds. A specific path is very difficult to predict, but the general pattern of movement is highly predictable. About one-third of all hurricanes travel directly from east to west without much latitudinal change. Most of the others begin on an east-west path and then curve prominently poleward, where they either dissipate over the adjacent continent or become enmeshed in the general flow of the midlatitude westerlies (see Figure 2–44).

Whichever path they follow, hurricanes do not last long. The average hurricane exists for only about 1 week, with a maximum life of 3 weeks. As soon as a hurricane leaves the ocean and moves over land, it begins to die, for its energy source (warm moist air) is cut off.

Hurricanes are best known for their destructive capabilities. Some of the destruction comes from high winds and torrential rain, but the overwhelming cause of damage and loss of life is from the high seas whipped up beneath the storm. Hurricane-tossed waves pound the coastline, damaging or destroying buildings, harbor facilities, moored boats, and other objects. However, destruction and tragedy are not the only legacies of hurricanes. Such regions as northwestern Mexico, northern Australia, and southeastern Asia rely on tropical cyclones for much of their water supply.

Thunderstorms

A **thunderstorm** is a relatively violent convectional storm accompanied by thunder and lightning. It is usually small and localized in extent and short-lived in duration. It is always associated with vertical air motion, considerable humidity, and instability, a situation that produces a towering cumulonimbus cloud and (nearly always) showery precipitation.

Thunderstorms sometimes occur as individual clouds, produced by nothing more complicated than thermal convection; such developments are commonplace in the tropics and during summer in much of the midlatitudes. They also frequently are found in conjunction with other kinds of storms or are associated with other mechanisms that can trigger unstable uplift.

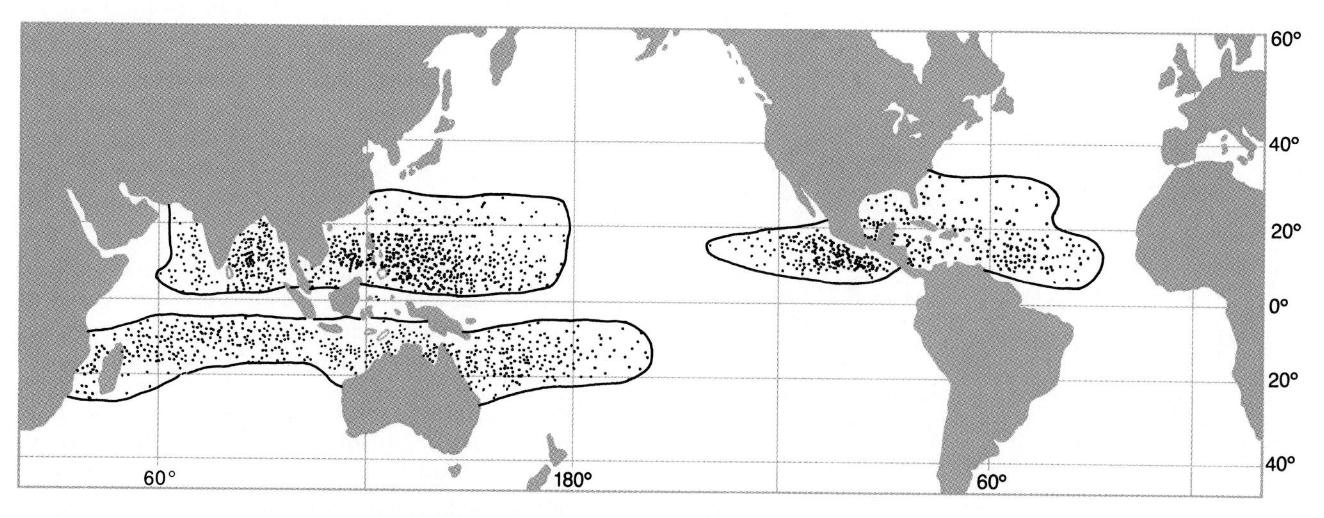

Figure 2–43. Location of origin points of tropical cyclones during the period 1952–1971. (After W. M. Gray, "Tropical Cyclone Genesis," *Atmospheric Science Paper 234*, Department of Atmospheric Science, Colorado State University, Fort Collins, 1975.)

Focus on Regions

One of the most basic terms in the study of geography is *region*. A region is a territory that exhibits a certain uniformity. Of all the things that can be found in any landscape, a geographer chooses only a few specific criteria and then maps those criteria to define a region. The criteria chosen may be either physical or cultural. Sometimes the criteria chosen are strictly descriptive. We call those *formal regions:* a region of mountains, for instance. Other regions can be defined on the basis of how they are organized. We call these *functional regions*. Examples are the suburban Denver region, which supplies commuters into downtown Denver, or any governmental jurisdiction.

The choice of criteria to define a region takes place in the geographer's mind. Therefore a region is a concept, an abstract idea. The geographer's ''region'' is comparable to the historian's ''period.'' Both time and space are continuous. How you divide them up into units of analysis depends on your purpose. Art historians define ''styles,'' sociologists define ''societies,'' and many other scholarly fields define their conceptual units of analysis.

Some regions may be demarcated clearly. Among regions defined on the basis of natural phenomena, for instance, watersheds can usually be mapped distinctly, as can a region defined as ''all territory over 1000 feet (305 m) above sea level.'' Among regions defined on the basis of cultural criteria, national borders are usually clearly marked.

In other cases, however, the regions that geographers define are less distinct in the landscape. Climatic regions are not sharply defined but imperceptibly merge into one another. The geographer's definition and trained discrimination ''draw a line'' between one climatic region and its neighbor. Regions defined by cultural phenomena often merge or overlap as well. The people living between two big cities, for instance, may listen to radio stations broadcasting from either city or from both cities. Therefore a geographer must carefully define the line between the market regions of the two stations: households that listen to station A or to station B more than 50 percent of the time, for example.

Sometimes people confuse regions based on natural phenomena with regions based on cultural criteria. Many people today refer to Africa as a region. In what sense is Africa a region? Africa is a continent, and some people might unconsciously accept that as sufficient to be considered a region. To the ancient Romans, however, northern Africa and southern Europe formed one region focused on the Mediterranean Sea, and that region was altogether distinct from sub-Saharan Africa, which was another region. The African continent is not in fact homogeneous by any criteria of physical environment or by the history or cultures of its many peoples. Whenever anyone refers to any territory as a region, we must be sure that we know and agree on exactly what territory the region includes and what the criteria of homogeneity are.

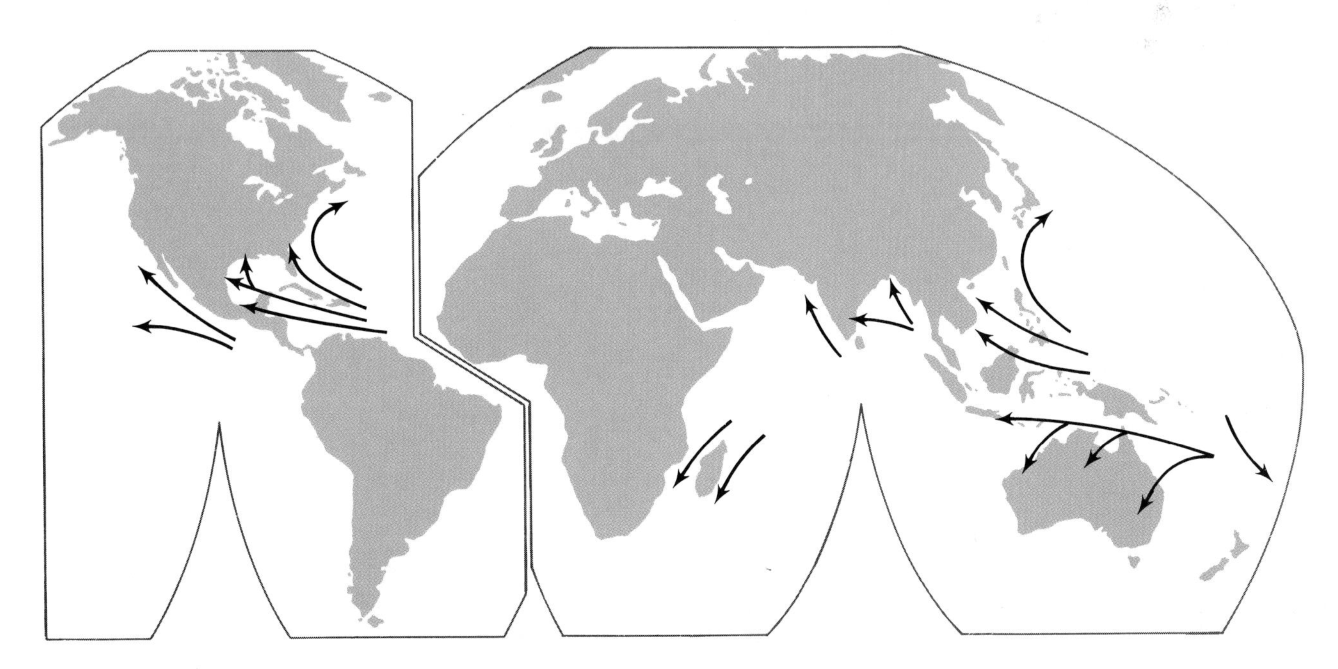

Figure 2–44. Major generalized tropical cyclone tracks.

Thus, thunderstorms often accompany hurricanes, tornadoes, fronts (especially cold fronts) in extratropical cyclones, and orographic lifting that forces unstable buoyancy.

Tornadoes

Although very small in size and localized in effect, the tornado is the most destructive of all atmospheric disturbances. A **tornado** is defined as a deep cyclonic low-pressure cell surrounded by a whirling cylinder of wind spinning so violently that the centrifugal force creates a partial vacuum within the funnel (see Figure 2–45). These are tiny storms, generally less than 0.25 mile (400 m) in diameter, but they have the most extreme pressure gradients known. These gradients produce winds of extraordinary velocity. Actual measurements are unknown because any tornado that has come close enough to an anemometer (a machine that records wind speeds) to be recorded has blown the instrument to bits. Scientists estimate, however, that winds can reach maximum velocities of 200 to 500 miles (320 to 800 km) per hour. Air sucked into the vortex also rises at an extreme rate, thought in some cases to exceed 100 miles (160 km) per hour.

Tornadoes are very erratic, occurring at random, jumping from place to place, and persisting only briefly. As with most storms, the exact mechanism of formation is not well understood. Tornadoes usually develop in warm, moist, unstable air associated with an extratropical cyclone or thunderstorm. Spring and early summer are favorable for tornado development because different air masses usually come into contact in the midlatitudes during those seasons. A tornado can form in any month, however.

Tornadoes occur in various portions of the middle latitudes and subtropics, but more than 90 percent of all such storms are reported from the United States. Between 800 and 1200 tornadoes are recorded annually in this country, but the actual total may be considerably higher than that because many small tornadoes that occur briefly in sparsely populated areas are not reported.

CLASSIFYING CLIMATES

The basic goal of the geographical study of climate, as of the geographical study of anything, is to understand its *spatial characteristics*—its distribution over the Earth. This is exceedingly difficult to do because climate is the product of the interaction of a number of different elements that vary from time to time and place to place. To cope with the great diversity of information encom-

Figure 2–45. The awesome funnel of a tornado reaches down from cloud to ground. (Courtesy of Howard Bluestein/Photo Researchers.)

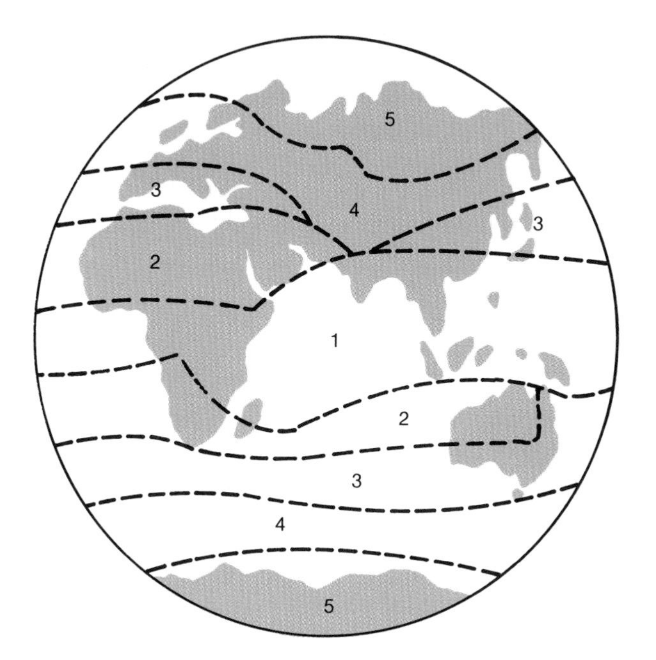

Figure 2–46. The major generalized climatic zones of the Eastern Hemisphere, as recognized today: (1) Equatorial warm-wet, (2) Tropical hot-dry, (3) Subtropical warm temperate, (4) Midlatitude cool temperate, (5) High-latitude cold.

passed by the concept of climate, we must identify and classify the most meaningful components of this concept. Scientists generally have chosen temperature and precipitation as the most significant and understandable features of climate, as well as the most available, to serve as the framework of their classifications.

Today we recognize five basic climates in the world: (1) equatorial warm-wet, (2) tropical hot-dry, (3) subtropical warm temperate, (4) midlatitude cool temperate, and (5) high-latitude cold (see Figure 2–46). Many climatologists have tried to refine the distinction among these five basic climatic zones and to subdivide the zones in various ways. Most efforts have relied primarily on natural vegetation as an indicator of climate.

The climatic classification that is by far the most widely used by geographers was developed by Wladimir Köppen (1846–1940), a Russian-born German climatologist who was also an amateur botanist. The Köppen system identifies five climate groups. The entire system uses as a database only the mean annual and monthly values of temperature and precipitation, combined and compared in a variety of ways.

Köppen defined four of his five major climatic groups by temperature characteristics, the fifth on the basis of moisture. Each group was then subdivided according to various temperature and precipitation relations. The system's distinctive feature is the use of a symbolic nomenclature (system of names) to designate the various climatic types. This nomenclature consists of a combination of letters, with each letter having a precisely defined meaning.

The Köppen system presented here is modified a bit from Köppen's "final" version (published in 1936), primarily for simplification (see Figure 2–47). Details of the system are shown in Table 2–3 on page 56, and each of the major types is described briefly below.

Tropical Humid Climates (Group A)

The A climates occupy almost all the land area of the Earth within 15° to 20° of the equator in both the Northern and Southern hemispheres (see Figure 2–48). This belt of tropical humid climates encircles the globe, interrupted occasionally by mountains or small dry zones. It dominates the equatorial regions and extends poleward to beyond the 25° parallel in some windward coastal lowlands.

The A climates are the only truly winterless climates of the world. The sun is high in the sky every day of the year, and even the shortest days are not appreciably shorter than the longest ones. These are climates of perpetual warmth, although they do not experience the world's highest temperatures. The fundamental character of the A climates, then, is molded by their proximity to the equator.

Another basic characteristic of the tropical humid climates is the prevalence of moisture. Warm, moist, unstable air masses frequent the oceans of these latitudes, and the Intertropical Convergence Zone persists for much of the year. Moreover, onshore winds and thermal convection are common. Thus, the A climate zone not only has abundant sources of moisture but also has mechanisms for uplift. Not surprisingly, then, these regions experience high humidity and considerable rainfall.

The tropical humid climates are classified into three types on the basis of the quantity and regime of annual rainfall. The *tropical wet* type (*Af* in the modified Köppen system) experiences relatively abundant rainfall in every month of the year. The *tropical monsoonal* type (*Am*) has a short dry season but a very rainy wet season. The *tropical savanna* type (*Aw*) is characterized by a longer dry season and a prominent but not extraordinary wet season.

Dry Climates (Group B)

The dry climates are even more extensive than the tropical humid climates (see Figure 2–49). They cover about 30 percent of the Earth's land area, which is more than any other climatic group. Although at first glance their distribution pattern may appear to be erratic and complex, it actually has a considerable degree of predictability. The arid regions of the world (other than in the Arctic) generally are the result of the lack of uplift of the air rather than the lack of moisture in the air. Vertical motion is suppressed by persistent stability, which is mostly due to the subsidence associated with subtropical high-pressure cells and secondarily to subsidence on the downwind side of mountain barriers.

The largest expanses of dry areas are in subtropical latitudes where thermodynamic subsidence is wide-

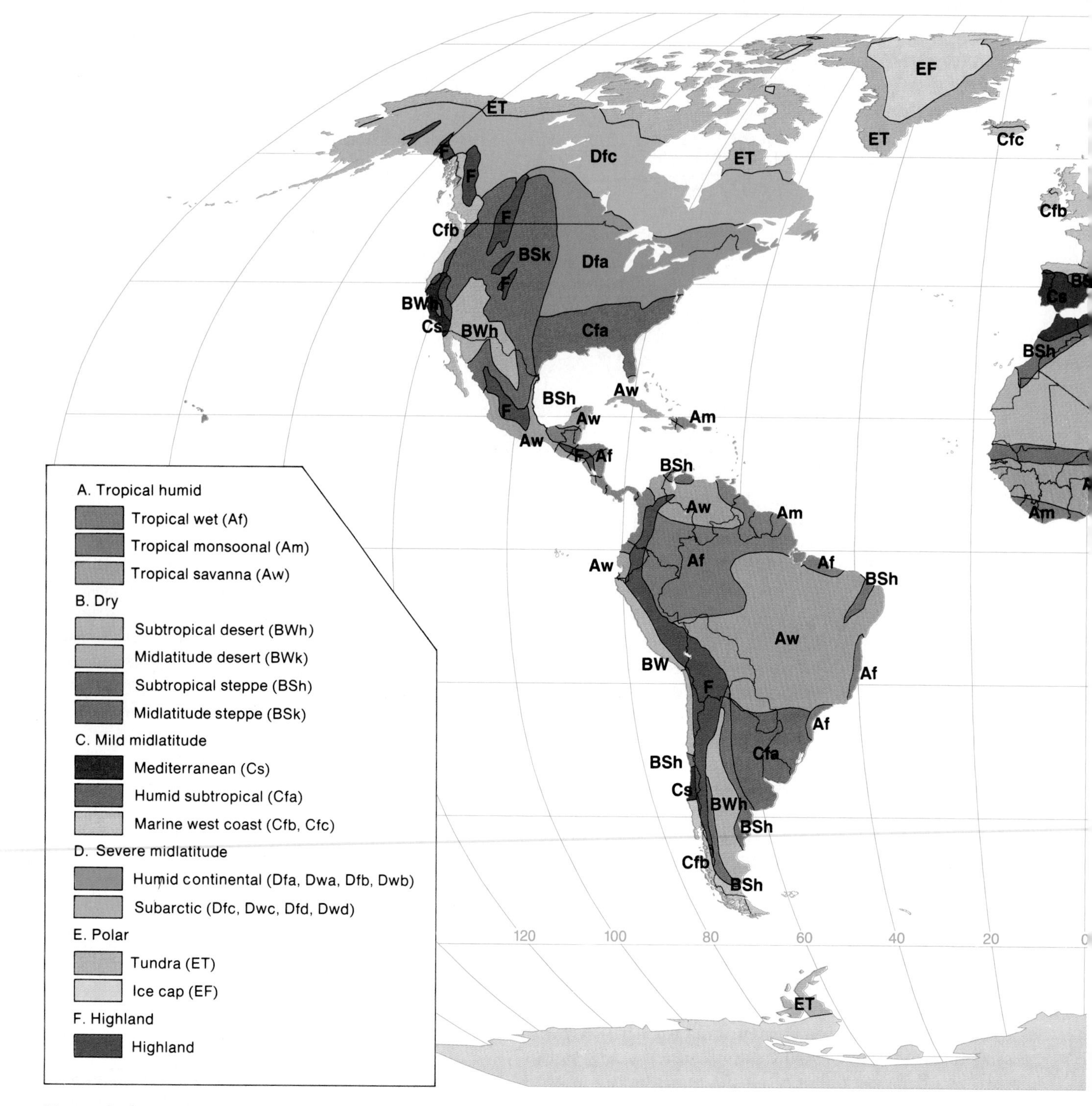

Figure 2–47. Climatic regions (modified Köppen system).

spread. These conditions are found in the western and central portions of continents, but they also occur over extensive ocean areas. In the midlatitudes, particularly in central Asia, the B climates are found in extreme continental locations, which means areas that are remote from sources of moisture, either because of distance or topographic barriers. In Asia and North America the subtropical and midlatitude dry lands merge to form continuous regions of moisture deficiency.

The concept of a "dry climate" is not as simple as it might sound. It involves, among other things, the balance between incoming moisture (precipitation) and outgoing (*evapotranspiration*, that is, the transfer of moisture to the atmosphere by transpiration from plants and evaporation from soil and plants). Climatic dryness depends not only on rainfall but also on temperature. The basic generalization is that higher temperature engenders greater potential evapotranspiration, so hot

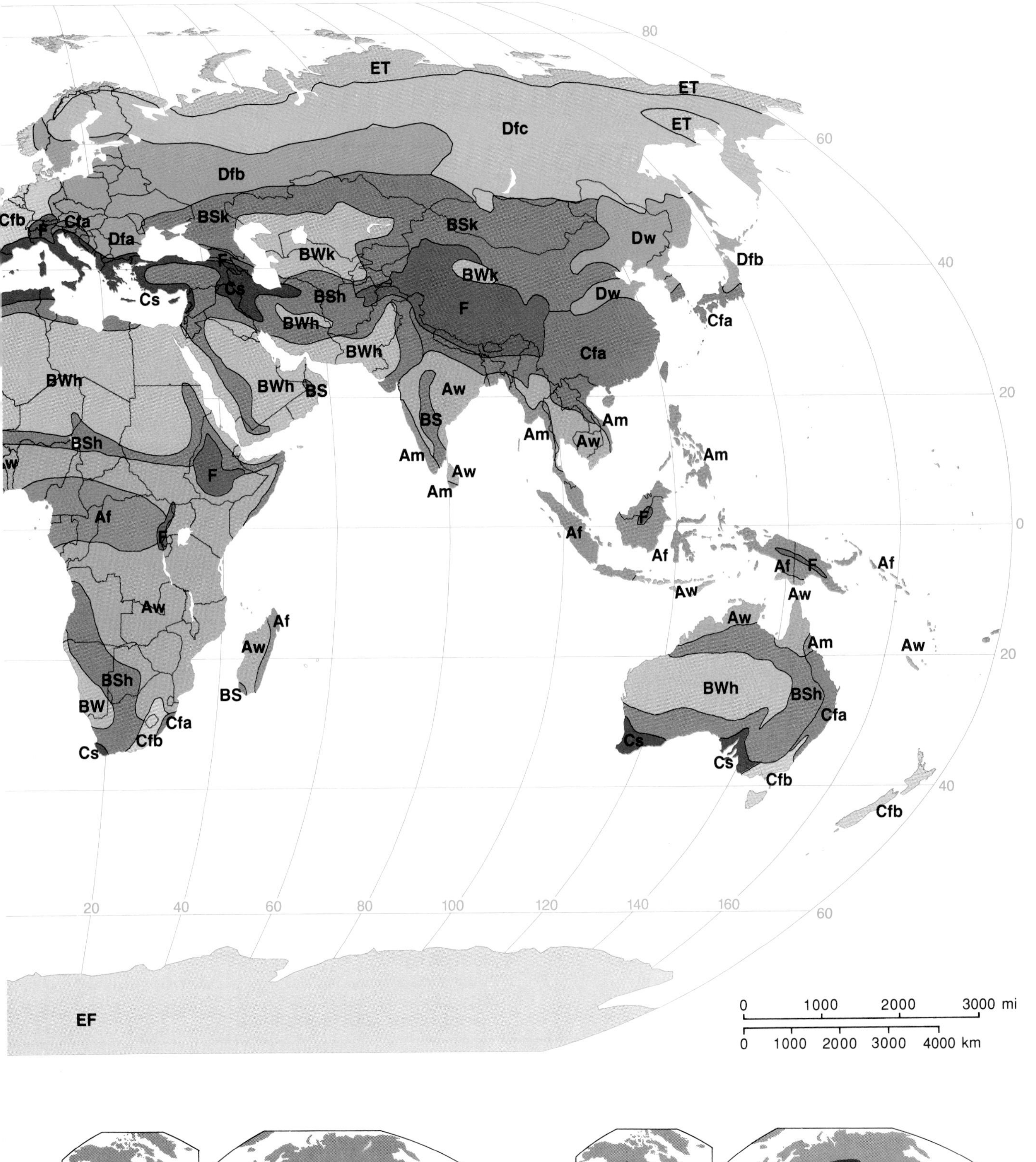

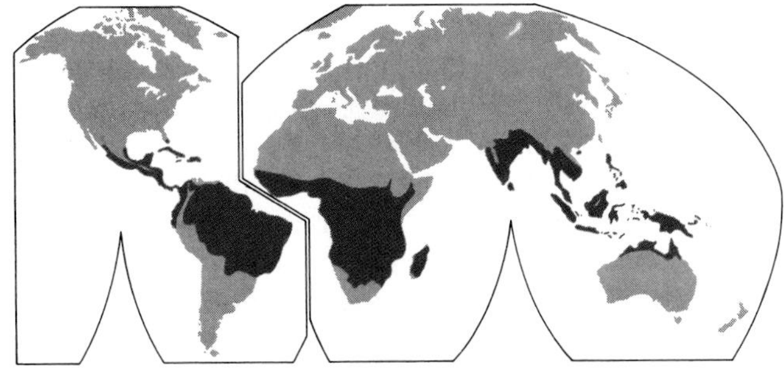

Figure 2–48. World distribution of A climates.

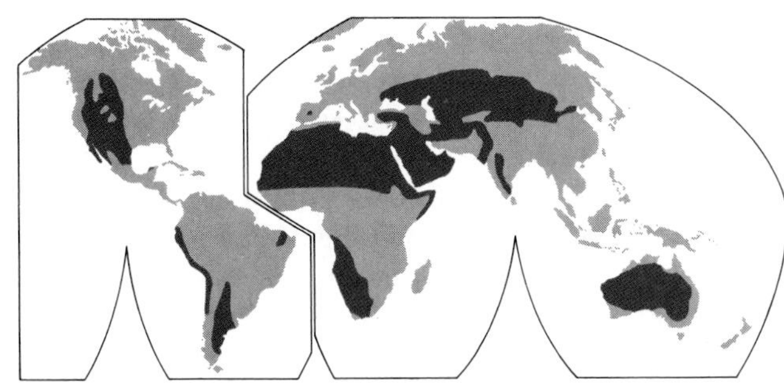

Figure 2–49. World distribution of B climates.

TABLE 2–3 ***Modified Köppen Climatic Classification***

Letters*						
1st	2nd	3rd	Derivation	Description	Definition	Types
A			Alphabetical	Low-latitude humid climates	Average temperature of each month above 64°F (18°C)	Tropical wet (Af) Tropical monsoonal (Am) Tropical savanna (Aw)
	f		German: *feucht* ("moist")	No dry season	Average rainfall of each month at least 2½ in. (6 cm)	
	m		Monsoon	Monsoonal; short dry season compensated by heavy rains in other months	1 to 3 months with average rainfall less than 2½ in. (6 cm)	
	w		Winter dry	Dry season in "winter" (low sun season)	3 to 6 months with average rainfall less than 2½ in. (6 cm)	
B			Alphabetical	Dry climates; evaporation exceeds precipitation		Subtropical desert (BWh) Subtropical steppe (BSh) Midlatitude desert (BWk) Midlatitude steppe (BSk)
	W		German: *Wuste* ("desert")	Arid climates; "true deserts"	Average annual precipitation less than approx. 15 in. (38 cm) in low latitudes; 10 in. (25 cm) in midlatitudes	
	S		Steppe, or semiarid	Semiarid climates; steppe	Average annual precipitation between about 15 in. (38 cm) and 30 in. (76 cm) in low latitudes; between about 10 in. (25 cm) and 25 in. (64 cm) in midlatitudes; *without* pronounced seasonal concentration	
		h	German: *heiss* ("hot")	Low-latitude dry climate	Average annual temperature more than 64°F (18°C)	
		k	German: *kalt* ("cold")	Midlatitude dry climate	Average annual temperture less than 64°F (18°C)	
C			Alphabetical	Mild midlatitude climates	Average temperature of coldest month is between 64°F (18°C) and 27°F (−3°C); average temperature of warmest month is above 50°F (10°C)	Mediterranean (Csa, Csb) Humid subtropical (Cfa, Cwa)
	s		Summer dry	Dry summer	Driest summer month has less than ⅓ the average precipitation of wettest winter month	Marine west coast (Cfb, Cfc)

regions can receive more precipitation than cool ones and still be "drier."

Geographers distinguish between *deserts*, which are arid, and *steppes*, which are semiarid. Deserts normally are large core areas of aridity surrounded by a transitional fringe of slightly less dry steppe. This arrangement is particularly marked in the subtropical dry lands, but it is less apparent in the midlatitudes, where the steppes may be more expansive than the deserts.

The B climates are subdivided into four types: ***subtropical desert*** (***BWh***), ***subtropical steppe*** (***BSh***), ***midlatitude desert*** (***BWk***), and ***midlatitude steppe*** (***BSk***).

Mild Midlatitude Climates (Group C)

We have noted previously that the middle latitudes have the greatest weather variability on a short-run basis. Seasonal contrasts also are marked in these latitudes. The midlatitudes lack the constant heat of the tropics and the almost-continual cold of the polar regions. This is a zone of contrasting air masses, with frequent alternating incursions of tropical and polar air producing more convergence than is found anywhere else except in the immediate equatorial zone. This air-mass conflict creates a kaleidoscope of atmospheric disturbances and weather

Letters*						
1st	**2nd**	**3rd**	**Derivation**	**Description**	**Definition**	**Types**
	w		Winter dry	Dry winter	Driest winter month has less than $\frac{1}{10}$ the average precipitation of wettest summer month	
	f		German: *feucht* (''moist'')	No dry season	Does not fit either s or w above	
		a	Alphabetical	Hot summers	Average temperature of warmest month more than 72°F (22°C)	
		b	Alphabetical	Warm summers	Average temperature of warmest month below 72°F (22°C); at least 4 months with average temperature above 50°F (10°C)	
		c	Alphabetical	Cool summers	Average temperature of warmest month below 72°F (22°C); less than 4 months with average temperature above 50°F (10°C)	
D			Alphabetical	Humid midlatitude climates with severe winters	4 to 8 months with average temperatures more than 50°F (10°C)	Humid continental (Dfa, Dfb, Dwa, Dwb) Subarctic (Dfc, Dfd, Dwc, Dwd)
	2nd and 3rd letters same as in C climates					
		d	Alphabetical	Very cold winters	Average temperature of coldest month less than −36°F (−38°C)	
E			Alphabetical	Polar climates; no true summer	No month with average temperature more than 50°F (10°C)	Tundra (ET) Ice cap (EF)
	T		Tundra	Tundra climates	At least one month with average temperature more than 32°F (0°C) but less than 50°F (10°C)	
	F		Frost	Ice cap climates	No month with average temperature more than 32°F (0°C)	
H			Highland	Highland climates	Significant climatic changes within short horizontal distances due to altitudinal variations	Highland (H)

* The code letters in the first column designate the various climatic types according to a modified Köppen system as described in the text.

changes. The seasonal rhythm of temperature is usually more prominent than that of precipitation. Whereas in the tropical climates the seasons are called "wet" and "dry," in the midlatitudes they clearly are "summer" and "winter."

The mild midlatitude climates occupy the equatorward margin of the middle latitudes, occasionally extending into the subtropics, and being elongated poleward in some west coastal situations (see Figure 2–50). They constitute a transition between the warmer tropical climates on one side and the severe midlatitude climates on the other.

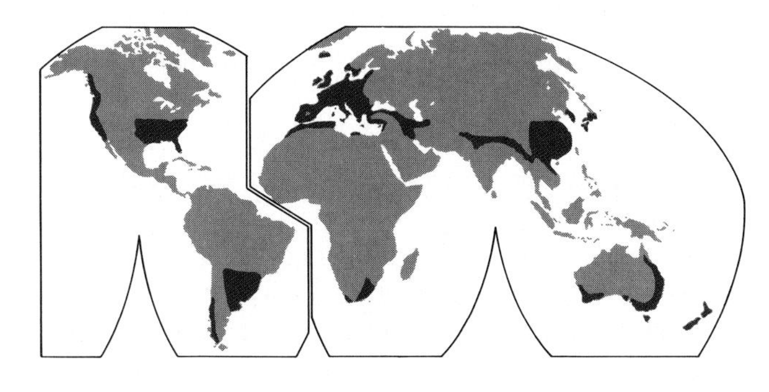

Figure 2–50. World distribution of C climates.

Focus on Desertification

The term *desertification* has come into prominent use since the 1960s to refer to the expansion of desert conditions into areas that were previously not deserts. The enlargement of deserts can come about through causes that are entirely natural, the most important being recurrent drought. It can also be caused by the activities of humans in degrading the land through overgrazing of livestock, imprudent agriculture, deforestation, and improvident use of water resources. The process of desertification is now generally accepted as a product of human-induced environmental degradation superimposed on a natural drought situation.

Desertification normally is associated with the margins of existing deserts and implies an expansion of the present desert into a spreading peripheral zone. However, this is not always the case. For example, "Dust Bowl" conditions in the North American Great Plains in the 1930s were not peripheral to any existing desert, yet they represented a clear and spectacular example of desertification long before the term was in common use.

The specter of desertification was brought to worldwide attention in the late 1960s by the onset of a cruel 6-year drought in the African Sahel. The Sahel is a subhumid-to-arid region on the southern margin of the Sahara Desert that occupies parts of ten countries, from the Atlantic Ocean on the west to the Ethiopian highlands on the east.

As the Sahelian drought intensified, from 1968 to 1974, vegetation disappeared, millions of livestock perished, tens of thousands of people died, and thousands of survivors migrated to an uncertain future in already overcrowded lands, particularly cities to the south of the drought-stricken region. Since 1974 the grip of the drought has been less intense, but almost every year has been "deficient" in rainfall, and the cumulative effect on existence in the Sahel has been overwhelming.

Traditional life in the Sahel was dominated by nomadic herding of goats, cattle, and camels. Herdsmen and their livestock followed the ITC rainbelt northward and southward in its seasonal migration. This generally kept human use of resources in balance with environmental conditions. Crop growing was mostly limited to favored river valleys. During the 1950s and 1960s, however, significant changes occurred. Human and livestock populations soared, due to improved medical conditions, above-average rainfall, and economic and political changes. Newly independent nations, however, closed international borders to traditional pastoral migration routes. Dry-land farming and some irrigation agriculture spread into traditional nomadic lands, tilling soil that should never have been broken. An enormous demand for firewood (the cooking and heating fuel for 80 percent of the Sahelian population) resulted in the elimination of most trees and large shrubs. Overgrazing removed most grasses and small shrubs. Then came year after year of searing drought. As a consequence of all these developments, land that formerly supported vegetation became desert.

Desertification is not limited to the Sahel. It is also being experienced on the northern margin of the Sahara, in southern Africa, in the Middle East, in India, in western China, in southern

Summers in the C climates are long and usually hot; winters are short and relatively mild. Precipitation conditions are highly variable, both as to total amount and seasonal distribution. Year-round moisture deficiency is not characteristic, but there are sometimes pronounced seasonal deficiencies.

The C climates are subdivided into three types, primarily on the basis of precipitation seasonality and secondarily on the basis of summer temperatures:

1. *Mediterranean* climates (*Csa*, *Csb*) have a pronounced winter-rain/summer-drought regime.
2. *Humid subtropical* climates (*Cfa*, *Cwa*) have hot summers and either no dry season or a winter dry season.
3. *Marine west coast* climates (*Cfb*, *Cfc*) have mild summers along with either no dry season or a limited summer dry season.

Severe Midlatitude Climates (Group D)

The severe midlatitude climates occur only in the Northern Hemisphere, because the Southern Hemisphere has no landmasses at the appropriate latitudes, 40° to 70°. This climatic group extends broadly across North America and Eurasia (see Figure 2–51).

Continentality is basic to the D climates. Landmasses are broader at these latitudes than anywhere else in the world. Even though these climates extend to the east coasts of the two continents, they experience little maritime influence because the general atmospheric circulation is westerly.

The most conspicuous result of continental dominance is the broad fluctuation of temperatures during the year. These climates have four clearly recognizable seasons: a long cold winter, a relatively short summer that varies from warm to hot, and transition periods in

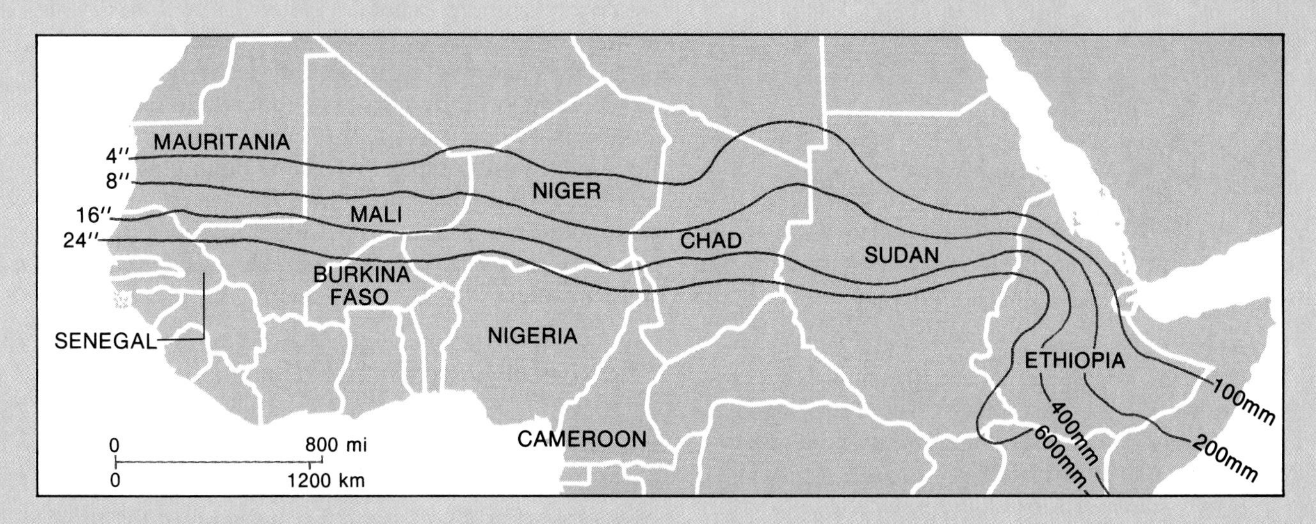

The Sahel Region of Africa. Isohyets of average annual rainfall are shown. Only the Sahelian countries are named.

Australia, in Chile and Peru, in northeastern Brazil, in the southwestern United States, in Mexico, and other places. The United Nations Environment Program estimates that 8100 square miles (21 million ha) of land is being desertified in the world each year.

But the Sahel is by far the largest and most overwhelming area of continuing desertification. Since the 1960s outside sources have contributed $10 billion to ease the Sahelian situation. Much of this has been for immediate famine relief and has had no positive long-term effect other than to prevent or postpone starvation for some people. There are no major success stories in the continuing, exhausting, complex struggle against Sahelian desertification.

What of the future? We know that the desert environment is fragile, but it is also resilient. The North American Dust Bowl recovered from its devastation of the 1930s. If human and livestock pressure on the Sahelian land could be relieved, and if a few years of "normal" rainfall would occur, we could expect striking improvement in the landscape. The burgeoning population, however, makes this an unlikely scenario. Indeed, drought-relief workers in the Sahel have suggested that either Mali or Mauritania could actually become wholly uninhabitable because of ecological catastrophe.

spring and fall. Annual temperature ranges are very large, particularly at more northerly locations where the winters are most severe. Precipitation is moderate, although throughout the D climates it exceeds the potential evapotranspiration. Summer is the time of maximum precipitation, but winter is by no means completely dry, and snow cover lasts for many weeks or months.

The severe midlatitude climates are subdivided into two types on the basis of temperature. The *humid continental* variety (*Dfa, Dfb, Dwa, Dwb*) has relatively long warm summers. The *subarctic* type (*Dfc, Dfd, Dwc, Dwd*) is characterized by short summers and very cold winters.

Polar Climates (Group E)

The polar climates are the most remote from heat (see Figure 2–52). Farthest removed from the equator, they receive inadequate insolation for any significant warming. By definition, no month has an average temperature of more than 50°F (10°C). If the wet tropics represent conditions of monotonous heat, the polar climates are known for their enduring cold. They have the coldest

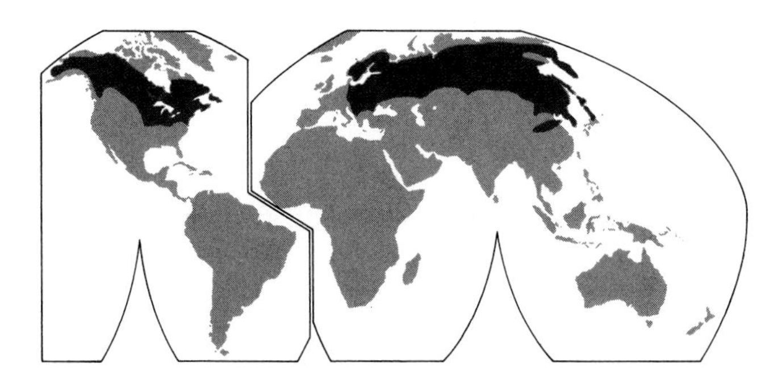

Figure 2–51. World distribution of D climates.

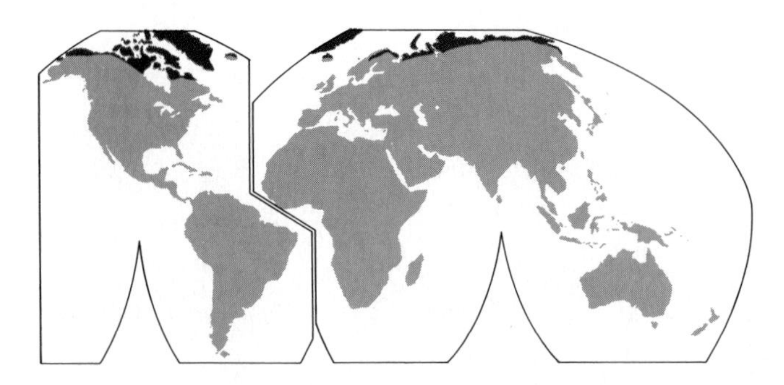

Figure 2–52. World distribution of E climates.

summers and the lowest annual and absolute temperatures in the world. They also receive very little precipitation, but evaporation is so minuscule that the group as a whole is classified as humid.

The two subcategories of the polar climates are distinguished by a difference in their low summer temperatures. The *tundra* climate (*ET*) has at least one month with an average temperature exceeding the freezing point. The **ice cap** climate (*EF*) does not.

Highland Climate (H)

Highland climate is not defined in the same sense as the other climatic types that have been discussed. Climatic conditions in mountainous areas have almost infinite variations from place to place, and many of the differences extend over very limited horizontal distances. Köppen did not recognize highland climate as a separate type, although most of his modifiers have added such a category. Highland climates are included (see Figure 2–53) in this book to identify relatively high uplands (mountains and plateaus) with extensive climatic variation within small areas.

Climates of highlands usually are closely related to those of the adjacent lowland regions, particularly with regard to seasonality of precipitation. Some aspects of highland climate, however, differ significantly from those of the surrounding lowlands. Latitude becomes less important as a climatic control than do altitude and exposure. The critical climatic controls on a mountain slope usually are relative elevation and angle of exposure to sun and wind.

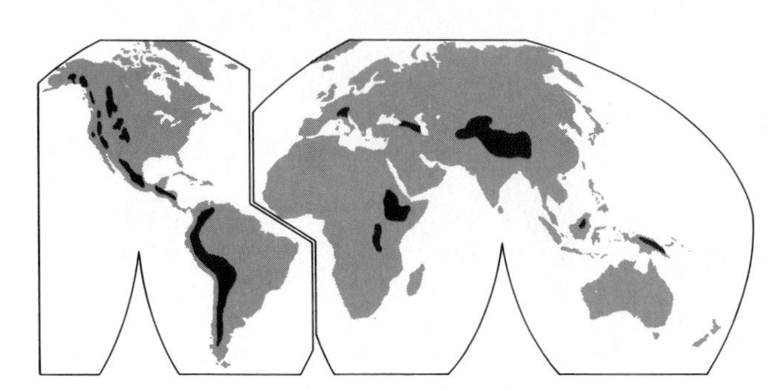

Figure 2–53. World distribution of H climates.

Changeability is perhaps the most conspicuous single characteristic of highland climate. The thin, dry air permits rapid influx of insolation by day and rapid loss of radiant energy at night, so daily temperature ranges are abnormally large. There is frequent and rapid oscillation between freeze and thaw conditions. Daytime upslope winds and convection cause rapid cloud development and abrupt storminess. Travelers in highland areas should be prepared for sudden changes from hot to cold, from wet to dry, from clear to cloudy, and from still to windy.

The Global Pattern Idealized

In the preceding pages, the basic characteristics and distribution of 15 major climatic types have been presented. From these data it is possible to summarize the global climatic pattern by constructing a model of the distribution of Köppen climatic types on a hypothetical continent of low and uniform elevation (see Figure 2–54). Its shape, with its great bulk in the Northern Hemisphere, corresponds roughly to that of the actual landmasses on Earth. This hypothetical continent illustrates how the climatic types would be distributed without the modifications and complications caused by the varying shapes, sizes, positions, and elevations of the Earth's actual landmasses.

Any such idealized, simplified representation is called a **model.** A good model simplifies reality enough to help us understand what our assumptions should lead us to expect. Then we can compare our model with reality, and if the real world deviates from the model in only minor ways or a few places, we may be able to

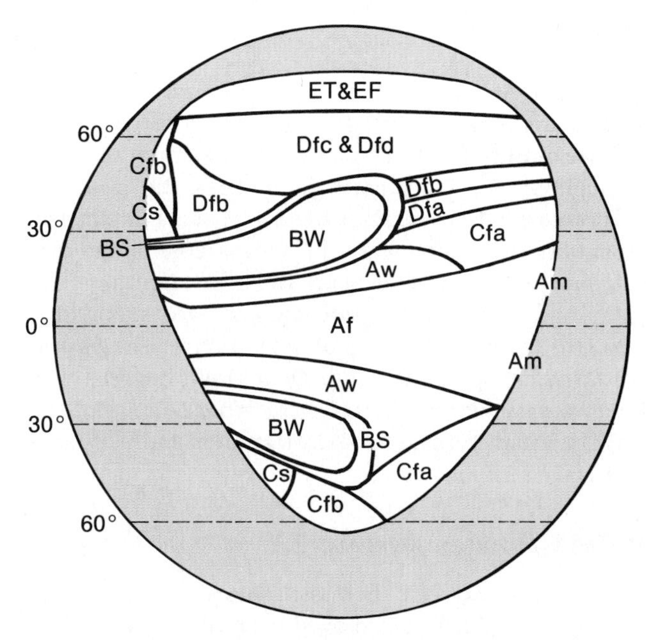

Figure 2–54. The presumed arrangement of Köppen climatic types on a hypothetical continent.

Focus on Deductive Geography

We have demonstrated the mapping of climates around the world and then simplified that map into a generalized model. We can now use that model to presume what the climate might be at any place. Reasoning with models introduces the differences between an inductive and a deductive approach in geography.

One reason we travel is that we think that other places are different from where we are. It is a travel agent's business to entice us to visit far-off places. "Visit exotic Marataria" trumpet the brochures, "and see things you've never seen before!" If you have already been to Marataria, how about a trip to lovely Jusopia, with its unique and special wonders and attractions? The travel agent must emphasize the uniqueness and specialness of each place because she or he wants us to buy tickets to all of them.

Geographers often take a similar view of the world, analyzing the ways in which specific elements of a place combine to give that place its individual character. The study of geography includes an important tradition of regional studies, which are in-depth analyses or syntheses of the attributes of different places. The content of regional geography courses usually has been an accumulation of many facts about individual places. The description of one place after another, usually according to a consistent set of criteria such as climate, resources, people, and economy, makes up a traditional world regional geography book, an encyclopedia of knowledge about the world's places. The popular *National Geographic* magazine similarly focuses on the individuality of each place.

Even the travel agent, however, would have to admit that places do have similarities. If you want to vacation on a tropical beach, you could choose any one of dozens in the Caribbean. Their similarity gives you an idea of what you are choosing when you select that category of vacation spot.

So it is with cities. Each large city has distinguishing characteristics, but it also has much in common with others. If you travel a lot or study many places, you soon begin to see many ways in which places are similar. Travelers begin to form categories, and each place they visit loses just a bit of its individuality in their mind as it becomes an example of a category. Urban planners look for similarities among cities in the hope of using one city's experiences to help solve similar problems in another city.

The two ways of looking at the world represented by the travel agent and the urban planner represent inductive and deductive ways of viewing reality. *Inductive thinking*, represented by the travel agent, begins with specific cases and leads to general conclusions. It proceeds stepwise, building up a picture of the world from a set of facts.

Deductive thinking, in contrast, starts from general assumptions about reality and then views each real-world example as a specimen. To the degree that any specimen varies from the abstraction, we try to explain the variation. If Buenos Aires, for example, which is in the abstract category of "big cities," does not have traffic jams, we try to explain why it deviates from that condition expected in that category. Deduction is a way of describing the uniqueness of places by explaining why they are not what we expect them to be.

To give just one example of deductive thinking, let us discuss the concept of distance. Distance is a powerful factor in our lives. We refer to the "friction of distance," and we try to overcome the cost and inconvenience of it. The distribution (the geography) of items in your home medicine cabinet demonstrates your attempt to overcome the friction of distance. You probably put the items you use most often on the closest, most convenient shelves, and the items you use less often farther away toward the top. Most people do, and we deduce that you do. If your cabinet demonstrates a significant deviation from this expected behavior, we would seek an explanation.

Köppen's model continent is an example of deductive geography. In Chapter 9 we will see how geographers Johann von Thünen and Walter Christaller used deductive notions to build theories of the use of land and of the distribution of cities across a landscape.

explain those local deviations due to local variations from our assumptions. This is an example of deductive reasoning. If, however, the real world bears little resemblance to our model, then we have to start over with new assumptions.

The model portrays the generalized distribution of all but the highland climate. Four of the climatic groups (A, C, D, and E) are defined by temperature, which means that their boundaries are strongly latitudinal because they are determined by insolation. The fifth climatic group (B) is defined by moisture conditions, and its distribution cuts across those of the thermally defined groups.

Such a model serves as a predictive tool. We can state, with some degree of assurance, that at a particular latitude and a general position (eastern, central, western) on a continent, a certain climatic type is likely to be found. Moreover, the locations of the climatic types relative to one another can be more clearly understood. The "real world" holds many refinements and modifications to this global pattern, but the schematic model shows the general alignments with considerable validity.

SUMMARY

The distinctive atmosphere of the Earth is a shallow encircling envelope of gases that makes life possible on our planet. The atmosphere is heated by a complex suite of processes that is set in motion by the arrival of radiant energy (insolation) from the sun. Some of this solar energy is absorbed directly by the atmosphere, but most of it is involved in a more complicated series of events that results in the direct transfer of energy to the atmosphere from the Earth's surface and only indirectly from the sun. Because of the high angle of the sun in low latitudes, there is a significant surplus of energy in the tropics and a corresponding deficit in the polar regions. However, the general circulation patterns of both atmosphere and oceans function to ameliorate this global temperature imbalance by transferring heat toward the poles and coldness toward the equator.

Atmospheric pressure is closely interdependent with density and temperature. Moreover, wind is an immediate response to differences in pressure. The direction of wind movement is determined by the balance among pressure gradient, the Coriolis effect, and friction. The general circulation of the atmosphere varies both seasonally and irregularly, but its major wind and pressure attributes are permanent in nature.

The amount of water vapor in the air, referred to as humidity, can be measured and expressed in several ways. Evaporation, condensation, and precipitation are the three basic processes that determine the role of moisture in the atmosphere. Precipitation is caused by complicated actions that usually are set in motion when large masses of air are forced to rise due to convection, orographic lifting, frontal uplift, or convergence.

Air masses and fronts are prominent components of major migratory pressure systems called extratropical cyclones and anticyclones that dominate midlatitude circulation. Other notable storms include tropical cyclones (hurricanes), thunderstorms, and tornadoes.

The Köppen climatic classification is widely used for pedagogic purposes. Based on temperature and precipitation characteristics, it recognizes six major groups of climatic types: tropical humid, dry, mild midlatitude, severe midlatitude, polar, and highland.

KEY TERMS

weather (p. 21)
climate (p. 21)
insolation (p. 21)
continental climate (p. 21)
maritime climate (p. 21)
currents (p. 23)
troposphere (p. 23)
average lapse rate (p. 23)
temperature inversion (p. 23)
isotherm (p. 24)
atmospheric pressure (p. 25)
barometer (p. 26)
millibar (p. 26)
isobar (p. 26)
updraft (p. 26)
downdraft (p. 26)
ascent (p. 26)
subsidence (p. 26)
wind (p. 26)
pressure gradient (p. 26)
Coriolis effect (p. 27)
friction (p. 27)
anticyclone (p. 27)
cyclone (p. 27)
subtropical high (STH) (p. 30)
trade winds (p. 30)
Intertropical Convergence Zone (ITC) (p. 30)
westerlies (p. 31)
jet stream (p. 31)
polar front jet stream (p. 31)
Rossby wave (p. 32)
polar high (p. 32)
polar easterlies (p. 32)
subpolar low (p. 32)
polar front (p. 32)
monsoon (p. 32)
water vapor (p. 33)
hydrologic cycle (p. 34)
humidity (p. 34)
absolute humidity (p. 34)
relative humidity (p. 34)
evaporation (p. 34)
condensation (p. 34)
dew point (p. 34)
condensation nuclei (p. 34)
adiabatic cooling (p. 35)
adiabatic warming (p. 35)
cirriform clouds (p. 35)
stratiform clouds (p. 36)
cumuliform clouds (p. 36)
fog (p. 36)
buoyancy (p. 36)
stability (p. 36)
rain (p. 37)
snow (p. 38)
sleet (p. 38)
hail (p. 38)
convective lifting (p. 38)
orographic lifting (p. 38)
frontal lifting (p. 39)
convergent lifting (p. 39)
isohyet (p. 39)
storm (p. 44)
air mass (p. 45)
front (p. 45)
warm front (p. 45)
cold front (p. 46)
stationary front (p. 46)
occluded front (p. 46)
extratropical cyclone (p. 47)
extratropical anticyclone (p. 48)
tropical cyclone (p. 50)
hurricane (p. 50)
thunderstorm (p. 50)
tornado (p. 52)
ice cap (p. 60)
model (p. 60)

QUESTIONS FOR INVESTIGATION AND DISCUSSION

1. In our study of physical geography, why do we concentrate primarily on the troposphere rather than on other zones of the atmosphere?

2. What is the importance of the so-called greenhouse effect to the heating of the Earth's surface?

3. Why do continental climates experience greater temperature extremes than maritime climates?

4. Pressure gradient is a pervasive force in the atmosphere. Why do winds not simply flow down the pressure gradient?

5. Why do trade winds cover such a large part of the globe?

6. How is atmospheric stability related to adiabatic temperature changes?

7. What are the principal harmful effects of acid rain?

8. Explain the differences between a warm front and a cold front. Why are there no fronts in an extratropical anticyclone? In a hurricane? In a tornado?

9. Plot the average monthly temperature and rainfall where you live. Which are the hottest and coldest months? In which climate type do you live?

10. Does your area suffer from acid rain? Does it export chemical pollutants to other regions?

ADDITIONAL READINGS

GEDZELMAN, S.D. *The Science and Wonders of the Atmosphere.* New York: John Wiley & Sons, Inc., 1980.

LUTGENS, F.K., and E.J. TARBUCK. *The Atmosphere: An Introduction to Meteorology*, 5th ed. Englewood Cliffs, NJ: Prentice Hall, 1992.

LYDOLPH, PAUL E. *The Climate of the Earth.* Totowa, NJ: Rowman & Allanheld, 1985.

MUSK, LESLIE, F. *Weather Systems.* Cambridge, England: Cambridge University Press, 1988.

PIELKE, ROGER A. *The Hurricane.* London: Routledge, 1990.

TREWARTHA, G.T., and L.H. HORN, *An Introduction to Climate*, 5th ed. New York: McGraw-Hill Book Company, 1980.

The Scientific Method

In the Middle Ages people thought that they had to prove things to know them. There are logical reasons, however, why it is impossible to prove almost anything.

For instance: Is there a hippopotamus in the room in which you are sitting right now? You probably cannot see one, hear one, smell one, or feel one, but how can you prove that one isn't there? How can you be absolutely sure that one isn't hiding? Perhaps the sandwich that you had for lunch has fermented into a drug that blinds you to hippos. Or maybe the Frisbee that knocked you on the head last week has affected your senses. Or maybe some other reason—no matter how improbable—prevents you from perceiving the hippopotamus. How can you know for certain? If you think about it long enough, you may go crazy, but you will not be able to prove that you are in a hippopotamusless room.

Early in the Renaissance, people decided that you really do not have to prove things. You can assume something is true until you are proven wrong. This way of thinking can help you accumulate knowledge faster.

One of the first people to articulate this idea was William of Ockham (or Occam), who lived from about 1285 until 1349. He cut away all the complications from medieval thinking with a much simpler theory, appropriately called *Ockham's razor*. "Entities must not unnecessarily be multiplied," he said. In other words, the simplest explanation that fits the facts is most probably true. If you are given a choice among possible explanations for any observation or phenomenon, you should prefer the simplest. If you cannot see, hear, smell, or touch a hippopotamus in your room, there probably isn't one there.

THE SCIENTIFIC METHOD DEFINED

Ockham's articulation of his razor eventually led to the step-by-step technique of investigating phenomena that we call the **scientific method.** This is a set of procedures used in the systematic pursuit of knowledge. If you want to learn about the world, follow these three steps: (1) Observe the world, (2) explain what you see as simply as possible, and (3) hold on to your tentative explanation until it is disproved.

Observe the World

At the time of the Renaissance many people believed that the statements given in ancient Greek or Latin texts or in the Bible should override their own observations. People were not confident that they were as smart as the ancients had been, and, in addition, they believed that much of what they saw in the world was the work of the devil. Therefore, if you believed your own senses you might be fooled and suffer eternal damnation.

The Bible records that Joshua said "Sun, stand thou still," and "the sun stood still." (Josh. 10.10) This passage was taken as proof that the sun goes around the Earth. Anyone who believed that the Earth goes around the sun was wrong and was considered heretical. Thus, when the Italian astronomer Galileo (1564–1642) wrote that the Earth goes around the sun, he was tried and punished. "I think," he wrote in his defense, "that in discussions of physical problems we ought to begin not from the authority of scriptural passages, but from sense experiences and necessary demonstrations. God is no less excellently revealed in Nature's actions than in the sacred statements of the Bible." A Church court nevertheless forced him to apologize for teaching that the Earth goes around the sun.

Explain What You See as Simply as Possible

Your explanation of your observations is called a hypothesis. A **hypothesis** is a proposition that is tentatively assumed in order to draw out all its consequences so that it can be tested against all the facts an investigator can gather. You must remember to cut away (with Ockham's razor) all unnecessary complications and explain what you see as simply as possible. If you cannot see, hear, smell, or touch a hippopotamus in your room, you should not hypothesize that one is hiding or that you have been drugged. Hypothesize that no hippopotamus is there.

It is not always easy to suggest a hypothesis. Sometimes it is hard to think of any possible explanation for an observed phenomenon, and it takes a brilliant insight to see a link between two facts that has never occurred to anyone else before. Biologist Sir Peter Medawar wrote: "Hypotheses appear in scientists' minds along uncharted byways of thought. . . . they are imaginative and inspirational in character. . . . they are adventures of the mind."

Hold on to Your Hypothesis Until It Is Disproved

If your hypothesis withstands test after test, then you may become confident in it. If I hold out my hand and let go of my pencil, I do not know for certain that it will fall to the floor. I know that that has happened before so regularly that people have gained considerable confidence in the notion of gravity. The force of gravity is a **theory,** which is a hypothesis that has been given probability by experimental evidence. I cannot, however, be absolutely certain that my pencil will fall if I let go. If I let go and my pencil hovers in the air, then the theory of gravity will have to be discarded, and we will have to hypothesize a new theory. Gravity is only a theory.

"Hypotheses appear in scientists' minds along uncharted byways of thought. . . . they are imaginative and inspirational in character. . . . they are adventures of the mind."

It should be noted that because the sun rises in the east and sets in the west, a geocentric (Earth-centered) theory that the sun goes around the Earth is actually the simplest possible explanation of the observed phenomenon, and it was held by thoughtful people, including Aristotle, for a long time. The geocentric theory cannot, however, be reconciled with additional observations of celestial phenomena. The Polish astronomer Copernicus (1473–1543) first proposed the heliocentric theory to replace the geocentric theory. The weight of observation supports the heliocentric theory. It explains all observed celestial phenomena better.

A theory that has withstood testing over a long time is often called a **law,** but this term is confusing and should be avoided. What does it mean to say that there is a "law of gravity"? Who passed it? Who enforces it?

The English philosopher and statesman Sir Francis Bacon (1561–1626) said, "We cannot command Nature except by obeying Her." In other words, we do not pass laws for nature to obey. We study nature's principles and try to understand them. Once we do understand them, however, we can harness them to work for us. Medieval magicians unsuccessfully waved magic wands and commanded nature to change one thing into another by reciting magical spells, but modern chemists can break up materials and reconstitute them into new products because the chemists understand and obey the principles of molecular bonding.

We do not always understand nature's principles as completely as we think we do. Theories that have been held for a long time and that seem "obvious" can be overturned overnight. For example, Sir Isaac Newton's theory that time is an absolute value was overturned by Albert Einstein's theory of relativity. Therefore, a true scientist will never say that he or she is certain of any theory. Scientists proceed on the basis of hypotheses. I only hypothesize that I am sitting on a chair, that I am wearing a blue shirt, and that I hear cars honking in the street outside. Who knows, maybe there was something about that sandwich I had for lunch.

GEOGRAPHIC COINCIDENCE AND CAUSE-AND-EFFECT

In geography, one use of the scientific method is to try to explain patterns. Geographers note when the occurrence of a phenomenon forms a pattern, or when two different phenomena occur in the same pattern. We read, for example, that certain climates often appear at corresponding latitudinal and continental locations around the world. This was a pattern. We also read that the patterns of certain types of climates and of certain types of soils were the same. This ability to recognize patterns may seem simple, but actually it is a learned skill that geographers exercise in common with art historians and others in the visual arts.

Once a pattern is discovered, then geographers hypothesize about what might be the cause of the pattern. When two phenomena exhibit the same pattern, they are called **coincidental,** and geographers hypothesize whether one of those phenomena might cause the other. There may be a cause-and-effect relationship between them, but we cannot immediately assume that one of the phenomena causes the other. We have to hypothesize a cause-and-effect relationship and then investigate it.

When scientists test a hypothesis, they experiment to see whether it is repeatable. Some scientists argue that the repeatability of results is actually what defines a science. If a chemist in Los Angeles performs an experiment and reports its results, a chemist in a laboratory in Paris must be able to repeat the experiment with exactly the same results.

If we geographers want to test a hypothesis about a cause-and-effect relationship between two phenomena, however, we cannot isolate the phenomena we want to test as easily as a chemist can isolate two chemicals in a test tube. The world is too complicated. An economic development policy that works in Thailand may not work in Argentina. This may be because of any one of a million factors that differentiate Thailand from Argentina, and none of these other factors may be under our control. To test geographical cause-and-effect hypotheses, we must compare as many different places as possible, and we must search for exceptions.

When beginning geographers make a few observations of a coincidental occurrence of two phenomena, they may hastily hypothesize that some factor *A* always causes some factor *B*: "Natural resources cause national wealth"; "A certain climate causes civilization"; "A certain religion causes social peace." The assumption that one factor alone is sufficient to explain any event is called a *single-factor hypothesis*. In many cases, however, we need to consider several factors.

Further research and study often reveal exceptions to our first observations. The world is rich in examples of different environments, of different ways in which human cultures have evolved and adapted to those environments, and of different ways in which various aspects of cultures have interacted. Therefore, we can test our hypothesis by hunting for exceptions to our first observations. If we cannot find exceptions or examples in the world today, we must explore the other dimension: time. We may find an exception or example at some place at some time in the past. The study of geography and the study of history are often complementary.

Some people tease geographers for being obsessed with finding unique cases of phenomena, places where things are different, even "odd." They say that geography is just a sophisticated game of *Trivial Pursuit*. The reasons that geographers collect examples of phenomena and exceptional cases, however, are to test cause-and-effect hypotheses and also to catalog the tremendously rich variety of phenomena on Earth. If a geographer finds just one example of a way of life, for example, on a tiny Pacific island or in a mountain valley in Switzerland, then our catalog of possibilities has been enlarged. The options among which we may choose to organize our own lives have increased.

Because geographers cannot always disentangle the interelationships among the phenomena we study, we must research cause-and-effect links by stating and restating certain questions very carefully. If any factor *A* (a certain climate, natural resources, or a certain religion) is coincidental with phenomenon *B* (a certain soil, national wealth, or environmental degradation) in one place, then we might hypothesize that phenomenon *A* is causing *B*. If we later find *B* someplace else where there is no *A*, then we might restate our hypothesis. We might hypothesize that *A* sometimes causes *B* but that *B* can also be caused by some other factor *C*. Then we have to guess what that factor *C* might be. Maybe *B* can be caused by a factor *D*, an *E*, or an *F*, as well. Sometimes many factors combine to explain the phenomenon under investigation.

Similarly, if we hypothesize that *A* is causing *B* at one place, and if we then observe *A* without *B* in another place, we might hypothesize that *A* has not yet had time to trigger *B* in that place.

We must carefully differentiate between factors that are *necessary* to cause an event and factors that are *sufficient* to cause an event. If *A* is necessary to cause *B*, then we cannot find *B* without *A*. If *A* is necessary but not sufficient to cause *B*, then it will be impossible to find *B* without *A* plus some other necessary factor. We must investigate what that other factor is. If *A* is both necessary and sufficient to cause *B*, then we will find *B* wherever there is *A*.

Maybe *A* and *B* are always found together, but there is actually no relationship between them. In common speech this is usually called "*mere* coincidence." Maybe *A* and *B* are always found together, but both are actually caused by some third factor *Q*. As we know, certain kinds of vegetation and certain soils are found together, but one does not cause the other. Rather, both are caused by climate. Still another hypothetical relationship is that *B* causes *A*.

How many potential geographical hypotheses can you count in the last four paragraphs?

Partial Explanation

The problem of soil classification illustrates that in geography one thing seldom explains another thing completely. Usually, one factor explains only part of another factor; it contributes to an explanation.

Long ago, soil scientists noted that soils vary according to the nature of the rock on which they form. Later, however, as observers gathered more information about the distribution of soils, they noted that the distribution of soil types actually varies more closely with climate than it does with bedrock. The various climates explain more of the variation in soil type than do the various bedrocks. Bedrock and climate are each partial explanations, and climate contributes a greater share of the explanation.

The science of statistics breaks down the degree of variation in one factor that can be attributed to any one of several other factors. When statisticians describe the variation of two measures together, called *covariation*, they use the word *explain* in a very special sense. To statisticians an explanation does not really answer the question of why something happens when something else happens. To them explanation means only that one factor varies with the other with a definable regularity. When a statistician says that "variation in factor *X* explains 80 percent of the variation in factor *Y*," the statistician is saying that they vary together 80 percent of the time. The statistician is not explaining anything in the normal sense of the word *explain*.

Our understanding of soil formation is not exact enough to analyze statistically the covariation of soils with either bedrock or with climate. That is, we cannot say that climate accounts for 65 percent and bedrock for 35 percent of the variation in soil types. There are simply too many other factors involved—windiness, slope, and so on.

In geography, as in everyday life, we can often list probable causes for many things . . . but we cannot always assign exact relative weights to the individual causes.

In geography, as in everyday life, we can often list probable causes for many things, and sometimes we can even rank those causes in order of their probable explanatory force, but we cannot always assign exact relative weights to the individual causes. We can say, for example, that the United States is a rich country because it has many natural resources, it has a history of relative political stability, and its economic system has encouraged people by allowing them to enjoy the rewards of their enterprise. These, however, are only a few of the relevant factors, and we cannot measure statistically the exact contribution of each of them.

In some cases geographers record that two things are found at the same place at the same time, but the degree of cause-and-effect relationship between them—if any—simply is not known. Any one or a combination of a tremendous number of other factors may be relevant. These are challenging questions for continuing research, and we will note a great number of such questions in this text.

KEY TERMS

scientific method (p. 64)
hypothesis (p. 64)
theory (p. 65)
law (p. 65)
coincidental (p. 65)

ADDITIONAL READING

BRONOWSKI, J., and BRUCE MAZLISH. *The Western Intellectual Tradition.* New York: Harper & Row, 1960.

HARVEY, DAVID. *Explanation in Geography.* London: Edward Arnold, 1969.

CHAPTER 3

Biota and Soils

A rich variety of types of vegetation flourishes in the mild climate of the Columbia River Gorge region in Oregon. (Courtesy of John M. Roberts/The Stock Market.)

Among the most important themes in geography are the interrelationships among its diverse elements. Climate is, in itself, a very significant environmental component, but it takes on added importance because of its notable effects on other components. In this chapter we focus on three particularly important relationships: climate and vegetation, climate and animal life, and climate and soils.

BIOTA

The term **biota** refers to the total complex of plant and animal life. Even the simplest of living organisms is extraordinarily complicated. When a student sets out to learn about an organism, whether it is an alga or an anteater, a tulip or a turtle, she or he embarks on a complicated quest for knowledge. An organism differs in many ways from other aspects of the environment, but most significantly in that it is alive, and its survival depends on an enormously intricate set of life processes. If our goal as geographers were to seek a relatively complete understanding of the world's organisms, we would need to live for centuries. As beneficial as such knowledge would be, however, a student of geography can settle for a less ambitious objective. With organisms, as with every other feature of the world, the geographer must focus on certain aspects rather than on a complete comprehension of the whole.

The geographical viewpoint is the viewpoint of broad understanding, whether dealing with plants and animals or anything else. This does not mean that the individual organism is to be ignored; rather, it implies that generalizations and patterns are to be sought and their significance assessed. Here, as elsewhere, the geographer is interested in distributions and interrelationships.

Biogeochemical Cycles

The web of life consists of a great variety of organisms coexisting in a diversity of associations. These organisms survive through a bewildering complex of systemic flows—flows of energy, water, and nutrients—that nourish them. The various chemical elements of these flows have been maintained by a cyclic passage through the tissues of plants and animals. In other words, these elements are absorbed by plants and animals as sustenance and are then returned to the air, water, or soil through decomposition.

These grand cycles, ultimately energized by solar insolation, sustain all life on our planet. They have continued for millennia, at rates and scales almost too vast to imagine. In recent years, however, the rapid growth of human population, accompanied by an extraordinary rate of exploitation of the Earth's resources, has adversely affected every one of these cycles. None of the damage is irreparable, but the danger is increasingly threatening. If the living things of our planet are to survive and prosper, the components of the biogeochemical flows must undergo continual recycling. This is to say that after these different elements are used, they must be converted into a reusable form. For some elements this can be accomplished in less than a decade; for others it may require hundreds of millions of years.

Flora

The natural vegetation, or **flora,** of the land surfaces is of particular interest to the geographer for three reasons:

1. Over much of the Earth, the terrestrial flora is the most significant visual component of the landscape. Plants often grow in such profusion that they hide or mask all other elements of the environment.
2. Vegetation is a sensitive indicator of other environmental attributes. By studying the characteristics of the flora of a particular region, we can often distinguish subtle variations in sunlight, temperature, precipitation, evaporation, drainage, slope, soil conditions, and other natural phenomena. Moreover, the vegetation frequently exerts a major effect on soil and animal life.
3. Vegetation often has a prominent and tangible influence on human settlement and activities. In some cases, it is a barrier or hindrance to human endeavors; in other cases, it provides an important resource to be exploited or developed.

Floristic Terminology Plants, like animals, belong to many different categories, or *taxa*. By familiarizing ourselves with some of the basic taxonomy, we can acquire important background information for our study of flora.

Taxonomic summation An important biological distinction can be made between "higher" and "lower" plants. Lower plants include two major groups: bryophytes and pteridophytes.

Bryophytes are mosses and liverworts. Presumably they have never in geologic history covered large areas of the Earth. **Pteridophytes** are ferns, horsetails, and club mosses, which are spore-bearing (rather than seed-bearing) plants. During much of geologic history, great forests of tree ferns, giant horsetails, and tall club mosses dominated continental vegetation. They are less widespread today.

The so-called higher plants reproduce by means of seeds. They are encompassed within two broad groups: gymnosperms and angiosperms.

The **gymnosperms** ("naked seeds") carry their seeds in cones. When the cones open, the seeds fall out. Gymnosperms were more important in the geologic past; today the only large surviving gymnosperms are cone-bearing trees such as pines. **Angiosperms** ("vessel seeds") are the flowering plants. They have seeds that are encased in some sort of a protective body, such as a fruit, nut, or seed pod. Most higher plants—including trees, shrubs, grasses, crops, weeds, and garden flowers—are angiosperms.

Descriptive terms Several other terms are commonly used to describe types of associations of vegetation. Some of the most important are discussed below.

1. The fundamental distinction among higher plants is often on the basis of the composition of the stem or trunk. *Woody plants* have stems composed of hard fibrous material, whereas *herbaceous plants* have soft stems. Woody plants are mostly trees and shrubs; herbaceous plants are mostly grasses, forbs (broad-leafed plants) and lichens.
2. An important distinguishing characteristic for trees is whether their leaves fall during particular seasons. An **evergreen tree** sheds its leaves on a sporadic or successive basis but at any given time appears to be fully leaved. A **deciduous tree,** in contrast, experiences an annual period in which all its leaves die and usually fall from the tree, due either to a cold season or a dry season.
3. Trees also are often described in terms of their leaf shapes. **Broadleaf trees** have leaves that are flat and expansive in shape, whereas **needle-leaf trees** are adorned with thin slivers of tough, leathery, waxy needles. Almost all needle-leaf trees are evergreen, and the great majority of all broadleaf trees are deciduous, except in the rainy tropics where all trees are evergreen.
4. **Hardwood** and **softwood** are two terms that are frequently used but are of doubtful value. Hardwoods are angiosperm trees that are usually broad-leaved and deciduous. Their wood has a relatively complicated structure, but it is not always hard. Softwoods are gymnosperms; nearly all are needle-leaf evergreens. Their wood has a simple cellular structure, but it is not always soft.

Climatic Climax Vegetation One of the difficulties in mapping world vegetation is that it is often changing. Even the most casual observer cannot help but notice such change in the local environment: Weeds invade the garden; grass pops up through cracks in the sidewalk; untended public parks grow wild; forests reclaim land that has been abandoned by farmers. Occasionally we read that the jungle has completely overgrown an ancient complex of temples somewhere in the tropics.

A full-grown jungle rich in varieties of species does not appear overnight but rather develops in slow stages. At any fresh spot the community of plants will naturally evolve through a series of increasingly complex stages of types of vegetation. The series that evolves at any spot differs, depending on the situation, but in each case the natural succession of plant types is called a **primary series,** or **prisere,** for that location. Imagine your classroom building being abandoned for hundreds of years. As it collapsed and crumbled, which types of plants would establish themselves in its cracks first? Which types would replace them? What would the area look like after 300 years?

A prisere has not developed to its final form if the variety of species that dominate in a place is still changing and if these plants are modifying certain aspects of the local environment, such as the soil beneath them, the conditions of light and shade, and the local drainage. In such cases, the plant species are in fact making the site more hospitable for a succession of plant communities, each of which will outcompete and thereby replace the existing community. Ultimately one community will evolve that can compete most successfully under the existing physical conditions. This community will establish itself and persist indefinitely, as long as the underlying physical conditions are relatively balanced. This stable association of vegetation in equilibrium with local soil and climate conditions is called **climatic climax vegetation.**

This concept enables us to think of the Earth's vegetation not as a static patchwork but as a dynamic system continually accommodating environmental changes and human interference. It is more useful than the term "natural vegetation," which is actually very difficult to define. What seems to be "natural" vegetation at any place and time may be in a state of flux.

Many places do not exhibit true climatic climax vegetation, and for many others we do not even know what the local climatic climax vegetation is because the flora is not yet fully mature but rather is still evolving. All land on Earth was once fresh and uninhabited by life forms—lava flows, recently uplifted sea floors, and so forth. Although vegetation covered some places millions of years ago, it has penetrated other places only recently. Areas of tropical rainforest in South America and Africa, for example, have been evolving without interruption for millions of years, but Northwest Europe and much of North America were scoured by glaciers just a few thousand years ago. The glaciers left a bare, inorganic surface, and the soils and vegetation located there today have had only a relatively short time to develop. This vegetation is still changing. *Cataclysmic* events such as floods, mudslides, or wildfires episodically clear substantial areas of their plant cover, but after each such event the vegetation resumes its slow evolution toward climax conditions.

Humans have interfered with the plant cover over much of the Earth, and in many places this has been occurring for so long that we do not know what plants could grow if humans would leave the area alone for a few centuries. Areas of China that today are grasslands are spotted with towns named "Wang," which means "oak tree." We assume that oaks must have grown there in the past, but we cannot be sure whether the climate has changed so that oaks can no longer grow there or whether human interference has destroyed the oaks.

Much of the Earth's surface is devoted to pastures or to crops, and people prevent it from reverting to any other state. An environment thus artificially maintained or arrested is called *plagioclimax*, or *false climax*. However, if a farmer's field is abandoned, priseral evolution recommences. The only true climax vegetation is climatic climax.

Fauna

Animals, known technically as **fauna,** occur in much greater variety than do plants. As objects of geographical study, however, they are less important than plants, for at least two reasons. First, animals are much less prominent in the landscape. Apart from extremely localized situations (see Figure 3–1), animals tend to be secretive and inconspicuous, whereas the vegetation is not only fixed in position, it also serves as a relatively complete ground cover wherever it has not been removed by human interference. In addition, animals do not provide the clear evidence of environmental interrelationships that plants do. This is due in apart to the inconspicuousness of wildlife, which renders it more difficult to study, but it is also due to the fact that animals are mobile and therefore more able to adjust to changes in the environment.

Food Chains The unending flows of energy, water, and nutrients through the Earth's biota are channeled in significant part by direct passage from one organism to another in a process referred to as **food chains.** A food chain is a simple concept: One organism eats another, thereby absorbing energy and nutrients; the second organism is consumed by a third, with similar results; the third organism is eaten by a fourth; and so on. In nature, however, the matter of who eats whom may be extraordinarily complex.

The fundamental unit in the food chain consists of plants that trap solar energy through **photosynthesis,** a process whereby plants utilize sunlight to produce stored chemical energy from water and carbon dioxide. The plants are then eaten by *herbivorous* (plant-eating) animals, which are considered the *primary consumers*. Herbivores in turn become food for other animals, called *secondary consumers*. There may be many levels of secondary consumers—for example, the plant-eating beetle is eaten by a frog, which is devoured by a snake, which is consumed by a hawk, and so on.

Evolution of Species Wherever the local environment remains stable enough to support evolution over long periods of time, fragile superstructures of species develop. *Mutation* within species seems to be random; that is, each generation of any given species of plant or animal will have a proportion of significantly new forms. If any new form survives and multiplies, the species has become modified, or a new species has evolved. Many new forms do not survive to multiply and establish themselves; others persist for long periods and then die out. Each species is a sort of natural experiment. Generally, the longer the period during which evolution has con-

Figure 3–1. Seals and sea lions are unusual animals in that they are highly gregarious. Most animals are secretive and inconspicuous and therefore are of less interest to geographers than are plants. This scene is near Point Año Nuevo on the central coast of California. (TLM photo)

ABC NEWS ***EPA Ozone Depletion More Serious Than Thought***

One of the major environmental problems currently threatening the flora and fauna on our planet is the depletion of the ozone layer. Ozone is a form of oxygen that is composed of three atoms (O_3) rather than two (O_2). Ozone molecules are created in the upper atmosphere, where they form a layer that protects plant and animal life from the harmful ultraviolet rays of the sun. Depletion of the ozone layer threatens the well-being of many species.

The only way to determine the extent of the damage to the ozone layer is through the use of satellite images. Without the use of satellites, we might not have known that this situation was occurring, until it was too late.

The depletion of the atmospheric ozone has been attributable to the increasing use of *chlorofluorocarbons* (CFCs), chemical compounds that contain chlorine, fluorine, and carbon atoms.

CFCs are most commonly used in air-conditioning and refrigeration units and in the manufacture of plastics. When these compounds are released, they rise into the upper atmosphere. There, the chlorine atoms break down ozone molecules into standard oxygen molecules. To prevent this, researchers are looking for other chemicals that can be substituted for CFCs.

The video listed above conveys the thoughts of a number of people concerning this problem.

1. Do you think that it will be possible to reverse the effects of ozone depletion, or are we only going to be able to stop the depletion at the present level?
2. In what ways can ozone depletion affect the food chains described in this chapter?

tinued uninterruptedly at a place, the more complex the community of species will be.

This explains why the oldest vegetative complexes on Earth, the great tropical rainforests, are so rich in species. They have uniquely high ***diversity indices***; that is, numbers of different species per unit of area. A 2.5-acre (1-ha) plot in the upper Amazon basin of Peru has been discovered to support 300 species of trees, which is almost one-half the number native to the entire United States.

Species of plants and animals may migrate. However, because of the seeming randomness of evolution, many of the species at any place may be unique to that place, and certainly the mix of species at any place is always unique. Each spot on Earth has a unique inventory of biological resources.

In our search for organizing principles through which to study world biota, two concepts are of particular value: *ecosystem* and *biome*.

ECOSYSTEMS

The term **ecosystem** is a contraction of the phrase *ecological system*. An ecosystem includes all the organisms in a given area, but it is more than simply a community of plants and animals living together. Rather, it involves the totality of interactions among the organisms and between the organisms and the nonliving portion of the environment in the area under consideration. An ecosystem, then, is fundamentally a biological community, or an association of plants and animals, expressed in functional terms; that is, it takes into account the interactions among the various components of the community. The concept is built around the flow of energy among the various components of the system, which is the essential determinant of how a biotic community functions (see Figure 3–2).

Although ecosystems might be easy to define, they are not always easy to identify. One problem in isolating an ecosystem is that of scale: An ecosystem can range in magnitude from a planetary system that encompasses all life on the planet to the ecosystem of a fallen log or even a drop of water. Therefore, if we are going to try to identify and understand broad distributional patterns of biota, we must focus only on ecosystems that can be recognized at a useful scale.

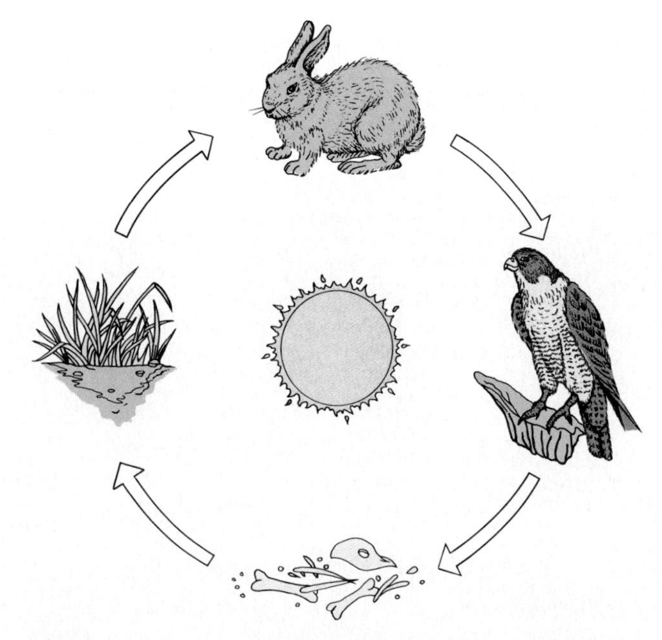

Figure 3–2. The flow of energy in a simple ecosystem. Energized by the sun, grass feeds a rabbit, which is eaten by a hawk, whose decaying remains furnish nutrients for the grass.

Biomes

Among terrestrial ecosystems, the one that provides the most appropriate scale for understanding world distribution patterns is called a **biome.** A biome is a large, recognizable assemblage of plants and animals in functional interaction with its environment. It is usually identified and named on the basis of its dominant vegetation association.

There is no universally recognized classification of the world's biomes, but scholars commonly accept about ten major biomes:

1. Tropical rainforest
2. Tropical deciduous forest
3. Tropical scrub
4. Tropical savanna
5. Desert
6. Mediterranean woodland and shrub
7. Midlatitude grassland
8. Midlatitude deciduous forest
9. Boreal forest
10. Tundra

A biome is composed of much more than merely the plant association that gives it its name. Many other kinds of vegetation usually grow among, under, and occasionally over the dominant association. Diverse animal species also occupy the area. Often, significant and even predictable relations are seen among the biota (particularly the flora) of the biome and the associated climate and soil types. The major biomes are summarized in the following section and are mapped in Figure 3–3.

Tropical Rainforest The **tropical rainforest** of the low latitudes, also called the **selva,** is probably the most complex of all terrestrial ecosystems. It contains a variety of trees growing close to one another, as portrayed in Figure 3–4 on page 76. Tropical flora consists primarily of tall, high-crowned, broad-leaved species that never experience a seasonal leaf fall because the climate is warm and moist throughout the year. Undergrowth is normally sparse in the selva because sunlight, which is essential to the survival of most green plants, cannot penetrate the dense forest covering, or *canopy*. Only where there are gaps in the canopy, as alongside a river, does light reach the ground, resulting in the dense undergrowth associated with a "jungle."

The interior of the selva, then, is a region of heavy shade, high humidity, windless air, continual warmth, and an aroma of mold and decomposition. As plant litter accumulates on the forest floor, it is acted on very rapidly by plant and animal decomposers, which find optimal conditions for their activities. The upper layers or "stories" of the forest are areas of high productivity, and there is a much greater concentration of nutrients in the vegetation than in the soil. Indeed, most selva soil is surprisingly infertile.

Rainforest fauna is largely **arboreal** (tree-dwelling) because the principal food sources are in the canopy rather than on the ground. Large animals are generally scarce on the forest floor, although there are vast numbers of **invertebrates,** animals without backbones, including insects. The animal life of this biome is typified by creepers, crawlers, climbers, and fliers. Common animal types include monkeys, arboreal rodents, birds, tree snakes and lizards, and countless species of invertebrates.

Rainforests occur in climatic regions characterized by consistent rainfall and relatively high temperatures. Thus, rainforests frequently are found in Af and Am climatic regions (see Figure 2–47 and Figure 3–5).

Tropical Deciduous Forest The selva and the tropical deciduous forest exhibit certain structural similarities, but they also exhibit several important differences. In the tropical deciduous forest, the canopy is less dense, and the trees are shorter, reflecting lower levels of precipitation. Due to a pronounced dry season that lasts for several weeks or months, many of the trees shed their leaves at the same time, allowing sunlight to penetrate to the forest floor. This produces an understory of lesser plants that often grow in such density as to produce classic "jungle" conditions.

The faunal assemblage of the tropical deciduous forest is generally similar to that of the rainforest. Although there are more ground-level **vertebrates** (animals with backbones) than in the selva, arboreal species are particularly conspicuous in both biomes.

This biome does not correspond directly to any specific climatic type, indicating greater complexity of environmental relationships (see Figure 3–6).

Tropical Scrub Extensive areas in the tropics and subtropics known as **scrub** are widespread in drier portions of the A climatic realm (Figure 3–7). This biome is dominated by a vegetation association that consists of low-growing, scraggly trees and tall bushes, usually with an extensive understory of grasses. The trees range from 10 to 30 feet (3 to 9 m) in height. They sometimes grow in close proximity to one another but are often spaced much more openly. Far fewer species are found in the scrubs than in the tropical forests; frequently just a few species will comprise the bulk of the taller growth over vast areas. In some areas a high proportion of the shrubs are thorny (see Figure 3–8).

The fauna of tropical scrub regions is notably different from that of the two biomes previously discussed. There is a moderately rich assemblage of ground-dwelling mammals and reptiles and of birds and insects.

Tropical Savanna **Savanna** lands are dominated by a plant cover that consists primarily of relatively tall grasses. Most savannas also contain a number of trees and shrubs. The savanna biome has a very pronounced

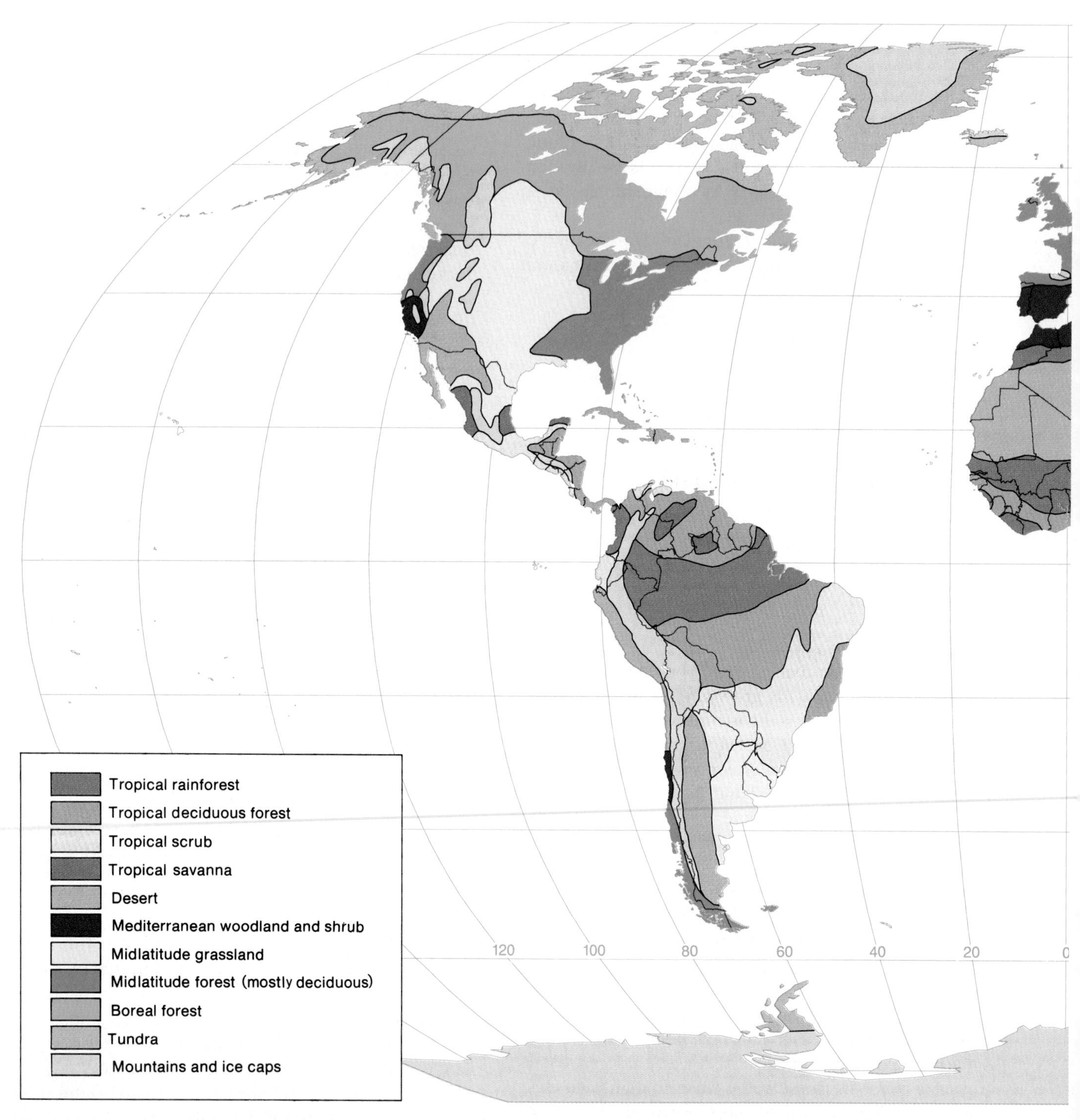

Figure 3–3. Major biomes of the world.

seasonal rhythm. During the wet season, the grass grows tall, green, and luxuriant. At the onset of the dry season, the grass begins to wither, and before long the above-ground portion is dead and brown. At this time, too, many of the trees and shrubs shed their leaves. The third "season" is the time of wildfires. The accumulation of dry grass provides abundant fuel, and most parts of the savannas experience natural burning every year or so. The recurrent grassfires actually benefit the ecosystem because they burn away the nonfertile portion of the grass without causing significant damage to shrubs and trees. When the rains of the next wet season arrive, the grasses spring into growth with renewed vigor.

Animal life within the savannas varies from continent to continent. The African savannas are the premier "big-game" lands of the world, with an unmatched richness of large animals—elephants, zebras, giraffes, and other species—but also including a remarkable diversity

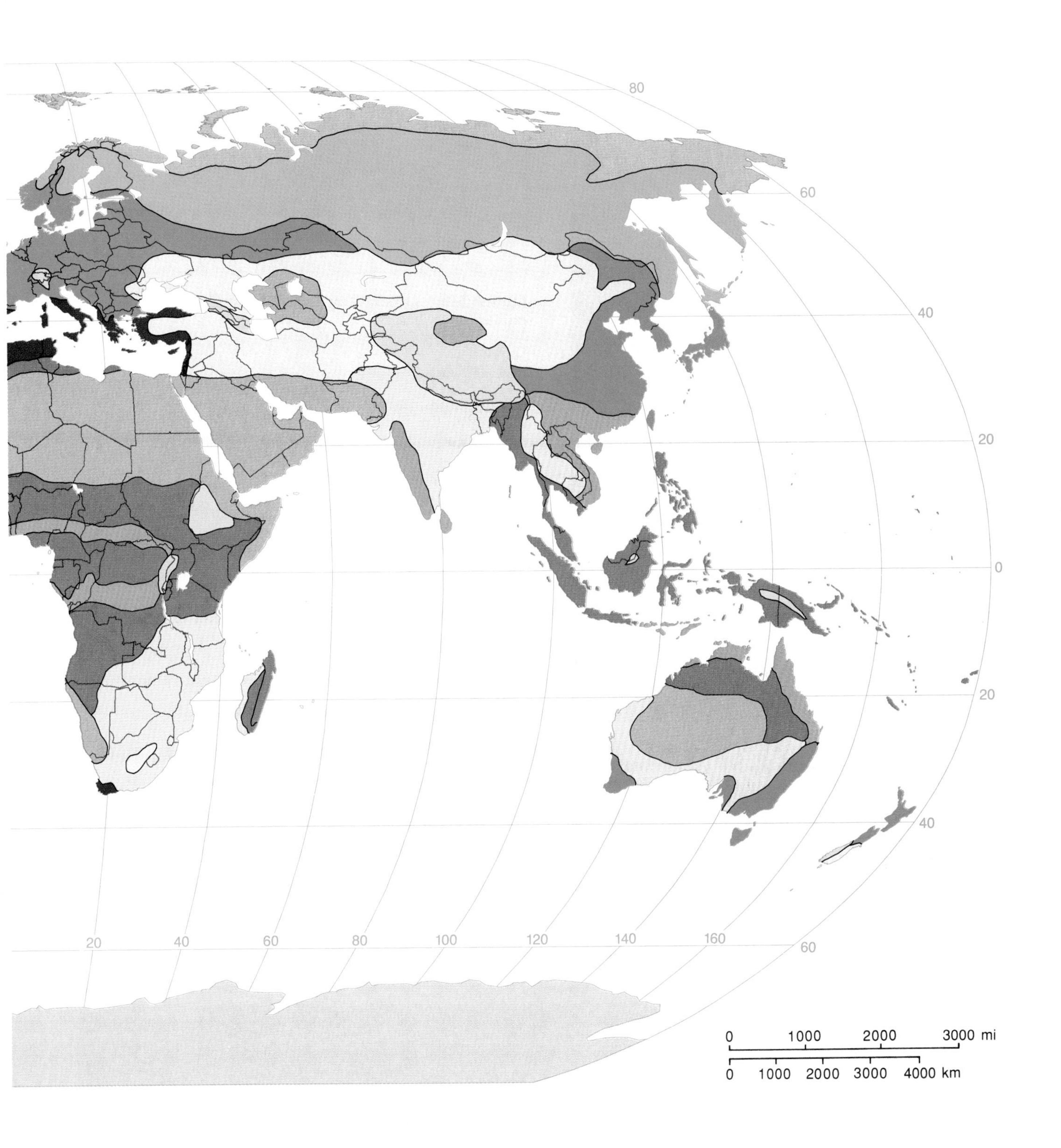

of lesser species (see Figure 3–9). The Latin American savannas, in contrast, have only a sparse population of large wildlife, and the Asian and Australian areas are intermediate between these two extremes.

There is an incomplete correlation between the distribution of the tropical savanna biome and that of the Aw (tropical savanna) climatic type. The correlation is most noticeable where contrasts in seasonal rainfall (wet season/dry season) are greatest, a condition particularly associated with the Intertropical Convergence Zone (see Figure 3–10, and compare it with Figures 2–17 and 2–47).

Desert In Chapter 2 we noted that precipitation diminishes as we move away from the equator. A parallel progression can be seen in the geographic distribution of biomes. Equatorial biomes such as the rainforests and deciduous forests have extensive vegetation. In contrast,

Figure 3–4. Rainforest vegetation is generally dense and multistoried. This scene is from the island of Viti Levu in Fiji. (TLM photo)

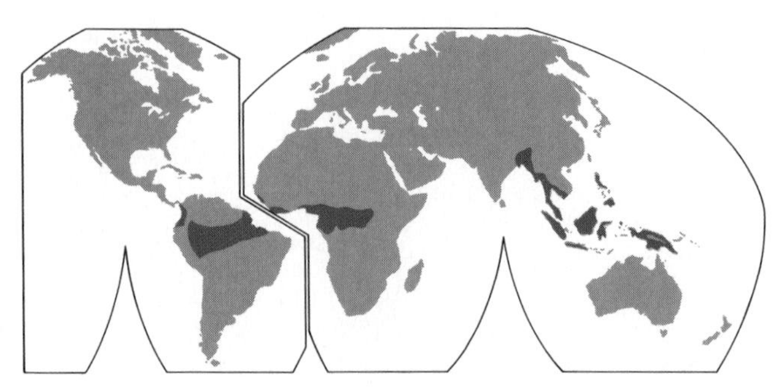

Figure 3–5. Generalized world distribution of tropical rainforest.

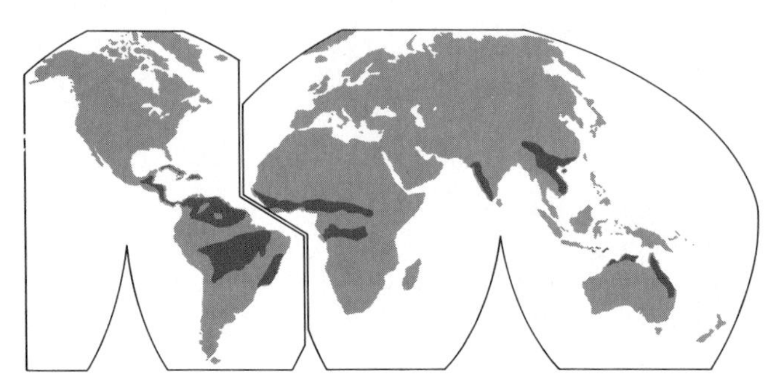

Figure 3–6. Generalized world distribution of tropical deciduous forest.

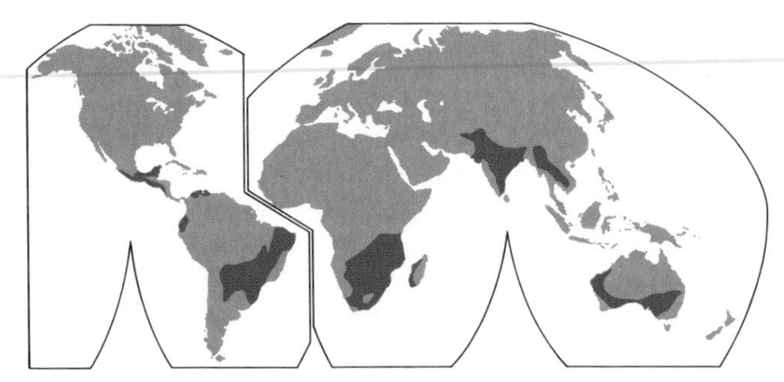

Figure 3–7. Generalized world distribution of tropical scrub.

Figure 3–8. A thorn scrub scene in northern Namibia. (TLM photo)

Figure 3–9. The savanna lands of Africa are famous for their immense herds of hoofed animals. In some localities these conditions still exist. The antelope here are topi, and the location is the northern Serengeti Plain of Kenya. (TLM photo)

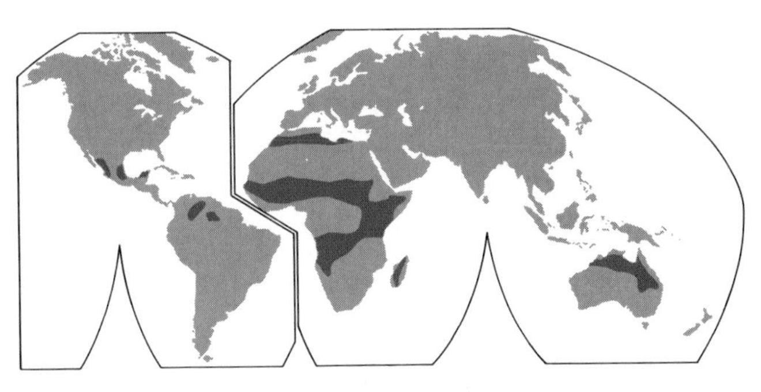

Figure 3–10. Generalized world distribution of tropical savanna.

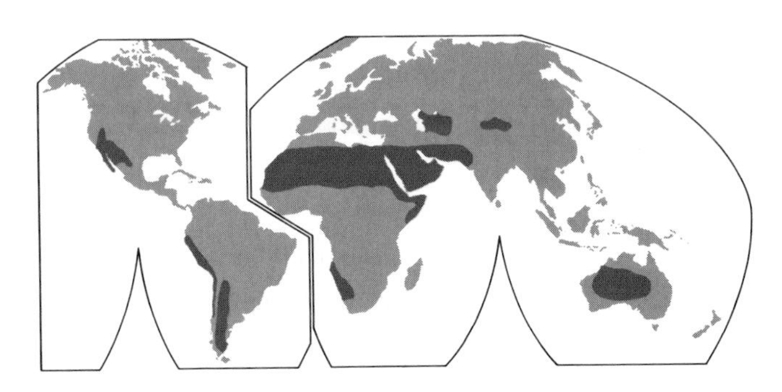

Figure 3–11. Generalized world distribution of desert.

Figure 3–12. Some deserts contain an abundance of plants, but they are usually spindly in form and are always drought-resistant in character. This scene is near Phoenix, Arizona. (TLM photo)

the **desert** biome occurs in the subtropics and in the midlatitude regions of Asia, North America, and South America (Figure 3–11).

Although we might think of deserts as largely devoid of plant life, desert vegetation is surprisingly variable. It consists largely of drought-resistant plants with structural modifications that allow them to conserve moisture, and drought-evading plants capable of hasty reproduction during brief rainy periods. Typically the plants are shrubs, which occur in considerable variety, each with its own mechanisms to combat the stresses of limited moisture availability (see Figure 3–12). Grasses and other herbaceous plants are widespread but sparse in desert areas. Despite the dryness, trees can be found sporadically in the desert, especially in Australia.

Animal life is exceedingly inconspicuous in most desert areas, leading to the erroneous idea that animals are nonexistent. In actuality, most deserts have a moderately diverse faunal assemblage, although the variety of large mammals is limited. Most desert animals avoid the principal periods of dessicating (drying) heat (day-

Focus on Rainforest Removal

Tropical rainforests comprise the climatic climax vegetation over an area of nearly 3 billion acres (1.2 billion ha), or about 8 percent of the Earth's total land surface. These remarkable forests are shared by 50 countries on 5 continents. This vegetative association represents a biome of extraordinary diversity. Biologists believe that rainforests are the home of perhaps half the world's biotic species, about five-sixths of which have not yet been described and named.

Throughout most of history, rainforests were considered to be remote, inaccessible, unpleasant places, and as a consequence they were little affected by human activities. In the present century, however, rainforests have been exploited and devastated at an accelerating pace. In the last decade or so, tropical deforestation has become one of the Earth's most serious environmental problems. The rate of deforestation is spectacular—51 acres (21 ha) per minute; 74,000 acres (30,000 ha) per day; 27 million acres (11 million ha) per year. More than half of the original African rainforest is now gone, about 45 percent of Asia's rainforest no longer exists, and the proportion in Latin America is approaching 40 percent.

The current situation varies in the five major rainforest regions:

1. The rate of deforestation is highest in southern and southeastern Asia, primarily as a result of commercial timber exploitation.
2. The current rate of deforestation is relatively low in central Africa.
3. Timber harvesting and agricultural expansion are responsible for a continuing high rate of forest clearing in West Africa. Nigeria has lost about 90 percent of its forests, and Ghana about 80 percent.
4. Deforestation of the Amazon region in South America has been moderate (about 5 percent of the total has been cleared), but it continues at an accelerating pace.
5. Very rapid deforestation persists in Central America, mostly due to expanded cattle ranching.

As the forest disappears, so does its animal life. In the early 1990s experts estimated that tropical deforestation was responsible for the extermination of one species per hour. Moreover, loss of the forests contributes to accelerated soil erosion, drought, flooding, declining agricultural productivity, and greater poverty for rural inhabitants. In addition, levels of atmospheric carbon dioxide increase because there are fewer trees to absorb it and because burning of trees for forest clearing releases more of it into the air.

The tragic irony of tropical deforestation is that the anticipated economic benefits are usually illusory. Much of the forest clearing, especially in Latin America, is in response to the social pressures generated by overcrowding and poverty in societies where most of the people are landless. The governments throw open ''new lands'' for settlement, and the settlers clear the land for crop growing or livestock raising. The result almost always is an initial increase in productivity, followed in only 2 or 3 years by a pronounced fertility decline due to nutrient loss, erosion, and infestation by weeds. Sustainable agriculture generally can be expected only with continuous heavy fertilization, a costly procedure.

The forests are renewable. If left alone by humans, they can regenerate, providing there are seed trees in the vicinity and the soil has not been stripped of all its nutrients. The loss of biotic

light in general and the hot season in particular) by resting in burrows or crevices during the day and prowling at night.

Mediterranean Woodland and Shrub In six widely scattered and relatively small areas of the midlatitudes, all of which experience the pronounced dry summer/wet winter precipitation regime of the mediterranean climatic type, is found a biome in which the dominant vegetation associations are physically similar but taxonomically varied; that is, they belong to different biological groups such as families and species (see Figure 3–13). The mediterranean biome is dominated by a dense growth of woody shrubs that is known as **chaparral** in North America but has other names in other areas.

The trees and shrubs are primarily broadleaf evergreens. Their leaves are mostly small and have a leathery texture or waxy coating that inhibits water loss during the long dry season. Moreover, most plants have deep roots.

Summer is a virtually rainless season, and summer fires are relatively common. Many of the plants are adapted to rapid recovery after a wildfire has swept the

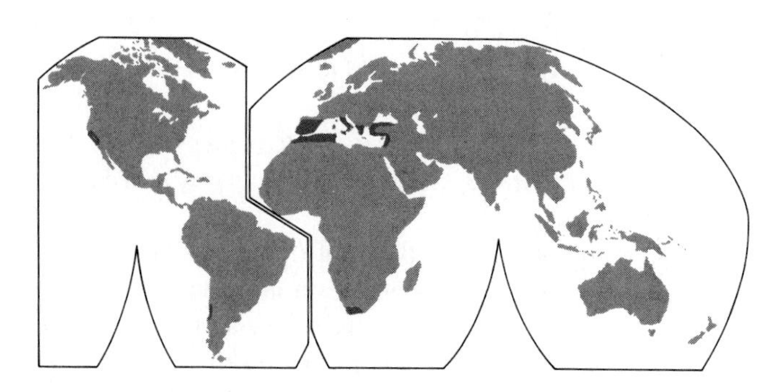

Figure 3–13. Generalized world distribution of mediterranean woodland and scrub.

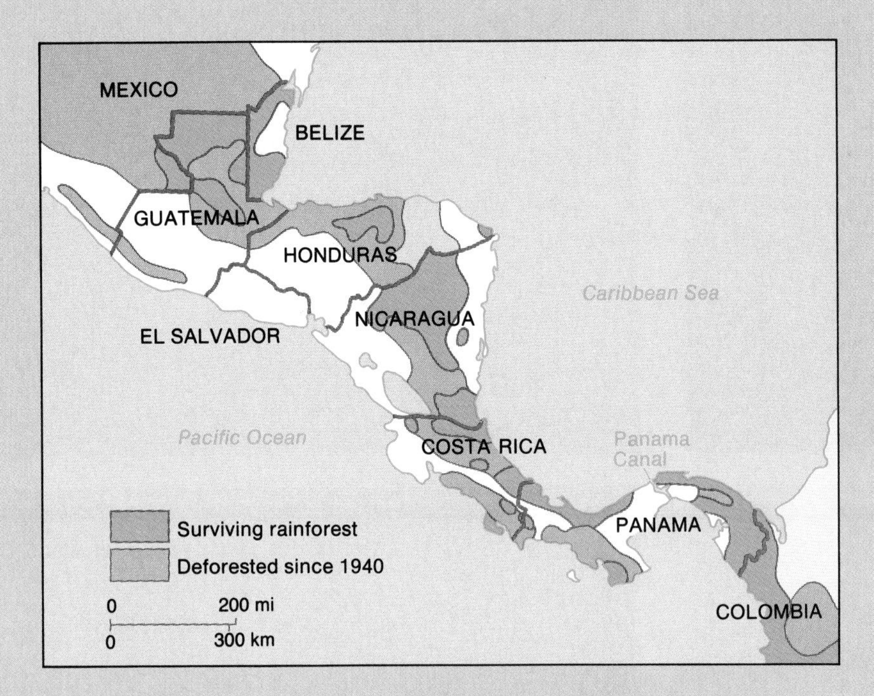

The shrinking Central American rainforest in the last half-century. This region has one of the fastest rates of deforestation in the world.

diversity, however, is much more serious. Extinction is an irreversible process. Valuable potential resources—pharmaceutical products, new food crops, natural insecticides, industrial materials—may disappear before they are even discovered. Natural plant and animal types that could be combined with agricultural crops and animals, respectively, to impart resistance to disease, insects, parasites, and other environmental stresses may also be lost. In the destruction of the tropical rainforest, even small, isolated, vulnerable groups of indigenous people may be wiped out.

Much concern has been expressed about tropical deforestation, and some concrete steps have been taken. The development of *agroforestry* (planting crops with trees, rather than cutting down the trees and replacing them with crops) is being fostered in many areas. In Brazil, which has by far the largest expanse of rainforest, some 46,000 square miles (119,000 km^2) of reserves have been set aside, and Brazilian law requires that any development in the Amazon region leave half the land in its natural state. In 1985 a comprehensive world plan, sponsored by the World Bank, the World Resources Institute, and the United Nations Development Programme, was proposed. It outlined concrete, country-by-country strategies to combat tropical deforestation. It is an $8-billion, 5-year project that deals with everything from scarcities of fuel woods to training forestry students. Unfortunately, its price tag makes its implementation unlikely, and it has yet to be undertaken. Meanwhile, the sounds of the axe, the chain saw, and the bulldozer continue to echo throughout the tropical forests.

area. Part of the seasonal rhythm of this biome is that winter floods sometimes follow summer fires, as slopes left unprotected by the burning away of grass and lower shrubs are susceptible to abrupt erosive runoff if the winter rains arrive before the vegetation has a chance to resprout (see Figure 3–14).

The fauna of this biome is not particularly distinctive. Seed-eating, burrowing rodents are common, as are some bird and reptile groups. There is a general overlap of animals between this biome and those adjacent.

Midlatitude Grassland Vast areas of relatively continuous grassland occur widely in the midlatitudes of all continents, as mapped in Figure 3–15. The relative absence of vegetation other than grasses reflects the lack of sufficient precipitation to support larger plant forms, or the frequency of fires (both natural and human-induced), which prevents the growth of tree or shrub seedlings. In the wetter areas, the grasses grow tall, and the term **prairie** is often applied. In drier regions, the grasses are shorter; such growth is often referred to as **steppe.**

Most of the grass species are perennials, lying dormant during winter and sprouting in summer. Trees are restricted largely to streamside locations, whereas shrubs and bushes occur sporadically on rocky sites. Wildfires are fairly common in summer, which helps to explain the relative scarcity of shrubs. The woody plants cannot tolerate fires and generally can survive only on dry slopes where there is little grass cover to fuel a fire.

These grasslands have provided extensive pasturage for grazing animals. Many of the smaller animals spend all or part of their lives underground, where they find some protection from heat, cold, and fire.

In the Northern Hemisphere the grassland biome occurs largely in regions with the steppe climatic type. In the Southern Hemisphere, grasslands are not closely associated with any particular climate.

Figure 3–14. Wildfires are a natural and recurrent part of the environment in many biomes, but the seasonal rhythm in mediterranean lands is particularly notable. These four photos record the annual regime in the Mount Lofty Ranges of South Australia. (1) Winter is moist, mild, and green. (2) Early summer is hot, dry, and brown. (3) Then comes the fire season. (4) Late summer is the black time. (TLM photos)

Midlatitude Deciduous Forest Extensive areas on all Northern Hemisphere continents, as well as more limited tracts in the Southern Hemisphere, were originally covered with forests of largely broad-leaved deciduous trees (see Figure 3–16). Except in hilly country, a large portion of this forest has been cleared for human use, meaning that very little of the original vegetation remains.

Tree species vary considerably from region to region, although most are broadleaf and deciduous. The principal exception is in eastern Australia, where the forest is composed almost entirely of varieties of euca-

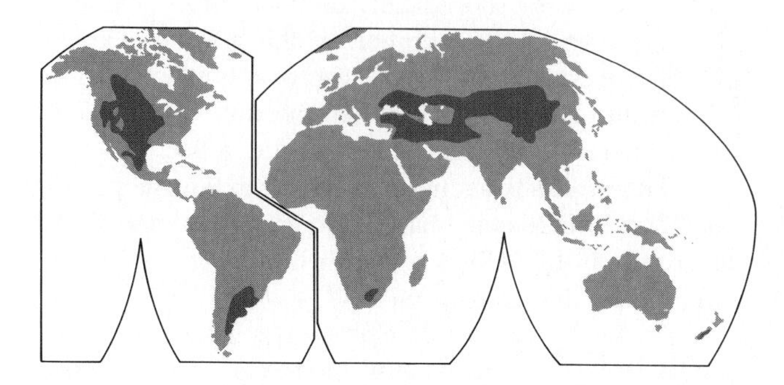

Figure 3–15. Generalized world distribution of midlatitude grassland.

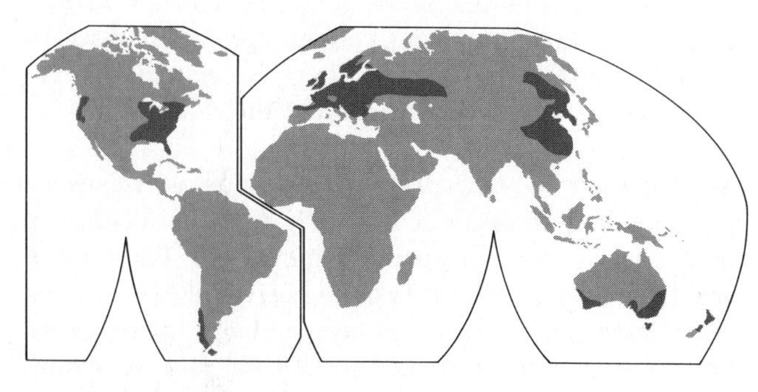

Figure 3–16. Generalized world distribution of midlatitude forest.

Figure 3–17. The original forests of Australia were composed almost entirely of species of eucalyptus. This scene is in eastern New South Wales, near Eden. (TLM photo)

lyptus, which are broadleaf evergreens (see Figure 3–17). In Northern Hemisphere regions, as you move northward, the deciduous forest gradually gives way to the needle-leaf evergreens. One exception occurs in the southeastern United States, in which extensive forests of pines (which are needle-leaf evergreens), rather than deciduous species, occupy most of the well-drained sites above the valley bottoms.

This biome generally has the richest assemblage of fauna to be found in the midlatitudes, although it does not have the diversity to match that of most tropical biomes. It has (or had) a considerable variety of birds and mammals, and in some areas reptiles and amphibians. Summer brings a diverse and active population of insects and other arthropods (animals with jointed legs and external skeletons, such as spiders). All animal life is less numerous (partly due to migrations and hibernation) and less conspicuous in winter.

Boreal Forest One of the most extensive biotic components of the world is the **boreal forest** or **taiga,** which occupies a vast expanse of northern North America and Eurasia (see Figure 3–18). This great northern forest contains perhaps the simplest assemblage of plants of any biome. Most of the trees are *coniferous* (cone-bearing), nearly all of which are needle-leaf evergreens. The variety of species is limited: Most are pines, firs, and spruces (see Figure 3–19). In some places the coniferous cover is interrupted by areas of deciduous trees, which often represent a subclimax situation following a forest fire. A close correlation exists between the location of the boreal forest biome and the subarctic climatic type. (Compare Figure 3–18 with Figure 2–47.)

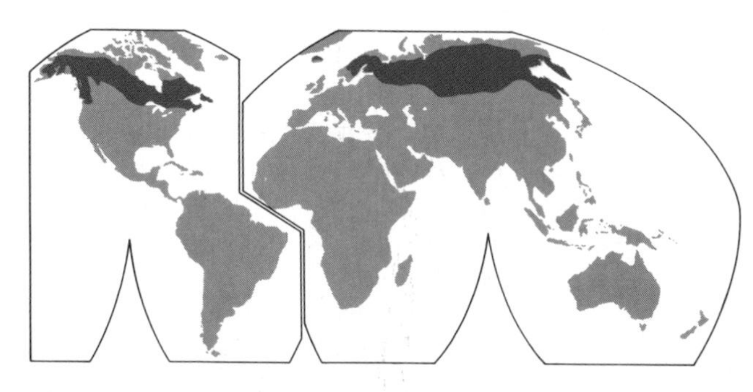

Figure 3–18. Generalized world distribution of boreal forest.

Figure 3–19. The boreal forest contains trees that generally are small, close-growing, and of uniform species composition. This spruce forest scene is in central Alaska, near Fairbanks. (TLM photo)

The trees grow taller and more densely near the southern margins of the taiga, where the summer growing season is longer and warmer. Near the northern margins, the trees are spindly, short, and more openly spaced. The ground usually is covered with a complete growth of mosses and lichens, with some grasses in the south, and with a considerable accumulation of decaying needles.

Poor drainage is typical in summer, due partially to permanently frozen subsoil, which prevents water from penetrating downward, and partially to the disruption of normal surface drainage by the action of glaciers during the recent ice age. Thus, bogs and swamps are numerous, and the ground generally has a spongy feel in summer. During the long winters, all is frozen.

The immensity of the boreal forest gives an impression of biotic productivity, but such is not the case. Harsh climate, floristic homogeneity, and slow plant growth produce only a limited food supply for animals. Species diversity among animals is limited, although the number of individuals of some species is impressive. Mammals are represented prominently by furbearers and by a few species of ungulates (hoofed animals), such as caribou and moose. Birds are numerous and fairly diverse in summer, but nearly all of them migrate to milder latitudes in winter. Insects are totally absent in winter but are superabundant during the brief summer.

Tundra The **tundra,** mapped in Figure 3–20, is essentially a cold desert in which moisture is scarce and summers are so short and cool that trees are unable to survive. The plant cover consists of a considerable mixture of species, many of them in dwarf forms. Included here are grasses, mosses, lichens, flowering herbs, and a scattering of low shrubs. The plants complete their annual cycles hastily during the brief summer, when the ground is often moist and waterlogged because of inadequate surface drainage and particularly inadequate subsurface drainage (see Figure 3–21).

Animal life is dominated by birds and insects during the summer. Extraordinary numbers of birds flock to the tundra for summer nesting and then migrate southward as winter approaches. Mosquitoes, flies, and other insects proliferate in vast numbers during the short warm season, laying eggs that can survive the bitter winter. Other forms of animal life are scarcer: a few species of mammals and freshwater fishes but almost no reptiles or amphibians.

Biomes and Climate One of the truly striking relationships in physical geography is the notable correlation between the global distribution of biomes and that

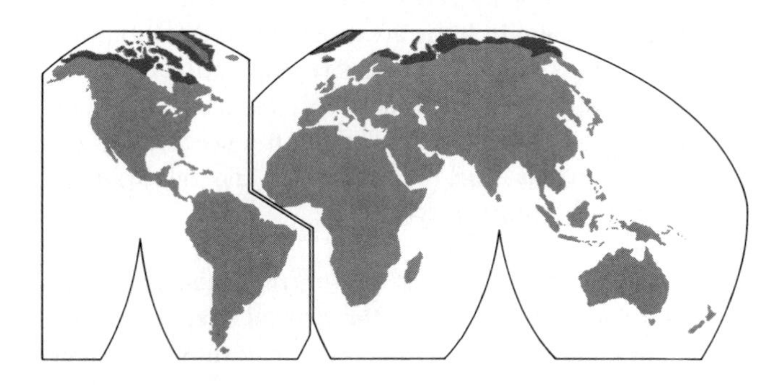

Figure 3–20. Generalized world distribution of tundra.

Figure 3–21. Tundra vegetation is found in high latitudes or high altitudes. This scene is at the 11,000-foot (3350-m) level in the Mummy Range of north-central Colorado. (TLM photo)

of major climatic types. This correlation is readily apparent when you compare the maps in Figure 2–47 and Figure 3–3. Although the interrelationships of climate, flora, and fauna are exceedingly complex, we can reasonably generalize that broad vegetation patterns are generally dependent on broad climatic patterns and that the distribution of animal life is significantly influenced by the climatic and vegetational milieu. To understand the world pattern of biomes we must consider many factors, but clearly a comprehension of climatic patterns is the starting point.

SOIL

Just as the sun is the source of energy for all life on Earth, so is the **soil** the essential medium in which all terrestrial life is nurtured. Almost all land plants sprout from this precious medium that is spread so thinly across the continental surfaces, with an average worldwide depth of only about 6 inches (15 cm).

Despite the implication of the well-known simile "as common as dirt," soil is one of the most complex features produced in nature. Soil is an infinitely varying mixture of weathered mineral particles, decaying organic matter, living organisms, gases, and liquid solutions. It is a dynamic mixture involving the slow but continuous disintegration of once-solid rock and the decomposition of once-living organisms. It is a locale of complicated chemical reactions involving a rich variety of microscopic and atomic particles.

Most important, however, soil is a zone of plant growth. In fact, soil can be conceptualized as a relatively thin surface layer of mineral matter that normally contains a considerable amount of organic material and is capable of supporting living plants. A key characteristic of soil is its ability to produce and store plant nutrients, an attribute that is made possible by the interactions of such diverse components as water, air, sunlight, rocks, plants, and animals.

The development of soil is initiated by the physical and chemical disintegration of rock that is exposed to the atmosphere and by the action of water percolating down from the surface. This disintegration is called **weathering.** The basic results of weathering are the weakening and breakdown of solid rock, the fragmentation of coherent rock masses, and the making of little rocks from big ones.

Thus, the soil is composed largely of finely fragmented mineral particles that are the ultimate product of weathering. Soil normally also contains an abundance

Figure 3–22. Soil develops through complex interactions of physical and biological processes.

of living plant roots, a variety of dead and rotting plant parts in varying stages of decomposition, a large quantity of microscopic plants and animals (both living and dead), and a variable amount of air and water. Soil is not the end product of a process. Rather, it is a stage in a never-ending continuum of physical, chemical, and biotic activities that at any given time and place, represents a dynamic life layer that has a unique set of physical and chemical properties (see Figure 3–22).

Soil-Forming Factors

Three kinds of variables interact in producing any particular soil: (1) the chemical composition of the rock that provides most of the bulk material, (2) the environmental conditions under which the rock is converted into soil, and (3) the length of time that the rock has experienced that environment. These variables in turn influence the five principal soil-forming factors, listed below (see Figure 3–23).

1. The *geologic factor* is the parent material (bedrock or loose rock fragments) that provides the bulk of most soils.
2. The *climatic factor* refers to the effects of temperature and particularly moisture on the organic and inorganic components of the soil.
3. The *topographic factor* refers to the configuration of the surface and is manifested primarily by aspects of slope and drainage.
4. The *biological factor* consists of both living plants and animals and dead plant and animal material that is incorporated into the soil.
5. The *chronological factor* refers to the length of time that the other four factors have been interacting in the formation of a particular soil.

Soil Properties

All soil types have certain identifying characteristics that help to distinguish them from one another. Some soil properties are easily recognized, but most can be ascertained only by precise measurement.

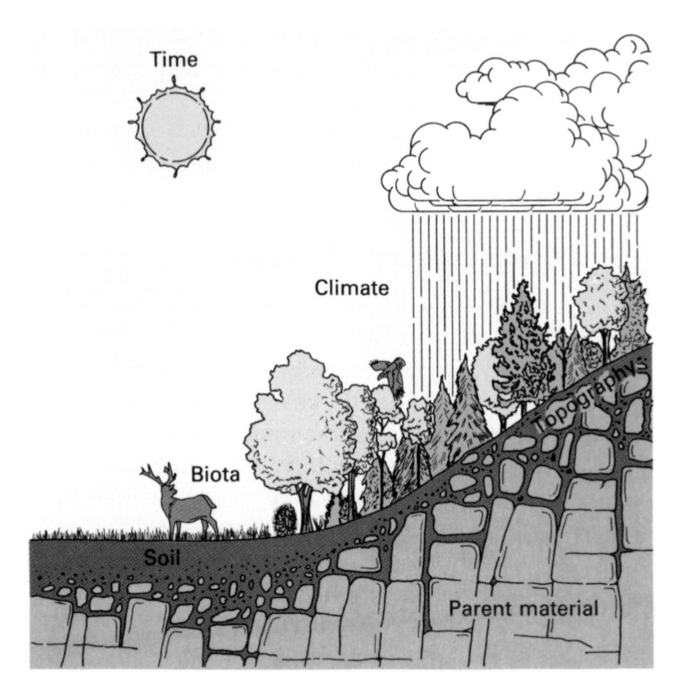

Figure 3–23. Five "factors" are interactive in soil formation. The geologic fundament and the topographic slope are acted upon by climatic and biologic agencies over a time continuum to produce a natural soil.

Color is the most obvious property. It is rarely definitive, although it can provide clues about the characteristics and capabilities of the soil. Black or dark brown colors usually indicate considerable organic content. This is a hint of high fertility, but it may also signify poor drainage. Reddish and yellowish colors generally indicate iron oxide stains on the outside of soil particles; these colors are most common in tropical and subtropical regions where many minerals are dissolved away, leaving insoluble iron compounds behind. Light-colored soils—gray or white—develop in varying environments. They may indicate lack of organic matter, an accumulation of salts, or (in humid areas) heavy **leaching,** that is, the dissolving away of minerals by percolation of ground water.

Texture is the size of the individual soil particles. The principal texture classes, ranked in order of size from largest to smallest, are sand, silt, and clay (see Figure 3–24). The best soils for farming usually have a mixture of these three components.

Structure is determined by the way that the individual particles tend to aggregate into larger masses or clumps. The size, shape, and stability of such aggregates have a marked influence on the ease of movement of water, air, roots, and organisms in the soil.

Soil Profiles

The development of any soil is expressed in two dimensions—depth and time—that have a close relationship with each other. The weathering of parent material, the addition of organic matter, and various chemical and biological reactions all combine to create new soil. As these soil-forming processes continue to operate—unless they are overbalanced by an even faster rate of soil loss through erosion—the soil becomes continually deeper. Along with deepening usually comes an increasing vertical variation in soil characteristics. Such properties as texture, structure, color, porosity, density, organic content, and chemical ingredients begin to vary with depth.

The vertical variation of soil properties is not random but rather reflects an ordered pattern. Soil tends to develop more or less distinct layers, called **soil horizons,** each with differing characteristics. A vertical cross section from the Earth's surface down through the soil layers into the parent material beneath is referred to as a **soil profile** (see Figure 3–25).

Time is an important factor in profile development, but the most critical factor is surface water that percolates into the soil. Descending water carries material from the surface downward and from the topsoil into the subsoil. Material from above is mostly deposited in zones a few inches or a few feet farther down. In the usual pattern, *topsoil* becomes a somewhat depleted horizon through **eluviation** (the physical removal of fine particles) and leaching, and *subsoil* develops as a layer of accumulation due to **illuviation** (the deposition of fine particles brought down from above).

Six soil horizons are usually recognized in well-developed profiles:

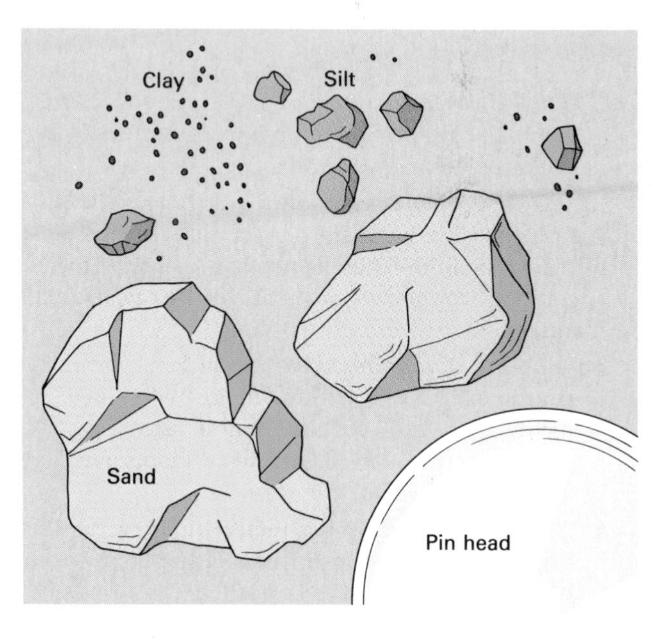

Figure 3–24. Comparative sizes of soil separates. This magnified example shows the size relationship of sand, silt, and clay. For scale, a portion of the head of a pin is shown in the lower right. (After Robert A. Muller and Theodore M. Oberlander, *Physical Geography Today: A Portrait of a Planet,* 2nd ed. New York: Random House, Inc., 1978, p. 260. © 1978 by Random House, Inc. By permission of McGraw-Hill, Inc.)

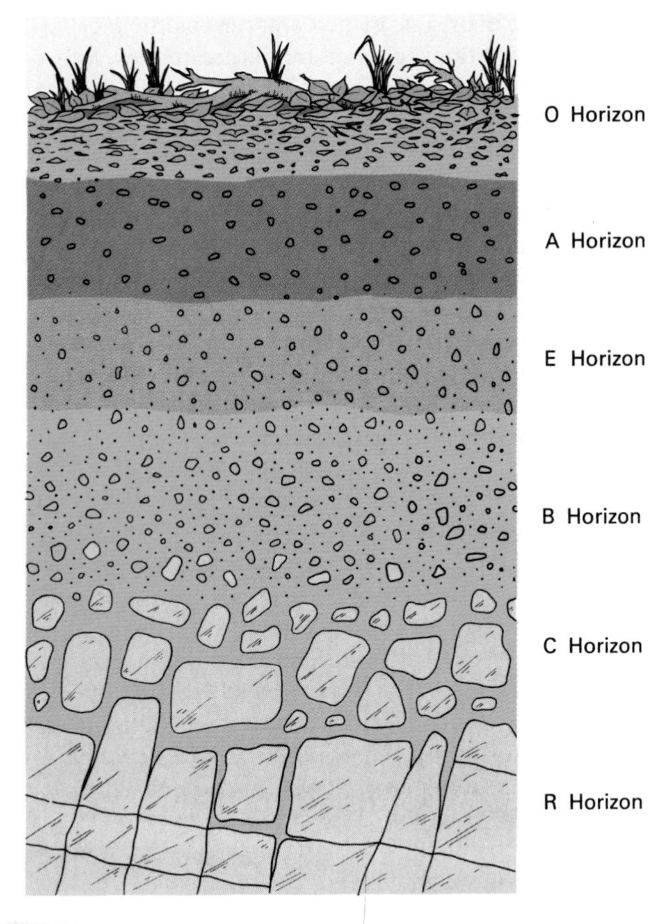

Figure 3–25. Idealized diagram of a soil profile. The true soil, or solum, consists of the O, A, E, and B horizons.

1. "O"—The *O horizon* is the immediate surface layer in which organic matter, both fresh and decaying, makes up most of the volume.
2. "A"—The *A horizon* is formed at or near the surface, usually as a mineral horizon that also contains considerable organic matter. It is usually dark in color due to the decomposition of the organic material.
3. "E"—The *E horizon* is an eluvial layer in which silicate clay, iron, and aluminum have been removed, leaving a concentration of highly resistant sand or silt particles, such as silica. It is usually light in color.
4. "B"—the *B horizon* is a mineral horizon of illuviation where most of the materials removed from higher layers have been deposited.
5. "C"—The *C horizon* is *unconsolidated* (loose) parent material beyond the reach of plant roots and most soil-forming processes except weathering.
6. "R"—The *R horizon* is *consolidated* (tightly connected) bedrock.

The profile is such an important indicator of the characteristics and capabilities of a soil that it is the principal diagnostic factor in soil classification. The almost infinite variety of soils in the world generally are grouped and classified on the basis of differences exhibited in their profiles.

Soil Fertility and Productivity

As geographers, we are particularly interested in the usefulness of soil to people. When speaking of a soil's innate **fertility,** we are referring only to the availability of nutrients for plant growth. A soil's **productivity,** however, is influenced by considerations of the soil's physical characteristics (such as structure), the needs of the particular crop being planted, details of the local climate, agronomic techniques, and many other variables. Fertile soils are not always productive, and the most productive soils are not always the most fertile.

Different soils react differently to being farmed, and they must be managed and husbanded differently. Techniques of plowing, planting, cultivating, fertilizing, draining, and irrigating are quite varied. Many crops will grow in soils that are only slightly acidic or alkaline, but others can withstand considerable acidity or alkalinity. Excess acidity normally can be reduced by the addition of an alkaline substance, such as pulverized limestone. An excess alkaline or saline content in arid soils usually can be removed by abundant irrigation water and good drainage.

Some soils yield abundantly if they are simply cleared, plowed, and seeded. Examples of such productive soils include those derived from limestone, alluvial soils deposited on floodplains, wind-deposited soils of fine texture called *loess* (discussed in Chapter 4), and certain volcanic soils.

Other soils may respond to treatment such as fertilization or crop rotation. The soils of the eastern United States illustrate this very well. They developed under an original forest cover in a humid climate. They have been thoroughly leached, their nutrient content is relatively low, and their natural fertility is limited. However, they generally have good structure, and local rainfall usually is dependable. These soils respond well to careful management, so with proper fertilization, drainage, and crop rotation, they can—and have—become productive.

Soil Classification

Soil classification is complicated because so many variables are involved, with regard to both soil formation and the characteristics that develop in the soil. Many classifications have been designed in the past, but the one that is by far the most important in the United States today was developed by the U.S. Department of Agriculture and is known simply as *Soil Taxonomy*.

Global Distribution of Major Soils At the highest level of the Soil Taxonomy system are ten orders of soils. Nine of these are arranged in a hierarchy in which each succeeding order represents an increased degree of weathering, particularly as expressed by mineral alteration and profile development (see Figure 3–26). The tenth order, histosols, is essentially an organic soil that lies outside the concept of the hierarchy. Figure 3–27 maps the global distribution of the orders of soils.

Entisols are the least developed of all soils. They have experienced little mineral alteration and are virtually without horizons. Characteristically they are either quite wet, quite dry, or quite rocky. They are widely distributed and are not closely associated with any particular climatic, floristic, or geologic conditions. Their immaturity usually means that they contain limited nutrients for plant growth and therefore are not very useful agriculturally, except where recently deposited floodplain sediments are involved.

Vertisols comprise a specialized type of soil with a large quantity of clay that becomes critical to the soil's development. It has an exceptional capacity for absorbing water and tends to swell when moistened and crack when dry. Vertisols are naturally fertile, but they are difficult to cultivate because their high clay content makes them sticky and plastic. Still, they have considerable agricultural potential in regions where power tools, fertilizers, and irrigation can be applied.

Inceptisols are immature soils with relatively faint distinctive characteristics. They are primarily eluvial in nature and exist in many differing environments, especially tundra and mountain areas. Their agricultural potential generally is quite limited.

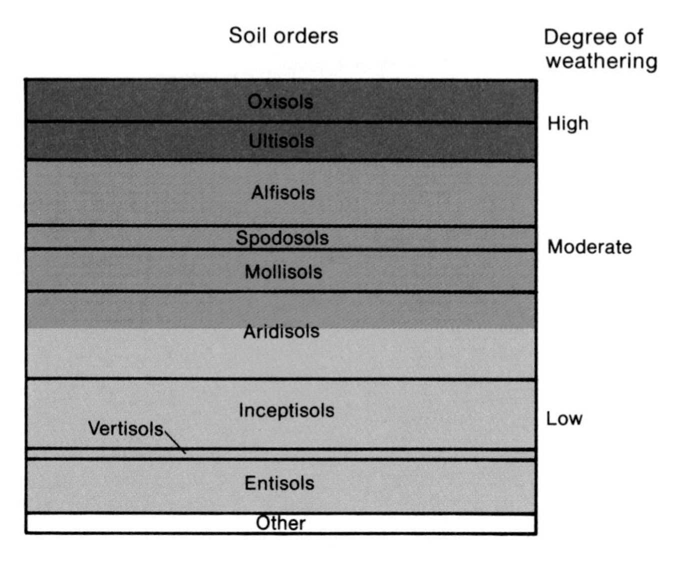

Figure 3–26. The relationship between soil orders and degree of weathering. The relative height of the band for each order is proportional to the approximate worldwide areal extent of that order. Nine of the ten soil orders fit this hierarchy; the base of the pyramid represents histosols and nonsoil surfaces, which have no relationship to degree of weathering.

Aridisols are mineral soils that are generally dry and low in organic matter. They cover nearly one-fifth of the land surface of the Earth and are found largely in desert and semidesert climate regions. The principal use of aridisols is for limited grazing of livestock although with irrigation they can be made useful for the intensive production of crops.

Mollisols are mineral soils with a thick, dark surface layer that is rich in organic matter. On the whole, mollisols are the most productive of the soil orders; thus, their agricultural potential is high. They are particularly common in midlatitude grassland areas.

Spodosols have a conspicuous subsurface horizon of organic accumulation, often with iron and aluminum. They are most widespread in areas of coniferous forest that have a humid continental or subarctic climate. Forestry has been the dominant economic activity on spodosols, which are not very fertile, but agriculture is possible if the land is carefully managed.

Alfisols are the most widespread of the maturely developed soils, occurring extensively in the low- and middle-latitude portions of the continents. They tend to be associated with transitional environments and are less common in regions that are particularly hot or cold, or wet or dry. They are distinguished by a subsurface clay horizon and a medium-to-generous supply of plant nutrients and moisture. As a soil order, alfisols rank second only to mollisols in agricultural productivity.

Ultisols are roughly similar to alfisols except that they are more thoroughly weathered and more completely leached of nutrients. They have experienced greater mineral alteration than any other soil in the midlatitudes, although they also occur in the low latitudes. Typically ultisols are reddish in color from the significant proportion of iron and aluminum in the A horizon. Ultisols are relatively infertile, but they may produce good crop or forage yields with careful management, particularly fertilization.

Oxisols are the most thoroughly weathered and leached of all soils. They evolved in warm, moist climates and invariably display a high degree of mineral alteration and profile development. Essentially they are tropical soils of low fertility.

Histosols represent the only soil order composed primarily of organic rather than mineral matter. They are usually saturated with water all or part of the year. They are black and acidic and are naturally fertile only for water-tolerant plants. Histosols occupy only a small fraction of the Earth's land surface, a much smaller area than any other order. They can be agriculturally useful, provided careful water management (especially drainage) is practiced.

Figure 3–27. Soils of the world.

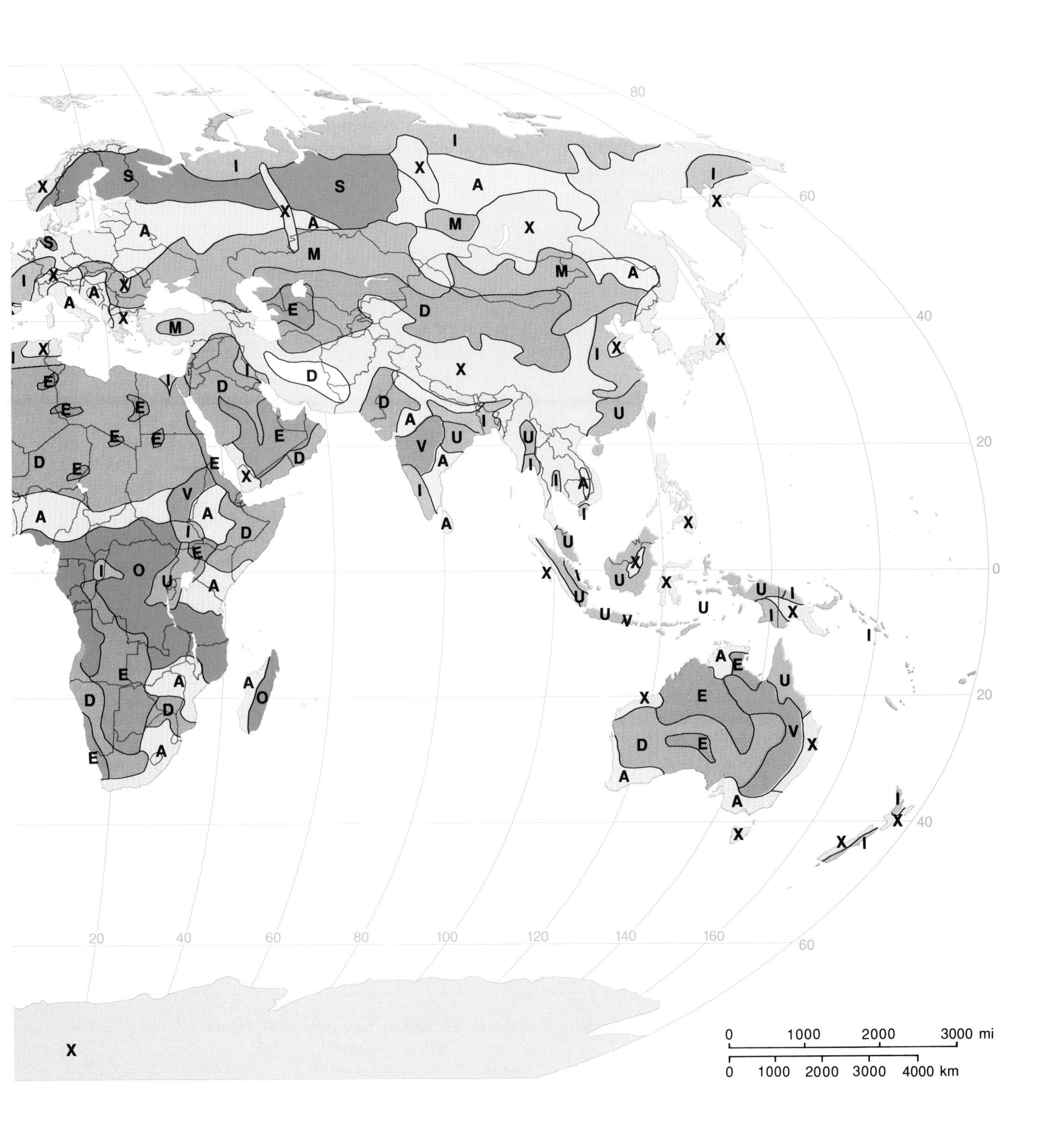
80
60
40
20
0
20
40
20
40
60
80
100
120
140
160
60
0 1000 2000 3000 mi
0 1000 2000 3000 4000 km

SUMMARY

Biota refers to the total complex of plant (flora) and animal (fauna) life. Useful organizing concepts for geographic study include ecosystem (the totality of interactions among organisms and their environment in a given area) and biome (a major assemblage of biota in functional interaction with its environment). The principal terrestrial biomes include tropical rainforest, tropical deciduous forest, tropical scrub, tropical savanna, desert, mediterranean woodland and shrub, midlatitude grassland, midlatitude deciduous forest, boreal forest, and tundra.

Soil is the essential medium in which all terrestrial life is nurtured. The five fundamental soil-forming factors (geologic, climatic, topographic, biological, and chronological) combine in numerous ways to produce an infinite variety of soil types. There is a multitude of soil classifications, but only one (called Soil Taxonomy) is now broadly accepted in the United States; it is logical, generic, and comprehensive.

KEY TERMS

biota (p. 69)
flora (p. 69)
bryophyte (p. 69)
pteridophyte (p. 69)
gymnosperm (p. 70)
angiosperm (p. 70)
evergreen tree (p. 70)
deciduous tree (p. 70)
broadleaf tree (p. 70)
needle-leaf tree (p. 70)
hardwood (p. 70)
softwood (p. 70)
primary series (p. 70)
prisere (p. 70)
climatic climax vegetation (p. 70)
fauna (p. 71)
food chain (p. 71)
photosynthesis (p. 71)
ecosystem (p. 72)
biome (p. 73)
tropical rainforest (p. 73)
selva (p. 73)
arboreal (p. 73)
invertebrate (p. 73)
vertebrate (p. 73)
scrub (p. 73)
savanna (p. 73)
desert (p. 75)
chaparral (p. 78)
prairie (p. 79)
steppe (p. 79)
boreal forest (p. 81)
taiga (p. 81)
tundra (p. 82)
soil (p. 83)
weathering (p. 83)
leaching (p. 85)
soil horizon (p. 85)
soil profile (p. 85)
eluviation (p. 85)
illuviation (p. 85)
fertility (p. 86)
productivity (p. 86)
entisol (p. 87)
vertisol (p. 87)
inceptisol (p. 87)
aridisol (p. 87)
mollisol (p. 87)
spodosol (p. 87)
alfisol (p. 87)
ultisol (p. 87)
oxisol (p. 87)
histosol (p. 87)

QUESTIONS FOR INVESTIGATION AND DISCUSSION

1. Why are plants generally of more interest than animals to geographers?
2. How does climatic climax vegetation become established?
3. Distinguish among, savanna, prairie, and steppe.
4. Discuss the distribution and characteristics of the selva.
5. Distinguish between texture and structure of a soil.
6. What is a soil profile?
7. Select one of the soil orders, and describe its distribution and characteristics.
8. What is the climatic climax vegetation in your area? Which plants might be in a prisere in your area?
9. In which biome do you live? Which soils are found in the area? If you live in a farming region, which crops are most common?

ADDITIONAL READINGS

COLLINSON, A.S. *Introduction to World Vegetation*, 2nd ed. London: Unwin Hyman, 1988.

FANNING, DELVIN S., and MARY C.B. FANNING. *Soil: Morphology, Genesis, and Classification.* New York: John Wiley & Sons, 1989.

MIEKLE, H.W. *Patterns of Life: Biogeography in a Changing World.* Boston: Unwin Hyman, 1989.

MORAIN, STANLEY A. *Systematic and Regional Biogeography.* New York: Van Nostrand Reinhold Co., 1984.

STEILA, DONALD, and THOMAS E. POND. *The Geography of Soils.* Totowa, NJ: Rowman & Littlefield, 1989.

CHAPTER 4

Topography

The jagged peaks and steep slopes of the Brenta Massif in the Italian Alps are spectacular land forms. (Courtesy of Robert Everts/Tony Stone Worldwide.)

The solid part of the Earth is an immense mass with a diameter of 8000 miles (12,800 km) and a circumference of 25,000 miles (40,000 km), but the focus of geographical inquiry is much more restricted. Geographers concentrate on the surface of the Earth because that is the zone of human habitation. As geographers we are interested in the Earth's interior only as it helps us to comprehend the nature and characteristics of the surface.

THE EARTH'S CRUST

As with all forms of matter, chemical elements occur in varying combinations to form the materials that compose the solid part of the earth. About 90 of these basic chemical substances are found in the Earth's crust, occasionally as discrete elements but usually bonded with one or more other elements to form **compounds.** These naturally formed compounds and elements of the crust are called **minerals,** solid substances having a specific chemical composition and a characteristic crystal structure. About 3000 different minerals are known.

Within the Earth is an unknown amount of molten mineral matter called **magma.** At or near the surface, however, all the fundament is a solid, generally known as **rock,** composed of aggregated mineral particles that occur in great variety and complexity. Solid rock sometimes is found right at the surface as an **outcrop,** but over most of the Earth's land area the **bedrock** is covered by a layer of broken and partly decomposed rock particles referred to as **regolith** (see Figure 4–1). Soil, when present, lies on the regolith. All rocks can be categorized into one of three fundamental classes—igneous, sedimentary, and metamorphic—depending on their genesis, or mode of origin (see Figure 4–2).

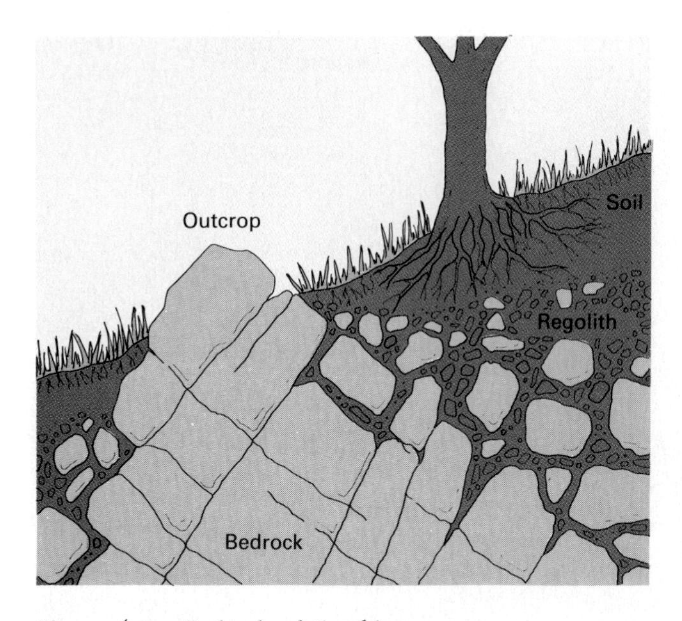

Figure 4–1. Bedrock relationships.

1. **Igneous rocks** are formed by the cooling and solidification of magma. They are crystalline in structure and are usually quite hard. Igneous rocks can be either extrusive or intrusive, depending on the conditions under which they solidified. **Extrusive** igneous rocks are spewed out onto the Earth's sur-

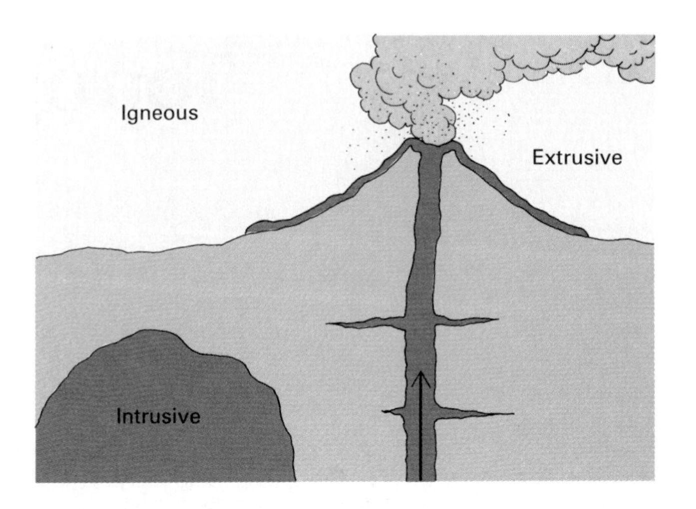

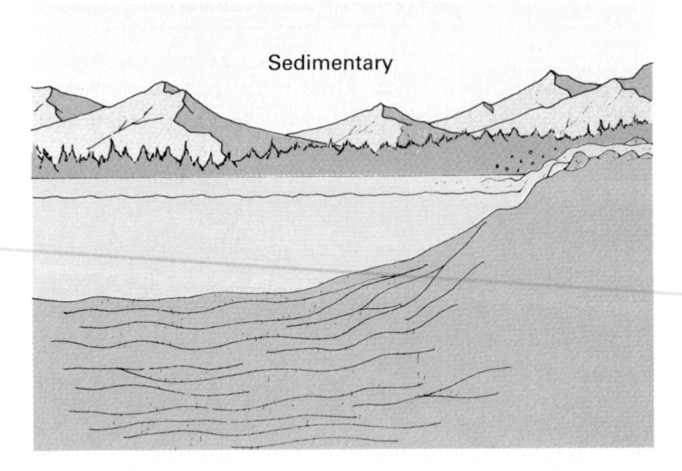

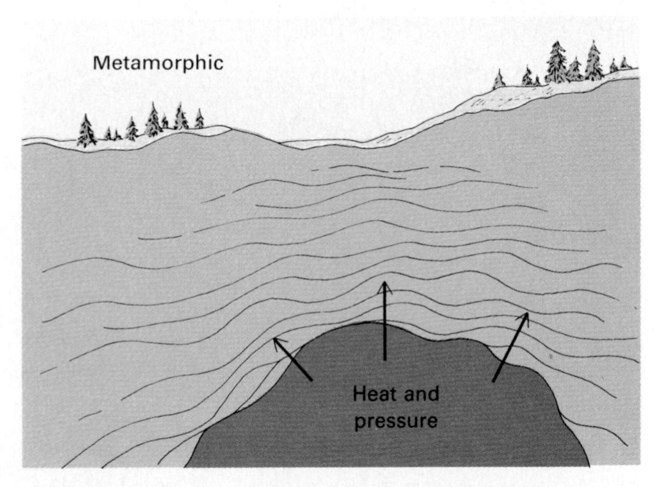

Figure 4–2. There are three basic types of rocks: (1) Igneous rocks are formed by the cooling of magma; (2) sedimentary rocks result from consolidation of deposited particles; (3) metamorphic rocks are produced when heat and pressure act upon preexisting rocks.

Figure 4–3. Nearly horizontal strata of interbedded limestone and shale in a road cut near Lyons, Colorado. (TLM photo)

face while still molten, solidifying quickly in the open air. **Intrusive** igneous rocks cool and solidify beneath the Earth's surface, where surrounding nonmagmatic material serves as insulation, which greatly retards the rate of cooling. The most common intrusive rock is granite, and the most widespread extrusive rock is basalt.

2. **Sedimentary rocks** are formed by the deposition of small rock fragments in a quiet body of water, particularly on the floor of an ocean. Such sedimentary deposits can be built to a great thickness, and the sheer weight of the massive overburden exerts an enormous pressure that causes adhesion and interlocking of the individual particles. Natural chemical cementation also normally occurs. This combination of pressure and cementation consolidates and transforms the sediments into sedimentary rock, of which the most common varieties are sandstone, limestone, and shale (see Figure 4–3).
3. **Metamorphic rocks** were originally either igneous or sedimentary rocks that have been drastically changed by massive forces of heat and pressure from within the Earth. By far the most common metamorphic rocks are schist and gneiss.

Geologic Time

Probably the most mind-boggling concept in all physical geography is the vastness of geologic time. In our puny human scale of time, we deal with such brief intervals as hours, months, and centuries, which does nothing to prepare us for the scale of the Earth's history. The colossal sweep of geologic time encompasses epochs of millions and hundreds of millions of years, periods that are extremely difficult for the human mind to comprehend. Figure 4–4 provides a diagram of the relative duration of geologic time intervals.

Topographic development involves processes that operate with excruciating slowness over the course of many centuries. Thus, only geologic history can provide a time frame vast enough to date and explain these processes.

INTERNAL GEOMORPHIC PROCESSES

The topography of the Earth's surface exists in infinite variety. This variety reflects the complexity of interactions between process and structure. In other words, the countless shapes and forms that make up the Earth's surface result from effects of long-term processes on the geologic foundation. We refer to these processes as **geomorphic** ("geo" meaning Earth and "morph" meaning form) because they determine the shape and form of the topography.

Geomorphic processes are relatively few in number, but they are extremely varied in their nature and operation. Scientists characterize these processes as either internal or external. The *internal* processes operate

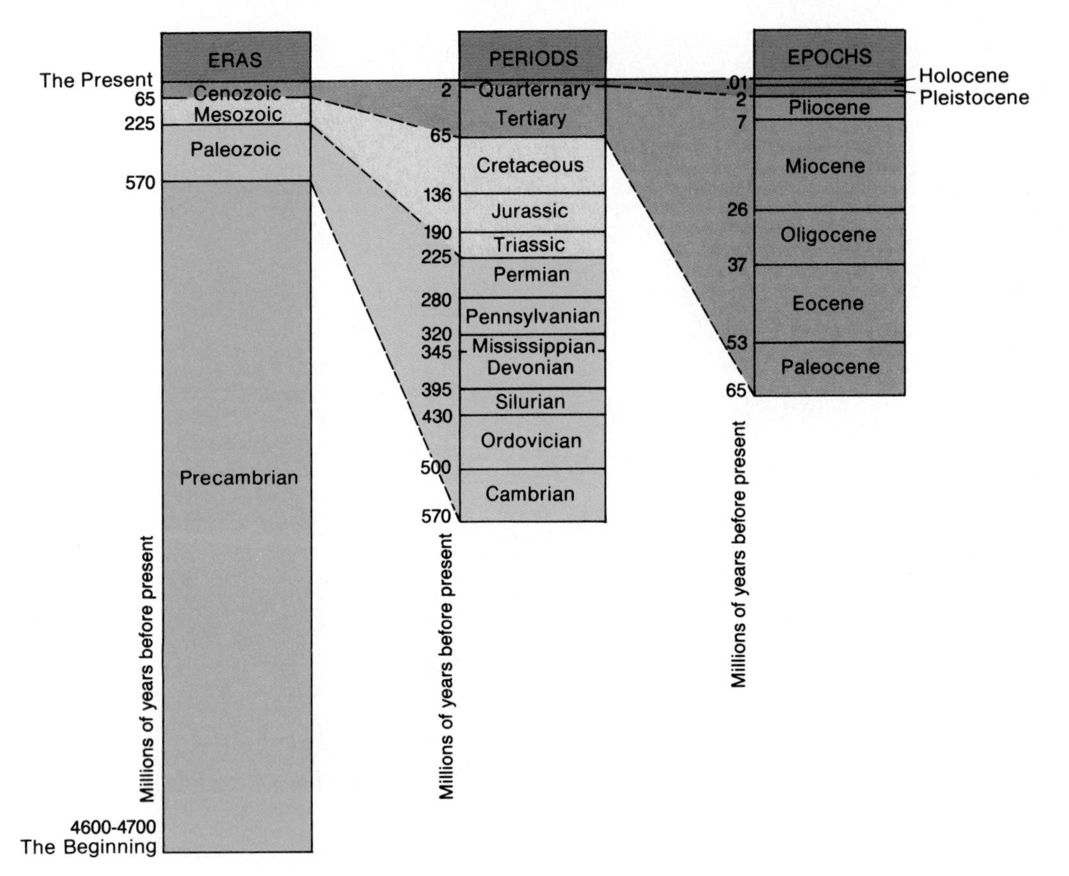

Figure 4–4. The relative duration of geologic time intervals. The height of the column is proportional to the length of the various units of geologic time. Geologic history is divided into four major units, called *eras*. The eras are subdivided into *periods*, and the more recent periods are further subdivided into *epochs*. The Precambrian era was seven or eight times longer than the other three eras combined.

from within the Earth, energized by massive and generally unpredictable forces that are imperfectly understood and that apparently operate independent of any surface or atmospheric influences. They result in *tectonic* activity, that is, various movements of the earth's crust. In general, they are constructive, uplifting, building forces that tend to increase the topographic irregularities of the land surface.

In contrast, the *external* processes are largely *subaerial*; that is, they operate at the base of the atmosphere and draw their energy mostly from sources in the atmosphere or the oceans. Unlike internal processes, external processes are well understood, and their behavior is often predictable. Moreover, their behavior may be significantly influenced by the characteristics of the preexisting topography, particularly its shape and the nature of the surface materials. The external processes generally may be thought of as wearing-down or destructive forces that eventually diminish topographic irregularities.

Internal and external processes thus work in more-or-less direct opposition to each other. Their battleground is the surface of the Earth, where this remarkable struggle has persisted for billions of years and may continue endlessly into the future.

In succeeding sections we will consider these various processes, but it may be useful to summarize them here so that they can be seen in totality. It should be noted that our classification is imperfect: Some items are clearly separate and discrete, whereas others overlap. This outline, then, represents a simple, logical way to approach a study of geomorphic processes, but is not necessarily the only or ultimate framework that could be used.

A. Internal processes
- **1.** Massive crustal rearrangement (plate tectonics)
- **2.** Diastrophism
  - **a.** Broad warping
  - **b.** Folding
  - **c.** Faulting
- **3.** Vulcanism
  - **a.** Extrusive
  - **b.** Intrusive

B. External processes
- **1.** Weathering
- **2.** Mass wasting
- **3.** Erosion/Deposition
  - **a.** Fluvial (running water)
  - **b.** Aeolian (wind)
  - **c.** Glacial (moving ice)
  - **d.** Solution (ground water)
  - **e.** Waves and currents (ocean/lake)

Massive Crustal Rearrangement

Prior to the twentieth century, most earth scientists assumed that the planet's crust was rigid, with continents and ocean basins fixed in position and significantly modified only by changes in sea level and periods of mountain building. The uneven shapes and irregular distribution of the continents were a matter of puzzlement, but it was generally accepted that the present arrangement was emplaced in some ancient age when the Earth's crust cooled from its original molten state.

The "rigid Earth" theory has been seriously called into question in recent years by a variety of discoveries and hypotheses. Prominent among these has been the recognition that the igneous rocks of the upper crust apparently occur in two layers that are well differentiated in several characteristics, especially density (see Figure 4–5). The general crustal structure appears to consist of continental masses of relatively lightweight *sial* "floating" on a foundation of denser *sima*. Thus, the continental masses are in continual, if extremely slow, motion.

Plate Tectonics The theory of **plate tectonics** is that the upper portion of the crust consists of a mosaic of rigid plates embedded in an underlying, somewhat plastic layer. The plates vary considerably in size: Some are almost hemispheric in size, whereas others are much smaller. The actual number of plates and the locations of some of their boundaries are as yet unclear. Current scientific thought suggests the existence of six major plates and perhaps twice that many smaller ones. All are thought to be about 60 miles (100 km) thick, and most consist of both continental (sial) and oceanic (sima) crust. These plates are coherent masses that move ever so slowly over the weaker underlying layer. These movements sometimes bring two plates together on a collision course. The plates are rigid and become significantly deformed only at the edges, where individual plates impinge on one another.

The process of *convection*, normally associated with fairly rapid flows of liquid or gas, apparently is at work in the interior of the Earth. A very sluggish thermal convection system appears to operate, slowly bringing

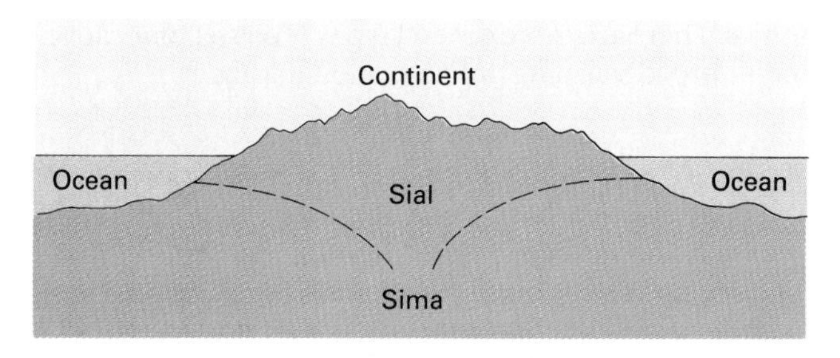

Figure 4–5. Igneous rocks on the upper crust of the Earth can be categorized as belonging to either sial or sima. Sial is an upper layer associated with the continents, whereas sima underlies both ocean basins and continental sial.

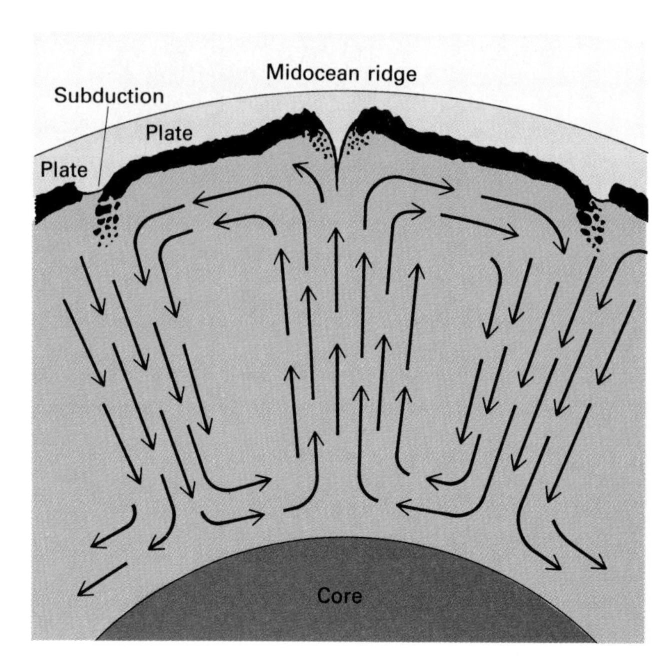

Figure 4–6. Schematic illustration of subsurface convection. Molten material reaches the surface of the midocean ridges and moves laterally across the ocean floor to be subducted downward at the edge of the plate, where it is melted and recycled into the system.

deep-seated (located well below the surface) molten material to the surface and, at the same time, pulling remelted crustal rocks into the depths (see Figure 4–6).

Systematic depth soundings have proved that running across the floors of all the oceans for some 40,000 miles (64,000 km) is a continuous system of large ridges. These ridges are usually located at some distance from the continents, often in the middle of the oceans (see Figure 4–7). The Mid-Atlantic Ridge, the world's mightiest mountain range, is the most prominent example of this phenomenon. Moreover, deep trenches occur at many places in the ocean floors, often around the margins of the ocean basins.

Scientists believe that oceanic ridges were formed by rising currents of deep-seated material, often accompanying volcanic eruptions, which spread laterally to form new ocean floors. Thus, new crustal material appears at the ridges and slowly moves outward, in an action called **sea-floor spreading.** At other places in the ocean basins, usually associated with trenches at the margins, older ocean-floor material descends into the interior in a process called **subduction,** where it is presumably melted and recycled into the convective system.

Continental drift The concept of plate tectonics provides us with a framework for understanding the massive crustal rearrangement that apparently has taken place during the relatively recent history of the Earth. Closely associated with plate tectonics is the concept of **continental drift,** which proposes that the present continents were originally connected as a single landmass, called *Pangaea.* About 200 million years ago Pangaea is

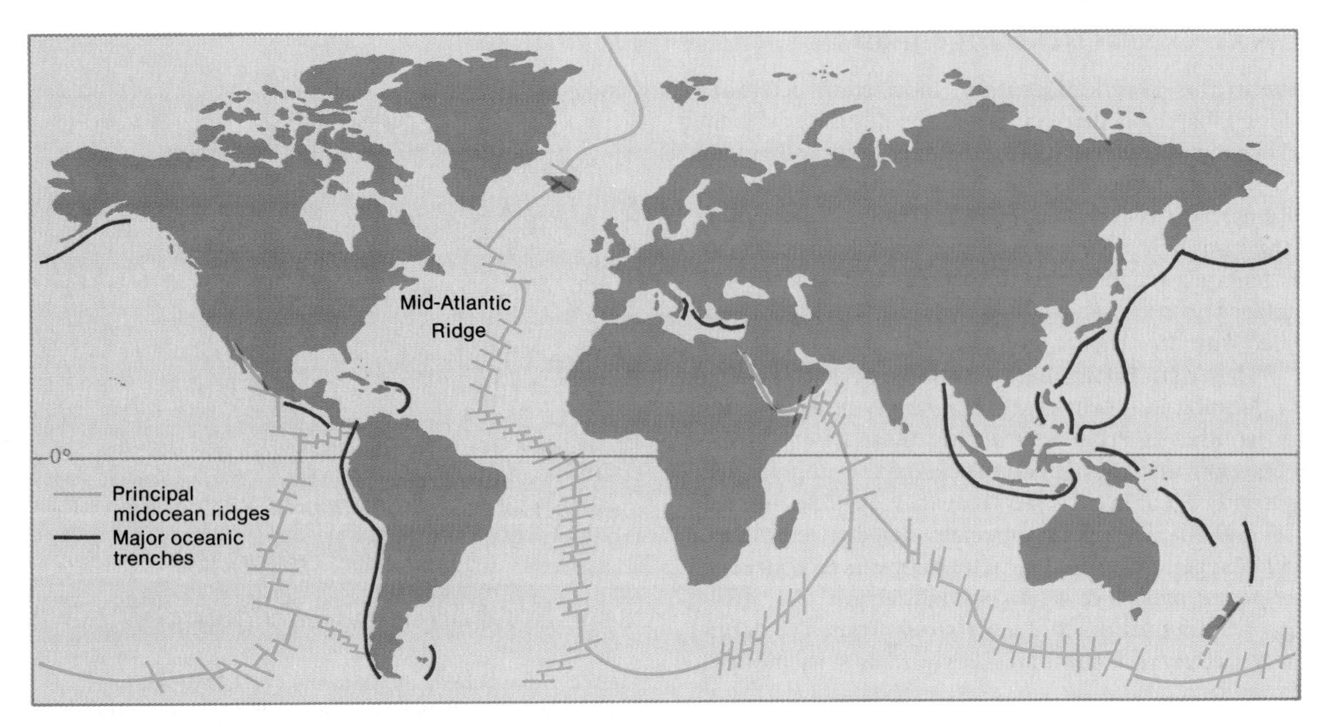

Figure 4–7. The principal midocean ridges form a continuous worldwide system, mostly far removed from any present continent. The principal oceanic trenches, on the other hand, are mostly situated close to continental margins.

believed to have broken apart into two massive landmasses, Laurasia in the Northern Hemisphere and Gondwana in the Southern Hemisphere (see Figure 4–8). The various crustal plates, with their attached continents and parts of continents, separated and drifted in various directions (see Figure 4–9). Their divergence was often associated with sea-floor spreading, and their convergence frequently involved collision, subduction, and mountain building. Over vast stretches of time, huge landmasses joined and separated, eventually forming the contemporary continents (see Figure 4–10). The drifting continues today, so that the present position of the continents is by no means their ultimate one.

Many unanswered questions remain concerning plate tectonics, not the least of which is the ultimate cause of plate movement. However, the present state of our knowledge is ample to provide a firm basis for understanding the broad patterns of most of the world's larger surface features—the size, shape, and distribution of the continents and ocean basins and many of the mountain ranges. To understand more detailed topographic features, we must turn our attention to less spectacular, but no less fundamental, internal processes, which often are directly associated with plate movements. Two principal groups of tectonic forces—diastrophism and vulcanism—are involved.

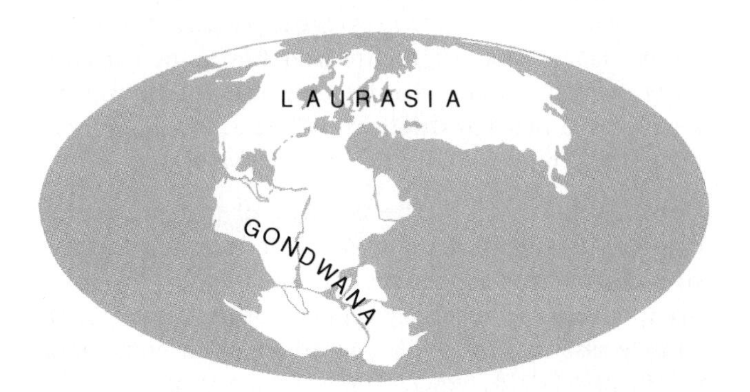

Figure 4–8. The presumed arrangement of the Pangaea supercontinent as of about 200 million years ago.

Diastrophism

Diastrophism is a general term that refers to the deformation of the Earth's crust. Crustal surfaces may be warped, and crustal rocks may be bent or broken in a variety of ways, in response to great pressures exerted from below or within the crust itself. Some of these vast pressures clearly result from movements of the crustal plates, but others appear to be unrelated to plate tectonics. There are three general types of diastrophic movement: broad warping, folding, and faulting.

Broad Warping Relatively extensive portions of the Earth's crust have been subjected to uplift or depression innumerable times throughout history. Areas that are now well above sea level are often covered with marine sediments, indicating that they were once in a submarine environment and therefore at a lower elevation than their present location. Conversely, many areas that are now covered by oceanic waters (for example, the North Sea) were once well above sea level.

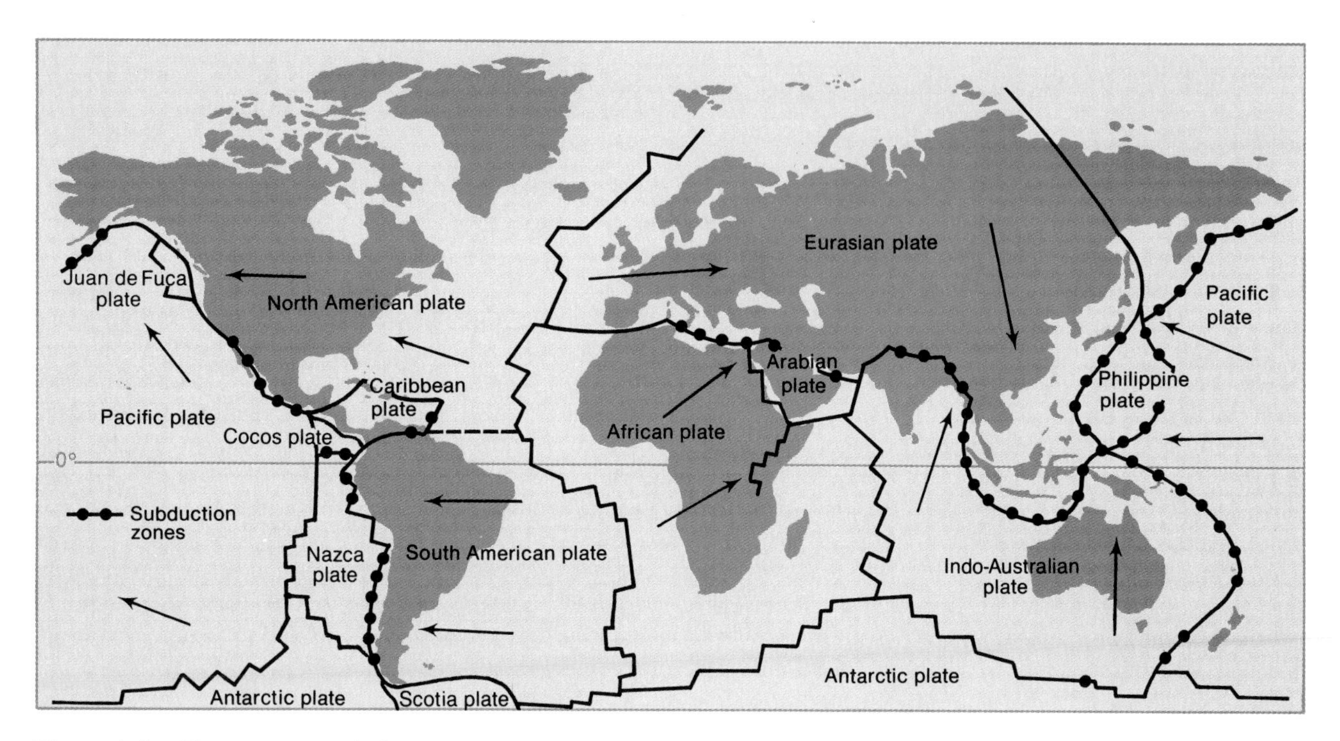

Figure 4–9. The major crustal plates and their generalized directions of drift.

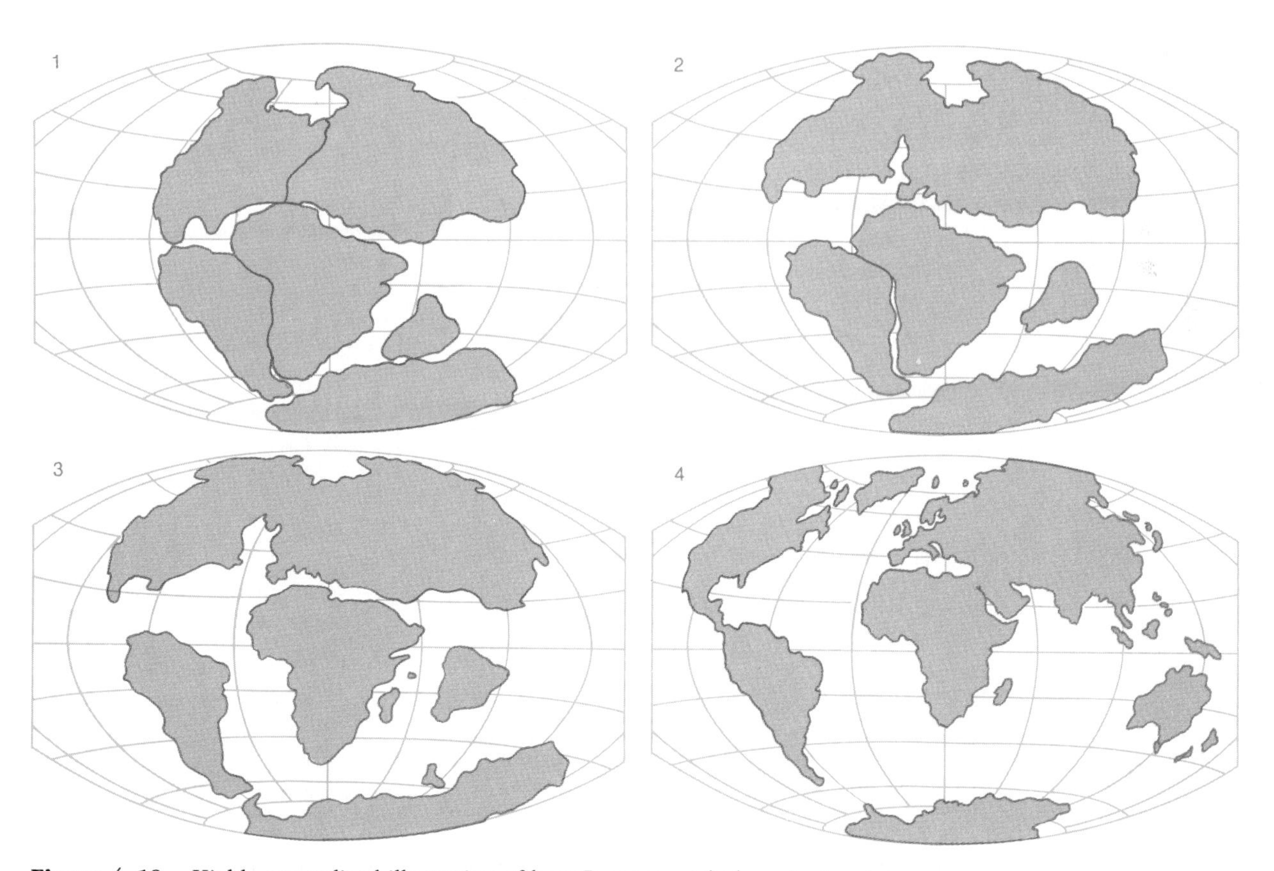

Figure 4–10. Highly generalized illustration of how Pangaea might have broken into separate continental masses. Each diagram represents a progression of about 65 million years from the preceding one, starting about 200 million years ago and concluding with the contemporary arrangement.

Focus on Earthquakes

Although many of the processes that operate in nature are slow and either benign or inoffensive to humanity, a few—such as tornados, hurricanes, volcanic eruptions, and earthquakes—are abrupt and capable of causing an enormous amount of destruction in a very short time. As shapers of terrain, earthquakes are of minor significance, but as potential producers of instantaneous havoc, they are overwhelming.

An earthquake is essentially a vibration in the crust that is produced by shock waves from a sudden displacement along a fault. The fault movement amounts to an abrupt release of energy from a long, slow accumulation of strain, which is an ongoing process in crustal deformation. The faulting may take place right at the surface, but it usually originates at considerable depth, extending downward as much as 400 miles (640 km) beneath the surface. The pent-up energy that is released moves through the lithosphere in several kinds of waves from the center of motion (called the *focus*). These *seismic waves* are transmitted outward in widening circles, almost like ripples produced when a rock is thrown into a pond, gradually losing momentum with increasing distance from the focus. The strongest shocks and greatest crustal vibration are usually felt directly above the focus on the surface, which is referred to as the *epicenter* of the earthquake.

Seismographs around the world record the arrival times and forces of seismic waves from earthquakes. Comparing records from different seismograph stations allows the focus of a quake to be pinpointed with great precision and its strength to be determined. Several different scales are used to indicate the violence of an earthquake, but by far the most widely used is the Richter scale of earthquake magnitudes, devised by California seismologist Charles F. Richter in 1935 to describe the amount of energy released in a single quake. The scale is logarithmic, so each successively higher number represents an energy release that is ten times greater than the preceding number. The scale numbers range from 0 to 9 but theoretically have no upper limit. Any earthquake with a Richter number of 8 or above is considered to be catastrophic, and lesser-numbered shocks can cause immense damage under certain conditions. The most violent known earthquakes have recorded 8.9 on the Richter scale. In comparison, the famous San Francisco quake of 1906 reached an estimated 8.3, the southern Alaska earthquake of 1964 recorded 8.5, the San Fernando Valley shock of 1971 attained only 6.6, and the San Francisco Bay area quake of 1989 recorded 6.9.

The effect of an earthquake on the topography is distinct from, although obviously related to, actual movements along a fault line. The most notable earthquake-caused terrain modifications are usually landslides, which may be triggered in hilly or mountainous terrain. The landslides themselves sometimes produce significant secondary effects such as blocking streams and thereby creating instant new lakes.

Another kind of hazard associated with earthquakes involves water move-

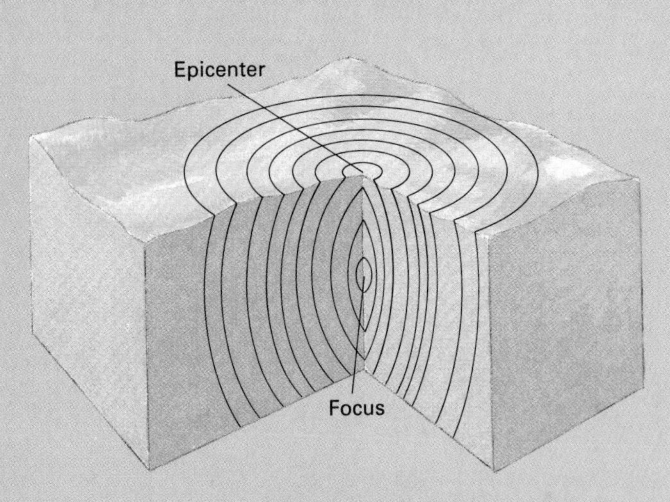

Schematic diagram to show the relationship of focus, epicenter, and seismic waves of an earthquake. The waves are indicated by the concentric circles.

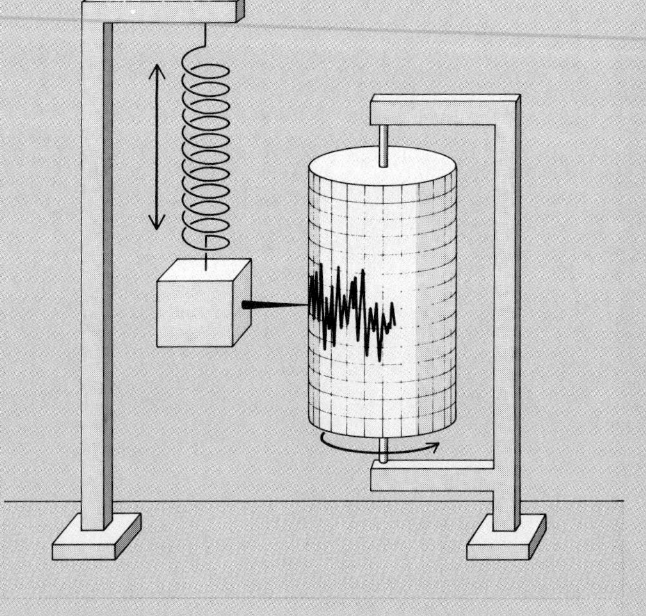

A simplified seismograph. The two posts, which are anchored in bedrock, pick up vibrations in the crust. The pendulum, suspended by a wire coil from one of the posts, traces the movement of the other post on the rotating drum.

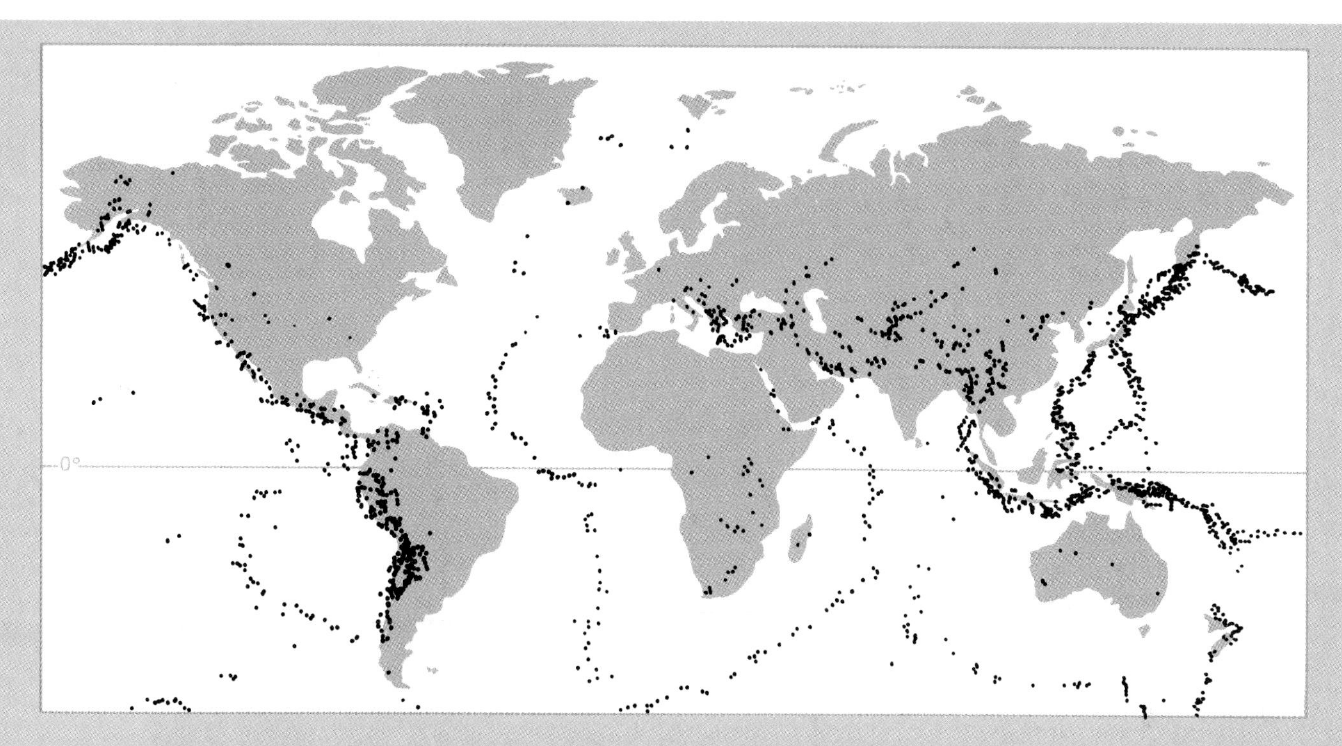

The distribution of epicenters for all earthquakes of at least 5.5 magnitude from 1963 through 1977. Their relationship to midocean ridges and oceanic trenches is striking.

ments in lakes and oceans. The abrupt crustal vibrations can set great waves in motion in lakes and reservoirs, causing them to overflow shorelines or dams in the same fashion that water can be sloshed out of a dishpan by shaking it. Much more significant, however, are great seismic sea waves, or *tsunamis*, which are sometimes generated by sea-floor movement associated with undersea earthquakes. These waves, sometimes occurring in a sequential train, move rapidly across the ocean. They are all but imperceptible in deep water; but when they reach shallow coastal waters, they sometimes build up to several feet or even tens of feet in height and may crash on the shoreline with devastating effect (where they are often incorrectly called *tidal waves*).

Although the effect of earthquakes on topography may be limited, it is sometimes cataclysmic for humankind. Most quakes are so slight as to occur without recognition, sometimes being mistaken as the rumbling of a passing truck. More severe earthquakes, however, can cause devastating damage within a few seconds. Rubble can be shaken off buildings, entire buildings may collapse, gas mains can be broken, igniting dangerous fires, and villages and even cities may be destroyed. Structural damage sometimes amounts to hundreds of millions of dollars, and loss of life can be counted in the thousands. Indeed, at least five earthquakes in history have killed more than 100,000 people each, and the long-term worldwide average is 10,000 quake deaths annually.

In any given year, tens of thousands of earthquakes occur somewhere in the crust, most of them followed by aftershocks. These aftershocks may number in the hundreds after a single quake and may continue for several days with diminishing intensity. "Significant" earthquakes occur on an average of between 60 and 70 times a year throughout the world. (To be classed as significant, the quake must have a magnitude of at least 6.5 on the Richter scale or cause casualties or create considerable damage.)

Earthquakes may occur anywhere, even in the middle of apparently very stable continental areas. Most, however, take place in association with the boundary zones of the great crustal plates, particularly along the midocean ridges and in the subduction areas of ocean margins. The greatest concentrations of earthquake epicenters are found around the rim of the Pacific Ocean.

Despite our expanding understanding of the nature and causes of earthquakes and the increasing sophistication of quake-detecting instruments, it is still quite impossible to predict a quake with any assurance. As one famous contemporary geophysicist states it, "At present we can't predict earthquakes any better than the ancients did."

Despite their awesome potential and occasionally devastating results, earthquakes represent only minor adjustments in the infinite continuum of events and process interactions that combine to maintain tectonic equilibrium in the Earth's crust—a scientific fact that may be of little comfort to a person who lives on the edge of the Pacific Ocean!

Figure 4–11. Tightly folded sedimentary strata in a road cut near Los Angeles. (TLM photo)

Folding Although we think of rocks as being rigid and brittle, crustal rocks often are deformed by being bent rather than breaking. When great pressure is applied for long periods, particularly in an enclosed, buried, subterranean environment, the result is often a slow plastic deformation that can produce folded structures of incredible complexity (see Figure 4–11). In some cases the folding is simple and symmetrical; elsewhere it may be extremely complex and totally asymmetrical. Indeed, the most severe crumpling of a dishrag can be duplicated in rocks by the actions of diastrophic folding.

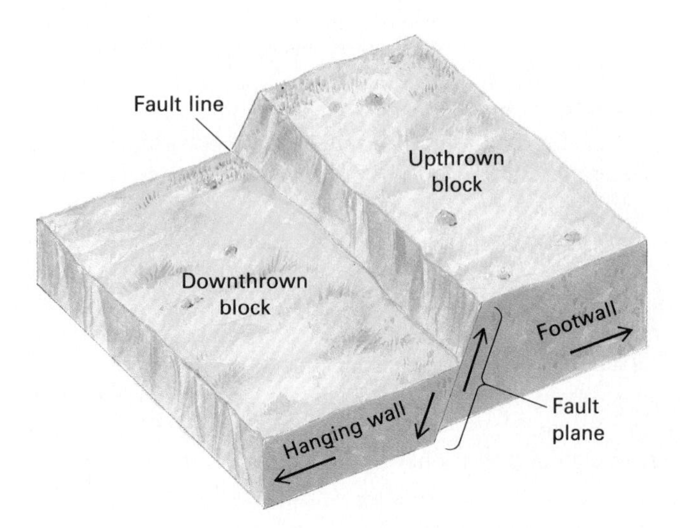

Figure 4–12. A simple fault structure.

Faulting Another prominent result of the various stresses in the crust is the breaking apart of rock material. When rock is forcefully broken with accompanying displacement (that is, an actual movement of the crust on one or both sides of the break), the action is called *faulting* (see Figure 4–12). The movement can be vertical, horizontal, or a combination of both (see Figure 4–13). Faulting usually takes place along zones of weakness in the crust; such an area is referred to as a **fault zone** or **fault plane,** and the intersection of that zone with the Earth's surface is called a **fault line.** Movement along a fault zone is sometimes very slow, but it can also occur as a sudden slippage, usually referred to as an **earthquake.**

Vulcanism

Vulcanism is a general term that refers to the movement of magma from the interior of the Earth to or near the surface. It sometimes consists of explosive volcanic eruptions that are among the most spectacular and terrifying events in all nature (see Figure 4–14). But it also involves much less violent phenomena, including the slow solidification of molten material below the surface. If the magma is expelled onto the Earth's surface while still in a molten condition, the activity is extrusive and represents volcanic activity. In contrast, if the molten material solidifies within the crust, it is an intrusive activity.

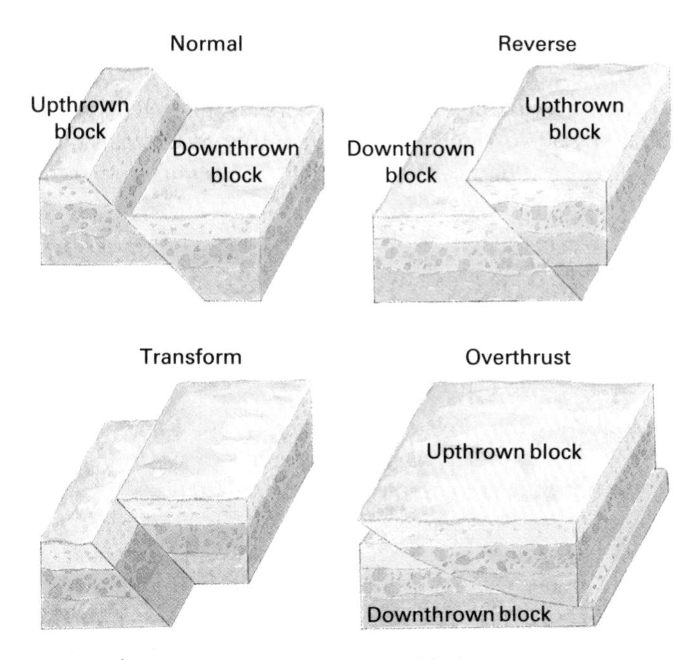

Figure 4–13. The principal types of faults.

Molten magma extruded onto the surface of the Earth, where it cools and solidifies, is designated as **lava.** The ejection of lava into the open air is sometimes volatile and explosive, devastating the area for miles around. In other cases, it is gentle and quiet, affecting the landscape more gradually. All instances, however, alter the landscape because the fiery lava is an inexorable force until it cools, even if it is expelled only slowly and in small quantity.

Although volcanoes are very conspicuous landscape features, many of the world's most extensive lava flows were not extruded from true volcanoes but rather issued quietly from great fissures that developed in the crustal surface. The mechanism that produces these great outpourings of lava is not clear, but apparently it involves both the midocean ridges, where upwelling magma fills the gaps between diverging plates, and more localized "hot spots," where magma is expelled through cracks in the plate itself. The lava that flows out of these fissures is nearly always basaltic and frequently comes forth in great volume. The term **flood basalt** is applied to the vast accumulations of lava that build up, layer upon layer, sometimes covering tens of thousands of square miles to depths of many hundreds of feet. Over the world as a whole, more lava has issued quietly from fissures than from the combined outpourings of all volcanoes.

When magma solidifies below the Earth's surface, it produces igneous rock. If this rock is pushed upward into the crust either before or after solidification, it is called an *igneous intrusion*. Most such intrusions have no effect on the surface landscape, but sometimes the igneous mass is raised high enough to deform the overlying material and change the shape of the surface. In many cases, the intrusion itself is exposed at the surface through uplift and/or erosion. When intrusions are thus exposed to the external processes, they often become conspicuous because they are usually resistant to erosion, and with the passage of time, they stand up relatively higher than the surrounding land. The intrusive process is usually a disturbing one for preexisting rock. Rising magma makes room for itself by a process called *stoping*, a mining term for ore removal by working upward. Stoping can involve the assimilation of invaded rock by the molten magma, heating the rock enough to make it flow out of the way, forcibly splitting rocks apart, or bending them upward. Adjacent to the new rock, the invaded rock may be physically and/or chemically changed due to exposure to the heat and pressure of the rising intrusion. Although igneous intrusions can assume an almost infinite variety of forms, the most common are batholith, stock, laccolith, dike, sill, and vein (see Figure 4–15).

Figure 4–14. Mexico's Paricutín volcano at the height of its activity. (Courtesy of K. Seger-Strom.)

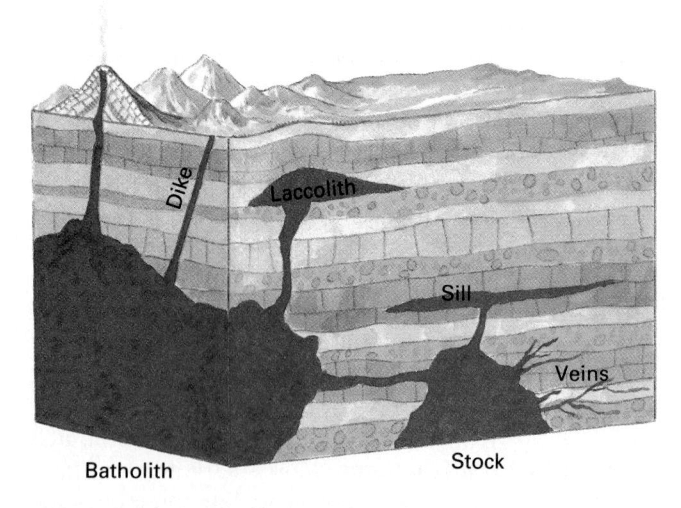

Figure 4–15. Some typical forms of igneous intrusions.

EXTERNAL GEOMORPHIC PROCESSES

At the same time these internal forces are occurring, various external forces are also at work in shaping the Earth's landscape. Although less dramatic and colorful than drifting continents and exploding volcanoes, these forces play a critical role in determining topographical features. Ultimately, the specific shape of the Earth's surface is sculptured by these external forces. The detailed configuration of peaks, slopes, valleys, and plains is molded by the work of gravity, water, wind, and ice, or some combination thereof.

The total effect of these actions—disintegration, wearing away, and removal of rock material—is generally encompassed by the term **denudation,** which implies an overall lowering of the surface of the continents (see Figure 4–16). Denudation is accomplished by the interaction of various agencies and forces, but for analytical purposes we can consider three types of activities—weathering, mass wasting, and erosion. Weathering is the breakdown of rock into smaller components by atmospheric and biotic agencies. Mass wasting involves the downslope movement of broken rock material, often as cohesive units, due to gravity, sometimes lubricated by water. Erosion consists of more massive and generally more distant removal of fragmented rock material.

Weathering

The first step in the shaping of the Earth's surface by external processes is accomplished by **weathering,** that is, the mechanical disintegration and/or chemical decomposition that destroys the coherence of bedrock and begins to fragment rock masses into progressively smaller components. Weathering is the aging process of rock surfaces, and all rock that is exposed at the surface will be weathered. Moreover, in some cases, weathering may reach as much as several thousand feet beneath the surface. This is made possible by open spaces within the rock bodies and even within the mineral grains. Subsurface weathering is initiated along these openings, which can be penetrated by agents from above (water, air, roots). As time passes, the weathering effects spread from the immediate vicinity of the openings into the denser rock beyond (see Figure 4–17).

Joints are the most common structural features of the crustal rocks. They are cracks that develop due to stress, but with no appreciable movement of the rock. These cracks are innumerable in all rock masses, serving to divide them into blocks of various sizes. Joints are almost universal and therefore are the most important of all rock openings in facilitating weathering.

The agents of weathering are relatively few in number, but they are complex in their interactions. Most, as implied by the term *weathering*, are atmospheric. Because it is in gaseous form, the atmosphere is able to penetrate readily into all cracks and crevices that extend down into the bedrock. From a chemical standpoint, oxygen, carbon dioxide, and water vapor are the three most important atmospheric components in rock weathering. Temperature changes are also significant. Most notable, however, is moisture in its liquid form, which can penetrate downward effectively into openings in the bedrock. Biotic agents also contribute to weathering, in part through the burrowing activities of animals and the rooting effects of plants, but especially by producing chemical substances that attack the rock more directly.

Mechanical Weathering For analytical purposes, it is convenient to recognize two major categories of weathering: mechanical and chemical. ***Mechanical***

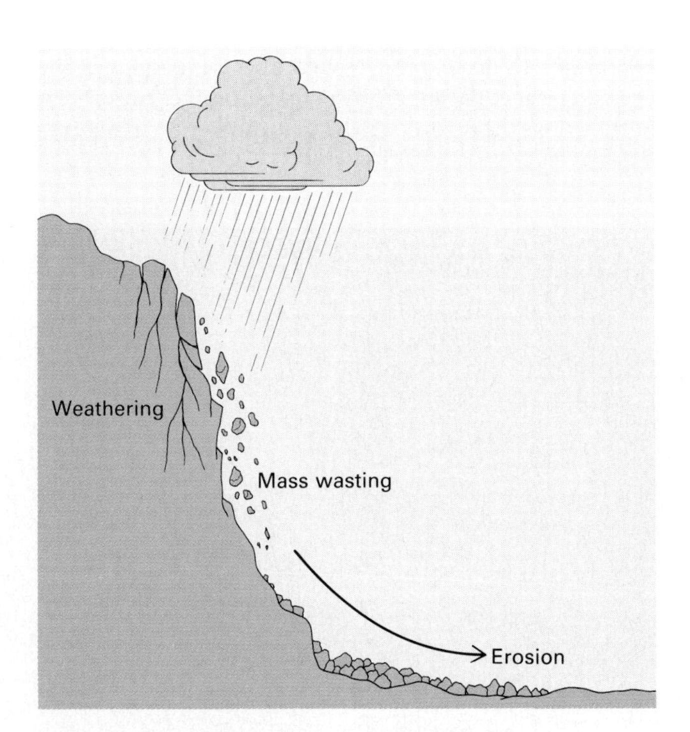

Figure 4–16. Denudation is accomplished by a combination of weathering, mass wasting, and erosion.

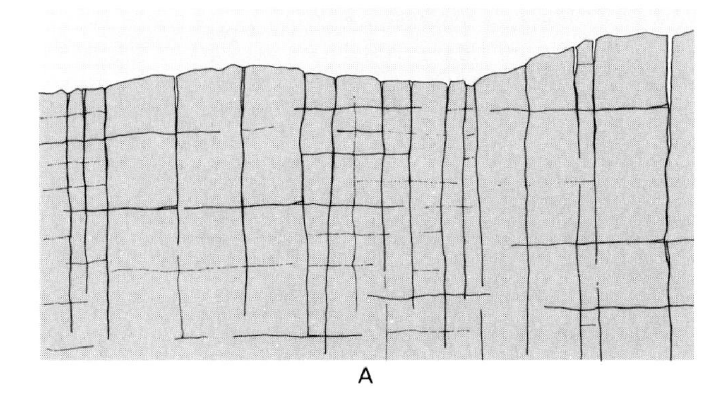

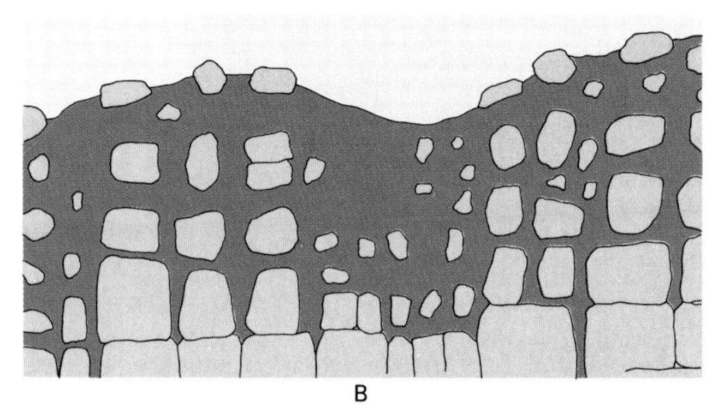

Figure 4–17. Schematic representation of the development of deep weathering in highly jointed rock.

weathering involves the physical disintegration of rock material. In essence, big rocks are made into little rocks by various stresses that fracture the rock into smaller fragments. Probably the most important single agent of mechanical weathering is the freeze/thaw action of water in open spaces in rock, known as *frost wedging*. When water freezes, it expands in volume by almost one-tenth, thus exerting enormous pressure on the surrounding rock (see Figure 4–18). Infinite repetition makes this frost wedging an irresistible force (see Figure 4–19). Temperature changes not involving freeze/thaw conditions also accomplish mechanical weathering, but much more gradually.

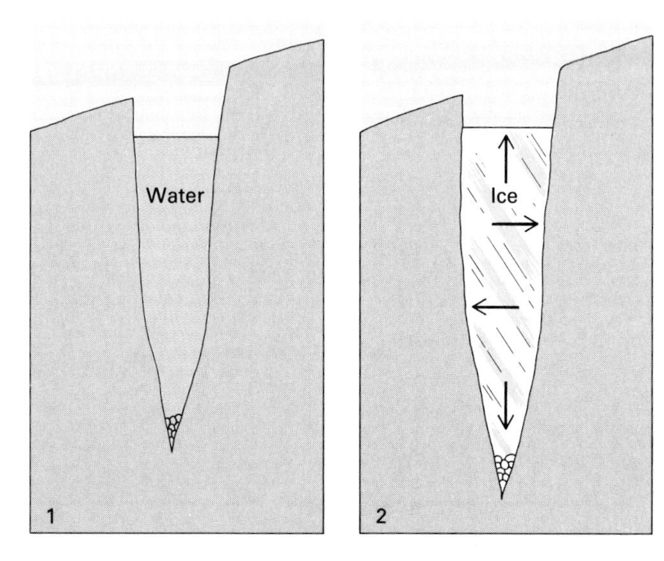

Figure 4–18. Schematic illustration of frost wedging. When water in a rock crack freezes, the ice expansion exerts a force that can deepen and widen the crack, especially if the process is repeated many times.

Chemical Weathering Mechanical weathering is usually accompanied by *chemical weathering*, which involves the decomposition of rock by the alteration of rock-forming minerals. Most rocks are composed of various minerals in combination and can be significantly affected by chemical weathering because the alteration of even a single important mineral constituent can lead to the eventual disintegration of the entire rock mass. Some of the chemical reactions that affect rocks are very complex, but others are simple and predictable. The principal reacting agents are the commonplace ones of water, carbon dioxide, and oxygen. The various chemical reactions may change both the appearance and the coherence of the affected rocks. Loose particles are produced at the surface, and beneath the surface the rock is chemically altered. The major eventual products of chemical weathering are clays, very common constituents of the surface that are produced solely by rock decomposition through chemical alteration.

Mechanical and chemical weathering normally occur together, but one is usually more important than the other. Mechanical weathering predominates in cold and dry regions, and chemical weathering in hot and wet regions.

Mass Wasting

The denudation of the Earth's surface is accomplished by a continuum of action that begins with weathering and ends with erosion. In between these two, there is often an intermediate stage in which weathered material is moved a relatively short distance downslope under the direct influence of gravity. This is called **mass wasting.**

Throughout our planet, gravity is inescapable; everywhere it pulls objects toward the center of the Earth. Where the land is flat, the influence of gravity on topographic development is minimal. On gentle slopes, minute effects are likely to be significant in the long run, and on steep slopes, the results often are immediate and conspicuous. Any loosened material will be impelled downslope by gravity, in most cases falling abruptly or rolling rapidly; in other instances, flowing or creeping with imperceptible gradualness.

All rock materials, from individual fragments to cohesive layers of soil, will lie at rest on a slope if undisturbed, unless the slope is very steep. The steepest angle that loose fragments can assume on a slope without moving downslope is called the *angle of repose*. The angle of repose represents a fine balance between the pull of

Figure 4–19. Frost wedging is an especially pervasive force on mountaintops above the treeline, as with these granite boulders on Australia's highest peak, Mount Kosciusko. (TLM photo)

gravity and the cohesion and friction of the rock material. The material that moves can be of any size, from gigantic boulders to tiny dust particles. Of particular importance, however, is the implication of "mass" in mass wasting. Large units of material, including fragmented rock, regolith, and soil, often are moved.

If water is added to the rock material through rainfall, melting snow, or subsurface flow, the mobility is usually increased, particularly if the fragments are of small size. Water is a lubricating medium, and it diminishes friction among the particles so that they can slide past one another more readily. It is clear, then, that mass movement will be particularly likely during and after heavy rains.

The presence of clay also contributes to mass wasting. The ability of clays to absorb water, combined with their extremely fine grained texture, produce a substance that is very slippery and mobile.

Although some types of mass wasting are rapid and conspicuous (*fall*, *slide*), others are slow and gradual (*flow*, *creep*). The principles involved are generally similar, but the extent of the activity and particularly the rates of movement are quite variable (see Figure 4–20).

The Fluvial Process

The work accomplished by running water is probably more significant in shaping landforms than that of all other external forces combined. This is not because running water is necessarily more forceful than the other agents. Indeed, such forces as moving ice and pounding waves often apply much greater amounts of energy per unit area than can be mustered by surface water. The most important reason that water is the dominant shaper of continental landforms is that it is found everywhere. Almost all parts of all continents except Antarctica experience occasional, frequent, or continuous movement of surface water. This is true even in desert areas where rains are rare and surface streams are virtually nonexistent. Other external forces, with the exception of wind, occur much less frequently and are therefore less significant. Running water is referred to as a **fluvial process** in this context. This encompasses both the general movement of surface water down the slope of the land surface, called *overland flow*, and the more precisely channeled movement of water along a valley bottom, or ***streamflow***.

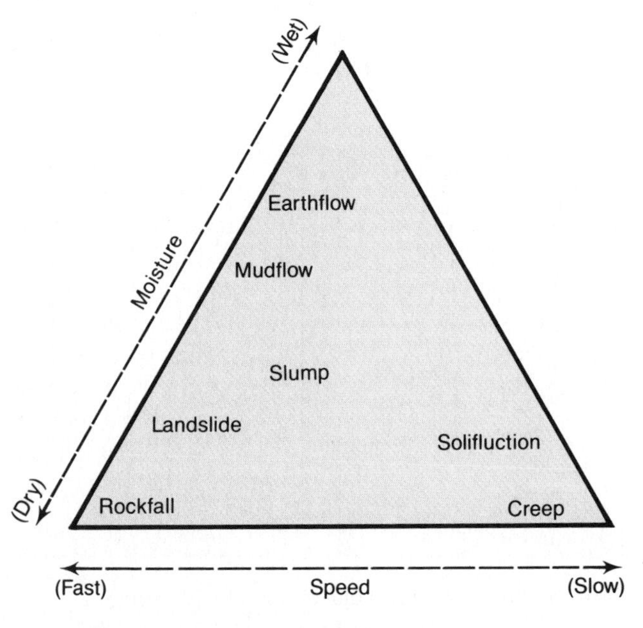

Figure 4–20. A mass-wasting schema.

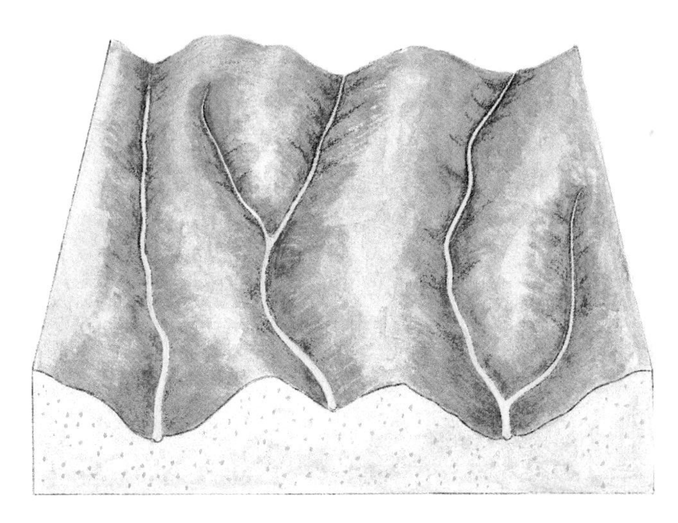

Figure 4–21. Valleys and interfluves. Valleys normally have clear-cut drainage systems; interfluves do not.

Valleys and Interfluves In a fundamental sense, the surfaces of the continents can be considered to consist of two topographic elements—valleys and interfluves. A **valley,** in the broad meaning of the term, encompasses that portion of the total terrain in which a drainage system is clearly established (see Figure 4–21). In some situations, the valley is narrow and elongated with a limited areal extent; in other cases, it may be extraordinarily broad and extensive. The outermost, or upper, limit of a "valley" in this context is not always readily apparent on the land, but it can be clearly conceptualized as a "lip" or "rim" at the top of the valley sides above which drainage channels are either absent or indistinct.

An **interfluve** is the higher land above the valley sides that separates adjacent valleys (see Figure 4–22).

Figure 4–22. The distinction between interfluve and valley is seen clearly here at a very large scale. The grass-covered area is an interfluve where runoff is unchanneled. When water trickles over the upper edge, or "lip" of the valley, it rapidly becomes channeled into the established drainage system. (TLM photo)

Interfluve means "between rivers." Some interfluves consist of ridgetops or mountain crests with precipitous slopes, but others are simply broad and flattish divides between drainage systems. Conceptually, all parts of the terrain not in a valley constitute a portion of an interfluve.

Erosion and Deposition

To grasp fully the process of denudation in general and the shaping of landforms in particular, we must first understand the relationship between erosion and deposition. All external forces perform both activities; they remove fragments of bedrock, regolith, and soil from their original positions in a process called **erosion,** and they relocate these materials in a process called **deposition.** The fluvial process produces one set of residual landforms by its erosional activities and another quite different set of landforms by its depositional activities.

The initial opportunity for fluvial erosion occurs when rain starts to fall. In the first few minutes of rainfall, much of the water penetrates into the soil, and there is little runoff. During heavy or continued rain, however, particularly if the land is sloping and there is a sparse vegetative cover, penetration diminishes, and most of the water proceeds downslope as overland flow.

Once the surface flow is channeled into a stream, its ability to erode is greatly increased by the enlarged volume of water. Erosion is accomplished in part by the direct power of the moving water, which has an impact and a dragging effect on material at the bottom and sides of the stream and can thus excavate and move considerable quantities of unconsolidated debris.

The erosive capability of streamflow is also significantly enhanced by the abrasive tools that it picks up and carries along with it. All sizes of rock fragments, from silt to boulders, exert a chipping and grinding effect as they are swirled and bounced or rolled downstream by the moving water. These "tools" break off more fragments from the bottom and sides of the channel, and they collide with one another, thus becoming both smaller in size and rounder in shape.

Whatever is picked up must eventually be set down, which means, from the standpoint of topographic development, that erosion inevitably is followed by deposition. Moving water, whether overland flow or streamflow, carries its load toward an ultimate destination either in an ocean, a lake, or some basin of interior drainage. The movement of the load is normally accomplished in spasmodic fashion; that is, debris is transported some distance and then dropped, only to be picked up later and carried farther along. Water moving fast and in large quantities can carry its debris a great distance, but sooner or later deposition will take place in response to a reduction of either velocity or volume of flow.

The general term applied to stream-deposited debris is **alluvium.** Alluvial particles tend to be sorted on the basis of size. Finer bits of silt, for example, can be carried greater distances by a smaller flow of water than can larger particles of gravel. Thus, the separation of different sizes of alluvial material is sometimes crude and indistinct, but it can also be clear-cut and precise.

Stream Channels Overland flow or surface runoff is a relatively simple process. It is affected by many factors, such as the nature of the rainfall, the vegetative cover, the character of the surface, and the shape of the slope, but its general characteristics are straightforward and relatively easily understood. Streamflow, in contrast, is much more complicated, in part because streams represent not only a process of denudation but also an element of topography.

A basic characteristic of streamflow, and one that further distinguishes it from the randomness of overland flow, is that it normally is confined to channels, which makes for a more complex flow pattern because the water can move up and down as well as forward or sideways. In any channel with even a slight gradient, gravitational pull overcomes frictional inertia to move the water down-channel. Except under very unusual circumstances, however, this movement is not straight and smooth and regular. Rather, it tends to be unsystematic and irregular, with many directional components involved other than a simple down-channel flow and with varying velocities in different parts of the channel.

A principal cause of this irregular flow is the retarding effect of friction along the bottom and sides of the channel, which causes the water to move slowest there and fastest in the center of the stream (see Figure 4–23). One effect of frictional retardation is to use up much of the stream's energy, decreasing the amount that is available for erosion and transportation.

Nearly every stream continually rearranges its sediment by *scouring* and *filling*, in response to variations in velocity and volume of flow (see Figure 4–24). During high-water periods of fast, voluminous flow, the stream is able to scour its bottom by detaching particles from its bed and shifting most or all sediment downstream. Conversely, during low-water periods, the flow is slowed, and sediment is more likely to settle to the bottom, which

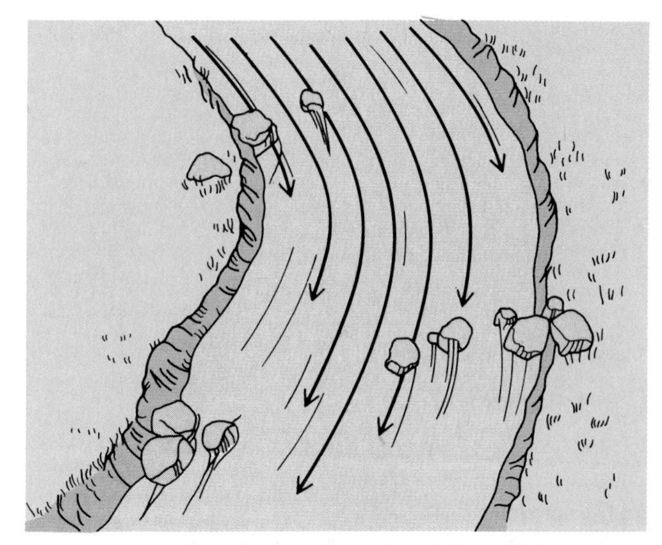

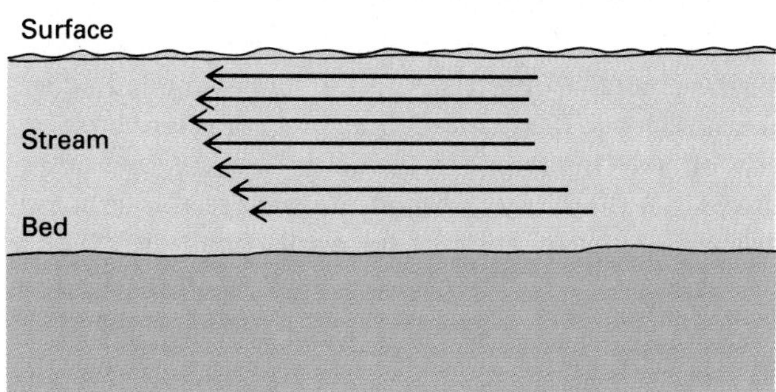

Figure 4–23. Friction retards velocity near the bottom and sides of a stream. The most rapid flow is near the center and slightly below the surface of the water, as shown in these hypothetical surface and cross-section views.

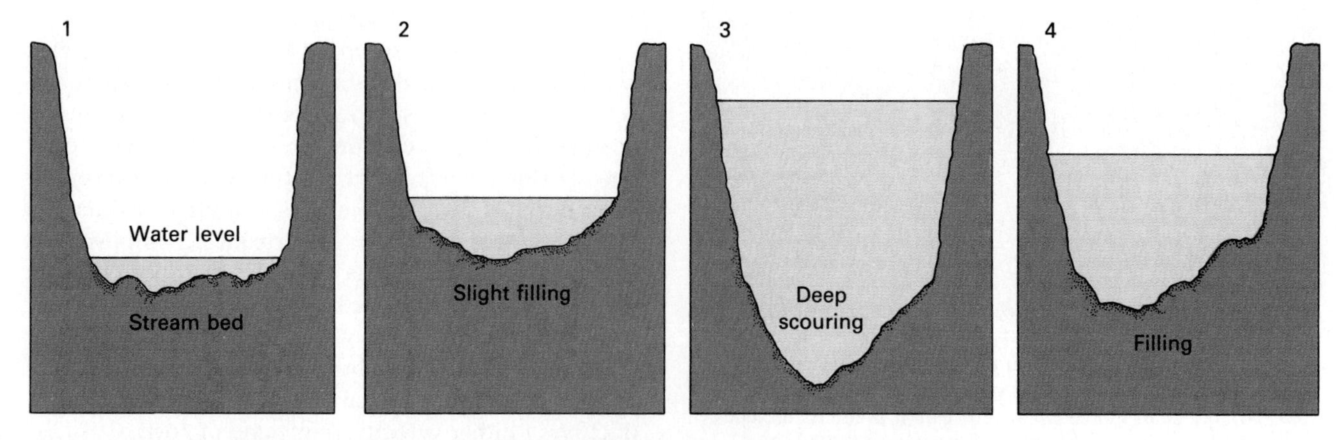

Figure 4–24. Schematic illustration of changing channel depth and shape during a flood: (1) The water flow is low prior to flooding. (2) As the volume of streamflow increases, the bed is raised slightly by filling. (3) Flood flow significantly deepens the channel by scouring. (4) As the flood recedes, considerable filling raises the channel bed again.

results in *aggrading*, the filling in (and thus the building up) of the channel bed.

Humans often construct an elaborate system of artificial embankments, or *levees*, alongside major river channels to move floodwaters downstream and thus protect the valley bottomland from inundation. There is a sequential cycle to levee building. If a levee is raised in an upstream area, then most downstream locales will require higher levees to pass the floodwaters on without overflow. The river tends to aggrade the channel, which raises the river bed above the level of the surrounding bottomland and invites disaster should the levee break. Thus, the river becomes higher than the adjacent valley floor, and the levee becomes the highest part of the local landscape.

The Shaping and Reshaping of Valleys The work of running water in shaping the terrain is accomplished in part by overland flow, but of much greater magnitude is the denudation that is associated with streamflow in the valleys. The shaping of valleys and their almost continual modification through time produces a changing sequence of landforms in most parts of the continental surfaces.

A stream excavates its own valley by eroding its channel bed, which is a major component of denudation. In addition, the actions of the streamflow frequently widen and lengthen the valley.

Valley deepening The process of valley deepening is simple and straightforward. A stream with either a rapid velocity or a large volume of flow will expend most of its energy in *downcutting*, that is, deepening its bed. Downcutting shows up most prominently in the upper reaches of a stream, where the gradient usually is steep and the valley narrow. The general effect of downcutting is to produce a deep valley with relatively steep sides and a V-shaped cross-sectional profile, because the only active site of water erosion is in the streambed itself, which is a relatively narrow area (see Figure 4–25).

Valley widening Where a stream gradient is steep, downcutting is the dominant activity, and valley widening is likely to be slow. Even at this stage, however, some widening will take place as the combined action of weathering, mass wasting, and overland flow removes material from the valley sides. In the valley bottom, downcutting will diminish with the passage of time and eventually will virtually cease as the stream develops a gentle profile. The stream's energy is then increasingly diverted into a meandering, side-to-side flow pattern, the reasons for which are not yet fully understood. As the stream begins to wander from side to side, *lateral erosion*

Figure 4–25. The Yellowstone River valley just below the Lower Falls of the Yellowstone is a narrow gorge of recent development. (TLM photo)

is initiated. In essence, this means that the principal current of the stream swings laterally from one bank to the other, eroding where the velocity is greatest and depositing where it is least. The water moves fastest on the outside of curves, and there it undercuts the bank, whereas on the inside of a bend, a sand bar is likely to accumulate.

The current often shifts its position, so that the undercutting is not concentrated in just a few locations. Rather, over a long period of time, most or all parts of the valley sides will experience it. The undercutting encourages the slumping of material into the stream as the process of lateral erosion continues (see Figure 4–26). Throughout this period of widening of the valley floor, mass movement and overland flow continually aid in the wearing down of the valley sides, and similar activities along tributary streams also contribute to the general widening of the main valley.

The frequent shifting of stream meanders produces an increasingly broader, flattish valley floor, which is largely or completely covered with deposits of alluvium. At any given time, a stream is likely to occupy only a small portion of the flatland, although during periods of flooding, the entire floor may be inundated. At these times the valley bottom is properly termed a **floodplain.** The outer edge of the floodplain usually is bounded on either

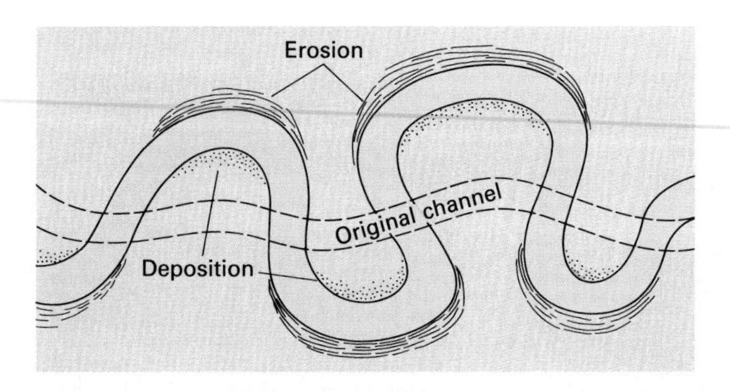

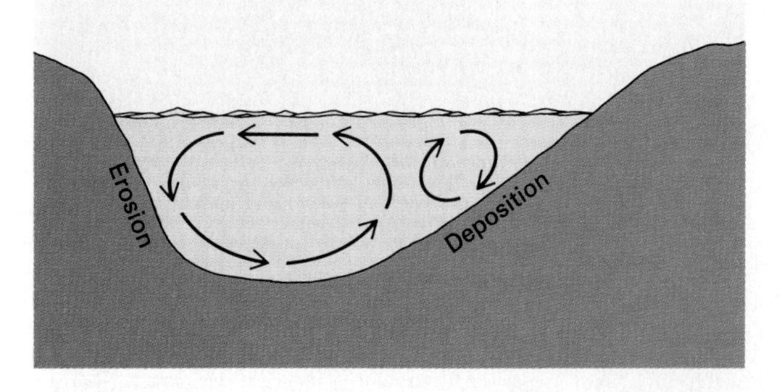

Figure 4–26. Valley widening is accomplished primarily by lateral erosion. The stream current shifts from side to side, eroding on the outside of curves and depositing on the inside. Continuation of this pattern produces an increasingly meandering channel. (After Figures 1.6 and 12.18 from *Essentials of Physical Geography*, Second Edition by Robert E. Gabler, Robert J. Sayer, Sheila Brazier, and Daniel Wise, copyright © 1982 by Holt, Rinehart and Winston, Inc. Reproduced by permission of the publisher.)

side by a clear-cut break in slope, marking the outer limit of lateral erosion and undercutting, where the flat terrain abruptly changes to a relatively steep slope, or a line of *bluffs*. This process of valley widening and floodplain development can extend to vast dimensions. In fact, the floodplains of many of the world's largest rivers are so broad that a person standing on the bluffs at one side cannot see the bluffs on the other side.

Valley lengthening The third dimension in the shaping of stream valleys is lengthening or extension. A stream may lengthen its valley by two different methods: (1) headward erosion at the upper end or (2) delta formation at the lower end. Both are commonplace occurrences, but the former is much more widespread and significant.

Headward Erosion The concept of *headward erosion* is critical to understanding how gullies and valleys can be created and extended. The upper perimeter of a valley is the line where the somewhat gentler slope of an interfluve changes to the steeper slope of a valley side. Overland flow from the interfluve surface drops more or less abruptly over this slope break, which tends to undercut the lip or rim of the perimeter, weakening it and often causing the collapse of a small amount of material (see Figure 4–27).

The result of this action is a net decrease in the area of the interfluve and a corresponding increase in the area of the valley (see Figure 4–28). As the overland flow of the interfluve becomes part of the channeled flow of the valley, there is a minute but distinct extension of gullies into the interfluve, comprising a headward extension of the valley. Although minuscule as an individual event, when multiplied by a thousand gullies and a million years, this action can lengthen a valley by tens of miles and expand a drainage basin by hundreds of square miles.

Delta Formation At the opposite, or seaward, end a valley can also be lengthened, in this case by deposition rather than by erosion. When a river flows into quiet body of water, such as a lake or ocean, its velocity is reduced, and its load deposited. Most of this debris is dropped right at the mouth of the river in a landform that has been termed for more than 2500 years a **delta,** after its supposed resemblance to the fourth letter of the Greek alphabet, Δ (see Figure 4–29). The classic triangular shape is maintained by some deltas, but it is severely modified in others in response to the balance between river-deposited sediments on the one hand and the removal of sediments by ocean waves and currents on the other. At some river mouths, ocean movements are so vigorous that no delta is formed at all.

Deposition in Valleys Thus far we have emphasized the prominence of erosion in the formation and

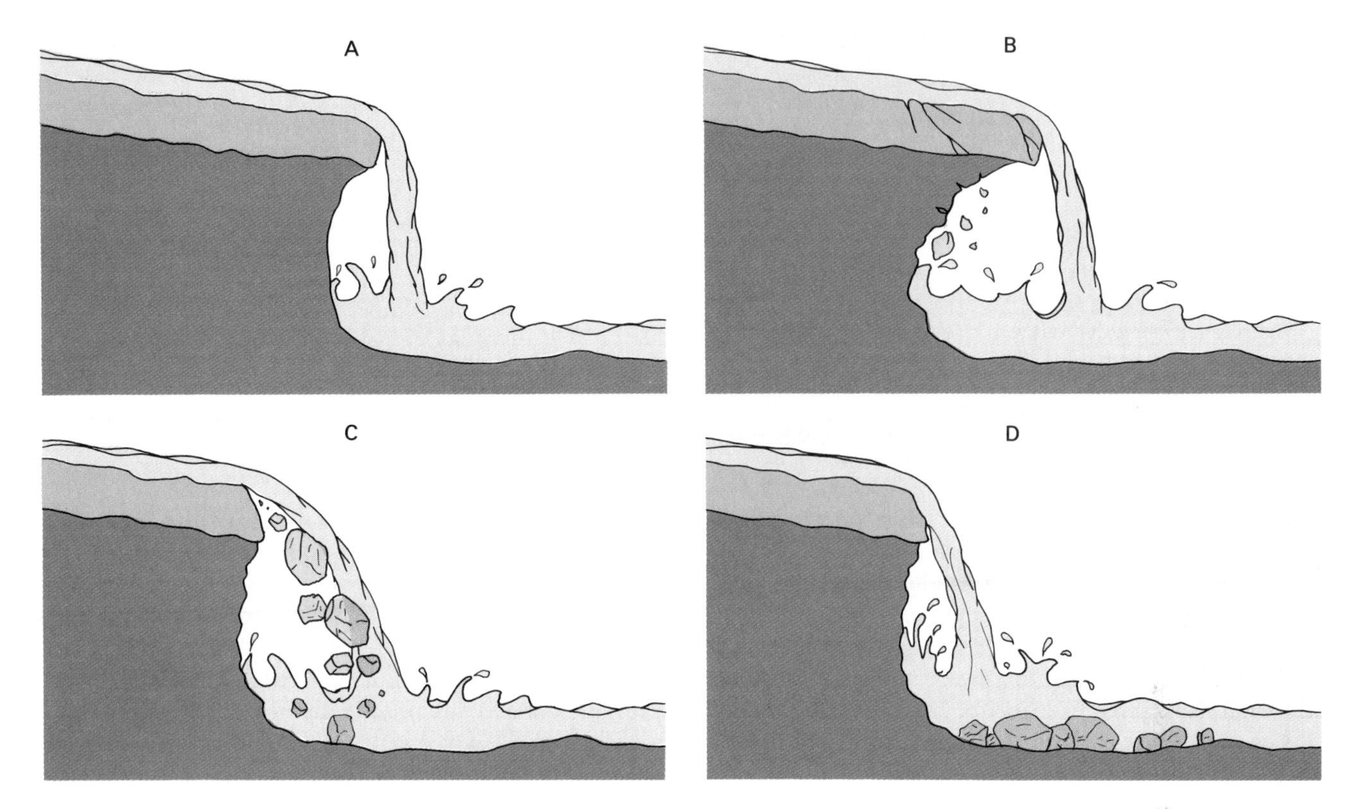

Figure 4–27. Headward erosion is accomplished by sheetflow pouring over the lip of a valley from an interfluve surface (A). The flow undercuts the lip (B) and causes collapse (C). This extends the valley headward at the expense of the interfluve (D).

Figure 4–28. This Wisconsin scene illustrates headward erosion. The grassy area being enjoyed by the horses represents an interfluve that is being cut into by headward erosion of the irregular, gorgelike valley in the foreground. (TLM photo)

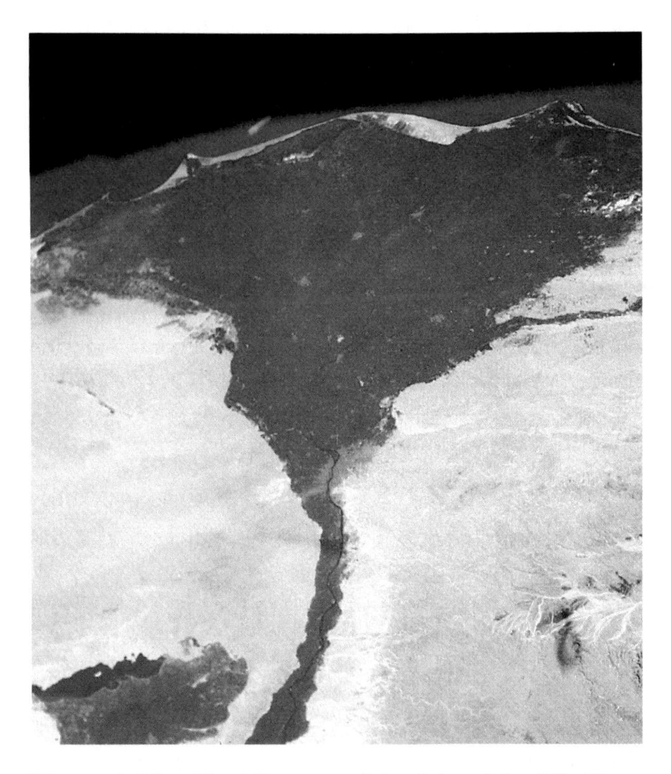

Figure 4–29. The full extent of the delta of the Nile River is shown clearly in this high-altitude (*STS-4*) image because almost its entire area is devoted to irrigation farming. The photo looks northward, with the Nile River entering from the bottom and the Mediterranean Sea at the top. (Courtesy of NASA.)

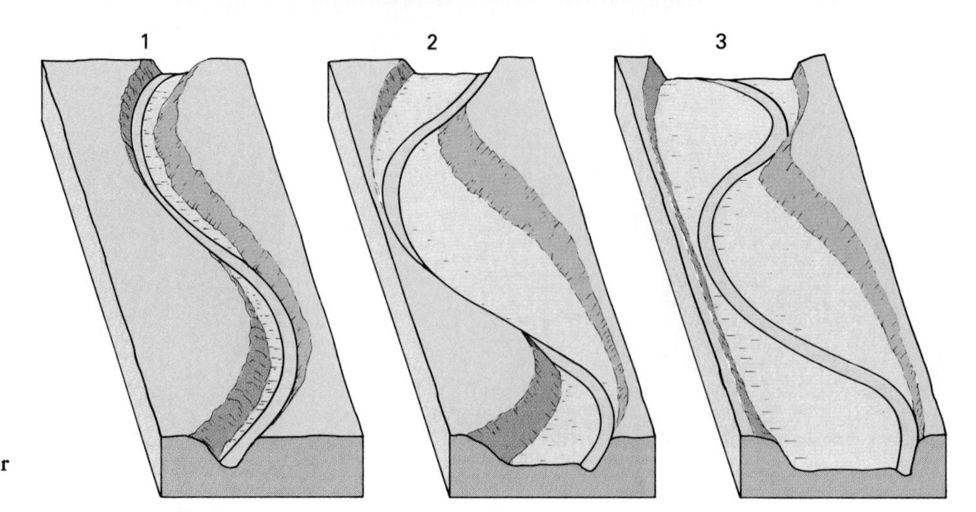

Figure 4–30. A stream widens its valley by lateral erosion, producing an increasingly broader valley floor by undercutting the bluffs on the sides of the valley. The flattish valley floor develops into a floodplain.

shaping of valleys. Deposition, too, has a role to play, although a more limited one in terms of overall valley configuration. Certain distinctive landforms are due entirely to deposition.

Fluvial deposits may include all sizes of rock debris, but smaller particles constitute by far the greatest bulk of the total. This is true primarily because the continuing collisions among particles break down larger particles into several smaller ones. It is no accident that most of the load carried by almost all streams in their lower courses consists of these smaller particles. The term alluvium can be applied to any stream-deposited sedimentary material, but in fact it mostly refers to sand, silt, and clay.

Although the principle of fluvial deposition is very clear—deposition takes place wherever the stream velocity is inadequate to transport the load—fluctuations in flow and variations in turbulence are such that alluvial deposits can occur almost anywhere in a valley bottom.

The most prominent of all depositional landscapes is the floodplain, made possible by the lateral erosion of a meandering stream, which produces a broad flattish valley floor (see Figure 4–30). This valley floor is inundated by overflow from the stream channel during floods, which occurs either regularly or sporadically. The floods leave broad and sometimes deep deposits of alluvium over the entire floor, which then comprises the floodplain.

The most conspicuous feature of a floodplain is the meandering channel of the river, which frequently changes its course with the vagaries of flow in such flat terrain. Meanders often develop narrow necks that are easily cut through by the stream, leaving abandoned *cutoff meanders*. Cutoff meanders initially hold water as *oxbow lakes* but gradually fill with sediment and vegetation to become *oxbow swamps* and eventually retain their identity as *meander scars*.

The Work of the Wind

Although the wind is a relatively constant and irrepressible force in many desert and coastal areas, it has only a limited effect on the landscape, with the important exception of such relatively impermanent features as sand dunes. **Aeolian processes** are those related to wind action. (Aeolus was the Greek god of the winds.) They are most pronounced, widespread, and effective in desert areas, although they are not restricted to dry lands. Wherever fine-grained, unconsolidated material is exposed to the atmosphere without benefit of vegetation, moisture, or some other form of protection, it is susceptible to wind action.

Aeolian Erosion The erosive effect of wind movement can be divided into two categories: deflation and abrasion. **Deflation** consists of shifting loose particles by blowing them into the air or rolling them along the ground. The wind is not strong and buoyant enough, except under extraordinary circumstances, to move anything more than dust and small sand grains, so no significant landforms are created by deflation.

Wind **abrasion** is analogous to fluvial abrasion discussed previously except that it is much less effective. Whereas deflation is accomplished entirely by air currents, abrasion requires the use of tools. The wind drives sand and dust particles against rock and soil surfaces in a form of natural sandblasting. Wind abrasion does not construct or even significantly shape a landform; it merely sculptures those already in existence (see Figure 4–31). The principal results of aeolian abrasion are the pitting, etching, and polishing of exposed rock surfaces and the further fragmenting of rocks.

Aeolian Transportation Rock materials are transported by wind in much the same fashion as they are moved by water, but less effectively. The finest particles are carried in suspension as dust. Strong, turbulent winds can lift and carry thousands of tons of suspended dust. In fact, some dust storms extend for thousands of feet above the Earth's surface and may move material over more than 1000 miles (1600 km) of horizontal distance.

A true sandstorm is a cloud of generally horizontally moving sand that extends for only a few inches or feet above the surface. People standing in its path would have

Figure 4–31. The wind sometimes is prominent in shaping surfaces that are covered with loose particles. This scene is near Barrow Creek in the Northern Territory of Australia. (TLM photo)

their legs peppered by sand grains, but their heads would probably be above the level of the sand cloud. The abrasive impact of a sandstorm, while having little erosive effect on the terrain, may be quite significant for the works of humans (such as posts, poles, and automobiles) near ground level.

Aeolian Deposition Sand and dust moved by the wind eventually are deposited when the wind dies down. The finer material, which may be carried long distances, usually is laid down as a thin coating of silt and has little or no landform significance. The coarser sand sometimes is spread across the landscape as an amorphous (formless) sheet, often called a *sandplain*. The most notable of all aeolian deposits, however, is the **sand dune,** in which loose, windblown sand is heaped into a mound or low hill.

Desert sand dunes Dune topography comprises one of the world's most distinctive landscapes. Dunes can assume an almost infinite variety of patterns. Unanchored dunes are deformable obstructions to air flow. They can move, divide, grow, or shrink. They are not dependent on a fixed obstruction for their continuance. They develop sheltered air pockets on their leeward sides that slow down the wind, so that deposition is promoted there.

Some dunefields are astonishingly extensive. One area of unbroken dunes in the Sahara Desert, for example, is as large as Colorado. In many cases, however, the expanse of dunes is interrupted by such nondune features as dry lake beds or rocky outcrops.

Several characteristic dune forms are widespread in the world's deserts. Among the most common are barchans, transverse dunes, and seifs (see Figure 4–32). The *barchan* usually occurs as an individual dune migrating across a nonsandy surface, although barchans may also be found in groups. A barchan is crescent-shaped, with the *horns* or *cusps* of the crescent pointing downwind. Sand movement is both over the crest from the windward to the leeward side and around the edges of the crescent to extend the horns. Barchans are most widespread in the deserts of central Asia and in parts of the Sahara.

Transverse dunes are also crescent-shaped, but their shape is less uniform than that of the barchans. They occur where the supply of sand is much greater; normally the entire landscape is sand-covered. Like barchans, their convex sides face the prevailing direction of wind.

The *seif* or *longitudinal dune* is rare in American deserts, but it might be the most common dune form in other parts of the world. Seifs are long, narrow dunes that usually occur in large numbers and in generally parallel arrangements. Typically they are a few dozen to a few hundred feet in height, a few tens of yards in width, and have a length that is measured in miles or even tens of miles.

Figure 4–32. Common dune forms.

Coastal dunes Winds also are active in dune formation along many stretches of ocean and lake coasts, whether the climate is dry or moist. On almost all flattish coastlines, waves deposit sand along the beach, where it is exposed to the elements, except perhaps at high tide (see Figure 4–33). A prominent onshore wind can blow some of the sand inland, often forming dunes. In some areas, particularly if vegetative growth is inhibited, the dunes slowly migrate inland, occasionally inundating forests, fields, roads, and even buildings.

Loess A specialized form of aeolian deposit that is *not* associated with dry lands is loess. **Loess** (pronounced *luhss*) is a wind-deposited silt that is fine-grained and usually buff-colored. Its most distinctive characteristic is its great vertical durability, which results from its fine grain size, high porosity, and vertical, joint-like cleavage planes. The tiny grains have great molecular attraction for one another, making the particles very cohesive. Moreover, the particles are angular, which makes them more porous. Thus, loess accepts and holds considerable amounts of water. Although relatively soft and unconsolidated, when exposed to erosion, loess maintains almost vertical slopes because of the structural characteristics outlined above, as though it were firmly cemented rock (see Figure 4–34). Prominent bluffs are often produced as erosional surfaces in loess deposits.

The formational history of loess is varied and incompletely understood. Its immediate origin is clearly aeolian deposition, but the materials apparently can be derived from a variety of sources, most notably from Pleistocene glaciation, which is discussed later in this chapter.

Most deposits of loess are in the midlatitudes, where some are very extensive, particularly in the United States, Russia, Ukraine, Kazakhstan, China, and Argentina. These deposits provide fertile possibilities for agriculture. Also, in China particularly, numerous cave dwellings have been excavated in loess because of its remarkable capability of standing in structurally rigid vertical walls. Loess can, however, be a major source of air pollution. Beijing, for instance, is afflicted by notorious loess windstorms.

Coastal Processes

The coastlines of oceans and large lakes are shaped significantly by the agitated edge of these waters, with the result that the terrain of the coastal area is often quite different from that of areas located just slightly inland. Coastal topography is produced in part by the "normal" geomorphic processes dealt with elsewhere in this book. In addition, seven "specialized" processes contribute to the distinctive shaping of coastal features. These include:

1. Long-term changes in water level
2. Tidal movements
3. Waves
4. Currents
5. Stream outflow
6. Ice push
7. Organic secretions, particularly coral

Erosive Action The most notable erosion, by far, along coastlines is accomplished by the action of waves. The incessant pounding of even small waves is a potent force in wearing away the shore, and the enormous power of storm waves almost defies comprehension. Waves break with abrupt and dramatic impact, hurling water and debris (and air) in a thunderous crash onto the shore (see Figure 4–35). Moreover, they often carry tools that make their erosive effectiveness even greater.

Figure 4–33. Coastal sand dunes on Hilton Head Island, South Carolina. A sparse growth of sea oats has partially anchored these dunes. (TLM photo)

Figure 4–34. Loess has a remarkable capability of standing in vertical cliffs, as seen in this road cut near Maryville, Missouri. (TLM photo)

Along with the water, sand, pebbles, and boulders may be hurled shoreward by the breaking wave, adding an abrasive effect of pounding and grinding. There is also a pneumatic action of air being forced into cracks and joints, and some dissolving of rock chemicals by the water.

All these wave actions tend to wear away the exposed coastal bedrock at the foot of steep sea cliffs. The most effective erosion is just at or slightly above sea level, so that a notch is cut in the base of the cliff. The cliff face then retreats due to collapse of the slope above the undercutting (see Figure 4–36). The resulting debris is broken, smoothed, and reduced in size by further wave action, and eventually most of it is carried seaward.

Where a shoreline is composed of sand or other unconsolidated material, currents and tides as well as

Figure 4–35. The continual pounding of waves can erode even the most resistant of coastal rocks. This scene is near Port Edward on the Natal coast of South Africa. (TLM photo)

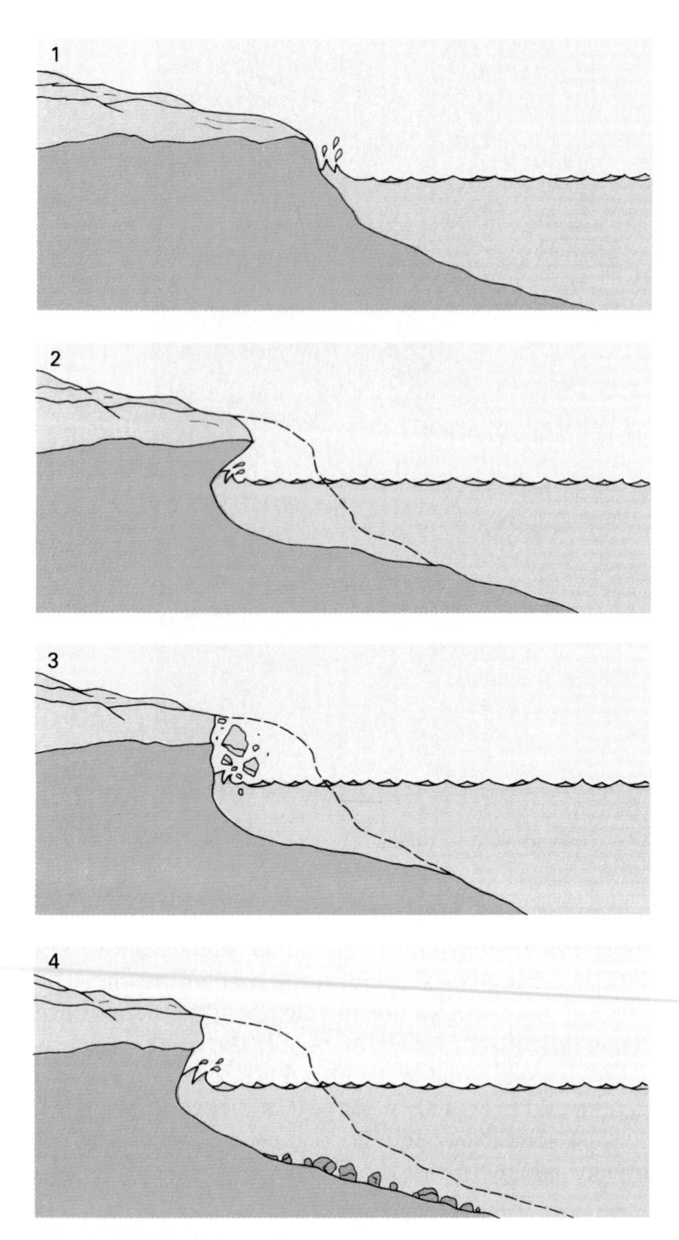

Figure 4–36. Waves pounding on exposed headlands accomplish the most effective erosion at water level, so a notch may be cut in the base of the slope. This tends to undermine the higher portion of the headland, which may subsequently collapse, producing a steep cliff. The notching/undercutting/collapse sequence may be repeated many times, causing a retreat of the cliff face.

waves will cause rapid erosion. Stormy conditions greatly accelerate the erosion of sandy shores. A violent storm can remove an entire beach in just a few hours, cutting it right down to the underlying bedrock.

Coastal Deposition Although the restless waters of coastal areas accomplish notable erosion and transportation, in many cases the most conspicuous topographic features of a shoreline are formed by deposition—including beaches, offshore bars, and spits (see Figure 4–37). As with other external forces, deposition occurs wherever the energy of the moving waters is diminished. Where wave and current action decreases, sediment is permitted to sink and come to rest.

Maritime depositional features along coastlines tend to be short-lived. This is due primarily to their composition, which is typically of relatively fine particles (sand and gravel), and the fact that the sand is not stabilized by a vegetation cover. Most of these features are under a constant onslaught by agitated waters, which can rapidly wash away portions of the sediment. Consequently, the sediment budget must be in some sort of balance for the feature to persist; in other words, removal of sand must be offset by addition of sand. Most marine depositional forms have a continuing sediment flux, with debris arriving at one end and departing at the other. During stormy periods the balance is often upset, and the feature is either significantly reshaped or totally removed.

Beaches The most widespread marine depositional feature is the **beach,** which is an exposed deposit of loose sediment that can range in size from fine sand to large cobbles. Beaches occupy the transition zone between land and water, sometimes extending well above the normal sea level into elevations reached only by the highest storm waves. Beaches sometimes extend for dozens of miles along straight coastlines, particularly if the relief of the land is slight and the bedrock is unresistant. Along irregular shorelines, beach development may be restricted largely or entirely to embayments (indentations in the coastline), frequently with an alternation of rocky headlands and bayhead beaches. Normally beaches are built up during periods of quiet weather and are removed rapidly during storms. Most beaches are broader and more extensive in summer and are worn away to become much smaller in winter.

Offshore bars Another prominent form of coastal deposition is the **offshore bar,** also called **barrier bar.** This is a long, narrow sandbar built up in shallow offshore waters, sometimes only a few hundred yards from the coast but often several miles at sea. Such bars are oriented approximately parallel to the shoreline, are extremely narrow in comparison with their length, and rise only a few feet above sea level at their highest points. They are believed to originate by the heaping up of debris where large waves (particularly storm waves) first begin to break in the shallow nearshore waters, although many of the larger offshore bars may have more complicated histories linked to the lowered sea level of the Pleistocene epoch. These bars are dangerous to ships, which may run aground on them. More disasters occur on these coastal bars than out at sea.

An extensive barrier bar isolates the water between it and the shoreline, forming a body of quiet salt or brackish water called a **lagoon.** The isolation of the lagoon contributes to its eventual disappearance, as it becomes increasingly filled with water-deposited sedi-

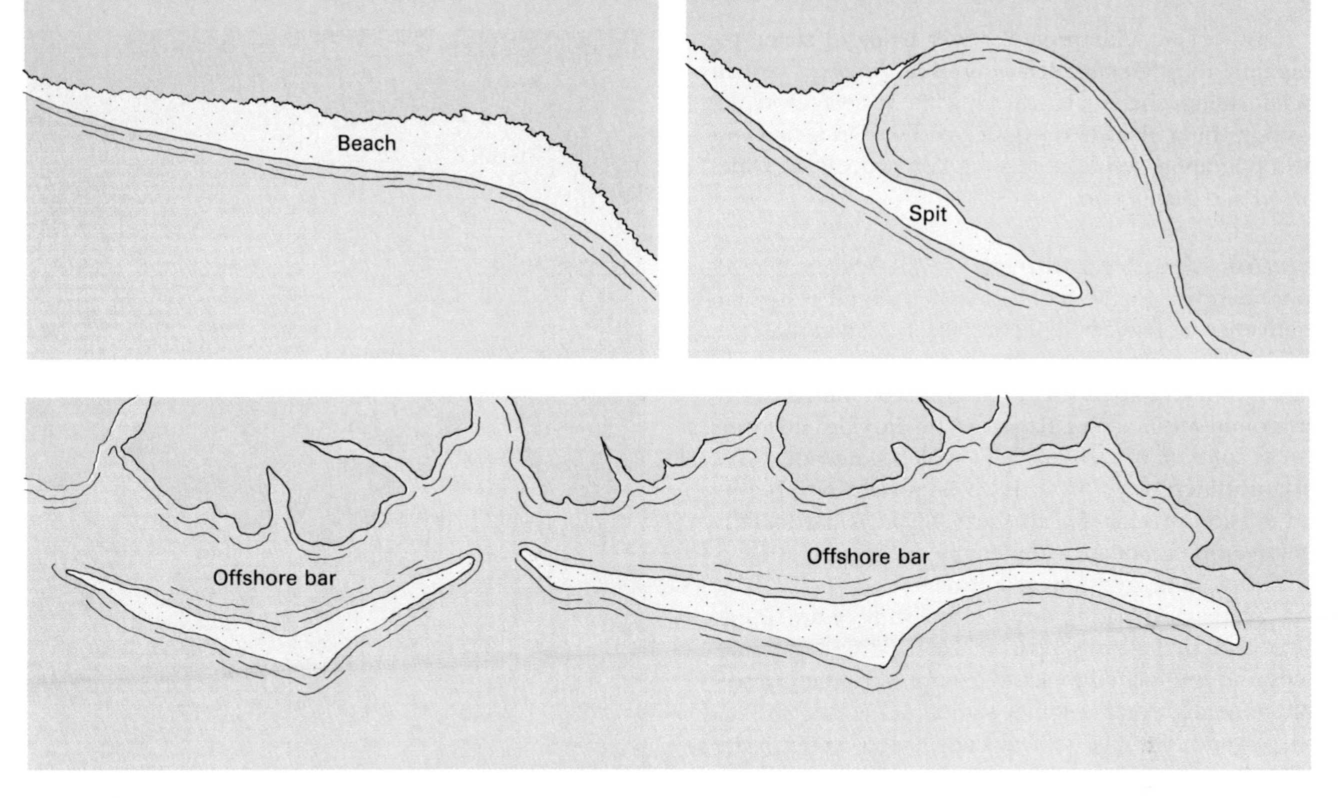

Figure 4–37. Deposition along coastlines takes many forms. Most widespread are beaches, offshore (barrier) bars, and spits.

ment from coastal streams and wind-deposited sand from the bar. Unless inlets across the bar permit vigorous tides or currents to carry lagoon debris seaward, most lagoons are slowly transformed into marshes and then into meadows by the long-term accumulation of sediment and by the growth of vegetation.

Spits A third form of coastal deposition consists of **spits**—longitudinal features formed under certain conditions of current flow and bottom configuration. Currents moving roughly parallel to the coast carry fine-grained material into the relatively deeper waters of bays, where deposition takes place. As the current brings in more sediment, the embankment becomes longer, wider, and higher. It may eventually reach the surface of the water to form a low, sandy projection that grows down-current.

Specialized Coastlines Two specialized types of coastal landform assemblages are particularly notable: fjorded coasts and coral coasts.

Fjorded coasts The most spectacular coastlines occur where a coastal terrain consisting of relatively steep slopes and considerable differences in elevation has undergone extensive glaciation. Glacial troughs in hilly coastal country can be cut far below the present sea level by submarine erosion by the ice or by subsequent drowning of subaerial valleys. In either case, once the ice has melted, a **fjord** is produced. In some localities these deep, sheer-walled coastal indentations occur in vast numbers, creating an extraordinarily irregular coastline, often with long narrow fingers of salt water reaching more than 100 miles (160 km) inland. The most extensive and spectacular fjorded coasts are found in Norway, western Canada, Alaska, southern Chile, the South Island of New Zealand, Greenland, and Antarctica.

Coral coasts In warm-water tropical oceans, nearly all continents and islands are fringed (continuously in some places, irregularly in others) with coralline formations. These are usually referred to as **coral reefs,** although the term *reef* implies a ridgelike development, and many coral accumulations bear no resemblance to such a form. In many cases coral formations are found with no continent or island nearby. This is usually an indication that a previous landmass has been slowly submerged, with the coral accumulations (which can live only in shallow water) gradually building the "reef" higher and higher.

Solution Processes

Underground water functions in a much more restricted fashion than does surface water. It is confined, it is largely unchanneled and therefore generally diffused, and it moves very slowly for the most part. Consequently, it is almost totally ineffective in terms of hydraulic power, corrosion, and other kinds of mechanical erosion.

However, underground water helps to shape topography through chemical actions. It dissolves certain rock-forming chemicals, carrying them away and depositing them elsewhere. Under certain circumstances the topographic results of this solution process are widespread and distinctive.

Solution and Precipitation The chemical reactions involved in the work of underground water are relatively simple. Although pure water is a poor solvent, almost all underground water is laced with enough chemical impurities to dissolve a few common minerals. Underground water accumulates carbon dioxide and forms a weak solution of carbonic acid, which slowly dissolves certain minerals. *Solution* is an important weathering and erosion process for all rocks, but it is particularly effective on carbonate sedimentary rocks such as limestone, dolomite, gypsum, and chalk.

Water percolating down into a limy bedrock dissolves and carries away a part of the rock mass. Because limestone and related rocks are composed largely of soluble minerals, great volumes sometimes are taken into solution and removed, leaving conspicuous voids in the bedrock.

Complementary to the removal of lime in solution is the precipitation (deposition) of lime from solution. *Precipitation* in this sense refers to a chemical process in which a solid (called a "precipitate") separates from a solution, usually falling to the bottom. In the case of underground water, mineralized water may trickle in along a cavern roof or wall. The reduced pressure of the open cavern, sometimes enhanced by evaporation, induces precipitation of whatever calcium minerals the water is carrying.

Caverns and Related Features Some of the most spectacular landforms produced by solution action are not visible at the Earth's surface. Solution action along joints and bedding planes in limy rocks beneath the surface often creates holes, sometimes large and extensive, in the form of vertical shafts or **caverns** or cave systems.

Caves are found in most areas of the world where there is a fairly massive limestone deposit at or near the surface. Some caverns are very extensive, with an elaborate network of galleries and passageways, usually very irregular in shape, and sometimes including massive openings ("rooms") scattered along the galleries.

Almost all limestone caverns are decorated with a wondrous variety of mineral deposits called **speleothems** (see Figure 4–38). Speleothems are formed when water leaves behind the minerals it was carrying in solution, due to pressure reduction and evaporation. The most striking speleothems are **stalactites,** which form on the cavern roof, and **stalagmites,** which form on the floor.

Figure 4–38. Conspicuous speleothems show clearly in this Carlsbad Caverns scene. (Courtesy of Bruce Coleman, Inc., copyright Nancy Simmerman.)

Karst Topography In many areas of limestone bedrock, the solution process has been so widespread and effective that a distinctive landform assemblage has developed at the surface, in addition to whatever caves may exist underground. The term *karst* (after a region in Slovenia, Croatia, and Yugoslavia) is applied to such solution topography, wherever it may be, despite the great variety in the terrains of different areas.

The most common surface features of karst landscapes are **sinkholes,** or **dolines;** these occur by the hundreds and sometimes by the thousands (see Figure 4–39). They are usually relatively small, rounded depressions formed by the dissolution of surface limestone. Sinks that result from the collapse of the roof of a subsurface cavern are called *collapse dolines.*

In many ways the most notable characteristic of karst regions is the absence of water flowing as streams on the surface. Most rainfall and melted snow seep downward along joints and bedding planes, enlarging them by solution in the process. Surface runoff that does become channeled usually does not go far before it disappears into a sink or an enlarged joint crack. An appropriate generalization concerning surface drainage in karst regions is that valleys are relatively scarce and mostly dry.

Figure 4–39. A karst landscape dominated by sinkholes and collapse dolines.

Glacial Modification of Terrain

In the long history of our planet, ice ages have occurred an unknown number of times. However, nearly all evidence of past glacial periods has been eradicated by subsequent geomorphic events. Only the last of the ice ages was recent enough to have exerted a significant impact on contemporary topography. Consequently, when referring to "the Ice Age," we usually mean the most recent ice age, which occurred during the epoch known as the Pleistocene. The Pleistocene epoch was only "yesterday" in terms of geologic time. In other words, it began at least 1.5 million years ago and ended less than 10,000 years ago.

Glaciations Past and Present The amount of ice that has built up on the surface of the Earth has varied remarkably over the last few million years. Periods of accumulation have been interspersed with periods of melting. Times of ice advance have alternated with times of ice retreat.

Pleistocene glaciation The dominant environmental characteristic of the Pleistocene was the dramatic cooling of high-latitude and high-altitude portions of the Earth, so that a vast amount of ice accumulated in many places. However, the epoch was by no means universally icy. In broad terms, the Pleistocene consisted of an alternation of *glacial* (times of ice accumulation) and *interglacial* (times of ice retreat) periods. North American (largely Canadian) evidence suggests that we still do not know the exact number of Pleistocene glacials and interglacials. Scientists today have identified at least 5 major glacial episodes, but as many as 11 different stages are indicated in some areas. Moreover, there were lesser variations within this broad pattern. Each of these stages experienced several briefer intervals of ice advance and retreat, presumably reflecting minor climatic changes.

At its maximum Pleistocene extent, ice covered just about one-third of the total land area of the Earth—nearly 19 million square miles (47 million km^2). Figure 4–40 highlights those regions that were covered with ice. North America experienced the greatest total area of ice coverage. The Laurentide ice sheet, which covered most of Canada and a considerable portion of northeastern United States, was the most extensive of all Pleistocene ice masses. It extended southward into the United States to a position approximating the present locations of Long Island, the Ohio River, and the Missouri River. Most of western Canada and much of Alaska were covered by an interconnecting mass of smaller ice sheets and glaciers. In western United States there were about 75 separate areas of highland ice accumulation, many of which extended outward into surrounding lowlands.

More than half of Europe was overlain by ice during the Pleistocene. Asia was less extensively covered, presumably because much of its subarctic portion received inadequate precipitation for ice to persist. In the Southern Hemisphere, the Antarctic ice cap was only slightly larger than it is today. Southernmost South America and the South Island of New Zealand were largely covered with ice. Other high mountain areas all over the world—in central Africa, New Guinea, Hawaii, and elsewhere—experienced more limited glaciation.

Contemporary glaciation In marked contrast to maximum Pleistocene glaciation, the extent of ice covering the continental surfaces today is very limited, with the notable exceptions of Antarctica and Greenland. About 10 percent of the Earth's land surface—amounting to some 6 million square miles (15 million km^2)—today is covered with ice, but more than 96 percent of that total is included in the Antarctic and Greenland ice caps. Many of the higher mountains of the world contain glaciers, although most are quite small. The distribution of ice sheets and glaciers in the Northern Hemisphere is shown in Figure 4–41.

Types of Glaciers A glacier is more than an ice-filled valley; it is a finely tuned environmental system with a nourishment budget (the equilibrium between accumulation and melting of ice) that is delicately balanced between growing and shrinking. Glacial systems can be classified into three categories: continental ice sheets, highland ice fields, and alpine glaciers.

Continental ice sheets By far the most significant glaciers, because of their immense size, are the continental ice sheets. Only two true ice sheets exist today, those of Antarctica and Greenland. These are vast blankets of ice that completely inundate the underlying terrain to depths of hundreds or thousands of feet.

Highland ice fields In a few high mountain areas, ice accumulates in a relatively unconfined sheet, called an ice field, that may cover a few hundred or a few

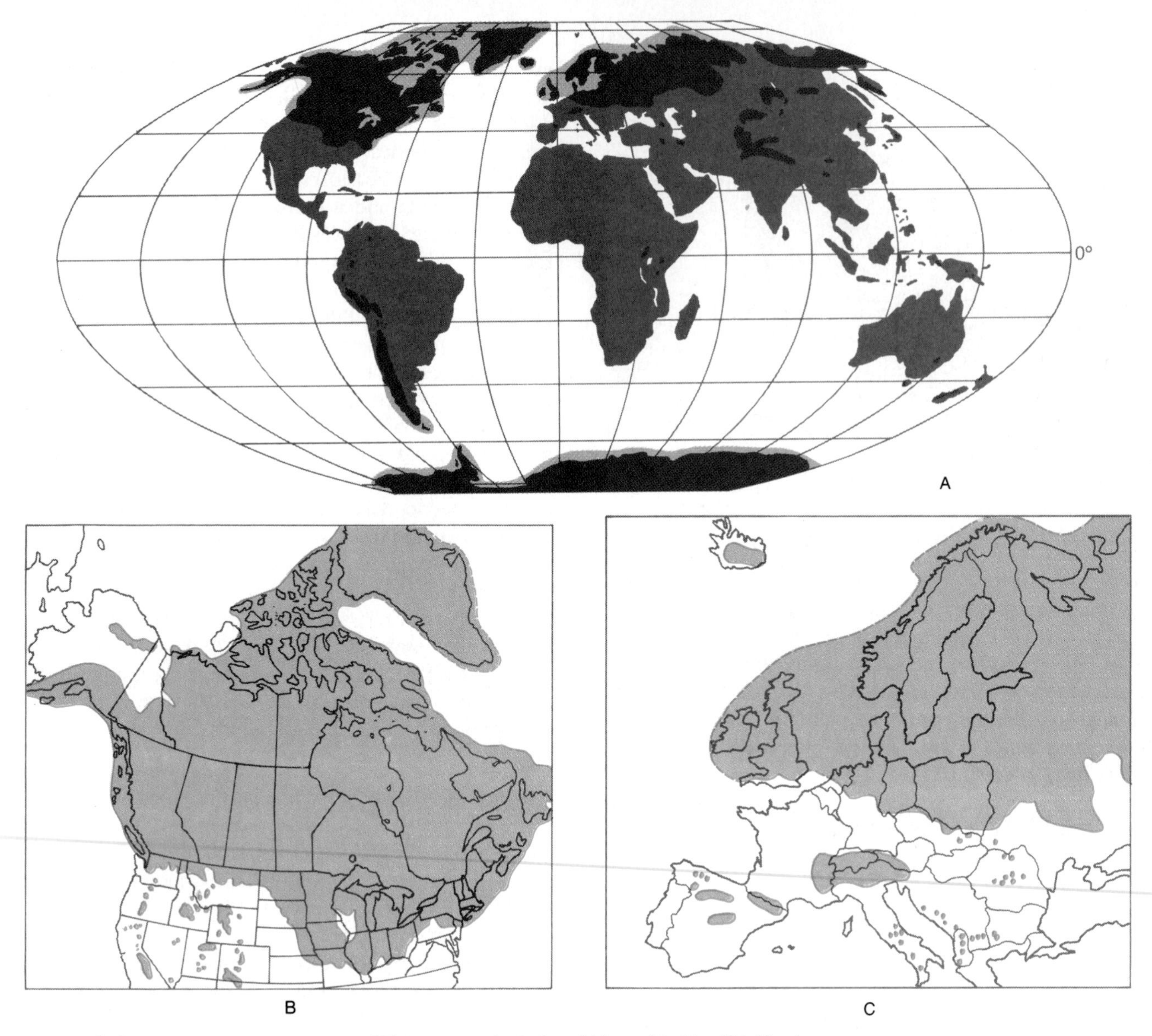

Figure 4–40. The maximum extent of Pleistocene glaciation: (A) worldwide, (B) North America, and (C) Europe.

thousand square miles, submerging all the underlying topography except perhaps for some rocky pinnacles (called *nunataks*) that protrude above the ice (see Figure 4–42).

Alpine glaciers *Alpine glaciers* are individual mountain glaciers that develop at the heads of valleys and normally move down-valley for greater or lesser distances (see Figure 4–43). Very small glaciers confined to the basins where they originate are called *cirque glaciers*. Normally, however, glaciers spill out of their originating basins and flow down-valley as long, narrow features that resemble rivers of ice; these are known as *valley glaciers*. Occasionally they extend to the mouth

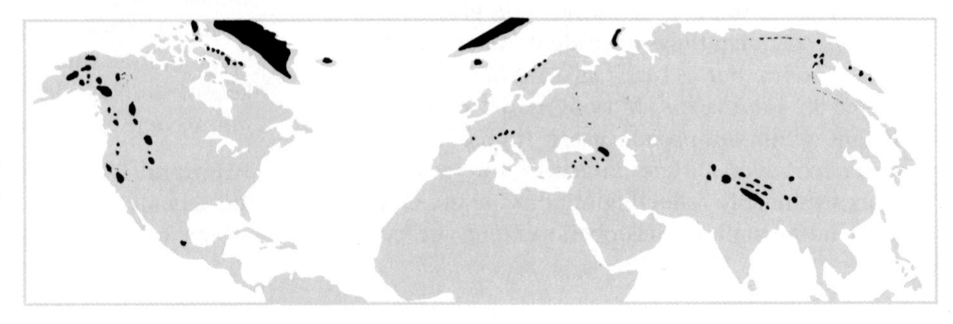

Figure 4–41. The distribution of contemporary ice sheets and glaciers in the Northern Hemisphere. Many small glaciers are not shown because of the small scale of the map.

Figure 4–42. Alaska's Juneau ice field covers an area of some 4000 square miles (10,000 km^2). Many nunataks rise above the general level of the ice. (TLM photo)

of the valley, where the ice then spreads out broadly over flatter land, forming a *piedmont glacier* (see Figure 4–44).

Regimen of Glaciers Snow falls and ice accumulates in many parts of the world, but glaciers do not always develop from these events. Rather, glaciers form only under certain circumstances. Although glaciers often are vast in size, they also are temporary and fragile, requiring just the right combination of temperature and moisture to survive. A slight warming or drying trend for a few decades can cause even the most extensive ice sheet to disappear. The persistence of any glacier depends on the balance between *accumulation* (addition of ice by incorporation of snow) and *ablation* (wastage of ice through melting and sublimation).

Formation of a glacier Glaciers can develop wherever more snow collects in winter than melts in summer, over a period of time. This occurs in certain high-latitude and high-altitude areas, but in an irregular and not always predictable pattern.

Snow is moisture that has crystallized directly from water vapor in the atmosphere. It floats to Earth as ice crystals that are only about one-tenth as dense as water. With the passage of time and compression from overlying snowfall, the crystalline snow is compressed into granular form that achieves a density about half as great as water. This material is called *névé* or *firn*. As time passes, the metamorphosis from snow to ice continues, with more air being forced out and the ice crystals increasing in size.

Glacial movement Glaciers are different from all other types of ice in that they flow outward or downslope from their area of accumulation. Surface ice may be brittle, but ice under considerable confining pressure, as below the surface of a glacier, can experience both internal deformation and movement.

When a mass of ice attains a thickness of several dozens of yards, it begins to flow in response to the

Figure 4–43. An alpine glacier scene in southeastern Alaska. This is the Davidson Glacier near Skagway. (TLM photo)

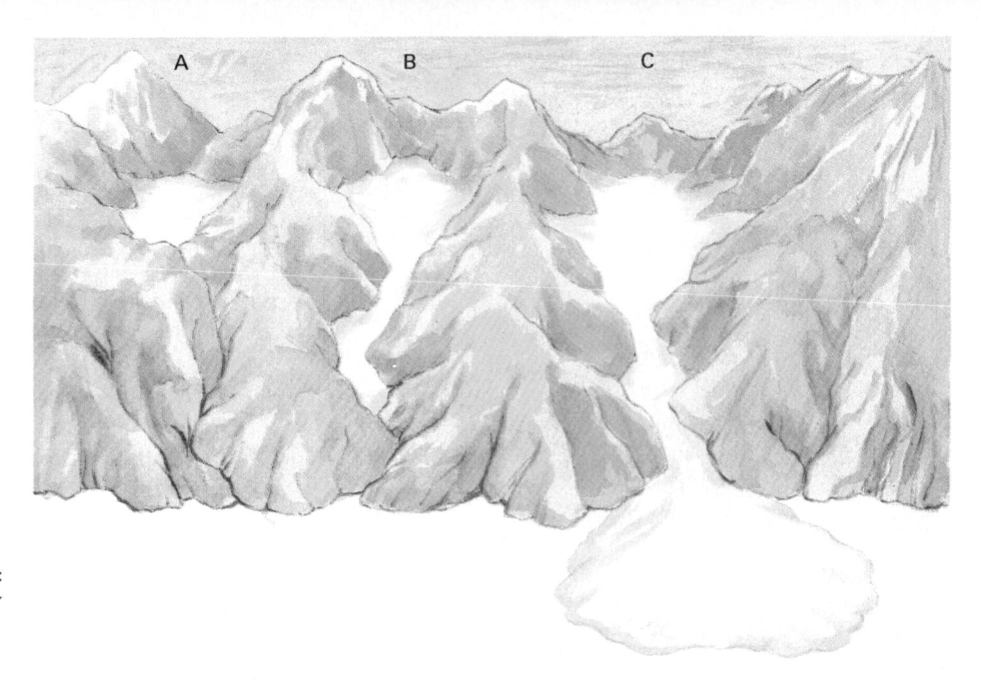

Figure 4–44. The three basic types of alpine glaciers: (A) cirque glacier, (B) valley glacier, and (C) piedmont glacier.

overlying weight. The entire mass does not move; rather, there is an oozing outward from around the edge of an ice sheet or down-valley from the toe of an alpine glacier. The glacier more or less molds itself to the shape of the terrain over which it is riding, although it simultaneously reshapes the terrain significantly through erosion.

Glaciers normally move very slowly, and the flow is often erratic, with irregular pulsations and surges over a short span of time. One of the most important principles of glacial movement is that as long as a glacier exists, it continues to move forward. This does not necessarily mean that the outer edge of the ice is advancing; this depends on the balance between accumulation and wastage of the entire ice mass (see Figure 4–45). Thus, a glacier can be thought of as a sort of conveyor belt that is always moving forward, but the movement does not affect the end of the conveyor belt. Even a "retreating" glacier is likely to be advancing throughout: It is retreating simply because its rate of wastage at the outer margin is more rapid than the rate of advance.

Glacial erosion Despite their spectacular nature, glaciers do not create landscapes; they merely modify existing ones by erosion and deposition. The very nature of a large moving mass of ice, however, is dramatic, and the amount of work that it can accomplish in a geologically short time is impressive.

As with streams, volume and velocity determine the effectiveness of glacial erosion. The depth of the erosion is limited in part by the structure and texture of the bedrock and in part by the relief of the terrain. Relief refers to the difference between the highest and lowest parts of the terrain. Glacial erosion is inhibited by low relief and is enhanced by high relief. On flat land the induced changes are relatively minor, but in mountainous areas the modifications may be striking.

As the slowly moving ice collides violently with bedrock, the frictional contact induces melting. Melted ice, however, can refreeze around rocky protrusions. As it does so, it frequently pries out rock particles located beneath the ice and drags them along in the general flow. Known as *plucking* or *quarrying*, this activity is probably the most important erosive work performed by glaciers.

Glaciers also erode by ***abrasion***, in which the bedrock is scoured or worn down by the rock debris dragged along in the moving ice. The "sandpaper" effect of a glacier is more effective than that of a river because glaciers can move much larger chunks of rock than can streams. Abrasion usually produces minor features, such as polished bedrock surfaces, striations (fine parallel indentations), and glacial grooves (deeper and larger indentations). Whereas plucking tends to roughen the underlying surface and provide the glacier with abrasive tools, abrasion tends to smooth the subglacial surface and dig striations and grooves.

In plains areas, the topography produced by glacial erosion is relatively inconspicuous. Prominences are smoothed, and small hollows can be excavated, but the general appearance of the terrain changes little. In hilly areas, however, the effects of glacial erosion are much more notable. Mountains and ridges are sharpened, valleys are deepened and made more linear, and the entire landscape becomes more angular and rugged.

Transportation by glaciers Although their movement is slow, glaciers are extremely competent, as well as indiscriminate, in their ability to transport rock debris. Because they are solid bodies, they are able to move immense blocks of rock, literally the size of houses. Moreover, they may transport these gigantic pieces for dozens or even hundreds of miles. Most of a glacier's load, however, is a diverse collection of particles of all sizes. Perhaps the most typical component of the load is *glacial flour*, which is rock material that has been ground to the texture of very fine talcum powder.

Glacial transportation moves the load outward or

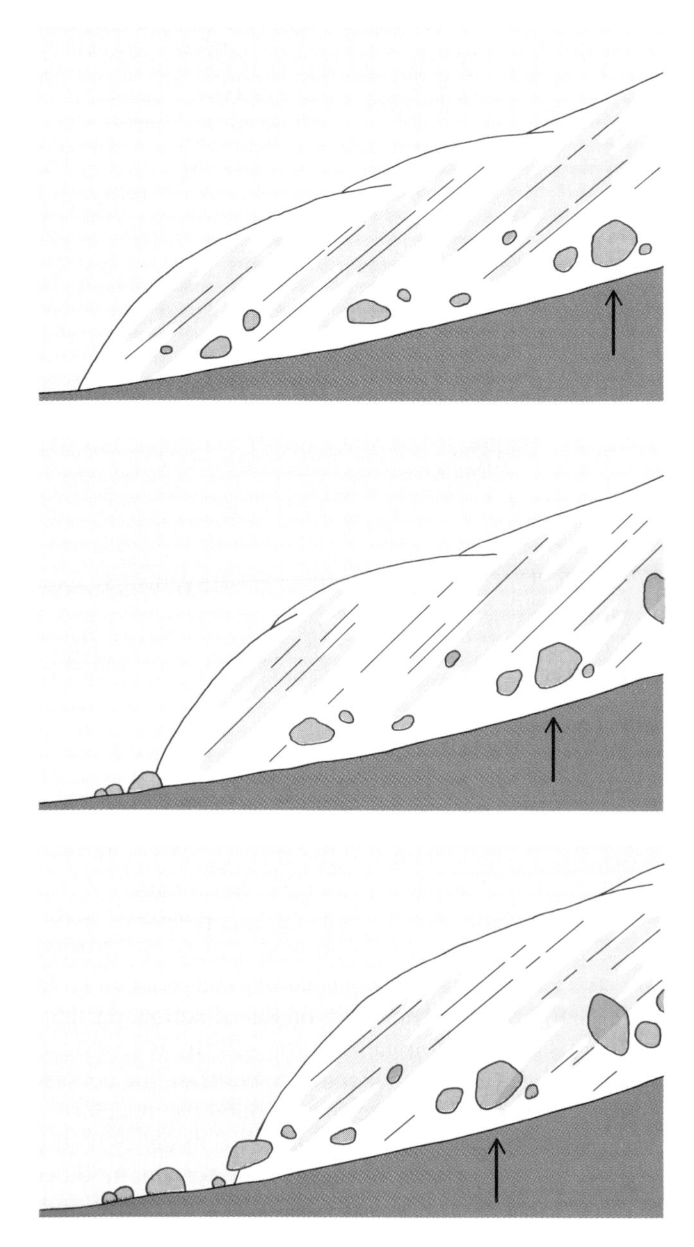

Figure 4–45. A flowing glacier is not necessarily an advancing glacier. In this sequential illustration, the front of the glacier is clearly retreating, but the ice itself continues to flow forward like a conveyer belt. The boulder marked by an arrow within the ice illustrates the principle.

down-valley at a variable speed. The "conveyor belt" action of a glacier continues throughout its existence. Even if the ice margin is retreating because of excessive wastage, there is still forward transport of debris by the advancing ice throughout the general extent of the glacier. The transport function of a glacier persists indefinitely unless and until the ice becomes so thin that a subglacial obstacle, like a hill, prevents further flow.

Glacial deposition Probably the major role of glaciers in landscape modification is to remove rock material from one area and deposit it in some distant region, where it is left in a fragmented and vastly changed form. This process is clearly displayed in North America, where an extensive portion of central Canada has been glacially scoured of its soil, regolith, and much of its surface bedrock, leaving a relatively barren, rocky surface dotted with bodies of water. Much of the material that was removed was taken southward and deposited in the midwestern states, producing an extensive plains area of remarkably fertile soil. Thus the legacy of Pleistocene ice sheets for the U.S. Midwest was the evolution of one of the largest areas of productive soils known, at the expense of central Canada, which was left impoverished of soil by those same glaciers. (Conversely, many valuable Canadian mineral deposits were exposed by the glacial removal of soil and regolith.)

The debris that is deposited by glaciers occurs in all sizes and shapes. The general term that is applied to all such material is ***drift***, a misnomer coined in the eighteenth century when it was believed that the vast debris deposits of the Northern Hemisphere were leftovers from biblical floods.

Rock debris deposited directly by moving or melting ice, with no meltwater flow or redeposition involved, is given the more distinctive name of **till** (see Figure 4–46). Till is an unsorted and unstratified agglomeration of fragmented rock material of all sizes. Most of the fragments are angular or subangular in form because they have been held in a relatively fixed position while carried in the ice and have had little opportunity to become rounded by frequent impact, as happens to pebbles in a stream.

Figure 4–46. A road cut through a deposit of glacial till near Lake Tahoe, California. Till consists of totally unsorted debris, ranging in size from tiny to huge. (TLM photo)

Figure 4–47. An alpine glacier issuing from a highland ice field in the Canadian Rockies. This is Athabaska Glacier in Jasper National Park. The glacier is about 10 miles (16 km) long and has heaped up conspicuous moraines on each side. (TLM photo)

Glaciofluvial deposition Much of the debris carried along by glaciers eventually is deposited or redeposited by glacial meltwater. In some cases this process is accomplished by subglacial streams issuing directly from the ice and carrying sedimentary material washed from positions in, on, or beneath the glacier. Much meltwater deposition, however, involves debris that was originally deposited by ice and subsequently picked up and redeposited by the meltwater well beyond the outer margin of the ice. Such **glaciofluvial deposition** occurs around the margins of all glaciers. The most important topographical features of glaciofluvial deposition are called **moraines.**

The results Apart from the oceans and the continents themselves, continental ice sheets are the most extensive features ever to appear on the face of the Earth. Their actions during the Pleistocene significantly reshaped both the terrain and the drainage of nearly one-fifth of the total surface area of the continents.

Most of the world's high mountain regions experienced massive Pleistocene glaciation, and many of them are still undergoing active glaciation on a reduced scale (see Figure 4–47). The effect of glacial action, particularly erosion, on mountainous topography is to create steeper slopes and greater relief. In contrast, the actions of ice sheets tend to smooth and round the terrain.

Glaciation, then, is a remarkable modifier of the landscape. In lowlands it smooths the terrain, modifies the drainage patterns, absorbs a great many standing bodies of water, and either removes the soil (through erosion) or significantly adds to the amount of soil-forming material (through deposition). In highlands it makes the interfluves more jagged, enlarges the valleys, and produces spectacular scenery.

SUMMARY

The basic composition of the Earth's crustal surface consists of a variety of minerals that form many kinds of rocks but that can be classified into three fundamental types—igneous, sedimentary, and metamorphic. Massive forces within the Earth are instrumental in reshaping the configuration of the crustal surface. The big picture of crustal shaping and reshaping has become much clearer in recent years as we become more familiar with the dynamics of plate tectonics, which are responsible for the broad topographic patterns over most of the planet.

The roles of weathering and mass wasting in the continuing drama of landform-shaping are fundamental but preparatory to the principal denudation activities, which are accomplished by the agents of erosion and deposition. Water is clearly the predominant external factor in the shaping of the terrain of the continents. Erosion, transportation, and deposition by both overland flow and streamflow produce a wide variety of landforms. Winds represent a minor force overall, although they continually reshape the surface configuration of depositional landforms in desert areas and along coastlines. The erosional and depositional effects of wave and current action and of sea-level fluctuations are particularly influential in shaping coastal terrain. Solution activity of percolating groundwater is capable of producing numerous subsurface caves in limestone bedrock, and solution processes sometimes are so widespread and effective that prominent surface features are created. About one-third of the Earth's land area was overlain by ice during the Pleistocene epoch (the most recent "ice age"), and the erosional and depositional effects of glacial action thoroughly reshaped the terrain.

KEY TERMS

compound (p. 92)
mineral (p. 92)
magma (p. 92)
rock (p. 92)
outcrop (p. 92)
bedrock (p. 92)
regolith (p. 92)
igneous rock (p. 92)
extrusive (p. 92)
intrusive (p. 93)
sedimentary rock (p. 93)
metamorphic rock (p. 93)
geomorphic (p. 93)
plate tectonics (p. 95)
sea-floor spreading (p. 95)
subduction (p. 95)
continental drift (p. 95)
diastrophism (p. 96)
fault zone (p. 100)
fault plane (p. 100)
fault line (p. 100)
earthquake (p. 100)
vulcanism (p. 100)
lava (p. 101)
flood basalt (p. 101)
denudation (p. 102)
weathering (p. 102)
joints (p. 102)
mass wasting (p. 103)
fluvial process (p. 104)
valley (p. 105)
interfluve (p. 105)
erosion (p. 105)
deposition (p. 105)
alluvium (p. 105)
floodplain (p. 108)
delta (p. 108)
aeolian process (p. 110)
deflation (p. 110)
abrasion (p. 110)
sand dune (p. 111)
loess (p. 112)
beach (p. 114)
offshore bar (barrier bar) (p. 114)
lagoon (p. 114)
spit (p. 115)
fjord (p. 115)
coral reef (p. 115)
cavern (p. 116)
speleothem (p. 116)
stalactite, stalagmite (p. 116)
sinkhole (doline) (p. 116)
till (p. 121)
glaciofluvial deposition (p. 122)
moraine (p. 122)

QUESTIONS FOR INVESTIGATION AND DISCUSSION

1. Distinguish between internal and external processes that shape the Earth's surface.

2. Explain how the presence of the Mid-Atlantic Ridge supports the theory of plate tectonics.

3. Explain the difference between folding and faulting.

4. What determines the erosive effectiveness of streamflow?

5. Why don't all rivers form deltas?

6. How important is wind in the sculpting of desert landforms?

7. What causes beaches to change shape and size with some frequency?

8. How is it possible for percolating groundwater both to erode and to deposit?

9. Discuss the mechanics of glacial movement.

10. Why are mountain areas that have experienced glaciation usually quite rugged?

11. Why are regions that have experienced continental glaciation usually poorly drained?

12. How is it possible for glacial ice to be advancing while the margin of the glacier is retreating?

13. Identify examples of different rock types found in your area or used in construction of local buildings.

14. Find small-scale examples on your campus of weathering, mass wasting, erosion, and deposition.

REFERENCES

BRIDGES, E.M. *World Geomorphology*. New York: Cambridge University Press, 1990.

CATES, DONALD R., ed. *Coastal Geomorphology*. London: George Allen & Unwin, 1980.

CHORLEY, R.J., ed. *Introduction to Fluvial Processes*. London: Methuen & Co., Ltd., 1971.

MOORES, ELDRIDGE, M., ed. *Shaping the Earth: Tectonics of Continents and Oceans*. New York: W.H. Freeman and Co., 1990.

RICE, R.I. *Fundamentals of Geomorphology*. New York: John Wiley & Sons, 1988.

SCHUMM, S.A. *The Fluvial System*. New York: John Wiley & Sons, Inc., 1977.

SHARP, ROBERT P. *Living Ice: Understanding Glaciers and Glaciation*. New York: Cambridge University Press, 1988.

The Influence of the Environment on Human Life

Many human activities are altering the Earth's environment and may, in turn, actually threaten humankind's own existence. The human organism has survived on Earth so far, however, so we can conclude that it has evolved successfully.

BIOLOGICAL EVOLUTION AND TECHNOLOGICAL EVOLUTION

For human beings the word *evolve* has a special meaning. Technological evolution long ago superseded biological evolution. As plants, insects, fish, birds, and other animals developed and spread around the globe, they all divided into a great number of species. In other words, they were no longer able to interbreed. Each species evolved distinctive biological capabilities and so survived to fill an **ecological niche,** which is a particular combination of physical, chemical, and biological factors that a species needs in order to thrive. Species evolved or else they perished. They did *not* evolve *in order to* survive; this is an error in logic called *teleology*, which assumes that a given process moves or is directed toward a particular end. The rejection of teleological explanations does not preclude the existence of a divine plan that cannot be understood by humankind.

Human beings are one single species. They can all interbreed, although in covering the globe they have evolved certain secondary physical characteristics, such as skin color and body shape. For the most part, the species has adapted itself to widely varying conditions on Earth through cultural, rather than physical, differentiation. Humans survive Minnesota winters, for example, by making warm clothes, not by evolving furry bodies. Through culture and technology, humankind has assumed responsibility for its own evolution. Genetic engineering offers the possibility that we might assume responsibility for our own biological evolution, as we have for that of domesticated animals. Some people fear that this capability might be misused.

The physical environment nevertheless affects or limits us, and the secondary physical characteristics have evolved in different physical environments. Human body bulk and basal metabolism, nose shape and skin color, lung capacity and the ratio of red corpuscles in the blood, and many more physiological responses to the physical environment may influence human performance and behavior. These should not be exaggerated, but neither should they be denied.

Interest in the influence of the environment on human affairs dates at least to the ancient Greeks. Many Greek writers believed that the principal determinant of a people's culture was the way they made their living and that this, in turn, was determined largely by their physical environment. This may have been true in the past, when peoples were almost entirely dependent on their local environments. Today, however, transportation and communication have released peoples from the constraints of their local environments.

The Greek physician Hippocrates (c.460–c.377 B.C.) argued in his book *Airs, Waters, Places* that civilization flourishes only under certain climatic conditions. Those just happened to be the conditions under which Hippocrates lived. Most people are prejudiced in favor of their home environments.

Historian Arnold Toynbee (1889–1975) turned Hippocrates's idea that a hospitable climate nourishes civilization upside down with his own *challenge-response theory*. Toynbee argued that people need the challenge of a difficult environment to put forth their best effort and to build a civilization. A rich environment encourages only sluggishness. We hear the challenge-response theory whenever anyone expresses a preference for an environment of seasonal change, as in the northern United States, over the almost tediously fine weather of southern California.

Both Hippocrates's and Toynbee's theories are intriguing, but they are diametrically opposed. Yet we can find evidence for both and exceptions to both. This should warn us that neither theory completely explains the relationship between climate and technological development.

CULTURAL ECOLOGY AND ENVIRONMENTAL DETERMINISM

The study of the ways societies adapt to environments is called **cultural ecology.** In contrast, the simplistic belief that human events can be explained entirely as the result of the physical environment is called **environmental determinism.** Any theory that promises to explain human affairs so simply is attractive, but it is also false. Human affairs are not simple, and when we examine any world situation or historic event carefully, environmental determinism, or any other single-factor explanation, proves insufficient.

Typical examples of environmental determinism state that geography determined that England would rule the seas, that the United States would expand to reach the Pacific Ocean, or that Russia would become a great power. Careful study of the history of each of these three countries, however, teaches that their geographical features may have contained possibilities for power or greatness, but that actual events resulted from people's seeing (or missing) opportunities, making decisions, being caught up in a web of social, economic, and political developments, and, more often than we might admit, being blessed or cursed by luck.

A milder form of environmental rhetoric personalizes nature by giving it the ability to think or to act. John Ruskin (1819–1900) labeled this style the *pathetic fallacy*. In typical examples, nature "offers" certain resources to humankind, or certain environments "discourage" settlement. In the earlier discussion, for example, we noted that Hippocrates argued that "a hospitable climate *nourishes* civilization," whereas Toynbee wrote that "a rich environment *encourages* only sluggishness." Both are examples of pathetic fallacies. In other cases of the pathetic fallacy, nature has feelings or encourages certain feelings in people (see Figure 1).

Nineteenth-century French geographers proposed the theory of **possibilism** as an antidote to all styles of environmental determinism. Possibilism propounds that the physical environment itself neither suggests nor determines what people will attempt, but it may limit what people achieve profitably. The choices and constraints involved in utilizing the natural environment are often as much cultural, economic, political, and social as they are technological. We do not grow bananas at the North Pole because it is easier and cheaper to grow bananas elsewhere.

Human affairs are not simple, and when we examine any world situation or historic event carefully, environmental determinism, or any other single-factor explanation, proves insufficient.

All forms of environmental determinism may seem to be harmless figures of speech, but they are intellectually dangerous. They predispose conclusions and close the mind to alternative explanations. "Research" becomes only a search for correlation between preselected environmental causes and the events that are to be explained. Watch carefully for these simplistic traps; you will find them in newspapers or in television commentaries every day.

OUR PERCEPTION OF THE ENVIRONMENT

The political commentator Walter Lippmann (1889–1974) differentiated "the world outside" from "the pictures in our heads." The world outside is the way things really are, but the pictures in our heads may be based on preconceptions, misperceptions, or incomplete understanding. Geographers call these pictures in our heads **mental maps.** The psychological theory of **cognitive behavioralism** argues that people react to their envi-

Figure 1. Rembrandt's "The Mill" (c. 1650) is described in the official guide to America's National Gallery of Art as "melancholy, conveying a mood of sublime sadness . . . through the stark simplicity of a windmill silhouetted in the fading light against the mist-filled sky. . . . incredibly moving." Can a landscape have feelings? Does it convey these feelings to you? Does any landscape necessarily evoke the same emotional response from all viewers, no matter how different the viewers' backgrounds or experiences? (Courtesy of the National Gallery of Art.)

ronment as they perceive it. In other words, people make decisions on the basis of the pictures in their heads, but they must act in the world outside. If there is a difference between the way a situation is and the way that we think it is, then we may get into difficulty by trying to do something impossible or by failing to take advantage of some opportunity.

We cannot understand why people make the decisions they do or act as they do by studying only the real environment in which they acted. We must discover what was in their heads when they made their decisions. In 1898, for example, Senator George Hoar of Massachusetts cast a key vote for the annexation of Hawaii because, according to his personal papers, he drew a line on a map from Alaska to southern California, he saw that Hawaii fell inside that line, and so he concluded that its possession was necessary for the country's defense. No one has been able to discover what sort of map he was looking at (perhaps one with an inset map of Hawaii?), but his misperception affected history.

BEHAVIORAL GEOGRAPHY

Behavioral geography is the study of our perception of the environment and of how our perception influences our behavior. Where we choose to live, shop, or visit depends on our feelings about places—whether we think certain places are good or bad, beautiful or ugly, safe or

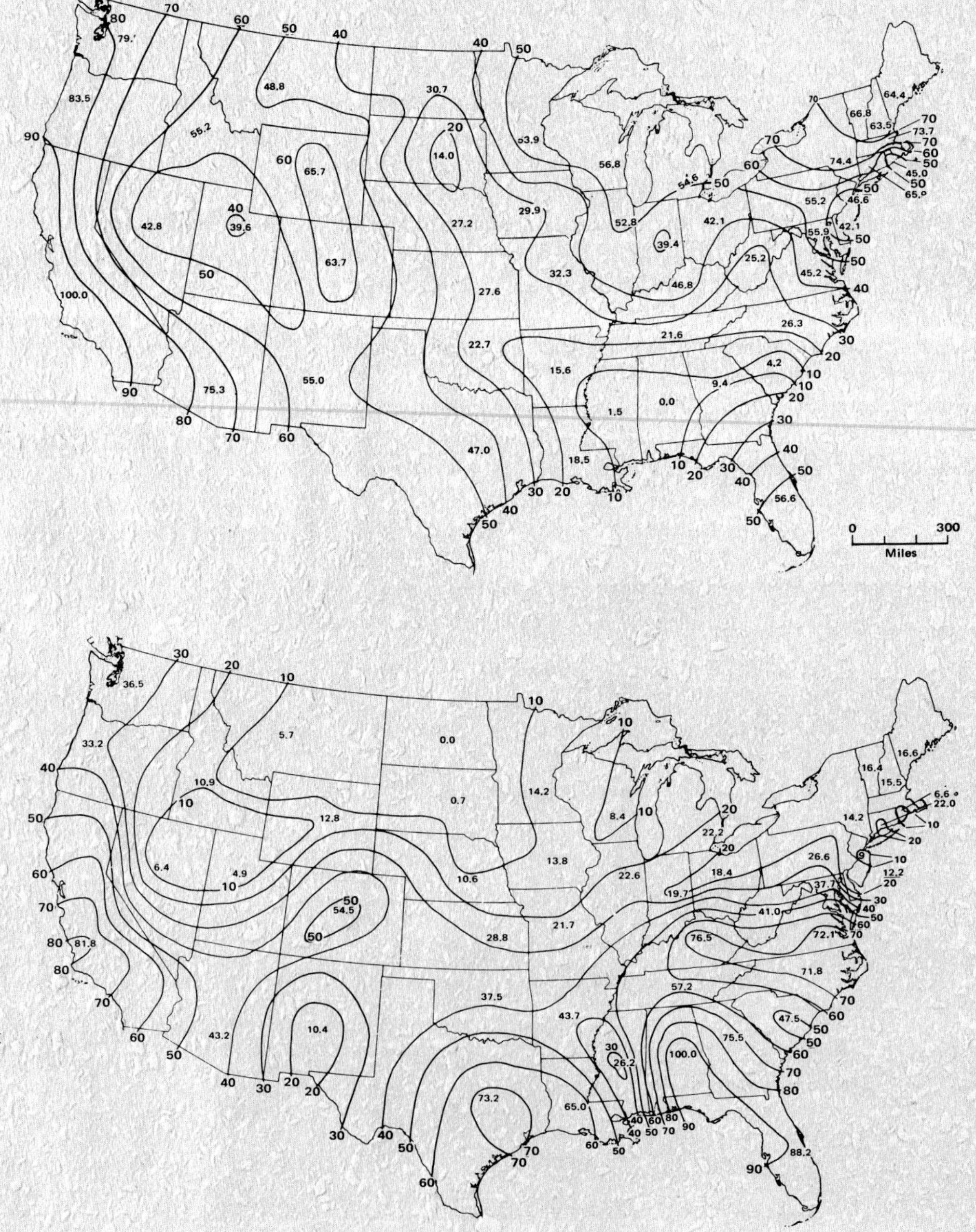

Figure 2. These mental maps reflect the residential desirability of the different states according to the opinions of students at the University of California at Berkeley (above) and the University of Alabama (below) in 1966. High numbers indicate a positive image in the students' minds. Students at both places seemed to like where they were, but the Alabamans had a better opinion of California than the Californians did of Alabama. Both seemed to have positive images of Colorado and low images of the Dakotas. These preferences are not necessarily based on personal experience. We all have opinions of places we have never visited, and a visit might surprise us. (From Peter R. Gould, "On Mental Maps," in *Michigan Interuniversity Community of Mathematical Geographers, Discussion Paper #9*, 1966. Reprinted with permission.

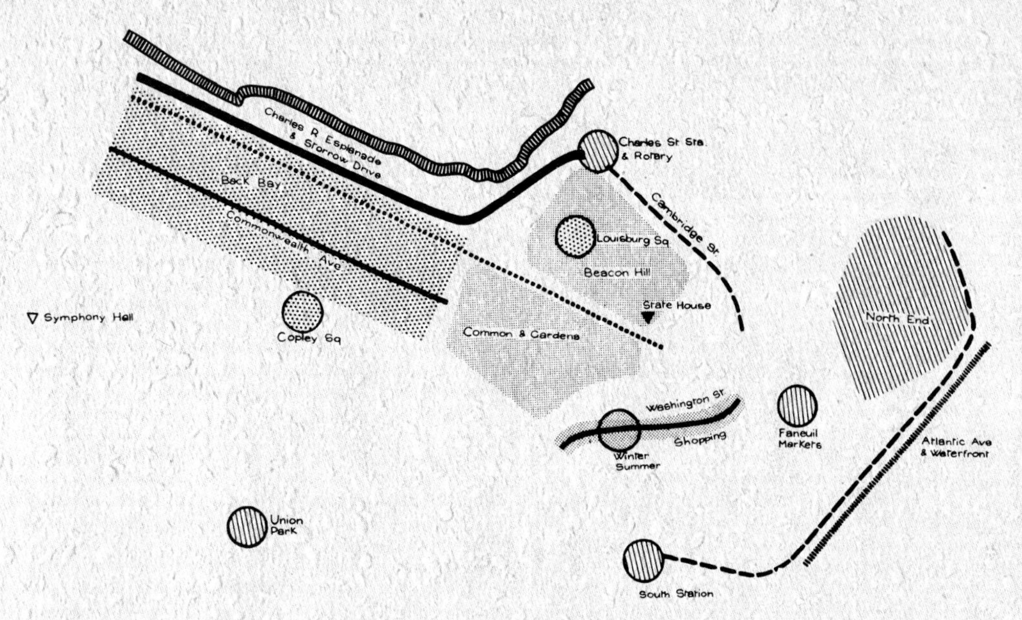

Figure 3. This composite mental map shows Boston's distinctive places according to a scheme developed by urban planner Kevin Lynch. Lynch discovered that people's mental maps seldom cover territories as regular grids do. People map space in terms of paths or routes; edges, such as the banks of rivers; nodes; districts; and landmarks. (Reproduced from Kevin Lynch, *The Image of the City*. Cambridge, Ma: MIT Press, 1960, p. 147. © 1960 by the Massachusetts Institute of Technology and the President and Fellows of Harvard College.)

dangerous (see Figures 2 and 3). Studies have investigated how we perceive environmental threats, either natural (earthquakes) or human-made (chemical or nuclear installations). How we perceive these threats affects our actions or reactions toward them.

Our individual notions of usefulness or of aesthetic value affect our evaluation of landscapes. Most urbanites, for example, agree that the Painted Desert in Arizona is beautiful, and it is a major tourist attraction. A subsistence farmer, however, might be appalled at such a landscape. We evaluate landscapes according to our individual criteria.

Aesthetic perception influences both our feelings about environmental preservation and also our attitudes toward our built environment—our homes and cities. We know that people's attitudes toward environments were different in the past, and our own attitudes toward the past influence our feelings about touring historic sites and preserving landmarks.

Studies that have grouped men, women, children, the elderly, the disabled, and people categorized in many other ways have taught us that each of these groups often perceives the environment in significantly different ways. Therefore, they may almost be said to live in different environments. This knowledge has proved useful in the design of new environments, such as housing. For example, Table 1 lists certain special considerations for constructing housing for elderly, hearing-impaired people.

Differences in perception may also be studied among different cultures. Human behavior is rooted in biology and physiology, but it is filtered through culture, the way of life of a group of people. Cross-cultural studies have revealed differences among what different cultures assume to be natural or unnatural, desirable or undesirable. **Proxemics** is the cross-cultural study of the use of space. People of different cultures actually rely on their eyes, their ears, and their noses differently, even if they have the same physical sensory capacity. They do not agree, for instance, on crowding. In a room in which most Americans would feel uncomfortably crowded, most Middle Easterners would not. Proxemics affects human behavior at all levels from polite interpersonal behavior to the design of buildings or even whole cities.

Human behavior is rooted in biology and physiology, but it is filtered through culture, the way of life of a group of people.

Some scientists believe that humans exhibit *territoriality*. Many animals lay claim to territory and defend it against members of their own species. Most aspects of territorial behavior among animals have to do with spacing, protecting against overexploitation of that part of the local environment on which the species depends for its living. Applying these studies to human behavior is complicated by the interplay between human biology and culture. People defend their standing space in a crowd (they bump back), urban gangs defend their "turf," and nations defend their land, but it remains unclear whether this behavior is rooted in biologically determined human territoriality.

HISTORICAL MATERIALISM

Human technology has given humankind greater control over the environment. **Historical materialism** is a school of thought that tries to write a plot for human history based on this idea. This understanding might then be used to predict the future and thus to make history scientific. Karl Marx (1818–1883) was the founder of

TABLE 1 ***Design Statements for the Hearing-Impaired Elderly***

Environmental concepts				
Communication	**Safety/ Security**	**Social interaction**	**Design statement**	**Comments**
			THE LIVING UNIT	
X		X	Spatial layout should provide for visual communication to occur between rooms, while not compromising visual privacy needs, e.g. between bedroom and living room.	A visual connection will facilitate lipreading or signing. Visual communication should be emphasized between "high" social interaction areas such as the kitchen-dining room, the bedroom-living room, the kitchen-outdoor space. Visual access can be controlled by the occupant through the use of shades-shutters or other devices.
X	X	X	General Communication systems should emphasize auditory as well as visual abilities of the population.	A visually oriented communication system enables the hearing-impaired to communicate effectively with the immediate environment. Entrance doors should provide visual observation outside the door to identify people prior to entry. Units should be wired for video-connections to the lobby, main office, and other units. Telephone adapters should adjust for auditory as well as pitch decrements.
	X		Emergency communications systems should reflect auditory as well as visual impairment and should be provided in several locations of the living unit.	Presbycusis places physical limitations on an individual's interaction with and awareness of the surrounding environment. Several, if not all rooms, of a unit should be equipped with: **a.** visual smoke alarms **b.** visual general building fire alarms **c.** an emergency call system.
		X	Criteria regarding sound transmission coefficients for interior and exterior walls should be adjusted to maintain acoustical privacy as well as minimize competing noises.	Hearing-impaired individuals may raise the volume of the television or radio. Standard sound transmission coefficients used in building design may not be sufficient to prevent unwanted noise. Similarly, windows should be double-glazed and weatherstripped. Doors should be soundstripped and of solid core construction.

Source: Adapted from Table 6 in A. Abend and A. Chen, "Developing Residential Design Statements for the Hearing-Impaired Elderly," *Environment and Behavior* 17(4):491. Reprinted with permission.

historical materialism. His friend and collaborator Friedrich Engels (1820–1895) said that Marx did for history what Darwin did for biology, that is, identified the underlying progressive force.

Historical materialists ask: What is the "story" of human history? What thread or theme can we trace in it through time? If we focus on politics, we notice that empires and entire civilizations rise and fall. Some periods are peaceful; others are plagued by wars. Some rulers have been good, but others bad throughout recorded history with no apparent evolution. In what aspect of human affairs can we demonstrate evolution?

Marx said that evolution is found in humankind's conquest of the physical environment to meet material needs. A contemporary ruler may or may not be wiser than one in the past, but a modern tractor can do more work than a horse dragging a wooden plow. Therefore, historical materialists insist, the advance of technology over the environment has to be history's plot, and it is this improvement of productive technology that encourages social evolution. Marx wrote:

> In the social production of their livelihood men enter into definite relations that are necessary and independent of their wills; these relations of production correspond to a definite stage in the development of their material forces of production. The sum total of these relations of production constitutes the economic structure of society, the real basis on which is reared a legal and political superstructure and to which correspond definite forms of social consciousness.

Thus, the relations of production (the economic system) and even the legal, political, and social aspects of society are determined by the degree of control a society has over its environment (the "stage" in its development). The degree of control over the environment is a matter of science and technology (see Figure 4).

From this interpretation of the past, Marx went on to predict the future. Eventually, Marx predicted, technological capability would produce such material abundance that the distributive systems of socialism and communism would be possible. ("From each according to his abilities; to each according to his needs.") Improving technology loosens nature's constraints on people. "The ascent of man," Engels wrote, is inevitably "from the kingdom of necessity to the kingdom of freedom."

Some of what Marx wrote has turned out to be misguided guesses, and some of it is simply false, but much of what Marx wrote is generally accepted. Today no one doubts that science progresses almost inexorably and that scientific progress often upsets society's economic, political, and even philosophical assumptions. For example, the invention of new birth control methods and the emergence of prenatal testing have challenged our laws and even our ideas of life.

Many of the political movements that have used Marx's name, however, disgrace his memory. The Communist parties that seized power in Russia in 1917 and that seized power or have been imposed on other peoples in the twentieth century bear no relationship to the humanitarian society that Marx idealistically envisioned being made possible by material abundance. Long after the collapse of the monstrous tyrannies imposed around the world in the name of communism, the idea of a prosperous society sharing its abundance will inspire people.

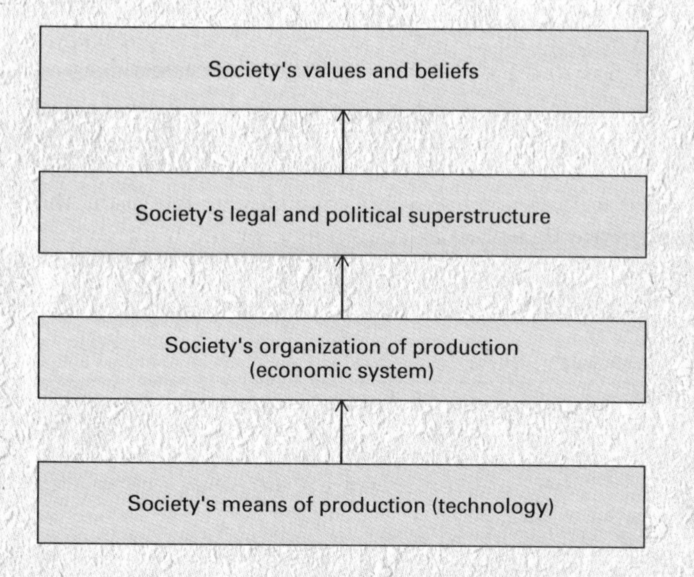

Figure 4. This chart outlines historical materialism's view that any society's values and beliefs, its legal and political systems, and its economic system rest on the foundation of its productive technology. Therefore, as that productive technology improves, all other aspects of the society will also have to evolve.

CULTURE AND ENVIRONMENT IN U.S. HISTORY

Each of the theories of the influence of the environment on human affairs contains some truth, but none of them can explain all events in history or in the present. The history of the United States demonstrates this. It often used to be told as the story of Europeans pressing westward across the continent and adapting their culture to new environmental circumstances. This approach neglected the contributions of Native Americans in teaching the Europeans how to adapt to the various North American environments, as well as the many contributions of African Americans and various Asian peoples. Even in analyzing the adaptations of European culture, however, U.S. history teaches that culture can be very conservative and that preconceptions and misperceptions can blind settlers to opportunities and difficulties in new environments and slow the process of cultural adaptation.

The English thought that the environment of eastern North America was similar to that of England, and so they anticipated no problems in the transfer of grains or livestock. They often failed to utilize superior native plants and animals. In eastern North America the summers are hotter, the winters colder, the sun brighter and hotter, the rainfall harder, and the environment generally suited to a much greater variety of vegetation than in England. The colonists nevertheless attempted to reproduce their lifestyle from home.

The settlers eventually learned farming and fishing techniques from friendly Native Americans, and they only slowly evolved a material culture adapted to the eastern seaboard environment. Wood, for instance, was more available than it was in the deforested British Isles, and so Americans came to rely on wood for a fuel, building material, and raw material for making tools and machin-

Figure 5. In 1806 explorer Stephen H. Long looked across these treeless Nebraska plains, thought them unfit for agriculture, and labeled them "The Great American Desert." This misconception delayed their settlement for decades. (Courtesy of Nebraska Department of Agriculture.)

ery. They also arduously learned and devised new techniques for utilizing the forest soils.

When U.S. expansion reached the semiarid and treeless Great Plains, early observers assumed that they faced a wasteland (see Figure 5). Major Stephen Long, the leader of an 1806 exploratory expedition, described the Plains as "wholly unfit for cultivation, and of course uninhabitable by a people depending upon agriculture." He labeled the Plains "The Great American Desert," and settlers avoided the region for the next 50 years.

Pioneers eventually devised new agricultural techniques, however. Because the settlers had not seen such soil in Western Europe and because there were no trees, the settlers first assumed that the soil must be poor. New techniques of turning and plowing the rich prairie soils evolved through laborious experimentation, and the most suitable seed grains were eventually brought over by Eastern European immigrants. Barbed wire substituted for wooden fencing, and windmills drew up precious water. Most national lawmakers came from the relatively well-watered East, and they failed to understand that the settlement of a drier area would necessitate different laws regarding homestead size and water rights. Thus, environmental misperceptions hindered even the formulation of a suitable framework of law.

KEY TERMS

ecological niche (p. 124)
cultural ecology (p. 124)
environmental determinism (p. 124)
possibilism (p. 125)
mental maps (p. 125)
cognitive behavioralism (p. 125)
behavioral geography (p. 126)
proxemics (p. 127)
historical materialism (p. 127)

QUESTIONS FOR INVESTIGATION AND DISCUSSION

1. Look for examples of each type of environmental determinism in newspapers or books.

2. Find in a library a copy of an old human geography book with a table of contents organized by environmental regions. Look for examples of environmental determinism.

3. Draw your own mental map of your town. Compare it with maps drawn by other students, nonstudents, visitors, children, and other groups. Compare your mental map with the distribution of local real estate values.

4. Think of an example of a scientific development, and trace how it has affected both economic and political life. Which assumptions about ourselves and our society has it challenged? Possible examples are television, the electric light, the air conditioner, or the birth control pill.

ADDITIONAL READINGS

Boal, Frederick W., and David N. Livingstone, eds. *The Behavioral Environment.* New York: Routledge, 1989.

Gould, Peter, and Rodney White. *Mental Maps.* Boston: Allen & Unwin, 1986.

Hall, Edward T. *The Hidden Dimension.* Garden City, N.Y.: Doubleday, 1966.

Marx, Karl. *Capital.* New York: Random House, 1977.

Sack, David. *Human Territoriality: Its Theory and History.* Cambridge, England: Cambridge University Press, 1986.

Sprout, Harold H., and Margaret Sprout. *The Ecological Perspective on Human Affairs.* Princeton, N.J.: Princeton University Press, 1965.

Toynbee, Arnold. *A Study of History.* New York: Oxford University Press, 1954.

CHAPTER 5

The Principles of Cultural Geography

This desert caravan illustrates how people adapt their culture to their environment, but they also expand their cultural possibilities through trade. (Courtesy of Tom Hollyman/Photo Researchers.)

Imagine that deep in the Brazilian rainforest there is a human community that is completely isolated. The people carry on their lives in total ignorance of the rest of us here on the planet, knowing nothing of international trade, of international political affairs, or even of the government of Brazil, which exercises no effective jurisdiction over them. Certainly they are affected by the complicated affairs of the rest of us. They see our airplanes in the sky, and their local environment is changing because the rest of us are polluting the atmosphere, but they explain these things either as natural phenomena or as the work of spirits. These isolated people—let's call them the "Happy" people, because perhaps they are—remain convinced that they are the only humans on the Earth (see Figure 5–1).

An isolated society such as this must depend entirely on its local environment for all its needs, including food, clothing, and shelter. Thus, for example, the Happy people would have learned over time which local plants and animals can be eaten and which local animals can be trained to work for humans. If these people receive no imports from the outside world, neither do they produce anything for others. Everything produced locally is consumed locally, and any surplus food or goods are stored for emergencies or else just waste away. Activities are probably closely related to the seasons.

As the Happy people use their local environment, they can also transform it. They can clear forests or plant new ones, drain swamps or dig irrigation systems, fertilize the soil, or deplete and waste it through poor agricultural practices. If they burn local vegetation, they make their own contribution to world atmospheric pollution. They can even redesign local landforms, as many societies have reduced steep mountainsides to stepped terraces for agriculture. People are never entirely passive; rather, they interact with their local environment.

THE DEVELOPMENT OF CULTURES IN ISOLATION

The Happy people develop in isolation their own distinctive **culture**. This word is often used to mean only fine paintings or symphonic music, but to social scientists it means everything about the way a people live: what sort of clothes they wear (if any); how they gather or raise food; whether they recognize marriages and if they do, whom (and how many spouses) they think it proper to marry and how they celebrate such ceremonies; what sorts of shelters they build for themselves; which languages they speak; which religions they practice; and whether they keep any animals for pets. Human cultures vary significantly in all these aspects.

As people modify their local landscape, we say that the landscape is transformed from a **natural landscape**, one without evidence of human activity, into a **cultural landscape**, which reveals the many ways people modify their local environment (see Figure 5–2). Aspects of cultural landscapes include both the treatment of the natural environment and also houses and other parts of the **built environment**. All the objects that people use are together referred to as that people's **material culture**. In addition, each culture includes a set of beliefs and practices.

Cultures constantly evolve and change through time, and, as we shall see later in this chapter, these changes may result either from local initiatives and developments or from foreign influences. In the past, human societies developed in greater isolation from one another than today, and the extraordinary diversity of human cultures testifies to human ingenuity. Different peoples living in very similar environments but isolated from one another developed astonishingly different lifestyles. Conversely, some aspects of cultures that have developed in different physical environments are startlingly similar.

Figure 5–1. This photograph shows the first contact between these formerly isolated people and the rest of humankind. On August 4, 1938, a U.S. scientific expedition entered the supposedly uninhabited Grand Valley of the Balim River in western New Guinea. They found over 50,000 people living in Stone Age circumstances totally unaware of the existence of anyone else on Earth. We cannot say for certain whether any totally isolated groups live on Earth today, but it becomes less probable each year. (Courtesy of Department Library Services, American Museum of Natural History.)

Figure 5–2. This aerial photograph of the Grand Valley of the Balim was taken a few weeks before the expedition actually contacted the valley's inhabitants. It must have been a surprise to look down into a supposedly uninhabited valley and to see these neat clearings and settlements outlined by irrigation ditches. This was clearly a cultural landscape, one modified by considerable human effort. (Courtesy of Department Library Services, American Museum of Natural History.)

No direct cause-and-effect relationship between a physical environment and any aspect of culture can be assumed.

CULTURE REALMS AND CULTURAL DIFFUSION

The place where a distinctive culture originates is called the **hearth area** of that culture, or **culture hearth**. Throughout history, various aspects of cultures such as clothing styles, diet, language, music, and architecture have spread out from their hearths and been adopted by other peoples. This process is called **cultural diffusion**, and the process of adopting some aspect of another culture is called **acculturation**. The entire region throughout which a culture prevails is called a **culture realm**. The use of the phrase "the Christian world," for instance, assumes that the prevalence of Christianity across a large region unites the peoples of that region in ways that might significantly be contrasted with, for instance, "the Islamic world." Each is a great culture realm, and as we shall see in Chapter 8, the prevalence of each of those religions encourages other observed similarities within those realms.

Smaller culture realms can be differentiated even within the United States. The North and the South were clearly self-defined and differentiated during the Civil War, but even today a number of regions can be identified and mapped across the country (see Figure 5–3).

The presence or impact of any cultural attribute may diminish away from its hearth area, just as the volume of a sound diminishes with distance. This phenomenon is called the **distance decay** concept. The farther away we travel from any big city, for example, the smaller the percentage of people who read that city's newspapers, tune in to its radio and television stations, and rely on its other services. We might try to define and map the territories of influence of the cities of Los Angeles and of San Diego, for instance, as culture regions. To the immediate south of Los Angeles, people probably subscribe to Los Angeles newspapers, watch Los Angeles television stations, and rely on service from the Los Angeles airport. Farther to the south, however, the influence of Los Angeles diminishes and that of San Diego becomes predominant.

Some aspects of culture may develop variations as people who carry that culture diffuse outward and away from one another. For example, different variations of a language called **dialects** develop as speakers move away from the hearth and toward the periphery of language regions.

Different cultures create noticeably different cultural landscapes. If we travel across a landscape with steeples with crosses on top, we know we are in the Christian culture realm. Minarets—towers attached to mosques—define the Islamic world (see Figure 5–4).

Local place names are an important feature of any cultural landscape. If the cities we visit bear names like St. Paul, we are in the Christian world; conversely, Islamabad ("the place of Islam," the capital of Pakistan) indicates the Islamic realm. The study of place names, called **toponymy**, often reveals a great deal about what people in a particular location do or believe. Take, for example, the historic city of St. Petersburg in Russia. Following the Russian Revolution of 1917, its name was changed to Leningrad. In 1991, with the demise of the Soviet Union, the city's name reverted to St. Petersburg. As this book is being written, places all across Eastern Europe are reverting to pre-Communist names (see Figure 5–5).

Toponymy can also reveal history when other evidence has been erased. It may tell us what people first

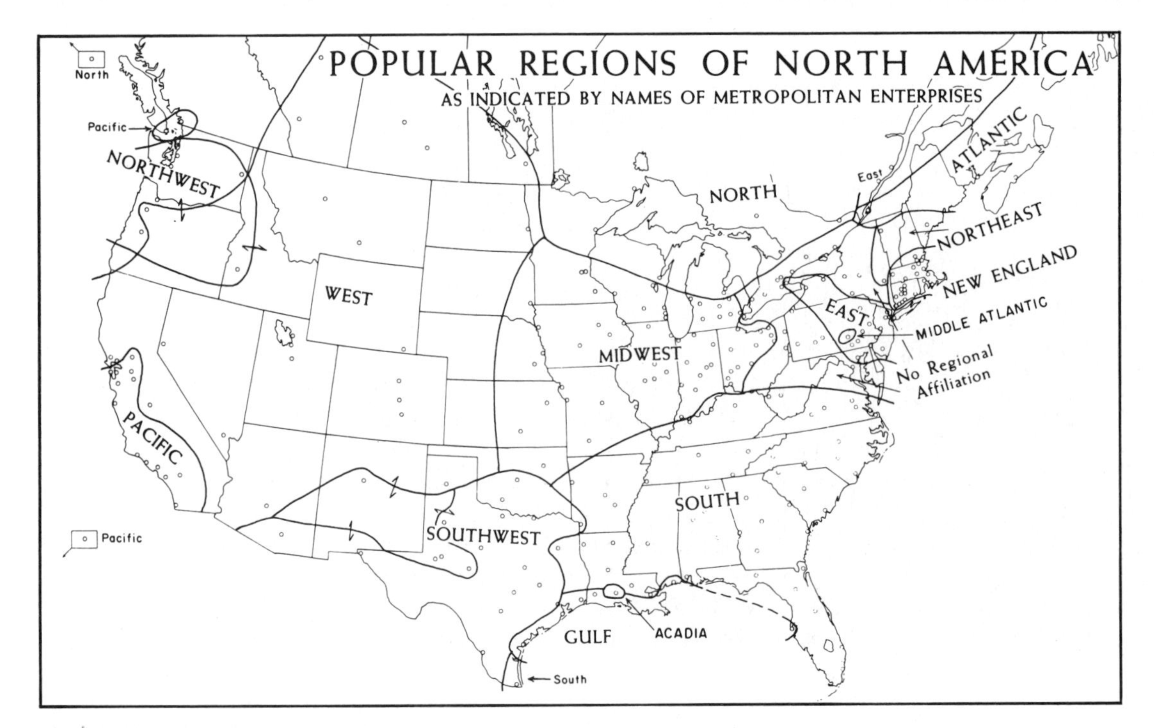

Figure 5–3. The United States is significantly homogeneous as a culture realm, but still there are many ways of mapping local cultures. The regions on this map are based on one single criterion: telephone directory listings of local metropolitan enterprises. Thus, they reflect regions as their inhabitants define them self-consciously. Outsiders to each of these areas, however, might have mental maps that differ from those of the local citizens. An Easterner, for instance, might consider Kansas City to be in the West, but to a Kansas Citian, home is in the Midwest. Does this map concur with your mental map of the United States? (After Wilbur Zelinsky, "North America's Vernacular Regions," *Annals,* Association of American Geographers, vol. 70, March 1980, p. 14. By permission of the Association of American Geographers.)

Figure 5–4. The modern skyscrapers in the background could be almost anywhere, but the grand mosque in the foreground suggests that this photograph was taken in the Islamic realm. The city is, in fact, Cairo, Egypt. (Courtesy of Roger Peters.)

Communist era name	Previous and now restored name
Andropov	Rybinsk
Brezhnev	Naberezhnye Chelny
Chernenko	Sharypovo
Frunze	Bishkek
Georgiu-Dezh	Lisky
Gorky	Nizhny Novgorod
Gotvald	Zmiev
Kalinin	Tver
Kuibyshev	Samara
Kirovbad	Gyanja
Leninabad	Khodjent
Leningrad	St. Petersburg
Mayakovsky	Bagdati
Ordzhonikidze	Vladikavkaz
Sverdlovsk	Yekaterinburg
Voroshilovgrad	Lugansk
Zhdanov	Mariupol

Figure 5–5. During the period of Communist government in the Soviet Union, many historic city names were changed to honor Communist leaders. Today many of these cities are reassuming their earlier names.

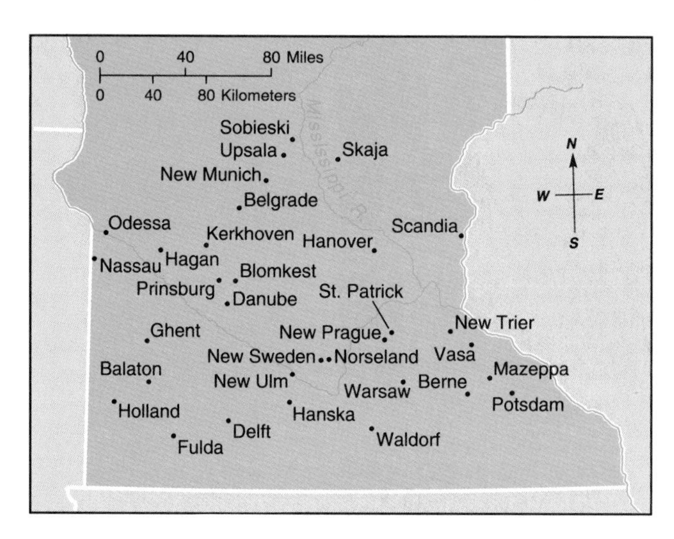

Figure 5–6. Some North American cities retain the names of explorers who just passed by, but these Minnesota place names supply a good clue to the origins of the immigrant settlers in the area. Each is a city in Europe or a hero to some national group. How many can you identify?

settled a place (see Figure 5–6). It may also indicate what type of environment they found, for example, as we discussed in Chapter 3 the many towns named "Wang" in the North China plain.

Culture realms may be demarcated by a great number of other features of the landscape. People usually rely on local materials for building, and so in one place stone may be the traditional building material, in another place brick, and in another wood (see Figure 5–7). Innovations in the uses of these materials may diffuse across cultures.

Individual styles of architecture often represent an adaptation to climatic conditions, and so a particular style may be adopted in different regions with similar climates, as shown in Figure 5–8. Cultural preferences are sometimes so powerful, however, that an architectural style may diffuse beyond the limits of where its building materials can be found and even beyond the architecture's physical comfort range. The style of the Italian Renaissance architect Andrea Palladio (1508–1580), for example, spread to England because of aesthetic preferences, despite the fact that Palladian buildings are uncomfortable in England's damp, cool climate. From England the preference diffused to America, even to areas that lacked both the appropriate climate and the necessary building materials. As Europe and the United States came to dominate other regions, Palladian architecture diffused throughout the world (see Figure 5–9).

Geographers who have investigated house types across the United States have concluded that diffusion paths conform to the migrations of various national and ethnic groups. Barns and other structures also frequently exhibit distinct architectural styles that reveal the origins of their builders.

Not only the design of individual buildings but the designs of whole settlements reflect cultural differences. Some rural societies cluster housing settlements, others string out settlements along transportation routes, and still others isolate settlements in individual farmsteads. The look and layout of whole towns and cities reveal the cultural backgrounds of their builders to a trained observer.

Cultural diffusion can be active and deliberate, as when an outside power conquers a region and tries to

Figure 5–7. The mosque at San, in the West African country of Mali, demonstrates that when people have mud and very little wood, they can still build extraordinary structures. This building is in the distinctive Dyula architectural style, named for a trading people who diffused the style as they moved around West Africa. Different building materials are available and suitable for different environments. (Courtesy of Professor J. Markusse, University of Amsterdam.)

Figure 5–8. The building in the left foreground is a bungalow, which is a Hindi word for a low-sweeping, single-story house with a roof extending out over a veranda. Such houses were first built in the mountain foothills of northern India, but the style was copied throughout the British Empire and eventually the United States. Today in English the word *bungalow* means almost any sort of small house.

This photograph shows Simla in the 1880s. Simla was the summer capital of the British Indian Empire between 1864 and 1947. It was cooler up at Simla (7100 feet—notice the vegetation) than down at New Delhi in the plains about 170 miles (275 km) to the south. The pseudomedieval English cathedral in the background seems an odd presence in northern India, but wherever the British went, they took their culture and traditions, including architecture. (Courtesy of the Library of Congress.)

Figure 5–9. The Morris-Jumel Mansion. Built in 1765 in Manhattan, 10 miles north of Wall Street, it is in the Palladian style, appropriate for the Italian climate but not for New York. It would have been impossible to keep warm. In Italy it would have been built of stone. Here, virtually on the frontier of Western civilization, it was built entirely of wood masquerading as stone. The choice of style demonstrates the power of prevailing aesthetic taste, even over comfort. (Courtesy of Washington Headquarters Association, New York City.)

impose its way of life. Cultural elements can also spread when one group discovers and adopts some aspect of a different culture that it considers superior to its own.

Diffusion Paths and Speeds

Tracing a course of diffusion may teach us a great deal about how peoples and cultures interact and influence one another. The paths of cultural diffusion are seldom so simple as the concentric ripples spreading out from a stone dropped into a pool. The patterns and speed of diffusion provide measures of cultural interactions. There are two basic types of diffusion: contiguous or contagious diffusion and hierarchical diffusion.

Contiguous diffusion occurs from one place directly to a neighboring place. Contiguous diffusion in the past is often revealed by the diffusion of artistic styles. For example, Italian Renaissance art quickly spread to the Low Countries of Northern Europe (today's Belgium and the Netherlands), a path that reflected trading and financial links. As long ago as the fourth century B.C., Greek art was carried far into Asia during Alexander the Great's conquests. Central Asian statues of Buddha sculpted during the centuries immediately following these conquests resemble Greek gods (see Figure 5–10). (The fact that we use B.C. here reveals that we live in the Christian realm; if we lived in the Islamic realm we would have used the Muslim calendar.)

Sometimes cultural diffusion does not occur contiguously across space but downward or upward in a hierarchy of organization. When such **hierarchical diffusion** is mapped, it shows up as a network of spots rather than as an ink blot spreading across a map. The Roman Catholic church illustrates both an organizational hierarchy and a geographical hierarchy. Each parish priest answers to his bishop, who from his cathedral presides over a diocese, which includes many parishes. The bishops, in turn, are under the authority of the Vatican (see Figure 5–11). An announcement from the Vatican diffuses to the cathedrals and then down to the parishes around the world. In hierarchies more information generally travels up and down the hierarchy than across any level of the hierarchy. Therefore, news from a parish about an innovation in a community service or church ceremony might reach the cathedral and even the Vatican before it is heard in neighboring parishes.

The "hub-and-spoke" flight patterns of the major U.S. airlines provide another example of hierarchical diffusion. Carriers collect passengers from many surrounding cities at hub airports and then fly them to their destinations in other major cities. To fly from one small city to another, a passenger may have to fly first to a nearby big city, then to another big city near the ultimate destination, and finally from that big city to the small city. This is flying up and down a hierarchy.

Advanced societies usually exhibit well-developed hierarchies of cities, so we can study how the hierarchy evolved through time (see Figure 5–12). Once a hierarchy is developed, the diffusion of any specific phenomenon can be traced from the few biggest cities "down" to the many more smaller cities and then further down to the even more numerous small towns and villages until the landscape is covered (see Figure 5–13).

One recent attempt to map the hierarchical diffusion of information in the United States focused on the flow of Federal Express Corporation overnight parcels. New York City was found to account for nearly 14 per-

Figure 5–10. This richly dressed young man is the Indian Prince Siddhartha before he became the Enlightened One ("Buddha") and gave up his worldly wealth. The statue exhibits a mix of styles: the curly hair, stocky physique, relaxed stance, and deeply carved drapery are Greek or Roman characteristics, but the facial expression is typical of Indian mysticism. The nimbus around the prince's head originated with Iranian statues of sun gods.

Where could all these cultural influences have come together and this extraordinary figure have been carved? The answer is the region of Peshawar in today's Pakistan, the crossroads of Eurasian travel routes where the Khyber Pass opens from Central Asia into the plains of the Indian subcontinent. It dates from the late second or early third century A.D. (Courtesy of the Seattle Art Museum.)

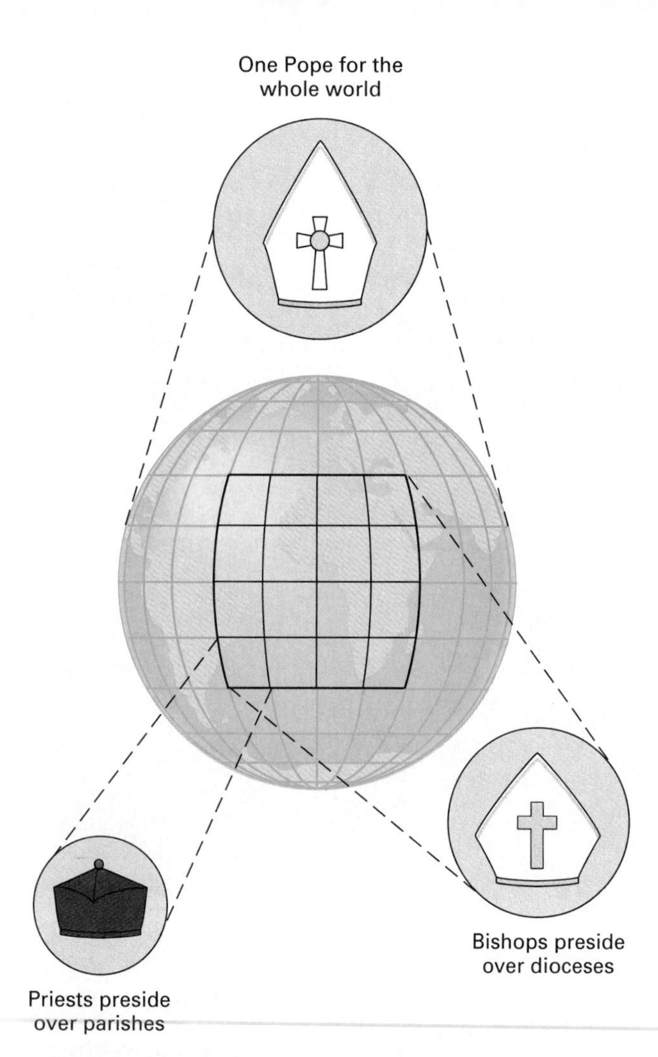

Figure 5–11. The Roman Catholic church illustrates both territorial and organizational hierarchy. The pope presides over the entire Church, but the Church is subdivided into dioceses, each presided over by a bishop, and the dioceses are subdivided into parishes.

cent of all parcels shipped among major cities. New York's most important links were with four regional information capitals: Los Angeles, Dallas-Fort Worth, Chicago, and Atlanta. The higher a city was found to be in the hierarchy, the more parcels it sent greater distances. As we might expect, the flow of information was chiefly up and down the hierarchy, not among the cities at any given level. The map of America's top domestic airline routes also reflects the national hierarchy of cities (see Figure 5–14).

Many phenomena diffuse hierarchically around the world today. The spread of a disease or of a cultural innovation such as a clothing fashion or a music hit can often be traced among the world's principal metropolises before it reaches down into the smaller cities in each country. This path of diffusion reveals the world's interconnectedness.

Barriers to Diffusion There are many barriers to cultural diffusion. Oceans and mountain ranges are examples. Historically, narrow bodies of water have often been easier to cross than equivalent land distances. Thus, Roman civilization diffused around the shores of the Mediterranean Sea, but the difficulty of crossing the Sahara Desert impeded its diffusion to tropical Africa.

In addition to deserts, topographical features such as mountains and valleys historically have blocked human communication and contributed to cultural isolation. In Asia the deeply cut valleys of the Tongtian, Mekong, Nu, and Brahmaputra rivers make overland travel between the Chinese cultural realm and the Indian subcontinent extremely difficult, so these two realms and great concentrations of population are surprisingly isolated from one another (see Figure 5–15, and look ahead to Figure 6–1). In Europe the Alps have impeded the flow of human

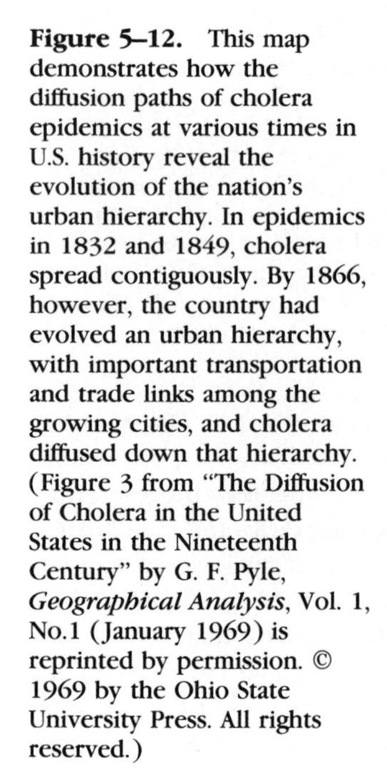

Figure 5–12. This map demonstrates how the diffusion paths of cholera epidemics at various times in U.S. history reveal the evolution of the nation's urban hierarchy. In epidemics in 1832 and 1849, cholera spread contiguously. By 1866, however, the country had evolved an urban hierarchy, with important transportation and trade links among the growing cities, and cholera diffused down that hierarchy. (Figure 3 from "The Diffusion of Cholera in the United States in the Nineteenth Century" by G. F. Pyle, *Geographical Analysis*, Vol. 1, No.1 (January 1969) is reprinted by permission. © 1969 by the Ohio State University Press. All rights reserved.)

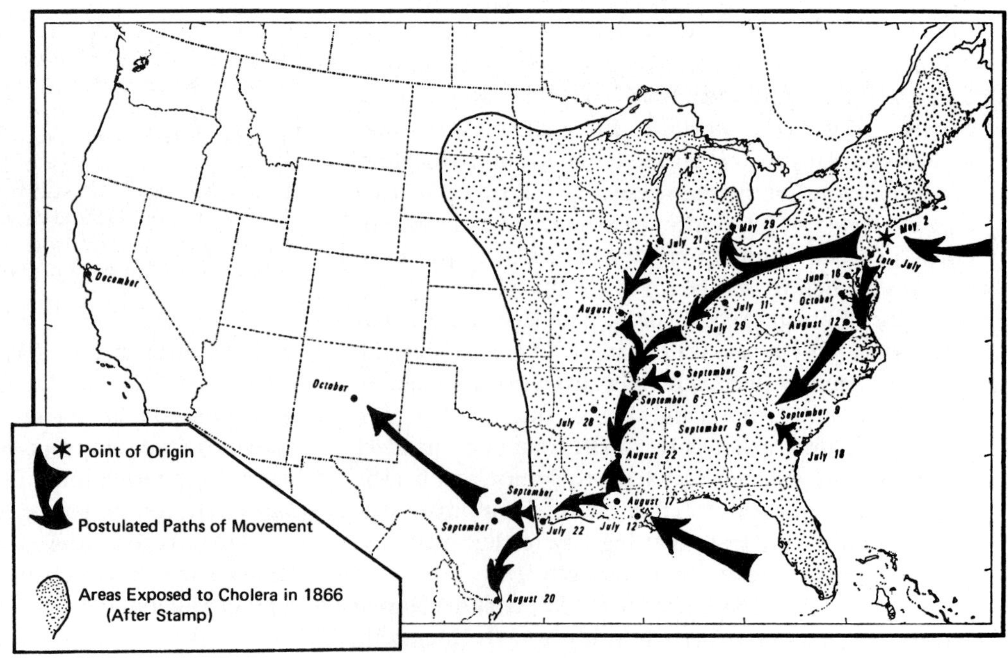

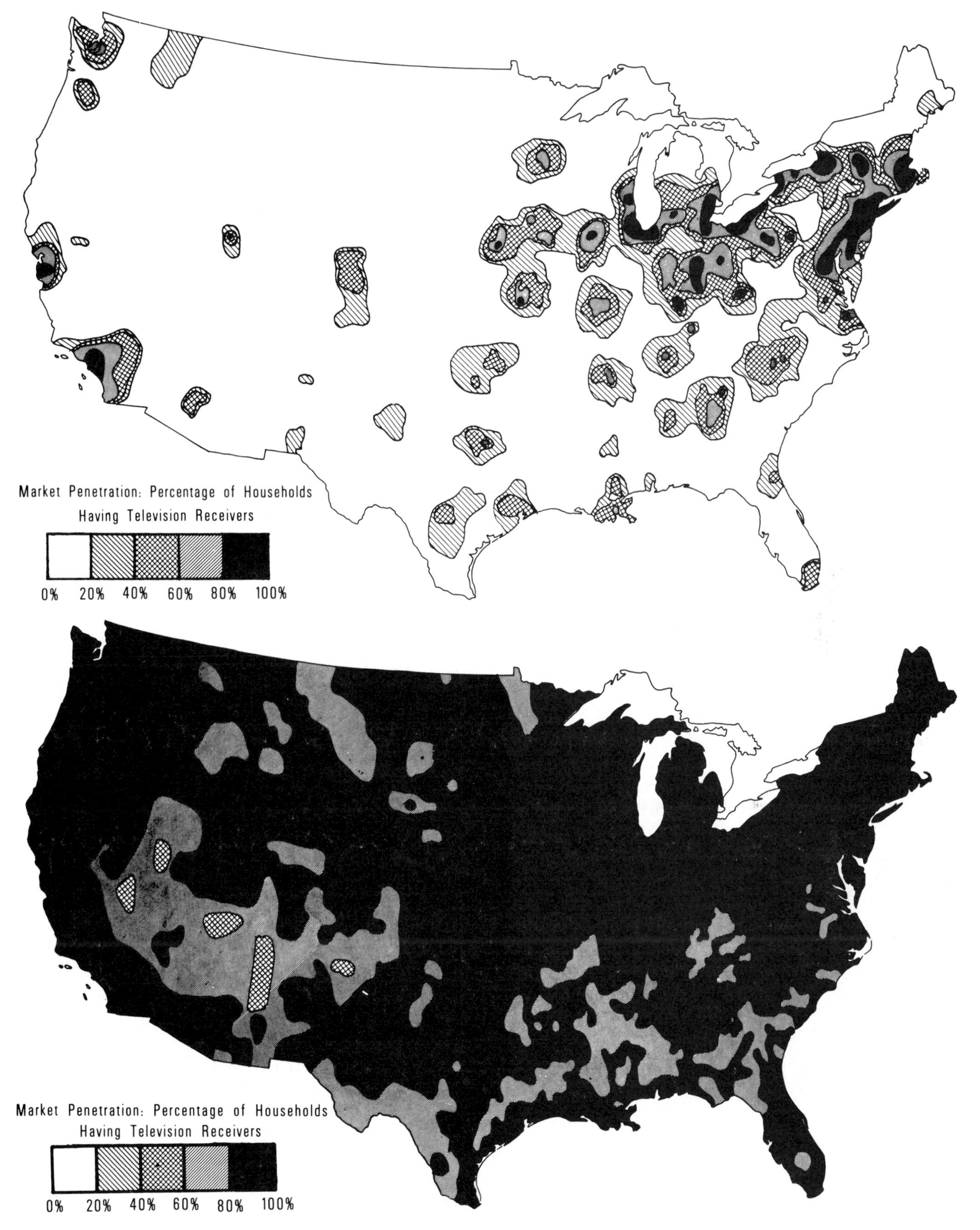

Figure 5–13. These maps show the increasing market penetration, or percentage acceptance of television sets in U.S. households from 1953 to 1965. This was before cable television or satellite dishes, and so viewers had to rely on receiving transmissions from metropolitan centers. How would the pattern of diffusion differ with today's technology? [Reprinted with permission from Brian J. L. Berry, "The Geography of the United States in the year 2000," *Transactions of the Institute of British Geographers*, LI (November 1970), 21–53.]

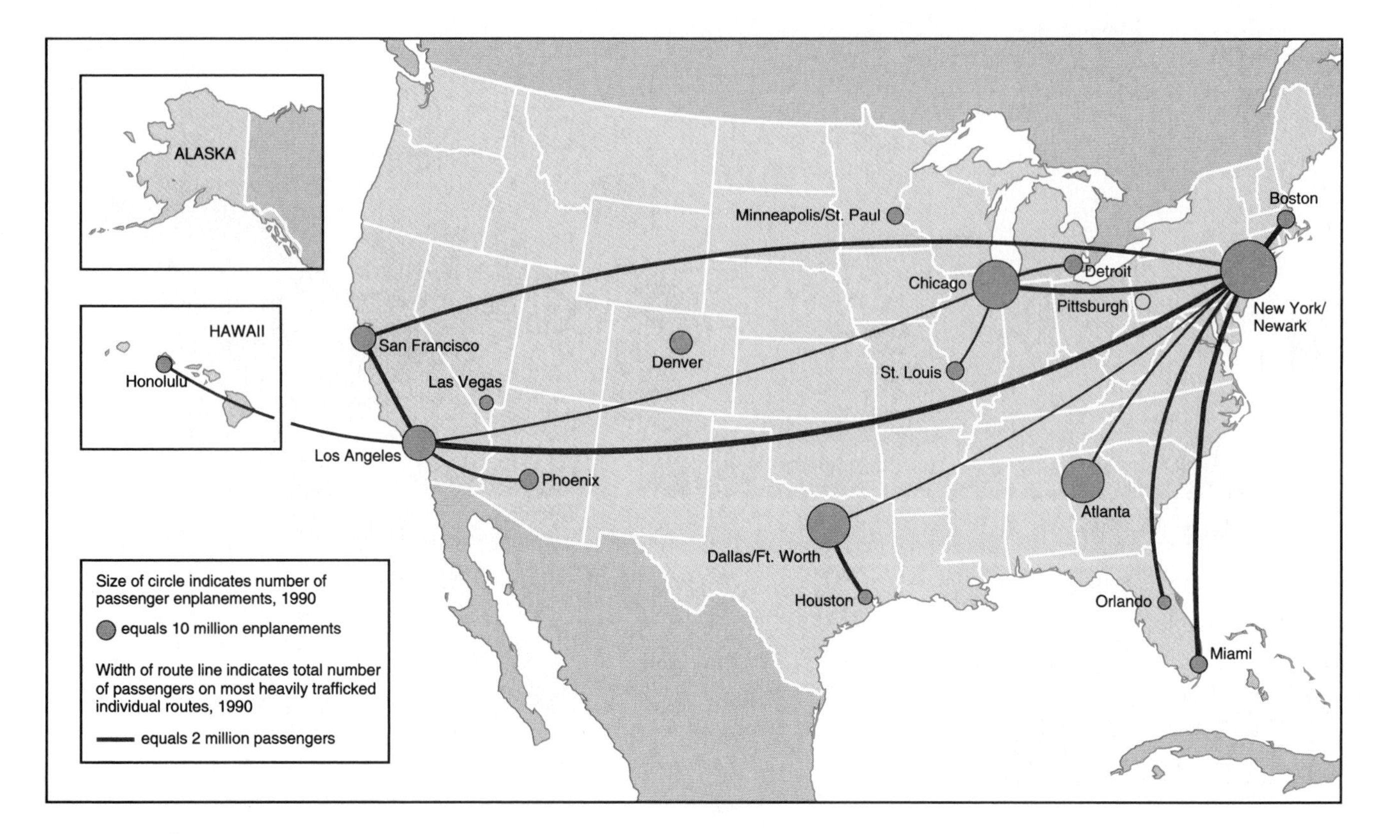

Figure 5–14. This map is a good indicator of the urban hierarchy in the United States.

populations, thereby contributing to cultural differentiation.

Other barriers to cultural diffusion include political boundaries and even the boundaries between two culture realms. Hostile misunderstanding, distrust, and competition between two culture groups can hinder communication and exchange between them.

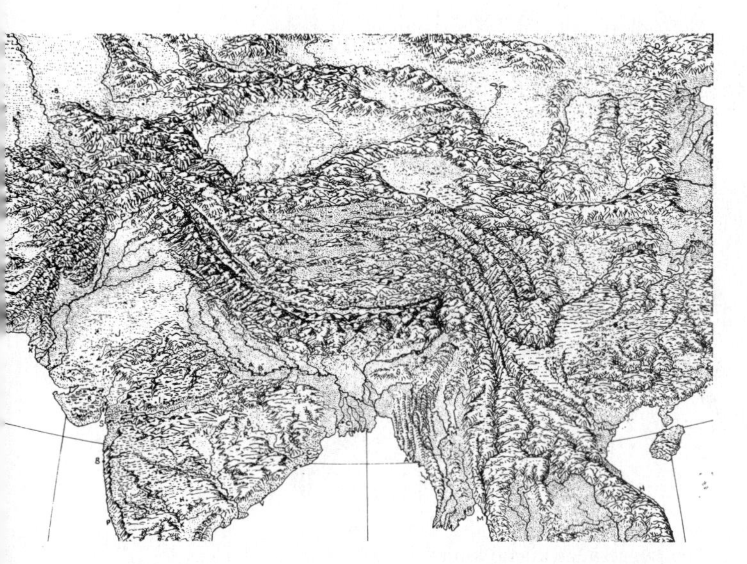

Figure 5–15. These parallel, deeply cut valleys have impeded transportation and communication between the Indian subcontinent and the Chinese cultural hearth area, contributing to the differentiation of the cultures. (Map by A. K. Lobeck; reprinted by permission of Hammond, Inc.)

Diffusionism Diffusion does not explain the distribution of all cultural phenomena. Sometimes the same phenomenon does occur spontaneously and independently at two or more places. At one time many scholars embraced **diffusionism,** the argument that there has been little independent invention of the components of civilization in different parts of the world. According to diffusionists, most civilizations owed their achievements to others. Diffusionist arguments have been used to minimize or underestimate the ingenuity of various peoples, but today diffusionism is repudiated by most anthropologists and by archaeological evidence.

In the history of mathematics, for example, the idea of the zero and its use as the basis of a numerical place system was conceived in two places: among the Maya, an Indian people of Central America, and among the ancient Hindus or perhaps the Babylonians before them. Diffusionists once argued that the Maya must have learned how to use the zero from Phoenicians who sailed across the Atlantic, but there is no evidence for this. We should never assume that any aspect of culture has diffused just because similar cultural artifacts are found in two places. We must be able to demonstrate the path of diffusion from one culture to another.

Diffusion Challenges Stability

The isolation in which the Happy people live is so different from the way most of us live today that we could almost be said to live in a different world. Modern people

Focus on Ethnocentrism

Ethnocentrism is a tendency to judge foreign cultures by the standards and practices of one's own, and usually to judge them unfavorably. Practices in other cultures that may at first seem strange to us, however, may in fact be sensible and rational. Conversely, some aspects of our own culture may seem strange or even offensive to others. One goal of studying cultural geography is to understand why other people act the way they do. Most Americans, for example, assume that a man should have one wife and a woman one husband at a time. However, due to death or divorce, a person may have a series of spouses throughout the course of his or her lifetime. This shocks some people, admittedly some Americans as well as people of other cultures. In contrast, Tibetans assume that a woman should marry all her husband's brothers. This is called *fraternal polyandry*. Through thousands of years fraternal polyandry has prevented unsustainable population increases and the fragmentation of landholdings in poor mountain valleys. The social ramifications of fraternal polyandry would upset most Americans. An American might ask how you identify the father of any given child, but to a Tibetan it makes no difference, and a Tibetan would consider the question to be in bad taste.

Any geography book will contain many examples of ways of life that contrast with your own. None is necessarily "right" or the best for everybody. All people have to overcome the assumption that "different from" the way we do things is "worse than" our way, and people everywhere must learn to appreciate and respect the integrity of other peoples' behaviors. Only by studying the extraordinary diversity of cultures can we truly understand even our own. Contrast helps us notice those things that we take for granted in our own culture.

Learning that other peoples do things differently from the way we do and coming to accept that fact is a minimal requirement for getting along in a diverse world, but the shrewd have always taken a step beyond that. They consider whether somebody else's way is superior to their own, and if it is, they adopt it.

no longer exploit their own local environments and develop their cultures in isolation. They are interconnected by transportation and communication of goods, people, ideas, and capital. All this movement we combine into the term **circulation**.

In the past, travel and transportation were more difficult and expensive than they are today, whether we measure the cost in time, money, or any other unit. This cost of transportation, which we call the **friction of distance,** was so high that only a few things moved far, and these things moved slowly. The place of origin and the path of diffusion of any innovation could be fairly well traced. Over the past 200 years, however, technology has reduced the friction of distance and accelerated the diffusion of cultural elements. We often hear that these developments in transportation and communication have "shrunk" the world.

Developments in electronics, for example, have disengaged communication from transportation. Originally information could move only as fast as a person could carry it, but in 1844 Samuel Morse demonstrated the first intercity telegraph line. This passed information almost instantaneously between Baltimore and Washington, D.C. Baltimore newspapers had previously carried news of events that had occurred in Washington a day or two before; Baltimore had been "a day or two away from Washington," but after Morse first demonstrated the telegraph, a Baltimore evening paper printed that afternoon's Washington political news. The paper concluded, "This is indeed the annihilation of space" (see Figure 5–16).

Today the global news industry exists specifically to ferret out and diffuse innovation. Ideas spread around the globe overnight. A new U.S. hit song is playing on Brazilian, Australian, and Kenyan radio stations within perhaps 24 hours. What we do not adopt, we at least know about.

Although we can have access to news and cultural elements from all parts of the world, we continue to be selective in what we pay attention to (see Figure 5–17). Our backgrounds, our education, our perceptions and prejudices, and the media to which we are exposed affect our understanding of other peoples and places and how we rank their relative importance. For a great variety of reasons we are more knowledgeable about or more interested in some places than others. These reasons are not necessarily related to the size or population of a particular region.

Today no pattern of culture realms is stable. Some religions are winning new converts, expanding geographically, and assuming new influence in world affairs, whereas others are withering away. Some languages are demonstrating the flexibility to adapt to new communication technology and are therefore gaining users at the expense of others. The relative rise and fall of the influence of various nations in world political or economic affairs wins or loses new adherents to their products, cultural artifacts, and lifestyles.

If the Happy people were to be discovered today, people in other nations would most likely adopt some aspect of Happy culture—perhaps a food or medicinal crop, or Happy music (see Figure 5–18). At the same

Figure 5–16. These "Men of Progress" were painted as if in conversation in 1857 by Christian Schüssele. Several of their inventions had helped shrink or settle the great distances across the United States. They are, from left, with their inventions: Dr. W. Morton (anesthetic ether); J. Bogardus (cast-iron building); S. Colt (revolver); C. McCormick (reaper); J. Saxton (fountain pen); C. Goodyear (vulcanization of rubber); P. Cooper (first American locomotive); J. Mott (the coal-burning stove); J. Henry (experiments with electricity); E. Nott (boilers); J. Ericsson (ironclad ship the "Monitor"); F. Estickel (steam engines); S. F. B. Morse (telegraph); H. Burden (many iron-working devices); R. Hoe (cylinder press and type-revolving printer); E. T. Bigelow (power loom); I. Jennings (many machines); T. Blanchard (machine tools); E. Howe (sewing machine). Whom would you put in such a picture today? (Courtesy of the National Portrait Gallery.)

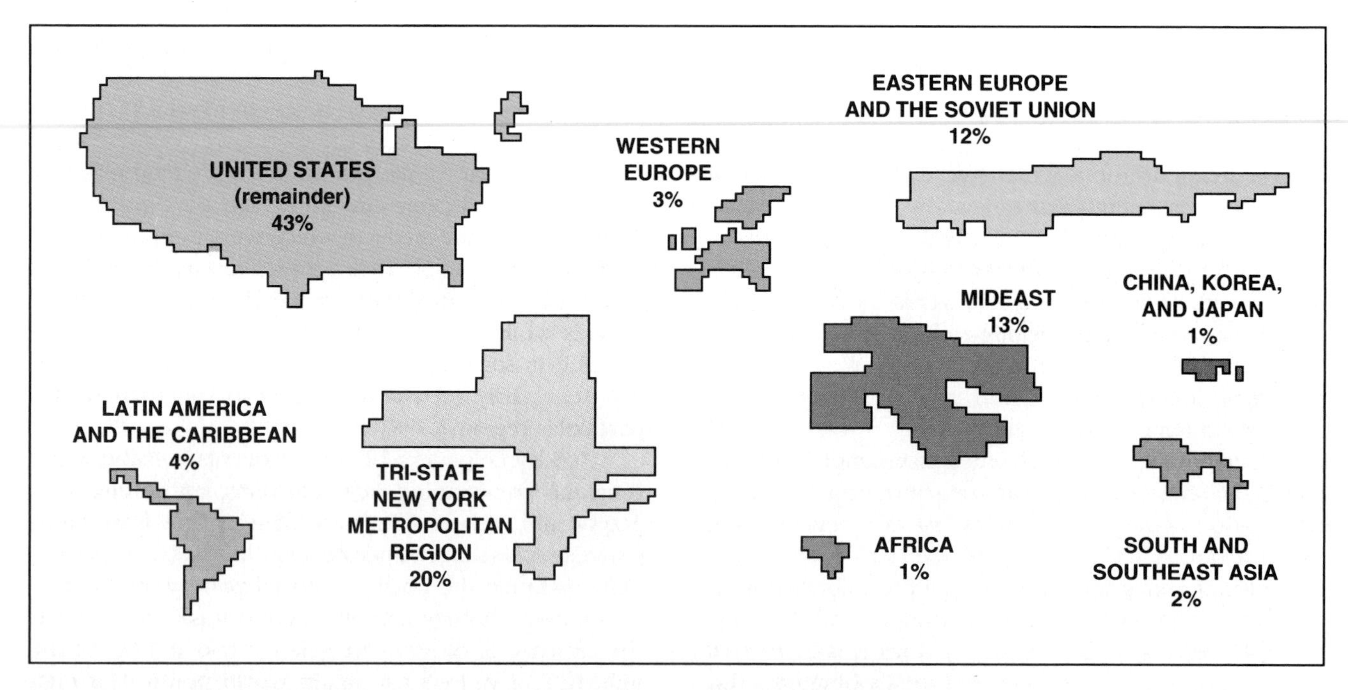

Figure 5–17. The relative sizes of world regions on this illustration correspond to the coverage devoted to each on the front page of *The New York Times* for 2 months in 1991. New York has a large and influential Jewish population, and this is reflected in the particularly high coverage of news from the Mideast. Miami has a large and influential Latin American population, and Miami has important trade links with Latin America. Considering those facts, how do you think "the view from Miami" would differ? How about Los Angeles? How about the city in which you live?

Figure 5–18. Dr. Michael Balick of the New York Botanical Garden learns plant uses from this traditional healer of the Kekchi Maya people in Belize. The knowledge that "primitive" cultures have accumulated of local plant uses, called ethnobotany, may hold keys to the cures of many diseases. Dr. Balick is researching potential anti-AIDS and anti-cancer agents for the National Cancer Institute. The healer's clothing reflects how fast these people are being acculturated, and their knowledge is being lost. (Courtesy of Dr. Michael Balick.)

time, the Happy way of life would be transformed by the infusion of products and ideas from the outside world.

Forces Fixing the Pattern of Cultures

Despite the force of diffusion, a number of factors tend to fix the geography of culture realms. Culture leaves its mark on the landscape. The fixed pattern of activities, land uses, transport routes, and even individual buildings represents investment, and it often guides, restricts, or predisposes future patterns and activities. We call this **inertia**, the force that keeps things stable. All of a people's assets fixed in place—railroads, pipelines, highways, airports, housing, and more—we call the **infrastructure**.

Culture itself is basically a set of values and ways of doing things, and culture groups seldom get displaced or eliminated entirely. Culture is learned behavior, and cultural norms are handed down through generations with a surprising integrity. Groups may be quick to take up new techniques or products, but tradition is a powerful force in human life, and imported ideas or lifestyles only slowly transform a people's entire cultural inheritance. This inertia is reinforced by the existence of territorially sovereign states and the enormous power that governments exert over their citizens to teach and to enforce norms of behavior. Lee Kuan Yew, prime minister of Singapore from 1965 through 1990, emphasized the role of culture in various Asian countries: "Culture is very deep-rooted; it's not tangible, but it's very real: the values and perceptions, attitudes, reference points, a map up here, in the mind."

Each local culture is today a unique mix between what originated locally and what has been imported, and each unique culture is a local resource just as surely as the minerals under the soil and the crops in the fields.

Most peoples value their culture, and they usually try to preserve key aspects of it. This will influence the way they interact with other peoples. A people's self-consciousness as a culture may be codified in their religion, as it is most notably with the Jews, or in their sense of their own history, called **historical consciousness**. Many Americans have difficulty understanding other peoples' historical consciousness, because one aspect of U.S. culture is an optimistic denial that history can shackle future opportunity.

Other peoples, however, nurture their traditional cultures and their historical consciousness. Many people act the way they do because that is the way their ancestors acted or to right wrongs or to account for deeds or misdeeds that took place long ago. It is essential for their own personal integrity that they do so.

Furthermore, some peoples take a long-term view of historical events that differs from popular U.S. attitudes. The teaching of history in many Muslim countries, for instance, recounts that Iberia was at least partly Muslim-ruled from 711 until 1492 (781 years), whereas it has been exclusively Christian since only 1492 (500 years): Some Muslims think it may return to Muslim rule. Traditional Chinese who are educated in the antiquity of their civilization may unself-consciously speak of units of time of hundreds of years even in casual conversation. To such people, Westerners are tremendously impatient, whereas to many Westerners, people who dwell on historical facts are stubborn, living life while looking in the rearview mirror.

The presence or absence of historical consciousness strongly influences current international affairs. Not all Americans recall that the United States conquered the Philippines early in the twentieth century and granted it full independence only in 1946. They may not see why this was relevant to negotiations between the Philippines and the United States about the continuation in the Philippines of U.S. military bases. Among the Filipinos, however, their former colonial status is a vivid historical memory. Many feel that they are owed reparations.

This chapter demonstrates that one major theme in geography is the tension between forces of change and forces for stability. Cultures diffuse, cultures change, and peoples can transform themselves and their behavior, but cultures and culture realms also have elements of stability. Any cultural pattern or distribution maps a current balance between those forces.

WHAT IS CULTURAL GEOGRAPHY?

By reading just this far you have already learned much about cultural geography and the geographical approach to studying cultures. Cultural geography maps the locations and distributions of human cultures and activities, and then it investigates the reasons for these distributions. If two phenomena have the same distributions, then geographers might hypothesize a cause-and-effect relationship. In each location, geographers also analyze the synthesis of the various cultural elements that makes each location unique. Geographers are always asking, Why there?

Endogenous Factors

Geographical research focuses on both endogenous and exogenous factors. **Endogenous factors** (generated within), or **site factors**, are elements of the specific local environment or of local cultural history and development. A study of endogenous factors can be summarized as *X* is there or is happening there because *Y* is also there. A map of the global distribution of swine, for instance, reveals that few are found in the Near East (see Figure 5–19). Why? What other local attributes of the Near East might explain this lack of swine? It cannot be any attribute of the local physical environment, which could support swine. The absence of swine can be explained by attributes of local culture. The local population is overwhelmingly Jewish and Muslim, and both Judaism and Islam forbid the consumption of pork.

Geographers traditionally begin inventories of specific places with the local physical geography: climate, landforms, soil, vegetation, and so forth. Each of these attributes is itself a product of many factors, and together they create a unique local physical environment. Human activities are carried out on this stage, and human activities interact with the local physical environment in creating a cultural landscape. For example, the climate of a place affects its soil, the soil affects local agriculture, and the nature of the agricultural system in turn affects local political life. In turn, however, political life affects agriculture, which in turn affects the soil. Cause and effect may be circular and almost infinitely complicated. Specific regional studies make up one of the finest traditions of geographical studies.

Exogenous Factors

In contrast to endogenous factors, **exogenous factors** (generated outside), or **situation factors**, refer to the way a particular place interacts with other places. Situation means specifically where a place is, relative to other places. The map of the world distribution of swine reveals a significant concentration in Iowa—more than the people there could possibly eat. Why there? The answer is that Iowans raise pigs and export pork for consumption elsewhere. In turn they import other foods. The Iowan economy is not isolated and self-sufficient but rather depends on exchange.

This principle of local specialization and exchange explains why there might be a concentration of pigs

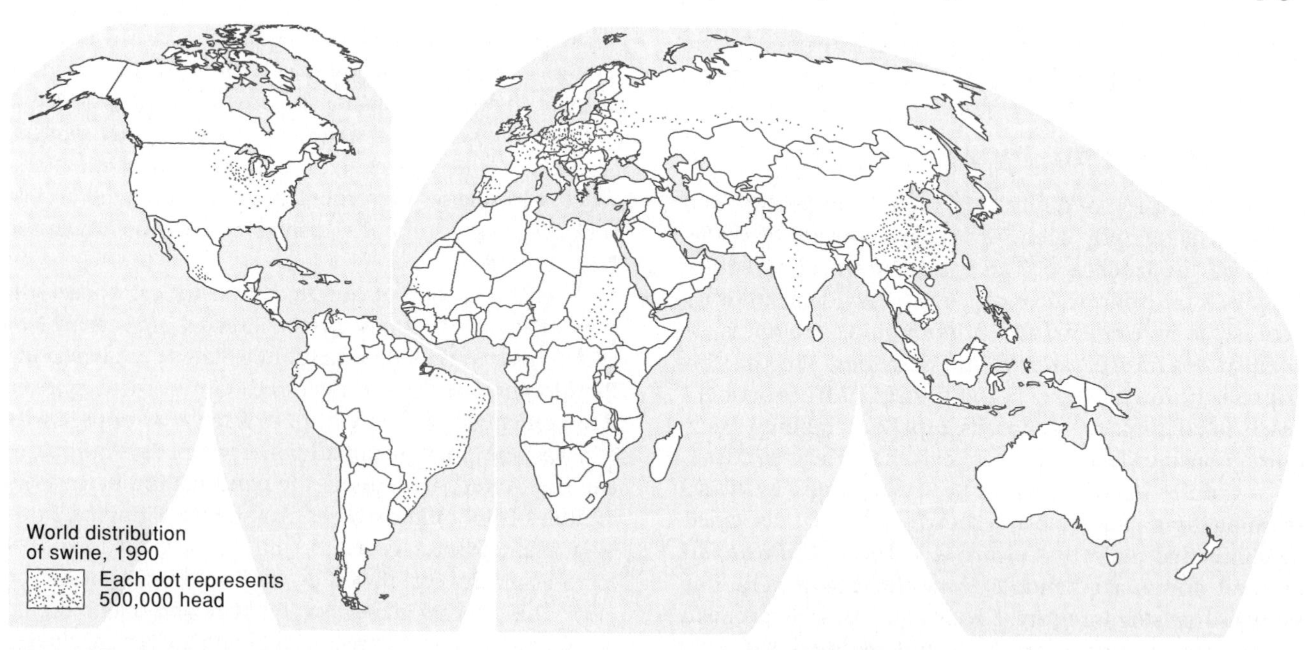

Figure 5–19. Today about 41 percent of the world's pigs are raised in China, 9 percent in Russia and Ukraine, 6 percent in the United States, 4 percent in Brazil, and another 4 percent in Germany. Some countries raise none, including countries that would be better off if they could increase their protein intake. What factors account for this distribution? (Data from World Resources Institute, *World Resources 1992–93*. New York: Oxford University Press, 1992, pp. 276–77.)

somewhere on earth, but we still have not answered the specific question, Why in Iowa? Swine were not first domesticated there; they are not even native to Iowa. Swine production is concentrated there today because a great supply of corn is available to feed the swine, markets are convenient, and the industry settled there years ago. Endogenous and exogenous factors of explanation are not contradictory. They complement each other.

As global communication and transportation have increased, the balance of factors that explain the local activities and culture at any place has tipped steadily away from endogenous factors and toward exogenous ones. In other words, what happens *at* places depends more and more on what happens *among* places. The patterns on maps can be understood only in terms of the patterns of movement that create them. Therefore, geographers must study not just *stasis*, which is motionlessness, but also *kinesis*, movement. Geography doesn't just *exist;* it *happens*.

AUTARKY AND TRADE

The study of how various peoples make their living and what they trade with other peoples is **economic geography**, which is an important component of cultural geography. When the Happy people were isolated, they neither imported nor exported any goods. They were totally self-sufficient. Such economic independence is called **autarky**.

As peoples and regions come into contact with others, however, they begin to exchange goods. At first people export only whatever they have in surplus and have no use for, and they view imports as luxuries unnecessary to their way of life. Trade, however, entangles people. People see how imports can raise their standard of living, and to pay for the imports they want, they dedicate more local effort to producing items for trade. That dedication pays off, because when people specialize in producing an item, their skill in producing that item usually increases, and the quantities produced can increase rapidly.

Improvements in transportation and communication facilitate exchange, and as trade multiplies, people produce more for consumption elsewhere, and an increasing share of the things they use are produced elsewhere. Eventually, if the terms of trade are favorable, people dedicate most of their efforts to exports, and they rely on imports not only for luxuries but even for their necessities. They surrender their self-sufficiency, and they become dependent on trade.

How Geographers Study Trade

Geographers ask three questions about trading systems. First, how do the people at any specific place make their living? Second, where do specific economic activities locate? That is, what are each economic activity's **locational determinants**? Finally, how do these factors affect other aspects of cultural geography? The first two questions are complementary—they are almost the same question asked from two perspectives.

Geographers often approach the first two questions by mapping the production of various items. These maps indicate the quantity of the item produced in each country and may also indicate each country's share of world production. In 1989, for example, about 778 million metric tons of crude steel were produced in the whole world. The Soviet Union produced 21 percent of that, Japan 14 percent, the United States 12 percent, China 8 percent, West Germany 5 percent, and so on.

Production figures alone, however, do not reveal why any place is producing what it is producing, nor do they reveal where any country's production is being used. Countries may import or export steel, and others, such as the United States, may import and export different types of steel, or they may even find it profitable to import steel in one part of the country and export steel from another region. Therefore, after we have gathered production figures of any item, we also need to examine the balance of imports and exports to understand the distribution of the final use of any product. Geographers' statistics and maps of production must always be supplemented with statistics about trade and maps of trade.

Some places obviously produce an item for export. Brazil is the world's second largest producer of iron ore, but it uses only one-third of what it produces. The remaining two-thirds are exported in exchange for machinery, fuels, and other goods.

Through the late 1980s the Ivory Coast produced about one-third of the world's cocoa beans. Why? The people there did not eat the beans. Rather, the terms of trade made it profitable for the people of the Ivory Coast to export them in exchange for other goods, including even food. This is, to a great degree, how the people of the Ivory Coast made their living.

What explains this extraordinary concentration of cocoa beans in the Ivory Coast? What are the locational determinants for cocoa-bean production? Did the cultivation of cocoa beans originate there? No, it originated in Mexico many centuries ago. Could the concentration of cocoa-bean production in the Ivory Coast have been due to some unique local environmental advantage? For example, could the water be special? Obviously not. The climate and the soils of the Ivory Coast are fine for the cultivation of cocoa beans, but so are the climates and soils of many other places. The concentration of cocoa beans in the Ivory Coast resulted from political and economic decisions to specialize production. These decisions were made within a context of global production, marketing, and consumption trading patterns. Those decisions and choices may be called the economic, political, and diplomatic environments of cocoa-bean production, and they rival or even supercede the physical environment as locational determinants.

Regions or peoples have not always freely chosen to enter into the system of production and exchange. Colonialism forced specialized production on vast areas. In some areas the peoples were forced to buy goods, and they had to develop export specialties to pay for these goods. The British actually went to war to force the Chinese to buy opium from British India (1839–1842). Opium is an addictive drug, but many desirable goods often stimulate peoples' wishes to enjoy them. Thus, new **felt needs** are created, which trigger a **revolution of rising expectations**. People begin to assume the availability of desirable goods, and in that way they become more deeply involved in the web of trade and circulation.

The share of any country's territory that is devoted to export production may be small, but the shares of the national population and the national income that are involved are rising everywhere. Even in a country as large and as rich as the United States, the share of the nation's economic life that was accounted for by foreign trade tripled between 1955 and 1985, from 10 percent to 30 percent of total national output. Furthermore, even within individual countries the growth of cities is creating markets for food. This is drawing farmers out of subsistence autarky and into production for national urban markets. Very few people today are self-sufficient. Find a picture of the most isolated people you can, and chances are high that they will be wearing or carrying something that they did not make themselves.

The Impact of Trade on Other Aspects of Culture

A local culture influences the degree to which people participate in trade. A culture can either limit circulation or encourage it. Some peoples, such as the Dutch, traditionally have embraced circulation and exchange (Figure 5–20). Others, such as the Myanmarese (formerly Burmese) or the Albanians from 1945 to 1990, prefer isolation. The Amish in the United States compromise: They specialize their farm production and market their goods, but they severely curtail which products they will accept in return.

When a local economy is transformed from self-sufficiency to specialized production for export, a process usually accompanied by increasing dependence on imports, every aspect of local culture is affected. The way people make their living is in itself a key aspect of their culture, and all aspects of culture interact, so an economic change will affect such cultural elements as religion, family, and politics.

Trade As Diffusion Economic exchange is a form of cultural diffusion. Every object in trade is an artifact of the culture that originates it, and so the introduction of a new item into a different culture—a sale—reduces the buyer's cultural uniqueness.

Both production and consumption are more than simply material aspects of existence. Goods are building blocks of lifestyles, of social and cultural identity, and of self-consciousness. The old saying "You are what you eat" is a simplistic ecological view of humankind. You are also what you have, what you make, and what you do. You become what you wear; what you live in. Choosing goods is an act of self-definition. Changes in production and in consumption change culture.

One example of a threat to culture from international trade is the construction of EuroDisneyland in France and the distribution of Disney-related consumer goods that will result from this enterprise. French intellectuals fear its impact because they are proud and protective of their culture. The French minister of culture, who controls the second largest budget in the national government, refused to attend the opening of EuroDisneyland, which French intellectuals have dubbed "a cultural Chernobyl" after the nuclear facility that spewed poisonous radiation across the landscape of Ukraine.

Holland has a 400-year head start gearing up for global strategies.

The Netherlands' commitment to international commerce dates back centuries. It's a vital part of the Dutch history and culture and people.

It's also vital to the Dutch economy.

You can see it in the Dutch telecom, logistical and industrial infrastructures. You can see it in the Netherlands' research agenda, in the orientation of its professional services, in the multilingual Dutch workforce, and in industrial policies and other traits that most other European countries are still just exploring.

More than 1,500 North American companies are already taking advantage of the head start the Netherlands gives them.

Just in the last year, companies like AFA-Polytek, Apple, Bruce Foods, Compaq, General Mills, Halliburton, Hercules, Hewlett-Packard, Rank Xerox and Westinghouse have announced major new projects in the Netherlands.

And when you look at the rate of return for foreign companies investing in the Netherlands, you'll see that a head start can translate into heady rewards.

Where better to anchor the European portion of your company's global activities?

Netherlands Foreign Investment Agency

NEW YORK (212) 246-1434 • SAN MATEO (415) 349-8848 • LOS ANGELES (310) 477-8288 • OTTAWA (613) 237-5030

DIST

Please send me literature on the Netherlands: Europe's Gateway to 1992.

Complete this coupon and mail it to: Mr. Irwin de Jong, Executive Director
Netherlands Foreign Investment Agency, One Rockefeller Plaza, New York, NY 10020.

Name ______ Title ______

Company ______ Telephone ______

Address ______

City ______ State ______ Zip ______

HIT THE GROUND RUNNING

This material is published by Ogilvy Public Relations Group, which is registered as an agent of the Government of the Netherlands. It is filed with the Department of Justice where the required registration statement is available for public inspection. Registration does not indicate approval of the contents by the United States Government.

Figure 5–20. This advertisement, which was widely placed in international business magazines, demonstrates how aggressively the Dutch seek interaction and trade with other peoples. (Courtesy of Netherlands Foreign Investment Agency.)

Today cultural diffusion through the exchange of consumer goods is swamping the world. The cheapest way to manufacture and distribute consumer goods is in large quantities. This might mean blanketing several different culture realms with one product. The integrity of each culture realm, however, resists homogenization. Therefore, cross-cultural advertising and marketing of consumer goods present fascinating cultural-geographical questions. Why do some products have worldwide appeal, whereas others are successful only in geographically restricted markets? What patterns of culture realms and markets does this reveal? Can salespeople deliberately break down cultural differences? A few consumer products have already achieved almost global diffusion: Coca-Cola, Sony, Kodak, and Levi's. Why do people around the world dress differently, eat different foods, and speak different languages, but almost everywhere emulate the Marlboro Man? Is there any place where people refuse cola drinks? (See Figure 5–21) (Yes, Utah. Why there? Mormons do not drink caffeine.)

Mass manufacturing and marketing can lower the cost of goods and raise standards of living, but at some sacrifice of cultural identity. Cultural diffusion through consumer goods is arguably the most significant method of cultural diffusion today. We must remember, however, that consumer goods are only one aspect of culture. In some places people wear blue jeans to political demonstrations against the United States. They like blue jeans, but that does not mean that they like everything about U.S. culture.

Trade and Lifestyle Changes As countries become dependent on trade, their standard of living becomes dependent on their competitiveness. Today nations alter their way of life to compete in world trade. International trade talks take up more and more issues of domestic life: pollution control, human rights, farm and industrial subsidies, patents and trademarks, government purchasing, industrial standards, and investment policies, including national and regional plans. When one country imports another's products, it is also importing the social, economic, and environmental standards embedded in those products. Trade triggers pressures to "harmonize" domestic conditions.

When an economy is tied into trade, even the selection of items that are produced for local use becomes dependent on whether those items are exportable. Items of local culture that attract an international market may be changed to increase exports. Local items that cannot be exported may no longer be produced, even for local consumption. In that way people slowly surrender their traditional material culture, and they may not even be aware of it.

Countries alter their domestic lifestyles to improve their competitive position in world trade and also to attract international investment. **Capital** (that is, "extra" money available for investment) is in limited supply. It migrates (gets invested) wherever someone is willing to pay the most to borrow it (interest rates are highest), and it gets invested in places and activities that offer the greatest potential for profit. As we enter the 1990s, Germany and Japan have a lot of capital, but Germany will probably use its capital to develop its eastern states (the former East Germany), and Japan will invest its capital in new infrastructure at home and in new factories overseas. The United States, Eastern Europe, the Commonwealth of Independent States (formerly the Soviet Union), Africa, and Latin America will compete to attract capital. Both Kuwait and Iraq need to be rebuilt after the devastating 1991 war, and Eastern Europe's polluted environment needs to be cleaned up. These competing global demands for capital mean that U.S. college students will have to pay higher interest rates on their tuition loans, and U.S. homebuyers on their mortgages. The world suddenly seems to be a very small place.

Figure 5–21. Some consumer goods practically blanket the Earth. (Courtesy of Nancy and Douglas Paley.)

LOCATION

Location is a unique attribute of any place. Geographers distinguish between **absolute location** and **relative location**. We discussed in Chapter 1 how absolute location pinpoints a spot on the globe—its latitude and longitude.

The relative location of a place, by contrast, helps us understand how it interacts with the rest of the world, and this can be as important as its absolute location or site. Relative location, or situation, determines *accessibility*, indicated by such terms as nearer and farther, easier or more difficult to reach, between, and on the way or out of the way. Accessibility can be a resource as valuable as mineral deposits or fertile soil. A convenient relative location between or among places can help a place develop as a trade center.

The relative location and accessibility of any place can change (see Figure 5–22 on page 150). The opening of a new highway stimulates the development of new housing and new roadside services such as shopping malls. At the same time it can choke older shopping malls by rerouting traffic past them. Similarly, at a larger scale,

CRITICAL THINKING: *Tombouctoo*

In modern slang "Timbuktoo" is a synonym for "nowhere," and few Americans even know that there really is such a place. In fact, Tombouctoo is a city in Mali at absolute location 16° 46′ N, 3° 01′ W. It plays no major role in today's world, but in past centuries it was a major urban center. Tombouctoo is situated on the Niger River where the river bends farthest north into the Sahara Desert. This puts Tombouctoo on a cultural border between nomadic Arab peoples to its north and the settled Black peoples to its south. The city was settled by the Taureg people from the North in 1087, and as a major contact point between north and south, it prospered as a market for slaves, gold, and salt and as a point of departure for trans-Saharan caravans to North African coastal cities. From these cities Tombouctoo's fame as a center of almost mythical riches spread around the Mediterranean. It was a center of Islam in West Africa, and it boasted a great university.

In the fifteenth century, however, Portuguese expeditions sailing around West Africa outflanked the trade routes on which Tombouctoo depended. The city declined rapidly. It was captured by the Moroccans in 1591, and by the time the French captured it in 1893, it was little more than an extensive ruin. Today it is a small city of 10,000 people who live surrounded by few remnants of the city's former greatness.

Questions

Compare Tombouctoo's location with that of Trabzon on the Black Sea in modern-day Turkey. Among which great empires was Trabzon once a major contact and trading point? How important is it today? Do you know of any other once-great cities that have declined as trade routes bypassed them? Conversely, which great ports have developed on the west coast of North America as trade with Asia has expanded during the twentieth century?

The geographer Ellen Churchill Semple opened her book *American History and Its Geographic Conditions* with the sentence: "The most important geographical fact in the past history of the United States has been their location on the Atlantic opposite Europe; and the most important geographical fact in lending a distinctive character to their future history will probably be their location on the Pacific opposite Asia." This was written in 1903. Has the twentieth century proved her correct?

The fabulously rich medieval king of Mali is shown here enthroned by the city of Tombouctoo. This map is from the Catalan Atlas (c. 1375), which was owned by the king of France. A search for the gold of Mali was one impetus for the ensuing European exploration and colonization of Africa. (Courtesy of Bibliotech Nationale, Paris.)

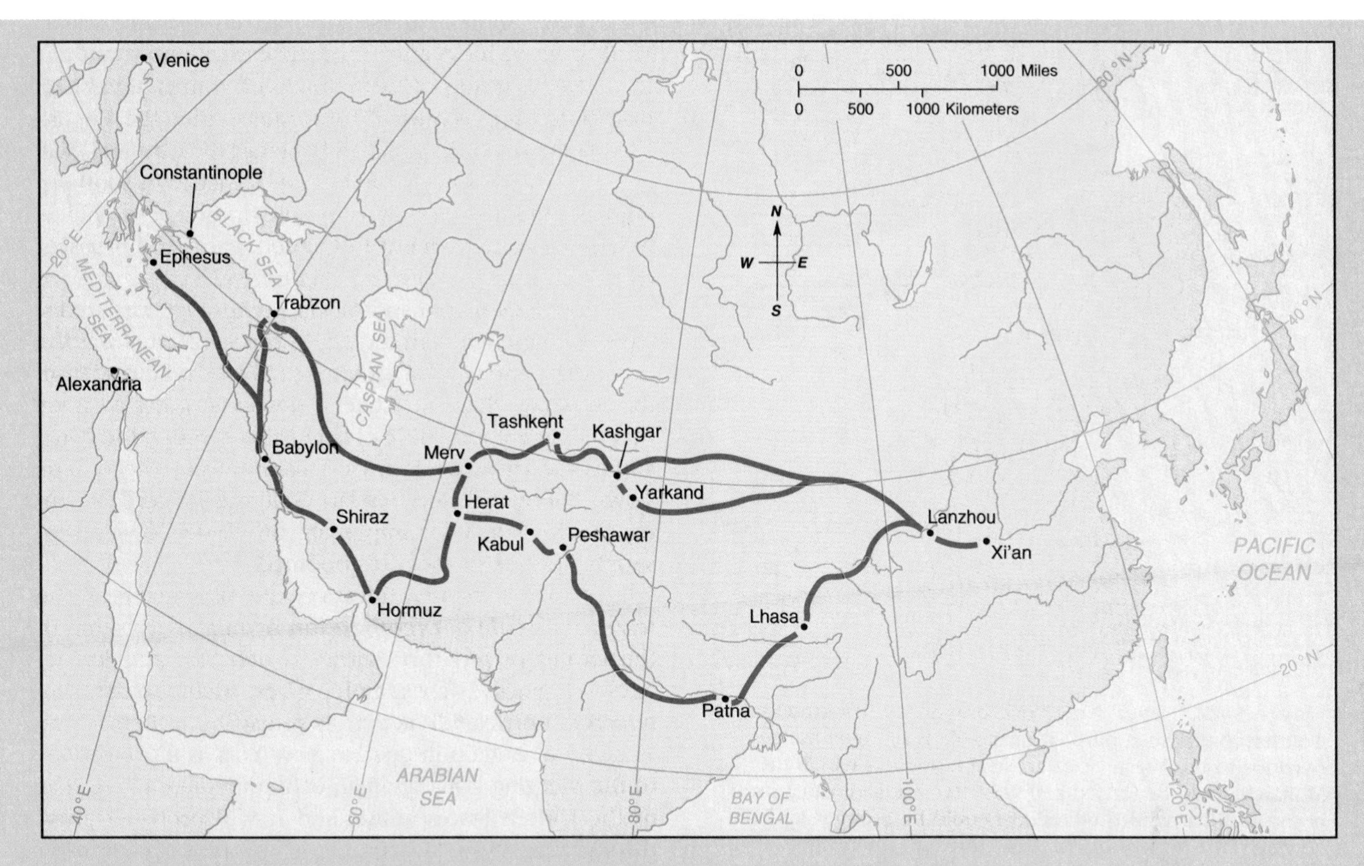

Venice, Trabzon, Kashgar, and Merv were great cities on major trade routes in the fifteenth century, but when sea routes opened around Africa and Asia, these and many other cities were suddenly "on the back road." Few ever recovered their former economic strength.

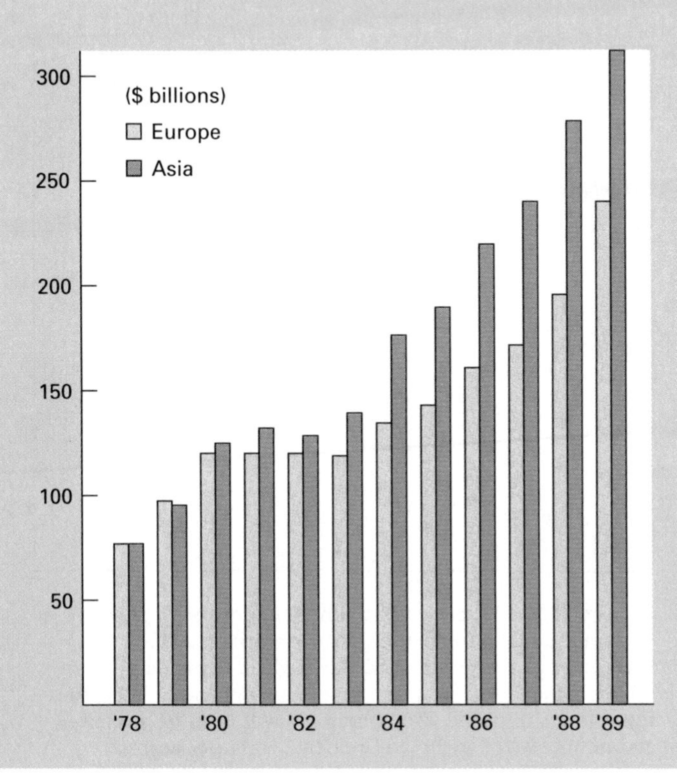

The total U.S. trade with Europe and Asia. The slow but steady rise of U.S. trade with Asia is just one piece of evidence confirming Ellen Churchill Semple's prediction of 1903. (Courtesy U.S. Dept. of Defense, as reprinted in *Forbes*, October 1, 1990.)

Figure 5–22. Utica, New York, originally prospered as a frontier commercial outpost in the Mohawk River valley corridor to the West. Utica arose at a point at which the Mohawk could be forded and was later bridged. Utica itself became industrialized when coal could be brought up from Pennsylvania by way of the 1837 Erie and Chenango Canal, whose route was later paralleled by the tracks of the New York, Ontario & Western Railroad visible on the map. With the opening of the Erie Canal and later the New York Central Railroad, industry developed around Utica Harbor and in the big buildings visible on the map downtown along the railroad. In the nineteenth century workers lived close to these in-town factories, and downtown Utica was a busy place. Virtually all new industrial development since World War II has spread out into the suburbs, taking advantage of the spaciousness and the accessibility provided by modern interstate and local highways. New buildings can be seen to the north across the river from the old downtown. Housing has also spread into the suburbs, and shopping and office malls have pulled more activities out of the downtown area. At any geographic scale of analysis, whether one is studying the location of a city in a country or focusing on locations within that city, new means of transportation redistribute accessibility. They open new areas to development and may choke off areas that once were favored sites.

great cities rise, and countries prosper if they are advantageously situated within national or international patterns of trade. Isolated areas usually fall behind and stagnate. When the Europeans learned how to sail around Africa to reach Asia, they brought about one of history's greatest redefinitions of relative locations. Europeans drew to the African and Asian seacoasts much of the commerce that had previously moved only by long and difficult overland caravan journeys. New seacoast cities sprang up, such as Bombay in India, Rangoon in Burma (today known as Yangon in Myanmar), and Hong Kong on the coast of China. Major cities in the interior of the Eurasian and African continents were suddenly "on the back road." Some of these, such as Kashgar and Samarkand, have never recovered their earlier importance.

The opening of new transport routes always redistributes accessibility. The tunnel under the English Channel between England and France (the Chunnel) will redistribute land values and many activities in southern England and northern France. Lille, in northern France, will be more convenient to London, Paris, and Brussels, and it will surely grow (see Figure 5–23).

Changes in transportation technology also redistribute accessibility. In the 1950s the Suez Canal was the principal route of oil shipments to Western Europe from the Mideast. New oil tankers, however, grew so large that they were incapable of squeezing through the Canal and had to sail around Africa. This rerouting placed Cape Town "between" the Persian Gulf and Rotterdam, Europe's chief port for importing oil. Cape Town's ship supply and repair facilities boomed.

A place's relative location can also change if the *territorial scale of organization* of an activity—that is, the extent of territory within which that activity occurs—changes. For example, when trade barriers between countries fall, activities redistribute themselves. The city of Buffalo in upstate New York is more central to the merging U.S.-Canadian economy than it is to that of the United States alone, and it will probably grow through the 1990s. All activities in Europe are readjusting from a national scale of organization to a supranational scale represented by the European Community.

Conversely, when territorial organization fragments, accessibility can be reduced. Calcutta was a pros-

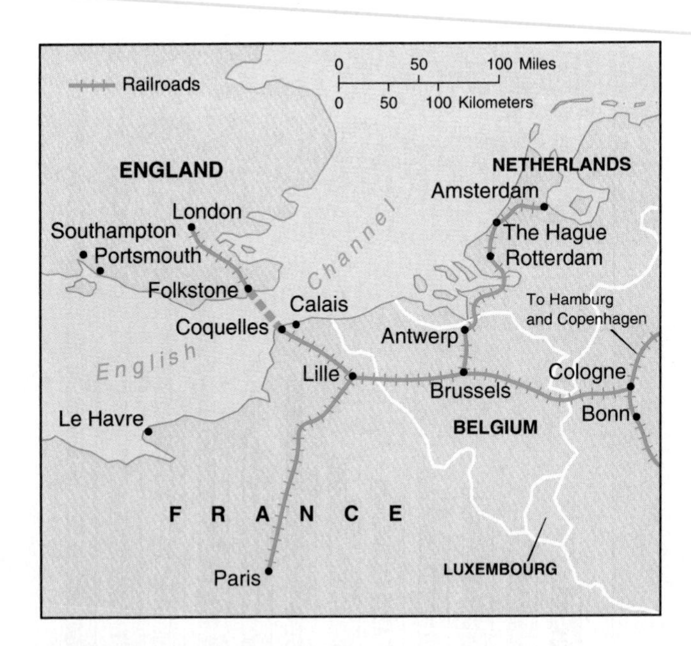

Figure 5–23. The idea of a tunnel under the English Channel has long been alternatively a dream and a nightmare to the French and the English, depending upon whether they were at peace or at war with each other. In these times of peace, the "Chunnel" will soon be a reality, and new flows of traffic will relocate activities across northern Europe.

perous city when Bangladesh and India were united as one colonial region. Since Calcutta's service area—the state of Bengal—was split into two countries, however, Calcutta has experienced a decline in access that has resulted in devastating poverty.

HUMAN ACTIVITIES DEFINE PATTERNS

All human activities find territorial expression, and maps of culture realms, of regions of economic specialization, or of political jurisdictions reveal current distributions of human activities. Human activities, however, are not static. They are dynamic, continuously organizing and reorganizing, forming and re-forming. **Historical geography** is the subfield within geography that studies the geography of the past and how geographical distributions have changed through time. The organization and use of territory is dynamic; no pattern is stable. Distributions and patterns are disrupted and reshaped repeatedly, just as the patterns in a kaleidoscope are. A map of human activities at any time is like a snapshot of the dancers in an intricate minuet. We cannot understand it if we do not know the principles of its choreorgaphy or even hear the music. For geographers to answer the question, Why there? requires the study of processes, the music, and the choreography.

Geographers always need to understand what makes things move and redistribute themselves. Some force is always required to overcome inertia. In Chapter 2 we saw that variations in air pressure cause the winds to blow. Air flows from points of high pressure to low pressure. Gravity causes water to flow to the sea. In studies of human activities, variations in job opportunities or in political liberties cause people to migrate. Variations in interest rates or in profitability cause capital to flow, and the places where capital is invested enjoy economic development. People travel to holy sites by the force of their religious beliefs. Ideas may flow under the power of conversion. Variations in places' natural endowments trigger trade flows. Produce flows to markets, raw materials to manufacturing sites, tourists to attractions. Variations in transport capability or freight rates direct flows of traffic. All these forces overcome geographic inertia.

Geographical inertia has also been reduced. The cost and time of moving almost anything—people, food, energy, raw materials, finished goods, capital, information—has steadily fallen, releasing many activities from the constraints of any given location (see Table 5–1). Few things are quite so fixed in place as they were in the past. If an activity can move or relocate freely, we call it a **footloose activity**. France's President Francois Mitterand likes to say that "History has accelerated." Geography has too.

Being familiar with the distributions of human activities is important, but it is not sufficient to understand geography. These distributions are constantly changing. If you learn the reasons why things get distributed the way they do, however, then you will have learned something useful for the rest of your life. You will always be able to go to a library to look up the latest reports and the most recent numbers. Get ready: The future does not look like the past.

THE GLOBAL DIFFUSION OF EUROPEAN CULTURE

Despite the rich variety of indigenous local cultures around the globe, the world is increasingly coming to look like one place. In consumer goods, architecture, industrial technology, education, and housing, the European model is pervasive. To ethnocentric Westerners and to westernized locals, this trend might seem natural. To them, this may be "modern" life, or "progress," or "development."

A dispassionate observer, however, might expect more diversity, more styles and models of development. Why are so many people around the world imitating Western examples and adopting aspects of Western cul-

TABLE 5–1 ***Transport and Communication Costs Expressed in 1990 Dollars***

Year	Average ocean freight and port charges per short ton of import and export cargo	Average air transport revenue per passenger mile	Cost of a 3-minute telephone call New York to London
1920	$95	na	na
1930	$60	$0.68	$244.65
1940	$63	$0.46	$188.51
1950	$34	$0.30	$ 53.20
1960	$27	$0.24	$ 45.86
1970	$27	$0.16	$ 31.58
1980	$24	$0.10	$ 4.80
1990	$29	$0.11	$ 3.32

Courtesy Institute for International Economics.
Sources: *Historical Statistics of the United States*; *Statistical Abstract of the United States*.

Figure 5–24. By studying this picture from a newsmagazine, can we tell who or where this man is? He is dressed as a modern, westernized man anywhere in the world today. The photo suggests that he knows something about modern computers; the one beside him is a U.S. brand, although many key parts are manufactured in Singapore. The buildings behind him are in a Roman classical or Palladian style. This picture seems entirely "Western." In fact he is Huzur Saran, an Indian computer scientist. He studied in the United States but works today at the Indian Institute of Technology. The photo was taken in New Delhi, the British-built, former imperial capital of India. (Courtesy of Dilip Mehta/ Contact Press.)

ture? The widespread adoption of the Western way of life is replacing both the positive and negative features of other cultures. It is the most pervasive example of cultural diffusion in world history (see Figure 5–24).

European Voyages of Contact

Europeans came to play a central role in world history and world geography because it was they who paved the way for the modern system of global interconnectedness. In the fifteenth century the great cultural centers of the world—the Inka and Aztec empires in the Western Hemisphere, the Mali and Songhai in Africa, the Mughal in India, Safavid Persia, the Ottoman Turks, the Chinese, and all the lesser empires and culture realms—were still largely isolated from one another (see Figure 5–25).

The European voyages of exploration and conquest connected the world (see Figure 5–26 on page 156). It was not inevitable that the Europeans would be the ones to do this. The Chinese were actually richer and more powerful than the Europeans, and earlier in the fifteenth century they had launched great fleets of exploration that had reached the east coast of Africa. We can hardly imagine how different world history and geography would be had the Chinese continued their initiatives and gone on to explore and conquer Africa and the Western Hemisphere. But they did not. Instead, for internal political reasons, the Chinese withdrew into themselves, and the initiative to draw the world together was left to the Europeans.

The first European initiative was the conquest of the city of Ceuta on the north coast of Africa in 1415 by Prince Henry of Portugal. He learned about trans-Saharan caravan routes and the riches of West Africa, which inspired him to found a naval academy at Sagres (see Figure 5–27 on page 156). Improvements in sailing technology allowed the Portuguese to sail down the west coast of Africa to reach beyond the Sahara Desert. Prince Henry, known as "The Navigator," launched the era of European seaborne colonial empires. Soon European powers were racing for colonies.

At the same time, the Russians forged eastward overland from their homeland in Europe west of the Ural Mountains. The Russian Empire pushed across Siberia to the Bering Strait, crossed over into Alaska, and eventually established colonies down the North American west coast as far south as today's California. This European outreach triggered the **Commercial Revolution,** between about 1650 and 1750. A tremendous expansion of trade followed the development of the first ocean-going freighters that could carry heavy payloads over long distances. The evolution of superior ships was paralleled by the evolution of superior naval gunnery and of additional useful technologies such as clocks, which were perfected to determine longitude at sea.

This first era of outreach ended with the death of the British explorer Captain James Cook in 1779. By then, for the first time in world history, Europeans could draw a fairly accurate outline map of all the world's continents and islands. The world was comprehended as it never had been before.

Expansion and Cultural Diffusion From that time on, Europe did not actually originate every "modern" idea and then impose it on the rest of the world. Europeans learned from others, too, and then transplanted around the world the ideas they had adopted from elsewhere. "Knowledge is power," said Sir Francis Bacon, and because the Europeans, and not the Chinese or any other group, had contacted all other civilizations, Europe became the clearinghouse of world information and products. Global diffusion was fixed hierarchically, with Europe as the apex. Europe became cosmopolitan, that is, familiar with many parts of the world. Other peoples, no matter how great their native civilizations were, remained more localized in the world Europe was creating. In a hierarchical system of diffusion, as we have seen, more information flows up and down the hierarchy than among the places at lower levels.

The redistribution of goods Consider, for example, what Europeans did with agricultural products. They took sugar cane from Asia and planted it in the Caribbean region. They replanted bananas from Southeast Asia to South America, cocoa from Mexico to Africa, and coffee from Arabia to Central America. All these foods remain major products of international trade and important factors in the export economies of many countries.

Europeans not only relocated the production of many goods around the world, they also introduced many products into world trade, both for the European home market and for other overseas markets that they pioneered. Europeans introduced many Indian goods, for instance, into China, and South American products into Africa and Asia. An Englishman stole tea plants from China and cultivated them in Calcutta, leading to the development of a tea industry in India and Ceylon (today known as Sri Lanka). Europeans created world markets, and they profited by controlling every stage: production, transportation, and marketing.

The redistribution of people and ideas Europeans also redistributed people and ideas. The culture of North America today is considered Western, and yet it draws on a diverse blend of contributions. The Native American cultures were overlain by several European and African traditions, with additional significant inputs by various Asian peoples. The story of Coca-Cola, one of the world's most familiar consumer products, exemplifies cultural blending. The popular drink was formulated in Atlanta, Georgia, and today symbolizes westernization so powerfully that the large-scale infusion of Western products into the non-Western world is often referred to as the "Coca-colonization of the world." The two original ingredients from which Coca-Cola takes its name, however, are coca, a Quechua Native American word for the South American tree whose leaves supply a stimulating drug (used today to make cocaine and crack), and cola, a Mandingo word for the West African nut that supplied the other original stimulating drug ingredient. Westerners borrowed the knowledge of both ingredients from their far-flung hearth areas, combined them, and marketed the drink worldwide.

Westernization has come to be synonymous with modernization, but modern world culture is by no means exclusively a Western product. For Westerners to think that it is, is presumptuous. For non-Westerners to think that it is, is to miss the many contributions of non-Western peoples. In a system of hierarchical diffusion, however, the direct source of much information for most non-Western places is the West, and this causes resentment.

The Industrial Revolution

Europe pulled ahead of the rest of the world economically as it underwent the tremendous transformation of the **Industrial Revolution**. Between about 1750 and 1850, Europe evolved from an agricultural and commercial society to an industrial society relying on inanimate power and complex machinery. Again, we cannot fully explain why Europe rather than any other part of the world first experienced this transformation, but we can list several factors that enabled Europe to industrialize. The voyages of discovery and conquest caused an influx of precious metals and other sources of wealth, stimulating industry and a money economy. The expansion of trade encouraged the rise of new institutions of finance and credit. In the mid-sixteenth century the joint stock company was developed. This was a way many investors could share both potential profit and risk in new enterprises. The creation of markets where stocks could readily be bought and sold granted capital new *liquidity*, which means easy conversion from one form of asset to another. This created, in the words of English writer Daniel Defoe, "strange unheard-of Engines of Interest, Discounts, Transfers, Tallies, Debentures, Shares, Projects."

In 1709 Abraham Darby first smelted iron ore using coke instead of charcoal, and in 1769 James Watt designed the steam engine, which multiplied the energy available to do work. Subsequent inventions and technical innovations in manufacturing, applied first to textiles and then across a broad spectrum of goods, dramatically increased productivity. Factory and industrial towns sprang up, canals and roads were built, and later the railway and the steamship expanded the capacity both to transport raw materials and to send manufactured goods to markets (see Figure 5–28 on page 157). New methods of manufacturing steel, chemicals, and machines played important parts in the vast changes. These innovations occurred first in Great Britain and subsequently spread to the continent of Europe and to the United States.

Beginning in the eighteenth century, Europe also first experienced an **Agricultural Revolution**. This de-

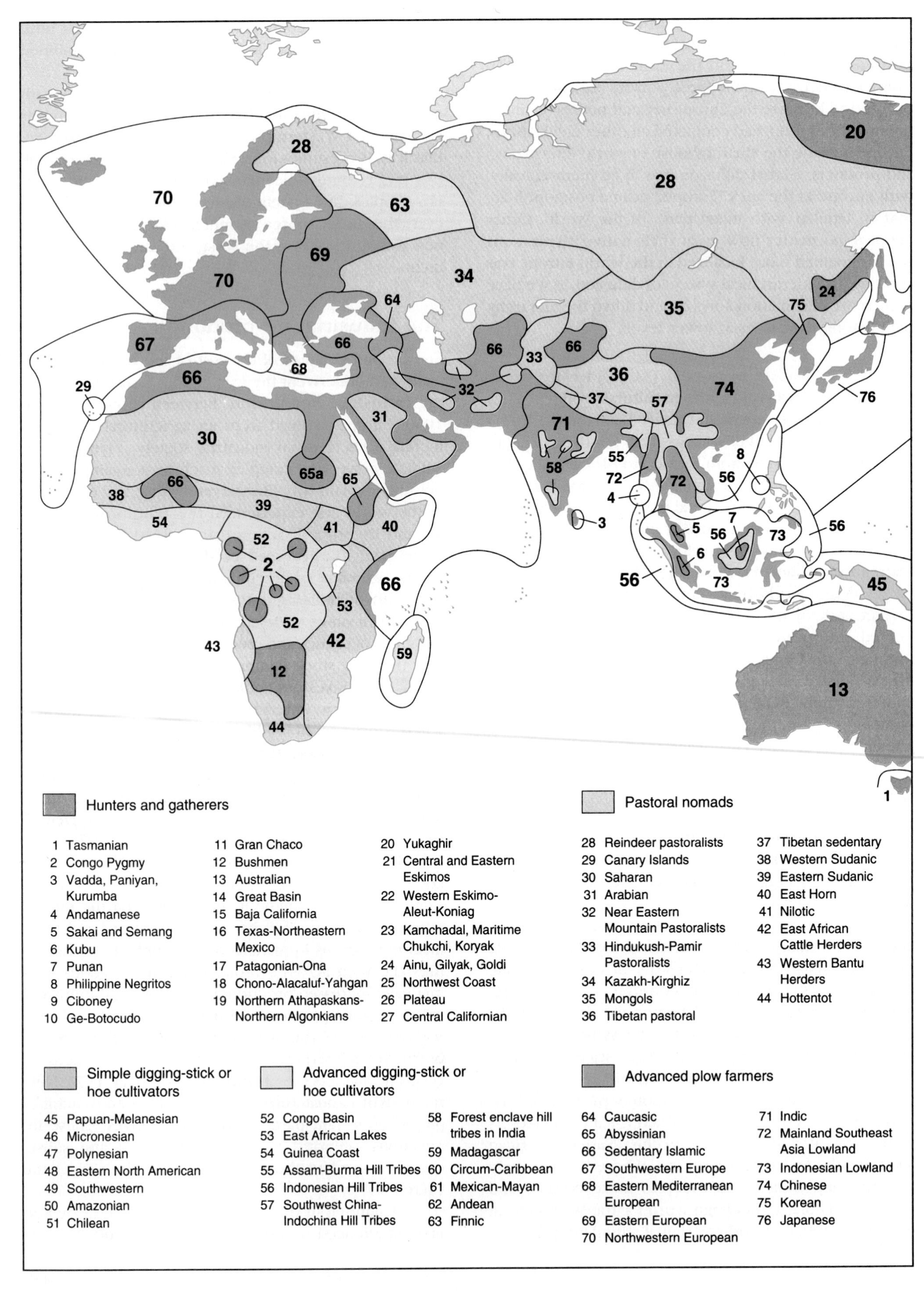

Hunters and gatherers
1 Tasmanian
2 Congo Pygmy
3 Vadda, Paniyan, Kurumba
4 Andamanese
5 Sakai and Semang
6 Kubu
7 Punan
8 Philippine Negritos
9 Ciboney
10 Ge-Botocudo
11 Gran Chaco
12 Bushmen
13 Australian
14 Great Basin
15 Baja California
16 Texas-Northeastern Mexico
17 Patagonian-Ona
18 Chono-Alacaluf-Yahgan
19 Northern Athapaskans-Northern Algonkians
20 Yukaghir
21 Central and Eastern Eskimos
22 Western Eskimo-Aleut-Koniag
23 Kamchadal, Maritime Chukchi, Koryak
24 Ainu, Gilyak, Goldi
25 Northwest Coast
26 Plateau
27 Central Californian
Pastoral nomads
28 Reindeer pastoralists
29 Canary Islands
30 Saharan
31 Arabian
32 Near Eastern Mountain Pastoralists
33 Hindukush-Pamir Pastoralists
34 Kazakh-Kirghiz
35 Mongols
36 Tibetan pastoral
37 Tibetan sedentary
38 Western Sudanic
39 Eastern Sudanic
40 East Horn
41 Nilotic
42 East African Cattle Herders
43 Western Bantu Herders
44 Hottentot
Simple digging-stick or hoe cultivators
45 Papuan-Melanesian
46 Micronesian
47 Polynesian
48 Eastern North American
49 Southwestern
50 Amazonian
51 Chilean
Advanced digging-stick or hoe cultivators
52 Congo Basin
53 East African Lakes
54 Guinea Coast
55 Assam-Burma Hill Tribes
56 Indonesian Hill Tribes
57 Southwest China-Indochina Hill Tribes
58 Forest enclave hill tribes in India
59 Madagascar
60 Circum-Caribbean
61 Mexican-Mayan
62 Andean
63 Finnic
Advanced plow farmers
64 Caucasic
65 Abyssinian
66 Sedentary Islamic
67 Southwestern Europe
68 Eastern Mediterranean European
69 Eastern European
70 Northwestern European
71 Indic
72 Mainland Southeast Asia Lowland
73 Indonesian Lowland
74 Chinese
75 Korean
76 Japanese

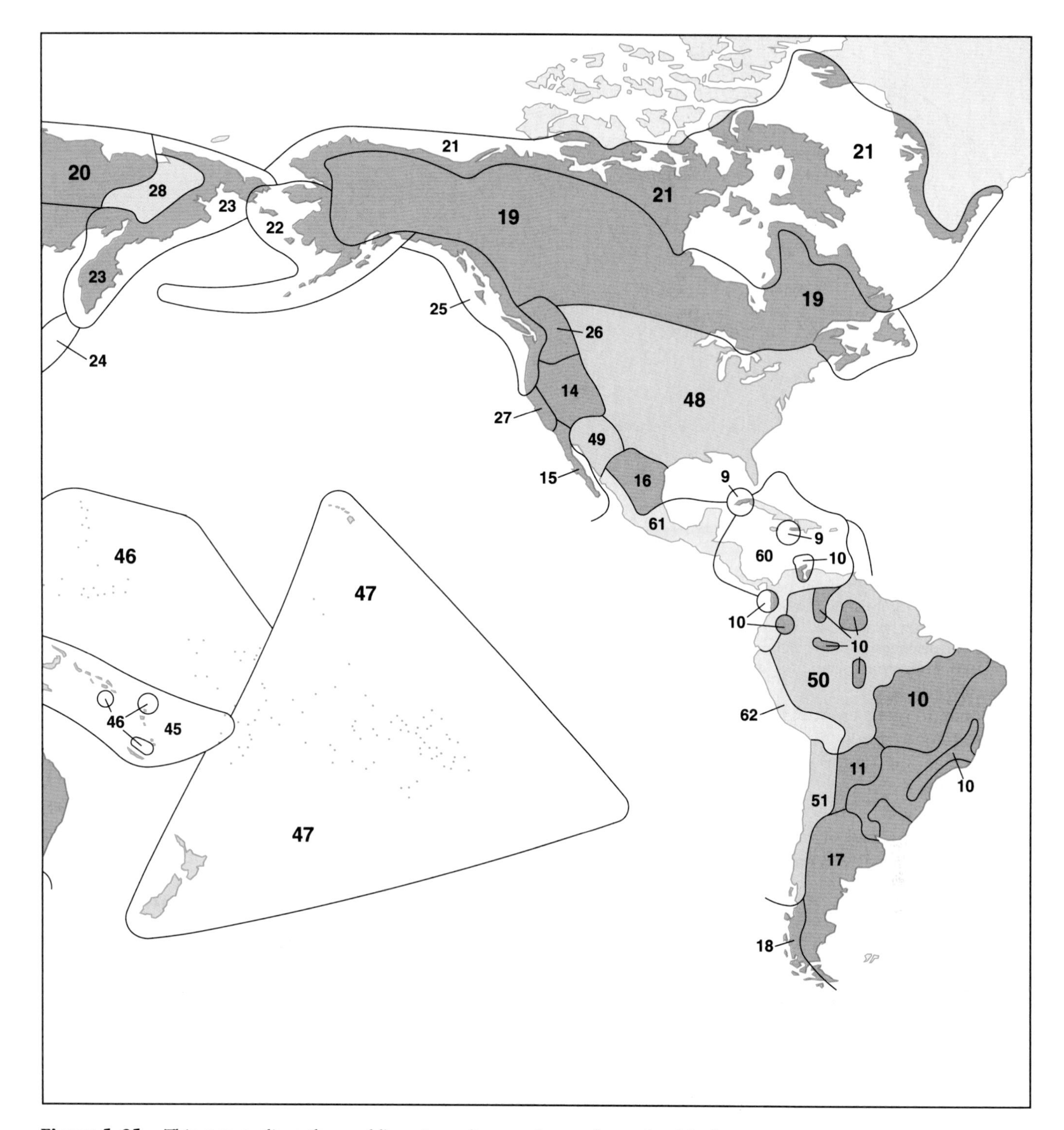

Figure 5–25. This map outlines the world's major culture realms as they existed in the year 1500. There was some interaction among them, but their isolation from one another was significant enough to allow each a distinct cultural individuality. The European voyages of exploration and conquest would soon scramble this map. (Adapted from G. W. Hewes, "A Conspectus of the World's Cultures in 1500 A.D.," *University of Colorado Studies*, no. 4, 1954.)

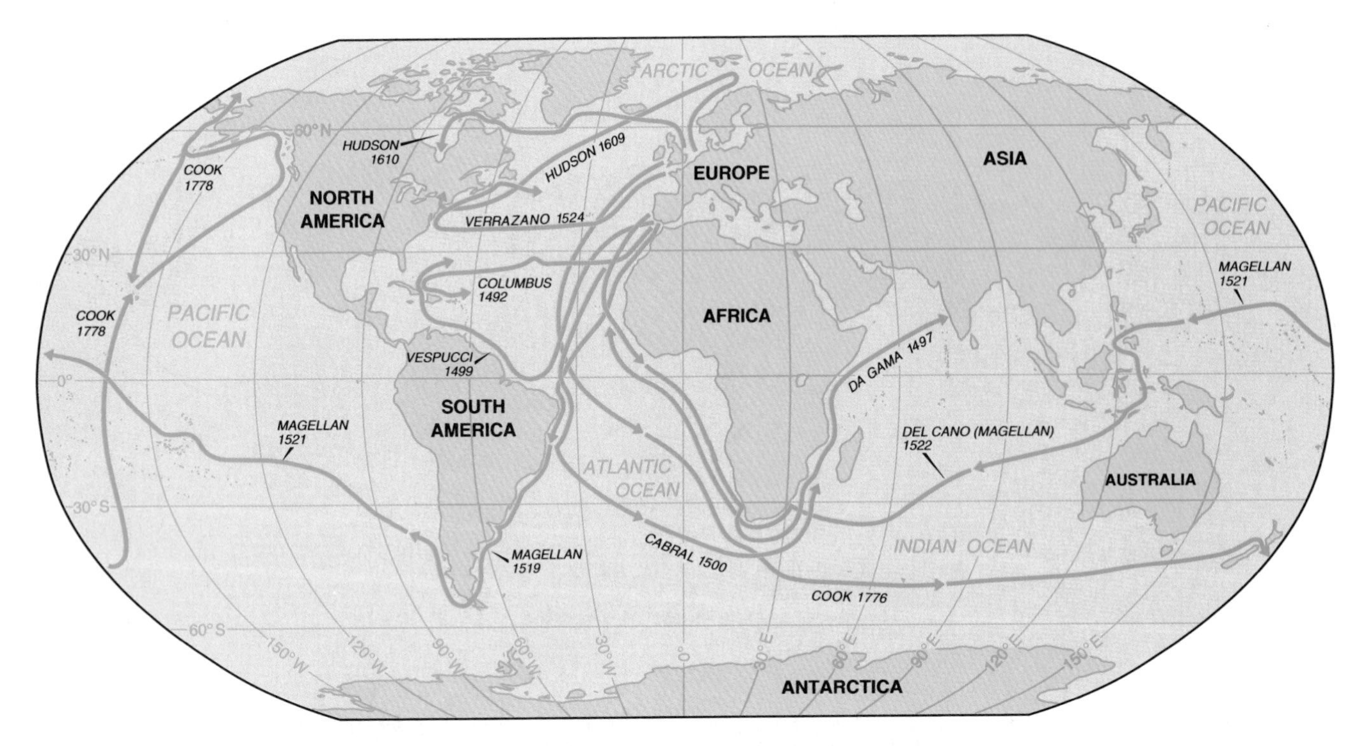

Figure 5–26. Each of the European voyages of exploration was a daring enterprise. The Portuguese first established a string of bases around the coasts of Africa and Asia across to Japan, where they established a trading post in 1543. Other European nations followed, and the race for discovery of new riches, new converts to Christianity, and help against Christianity's powerful Islamic foe was launched.

velopment, to be examined in detail in Chapter 7, both increased food production and released agricultural workers from the land, thereby creating a supply of industrial labor.

Europe and European settlements around the world drew far ahead of any other places and peoples in productive capacity. As recently as 1800 the per capita incomes of the various regions of the world were close. If we index the Western European per capita income in the year 1800 as 100 units of wealth, then estimated per capita income in North America was 125, in China 107, and in the rest of the non-European world 94. By 1900, however, European and North American incomes were several times those of non-Western peoples.

Figure 5–27. These Ebrie women of the Ivory Coast are wearing contemporary gold items like those that first enticed the Europeans to sail around the Sahara to explore new routes to West Africa. (Photo by Eliot Elisofon, National Museum of African Art, Eliot Elisofon Archives, Smithsonian Institution.)

Commercial Contacts and Economies

At the beginning of the age of the European voyages, European demand for foreign products such as spices, sugar, fruits, and North American furs grew rapidly. Soon the Europeans were no longer content to trade with native peoples for these goods, and the Europeans themselves established overseas estates and plantations and applied large-scale techniques to specialized production.

European commercial plantations at first concentrated along the coasts, but in the nineteenth century the railroad allowed penetration of the continental interiors to superior agricultural lands or, later, as Europe industrialized, to mineral deposits. The world's railway network expanded from 125,000 miles (200,000 km) in 1870 to over 625,000 miles (1 million km) by 1900. What had been European treaty ports and coastal enclaves became vast inland empires (see Figure 5–29). The steamship also greatly increased the possibility of transporting African minerals, and new quantities of minerals supplied Europe's multiplying factories. Between 1840 and 1870 the world's merchant shipping rose from

Figure 5–28. This is the world's first iron bridge, built over England's Severn River in 1779. It demonstrated iron's strength, and the bridge's bold design invited other uses for the new material. Its builder, John Wilkinson, launched the first iron boat in 1787, and when he died, he had himself buried in an iron coffin. (Courtesy of the British Tourist Authority.)

10 million tons to 16 million tons, and then it doubled in the next 40 years.

New cities emerged in the non-European world as coordinating centers for these commercial activities; new ports sprang up along the seacoasts. The major seaport cities from the Strait of Gibraltar, around Africa, across the Indian Ocean, and throughout South Asia are the products of European contact (see Figure 5–30). The same is true for most of the port cities of the Western Hemisphere. The railroads and associated commercial economies at first affected only a small percentage of the population, but over time an increasing share of the population and territory were drawn into the emerging global economy. In some countries, however, the modern commercial economy still overlies a traditional subsistence economy. This economic dichotomy is called **dualism**. There may be little exchange between the two economies unless refugees from the collapse of the traditional economy flee to the slums of the modern cities.

Political Conquest

In two waves of exploration and conquest—the first extending from 1415 to 1779 and the second occurring at the end of the nineteenth century—Europe (and, in the second period, the United States) conquered most of the rest of the world. One of the original reasons for this conquest was the wish on the part of European nations to protect their investments in foreign lands and to control these lands as markets for themselves. European countries divided up the rest of the world to regulate their own rivalry. Superior military power guaranteed their ascendancy over the natives. The Europeans' confidence in their military superiority was expressed by British satirist Hilaire Belloc, who wrote in honor of the newly invented machine gun: "Whatever happens we have got/ The Maxim gun, and they have not."

The United States and most of Latin America won independence between 1775 and 1825, but between 1875 and 1915 about one-quarter of the Earth's land surface was distributed or redistributed as colonies among a half-dozen imperialist states. Of all the countries in the world today, the only ones never ruled by Europeans or by the United States are Turkey, Japan (although occupied by the United States from 1945–1952 and its Constitution imposed), Korea (which was ruled by Japan from 1910 to 1945 and remains divided today), Thailand (left as a *buffer state* between the French and English empires), Afghanistan (a buffer between the English and Russian empires), China (which was nevertheless divided into foreign "spheres of influence"), and Mongolia (ruled by China). The peoples of every other country on the world map experienced European or American imperial rule.

European political domination imposed European concepts of law that drastically changed native societies and still linger today. Among the most important European ideas were those of property rights and landownership. Before the Europeans came, land was generally considered a good held in common for all members of the community. Local political leaders apportioned and adjudicated land use and tenure by customs that brought the community together. Neither the leaders nor anyone else owned land. The idea that any individual could own land and determine alone how to use it was largely unknown. Europeans introduced their idea of "ownership" of an "estate" that could be bought, sold, or mortgaged by individual contract. European-style ownership introduces liquidity, but it dissolves social cohesion. Land was no longer a common good, and the regulation of its use was no longer a shared community affair.

Native-American tribal leaders, for instance, did not by their own customs have the right to transfer land out of tribal control, and they frequently did not actually understand what Europeans wanted. In the history of the United States innumerable wars were sparked when Native Americans returned to hunt or harvest unoccupied land that the Europeans insisted the Natives had sold. Throughout Africa and Asia the Europeans often simply

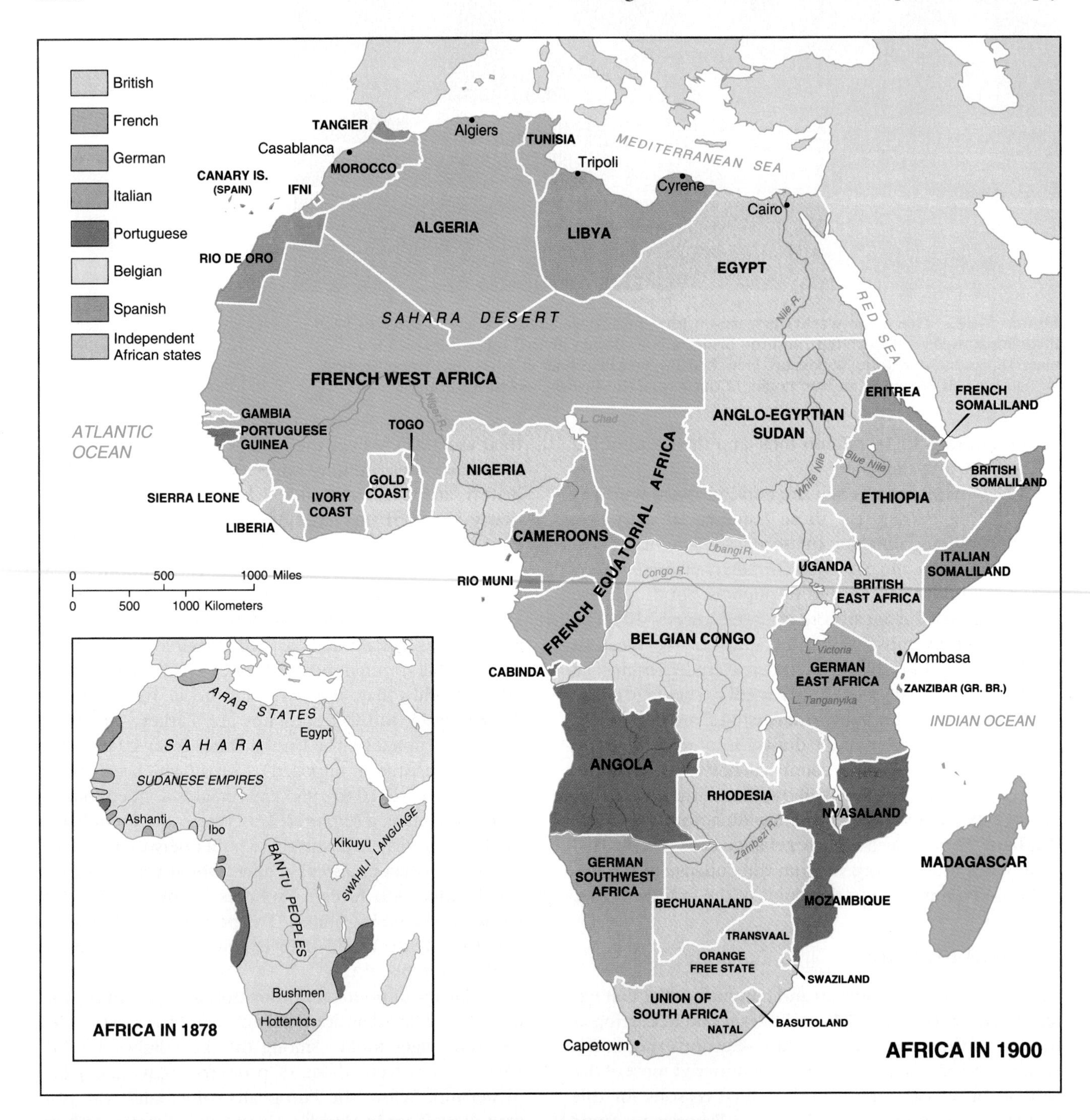

Figure 5–29. European treaty ports and trading stations along the African coast were transformed into inland empires when the Europeans decided to parcel out the continent to prevent competitive war among themselves. Industrial Europe demanded African raw materials, and the railroad allowed the Europeans to draw them out of the African interior. The native African peoples were not consulted in the political reapportioning, which was completed at a conference in Berlin in 1884–85.

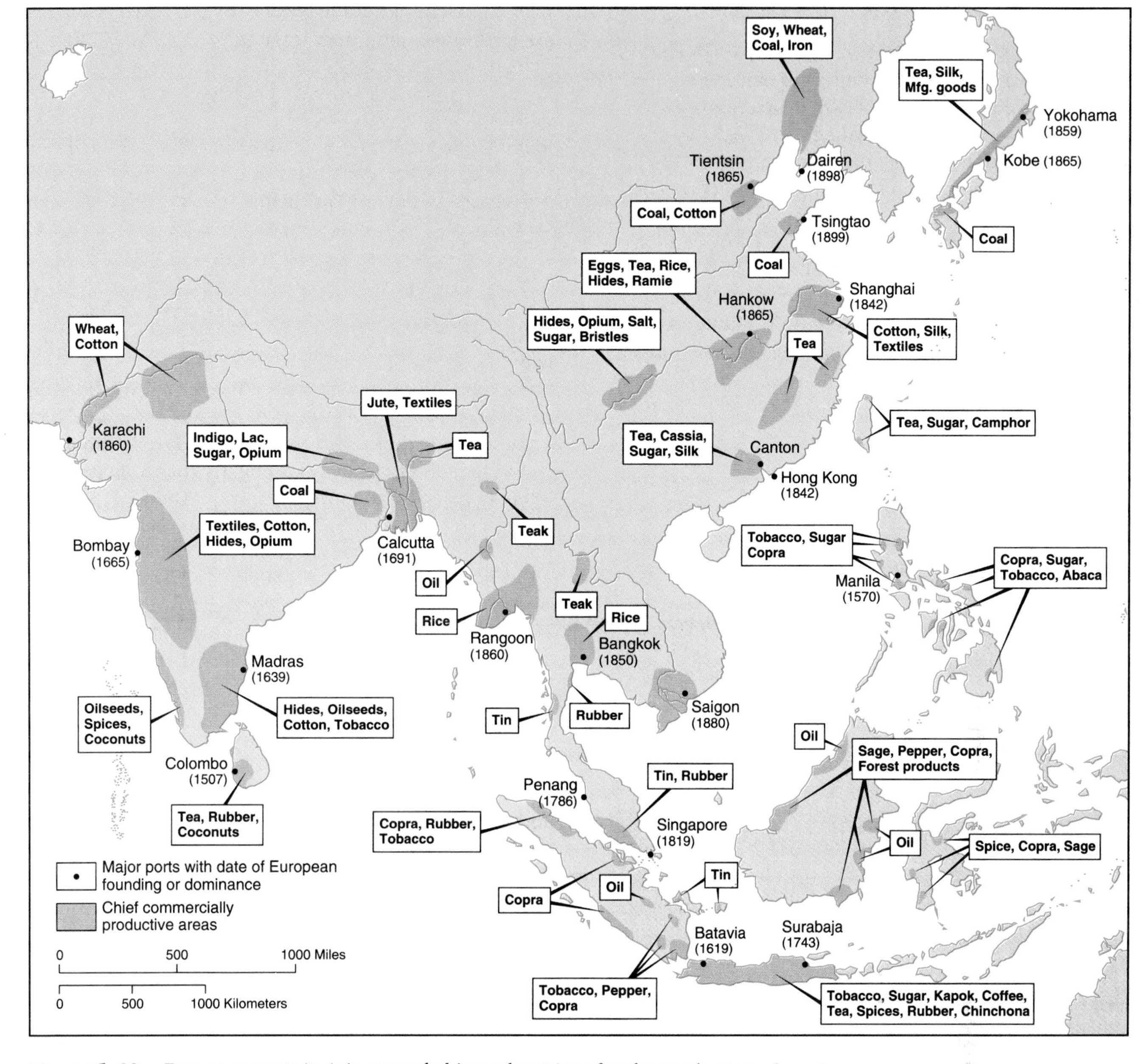

Figure 5–30. European ports in Asia expanded into plantations for the production of goods valuable in international trade. The plantation regions then expanded into political colonies, as in Africa.

assigned ownership to the local politicial leaders. This ended traditional egalitarian systems and created new classes of rich and poor. The descendants of many of these newly enriched leaders remain great landholders today. Many traditional societies crumbled under the transformation of land into a private commodity.

European law also tended to transform labor into a commodity. Sociologists differentiate a group of people held together by traditional networks of rights and responsibilities, called a **community**, from a group interacting as more self-interested individuals, called a **society**. Traditional communities were not idyllic; they restricted individual liberties, and slavery and serfdom were not unknown. These constraints, however, were often balanced by strong webs of responsibilities and rights that usually kept anyone from being entirely outcast and starving. The European idea of a self-regulating market for individual labor is more abstract and impersonal. It cut traditional ties of both rights and responsibilities.

In addition, Europeans required the use of money as a universal measure of value. Natives were forced to work for wages or to sell goods for money to pay taxes.

Europeans also brought their forms of administration, government, centralized state authority, written arrangements, uniformity, secularization, economic planning, public accounting and treasury control, central administration, and decision-making. In many cases the civil services that the Europeans left behind remain the pride of new nations, as in India.

Cultural Imperialism

European rule was marked by **cultural imperialism**, which is the substitution of one set of cultural traditions for another either by force or by degrading those who fail to acculturate and rewarding those who do. Europeans seldom doubted that native cultures were inferior and that native peoples needed "enlightenment." Therefore, the Europeans destroyed other ways of life, including religious and political traditions as well as physical artifacts such as art and architecture (see Figure 5–31).

One reason for this policy was the nature of Christianity. Christianity is a proselyting religion, which means that its adherents try to convert others to their faith. Many other religions accept that there is truth in all religions, and their tolerance helps explain the inroads made by Western culture.

Additionally, European military and technological superiority presumed European superiority in all other aspects of life. The conquerors often failed to appreciate the value of the civilizations they conquered, or even their science and technology, and so the Europeans did not learn as much from others as they might have.

European cultural imperialism began with the systematic training of local elites. The missionary schools produced converts who proselyted among their own people, helping to eradicate the local culture. Later the government schools turned out bureaucrats and military officers who helped govern their own people.

Figure 5–31. This page is from a book called the *Grolier Codex,* one of only four books known to survive from the considerable libraries of Central America's Mayan civilization. The rest were burned by the Spanish. This book contains astronomical records regarding the recurrence of the planet Venus. The vertical columns of markings along the left are counts of days. The warrior, perhaps mythological, stands before an incense burner. (Courtesy of Justin Kerr.)

A second channel of transmission was reference-group behavior. People who wish to belong to, or be identified with, a dominant group will often abandon their traditions to adopt those of the dominant group. The returned slaves who carried the first wave of westernization in West Africa wore black woollen suits and starched collars in the tropical heat. In India the native officer corps imitated English officers, complete with waxed mustaches. Premier Lee Kuan Yew of Singapore, a Chinese, even boasted of being "The last true Englishman east of Suez."

Local elites also adopted Western ways because they were made to feel ashamed of their color and their culture. Given the racism and cultural imperialism of the rulers, natives could succeed only by succeeding on the whites' terms, by adopting their ways. Colonial rule was an experience in racial humiliation, as the autobiographies of the new national leaders of every African and Asian country recount (see Figure 5–32). The colonial school systems implanted in children's minds an image of the power and beneficence of the "Mother Country." The children in the Congo, for instance, knew more about Belgium than about their own land and peoples. History texts used in French African colonies opened "Our ancestors the Gauls. . . ." Schooling focused on each colony's ruler, and so the subjects scarcely knew that other countries existed. Residents of the Congo, for example, referred to all whites as Belgians.

Self-westernization By the end of the nineteenth century, the elites of the entire non-Western world were taking the Europeans as their reference group. Only three major nations were never colonized—the Turks, the Chinese, and the Japanese—but these were all militarily humiliated, and this experience was traumatic for them. All three had had empires of their own, and their defeats forced them to reconsider all the assumptions on which their institutions and the peoples' lives were based. All three preempted or coopted Western civilization, and to some degree this saved them from Western political rule.

Mustafa Kemal (1881–1938) seized power in Turkey and forced the country to undergo self-westernization (see Figure 5–33). In China the Boxer Rebellion, an uprising that occurred in 1900, challenged Western influence but failed to get rid of it. The Republican Revolution of 1911 attempted to modernize the country, but China's subjection to the West continued until another leader, Mao Zedong, applied an alternative brand of westernization, namely, Communism, although in a uniquely Chinese adaptation. The Japanese were forced to open their society to Western trade in 1853, after centuries of near-total isolation. They decided to become thoroughly Western but to retain control of the process.

Figure 5–32. In the 1935 film *Saunders of the River*, Commissioner Saunders supervises and keeps the peace among tribes treated as dangerous children in a British African territory. The movie is racist and illiberal, and in this key scene Saunders prods a submissive native in the chest with his cane while lecturing on the superiority of British civilization. That young African actor was Jomo Kenyatta, who would later serve as president of newly independent Kenya. (Courtesy of the Kobal Collection.)

The Japanese escaped the domination of the West by embracing the West. The Meiji Restoration in 1867 launched the westernization of Japan with astonishing speed. Japan sent students to the United States and Europe and adopted Western science, technology, and even many cultural traits so successfully that by 1904 the country was able to defeat a Western power—Russia—in war (Figure 5–34).

Although in World War II Japan was defeated in its attempt to become a colonial power, its early military successes discredited the Western powers and helped encourage non-Western peoples, particularly in Asia but later in Africa as well, to pursue national independence after 1945. Despite Japan's military defeat, it later developed into one of the world's dominant economic powers.

The period of World War II actually created in the minds of non-European peoples a new definition of themselves and of their relationship to Europe. Japan's ostensible purpose in the war had been to liberate Asia from the whites. "Asia for the Asians" was a Japanese rallying cry intended to mobilize all Asian peoples against Europeans. Most peoples of Asia had never thought of themselves as "Asians," let alone as a distinct group in opposition to Europeans. "Asia" is a European geographical term previously unknown to the people it categorizes. Asia had never experienced the cultural cohesion that Christianity gave Europe. Koreans and Cambodians, Japanese and Malays, Burmese and Manchu are culturally more different from one another than are Italians from Germans, Spaniards from Scots, and Portuguese from Poles. "Africa" too is a European geographical term. Only in recent decades have some peoples of that continent begun to think of themselves as having common "African" interests, and that only in opposition to outsiders.

The period of European relations with Africa and Asia that began almost 500 years ago is ending. There

Figure 5–33. Mustafa Kemal "Ataturk," which means "father of the Turks," recognized the technological superiority of the West, and he consciously turned his people toward westernization. One of his changes was the transformation of the writing of Turkish from the Arabic into the Roman alphabet. Here he teaches a class in an Istanbul public park. (Courtesy of the Turkish Government Tourist Office.)

Figure 5–34. This is a Japanese woodblock cutaway view of a German battleship that visited Yokohama in 1873. This woodcut would later provide a virtual blueprint for the technologically backward Japanese Navy to copy. (Daval Foundation: from the Collection of Ambassador and Mrs. William Leonhart. Courtesy of Arthur M. Sackler Gallery, Smithsonian Institution, Washington, DC.)

are no longer any European colonies on the African continent, and the Asian empires of the Dutch, English, Portuguese, and French have been surrendered except for a few islands. Great Britain is scheduled to yield Hong Kong to China in 1997, and in 1999 Portugal will return the Chinese territory of Macao, which it has held since 1557. By 2000 only Russian territories and U.S. military bases and territories will remain as enclaves of non-Asians in Asia.

The West and Non-West Since Independence The fixation with the West among the elites in the non-European world did not end when these countries won political independence. Those who assumed power in Africa and Asia were descendants of the Western-educated class who had become disappointed with their rulers, and their demands for independence included quotations from Western political writers whom they had read at Western universities.

Few of these Western-educated elites developed indigenous models of development because they were not themselves members of the traditional elites, such as chiefs or religious precolonial leaders. They were themselves more Western, and they wanted to transform their new countries into Western-style societies as quickly as possible. They started building in the middle of their commercial cities and went on building outward. The technical term for the favored cities is **growth poles**. The leaders hoped to cover their national territories with modern economies and societies, but they did not realize how long it would take for the majority of their people to benefit from these developments.

The departure of the colonial rulers caused a local status vacuum, because the traditional sources of status, such as religion, family, and customs, had faded. The new power groups—politicians, bureaucrats, and businessmen—defined their status in the only way they knew, which was to exhibit Western material goods. Their housing, clothing, and means of transportation all mimicked Western status symbols.

In many countries the new rulers have practiced a sort of internal colonialism on their own people. For example, the rulers who assumed power in Latin America following the independence movements of the early nineteenth century were of European stock, and their descendants still dominate this region. Peoples of Native-American and African backgrounds have been treated as subject groups, forced to adopt European culture, religion, and language, and subjected to discrimination. The situation is not very different in Africa and Asia. The "colonizers" are generally the westernized elite; the "colonized" are all those who do not belong to this group, which is often the majority of the population. Government by an elite, privileged clique is called an **oligarchy**.

This elitism plays a role in keeping many of these countries poor. The elite use their capital to buy foreign goods. They also invest it abroad rather than in their own countries. The oligarchs fear domestic social upheaval, but by investing abroad rather than in their own countries they make upheaval more probable. The amounts of capital accumulated abroad by rich citizens of the poor countries often exceed the debts of those poor countries. In 1988 some countries' debt payments as a percentage of their total exports were as high as 24 percent for Morocco, Zimbabwe, and Pakistan; 25 percent for Indonesia; 30 percent for Mexico; 34 percent for the Philippines; and 39 percent for Madagascar. National debts are usually owed by the government, but in fact the assets have been taken by private individuals, often corrupt politicians, and deposited or invested abroad. The presidents or rulers of many of the poorest countries in Latin America, Africa, and Asia are regularly counted among the world's richest individuals.

WESTERNIZATION TODAY

The spread of Western culture continues today. Western culture diffuses from the top of societies, from the examples and activities of the local elites. Young people also diffuse it by adopting Western dress and lifestyles as status symbols. International studies reveal that these two groups—the rich and the young—are everywhere the most cosmopolitan consumers. Even the schools have become instruments of westernization. Their syllabuses emphasize modern, urban activities and values. The young sometimes emerge oblivious to their traditional culture, or even despising it.

The adult media reinforce the message. Western television programs, movies, advertisements, and videos penetrate millions of homes and implant Western values. Night after night, on television screens around the world, the images of the good life are images of the life among the wealthy in the United States. This imagery has dramatically changed behavior. In traditional societies, for instance, sexuality is associated with childbearing, but in modern lifestyles it is not. Some governments have hypothesized that the broadcast in their countries of the new behavior model is responsible for lowering their national birth rates. Consumption patterns also change because people want to copy what they see. This occurs regardless of any specific advertising (see Figure 5–35).

Tourism provides still another channel of westernization. Westerners are attracted to "different" and "unspoiled" places, but the presence of tourists changes the places they visit. National cultures can degenerate into commercialized spectacles and shoddy souvenirs. Many local people abandon their own material culture and adopt that of the visitors.

Global flows of professionals and of professional education are another powerful force. The rich Western countries export professional services, and people from around the world attend Western schools for professional education. A substantial share of the elite and of professionals in most countries have been educated in Europe or the United States. These graduates, who will presumably assume influential roles in their societies, have been acculturated to Western ways.

The diffusion of Western standards seems to carry an especially powerful impact in design professions. Western architects, civil engineers, and urban planners—or non-Western individuals educated in Western schools—are transforming the built environment around the world. The cultural landscapes of the poor countries are increasingly "modern," and there is only one global "cultural landscape of modernity." There is no prototype other than the Western. The cultural landscape, in turn, influences and conditions the lives and even the thinking of the area's inhabitants.

World flows of capital overwhelm the poor countries. Many cultures were transformed in the 1980s by capital investment according to the standards of the cultures that had exported the capital.

Under this barrage of westernization, many traditional cultures and social structures are being radically transformed, and these transformations are not always in accord with the people's conscious wishes. In some cases whole cultures disappear, and their disappearance reduces cultural diversity and impoverishes all of us.

Contemporary Reevaluations

In 1898 Winston Churchill, future prime minister of Great Britain, was a 25-year-old journalist. After witnessing the Battle of Omdurman, which eliminated the last Sudanese resistance to British imperialism, Churchill wrote: "These extraordinary foreign figures . . . march up one by one from the darkness of barbarism to the footlights of civilization . . . and their conquerors, taking their possessions, forget even their names. Nor will history record such trash."

Figure 5–35. Television, particularly U.S. television programing, penetrates practically every region. How do these people interpret what they see? How does it transform their cultural inheritance? The truth is that we do not know the answers to these questions. (Courtesy of Raghu Rai/Magnum.)

Figure 5–36. The extraordinary 1981 international hit movie *Lion of the Desert* was financed by the government of Libya to teach Libya's history from Libya's point of view, rather than Hollywood's. Anthony Quinn starred as Omar Mukhtar, who defeated Italy's attempted conquest of Libya from 1911 to 1931. Not all countries have been able to seize control over telling their own versions of history this way, either to international audiences or even to their own people. (Courtesy of Film Stills Archive, Museum of Modern Art.)

History is written by the winners, and if the winners devalue all cultures but their own as "barbarism" and "trash," then much of value is destroyed, and the losers, in effect, have their history stolen from them. After three-quarters of a century of British rule and cultural imperialism, the descendants of the Sudanese warriors would in fact have little knowledge of their ancestors' preconquest culture and civilization. That is why to many people the study of their history is a fierce reclaiming of their history, of their independent identity, of their self-respect (Figure 5–36).

The second generation after independence of many non-European peoples has attempted to revive their own cultural history and values. Movements of cultural reaction have swept the world, and many of these are angry. The title of an Iranian book written in 1962 disdains intellectual subjection to the West as *Weststruckness*. The experience of Iran since the fall of the Shah, who tried to westernize that country, and his replacement by fundamentalist clerics in 1979 has shown that restoration of the old ways may be no better than their eradication. Some features of traditional life, such as the low status of women, conflict with modern ideas of justice and equity. The West has been guilty of sexism, racism, and imperialism, but that was common behavior through recorded history. Western culture is unique among all civilizations, however, in having recognized, named, and tried to remedy these historic diseases.

All modernizers around the world today face a problem. Reactionary revival, that is, re-creation of some idealized model of what their society was like in the past, is impossible. But many modernizers also reject total conversion to Western ideas and standards. They must therefore produce some synthesis of civilizations.

Many of the problems that countries face today—problems such as urbanization, pollution, and cultural confusion—are problems that Europe faced first but that European culture has been unable to solve. Some observers have spoken of the "exhaustion" of Western modernity and the opening of the postmodern world. Eu-

Figure 5–37. The strong, bold forms of this sculpture by Jacques Lipchitz demonstrate that African art was one of the main influences on the twentieth century western art movement called cubism. Lipchitz was one of the founders of cubism, along with Pablo Picasso and Georges Braques, all of whom acknowledged their artistic debt. (Courtesy of Hirshhorn Museum and Sculpture Garden, Smithsonian Institution, Gift of Joseph H. Hirshhorn, 1966.)

rope may not be so central as it once was (Figure 5–37). More and more people everywhere are today involved with more than one culture. Latin America is reaching out directly to Africa, Asians to Latin America. Perhaps world cultural diffusion will be less hierarchical in the future.

In our search for solutions to world problems we might look to the other world cultures. The worldwide revival of interest in non-Western cultures is both a search for new cultural solutions to world problems and a revival of self-respect on the part of peoples whose cultures were overlain by European culture. The United States, the self-proclaimed "melting pot," demonstrates this in a microcosm. African Americans are reexamining their heritage. We are all rediscovering that Native-American culture has much to teach us in its respect for ecology, and new Hispanic and Asian cultural voices are also contributing.

SUMMARY

An isolated society depends entirely on its local environment for all of its needs. The people receive no imports from the outside world, nor do they produce anything for others. They develop their own distinctive culture. Human cultures vary significantly. People make use of their local landscape, but they also modify it and thus transform it from a natural landscape into a cultural landscape. Different cultures will create noticeably different cultural landscapes. Culture realms may be demarcated by land use, place names, building materials, styles of architecture, and even the designs of whole settlements.

The place where a culture originates is called the culture hearth. Various aspects of cultures may spread out and be adopted by other peoples in a process called cultural diffusion. Cultural diffusion can be active and deliberate, or else one group may discover and adopt some aspect of a different culture that it considers superior to its own. The impact of any cultural attribute may diminish away from its hearth area; this phenomenon is called the distance decay concept. Some aspects of culture may also develop variations as people who carry that culture diffuse outward and away from one another. Tracing a course of diffusion may teach us a great deal about how peoples and cultures interact. Diffusion processes may be contiguous or hierarchical. Barriers to cultural diffusion may be topographic, political, or even cultural. Diffusion does not explain the distribution of all cultural phenomena. Sometimes the same phenomenon occurs spontaneously and independently at two or more places.

Cultures diffuse, cultures change, and peoples can transform themselves and their behavior, but cultures and culture realms also have elements of stability. Any cultural pattern or distribution maps a current balance between the forces of change and of stability. Cultures are constantly evolving and changing through time, and these changes may result either from local initiatives and developments or else as the result of influences from other places. Each local culture is a mix between what originated locally and what has been imported. Endogenous factors, or site factors, are elements of the specific local environment or of local cultural history and development. Exogenous factors or situation factors refer to the way a particular place interacts with other places. Endogenous factors of explanation and exogenous factors complement each other. As global communication and transportation have increased, the balance of factors that explain the local activities and culture at any place has tipped steadily away from endogenous factors and toward exogenous ones. What happens at places depends more and more on what happens among places.

In studying economies and trade, geographers note how the people at any specific place make their living, where specific economic activities locate, and how economic activities and trade affect other aspects of cultural geography. The role any place plays in systems of trade may depend upon its situation or relative location. Accessibility can be a resource, and accessibility can change with changes in transport routes, technology, or the territorial scale of organization of any activity.

All human activities find territorial expression, and maps of culture realms, regions of economic specialization, or political jurisdictions reveal current distributions of human activities. These activities, however, are dynamic, continuously organizing and reorganizing, forming and reforming. Therefore their distributions and patterns are disrupted and reshaped repeatedly. Geographers need to understand what forces make things move and redistribute themselves.

Despite the rich variety of indigenous local cultures around the globe, the European cultural model is widespread. Europeans came to play a central role in world history and geography because they paved the way for the modern system of global interconnectedness. Global diffusion was fixed hierarchically, with Europe as the apex. Europe conquered most of the rest of the world, and European political domination imposed European concepts of government, law, property, and other aspects of culture. The spread of Western culture continues today, and many traditional cultures and social structures are being radically transformed. In some cases whole cultures disappear.

The second generation after independence of many non-European peoples has attempted to revive their own cultural history and values. Perhaps world cultural diffusion will be less hierarchical in the future.

KEY TERMS

culture (p. 132)
natural landscape (p. 132)
cultural landscape (p. 132)
built environment (p. 132)
material culture (p. 132)
hearth area (p. 133)
culture hearth (p. 133)
cultural diffusion (p. 133)
acculturation (p. 133)
culture realm (p. 133)
distance decay (p. 133)
dialects (p. 133)
toponymy (p. 133)
contiguous diffusion (p. 137)
hierarchical diffusion (p. 137)
diffusionism (p. 140)
circulation (p. 141)
friction of distance (p. 141)
ethnocentrism (p. 141)
inertia (p. 143)
infrastructure (p. 143)
historical consciousness (p. 143)
endogenous factors (p. 144)
site factors (p. 144)
exogenous factors (p. 144)
situation factors (p. 144)
economic geography (p. 145)
autarky (p. 145)
locational determinants (p. 145)
felt needs (p. 146)
revolution of rising expectations (p. 146)
capital (p. 147)
absolute location (p. 147)
relative location (p. 147)
historical geography (p. 151)
footloose activity (p. 151)
Commercial Revolution (p. 152)
Industrial Revolution (p. 153)
Agricultural Revolution (p. 153)
dualism (p. 157)
community (p. 159)
society (p. 159)
cultural imperialism (p. 160)
growth pole (p. 162)
oligarchy (p. 162)

QUESTIONS FOR INVESTIGATION AND DISCUSSION

1. What were a few distinctive cultural products of your region 50 years ago? Religious observances? Food products? Clothing or costumes? Architectural styles? Games? Have they been exported from the region? Are they still typically produced? Attend a local street fair or celebration. Which aspects of that festival are different from such a festival 100 miles away?

2. For examples of distance decay, map the homes of faculty or students at your college or in your class or of customers of a local store.

3. Take a photo of a local landscape, and then analyze how different aspects of it reveal where it is—the dress of the inhabitants, vegetation, languages on signs, and so forth.

4. What do local place names reveal?

5. How many cities are serviced by nonstop flights from your town?

ADDITIONAL READINGS

EYLES, JOHN, ed. *Research in Human Geography.* New York: Basil Blackwell, 1988.

KEAY, JOHN, ed. *The Royal Geographical Society History of World Exploration.* New York: Mallard Press; London: Paul Hamlyn Publishing Co., 1991.

MCNEILL, WILLIAM. *The Rise of the West.* Chicago: University of Chicago Press, 1963.

NEHRU, JAWAHARAL. *Glimpses of World History*, centenary ed. Oxford, England: Oxford University Press, 1989. Originally published in two vols., 1934–1935.

READER, JOHN. *Man on Earth: A Celebration of Mankind: Portraits of Human Culture in a Multitude of Environments.* Austin: University of Texas Press, 1988. Reprint. New York: Harper & Row.

THOMAS, WILLIAM, ed. *Man's Role in Changing the Face of the Earth.* Chicago: University of Chicago Press, 1956.

CHAPTER 6

Population, Population Increase, and Migration

This crowded street is in a city in India. Meanwhile, large areas of the Earth remain uninhabited. Population geographers map and try to explain the distribution of humankind. (Courtesy of Frederica Georgia/Photo Researchers.)

This chapter examines the world's **population geography,** that is, the distribution of people across the Earth. The relative distribution of humanity is constantly changing because of two factors. One is the fact that different countries have different internal population dynamics, such as the numbers of births and deaths, which produce different rates of increase or decrease. The second factor is human **emigration** out of some places and **immigration** into others. We shall examine how migration has affected the distribution of the world's peoples in the past and how continuing migration affects world affairs today.

In studying population geography we shall occasionally include ideas and information about **demography,** which is the study of individual populations in terms of specific group characteristics such as the distribution of ages within the group, the ratios between the sexes, and income levels. Demography means "describing people."

THE DISTRIBUTION OF THE HUMAN POPULATION

In 1990 the population of the Earth was about 5.3 billion, but people are not evenly distributed across the landscape (see Figure 6–1). About 90 percent of the population is concentrated on less than 20 percent of the land area.

The map reveals three major concentrations of population: (1) East Asia, where eastern China, the Koreas, and Japan total well over 1 billion people; (2) South Asia, where India, Pakistan, and Bangladesh together account for another billion; and (3) Europe from the Atlantic to the Urals, with almost another billion. Southeast Asia forms a secondary concentration, with just under 300 million people; the eastern United States and Canada are home to another 175 million.

These five population concentrations, plus parts of West Africa, Mexico, and areas along the eastern and western coasts of South America, are densely populated, with more than 60 people per square mile (25 per square km). More than half of the Earth's land area, by contrast, has fewer than 2 people per square mile (1 per square km). A surprising percentage of the Earth's land surface is virtually uninhabited: central and northern Asia, northern and western North America, and the vast interiors of South America, Africa, and Australia. The countries that occupy these spaces are enormous in area, but they contain relatively few people. Mongolia, for instance, is 20 times as big as the state of Pennsylvania, but Mongolia's 2 million people would fit into the city of Philadelphia. The three African countries of Chad, Niger, and Mali together cover 35 times the area of South Korea, but their combined population is about the same as that of South Korea.

Figure 6–2 is a world map drawn so that the size of each country reflects its population, not its land surface area. The most populous countries are shown as the

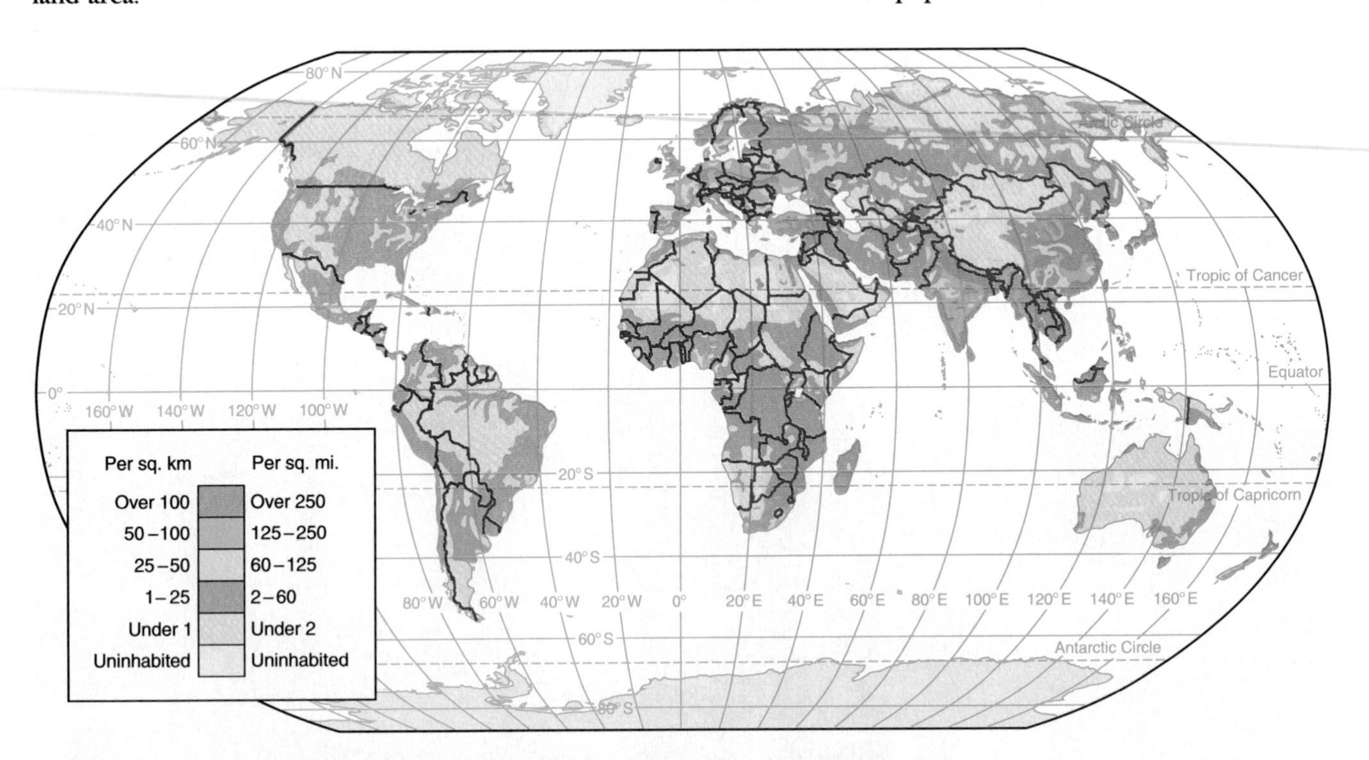

Figure 6–1. The aspect of this map that most surprises people is the amount of the earth's surface that is almost uninhabited. About 75 percent of the total human population lives in the Northern Hemisphere between 20° and 60° north latitude, and even great expanses of that area are sparsely populated.

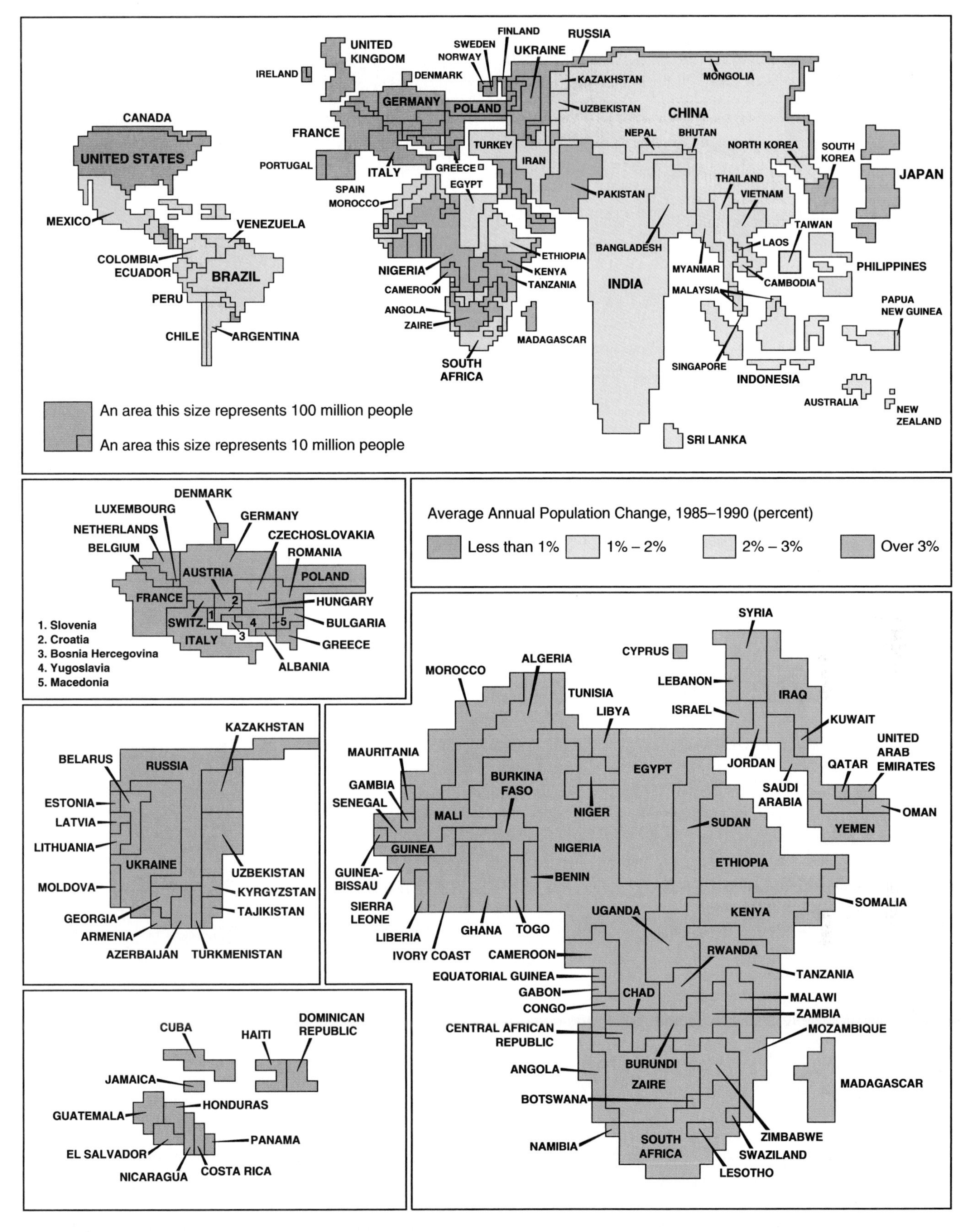

Figure 6–2. On this population cartogram the size of each country reflects its population, not its actual land area. On a population cartogram the whole African continent is smaller than India because it is less populous. South America also is very small. The color coding conveys additional information—the rates of natural population increase. Almost equal quarters of the human population live in countries in which the rate of increase is less than 1 percent; between 1 and 1.4 percent; between 1.5 and 2.1 percent; and over 2.2 percent. (Data from World Resources Institute, *World Resources 1992–93.* New York: Oxford University Press, 1992, pp. 246–47.)

Focus on Cartograms

Standard maps try to depict the Earth's features in their correct locational and size relationships. Sometimes, however, it may be more important to convey a visual impression of the magnitude of something than to convey a visual impression of exact spatial locations. To do this cartographers design special maps called cartograms. All cartograms replace physical distance with some other measure. The two main types of cartograms are *area cartograms* and *linear cartograms*. On an area cartogram, a region's area is drawn relative to some value other than its land surface area. In Figure 6–2 that value is population. The bigger a country is drawn on the cartogram, the greater is its population.

When a cartographer makes a conversion from physical space to something else, he or she tries to retain as many spatial attributes of the conventional map as possible. If recognizable shape, proximity, and continuity are preserved, it is easier for the viewer to compare the cartogram with a standard map. Each of these attributes of space can be retained, however, only by distorting one or more of the others.

Linear cartograms draw our attention to distance. The transit map, for example, is called a *time-distance* map. Most people use time-distance measurements in their daily lives without realizing it ("That's about 20 minutes from here."). Geographers also frequently measure the *cost-distance* between two places. Transportation over land is usually more expensive than transportation over water, and so two seaports may be closer to each other, in cost-distance, than either is to inland cities.

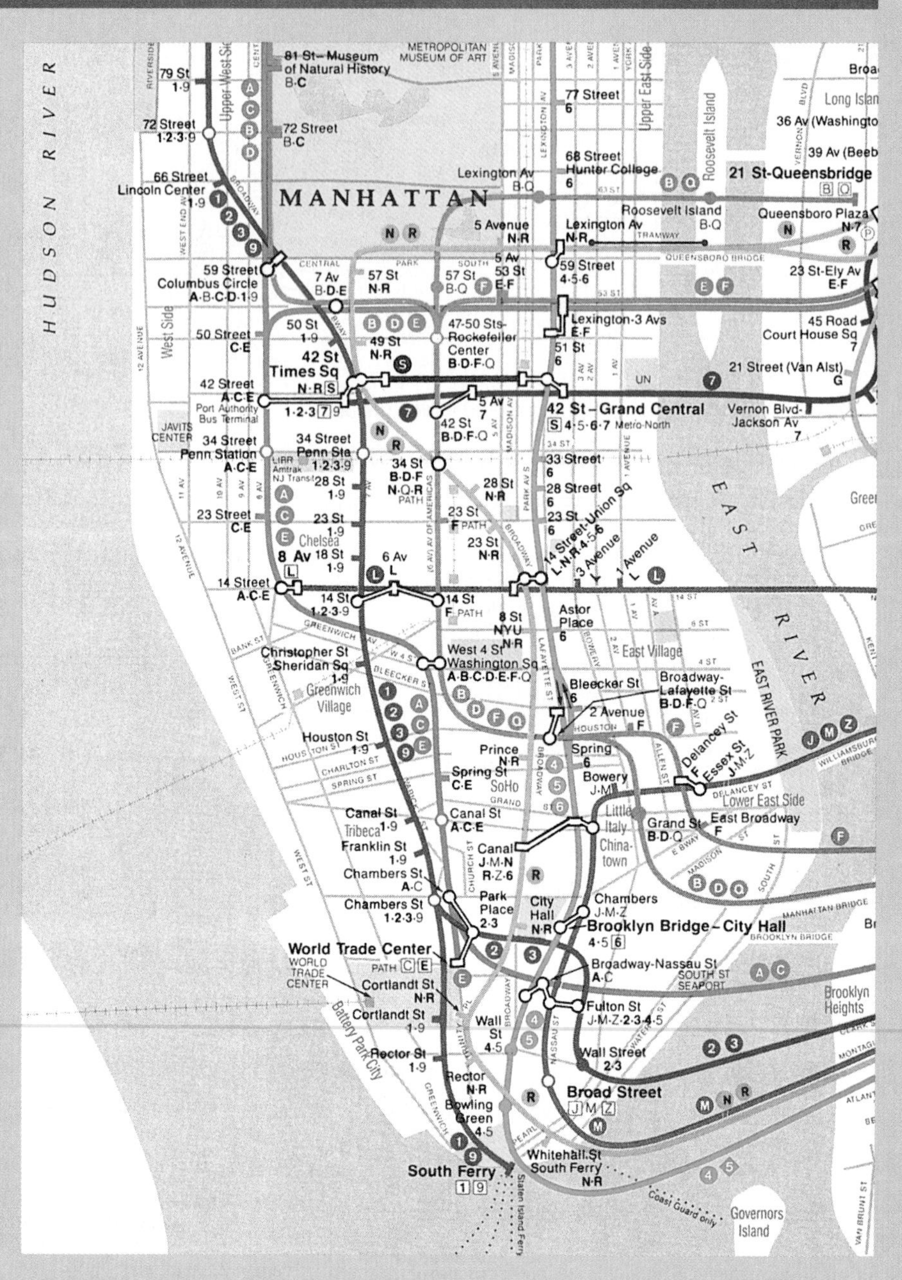

This subway map of Manhattan distorts the island's true shape in order to clarify the subway lines and stops. Subway riders think in terms of stops, not distances. Therefore, on many transit maps the distance between stops is shown as a uniform distance. (Reprinted with permission from the New York City Subway Map. © 1992 New York City Transit Authority.)

Cartograms may look strange to us because the scale depends on something other than the physical units we expect. Cartograms are not, however, lies or tricks. Their purpose is to portray some other important aspect of reality.

Area cartograms of such things as countries' populations (Figure 6–2) or economic output (Figure 11–4) visually convey the relative population or wealth of different countries better than standard land area maps do. The viewer must be certain to note when a figure is a cartogram. Most maps in this book are standard land-area maps, but the book also contains cartograms. These are especially useful when we want to illustrate human population dynamics or the conditions in which people live. A cartogram will show not only where people live in certain circumstances, but at the same time, it shows what share of the total world population lives in those circumstances.

biggest, and the least populous as the smallest, no matter how big or small their land surface areas are. Maps like this are called **cartograms.**

The world's ten most populous countries contain almost two-thirds of all humankind: China (1100 million people), India (850), the United States (250), Indonesia (180), Brazil (150), Russia (150; 280 million together with the other states of the Commonwealth of Independent States), Japan (125), Bangladesh (115), Nigeria (110—see the Focus box on p. 172) and Pakistan (110).

EXPLAINING THE POPULATION PATTERN

A number of factors combine to explain the distribution of the human population, including the physical environment, history, and local differences in rates of population increase.

The Environmental Factor

It is often thought that the population density of a place reflects the productivity of the local environment. Some scholars differentiate simple **arithmetic density**—the number of people per unit of area—from **physiological density**—the density of population per unit of arable land. Furthermore, we might assume that a fertile region could support a higher population density than a less fertile land; in other words, the fertile area has a high **carrying capacity.** In an infertile region the same arithmetic density could impose a greater strain upon local resources.

Neither measure of density, however, tells us anything about the local standard of living or allows us to judge whether any region is overpopulated. Trade and circulation free societies from the constraints of their local environments and allow them to draw resources from around the world. Local carrying capacity has become less important in determining how many people can live in a given area or how well they can live, and the physiological density of a highly industrialized country is irrelevant to its welfare. Some of the most densely populated areas in the world support some of the richest populations, whereas in others the people are poor. Conversely, people in some of the sparsely populated regions are rich, and in other sparsely populated regions the people are poor. No clear correlation between density and welfare can be drawn.

Climate The clearest relationship between population distribution and environmental factors is that population densities are low in most of the world's dry areas and cold areas (compare Figure 6–1 with Figures 2–29 and 2–47). The major exceptions to this rule are places where rivers flowing through dry areas provide water for irrigation, such as India, Pakistan, and Egypt (look back at Figure 4–29).

Figure 6–3. The construction and intensive cultivation of rice terraces such as these on the Pacific island of Bali supports a very high human population density. (Courtesy of Professor J. Markusse, University of Amsterdam.)

The Earth's coldest areas are inhabited only where mines or other special resources make it worthwhile to support populations at work there, as in parts of central Asia or high in the mountains of South America. People will settle in harsh areas if it is profitable to do so; the settlement in those areas demonstrates our earlier discussion of possibilism.

Some warm and wet equatorial regions are also sparsely populated, such as the Amazon Basin in South America, the Congo Basin in Africa, and the island of New Guinea. Despite the high biological productivity of these regions, conventional agriculture has not proved successful. This failure is attributed to nutrient-poor soil or to competition from other plants, but heat multiplies biological activity, and this provokes another hazard—the flourishing of life forms hostile to humans and their agriculture. Insects such as mosquitoes, tsetse flies, blackflies, and sandflies thrive, as do the diseases they carry. In addition, these environments support countless forms of parasites, microbes, and fungi that weaken and kill humans, wilt and blight their plants, eat crops alive in

Focus on National Censuses

Counting a country's population might seem to be a fairly easy task, but in fact throughout history—and in many countries today—it has been difficult or impossible. Most states counted their people only to record how many were on the tax rolls—or ought to have been. The Canadian census of the late seventeenth century was probably the world's first census undertaken solely to count the population. The U.S. Constitution required a decennial census (every 10 years) for the novel purpose of reapportioning seats in the House of Representatives (Table 6–1).

The United Nations annually publishes the *Demographic Yearbook*, but the figures in it are, in many cases, only estimates. Some countries lack the administrative apparatus to carry out a full census effectively. Portions of the country may be inaccessible, and the population may be mobile. Portions of the population may avoid being counted out of superstition, distrust of the central government, or the wish to avoid taxation. In April 1990, one-third of U.S. households threw their census forms into the wastebasket.

Nigeria exemplifies a country in which politics militates against an accurate census. The census of 1991 counted 88.5 million Nigerians, but counting is politicized because the balance of population among the country's 30 states determines how much money each state gets from the federal government. The numbers dictate the political, ethnic, and religious balance of the country, and no one wants to be part of a well-documented minority.

Nigeria's population is dominated by three broad ethnic groups: the Hausa-Fulani in the north, Yorubas to the west, and Ibos in the east. The last fairly reliable census, in 1952–1953, gave the largest share of the population to the Hausa-Fulani. Consequently, when Nigeria received its independence in 1960, northerners dominated its first government. Subsequent censuses were fiascos. The count of 1962 suggested that in the intervening decade the population of the east had increased by 71 percent, that of the west by 70 percent, and that of the north by only 30 percent. The government ordered a recount, and the north's increase magically rose to 67 percent. Most northern and western politicians agreed to accept that and to assume that the national population was 55.6 million, with 30 million in the north. A 1973 census demonstrated similar creative counting techniques, and thus the north has been dominant in every government. Southerners have accepted that, provided always that each region had a share of people in the nation's top jobs. The 1991 census counted 88.5 million Nigerians, of whom 47.2 million, or 53.7 percent, lived in the north. This census was the first deliberately to avoid questions of religious or tribal affiliation. It is bound to be contested, as it is far below earlier national and international estimates of 110 million. Even the U.S. census is politically contested.

the fields, or quietly feast on them in granaries or storerooms.

Heat also affects human productivity. Work generates heat, and the body has to lose excess heat to work efficiently. The productivity of manual workers decreases by as much as one-half when the temperature rises to about 95°F (35°C), as is quite common in the tropics. The invention of the air conditioner, however, has made it more comfortable for people to settle in the warm regions of the United States, and this has transformed the country's population geography. Virtually all of the country's population growth since 1970 has taken place in the southern "Sunbelt." This illustrates again how cultural and technical innovation can overcome the constraints that the theory of environmental determinism would suggest.

Tropical Asia is home to several great concentrations of people, but these concentrations are mostly in regions that are not wet year-round. Some of these, such as the island of Java or the Deccan Plateau of India, offer rich volcanic soils, and others are well adapted to the cultivation of rice paddies (see Figure 6–3).

Rice terraces are extremely productive, yielding the highest number of calories per acre of any known crop. Paddy flooding plus intensive supervision and care limit the growth of competing vegetation, and algae in the water supply nitrogen as a plant nutrient. Rice paddies can support increasing numbers of cultivators on a given unit of cultivated land. The capacity of terraces to respond to intensive care is amazing. It seems almost always possible to increase the yield of a rice terrace by working it just a little bit harder. In addition, rice terraces create a stable ecosystem. They do not break down the physical system.

In conclusion, most of the world's population is concentrated in areas of seasonal environments that are

TABLE 6–1 ***The First U.S. Census in 1790 Revealed that Virginia Had a Full 22 Percent of the U.S. Population***

DISTRICTS.	Free white males of sixteen years, and upwards, including heads of families.	Free white males under sixteen years.	Free white females, including heads of families.	All other free persons.	Slaves.	Total.
Vermont - -	22,435	22,328	40,505	252	16	85,539
New Hampshire -	36,086	34,851	70,160	630	158	141,885
{ Maine - - -	24,384	24,748	46,870	538	none.	96,540 }
{ Massachusetts -	95,453	87,289	190,582	5,463	none.	378,787 }
Rhode Island -	16,019	15,799	32,652	3,407	948	68,825
Connecticut -	60,523	54,403	117,448	2,808	2,764	237,946
New York - -	83,700	78,122	152,320	4,654	21,324	340,120
New Jersey - -	45,251	41,416	83,287	2,762	11,453	184,139
Pennsylvania -	110,788	106,948	206,363	6,537	3,737	434,373
Delaware - -	11,783	12,143	22,384	3,899	8,887	59,094
Maryland - -	55,915	51,339	101,395	8,043	103,036	319,728
{ Virginia - -	110,936	116,133	215,046	112,866	292,627	747,610 }
{ Kentucky - -	15,154	17,057	28,922	114	12,430	73,677 }
North Carolina -	69,988	77,306	140,710	4,975	100,572	393,751
South Carolina -						
Georgia - -	13,103	14,044	25,739	398	29,264	82,548

	Free white males of twenty years, and upwards, including heads of families.	Free males, under twenty-one years.	Free white females, including heads of families.	All other free persons.	Slaves.	Total.
Southwestern territory	6,271	10,277	15,363	361	3,417	35,691
Northwestern territory			15,365	361	3,417	

not too wet, too hot, too dry, or too cold. In fact, the plurality of the Earth's population lives in areas that are designated "C" climate (see Figure 2–47).

Topography Topography often affects population distribution, although its effect is less prominent than that of climate. People tend to settle on flat lands because of the ease of cultivation, construction, and transportation. Thus, most of the densest concentrations of population in the world are found on level terrain (see Figure 6–4).

Flatness alone, however, is insufficient to attract a large population. Usually there is an association of other environmental attributes, such as fertile soil, available water supply, and moderate climate, that helps to explain high population density. Thus, flat land in central Siberia, western Australia, and central Brazil may appear very empty on the population map because other factors, such as the nature of the climate or soils, diminish the area's productivity.

Conversely, sloping land (hills and mountains) usually has a sparser population density, although this is not always the case. High density is found in many mountainous portions of South America, Japan, New Guinea, Southeast Asia, Central Europe, and elsewhere. Specialized attractions, such as mineral resources, often help explain such concentrations. Moreover, in most mountainous areas the settlements actually are concentrated in valley bottoms, even though they may be small and constricted. Thus, the connection between topography and population density is too inconsistent to validate a direct relationship.

Soils People usually settle areas of fertile and potentially productive soil unless there are powerful negative factors (compare Figure 6–1 with Figure 3–27).

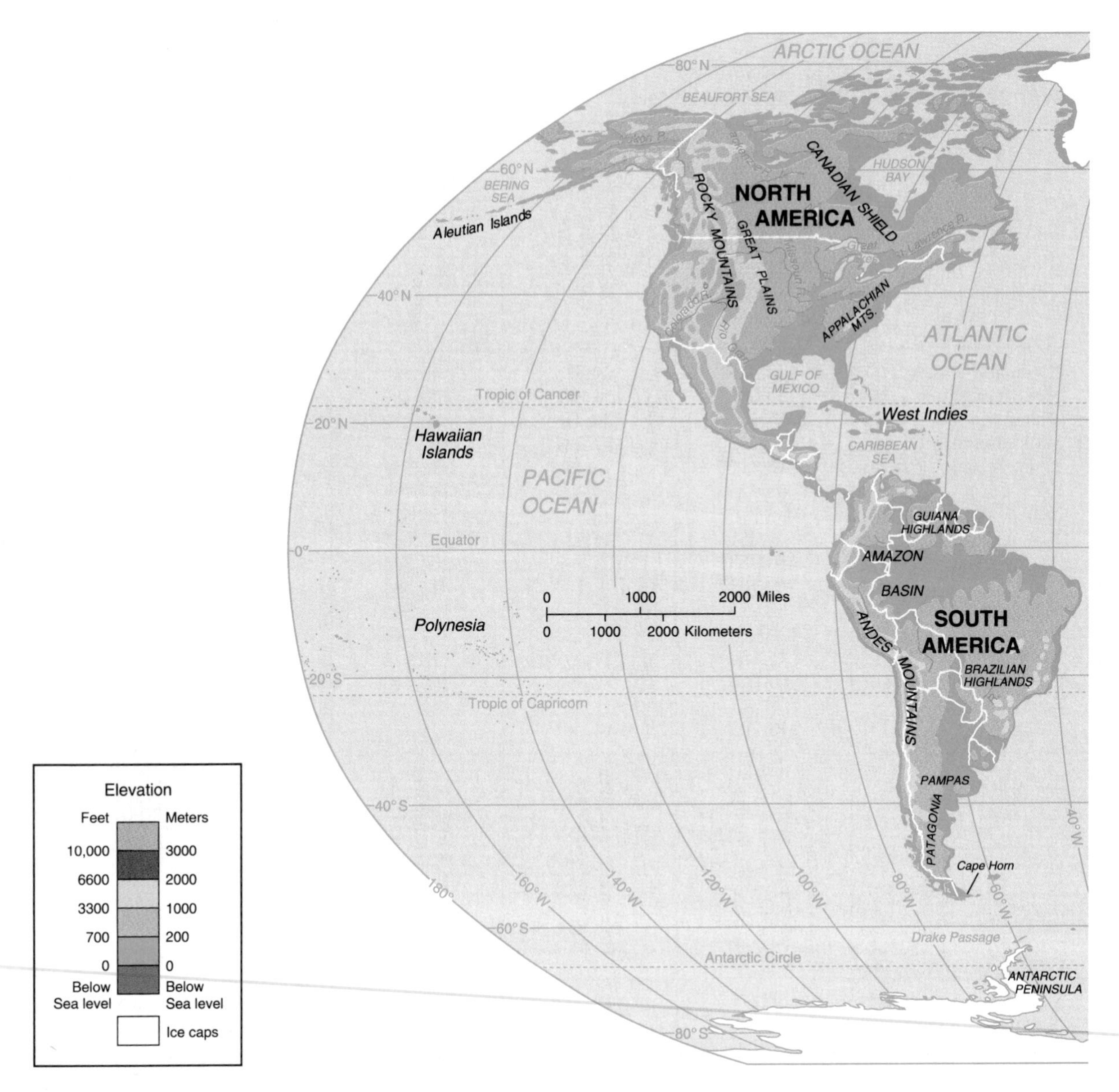

Figure 6–4. The topography of the Earth.

Thus, most river floodplains are densely populated unless they are located in extremely cold or dry areas. Floodplain soils are typically fine-textured and full of nutrients, so they can be used intensively for agriculture. People do not necessarily avoid areas of poor soils, however, because some poor soils can be enriched with fertilizers.

History

History helps explain patterns of human settlement. The populations of China and the Indian subcontinent achieved productive agriculture and relative political stability thousands of years ago. These peoples domesticated many plants and animals very early. Intensive cropping and irrigation yielded generous food supplies, which supported rising populations. The Western European population multiplied when the Europeans gained material wealth and improved their food supplies as the result of early world exploration and conquest. Later, during the Industrial and Agricultural revolutions, European productivity multiplied again. Migrants from Europe settled and brought European technological sophistication to more sparsely populated areas. Thus, some of today's secondary population concentrations grew as extensions of Europe. These include eastern North America, coastal South America, South Africa, and Australia and New Zealand. It has been estimated that Europeans and their descendants increased from 22 percent of the world's population to 35 percent in the period from 1800 to 1930.

When we look at the maps of current world population distribution, we should remember that considerable population reductions have occurred in some areas. The arrival of the Europeans in the Western Hemisphere triggered wholesale depopulation, as the Native Americans succumbed to disease and mistreat-

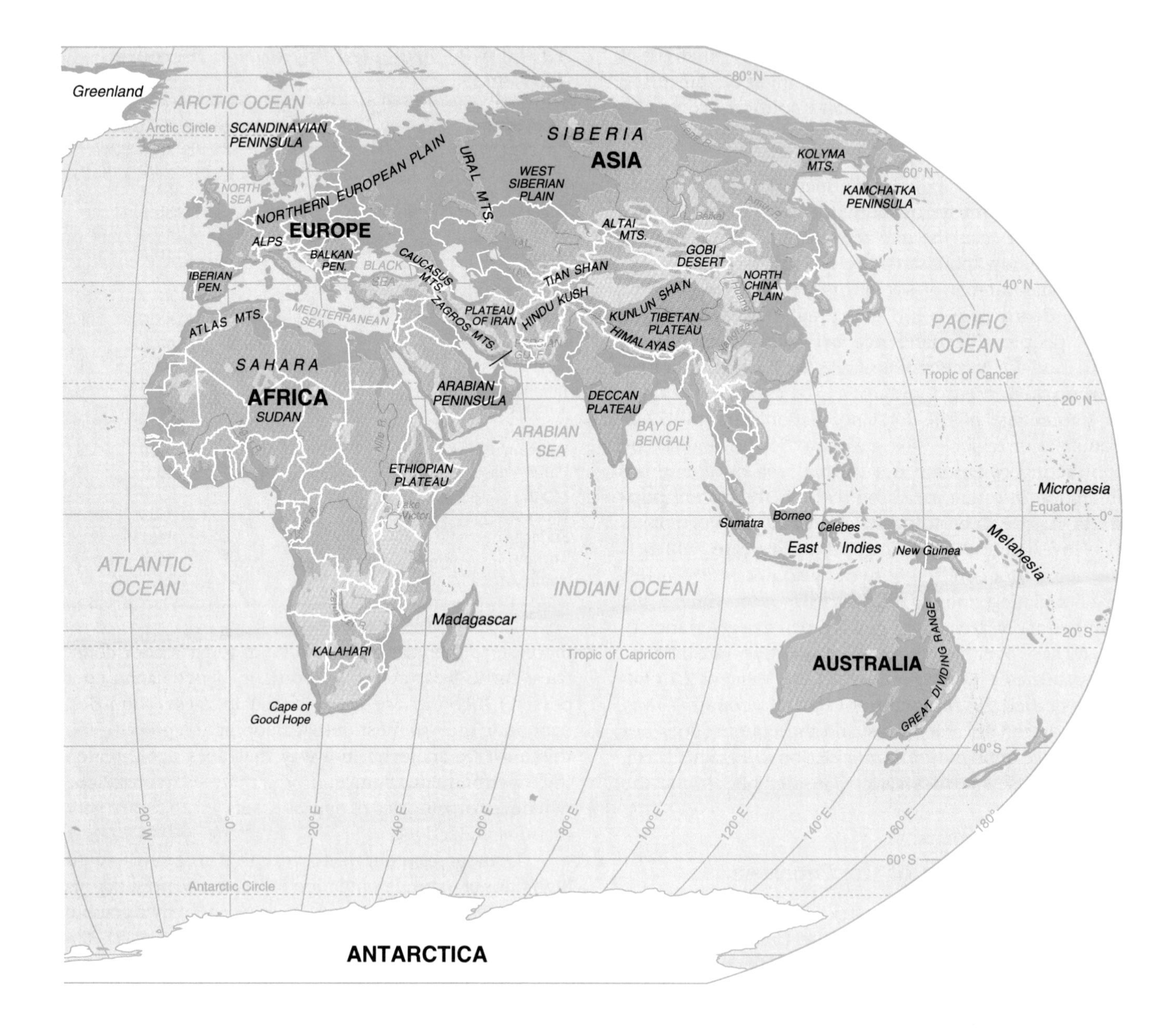

ment. We cannot calculate what the total population of the Western Hemisphere would be today, or its distribution, if the Native Americans had not suffered this terrible loss but had lived and multiplied. Similarly, we cannot compute what the population of sub-Saharan Africa would be today had millions of Africans not been drawn out as slaves.

Human population distribution can also be affected by the demarcation of cultural territories, particularly political territories. Two governments on opposite sides of an international border may have different environmental policies, and these different policies can drastically alter the carrying capacity of the environment. Furthermore, a country may manipulate the distribution of its population. It might subsidize or command the settlement of harsh territories to occupy them effectively, or it might establish settlements along its borders for defense.

WORLD POPULATION INCREASE

World population density is increasing overall. The Earth's population reached 1 billion around 1800, 2 billion by 1930, and 4 billion by 1975. Today there are over 5 billion people, and their numbers are increasing at a rate of about 87 million per year. It is doubtful, however, whether the present rate of increase will continue into the future. In the second half of the 1960s the rate of growth of the world's population began to slow down. The average annual population increase fell from 2.06 percent in the period 1965–1970 to 1.73 percent for the period 1985–1990.

A **population projection** is a prediction of the future assuming that current trends remain the same or else change in defined ways. Projecting population is an uncertain task, and the smallest rise or fall in the percentage of increase today would increase or decrease the total population in the next century by hundreds of millions of people. Nevertheless, it might be difficult to sustain the world's economy and political order with twice as many people as today, to say nothing of three times the present population.

The rate of population increase varies greatly among the countries of the world, as shown in Figure 6–2. In each case, the rate represents a balance among several demographic statistics. The **crude birth rate** is the annual number of live births per 1000 people. The **crude death rate** is the annual number of deaths per 1000 people. The difference between the number of births and the number of deaths is the **natural increase** or **natural decrease,** and that can be expressed as a rate or percentage of the total population. For individual countries or regions, this figure can then be modified by subtracting emigration out of that area or adding immigration into that area. The result is the overall **population growth** or **population decrease.** Geographers also investigate each country's **fertility rate,** which is the number of children born per year per 1000 females in a population, and the **total fertility rate,** which is the number of children an average woman in a given society would have over her lifetime. A total fertility rate of about 2.1 stabilizes a population, and so this value of 2.1 children is called the **replacement rate.** If a country's total fertility rate falls below this, and the country does not experience immigration, its population will gradually fall. Each of these statistics varies considerably around the globe.

The Distribution of the Increase

Figure 6–2 shows how the rates of population increase vary from country to country. The United Nations estimates that 70 percent of the total projected increase from 1989 to 2025 will occur in just the 20 countries listed in Table 6–2. Ninety-five percent of the population increase to the year 2025 will be in today's poor countries, those least able to support it. If per capita (per person) incomes are to be raised in these countries, economic growth must outpace population growth, and this must be achieved in a way that does not threaten the environment. Future productivity is jeopardized whenever inventories of natural assets left to future generations are reduced.

TABLE 6–2 ***Projected Population Increase, 1989–2025***

Country	Increase (in millions)
India	592.2
China	357.1
Nigeria	188.3
Pakistan	144.4
Bangladesh	119.4
Brazil	95.4
Indonesia	82.7
Ethiopia	65.6
Iran	65.6
Zaire	63.5
Mexico	61.5
Tanzania	57.5
Kenya	52.5
Vietnam	50.8
Philippines	49.0
Egypt	39.9
Uganda	36.8
Sudan	34.4
Turkey	34.0
South Africa	28.0

Demographic analysis reveals that in countries with high rates of natural population increase, the populations are young. This statistic is often represented by a graphic device called a **population pyramid** (Figure 6–5). Today about 33 percent of the world's population is under 15 years of age, and that figure will increase to about 40 percent by the year 2000. The median ages of the pop-

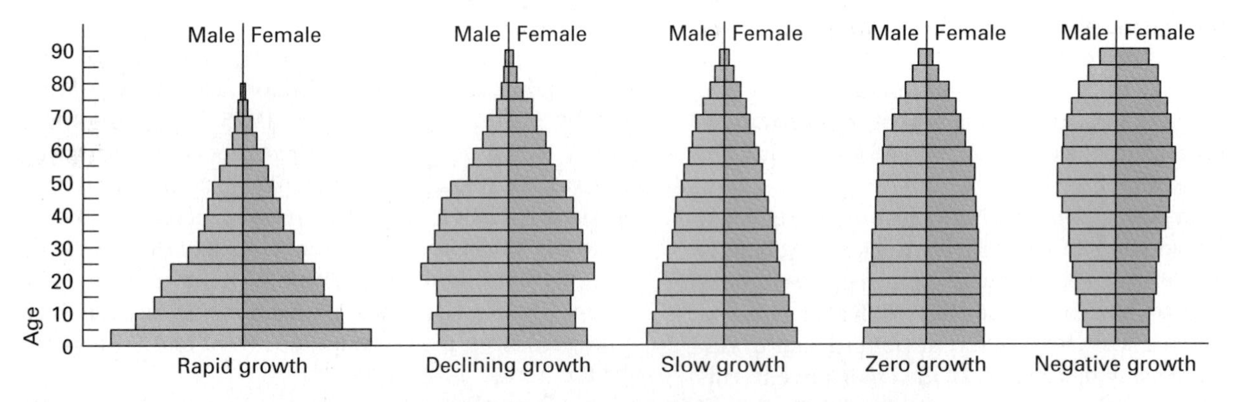

Figure 6–5. These are examples of a graphic device called a population pyramid, which shows the age and sex structure of a country's population. The population pyramids of the poor countries, where birth rates are high and life expectancies are limited, typically exhibit broad bases of many children tapering to narrow tops of fewer older people. The population pyramids of the rich countries, where both crude birth rates and crude death rates are low and life expectancies are long, more closely resemble columns.

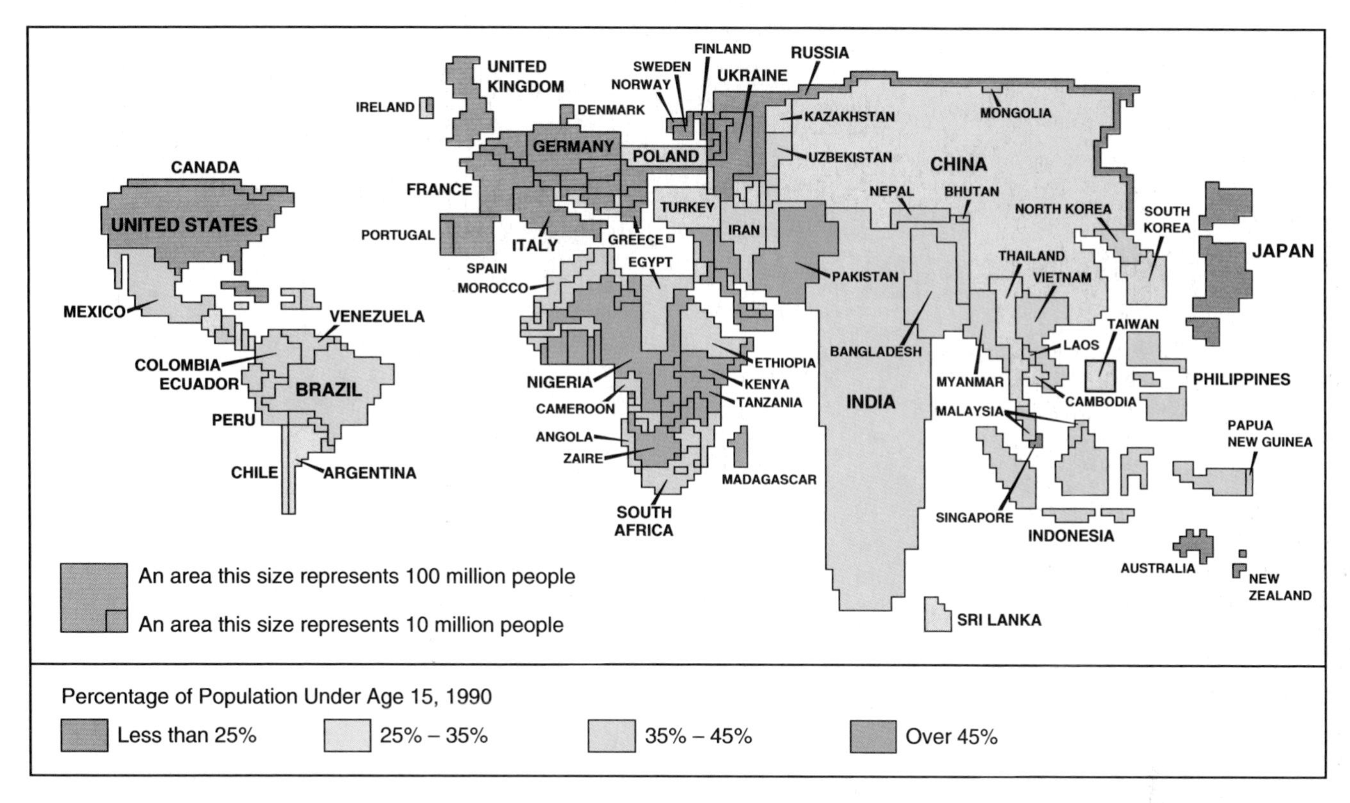

Figure 6–6. This cartogram is drawn on the same base outline as Figure 6–2. It maps where the national populations are young and also conveys a visual idea of what proportion of the world population is young. Comparison with Figure 6–2 suggests that populations are young where rates of population increase are high. (Data from World Resources Institute, *World Resources 1990–91.* New York: Oxford University Press, 1990, pp. 256–57.)

ulations in many countries in Latin America, Africa, and the Near East are between 15 and 18 years. Forty percent of Mexico's 86 million citizens are under 15 (see Figure 6–6).

In a country with a low median age, a high percentage of the total population is drawing on national resources but has not yet reached its most productive working years. The national budgets of these countries are consumed by the challenges of feeding, clothing, and schooling youngsters. As the youngsters reach their mature working age, the national economy must be able to provide productive jobs for them or else their frustration may break out in civil disorder. This scenario underscores the need for economic development.

In contrast, the median ages of populations of wealthy countries are higher. In Western Europe, for example, the median age is 34; in the United States it is nearly 33. These countries face a challenge that is the opposite of that faced by the poor countries: Increasing percentages of their populations are either approaching retirement or are already retired. National economies with shrinking labor forces must be able to support increasing numbers of pensioners.

The **dependency ratio** of a country suggests how many of its people are in their most productive years. It is defined as the ratio of the combined child population less than 15 years old and adult population over 65 years old to the population of those between 15 and 65 years of age. In the poor countries this ratio may be as high as 70, whereas in the rich countries it is usually about 50.

The Demographic Transition The history of population growth in today's rich countries reflects a phenomenon called the **demographic transition,** which involves three distinct stages of growth. In the first stage of this evolution, both the crude birth rate and the crude death rate are high, and so the population does not increase rapidly. All countries experienced this state in the past, when the human population was a fragile number in constant danger of actual extinction, locally if not globally, by periodic epidemics.

As incomes increase and as medical science develops, however, crude death rates drop dramatically (see Figure 6–7). The **infant mortality rate,** which is the number of infants per 1000 who die before reaching 1 year of age, usually falls almost immediately (see Figure 6–8). Another factor lowering death rates was the improvements in the quantity and quality of food that resulted from the Agricultural Revolution, discussed in the next chapter. Crude birth rates, however, remain high (see Figure 6–9). Thus, in this second stage of the demographic transition, the rate of natural increase is high. Some scholars compute each country's population **doubling time.** A number compounding at 3 percent annually doubles in only 23 years, but a number compounding at only 1 percent annually takes 70 years to double.

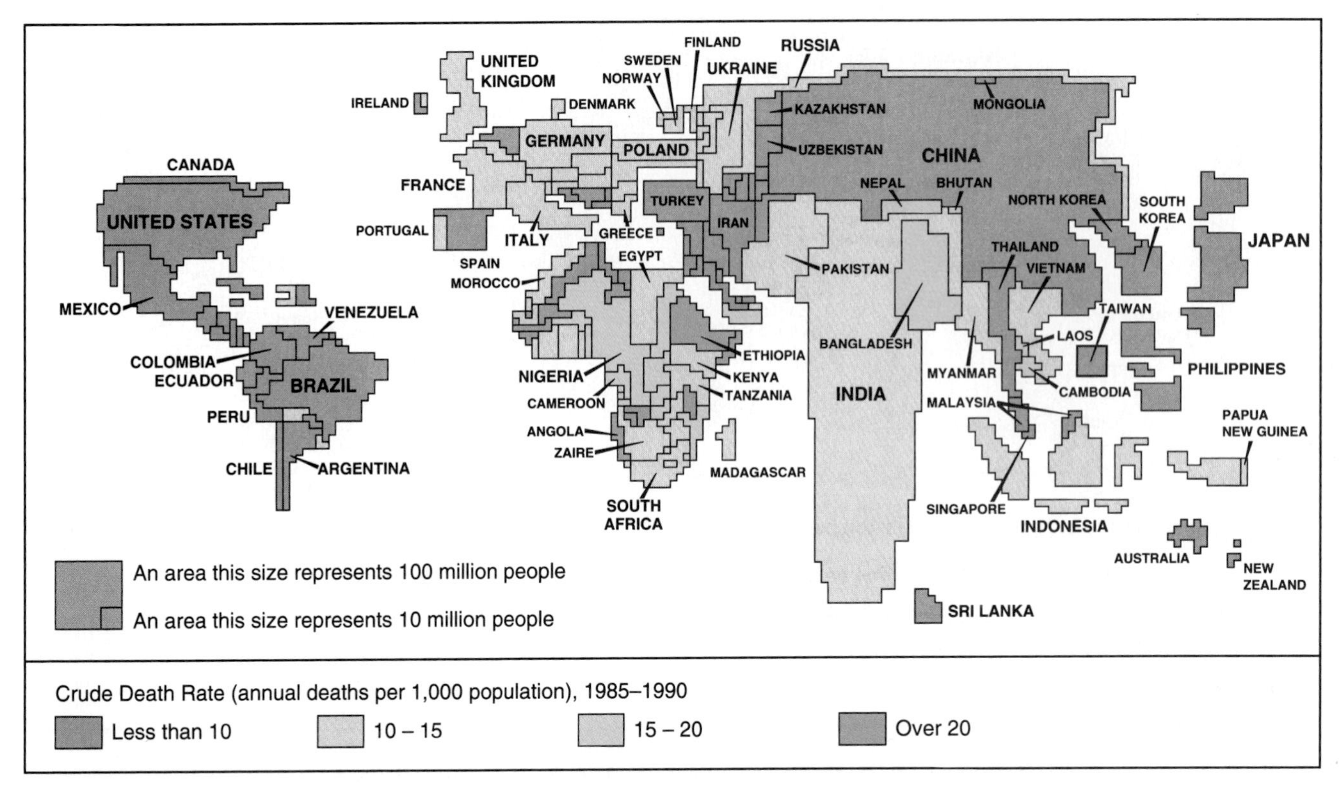

Figure 6–7. Crude death rates are falling almost everywhere, but this cartogram reveals that they remain high in some countries, which are home to large numbers of people. Today only environmental disasters, epidemics, or war can substantially increase a country's crude death rate. (Data from World Resources Institute, *World Resources 1992–93*. New York: Oxford University Press, 1992, pp. 250–51.)

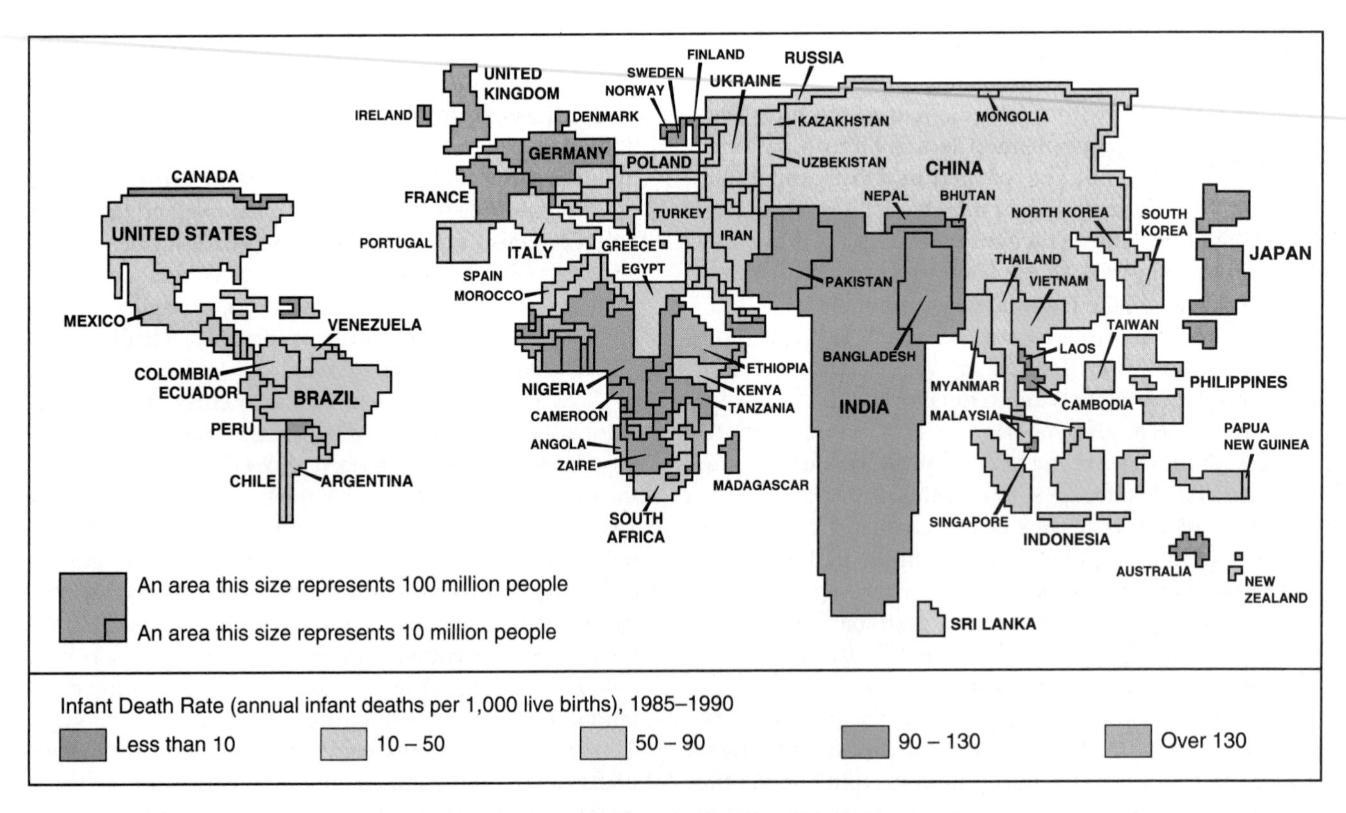

Figure 6–8. This cartogram shows that tragically high infant mortalities still plague many populous countries. (Data from World Resources Institute, *World Resources 1992–93*. New York: Oxford University Press, 1992, pp. 250–51.)

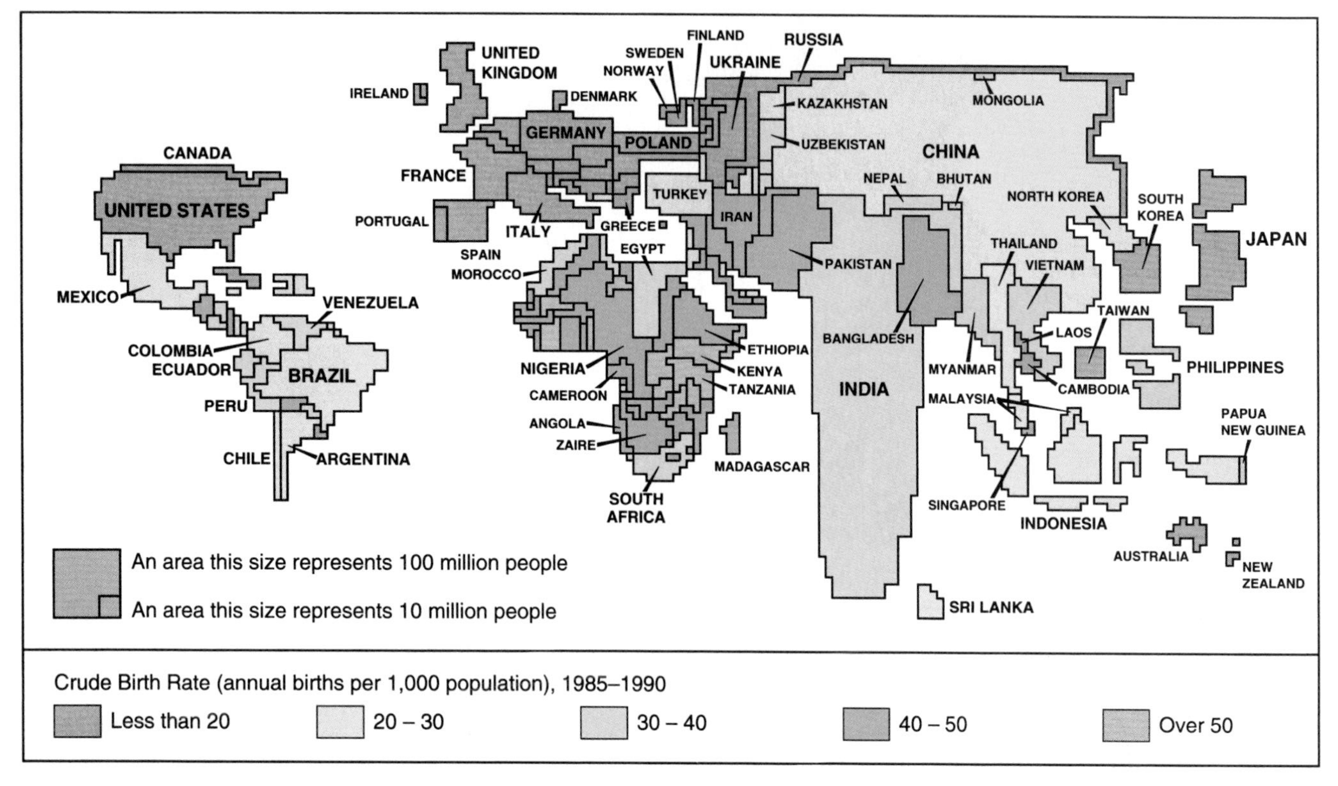

Figure 6–9. This cartogram illustrates that birth rates remain high among a large share of the earth's population. Today's rich countries generally have low rates, but the poor countries still show high rates. Compare this cartogram with Figure 6–6. Countries with high birth rates have young populations. (Data from World Resources Institute, *World Resources 1990–91*. New York: Oxford University Press, 1990, pp. 256–57.)

Several reasons explain the persistence of high birth rates during the second stage of the demographic transition. In traditional societies children may be economic assets. They provide more hands to help in the fields. Also, people traditionally expect their children to look after them in their own old age. If infant mortality rates are high, parents have many offspring to ensure that some survive into adulthood. Not until one or two generations have passed will adults accept the reality of lower infant mortality rates and start having fewer children.

The rich countries of the world today demonstrate a third stage of the demographic transition. Their crude death rates remain low, but their crude birth rates have also dropped. Total fertility rates are at or below the replacement level. This decline in childbearing occurred with economic growth, urbanization, and rising standards of living and education.

Today few rich countries have rates of natural population increase higher than 1 percent. In Western Europe only Ireland records a total fertility rate higher than the replacement rate. Once the current generation has passed in many rich countries, their populations might fall by anywhere from one-tenth to one-third in each generation. Italy, for example, has the lowest crude birth rate in the world, and at current rates its population could decline from more than 57 million in 1990 to 23 million by 2020. Reasons cited by Italians for having fewer children typify the social changes in the rich countries: Women are pursuing careers, day-care services are limited, urban housing is short, and abortion has been legalized.

Unlike the rich countries, many of the poor countries have not made the shift from the second to the third stage of the demographic transition. Birth rates remain high, as do rates of population increase. In even the world's poorest countries, antibiotics and immunization have helped many more babies to survive (see Figure 6–10). Infant mortality in many poor countries is today as low as it was quite recently in the rich countries.

Meanwhile, crude birth rates and total fertility rates remain high. The average total fertility rate for African countries, for example, is 6.1. This is partly because in many of these countries, having more children remains the economically rational choice for the poor—for all the same reasons as it was in the past for the populations of today's rich countries (see Figure 6–11). It can even be argued that some African countries face labor shortages in agriculture, and because they lack capital for agricultural machinery, increases in rural labor forces would increase food output.

Modern medicine also keeps people alive longer, and the combined effect of lowering infant mortalities and keeping people alive longer expands **life expectancy,** the average number of years that a newborn baby within a given population can expect to live (see Figure 6–12). Lengthening life expectancies is still another fac-

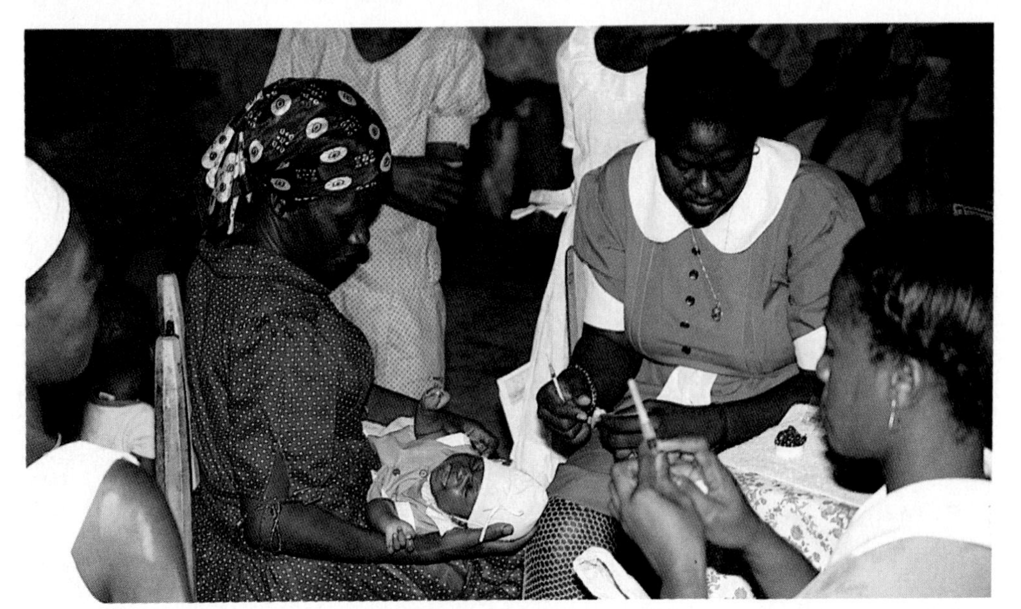

Figure 6–10. These Haitian children are getting protection against measles. The introduction of health measures such as this greatly reduces the mortality rates among children, and can trigger high population growth in countries in the second stage of the demographic transition. (Courtesy of the Agency for International Development/ John Metelsky.)

tor that can increase total populations. Today life expectancies in even the poor countries are lengthening faster than they did in the past in the countries that are today in the third stage of the demographic transition. For example, life expectancy in Mexico, a poor country in the second stage of the demographic transition, is already higher than life expectancy was in France as recently as 1950, and France is a rich country clearly in the third stage.

Total Projected Increases Even if worldwide crude birth rates and total fertility rates were to fall tomorrow, the total number of people on Earth would continue to grow. This is because the number of young women reaching childbearing age is larger than ever before. In Brazil, for example, the fertility rate has dropped 30 percent since 1965, yet the birth rate has dropped by only 19 percent, and the total number of births each year has risen from 3 million in the late 1950s to almost 4 million.

The Earth's population will go on rising, but the sooner the world's total fertility rate falls to the replacement rate, the lower the total population will be at which the numbers eventually level off. Many observers insist that the sooner the Earth reaches a constant population, **zero population growth** overall, the better. When will this happen? At what figure?

The United Nations offers a range of projections pertaining to this question. If the world's total fertility rate drops to the replacement rate as soon as 2015, then the world's population will stabilize at about 8 billion. If the world reaches replacement total fertility by the year 2035, then its population will stabilize toward the end of the 21st century at 10.2 billion. If, however, the total fertility rate does not fall to replacement level until 2055, then the population will not stabilize until it has reached 13 billion. The difference between the two extreme estimates, 5 billion people, is almost the Earth's present population.

ABC NEWS ***America's Infant Mortality Crisis***

High infant mortality rates are typically found in countries with low educational levels, unsanitary conditions, and inadequate access to health care. Thus, we probably would not think that a wealthy country like the United States would have a high infant mortality rate. In fact, the United States ranks only 20th in lowest infant mortality rates. Each year, 40,000 babies in the United States die before their first birthday. This problem is especially acute among the poor, especially the black poor. For example, a black baby born in Washington, D.C., is more likely to die before its first birthday than is a baby born in Cuba. Problems such as maternal drug use during pregnancy and lack of prenatal care (due to factors such as poverty and lack of education) are claiming the lives of thousands of U.S. infants. The same prolems arise in some of the country's more rural areas. In fact, Mississippi has the highest infant mortality rate of any state. In these areas the problem is frequently a shortage of doctors.

As you watch the video titled "America's Infant Mortality Crisis," consider the following questions.

1. What are the major reasons for the different rates of infant mortality from place to place and among different groups across the United States?
2. How does the infant mortality rate reflect on the national health-care delivery system?
3. What steps can be taken to help reduce the infant mortality rate?

Figure 6–11. Children can be an economic asset if they are quickly put to backbreaking labor in the fields or rice paddies, as in this photograph taken on the Indonesian island of Java. (Courtesy of the Agency for International Development.)

Slowing Projected Increases

There are two ways of slowing natural population increases: (1) increasing the death rate, and (2) lowering the birth rate. These two possibilities are discussed next.

The Crude Death Rate Nobody advocates increasing the crude death rate, but it could happen in the event of a pandemic of a new fatal disease. Influenza viruses change their antigenic (disease-causing) nature frequently, and they are difficult to battle. The last truly global outbreak of a fatal influenza occurred in 1918, during which some 22 million people died. Still more recently, acquired immune deficiency syndrome (AIDS), a viral disease that destroys the body's ability to fight infections, has spread around the world. People do not actually die of AIDS but of the opportunistic infections

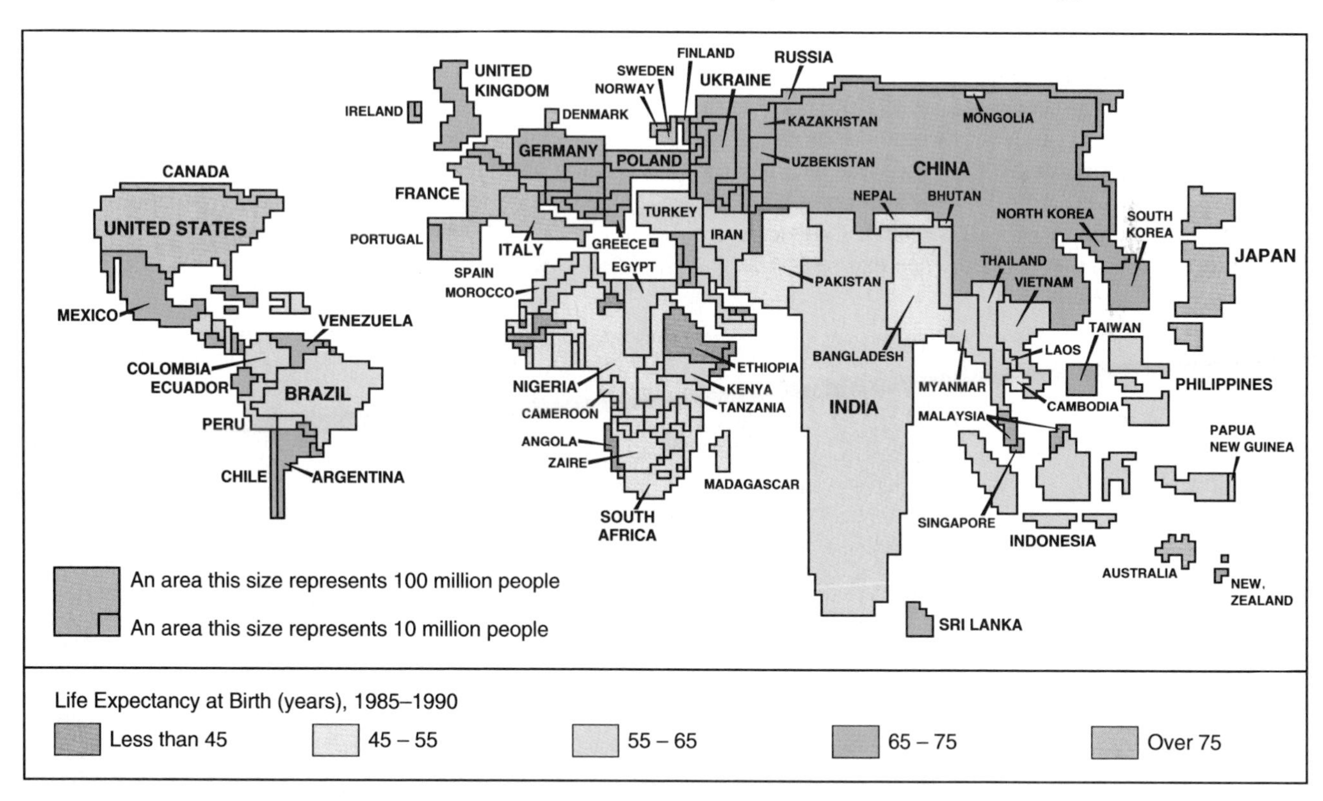

Figure 6–12. Life expectancies are lengthening everywhere, but much of the human population still cannot be expected to reach what we would call "old age." Compare this cartogram with Figures 6–7 and 6–8. Countries with high death rates have low life expectancies. (Data from World Resources Institute, *World Resources 1990–91*. New York: Oxford University Press, 1990, pp. 256–57.)

Figure 6–13. This young mother in Kenya stops to read the warnings about AIDS. The spread of this disease across Africa has cast doubt on all projections of population increase there. (Courtesy of Wendy Stone/Gamma Liaison.)

that follow it. In mid-1990 the World Health Organization (WHO) estimated that 8 million to 10 million people worldwide had contracted the AIDS virus, called the human immunodeficiency virus (HIV), and that about 700,000 had developed the actual disease. By the year 2000, about 4 million people will have died from AIDS-related illnesses, another 10 million people will be sick, and the total number infected may approach 40 million. In the United States over 100,000 people had died by early 1992, over 200,000 more had been diagnosed with the disease, and an estimated 1 million were infected. The disease has been spreading fastest in the poor countries, and by 2000, 90 percent of all infections will probably occur there. Already by 1990 the virus had infected one-third of the population in some parts of Africa, and in several African countries AIDS-related illnesses were the leading cause of adult death (see Figure 6–13).

Periodic outbreaks of infectious diseases can kill large numbers of people. For example, cholera, which is characteristic of crowded, unsanitary conditions, killed thousands of people in South America in the early 1990s. Natural disasters also claim tens of thousands of lives each year. A tropical storm in Bangladesh in May 1991 claimed as many as 200,000 lives (see Figure 6–14).

War also tragically increases the death rate. Iraq, a nation of some 19 million people, suffered the loss of an estimated 120,000 young male soldiers in its 1980–1988 war with Iran, and by some estimates tens of thousands more in its 1991 war with UN forces. Those losses will depress the nation's birthrate and affect its demographic balance for decades. The victims of wars also include noncombatant civilian populations.

The Crude Birth Rate Lowering the crude birth

Figure 6–14. The Indonesian island of Java has a high local death rate, a reality that is reflected in the prominence of a local coffin store. (Courtesy of the Agency for International Development.)

Figure 6–15. This Indonesian government billboard tries to reduce traditionally high family sizes by suggesting that smaller families can be healthier, happier, and better off. (Courtesy of the Agency for International Development.)

rate has become the focus of countries' efforts to reduce overall population increases. Today almost all poor countries are committed to family planning (see Figure 6–15). A traditional cultural preference for large families is strong in some areas, and, as emphasized above, having children may even be the economically rational choice for some families in poor countries. Surveys have shown, however, that sizable portions of the populations of most poor countries want smaller families than have traditionally been normal. Except in Africa and much of the Middle East, crude birth rates are now declining in almost every country faster than anybody foresaw 20 years ago.

Sterilization is the most popular form of birth control in the poor countries. Some 98 million women and 35 million men around the world had chosen that solution by 1990. The other current mainstay is abortion. About 28 million abortions are performed annually in the poor countries, another 26 million in the industrial world. WHO estimates that improperly performed abortions kill about 300,000 women around the world each year.

Modern technology has provided women with new means of birth control. One example is the drug RU486, which induces a miscarriage if used in the early stages of pregnancy. The French government, part owner of the patent, has referred to it as "the moral property of women." Still another recent invention is a steroid-filled capsule embedded in a woman's arm. It provides contraceptive protection for 5 years. Use of these methods is rapidly spreading.

Opposition to birth control Despite technological advances, family-planning programs face obstacles. The manufacture, distribution, and instruction in the use of contraceptive devices is expensive. Two-thirds of the users of condoms, diaphragms, and sponges live in the industrialized world, and attitudes about family planning in the rich countries determine whether these countries offer assistance to the poor countries—even for those poor countries that want it.

Religion can block birth control programs. Speaking in Mexico in May 1990, Pope John Paul II reiterated the Roman Catholic church's position: "If the possibility of conceiving a child is artificially eliminated in the conjugal act, couples shut themselves off from God and oppose His will." The Catholic church also adamantly opposes abortion. Some other religions also preach against specific forms of birth control and abortion.

Another obstacle to birth control is the low status of women. In numerous societies women lack political and economic rights, have limited access to education, and in general exercise little control over their own lives. The United Nations estimates that fully half of the married women in the poor countries do not want any more children, yet many have little or no access to methods of birth control. For example, in Pakistan, Bangladesh, and Arab countries where women suffer low status, crude birth rates remain high. In contrast, on the Indonesian island of Java, where women enjoy greater rights and better education than in many other Muslim regions, crude birth rates are falling.

In culture realms where male children are preferred to females, birth rates frequently remain high because if a couple's first child is female, they will continue having children until a male child is born. In these traditional societies, a girl is often regarded as just another mouth to feed, a temporary family member who will leave to serve her husband's kin. Also, in many societies a woman's parents must provide her with a *dowry*, a substantial financial settlement at the time of her marriage. A son, on the other hand, means more muscle for the farm work, someone to care for aged parents, and someone who can, in several Oriental and traditional African religions, ritually burn the necessary offerings to ancestors. A son carries on the family name. The financial preference for male children may be reduced in societies in which women are winning the right to work outside the home.

In societies in which men believe that the number of their children is a measure of their masculinity, family planning relies on contraceptive techniques that do not involve men's participation or even knowledge. In Mexico, for instance, an estimated 60 percent of women who receive government-sponsored birth control do so secretly.

Throughout the world, fertility rates are lower in urban than in rural areas. Moreover, the larger the city, the more likely women are to use contraception. Family-planning services may be easier to reach. Children cost more to raise in big cities, and space is at a premium. Roughly half of the world's people will live in cities by 2000, and this will probably lower total fertility rates.

Birth control programs: some examples The most dramatic and significant change in fertility has been in China, where the annual rate of natural population increase dropped from 2.9 percent to 1.4 percent between 1965 and 1990. Because China is home to about 22 percent of the human population, that decline alone accounts for much of the change in world trends. In 1979 China launched a one-family/one-child policy with the aim of limiting the population to 1.2 billion by 2000. The government offered incentives to reduce child-bearing, such as financial rewards and special privileges for small families, and penalties for exceeding the targets. Unfortunately, this policy perpetuates traditional prejudice in favor of males. Women are allowed to have a second child only if the first is a girl, but sterilization is obligatory after the second child. Women have been coerced into having abortions, and there have been reports of *female infanticide*, the killing of female children. China is determined to continue to reduce the crude birth rate, but the total population will nevertheless reach 1.27 billion by 2000.

India's population grew from 342 million at independence in 1947 to 850 million in 1991. With a growth rate of 2.2 percent per year, it will pass 1 billion by the end of the century, and India will overtake China as the world's most populous country by the middle of the next century. Variations in crude birth rates within India reflect those around the world. Family sizes are smallest in regions where the education levels and social status of women are high, and population growth is concentrated in the areas with the lowest levels of education, especially the northern Hindi belt, where literacy among women is as low as 5 percent.

In the mid-1970s a government-sponsored family-planning drive was given coercive powers, but after widespread reports of forced sterilizations, the government backed down. Today radio and television advertising and information have brought to all Indians the message that small families are healthier and happier. Sterilization accounts for about 90 percent of India's program.

To encourage lower birth rates, the Indian government also established a program that offers financial support to elderly people. This program removes the incentive for couples to have many children so they will be taken care of in their old age.

Recent signs from Zimbabwe, Kenya, and Botswana suggest that fertility may at last be starting to fall in tropical Africa. Those three countries have increased contraceptive use dramatically. In Botswana few couples used contraception 10 years ago; today around 27 percent do so. In Zimbabwe the figure is 40 percent. Kenya has long had one of the highest rates of natural population increase in the world, estimated at 4.1 percent per year, but even Kenya has recently reduced its crude birth rate. Key factors in Kenya seem to have been the government's promotion of family planning, a well-developed rural health system that can deliver contraceptive services, and the widening realization that families are financially better off with fewer children.

Brazil's total fertility rate has been cut in half in one generation. By the year 2000, Brazil will probably have 170 million people, 50 million fewer than demographers predicted in the 1970s. The falling fertility rate has been attributed to the spread of contraceptive devices and to access to television. Television transmits images, attitudes, and values of modern, urban, middle-class existence in which families are small, affluent, and consumer-oriented. Similar drops in fertility have occurred in Colombia and Mexico, both of which have government-sponsored family-planning programs.

A resurgence in total fertility rates? Since 1975 the decline in fertility seems to have leveled off in some poor countries. In India the total fertility rate is stuck at 4.8. In Indonesia it has dropped by more than a quarter since 1962, to 4.3, but it has stuck there, and so the population is still rising by 2 percent per year. Lack of women's rights and the preference for sons throughout Asia and Africa may prevent total fertility rates from ever declining to replacement levels.

Even in the rich countries, the projections that declining total fertility rates will lead to overall population decreases may be premature or exaggerated. In several advanced countries, rates have recently risen. For example, total fertility rates in Sweden sank to a low of 1.6 in 1978 and then rose to 2 in 1989 and 2.1 in 1990. Similar though smaller rises have been noted in Denmark, Norway, and Iceland. Even in the former West Germany the rate climbed from its record low of 1.28 in 1985 to 1.42 in 1989. In the United States too, the rate has been creeping upward, to 2.

The explanation for these trends could lie in the growing ability of women to earn their own living outside the home. Sweden offers the most advanced day care and parental leave in the world. In the United States some employers have adopted comparable packages, and women's salaries are rising. When child-care costs drop as a percentage of women's wages, the fertility rate rises. Also, in the United States fertility is about 25 percent higher for foreign-born than for native-born women. Because a growing percentage of Americans are foreign-born, the overall fertility rate has increased.

Cautions We might still be underestimating future population growth. The UN projections rest on the assumption that life expectancy in the poor countries in 2020 will be no higher than it was in North America in 1950. In fact, life expectancies in many of these countries are rising. Crude death rates, particularly infant mortality rates, will, it is hoped, continue to fall, and these falling death rates will trigger at least a temporary boost in rates of natural population increase (see Figure 6–16). It is hoped, however, that crude birth rates will also fall quickly enough to offset that increase.

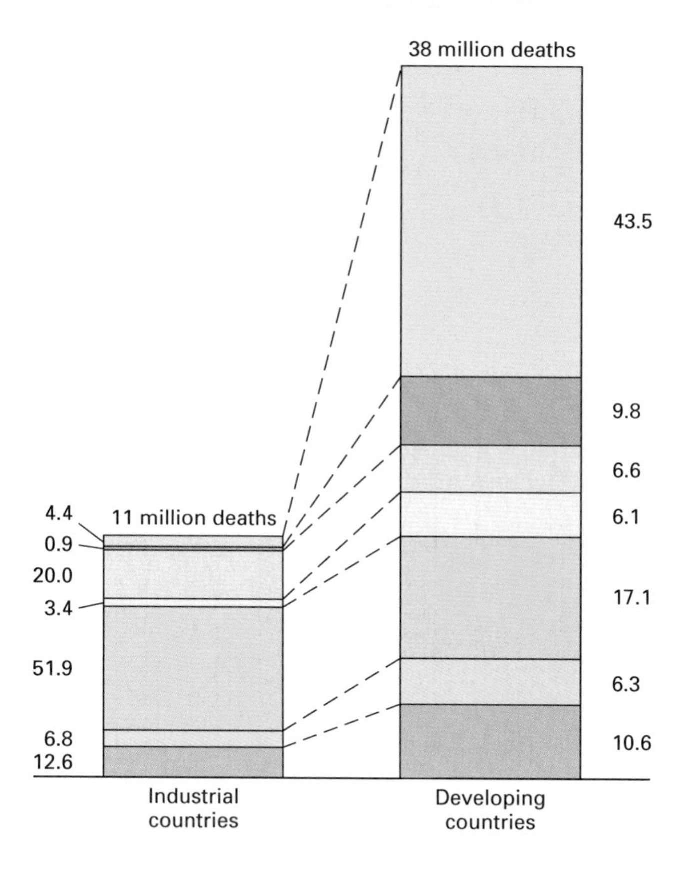

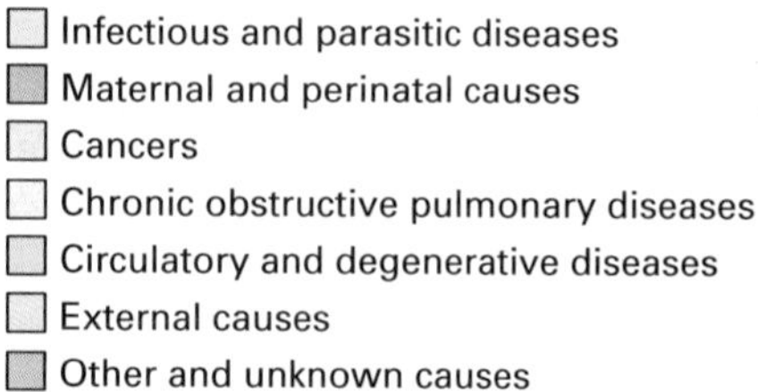

Figure 6–16. The principal causes of death are very different in the rich countries and in the poor countries. A great many deaths in the poor countries are due to causes that modern medicine can cure. Introducing these cures into the poor countries would increase total populations very rapidly. (Adapted with permission from The World Bank, *World Development Report 1991*, p. 62.)

The Aging of the Populations in Rich Countries

For the rich countries, substantial natural population increases seem to be over, and this presents new and different opportunities and problems. As people have fewer children and live longer, ratios of the old to the young rise rapidly (see Figure 6–17).

Unprecedented questions arise about the possibility of sustaining economic growth and about the equitable distribution of wealth among different generations. Increasing shares of national wealth and expenditures must be directed to health care and other problems of the elderly. Taxes on the working population may rise to support pensioners.

The settlement patterns of retired people can bring about a geographical redistribution of the national wealth. Pensioners in Florida and Arizona, for example, are spending money earned in Illinois and New York. In addition, the United States is the only country that makes social security payments to noncitizens who have retired after working in this country. Because many of these people retire to their native countries, these payments drain capital that otherwise could be used for investment in the United States. For some poor countries, however, these same payments represent a significant contribution to the balance of trade and payments. People in the Dominican Republic, for example, receive more than $1.5 million each month in U.S. social security benefits.

The leveling off and aging of a country's population also affects the long-term use of the land. Land-use decisions in the United States have always been based on the assumption that there would be future growth and population increase, but that may no longer be the case. The decisions that are being made today, for instance, and that will be made throughout your lifetime about the use of land, about the care of the environment, about wilderness and farmland preservation, and about urban-suburban development may be permanent.

Some people in the rich nations have urged increasing fertility. Stimulating fertility in the rich countries, however, might discourage the family-planning efforts in the poor countries. The renewed emphasis on increasing fertility is also often based in racist attitudes that there are not enough of "us" and too many of "them" (see Figure 6–18).

The Future The problems associated with rapidly aging populations are today restricted to the rich countries, but quite early in the next century countries now labeled young and poor will themselves be aging fast. Today's enormous numbers of young and poor people will grow older in countries in which the crude birth rate is dropping, and fewer young people are coming along behind them. Therefore, the median ages of those countries will rise sharply. By 2020 the median age of

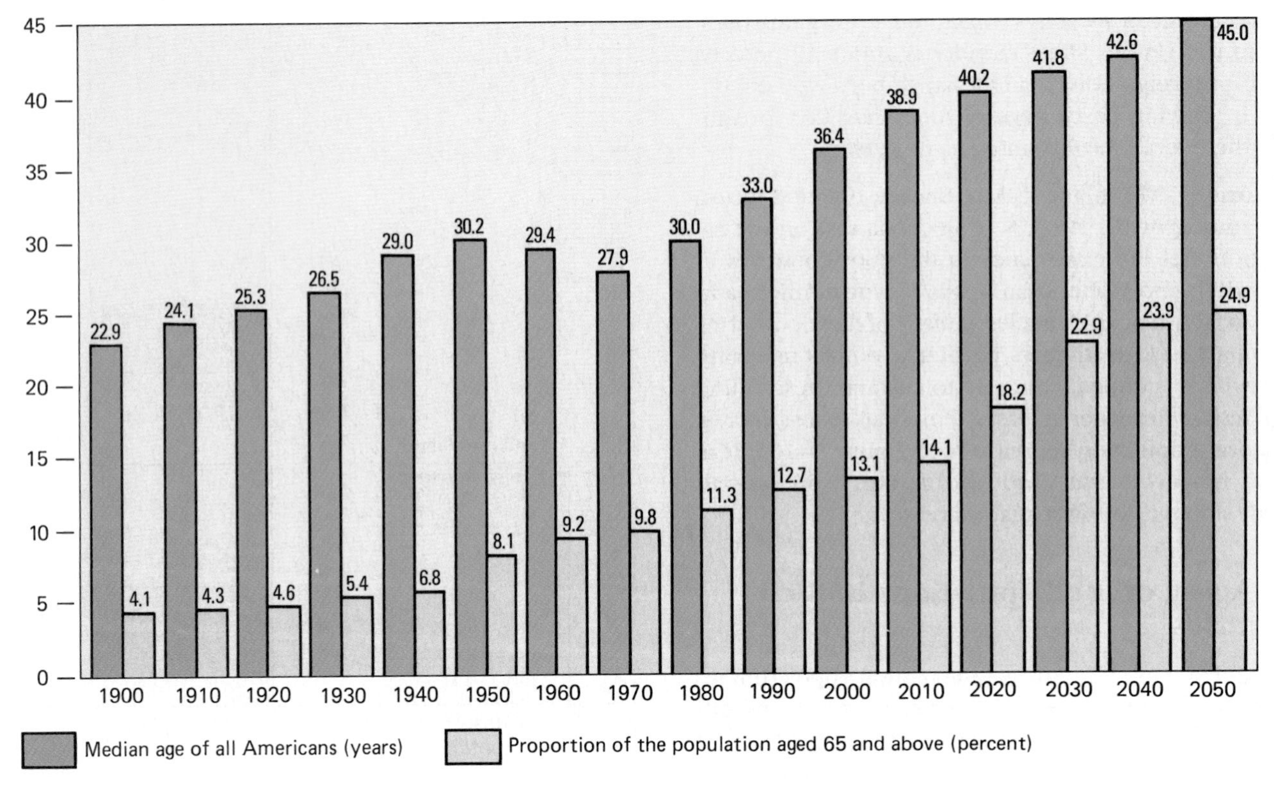

Figure 6–17. This chart projects the increasing percentage of the U.S. population that will be 65 years old or older through future decades. The aging of the population is already triggering political struggles over questions of taxation and social security benefits. In what year do you intend to retire, and what percentage of the population will be of retirement age then? (Reprinted with permission from John J. Macionis, *Sociology*, 3rd ed. Englewood Cliffs, N.J.: Prentice Hall, 1991, p. 370.)

Mexico's population, about 17 in 1990, will probably be 33.4 years. By that time, the percentage of people in China over 60 years of age will be the same as in Europe. Some poor countries will soon be aging faster than the rich countries are now. Those countries do not have national welfare or social security programs to replace lost traditions of families caring for the old. They have perhaps two generations to build up their national incomes to be able to look after their elderly.

Figure 6–18. So few native French citizens are having babies that this one is shown thinking "It seems that I'm a sociocultural phenomenon." France's low birth rate has alarmed the French government into proclaiming "France needs children." (Courtesy of N. Maceschal/The Image Bank.)

HUMAN MIGRATION

Human beings do not stay put; they never have. Wanderings and migrations of people around the world have distributed and redistributed populations throughout history and even prehistory, and significant redistributions are still going on today.

Geographers who analyze human movements divide the causes for those movements into push factors and pull factors. **Push factors** drive people away from wherever they are. Push factors include starvation and political and religious persecution. **Pull factors** attract people to new destinations. Pull factors include economic opportunity and the promise of religious and political liberty. Physical geography can also serve as a push

The thriving City of Eden as it appeared on Paper

The thriving City of Eden as it appeared in Fact

Figure 6–19. Charles Dickens's comic novel *Martin Chuzzlewit* (1844) recounts these gentlemen's dismay when they migrate to the U.S. frontier and find that it is much more primitive than land developers had suggested. Many of Dickens's U.S. readers took offense, but in fact such misrepresentations were common.

or pull factor. For example, the movement of people from colder to warmer regions has remapped the population geography of the United States.

Push and pull factors should be investigated with care. People choose a destination, for example, because of what they think they will find at their target, not because of the reality of conditions at the place. Therefore, to know why someone moved, we cannot study geographic factors alone; we must learn what was in the person's mind. Migrants can be surprised and disappointed in what they find at their new homes (see Figure 6–19). This demonstrates what we learned about cognitive behavioralism.

Migration has not always been voluntary. Some migrations have been forced, and millions of people suffered the tragedy of migrating in slavery, regardless of push or pull considerations of their own.

Movements of people can often be explained by studying flows of trade or by mapping the most convenient transport routes away from those places people want to leave. People also follow information. Successful migrants write home, and they pave the way for new arrivals by providing them with employment and financial assistance. Information often travels between formerly imperial powers and their former colonies, as between France and several African countries or between the United States and the Philippines. In these cases potential migrants are partially acculturated to their target before they leave home. Information flows can be astonishingly place-specific. In 1990 an estimated 80 percent of the Hispanics living in metropolitan Washington, D.C., were from the tiny Central American country of El Salvador—many of them undocumented. Nobody knows exactly how word first spread throughout El Salvador to trigger this migration.

Many migrants intend to stay in their new location only until they can save enough capital to return home to a higher standard of living. These **sojourners** are usually men who are either unmarried or have left their families in their home country. Many sojourners eventually decide to stay, at which point they send for their wives and families. The United States has always attracted great numbers of sojourners. In various years between 1890 and 1910 the number of Italians leaving for Italy was as high as 75 percent of the number of Italians arriving in the United States. Even today many Greeks, Irish, Caribbean peoples, and Africans shuttle between their homelands and the United States.

Some migrations have numbered in the millions of people. Some small migrations have been disproportionately significant because the migrants played key roles in transforming the government, language, economics, or social customs in their new homelands. As we shall see later in this chapter, the impact of the international migrations of Indians and Chinese, for example, or of Caribbean blacks into the United States is not fully realized by their numbers alone.

Human migration continues around the world today—of rural peasants everywhere into the cities (discussed in Chapter 9), of economic refugees from poor countries into the rich countries, and of political refugees fleeing tyranny and persecution.

Prehistoric Human Migrations

According to the theory of evolution, if one species disperses, its different dispersing groups encounter different environmental circumstances. If, through generations, different specific mutations thrive and multiply in each of the new environments, different **races** of the original species can be identified. If the races are different enough that they can no longer interbreed, then they must be recognized as distinct species. This has not happened in the case of humankind. Human beings are one single species, and all its members can successfully interbreed. In this biological community of humankind, subdivisions are insignificant.

Humankind has traditionally been divided into races, however, according to certain characteristics. The traditional criteria for these subdivisions have been external features, known as *secondary characteristics*, such as eyefolds or skin color. A number of different classification systems have been based on these criteria, the most common of which divides humankind into three races: Caucasoid ("white"), Mongoloid ("yellow"), and Negroid ("black").

Other classificatory systems are based on an internal criterion, the analysis of blood types. Studies and classifications of blood types have helped explain the distribution of certain diseases, and in that way they have helped direct researchers toward cures. Sickle-cell anemia, for example, a condition of the red blood cells, is very common in Equatorial Africa, and it is carried by 9 percent of African Americans. Research has shown, however, that the trait crosses "racial" lines. It is widespread in India, Greece, and Turkey.

Classification of people by blood type yields a different set of categories ("races") from classification of people by skin color or other external features. Individuals cross over categories, thus casting doubt on the significance of either system of categories.

Geneticists—scientists who study the passage of traits from parents to offspring—approach the question of human variation from a cellular or molecular level. They are currently working on global surveys of *genetic markers*—variations in proteins and enzymes that reflect differences in people's genetic makeup. The inheritance of genetic traits is controlled by the arrangement of molecules within an individual's deoxyribonucleic acid (DNA), which is contained within the person's chromosomes. The conclusions of researchers who study genetic markers do not always agree with those of any other classification system. Japanese, Koreans, and Inuits (Eskimos), for instance, have traditionally been included among Mongoloids, but genetics places them closer to Caucasoids.

Some geneticists believe that they have sketched a family tree of all humankind, depicting the divergence of the world's peoples from a single stock that lived in Africa as long as 200,000 years ago. From this one stock, known as *Homo sapiens*, a three-branch family tree evolved: North Eurasians, Southeast Asians, and black Africans. The black Africans stayed "home," and the others migrated away. The peoples who would become the Native Americans, for instance, seem to have walked over a land bridge from Asia where the Bering Strait is located today between 15,000 and 35,000 years ago.

All these events are deep in prehistory, but research and hypothesis formulation about the origins and diffusion of humankind continue in geography, biology, anthropology, archeology, and, as we shall see, linguistics. This is an exciting frontier of human self-knowledge.

In fact it is impossible to make unambiguous distinctions among races. How many races there are, which groups of people are in each, and even the very meaning of "race" are all open to question.

Nevertheless, attempts at racial classification continue. Although they frequently are based on scientific curiosity or methodologies, they have also been used to buttress racist ideologies. **Racism** is a belief in the inherent superiority of one race over another and the linking of human ability, potential, and behavior to racial inheritance. Racism is wrong, immoral, and injurious to both those who discriminate and those discriminated against. Racists' search for "racial purity" is nonsense in a world in which people have intermingled as they have throughout human history.

THE DIFFUSION OF PEOPLES IN THE MODERN WORLD

For purposes of understanding contemporary geography, we begin our study of human diffusion 500 years ago, before the European voyages of exploration and conquest. The most significant population transfers during those years were migrations of Europeans to carve enclaves of settlement around the world, migrations of blacks out of Africa, and migrations of other groups instigated by European expansion.

European Migration to the Americas

The substitution of populations in the Western Hemisphere was the greatest anywhere (Figure 6–20). The Native-American population fell victim to slavery, warfare, and, most important, to European diseases, including smallpox, influenza, measles, and typhus. These diseases do not seem to have occurred before in the Western Hemisphere, and so the natives had no immunity to them. The Native Americans, in turn, transmitted syphilis to the Europeans.

We have only estimates of the Native-American populations at the time of the European conquests, and they range from 30 million to 100 million. The numbers of deaths were appalling. There were between 12 million and 25 million Native Americans in Mexico before the

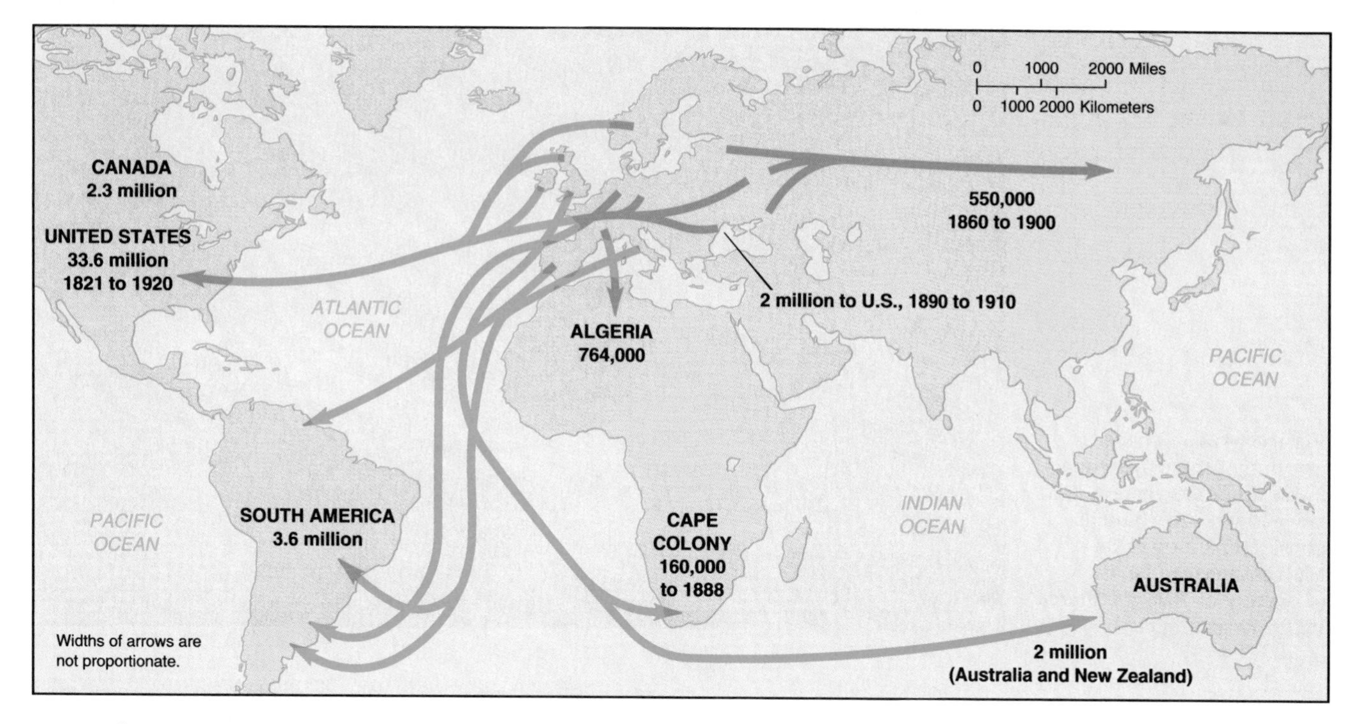

Figure 6–20. The nineteenth century saw the greatest migrations of Europeans. Europeans regarded the Western Hemisphere as a "New World" open to European settlement, and their massive migrations forever changed hemispheric demographics.

arrival of the Spaniards; by 1600 that number had declined to about 1.2 million. The population of the Inka Empire in South America plummeted from about 13 million in 1492 to 2 million by 1600. The estimated 5 million Native Americans in Brazil shrank to only 220,000 today. In what is presently the United States and Canada, the Native-American population fell from about 2 million in 1492 to a low of 530,000 in 1900.

North America received some 40 million Europeans. Today they and their descendants make up the majority of the population. The number of self-identified Native Americans in the United States tripled to 1.8 million between 1960 and 1990, perhaps reflecting growing cultural pride, but in 1990 they constituted less than 1 percent of the national population. Canada's 500,000 legally acknowledged "status Indians" constitute only 2 percent of that country's population.

Both the United States and Canada have mixed policies of both assimilating Native Americans and at the same time setting aside lands for the preservation of Native-American cultures. In both countries Native Americans retain preferred hunting and fishing rights over substantial areas in addition to their tribal lands, or reservations, over which they exercise more substantial, but never complete, control (see Figure 6–21). Canada has granted the Inuit people (Eskimos) political domain over roughly one-fifth of the country, the eastern half of the former Northwest Territories, now the territory of Nunavut. Continuing litigation of other extensive native claims embroils national politics. In the United States Native-American lands represent 2.2 percent of the total land area.

About 4.5 million Europeans, mostly from the Mediterranean area, migrated to "Latin America," and that region developed a complex racial structure. Ibero-American (descended from Spain and Portugal) societies were composed as a pyramid of varying proportions, with a large base of Native Americans and blacks, a lesser number of people of mixed race (a person of mixed white-black ancestry is usually called a ***mulatto***; one of mixed black-Native American a ***zambo***; and one of mixed white-Native American a ***mestizo***), and a minority of whites on top. Several Latin American colonies and later the successor independent states institutionalized racial status rankings. Native Americans lost their rights in their own homelands, even though they may have constituted the majority of the population. Only since the mid-1980s have some South American countries, including Peru, Venezuela, Colombia, and Brazil, set aside lands for Native Americans.

Native Americans still make up one-third to one-half or more of the populations of Mexico, Guatemala, Ecuador, Peru, and Bolivia. Native Americans have played the most conspicuous political role in Mexico, but they may be waking to power in Peru and in Ecuador. A Peruvian of Japanese ancestry won that country's presidency in 1990, explicitly allying himself with the non-white peoples. Among the southernmost midlatitude states Chile claims to be 95 percent mestizo, but the Native Americans were practically exterminated in Uruguay and Argentina.

In Central America most people are mestizos. Costa Rica is the only country in which whites form the majority.

Figure 6–21. This is a meeting of the tribal council of the Cherokee nation. U.S. law allows such councils considerable power on Native-American lands. (Courtesy of Nancy J. Pierce/ Photo Researchers.)

The African Diaspora

Black peoples have migrated out of Africa, either freely or in slavery, for centuries. These migrations are known, collectively, as the **African diaspora**, from the Greek word for "scattering," first used by the Jews to describe their worldwide migrations.

Racial slavery, as the modern world has come to know it, originated in medieval Islamic societies. Muhammad himself owned slaves, and Arabs and Persians invented the long-distance slave trade that transported millions of sub-Saharan captives out to a realm stretching from Islamic Spain across to India (Figure 6–22). By the fifteenth century, when Christian Europe launched its geographic expansion, a slave system based on African labor, including sugar plantations and a fully developed slave trade from Africa, was already in place across the Muslim world.

The Islamic slave system lasted over a period of 12 centuries, and in total numbers it probably surpassed the African slave trade to the Western Hemisphere. The trade diminished slowly in Islamic societies, lingering longest in the Arab countries. (It was finally banned in Saudi Arabia in 1962 and in Mauritania in 1980). The absence of a large population of black survivors across this enormous stretch today can be explained by the high mortality rate, by assimilation, and by the practice of castrating slaves. Even in central India today, however, there are communities of blacks who are the descendants of African slaves.

Figure 6–22. Malik Ambar was a great Indian military leader in the seventeenth century. A number of black Africans rose from slavery to prominence across the Islamic world, either through domestic or military service. (Ross-Coomaraswamy Collection. Courtesy, Museum of Fine Arts, Boston.)

The European Slave Trade European transportation of black slaves to the Western Hemisphere had a more dramatic impact on population geography. In the fifteenth century Europeans took a few hundred slaves from West Africa to Europe and the Atlantic Islands, and

Figure 6–23. Tens of thousands of Africans passed through this old slave market on Gorée Island off the coast of today's Senegal on their way to the Western Hemisphere. Today the government is restoring the island's historic fortications as a museum and raising money to build an international conference center (Courtesy of Brian Seed/TSW Chicago.).

in the 1520s Europeans began transporting slaves to the Western Hemisphere (see Figure 6–23). Black Africans were proposed as an alternative labor supply to the dying Native Americans. Some Europeans viewed the enslavement of blacks as a twofold humanitarian gesture: It would save Native-American lives and at the same time would convert the Africans to Christianity. Fray Bartolomé de las Casas (1474–1566), whose writings contain the strongest denunciations of brutality toward the Native Americans and who won the title Protector of the Indians, was one of the earliest advocates of black slavery as an alternative to forced Native-American labor. He lived to abhor the black slave trade.

From 1526 until the trans-Atlantic slave trade came to an end in 1870, about 10 million to 12 million slaves reached the Western Hemisphere from Africa (see Figure 6–24). Migration of blacks to the Western Hemisphere far exceeded migration of whites until the middle of the nineteenth century. Slavery was ended in the United States in 1865 and in Brazil in 1888.

The sources of the slaves reflected patterns of European colonialism in Africa. The Portuguese African colonies of Angola and Mozambique provided the slaves of Portuguese-ruled Brazil, for instance, whereas West Africa supplied most of the slaves for North America.

The pattern of resettlement in the Western Hemisphere reflected the uses to which the slave labor was put. Slaves were brought to tropical and semitropical plantations in the U.S. Southeast, to the islands of the West Indies, and, throughout Latin America, in the "tierra caliente" (hot land) at elevations of less than 3000 feet (2700 m). Mexico was an exception because there the slaves were used in the cities and mines. Mexico never had a significant rural black population, and eventually Mexico's black population died out or mixed with the other groups almost entirely. Mexico abolished slavery at independence in 1821.

Blacks in the Americas Today Today descendants of slaves constitute the majority of the population on most West Indian islands and in Belize. In South America, blacks still occupy, for the most part, the lowland areas. After centuries of intermarriage, tropical South American countries still have significant minority black, zambo, or mulatto populations.

In the United States, about 12 percent of the population is African American—27 million people—giving the country one of the world's largest black national populations. The black population was originally concentrated in the Southeast, but blacks began to migrate to the cities of the North just before World War I (see Table 6–3). Between 1910 and 1970 some 6.5 million African Americans headed northward, and the percentage living in the South fell from 70 percent to 50 percent. The civil rights struggles of the 1960s created new opportunities for blacks in the South, and partly for that reason and partly for reasons not fully understood, African Americans have been migrating back to the South since the mid-1970s. About 52 percent of African Americans lived in the South in 1980; 56 percent in 1990 (see Figure 6–25).

The twentieth century has also witnessed a steady migration of West Indian and Caribbean blacks into the United States. Between 1960 and 1990 alone, about 2 million Caribbean blacks migrated to the United States. A high percentage of these people brought capital or job skills, and they and their descendants have played a role in twentieth-century U.S. politics and culture disproportionate to their numbers. Statistical descriptions of America's foreign-born blacks and of their children reveal high educational attainment, high income levels, substantial representation in professional and managerial occupations, and low levels of unemployment. These statistics more closely match the statistical descriptions of America's highest-achievement whites, the Jews and the

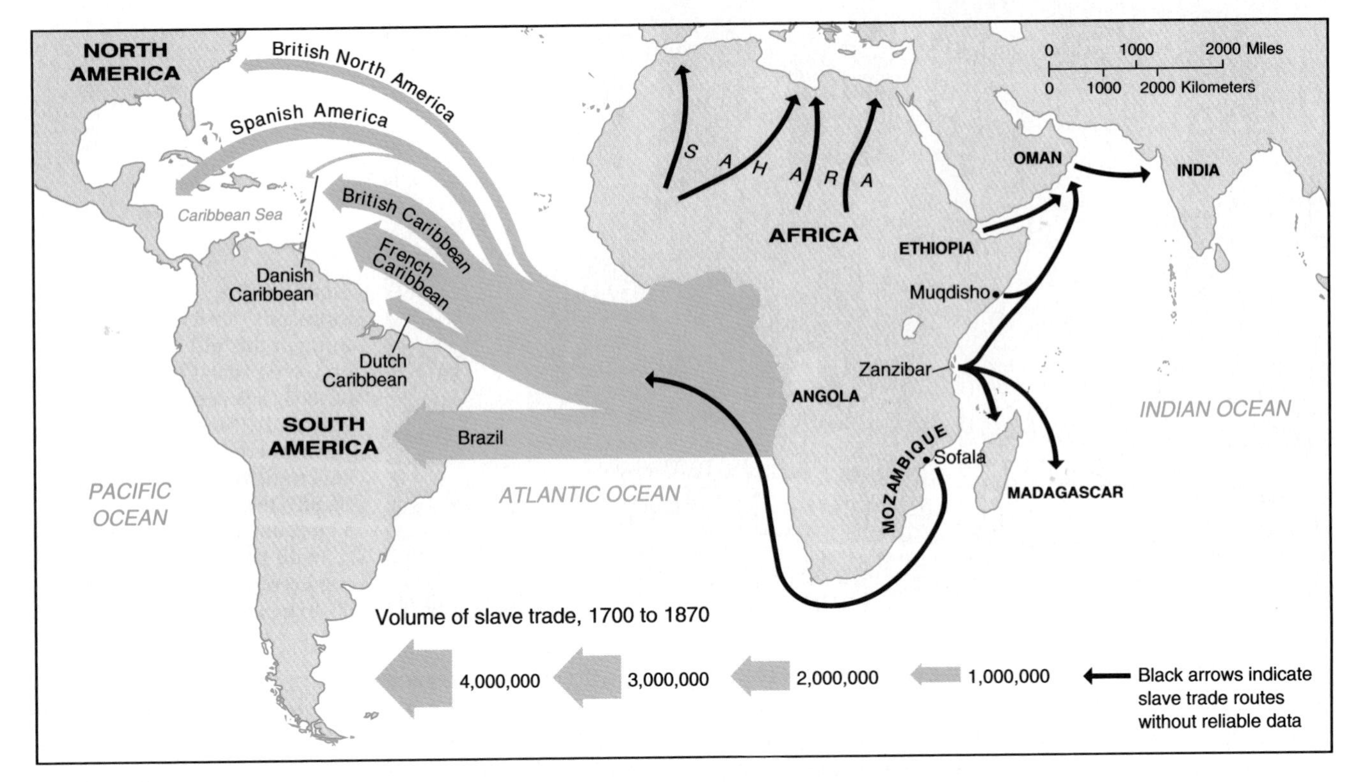

Figure 6–24. Arab slave dealers first transported black slaves north and east, throughout the entire region from Spain to India. Later European colonial interests in Africa determined the sources and destinations of slaves for the Western Hemisphere. The horrors of the African slave traffic were such that about four times as many people were captured in the interior of Africa as the 10 million to 12 million who reached the Western Hemisphere. Notice how relatively few were brought to today's United States.

Irish, than they match the description of the country's often underprivileged native blacks. The first African-American Chairman of the Joint Chiefs of Staff, General Colin Powell, was the son of immigrants from Jamaica. This flow of migrants and sojourners to the United States, combined with the flow of U.S. tourists to the Caribbean, has caused a degree of acculturation sometimes referred to as the "Americanization of the Caribbean."

The cultural and economic might of black Americans overall exerts a tremendous attraction in Black Africa. Black America's literature and consumer products blanket the continent, and black American civil rights leaders are heroes, a tradition begun by W. E. B. Du Bois (1868–1963), who settled in Ghana (Figure 6–26). Senegalese poet-president Léopold Senghor (1906–) singled out poet Countee Cullen (1903–1946) among African Americans whose works sustained

TABLE 6–3 ***Estimated Net Black Migration by Region, 1871–1980***

Intercensal period	South	Northeast	North central	North total	West
1871–80	−60	+24	+36	+60	(na)
1881–90	−70	+46	+24	+70	(na)
1891–1900	−168	+105	+63	+168	(na)
1901–10	−170	+95	+56	+151	+20
1911–20	−454	+182	+244	+426	+28
1921–30	−749	+349	+364	+713	+36
1931–40	−347	+171	+128	+299	+49
1941–50	−1,599	+463	+618	+1,081	+339
1951–60	−1,473	+496	+541	+1,037	+293
1961–70	−1,380	+612	+382	+994	+301
1971–75	+14	−64	−52	−116	+102
1976–80	+195	−175	−51	−226	+30

Note: Numbers in thousands. Plus sign (+) denotes net in-migration; minus sign (−) denotes out-migration.
Source: US Bureau of the Census 1981.

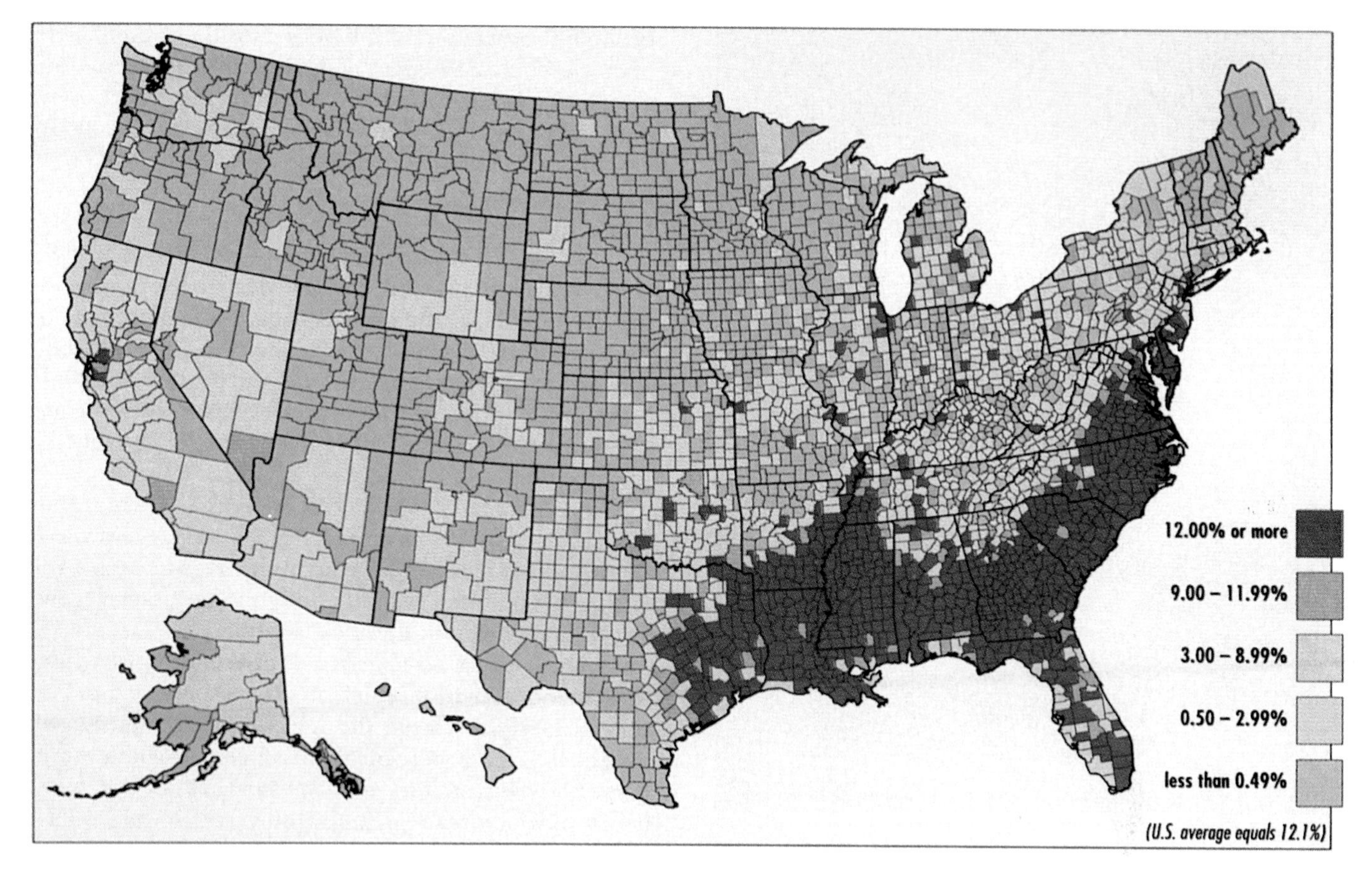

Figure 6–25. The counties with the highest percentages of African Americans among their populations form a solid arc through the U.S. Southeast, plus a few scattered urban concentrations in the North and West. (Reprinted with permission © *American Demographics* (July 1991):8.)

racial pride throughout Africa under colonialism. Because of America's black population and its many achievements, American culture has diffused throughout Africa to a surprising degree. It is displacing the cultural influence of Africa's European former rulers, despite the fact that the United States never had any African colonies.

European Migration to Asia

Although the Spanish, French, Dutch, Germans, Portuguese, British, and Americans all created empires in Asia, few people from these countries settled permanently in the Asian colonies. In no cases did any of these peoples come to make up significant proportions of mainland Asian populations. The most important European settlers in Asia can conveniently be divided into those who migrated by land—the Russians—and those who migrated by sea—settlers from the British Isles who went to Australia and New Zealand.

Russian Expansion Ever since the Russians rose up against their Mongol rulers in the sixteenth century, they have sent a steady stream of settlers to the East, over the Ural Mountains into Asia. Between 1860 and 1914 almost 1 million Russians resettled in the East, and this flow continued within the political framework of the Union of Soviet Socialist Republics. In 1990 the population of Kazakhstan, which was ostensibly a republic of the Kazakh people, was 38 percent ethnic Russian. Russians migrated particularly into the cities in the non-Russian republics. In 1990 Tashkent, the capital of the Uzbek Republic and the third largest city in the USSR, was primarily Russian.

The dissolution of the USSR may trigger mass migrations of Russians living in other republics into Russia. Many of these have not lived in Russia for generations. This problem is discussed at greater length in Chapter 12.

Australia and New Zealand Over 2 million Europeans, mostly from the British Isles, moved to Australia and New Zealand in the nineteenth century, and another wave of 3 million Europeans (half British) arrived between 1945 and 1973. They all but exterminated the native peoples. Latest censuses count only 145,000 Aborigines in Australia (of a total population of 17 million) and 295,000 Maoris among New Zealand's 3.3 million people. The Maoris may regain control over parts of the islands of New Zealand, pending outcome of judicial

Figure 6–26. U.S. civil rights leader W. E. B. Du Bois migrated to Ghana after that African country achieved independence in 1957. Ghana's president Kwame Nkrumah invited Du Bois to come over to edit a series of publications on black history. As a result, Du Bois earned fame throughout Africa. (Courtesy of AP/Wide World.)

review of the 1840 Treaty of Waitangi. Legal actions can be expected to last until the end of this century.

In both Australia and New Zealand, commercial farming and ranching thrived from the beginning of white settlement. Wool, wheat, and later frozen meat were sent to markets within the British Empire. In Australia, white settlement in the dry areas was hazardous and only intermittent until spectacular mineral discoveries attracted some people inland. In general, however, the white populations of both Australia and New Zealand have always been highly urbanized along or near the coasts.

European Migration to Africa

Europeans migrated to Africa and settled in substantial numbers only where environmental conditions made possible a Mediterranean or European style of life and commercial agriculture.

East Africa, Northwest Africa, and West Africa The East African highlands attracted white settlement after the opening of the Suez Canal in 1867. Europeans claimed the fertile lands in today's Kenya, Tanzania, and Uganda, establishing there what was for them a pleasant and prosperous lifestyle. The natives were crowded onto less fertile lands, and capital investment improved the economic situation of the whites, but not of the native peoples. Much of the black population ended up as workers on the white-owned farms or in white enterprises. Tens of thousands of whites remain scattered throughout East Africa, but their economically privileged position is now under pressure from the black-majority governments.

Northwest Africa is called in Arabic the ***Maghrib***, "the island," because the Mediterranean to the north and the vast Sahara to the south isolate this region, which itself supports settlement and agriculture. Here, too, European colonial governments reserved the best lands for European settlers. During French rule of Algeria from 1830 to 1962, some 1.2 million Europeans settled there—about as many as in all 26 countries of tropical Africa combined. When Algeria, Morocco, Tunisia, and Libya gained their independence, most white-owned farms were confiscated or bought by Arabs or Arab governments and often redistributed to the peasantry. Few Europeans remain in these countries.

West Africa never experienced substantial white settlement. The heat and humidity discouraged a European lifestyle, and so the region's population never exceeded 1 or 2 percent white, most of whom were colonial administrators and entrepreneurs. The incidence of fatal disease among whites caused the area to be called "The White Man's Graveyard." Today whites still come or serve as business executives, technicians, or even government advisors, but they still make up only insignificant percentages of West African populations.

South-Central and South Africa Southern Africa includes large areas that Europeans found well suited to settlement and creation of a European lifestyle. Several colonies did achieve considerable prosperity—for the whites—but always based on repression and exploitation of the native black majority. As these colonies won independence, white racist regimes yielded to black-dominated governments (see Figure 6–27).

The Republic of South Africa In 1671 the Dutch established Cape Town colony as a staging point for voyages to the East, and numerous Dutch farmers, called Boers, came to settle. Britain seized the strategic colony in 1806. The Boers fled north onto a high plateau, the Veld, where they established two new republics, the Orange Free State and the Transvaal. The British eventually overcame these republics in war and annexed them in 1902. In 1910 the British united these two republics with the British colonies of Cape Province and Natal to form the independent Union of South Africa. This state became a republic in 1961 and drew into increasing international seclusion in the face of opposition to its internal policy of racial segregation, called **apartheid.**

The white minority perpetuated a strictly segregated society, with each ethnic group holding a fixed status. Laws enacted by the government divided the population into four groups: whites (British and Boers com-

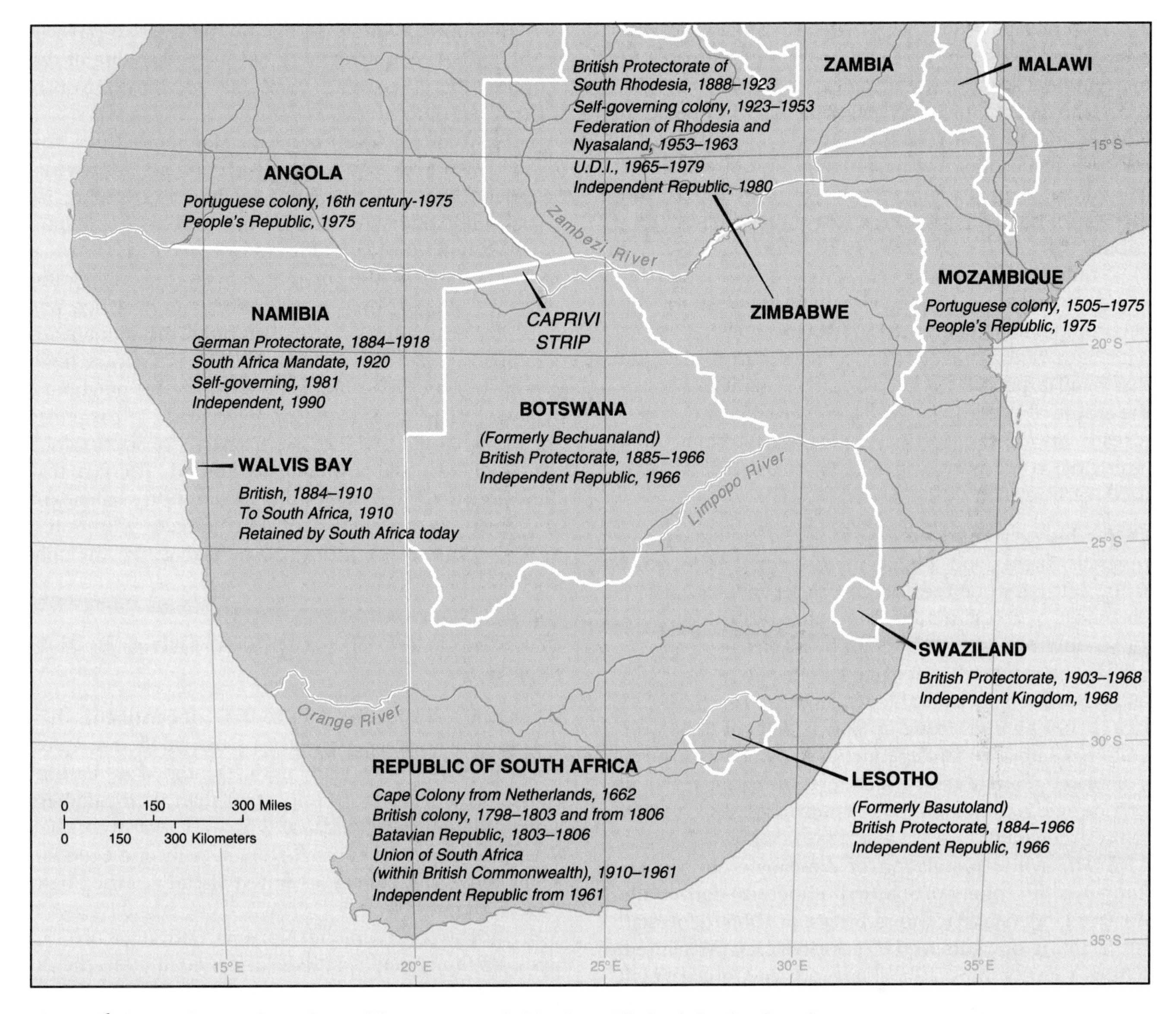

Figure 6–27. Substantial numbers of Europeans settled in the midlatitude lands of southern Africa when these were colonies, and today's independent countries are still working to achieve ways for the whites and majority blacks to live together.

bined, totaling 13 percent of the total population in 1987), Asians (2 percent), Mixed (or Colored, 9 percent), and blacks (75 percent).

The Colored, concentrated in Cape Province, are mulatto products of early mixing between Boers and Hottentots. Most Asians came to Natal to work on the sugar plantations, but eventually many rose to middle-class status or even achieved considerable wealth in the capital, Durban. The South African government stopped immigration of Asians in 1913, and it maintained for 50 years the legal fiction that they would someday return to Asia. It did this to restrict Asian political rights. Only in 1963 was the Asians' residency recognized as permanent.

The consequent constitution created a three-chamber Parliament, one white, one Asian, and one Mixed. Blacks were excluded from participation, and the government planned to move them into segregated *Bantustans*, on the basis of ethnic roots. Four such homelands were granted formal independence, but no other country in the world recognized them. Six other homelands were granted dependent status. The Bantustans covered only 13 percent of the territory of South Africa, rural areas of low agricultural potential and without other resources. The blacks obviously would never be able to support themselves in these lands; rather, they would have to sell their labor within South Africa. There they would be immigrant workers, with no political rights. The other 87 percent of the territory was reserved for whites. By 1987 the per capita incomes of the various groups were: whites, 14,880 rands; Asians, 4560 rands; Mixed, 3000 rands; and blacks, 1246 rands. The infant mortality rate for the whites was 9 per 1000; for the Asians, 16; for the Mixed, 41; and for the blacks, 63.

By the late 1980s the system of apartheid and Bantustans had begun to collapse under the pressures of international sanctions and internal rebellion. In 1991 the government finally repealed the 1950 Population Registration Act, which identified everybody by race. By the early 1990s a new governmental system was being negotiated. Democratic rule may yet be achieved, and the Bantustans will probably be reunited with the Republic.

Any new democratic government will face tremendous pressures. Blacks understandably want the life of conspicuous affluence that South African whites enjoy, but South Africa does not have the wealth to provide that for all its people. Democratic government also brings the threat of warfare among the several black peoples. Already during the black struggles against white rule, intertribal rivalries have resulted in massacres of hundreds of blacks by other blacks.

Other Southern African states The Portuguese colonies of Angola and Mozambique did not have many white settlers when they became independent in 1975, and most of those settlers fled, often destroying what they could not take with them. Botswana also had few white inhabitants when it gained independence in 1966, and when Namibia gained independence from South Africa in 1990 whites made up only 5 percent of its population. Zambia won independence in 1964 with a black-dominated government, which in 1975 nationalized all private landholdings and other enterprises. By 1991 few whites remained.

The white population of Zimbabwe (previously known as the colony of Southern Rhodesia) unilaterally declared independence from Britain in 1965 to forestall the granting of rights to the majority black population. Fifteen years of international boycotts and internal civil warfare followed until a new constitution was internationally accepted. It guaranteed the whites, who constituted 6 percent of the population, political protection in a system of racial representation. This system lasted until 1987, when the whites' privileged status was ended. By 1990, due to white emigration and a high black birth rate, whites made up less than 1 percent of the population. These few whites, however, still held title to the best agricultural land. In the 1990s the government may nationalize and redistribute the land, and the success of this policy may determine the country's future.

Future economic cooperation in southern Africa After years of white political and economic domination, southern Africa's future welfare may depend on cooperation. The ten majority-ruled states of southern Africa—Angola, Botswana, Mozambique, Namibia, Tanzania, Zambia, Zimbabwe, Lesotho, Malawi, and Swaziland—are united in a Southern African Development Coordination Conference (SADCC), and South Africa may join this group. Trade between South Africa and these other countries, however, has been hampered not only by political sanctions but by the relative poverty of the SADCC countries. In 1990 the total value of the output of their economies combined was only one-quarter that of South Africa.

New international acceptance for South Africa may actually inflict costs on these other countries. Botswana, Swaziland, and Lesotho have profited from being included in a South African customs union; Mozambique received outright financial assistance from South Africa; and workers from Mozambique, Swaziland, and Lesotho all found jobs in South Africa. When South Africa was isolated internationally, it had to bribe these countries for friendly relations. A democratic South Africa, however, might be less willing to subsidize its neighbors. Malawi, Botswana, Lesotho, and Swaziland in particular are all so closely tied into the economy of South Africa that their effective autonomy is limited. The fact that they are landlocked makes them even more dependent on South Africa. Namibia must cooperate with South Africa because South Africa retains Walvis Bay, the only port that Namibia can use.

The Migration of Indians to Other British Colonies

When the British ruled the Indian subcontinent, they encouraged migration to other parts of their empire. These migrants, historically undifferentiated as "East Indians," have been more important than their numbers alone might suggest, because wherever they went they formed a bourgeois infrastructure socially and economically above the natives, as noted already in the South African state of Natal.

Indians settled in both Myanmar and Singapore, but since Myanmar achieved independence in 1948 the Indians have been persecuted there. Over 200,000 left in the 1980s, but several hundred thousand remain. Singapore is about 6 percent Indian. Indians went also to the Pacific Island of Fiji, but continuing friction between Indians and Fijians has triggered Indian flight to Australia and New Zealand.

Indians still form key elements of the populations in several West Indian countries, even where their percentages are small. Belize is about 2 percent Indian and Jamaica about 3 percent, but Trinidad and Tobago is about 40 percent Indian, and Guyana on the South American mainland is about 50 percent Indian.

Indian migration to British East Africa surpassed white migration there. Indians generally made up less than 5 percent of the populations, but they dominated the commercial infrastructures of the new countries that emerged in the 1960s (Uganda, Kenya, Malawi, and Tanzania), and their high visibility made them targets of the majority blacks' ethnic hostility. Uganda even expelled over 80,000 Indians, who emigrated to Great Britain, Canada, and the United States. Several African countries have recently tried to entice the Indians to return. Over

70 percent of the economy of Kenya was said to be in Asian hands in 1990, and Uganda was offering economic privileges to returning Indians or their descendants. The Indians demonstrate that entrepreneurial skill is one of the most valuable resources in the world.

The Overseas Chinese

From early in the nineteenth century until about 1930, waves of South Chinese left their homeland to work in the British, French, and Dutch Asian empires (see Table 6–4). Their success mirrored that of the Indians. At the time the nations of Southeast Asia gained their independence, their economies were firmly in the hands of these small Chinese minorities, who were often linked across international borders by bonds of family, clan, and common home province.

The Chinese commercial supremacy and their international links still survive, so that today these 35 million Overseas Chinese constitute a formidable economic power throughout Southeast Asia. Even where the Overseas Chinese make up small minorities of the population, they exercise financial influence far beyond their numbers. The economies of several of the countries listed in Table 6–4 are virtually controlled by the Overseas Chinese. Furthermore, as the Japanese have made considerable investments in these countries, the Japanese have preferred Chinese business partners. This partnership has everywhere bolstered the economic dominance of the Overseas Chinese.

Just as the migrant Indians faced native ethnic hostility, so have the Chinese. For example, ethnic resentments dominate the history of Malaysia. When the British ruled Malaysia as a colony, they founded both Singapore and Kuala Lumpur. The populations of both cities came to be overwhelmingly Chinese, although the majority of the population of the entire colony was Malay. Malaysia was granted independence in 1957, but animosity was so great between Chinese and Malays that in 1965 Singapore broke off as an independent, Chinese-dominated city-state. Even in Malaysia, however, the Chinese minority still controls the economy, although the Malay-dominated government blocks Chinese and Indians from advancement in government positions and encourages Malay entrepreneurship. The population of Malaysia today is about 59 percent Malay, 32 percent Chinese, and 9 percent Indian, and the government is encouraging high fertility rates among Malays.

TABLE 6–4 ***Percentage of the Population of Southeast Asian Countries That Is Overseas Chinese, 1990***

Country	Percentage of ethnic Chinese
Hong Kong	98
Singapore	76
Malaysia	32
Thailand	14
Taiwan	14
Cambodia	5
Myanmar	3
Vietnam	3
Indonesia	3
Australia	3
Philippines	1.5

Source: CIA *World Factbook.*

In Indonesia the 5 million Chinese control as much as 80 percent of the country's private industry. Chicken of the Sea brand tuna is a familiar part of a Chinese-Indonesian multinational corporation. The Indonesian government directs investment by Chinese-owned corporations into businesses owned by other ethnic groups. A 1965 Communist coup attempt, supposedly backed by China, ignited anti-Chinese sentiment, and during the next year (*The Year of Living Dangerously* is a Hollywood film about the events) as many as 500,000 ethnic Chinese were slaughtered. The future status of Overseas Chinese in Indonesia remains unclear.

Overseas Chinese play a similar elite role in the Philippines, and Thailand's commercial infrastructure is almost entirely Chinese. Thailand's largest corporation, Charoen Pokphand, was founded by two brothers from South China in 1921, and today it is one of the world's largest producers and exporters of foods. The Bumble Bee brand of tuna popular in the United States is one tiny part of another Chinese-Thai multinational conglomerate. Notice the importance of Thailand as a food and fish producer and exporter in Tables 7–1, 7–2, and 13–4.

Ambitious Chinese also established themselves in New South Wales, Australia, along the west coast of the Western Hemisphere from British Columbia to Ecuador, and in small but significant numbers in the Caribbean.

Tens of thousands of Chinese migrated to California in the mid-nineteenth century, but they were victims of intense racism. A new wave of Chinese migration to the United States began in 1968 and is discussed later in this chapter.

Tens of thousands of Hong Kong Chinese are emigrating before that city reverts to Chinese rule in 1997. Many of them have headed for Australia, and others have made Vancouver, Canada, which in 1991 was 15 percent ethnic Chinese, a major financial center. Other Hong Kong Chinese purchased East German passports just before German unification in 1990, and over 50,000 more were attracted to Panama in the 1980s when Panama sold citizenship.

MIGRATION TODAY

Human migration has by no means come to an end. In 1990 the United Nations estimated that 30 million people lived outside the countries in which they were born, and large-scale migrations still make daily news. The UN *Universal Declaration of Human Rights* affirms every person's right to leave his or her homeland to seek a better life elsewhere, but it cannot guarantee that there will be any place willing to take him or her.

As in the past, the major push and pull factors behind contemporary migration are economic and political. People are trying to move from the poor countries to the rich countries and from politically repressive countries to more democratic countries. In addition, civil and international warfare inevitably create large numbers of refugees. Pressures for migration are growing, and in coming years they may constitute the world's greatest political and economic problem.

The peoples of the rich and democratic countries are beginning to exhibit "compassion fatigue," and doors are closing all around the world. Today no country in the world allows free immigration, although democratic governments are demonstrably incapable of coping with illegal immigration. Once immigrants are in, democratic countries find it difficult to muster the brutality necessary to evict them.

Some poor nations may be relieved to see their unemployable or troublesome people emigrate, especially if those people send money to their families back home. Several countries have come to depend on such **workers' remittances** for a substantial share of their income. In 1988 workers' remittances from abroad equaled 75 percent of the value of Egypt's merchandise exports, 60 percent of Bangladesh's, 46 percent of Pakistan's, and 31 percent of Greece's. Many rich countries today accept foreign workers to do the lowest-paid or lowest-status work in their country, but they do not grant these workers citizenship and all the rights that citizenship would bring. The host countries accept them only as sojourners, and it may be unclear whether the migrant workers themselves hope to stay.

Political Refugees

Many migrants claim to be political refugees, but it is often difficult to discern who is a political refugee and who is only an "economic refugee" seeking a higher material standard of living. As defined by a 1951 Geneva Convention, a **refugee** is someone with "a well-founded fear of being persecuted in his country of origin for reasons of race, religion, nationality, membership of a particular social group, or political opinion." Several countries have expanded the definition to include persecution due to sexual orientation. Worldwide, international wars and civil wars boosted the numbers of political refugees in the late 1980s. The United Nations estimated that in 1990 some 15 million people around the world were only "temporarily" settled wherever they were at that time. The 1991 war of the UN allies against Iraq may have added as many as 1 million or 2 million refugees to the world total (see Figure 6–28).

Other Major Migrations

Emigration From Eastern Europe The crumbling of Eastern Europe's Communist bloc has enabled many of its citizens to seek their fortunes in the West, and the joy with which Western Europe greeted the political development was quickly followed by alarm. Roughly 2 million people left Eastern Europe in 1990, of whom 400,000 were from the Soviet Union. If the economies of Eastern European countries deteriorate in the 1990s, these numbers could rise (see Figure 6–29).

In 1991 an official of the government of the Soviet Union estimated that 1.5 million to 2 million Soviet citizens might apply to leave for work in the West each year until 2000. This may have been a crude form of blackmail to extract economic aid from the West, but Western Europeans do not know how to cope with such a challenge. The United States may accept many Eastern Europeans.

The emigration of Jews from Russia and the other Commonwealth states is a special case. The United States long pressured the Soviet Union to allow Jewish emigration, and each year from 1978 through 1990 more Soviet Jews migrated to the United States than to Israel.

Israel has tried to attract Jewish immigrants. It has built new housing in territories it occupies but to which it does not hold internationally accepted title. This has become a major international issue.

Other Migration into Europe The migration of Europeans around the world has recently been mirrored by migrations of former colonial peoples into Europe. Large numbers have settled in Great Britain (in 1990 almost 9 percent foreign-born, mostly West and East Indian), in France (about 11 percent foreign-born, mostly North African Arab and African blacks), in Belgium (8 to 9 percent foreign-born, mostly Turkish and North African), in Spain (about 2 percent North African), and in the Netherlands (about 4 percent foreign-born). Some 1.5 million Turks lived in Germany in 1990.

The European countries have not been notably successful, economically or socially, at absorbing the immigrants who have already arrived, and exclusionary movements are on the rise. Many immigrants from the Near East and North Africa are Muslims, and, as is discussed in Chapter 8, their religion sometimes clashes with traditional European Christian culture. The immigrant peoples have higher birth rates than the Europeans do, and riots have swept immigrant ghettoes in several European capitals. Compared to the Europeans around them, these groups achieve lower levels of education, occupy substandard housing, and receive inferior services (see Figure 6–30).

Migration to Australia About 90 percent of Australia's 17 million population are of British and European ancestry, and only 4 percent are Asian. Australia has in-

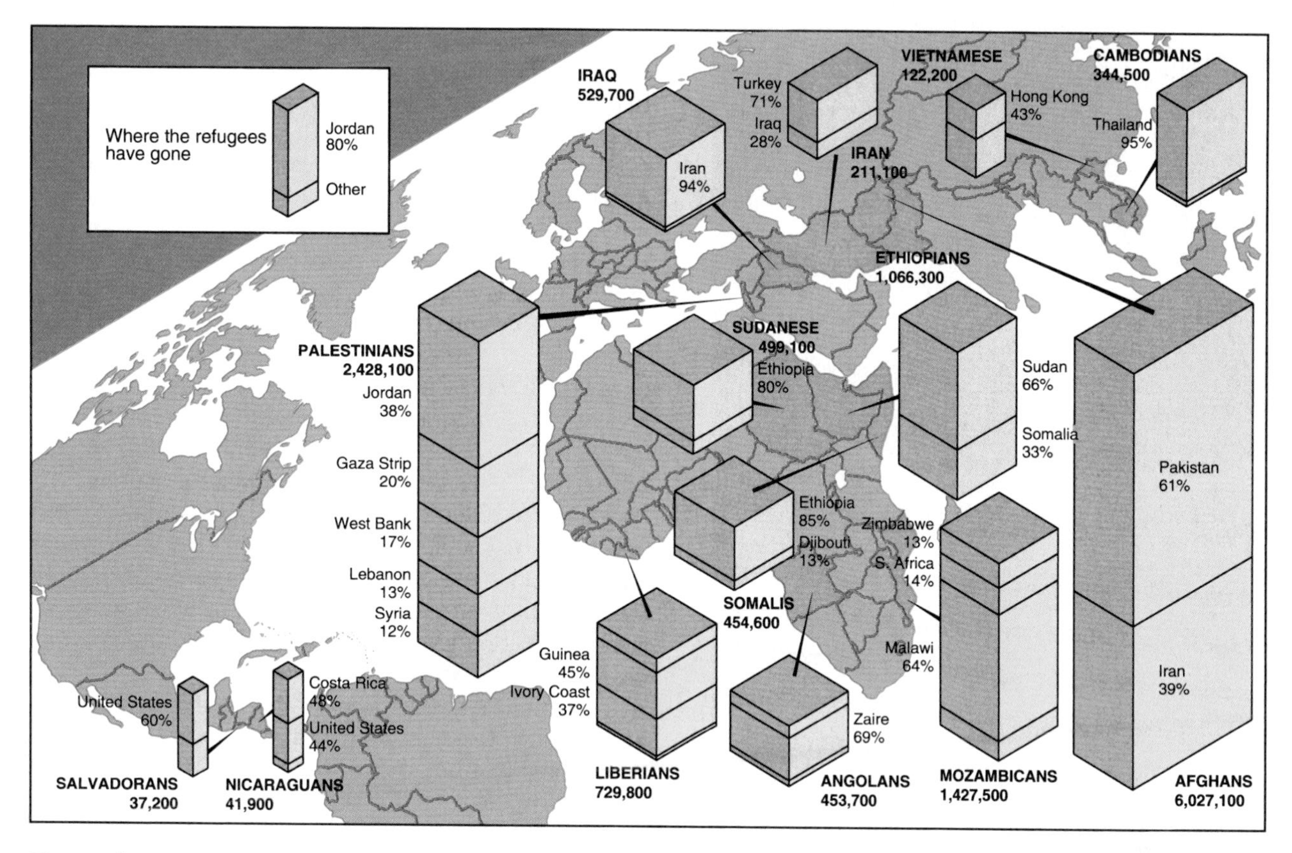

Figure 6–28. This map shows the world's most numerous refugee populations on December 31, 1990. It does not include the millions of people in countries around the world who are "internal refugees," that is, within their own home countries but not at their own homes because of civil wars or persecution.

dicated that it might absorb more settlers. The country abandoned an explicit "White Australia" policy in 1973, but it still has to decide on which basis it will open its borders: refugee status, merit, investment, family ties, or other criteria. Today about one-third of Australia's newest immigrants are Asian, and the proportion is rising.

Migrations of Asian Workers An estimated 4 million to 5 million Asians work abroad, and in many cases it is doubtful whether they will ever return to their homelands. Meanwhile, their remittances to their homelands are an important element in the economies of their home countries.

The absence of these workers lowers the local unemployment rates in their homelands, but it also constitutes a loss of important skills needed back home. Although the home country might not have the resources to reward these workers adequately, it frequently needs

Figure 6–29. A line of soldiers in the service of a rich country pushing back a crowd of desperate economic refugees from a poor one may be a frequent sight in coming years. In this case in 1991 the soldiers are Italian, the refugees Albanian. (Courtesy of G. Giansanti/Sygma.)

Figure 6–30. North African immigrants in Brussels offer a variety of services and goods to their compatriots. (Courtesy of Robert Mordant.)

their contributions. A young Filipino woman, for example, may be needed in her home town as a school teacher, but staying there she would earn $40 per month. If she is earning $150 per month as a maid in Singapore or $384 per month as a maid in Hong Kong, she is better off, but the Phillipines suffers. In 1990 the Filipino consulate in Hong Kong estimated that 52,000 Filipino women were working as maids in Hong Kong. Many were college graduates.

In the 1970s and the 1980s the oil wealth of the Arab states drew into them many Asian workers who seem little disposed ever to go home. About 1.5 million were in Iraq and Kuwait when Iraq invaded Kuwait in 1990. The 1991 Mideast War left these people stranded, and their future is in doubt. As long as the Middle East remains an international hotspot, and as long as Arab oil supplies are important to world prosperity, any destabilizing migration into Arab countries is of world concern.

Hundreds of thousands of Vietnamese fled their homeland after the triumph of Communism there in 1975. The United States and several other countries have accepted thousands of these people, but tens of thousands remain in camps throughout Southeast Asia. Many may be forcibly repatriated to Vietnam.

Migration to the United States The single largest migration flow for the past 150 years has been migration to the United States. In 1989 the United States welcomed 612,086 new immigrants and granted legal status to 478,814 residents who had previously been considered illegal aliens. These numbers do not include illegal immigrants, who could number several hundred thousand each year. Because the nation's overall birth rate is low, immigration is responsible for about 27 percent of U.S. population growth.

The 1990 census counted over 18 million foreign-born U.S. residents, the highest number ever. The source-areas of U.S. immigrants are changing, and this will bring long-term consequences (see Figure 6–31). In 1965 the proportion of European to non-European immigrants was 9 to 1; by 1985 that was reversed. In 1990 three-quarters of all Americans were white; by 2050 that percentage may fall below half. People who are 30 years old or older living in the United States today might have the experience of having grown up in one country and growing old in another without having moved.

In the U.S. census, people choose the racial or ethnic group with which they wish to identify themselves (see Table 6–5). Between 1980 and 1990 the percentage of whites in the total U.S. population decreased, and the percentage of Asians increased the fastest, 11 times the nation's overall population increase. The mix of the Asian-American population also greatly diversified. People of Hispanic origin, who may be of any race, also increased substantially. Over half of Hispanic Americans are of Mexican origin.

The United States was dubbed *the melting pot* in a 1914 novel of that title, but various groups have maintained their cultural identities. Perhaps *cultural mosaic*, which is what Canada proudly calls itself, is a better term.

Immigration into the United States was totally unrestricted until the late nineteenth century, when the government enacted rules explicitly designed to keep out the Chinese and other Asians (see Figure 6–32). Later, in the 1920s, the government issued immigration quotas for various nationality groups. The first quotas favored British, Irish, and Germans (see Table 6–6 on p. 203). By 1920, however, substantial numbers of Italians, Jews from Russian, and other Southern and Eastern Europeans had already immigrated, and the ethnic makeup of the country had already changed. After the 1920s a maximum quota was set on immigrants from the Eastern Hemisphere (in practice, Europe), but there was in theory no ceiling on immigrants from the Western Hemisphere.

In 1965 the United States changed the rules again, in a way that tended to favor Latin Americans and Asians (see Figure 6–33). The number of Asians in the United States jumped from 891,000 in 1960 to 7.3 million in 1990, almost 3 percent of the total national population.

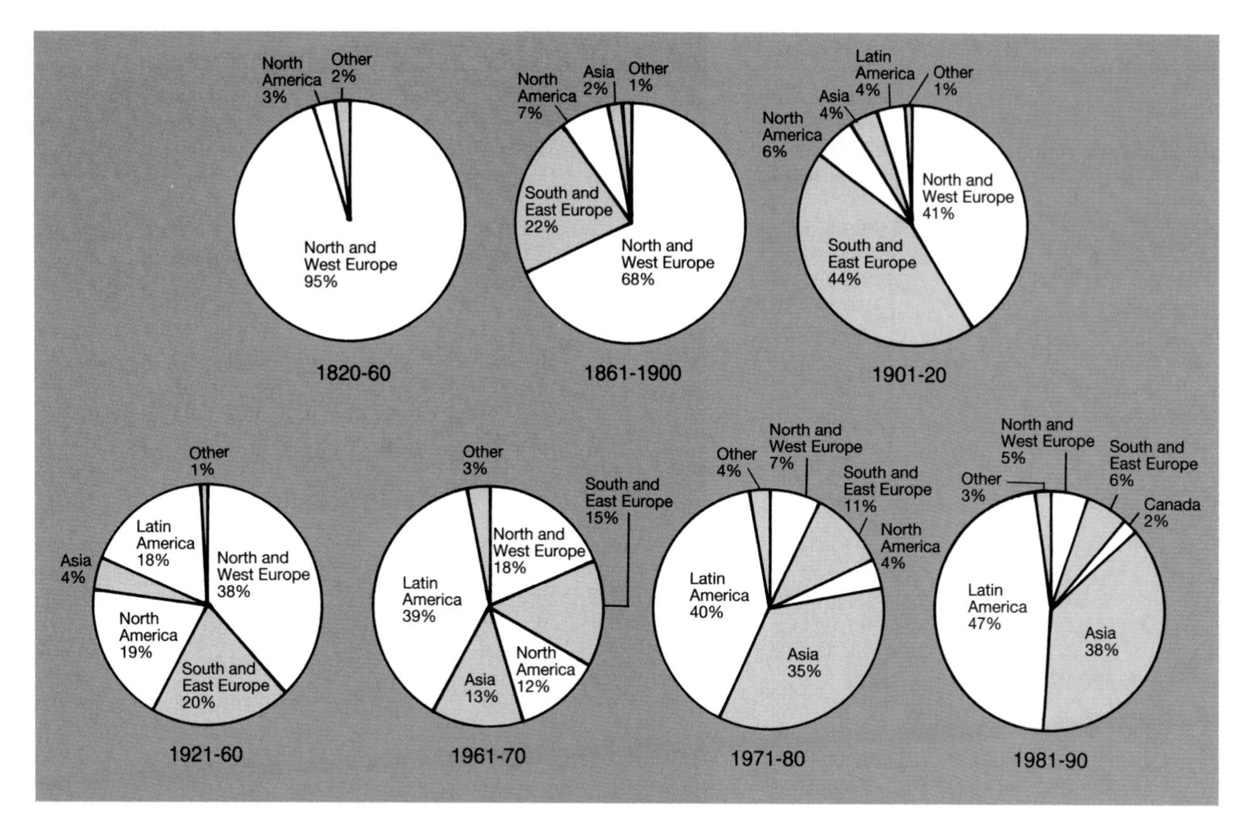

Figure 6–31. These graphs show how the source-areas of immigrants to the United States have changed through history. The percentages of immigrants from Northern and Western Europe have dropped steadily, as the percentages from Latin America and Asia have risen. These trends have affected the composition of the national population. (Based on data From U.S. Department of Justice, Immigration and Naturalization Service.) (Reprinted with permission from Tom L. McKnight, *Regional Geography of the United States and Canada.* Englewood Cliffs, N.J.: Prentice Hall, 1992, p. 41.)

CRITICAL THINKING: *Is Immigration a Substitute for Education?*

Many critics have long argued that the labor force provided by immigration has allowed the United States to slight the education and training of the nation's native work force. They have suggested that the country invest in educating and training its own workers rather than rely on immigrant labor. In his 1895 speech "Let Down Your Bucket," black educator Booker T. Washington (1856–1915) encouraged southern industrialists to seek their labor force around them in the willing blacks who had so long "tilled your fields, cleared your forests, built your railroads and cities," rather than "to look to the incoming of those of foreign birth and strange tongue and habits . . . to buy your surplus land, make blossom the waste places in your fields, and run your factories." This would require "helping and encouraging them . . . and education."

Today, 100 years later, some U.S. industrialists insist that the national labor force still has not been properly trained and that skilled immigrant labor is needed. The National Association of Manufacturers (NAM) defended the 1990 legislation that welcomed skilled immigrants. An NAM spokesperson said of the U.S. labor market: "The bottom line is that the talent isn't there. It certainly might be there in 20 years. But we are where we are today because the American educational system has failed us."

Questions

Which is cheaper, importing a trained labor force or educating native-born workers? In the short run? In the long run? Does the immigration of skilled people discourage a country from educating its own workers?

TABLE 6–5 ***The Racial Composition of the U.S. Population in 1990 and 1980***

Race and Hispanic origin	1990 Census		1980 Census		Number change	Percent change
	Number	Percent	Number	Percent		
RACE						
All persons	248,709,873	100.0%	226,545,805	100.0%	22,164,068	9.8%
White	199,686,070	80.3%	188,371,622	83.1%	11,314,448	6.0%
Black	29,986,060	12.1%	26,495,025	11.7%	3,491,035	13.2%
American Indian, Eskimo, or Aleut	1,959,234	0.8%	1,420,400	0.6%	538,834	37.9%
American Indian	1,878,285	0.8%	1,364,033	0.6%	514,252	37.7%
Eskimo	57,152	0.0%	42,162	0.0%	14,990	35.6%
Aleut	23,797	0.0%	14,205	0.0%	9,592	67.5%
Asian or Pacific Islander	7,273,662	2.9%	3,500,439	1.5%	3,773,223	107.8%
Chinese	1,645,472	0.7%	806,040	0.4%	839,432	104.1%
Filipino	1,406,770	0.6%	774,652	0.3%	632,118	81.6%
Japanese	847,562	0.3%	700,974	0.3%	146,588	20.9%
Asian Indian	815,447	0.3%	361,531	0.2%	453,916	125.6%
Korean	798,849	0.3%	354,593	0.2%	444,256	125.3%
Vietnamese	614,547	0.2%	261,729	0.1%	352,818	134.8%
Hawaiian	211,014	0.1%	166,814	0.1%	44,200	26.5%
Samoan	62,964	0.0%	41,948	0.0%	21,016	50.1%
Guamanian	49,345	0.0%	32,158	0.0%	17,187	53.4%
Other Asian or Pacific Islander	821,692	0.3%	NA	NA	NA	NA
Other race	9,804,847	3.9%	6,758,319	3.0%	3,046,528	45.1%
HISPANIC ORIGIN						
All persons	248,709,873	100.0%	226,545,805	100.0%	22,164,068	9.8%
Hispanic origin*	22,354,059	9.0%	14,608,673	6.4%	7,745,386	53.0%
Mexican	13,495,938	5.4%	8,740,439	3.9%	4,755,499	54.4%
Puerto Rican	2,727,754	1.1%	2,013,945	0.9%	713,809	35.4%
Cuban	1,043,932	0.4%	803,226	0.4%	240,706	30.0%
Other Hispanic	5,086,435	2.0%	3,051,063	1.3%	2,035,372	66.7%
Not of Hispanic origin	226,355,814	91.0%	211,937,132	93.6%	14,418,682	6.8%

* Persons of Hispanic origin may be of any race.

In 1990 the United States rewrote the national immigration legislation still again to give preference to immigrants who would bring either skills or money. This has been Canadian and Australian policy for years, but such a policy is new in U.S. history. The nation's new rules favor immigrants bringing job skills or knowledge in mathematics, engineering, and the sciences, thus inviting the world's skilled workers to come, and foreign students already here to stay. The legislation reduced the number of unskilled workers allowed to immigrate, but it lifted the overall ceiling from 530,000 per year to 700,000 in 1992–1994 and 675,000 thereafter.

The legislation also set aside a quota of 10,000 for millionaires able to invest capital in the U.S. economy. Millionaries do not have to demonstrate special skills or education, but 30 percent of the millionaires must invest either in rural areas or wherever the unemployment rate is 1.5 times the national average.

Figure 6–32. This cap pistol reflects the virulency of U.S. anti-Asian animosity in 1876. (Courtesy of the Smithsonian Institution.)

The Brain Drain

Many nations are losing their best-educated people and most skilled workers through emigration. This migration from less-developed to more-developed countries has been called the **brain drain**. In some cases a country's

TABLE 6–6 ***Immigration Quotas under the National Origins System***

Rank	Country or area	Quota	Rank	Country or area	Quota
1	Great Britain and Northern Ireland	65,721	15	Denmark	1,181
2	Germany	25,927	16	Hungary	869
3	Irish Free State	17,853	17	Yugoslavia	845
4	Poland	6,524	18	Finland	569
5	Italy	5,802	19	Portugal	440
6	Sweden	3,314	20	Lithuania	386
7	Netherlands	3,153	21	Greece	307
8	France	3,086	22	Rumania	295
9	Czechoslovakia	2,874	23	Spain	252
10	Russia	2,701	24	Latvia	236
11	Norway	2,377	25	Turkey	226
12	Switzerland	1,707	26	Syria and Lebanon	123
13	Austria	1,413	27	Estonia	116
14	Belgium	1,304		All others	100

Source: U.S. Department of State, *Admissions of Aliens into the United States* (Washington, D.C.: U.S. Government Printing Office, 1932), pp. 102–4.

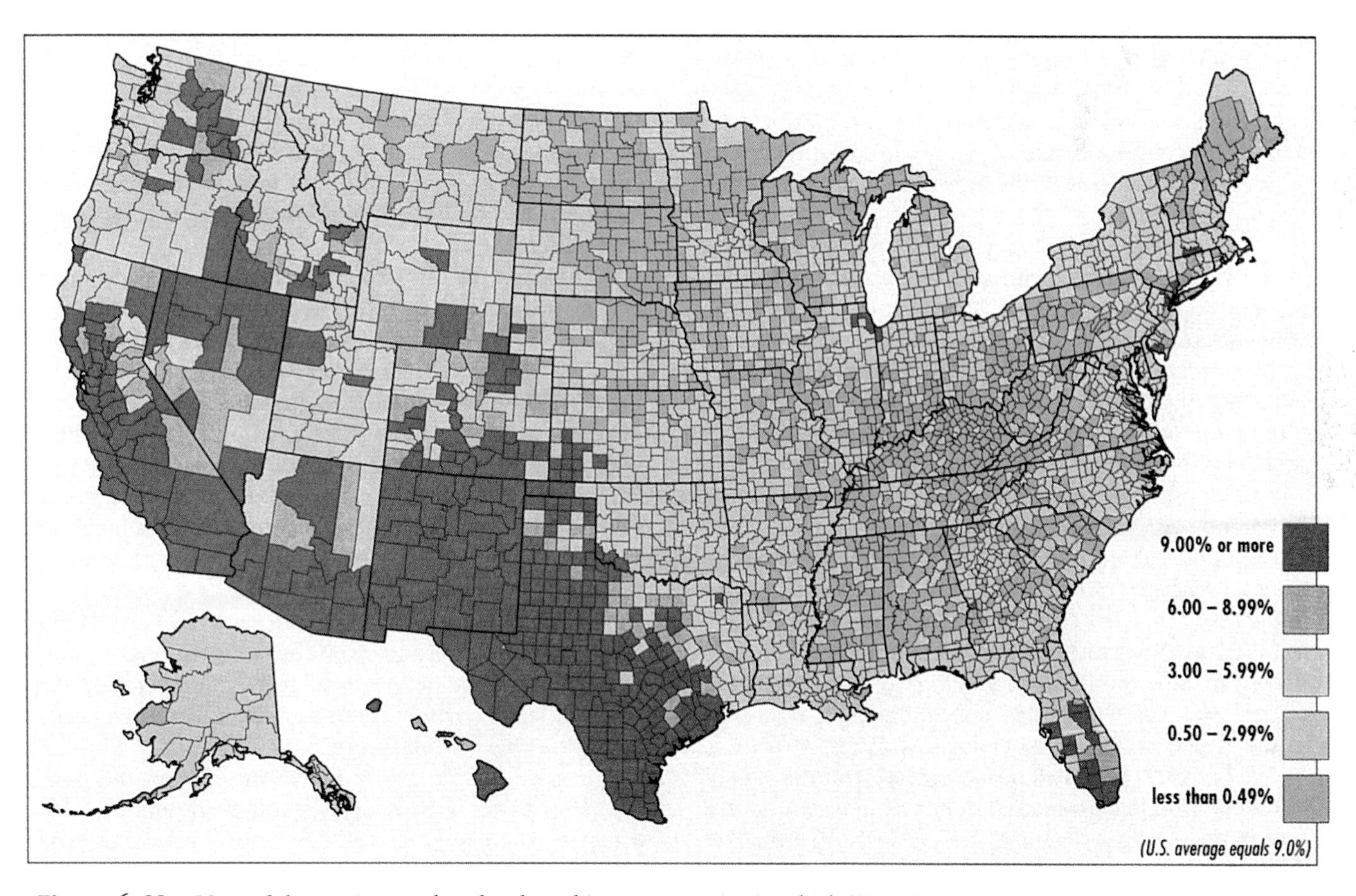

Figure 6–33. Most of the territory colored red on this map, counties in which Hispanic persons make up more than their 9 percent share of the overall national population, was taken from Mexico. Today, however, Mexicans boast that Los Angeles is the world's second largest "Mexican" city. In 1990 four states were home to 71 percent of U.S. Hispanics: California, Texas, Florida, and New York. In California and Texas, Mexicans make up over 75 percent of the Hispanic population. New York's population is evenly split between Puerto Ricans and other Hispanics, and Florida has the most diverse Hispanic population. (Reprinted with permission © *American Demographics* (July 1991):14)

schools are qualifying students for professional careers faster than the economy can absorb them. In other cases the skilled or educated seek political freedom, and in still other cases the country's best students go abroad for training to help their own people, but they stay abroad and never return home.

No country can afford a drain of its most highly trained people. Several countries have levied taxes on educated or skilled emigrants to prevent them from leaving, although the policy's ostensible goal is to recapture the government's investment in their education.

The United States profits most from the drain of skilled labor from other countries, and, in fact, the nation's scientific establishment would suffer without them. About one-third of the nation's Nobel Prize winners in the sciences have been foreign-born. Thousands of senior scientists, engineers, and other professionals migrate to the United States each year, and about half of the young foreigners who come for education never return home. Each year from 1960 to 1990 the percentage of U.S. Ph.D.'s in science and engineering granted to foreign students rose, whereas the number of Americans getting such degrees barely held steady. The percentage of non-citizens earning advanced degrees in all fields in the United States grew from 15 percent in 1972 to 26 percent in 1989. Foreigners earn more than half the degrees in technology and sciences. In 1990, 12 percent of the 1.2 million graduate students in the United States were not U.S. citizens.

The drain of scientists, engineers, and physicians from poor areas undermines the ability of these areas to develop. Any advanced country is reluctant to expel these skilled immigrants, who would surely just go to another advanced country anyway. How these skilled professionals are to be kept in their homelands, or attracted back, is a problem that must be solved if the gap between the world's rich and poor countries is not to widen further.

SUMMARY

The population of the Earth is about 5.3 billion people, but 90 percent of the population is concentrated on less than 20 percent of the land area. Major population concentrations are in East Asia, South Asia, and Europe. Secondary concentrations are in Southeast Asia and eastern North America. Much of the Earth's land surface is virtually uninhabited.

Factors that explain this distribution include the physical environment, history, and local differences in rates of population increase. Most of the population is concentrated in areas of C climate, but people will settle in harsh areas if it is profitable to do so. In some places local population densities may be high, but trade and circulation free societies from the constraints of their local environments.

The populations of China and the Indian subcontinent long ago achieved productive agriculture and relative political stability. Europeans multiplied when they gained material wealth and improved their food supplies as the result of world exploration and conquest, and later again during the Industrial and Agricultural revolutions. European migrants took technology to more sparsely populated areas. Population reductions have occurred in the Western Hemisphere and sub-Saharan Africa.

The world population is increasing, but the rate of growth is slowing. The rate of increase in each country represents a balance among the crude birth rate, the crude death rate, the fertility rate, the total fertility rate, and migration. Most of the increase is occurring in the poor countries. The history of population growth in today's rich countries traces a demographic transition. In the first stage both the crude birth rate and the crude death rate were high. In a second stage medical science improved, and so crude death rates dropped. Therefore, rates of population increase rose. In the third stage crude birth rates have dropped, and so rates of natural population increase are low.

Many of the poor countries are in the second stage, but we project that they will in the future enter the third stage. Even if worldwide crude birth rates and total fertility rates were to fall tomorrow, however, the total population would continue to grow because the number of young women reaching childbearing age is larger than ever before. The sooner the world's total fertility rate falls to the replacement rate, the lower the total population will be at which the projected numbers level off. Current projections range from a maximum of 8 billion to 13 billion people sometime in the 21st century.

Human migrations have redistributed populations throughout history, and significant migrations continue. Push factors drive people away from wherever they are, and pull factors attract them to new destinations. Some migrations have been forced.

Today people are trying to move from the poor countries to the rich countries and from politically repressive countries to more democratic countries. The brain drain of educated people and skilled workers from poor countries undermines their ability to develop.

KEY TERMS

population geography (p. 168)
emigration (p. 168)
immigration (p. 168)
demography (p. 168)
cartogram (p. 171)
arithmetic density (p. 171)

physiological density (p. 171)
carrying capacity (p. 171)
population projection (p. 175)
crude birth rate (p. 176)
crude death rate (p. 176)
natural increase (p. 176)
natural decrease (p. 176)
population growth (p. 176)
population decrease (p. 176)
fertility rate (p. 176)
total fertility rate (p. 176)
replacement rate (p. 176)
population pyramid (p. 176)
dependency ratio (p. 177)
demographic transition (p. 177)
infant mortality rate (p. 177)
doubling time (p. 177)
life expectancy (p. 179)
zero population growth (p. 180)
push factors (p. 186)
pull factors (p. 186)
sojourners (p. 187)
race (p. 188)
racism (p. 188)
African diaspora (p. 190)
apartheid (p. 194)
workers' remittances (p. 198)
refugee (p. 198)
brain drain (p. 202)

QUESTIONS FOR INVESTIGATION AND DISCUSSION

1. Investigate the problems faced by the U.S. Census Bureau in 1990.

2. If you can, investigate the numbers of children each generation in your family had through the generations.

3. Investigate the various churches' positions on family planning and birth control.

4. Investigate any country's family planning efforts and experience.

5. Do you know any elderly people who have retired and are now collecting pensions? Where did they spend their working years? Where are they living now? Can you relate your findings to migration trends within the United States?

6. Do you know how many generations of your family have moved either to this country or within this country? Which push and pull factors motivated them?

7. Which Native-American peoples inhabited your region prior to the coming of the whites? Were they exterminated? Did they move? Are any of their nation left? Are there any Native-American lands near where you are?

8. Has your local community seen net in-migration or net out-migration of its native black population in recent years? If in, where from? If out, where to? Are first- or second-generation Americans a significant share of your local black community? If so, where did they come from? Africa? The Caribbean?

9. How many different national backgrounds did the 1990 census reveal in your community? Have newcomers settled in distinct neighborhoods? Which push and pull factors motivated them?

10. If you thought that the population of your hometown were going to stabilize, how would you use whatever land is currently undeveloped? How might you think about redistributing jobs, housing, and shopping? Which lands would you set aside for recreation?

11. Compare the cartogram in Figure 6–2, countries drawn according to their population size, with Figure 11–4, countries drawn according to their economic output, and with a standard area world map. Do you think that any of these criteria—population, economy, or area—ought to determine the amount of time you devote to each country in your studies? Which one ought to determine how much time on the evening news is devoted to each country? Can you suggest any other measures of each country's "importance?"

12. If your local region were entirely dependent on local agriculture, what would be your local environment's carrying capacity?

13. In what year do you intend to retire? From Figure 6–17, calculate what percentage of the U.S. population will then be of retirement age. What percentage will be working to support you?

ADDITIONAL READINGS

JONES, HUW. *Population Geography.* London: Paul Chapman Publishing, 1990.

PAN, LYNN. *Sons of the Yellow Emperor: A History of the Chinese Diaspora.* Boston: Little, Brown & Co., 1991.

PHILLIPS, DAVID R. *Health and Health Care in the Third World.* Essex, England: Longman, 1990.

POPULATION REFERENCE BUREAU. *Population Bulletin.* Washington, D.C., quarterly.

TEITELBAUM, MICHAEL, S., and JAY M. WINTER, eds. *Population and Resources in Western Intellectual Traditions.* Cambridge, England: Cambridge University Press, 1989.

THERNSTROM, STEPHAN, ed. *Harvard Encyclopedia of Ethnic Groups.* Cambridge, MA: Harvard University Press, 1980.

U.S. COMMITTEE FOR REFUGEES. American Council for Nationalities Service. *World Refugee Survey.* Washington, D.C., annually.

CHAPTER 7

Food and Mineral Resources

International trade in food and raw materials has risen almost without interruption, whether trade is measured in tons, by value, or even as a percentage of all food and raw materials produced. (Courtesy of Four By Five/Superstock.)

The world population continues to rise, and each of the Earth's billions of people aspires to superior material welfare. These aspirations impose tremendous pressures on the Earth's supplies of food and mineral resources. This chapter examines the world distributions of these resources. In addition, the chapter investigates whether there are enough resources to satisfy the Earth's population and to provide rising standards of living for millions of people.

In general, inventories of resources depend on the definition of a resource. A **resource** may be defined as anything that can be consumed or put to use by humankind. Based on this definition, inventories of resources are continually expanding. Humans are constantly learning to produce more of some things and finding new ways of using things previously thought useless. This chapter focuses on world supplies of food, water, and mineral resources. The basic perspective of this chapter is that sufficient resources exist or can be supplied to meet basic human needs. At the same time, however, we must recognize that many people live in conditions of deprivation. That is humankind's tragedy.

FOOD SUPPLIES

The world production and distribution of food is the result of complex interaction among environmental, technological, political, and economic factors. The result is a paradox of oversupply in some places and insufficient supplies in others. Some countries are glutted with food. Their citizens might pay millions of dollars to store surpluses, or their governments might engage in **dumping,** that is, selling the food on the world market for less than it costs to produce. Even in many of those countries, however, some people go hungry. In other countries, overall national food supplies are insufficient to provide the entire population with a nutritious diet. In some of these countries the majority of the populations are peasant farmers surviving under subsistence conditions while the urban populations in those same countries eat subsidized imported food. These countries subsidize urban food supplies either to encourage urban economic growth or else to pacify urban populations that might otherwise riot.

We first examine the environmental and technological capability of the planet to produce enough food to supply the entire human population. This capability is secure today and in the future. We then examine the complex of factors that explains the tragic paradoxes in the maps of current world food supplies.

The World Population: Food Ratio

Thomas Malthus and the Malthusian Equation

Discussions of world population growth inevitably invoke the ideas of Thomas Robert Malthus (1766–1834), who asked whether humankind would always be able to feed itself. Malthus's statement of the relationship between population and food supply still demands our attention. An understanding of Malthus's intellectual milieu will also help us understand his argument.

Thomas's father, Daniel, was a disciple and personal friend of both the French philosopher Jean Jacques Rousseau (1712–1778) and of the English theorist William Godwin (1756–1836). Daniel Malthus and Godwin spread in England the new, optimistic "Enlightened" idea that progress is achievable through human action. Godwin's wife, Mary Wollstonecraft Godwin (1759–1797), was one of the first feminists, and her essay *A Vindication of the Rights of Women* (1792) raised a storm of opposition. William and Mary Godwin's daughter, also named Mary (1797–1851; wife of the poet Percy Bysshe Shelley), was less optimistic about the future of humankind. She later wrote the novel *Frankenstein* in 1818. In this classic, Dr. Frankenstein's confidence in science leads him to create a living creature that he hopes he can control but that turns on him and kills him. The idea that science can be a "Frankenstein's monster" that creates more serious problems than it solves, survives in our imagination.

In these debates between people who were optimistic about humankind's future and people who were pessimistic, Thomas Malthus was a pessimist. When his father challenged Thomas to put his ideas of the future into print, Thomas produced, anonymously, his *Essay on Population* (1798). Thomas was by that time a professor of political economy and also a clergyman. Stated most simply, Thomas Malthus argued that food production increases arithmetically: 1, 2, 3, 4, 5 . . . units of wheat. Population, however, increases geometrically: 2, 4, 8, 16, 32. . . . people. Thus, the amount of food available per person must decrease.

The human population can be kept in balance with food supplies only through checks on population increase. Malthus defined two types: positive and preventive checks. **Positive checks** refer to premature deaths of all types, such as those caused by war, famine, and disease.

The alternatives to these positive checks are human actions designed to limit population growth. Malthus called these behaviors **preventive checks.** An example of a preventive check is a decision by a young couple to delay marriage and childbearing. According to Malthus, couples should have a sense of responsibility for the economic welfare of the children they might produce, or at least couples ought to fear their own inevitable social and economic decline if they produce too many children. "There are perhaps few actions that tend so directly to diminish the general happiness as to marry without the means of supporting children," he wrote. "He who commits this act, therefore, clearly offends against the will of God."

Because Malthus did not have much faith in people's ability to restrain themselves, he did not believe that preventive checks could effectively control population growth. Convinced that the poorest people would have the most children, he predicted that the future of humankind would consist of endless cycles of war, pestilence, and famine.

Thomas Malthus's work had an immediate and powerful political impact. Prime Minister William Pitt the Younger read it and was convinced to stop supporting public assistance programs. Pitt was convinced that these programs were counterproductive because they encouraged the poor to have more children, thereby increasing the overall numbers of poor people. According to this philosophy, welfare programs were not only economically counterproductive, but they were also immoral. They severed the vital link between a person's action (procreation beyond ability to support) and the consequences of that action (starvation).

Developments since Malthus Since Malthus's theory was first published, the human population has increased from 1 billion to over 5 billion. The mass starvation predicted by Malthus has not occurred, however, for several reasons.

Increased production To begin with, disasters such as wars and epidemics have killed tens of millions of people, thereby restricting population growth. In addition, vast areas of the planet that were scarcely utilized during Malthus's lifetime have been opened to productive agriculture. The greatest agricultural surplus areas of both North and South America, as well as surplus regions in Australia and South Africa, have been developed since Malthus published his theories.

Since the middle of the last century alone, 3.5 million square miles (9 million km^2) of the Earth's surface have been converted to permanent croplands. The United Nations reports that just between the periods 1975–1977 and 1985–1987, the world cropland area increased by almost 3 percent. Most of these lands were opened by irrigation.

Not only have new croplands been planted, but many food crops have been transplanted to new areas where they have thrived, in some cases better than in their areas of origin (see Figures 7–1 and 7–2). Even before Malthus began writing, the Western Hemisphere had contributed important food crops to the Eastern

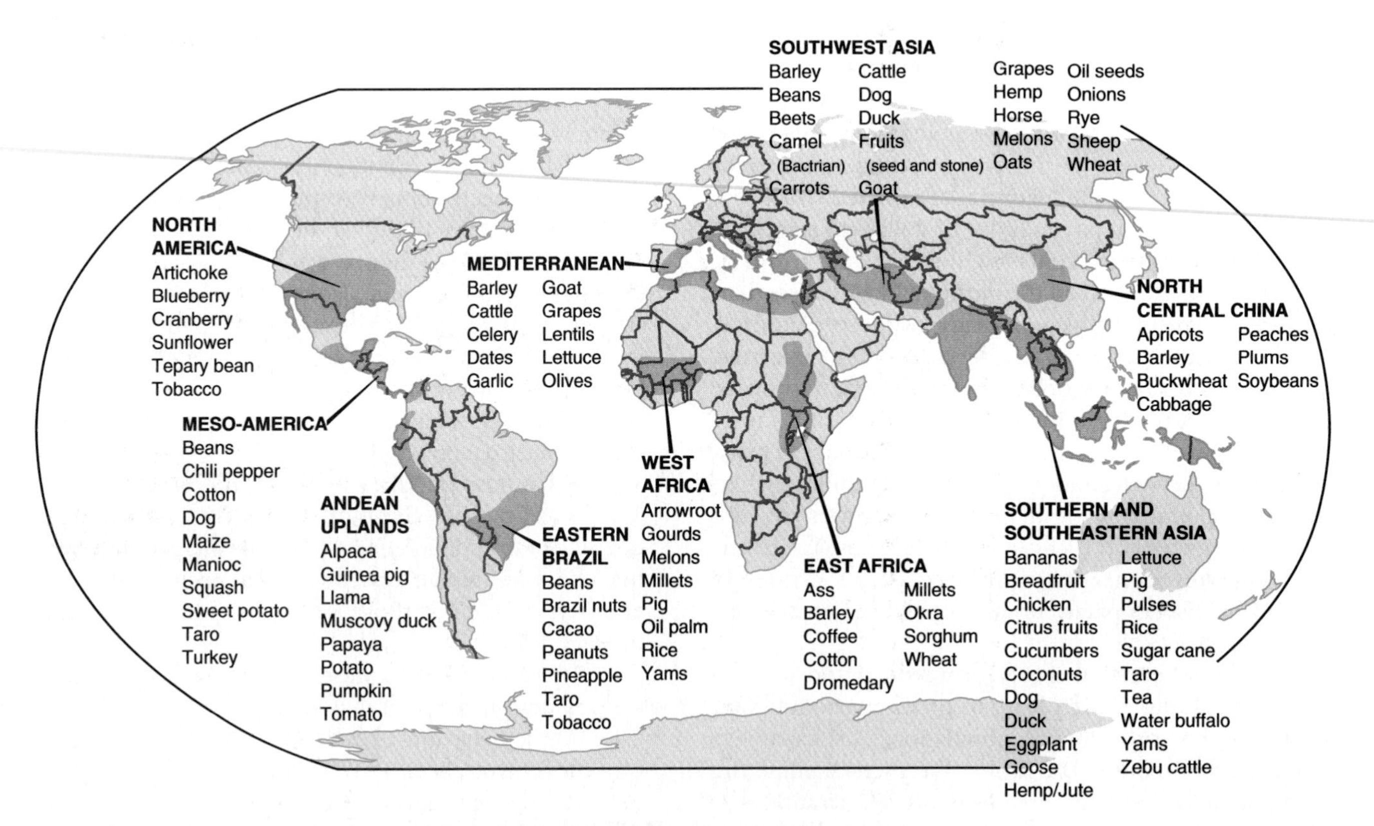

Figure 7–1. This map shows the origins of the world's food crops and domesticated animals. These plants and animals have been so widely redistributed, however, that today's leading producers of many of these are not the same as the areas in which they were first domesticated.

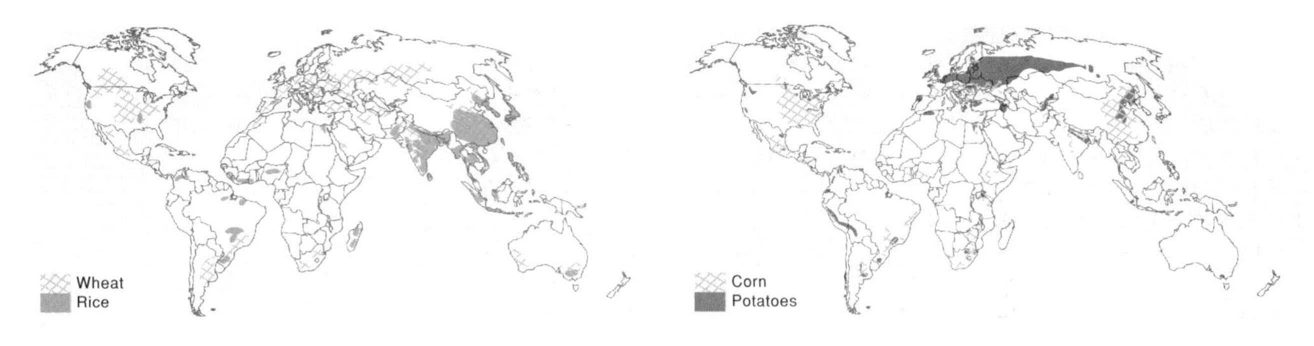

Figure 7–2. Distribution of current world production of wheat, rice, corn, and potatoes. (Data adapted from Rand McNally, *Goode's World Atlas*, 18th edition. © 1990 by Rand McNally R. L. 92-S-76.)

Hemisphere. For example, the potato, which yields the second highest number of calories per acre of any crop, is native to the Andes region of South America. By Malthus's day it had already become a major food in northern Europe, and today it is the world's fourth most important food crop (measured by total tonnage harvested). China has recently recognized the potato's versatility and is today, after Russia, the world's second largest producer. India is not far behind, and the crop is becoming a mainstay throughout Africa and Asia. New genetically engineered plants are being introduced to the tropics, where they are ready to harvest only 40 to 90 days after planting. Potato planting continues to spread, and the value of the world's potato crop increases each year. The true treasure of the Andes was not the gold that the Spanish conquerors sought, but the potatoes they trampled.

Corn (or "maize" or "Indian corn" as the Europeans call it) also was a Western Hemisphere crop, as was manioc, or cassava. In the sixteenth century Portuguese traders introduced manioc into Africa's Congo River delta area, and it rapidly became a dietary staple there. The introduction of manioc to Africa improved diets there so much that it is often cited as one reason why Africa's population did not decline during the period in which millions of Africans were taken off into slavery. Manioc is still the principal source of calories for Africans. Successful transplantation of additional crops continues around the world today.

Transportation and storage Improvements in worldwide transportation have allowed regional specialization in food production, and specialization can multiply productivity. Today railroads, trucks, and cargo ships—many of them refrigerated—move quantities of food that could not have been imagined in Malthus's day.

Improvements in transportation also allow the shifting of food from surplus to deficit areas. Thus, fewer people need die from local famines. In the past it was not uncommon for a food surplus to rot in the sun 100 miles from a starving population because the food could not be transported.

The technology of food storage has also continually improved, decreasing both spoilage and loss to pests (see Figure 7–3). Improvements have been continuous from the introduction of the silo in the nineteenth century to freeze drying—actually, the Inka did this 500 years ago, but the rest of the world learned only recently—and antiseptic packaging.

The green revolution The introduction of higher-yielding and hardier strains of crops through advances in botanical science is known as the **green revolution.** Specialists at Texas A&M University and at the International Rice Research Institute in the Philippines have produced new strains of rice that are more resistant to disease and pests, have shorter growing cycles, and have greater yields. New potato strains are being developed at the International Potato Center in Lima, Peru. Wheat yields have multiplied from new hybrids. India, where 1.5 million people died in a 1943 famine, became a grain exporter in 1977, even though its population doubled between those years.

Other technological advances The green revolution is just one aspect of a complex of factors known as the **scientific revolution in agriculture.** New pesticides save crops from insects that once wiped out entire harvests, and new fertilizers multiply yields per acre. Parallel scientific research has been directed toward livestock production (see Figure 7–4). As with crops, the developments include world redistribution, increases in total numbers, improved breeding, and even "engineering" for greater hardiness and higher yield from each animal. In 1950 the average U.S. dairy cow gave 618 gallons (2339 L) of milk per year; by 1990 that number had risen to 1703 gallons (6446 L). This level of productivity applies so far to only a fraction of the world's livestock herds.

Farm machinery invented since Malthus's day not only releases workers from the fields but also increases yields by improving the regularity of plant spacing and the efficiency of harvesting. Heaters rescue many crops threatened by freezes. During the past 200 years humans have literally re-formed the earth with drainage projects where there was too much water and with irrigation projects where there was not enough. Projects such as these are not new in theory, but they are achieved in scale beyond anything that Malthus could have foreseen.

Figure 7–3. In this picture an officer of the U.S. Agency for International Development (AID) oversees the construction of a silo in Peru. (Courtesy of the Agency for International Development.)

Diversifying our diets The green revolution focused attention on the improvement of just a few of humankind's most important crops, and the success in improving these crops has raised concern that humankind is becoming too reliant on too few crops (see Figure 7–5). Wheat responded to scientific yield enhancement so well that today humankind relies more on wheat than on any other single crop in the past. If a new wheat disease appeared today as suddenly as the potato blight appeared in the last century, it could destroy a significant percentage of humankind's total food supply. An example of such a disaster occurred in the U.S. corn crop in 1970. A new fungus suddenly appeared that was lethal to most of the nation's corn crop, and U.S. corn production fell 15 percent.

A diversity of crops offers protection against catastrophe in case any one crop should fail, but modern commercial agriculture is increasingly specialized. Only primitive agriculture, which in fact is very sophisticated (see Chapter 5), has preserved diversity. Therefore, paradoxically, new scientific appreciation of primitive agriculture may multiply world harvests. Ethnobotanists have reported more than 50 kinds of potato grown in a single village in the Andes. Scientists are just now re-

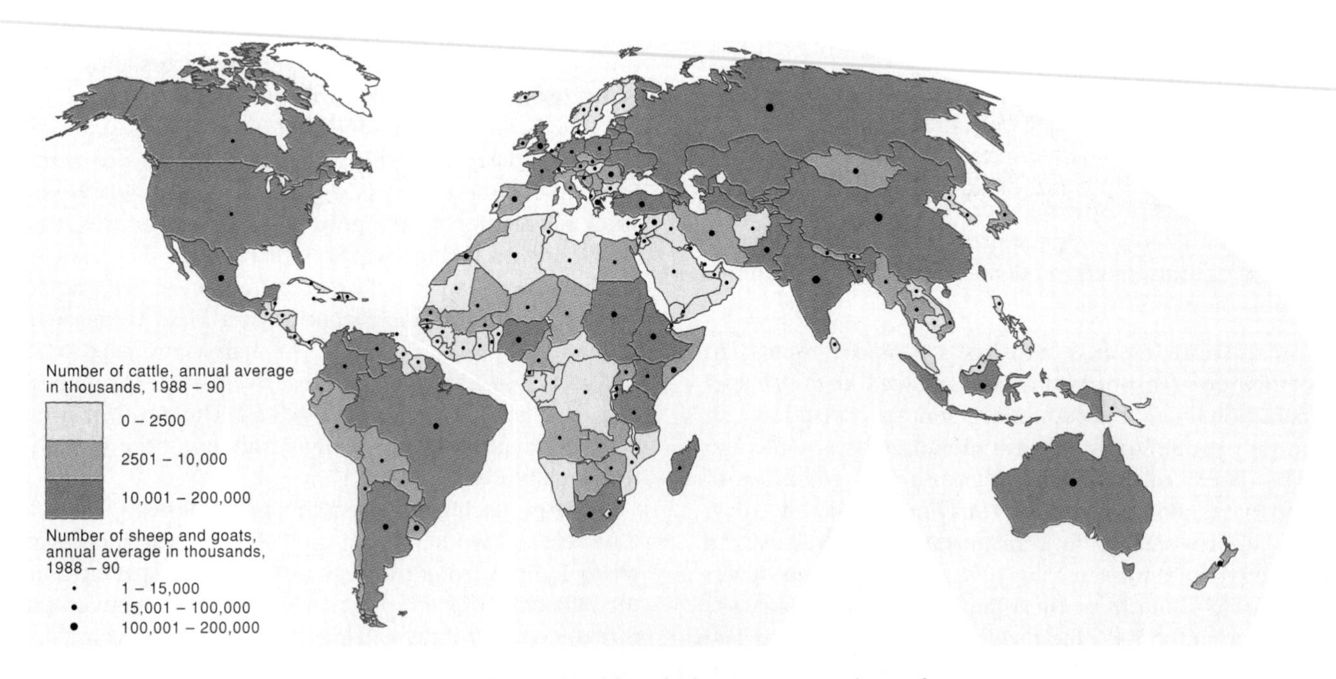

Figure 7–4. Cattle are widespread around the world, although the greatest number—almost 16 percent—are in India. The world distribution of sheep and goats is remarkable for their low populations in the United States and Canada. Can you hypothesize as to why? The answer does not lie in physical-environmental conditions. (Data from World Resources Institute, *World Resources 1992–93*. New York: Oxford University Press, 1992, pp. 276–77.)

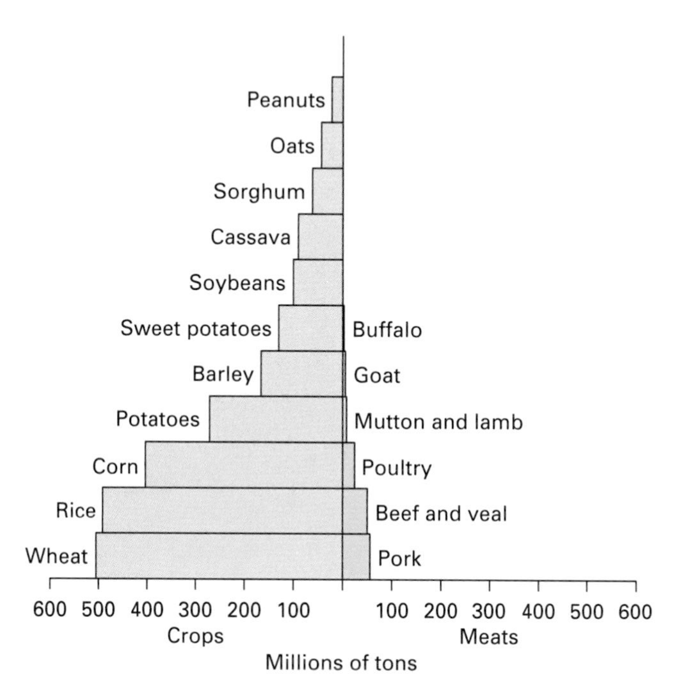

Figure 7–5. Today most of humankind is dependent on a handful of plant and animal species for its food. The 4 leading crops contribute a greater portion of the world total than the next 20 combined. Three-quarters of the human diet is based on only 8 crops.

discovering valuable food crops that were cultivated by the Inkas 500 years ago. When archaeologists returned some Bolivian fields from the most modern farming techniques to the methods used by the Tiahuanaco people over 1000 years ago, yields multiplied by 7.

Today humans rely on only 20 plant species for almost all our food, but in the course of human history about 7000 species have been utilized, and at least 75,000 species have edible parts. Many of these are demonstrably superior to the plants cultivated today. The New Guinea winged bean, for instance, is entirely edible—roots, seeds, leaves, stems, and flowers. It grows rapidly, up to 15 feet (4.5 m) in a few weeks, and it offers a nutritional value equal to that of soybeans. The American Academy of Sciences has recommended cultivating 36 other crops including amaranthus, buffalo gourd, tamarugo, guar, mangosteen, and soursop. Most of us have undoubtedly never heard of these, and yet each of them offers enormous potential for food.

The potential of raising food from more and different plants is a reason why the extinction of plant species on Earth threatens humankind's welfare. The steady loss of genetic diversity, even before botanists have had the opportunity to study each of the Earth's millions of species, is a loss of potential new crops or genes for interbreeding. Humankind's destruction of plant species is the equivalent of destroying whole libraries of unique books before the books have been read and their information inventoried.

Agricultural technology offers many alternative foods, but what people eat, or refuse to eat, is to an astonishing degree *cultural* (see Figure 7–6). Culinary imagination could probably make a wide variety of alternative crops appealing and palatable.

The Future

Until now all the factors just discussed have held off the specter of worldwide starvation. Is it possible that humankind is now, at last, at the end of its ability to increase food supplies?

The answer to this question is no. There is no reason to fear that humankind is, technologically, in danger. If demographers are correct in their projections of the Earth's future population, the population can be fed. Although farmers now utilize almost all potential cropland, and the amount of cropland per capita is falling, humankind has scarcely begun to maximize productivity even with present-day technology. The leading contemporary technology has been applied to only a small portion of the Earth, and newly emerging technology offers still greater possibilities (see Figure 7–7).

A 1983 study by the Food and Agriculture Organization (FAO, a UN agency) of the world's soil and climate determined that with basic fertilizers and pesticides, all cultivable land under food crops, and the most productive crops grown on at least half the land, the world in the year 2000 could feed four times its projected 2000 population (see Figure 7–8). (That study discounted any possible technological breakthroughs between 1983 and 2000.) If average farm yields rose just from the present 2 tons of grain equivalent per hectare

Figure 7–6. Peasants bring live iguanas into downtown Hanoi, Vietnam, for sale for Saturday night dinner. Many people around the world find nutrition in things Americans might not find very tempting. (Courtesy of Nancy and Douglas Paley.)

Figure 7–7. These two photographs illustrate the extremities of capital investment in agriculture. If the Senegalese peasant (a) had as much capital to invest in farm machinery, fertilizer, and improved seeds as the Nebraska farmers (b) do, who knows how much food he could raise. (Courtesy of (a) the Agency for International Development/Carl Purcell and (b) U.S. Department of Agriculture.)

(2.47 acres) to 5 tons, the world could support about 11.5 billion people (see Figure 7–9). Each person could enjoy "plant energy"—food, seed, and animal feed—of 6000 calories per day, the current global average. (North America currently uses about 15,000 calories per person per day, but most of that is consumed by animals, which are then eaten by people.)

In addition, humankind could improve its diet by concentrating on raising those domesticated animals that most efficiently transform grain into meat. Chickens are the most efficient. They yield 1 pound (0.46 kg) of edible meat for every 4 pounds (1.8 kg) of grain they consume. Pigs produce 1 pound for every 7 pounds of grain, and beef cattle 1 pound for every 15 pounds (6.8 kg) of grain. In addition, chickens reach maturity—and therefore can be consumed—much faster than do pigs and cattle. Therefore, a greater emphasis placed on raising chickens could immensely improve the human diet (see Figure 7–10). China is widely replacing swine with chicken farming, and several other countries have established programs to multiply their chicken populations. In many societies, however, cattle are viewed as a status symbol, and this preference delays the switch into more productive livestock.

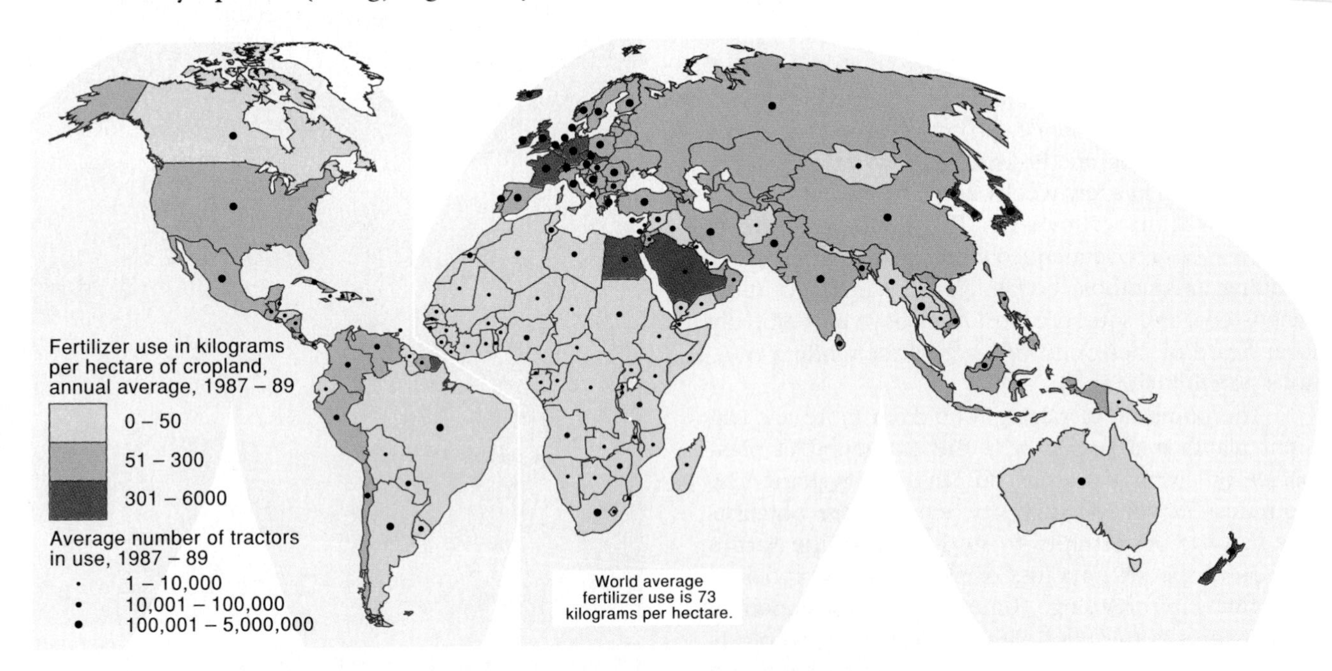

Figure 7–8. Such basic capital inputs in agriculture as the amount of fertilizer used and the number of tractors available vary tremendously around the world. (Data from World Resources Institute, *World Resources 1992–93*. New York: Oxford University Press, 1992, pp. 274–75.)

Figure 7–9. Yields per hectare vary partly as a result of natural physical conditions but also to a great degree according to the capital inputs in agriculture. (Data from World Resources Institute, *World Resources 1992–93*. New York: Oxford University Press, 1992, pp. 272–73.)

Biotechnology in Agriculture New scientific advances promise to multiply future food yields. Biotechnology offers genetically altered crops that can be custom designed to fit the environment, produce bountiful harvests, and resist plant diseases. One bacterial gene eliminates the need for chemicals to kill worms by producing a natural protein that disintegrates the worms' digestive system. Genetically engineered viruses can be used as pesticides. In 1988 scientists mapped the *genome* of rice—the set of 12 chromosomes that carries all the genetic characteristics of rice. This development could enable geneticists to produce improved strains of rice. Biotechnology can replace chemical pesticides and fertilizers, whose biological or even genetic impact on our own bodies is not fully understood.

Scientists have also conducted research on ***halophytes***, plants that thrive in salt water. Interbreeding halophytes with conventional crops has made these crops more salt-resistant, which means that they can grow in more diverse environments. Farmers are today harvesting lands in Egypt, Israel, India, and Pakistan once thought too salt-soaked to support crops. Conventional crops may someday be grown in salt water.

Mechanized fishing on technologically advanced

Figure 7–10. Chicken production in advanced countries has moved from the barnyard into virtual factories in which chickens are mass-produced. (Courtesy of U.S. Department of Agriculture.)

Figure 7–11. The fish in these tanks in Idaho are not "caught," but raised under controlled conditions. The latest technological innovations in aquaculture improve the recycling of the water and allow the harvest of fertilizer nutrients from the fish waste. (Courtesy of U.S. Department of Agriculture.)

ships has already multiplied yields from the sea, but humankind has scarcely begun the shift from hunting and gathering seafood (fishing and gathering a few aquatic plants) to **aquaculture,** which involves herding or domesticating aquatic animals and farming aquatic plants (see Figure 7–11). Humankind took this step with agriculture and livestock herding on land thousands of years ago. Presumably this food frontier will be expanded.

All these possibilities justify optimism. The economist Henry George (1839–1897) succinctly contrasted the rules of nature with the multiplication of resources through the application of human ingenuity. He said, "Both the jayhawk and the man eat chickens, but the more jayhawks, the fewer chickens, while the more men, the more chickens."

This principle is key to understanding and counting all resources, but technological solutions to problems can still trigger unexpected new problems. Overreliance on insecticides, for example, can lead to the poisoning of farm workers or the contamination of water supplies. The debate between the optimists and the pessimists continues.

WORLD DISTRIBUTION OF FOOD SUPPLIES AND PRODUCTION

The amount of nutrients needed per capita in each country varies greatly, depending, for the most part, on the climate and on the population's median age. A country with a low median age needs less food per capita. Figure 7–12 maps the actual available nutrition per capita. The food supplies in some countries fall short of per capita needs. Many people in those countries are well fed, but many others are hungry. Even in the countries in which overall per capita food supplies are adequate, some people are hungry.

The world could produce enough food for everyone, but a number of obstacles prevent food production and distribution where it is needed. No country is autarkic in food. Most countries both import and export food, and a few countries are net exporters of food despite the fact that portions of their own populations are undernourished. This may be because of injustice or civil strife in the country. This is a problem of economics, politics, and morality, not geography or technology. Tables 7–1 and 7–2 list the major importers and exporters of the major foods in world trade.

Problems in Increasing Food Production

Advanced technologies create the potential for increasing food production, but they do not guarantee that this will occur or that the food will be distributed where it is needed. Unfortunately, many practices and conditions actually work against increasing the food supply. One is that the most advanced technology is not spread around the whole world, nor is it distributed in the most economic way. The use of fertilizer exemplifies this.

Fertilizers: Overuse and Diminishing Returns Fertilizers can improve crop yields significantly when used in appropriate amounts. The quantities of fertilizer applied in the advanced countries, however, for many years have far exceeded the point of diminishing returns.

Diminishing returns exists when, in successively applying equal amounts of one factor of production to the remaining factors, an added application yields a smaller increase in production than the application just preceding. Table 7–3 on p. 217 demonstrates this concept with fertilizer use. In the table, the point of diminishing returns is 4 pounds (1.8 kg) of fertilizer. This is true because adding the second and third pounds of fertilizer increases the yield by 11 bushels, but adding the fourth increases the yield by only 10 bushels. At some point, then, the diminishing increases in crop yields no longer justify the use of greater amounts of fertilizer at one place. That fertilizer could more efficiently be applied somewhere else that has not reached the point of diminishing returns.

Today, U.S. agriculture uses 160 pounds (73 kg)

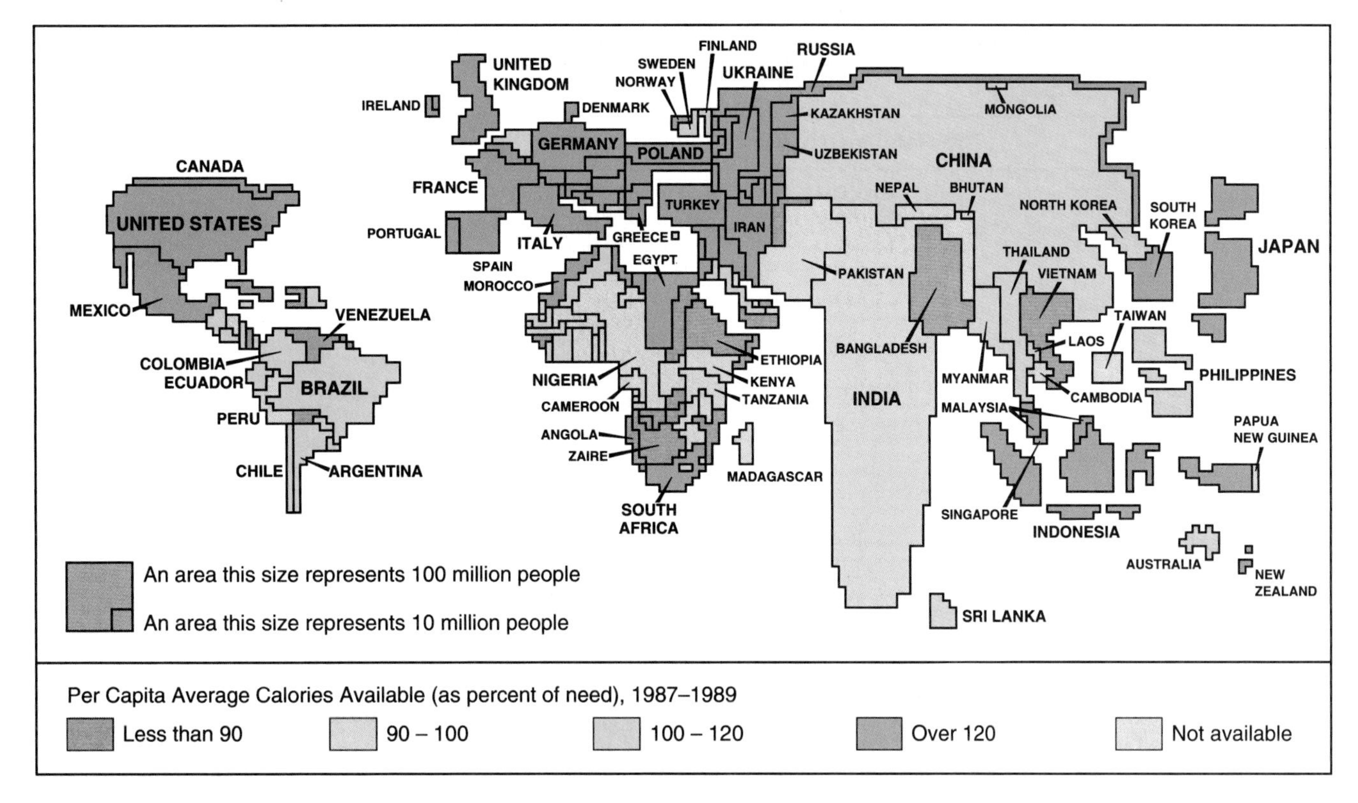

Figure 7–12. This population cartogram shows the calories that reach the population as a percentage of estimated need. (Data from World Resources Institute, *World Resources 1992–93*. New York: Oxford University Press, 1992, pp. 250–51.)

of fertilizer for each man, woman, and child in the country. If each U.S. farmer's last 10-pound (4.5-kg) bag of fertilizer were given to a farmer in a poor country, the world food supply would increase dramatically. In other words, the same fertilizer that yields only diminishing returns in the United States could be used much more efficiently and productively in a poor country.

Lack of Financial Incentives Agriculture almost always demonstrates the economic dictum that "the market produces the surplus." In other words, if a market exists for any surplus that farmers can produce, and if the political and economic system sufficiently rewards farmers for their effort, production of a surplus is all but assured (assuming adequate technology is available). In many countries, however, the farmers have little incentive to increase food production. Many governments tax farmers heavily in the form of artificially low prices for crops, export levies, high exchange rates, state-run marketing boards, and high import duties on needed tools and agricultural chemicals. This wealth has gone to urban civil service bureaucracies, military spending, or unprofitable state-run industries. Many farmers stay at the subsistence level because they have no incentive to produce more food. Meanwhile, peasants migrate to cities, increasing the demand for subsidized food.

Patterns of Landownership In many countries, systems of land ownership retard food production while increasing political and social inequalities. In Brazil in 1990, for example, 70 percent of the rural population lacked title to land, whereas 0.7 percent of the farms occupied 43 percent of the farmland. Many of the large landholdings were not worked to maximum productivity, and the concentration of landownership spurred the landless to attempt to farm new claims in the rainforest regions. This expansion destroys the environment and produces disappointing yields (see Chapter 3).

The collectivization of agriculture under communist regimes crippled farm production in several countries, but today private farming is being encouraged. When Poland ended food subsidies and restrictions on direct marketing in 1989, farmers' markets sprang up overnight, food shortages eased, and prices fell. The Soviet Union always allowed small private market gardens. These accounted for less than 2 percent of total Soviet farmland, but they yielded 60 percent of the country's potato crop and around 30 percent of its vegetables, meat, eggs, and milk. Today the successor states of the dissolved Soviet Union are slowly returning to private farming. China privatized its previously collectivized agriculture in 1978, and within 10 years farm output rose 138 percent. China turned from a net importer of food products into a net exporter. (Evidence suggests that a side effect of privatization in China is that it may increase the birth rate by intensifying the traditional urge to have sons to carry on the family farm.)

In many countries the landholding systems are holdovers from colonial periods. One widespread con-

TABLE 7–1 ***Wheat, Rice, and Corn: Top Ten Exporters and Importers, 1988***

Exporters		Importers	
Wheat	**(Thousand metric tons)**		**(Thousand metric tons)**
United States	40,500	Soviet Union	21,180
Canada	20,079	China	14,547
France	14,847	Japan	5,724
Australia	12,303	Egypt	5,267
Argentina	3,643	Italy	4,904
West Germany	3,185	South Korea	4,116
Saudia Arabia	2,058	Algeria	3,350
Turkey	1,993	Iran	3,200
United Kingdom	1,927	Iraq	2,800
Hungary	1,790	Netherlands	2,454
Rice			
Thailand	5,267	India	684
United States	2,260	Bangladesh	674
Pakistan	1,210	Iraq	603
China	699	Iran	600
Italy	510	Soviet Union	498
Australia	297	Hong Kong	364
Uruguay	273	Saudia Arabia	363
North Korea	236	China	310
India	200	Senegal	310
Spain	119	France	297
Corn			
United States	46,568	Japan	16,555
France	6,011	Soviet Union	11,426
Argentina	4,217	South Korea	5,051
China	3,917	Taiwan	4,459
Thailand	1,209	Mexico	3,303
South Africa	745	Spain	2,244
Spain	731	Netherlands	2,045
Belgium & Luxembourg	524	Belgium & Luxembourg	1,487
Greece	500	Italy	1,471
Soviet Union	365	United Kingdom	1,338

Source: *Statistical Abstract of the United States*, 1991, #1476–1477, pp. 856–57.

TABLE 7–2 ***Top Ten Importers and Exporters of Pulses (Lentils, Peas, and Other Dried Leguminous Vegetables), 1987–89***

Exporters		Importers	
	Metric tons (annual average)		**Metric tons (annual average)**
Turkey	804,485	Netherlands	680,494
France	608,287	Germany (combined)	677,455
United States	497,316	India	656,900
China	409,351	Italy	351,868
Canada	392,909	Belgium	345,556
Australia	384,349	Japan	177,653
Hungary	208,756	Pakistan	136,838
Thailand	199,556	Spain	128,522
United Kingdom	165,302	Cuba	125,593
Denmark	158,833	Algeria	109,936

Source: World Resources Institute, *World Resources, 1992–93*, pp. 278–79.

TABLE 7–3 ***The Diminishing Returns of Fertilizer Use***

Fertilizer (Pounds)	Total yield (Bushels)	Production increase (Bushels)
1	10	10
2	21	11
3	32	11
4	42	10
5	51	9
6	59	8

sequence of colonialism was the replacement of precolonial communal systems with private property. The distribution of the property was highly inequitable, creating one class of large-estate holders and another, much larger, class of landless poor. The former often moved to the cities, becoming absentee landlords unwilling to make the investments necessary to maximize the return from their agricultural holdings.

Individual farmers of small personal holdings are more apt to maximize their productivity. In the Philippines, for example, large blocks of land were granted to the Spanish colonial elite, and the natives on these land grants became the serfs or tenant farmers. After a few generations, the landlords were mostly mestizos, and their descendants own the best agricultural land. This land-grant system prevailed wherever the Spaniards colonized in Central and South America, and it is at least partly responsible for political unrest there today.

Indonesia presents a contrast to the former Spanish colonies. It is much larger than the Philippines but much like it in geography and population type. For more than 300 years under Dutch rule, however, only native Indonesians could own land, except for city lots. Dutch, Arabs, Chinese, and English could rent agricultural land, but they could not acquire title to it. Thus, landownership among Indonesian peasants remained much more widespread than it was in the Philippines. Indonesian peasants who have inherited their land and expect to leave it to their children are much less likely to join peasant guerrilla movements than landless Filipinos are.

When land is leased to those who actually work it, the conditions of tenancy and of payment determine whether the farmers are encouraged to produce significant surpluses. If the landlord takes too large a share, surpluses will be small.

Unsuccessful land redistribution: Mexico Land redistribution may increase incentives for growing more food, but returning land to communal ownership does not necessarily increase productivity. This is because another factor in successful modern farming is capital investment in irrigation, chemicals, improved seeds, and machinery. Farmers can raise capital in the private sector—borrow it if necessary—only if their land can serve as *collateral*, that is, a thing of value that can be seized by a creditor in case of nonpayment of a debt. Peasants cannot borrow money to invest in farming if the ownership of the land is frozen in communal holdings. Few governments have been willing or able to extend loans to their farmers.

Mexico demonstrates this phenomenon. More than one-half of the country's arable land is held in *ejidos*, a form of land tenure in which a peasant community collectively owns a piece of land and the natural resources and houses on it (see Figure 7–13). Mexican law states that ejidos are "inalienable, nontransferable and nonattachable." They cannot be used as collateral, and this illiquidity discourages people from lending money to them. This system, combined with a government tradition of paying farmers low prices for their crops while at the same time subsidizing food for urban consumers, has caused a disaster in the countryside that has in turn intensified migrations of peasants to the cities. Mexico must import 35 percent of its total food supplies, including nearly half of its staples of corn, wheat, and beans from the United States. Mexico is also one of the world's largest importers of dairy products.

In 1991 the Mexican government launched an effort to overhaul the ejido system. Official recognition of ejido ownership was, however, a principal issue of the Mexican Revolution (1910–1920), in which over 1 million Mexicans lost their lives, and many Mexicans remain attached to the idea. Changing the system will require an emotional political battle, but the government hopes it will substantially increase food production.

Figure 7–13. Ejidos are a traditional Mexican form of communal landholding. As the rural population has risen, the amount of land per farmer has dropped, and capital inputs are meager. (Courtesy of Carl Frank/Photo Researchers.)

Figure 7–14. Agricultural colleges, such as this one in Tanzania, instruct local farmers in the latest farming techniques. Most agricultural research, however, has been done and continues to be done in the areas that are already rich and advanced. More is known about the physical conditions for agriculture in Iowa than in Tanzania. Research remains to be done into the physical conditions of farming, such as soils, in many underdeveloped countries. (Courtesy of the Agency for International Development.)

Barriers to increasing production: Africa The return of land to communal holdings in some African countries has mirrored the Mexican experience. The good farmers cannot borrow money for investment. Some land is poorly used by some farmers, and the good farmers cannot expand their holdings. All these conditions promote increased migration to the cities. As long as tenure depends on occupancy, migrants to the city leave their families in the villages. In some parts of rural Zimbabwe, 40 percent of the households have lost the father to the town. The overworked women, left to tend children as well as crops, can scarcely rise above subsistence. The United Nations Children's Fund (UNICEF) estimates that women grow 80 percent of Africa's food (see Figure 7–14).

Civil wars have reduced food production in some areas of the world, again particularly in Africa. The continent fed itself in 1960, but by 1990 Africa was importing 40 percent of its food supply. In 1976 the U.S. Central Intelligence Agency (CIA) estimated that Angola could produce enough food for 250 million people, but in 1989 that country had to import half of the food for its 9 million people.

We cannot predict what future technology will bring, but given all the complicated factors that inhibit food production in many countries, and given the fact that food is available in surplus in world markets for countries that can afford to import it and choose to do so, it is difficult to map potential production and supplies meaningfully. The world does have, however, a food distribution problem.

The United Nations estimated that about 500 million people were undernourished in 1991, about 9 percent of the Earth's population. Many of these people could be fed if their homelands could stop discriminating against farmers (particularly women), end their civil wars, stop subsidizing urban populations at the expense of farmers, let their farmers import the tools and chemicals they need, and reform landholding systems. If any country chooses not to raise its own food but to develop its economy by doing something else and to buy food on world markets, large amounts of food are available at low and falling prices.

Subsidized Food Production and Export

The system of world trade exacerbates the problems in the patterns of world production and availability of food. The percentage of the world's food that is traded internationally is rising, and so the terms of trade in food are increasingly important.

International trade in manufactured goods has been subject to international regulation since 1945. Trade in agricultural production, however, which is approaching $500 billion per year, is not. The rich countries have erected tariff walls to close their markets to food imports from the poor countries, and they have subsidized their own farmers so generously that they can export or dump subsidized food. The Organization for Economic Cooperation and Development (OECD) estimated that overall government subsidies as a percentage of agricultural output in 1991 totaled 30 percent in the United States, 49 percent in Europe, and 66 percent in Japan.

This system distorts the world geography of agriculture. It contributes to production of agricultural surpluses in the rich countries. Furthermore, it discourages increasing production in the poor countries by dumping food on their urban markets, thereby minimizing the urgency of investing in their own rural areas.

Why Is Agriculture Subsidized? The rich countries subsidize their farmers and erect import barriers for several reasons. First, some countries pursue self-sufficiency in food production as a national security measure. The governments want some defense against grain embargos and crop failures in those countries that normally have surpluses to market. Subsidizing a country's own farmers offers some protection against these threats. Following U.S. threats to halt grain shipments to the Mideast in the 1970s, for example, Saudi Arabia spent heavily to improve its agriculture (see Figure 7–15). Iraq failed to do so, and by the time it occupied Kuwait in 1990, Iraq depended on imports for almost 80 percent of its nutritional needs.

Second, many rich countries, particularly in Europe, subsidize agriculture to keep their farmers from migrating to the cities in search of work. Rural to urban migration can overwhelm the ability of cities to absorb the new labor force, and it threatens the quality of life in the cities.

Third, rich countries may subsidize farming as part of national land-use plans. To them it is valuable in itself to have a farming countryside and to preserve green areas around cities, called **greenbelts.** Their urban citizens enjoy driving through or visiting convenient rural landscapes.

Fourth, many rich countries subsidize their farmers because their systems of political representation favor farmers. The farmers in the United States, Japan, and Europe constitute a decreasing percentage of the national populations, and yet in Europe and Japan particularly the electoral districts have never been readjusted for the relative depopulation of the countryside and the urbanization of the national populations. Thus, a rural vote may heavily outweigh an urban vote, giving farmers a disproportionate role in politics while effectively muffling the voices of urban consumers. Most urban consumers remain unaware that they are paying premium prices as subsidies to their nation's farmers because the cost of food as a percentage of their incomes is falling anyway. Thus, electoral reform in the rich countries is necessary to encourage food production in the poor countries.

The disproportionate political power and economic subsidization enjoyed by farmers in the rich countries contrasts with the situation in the poor countries. In those countries national governments overtax farmers and subsidize the urban populations.

Agricultural subsidies in the wealthy countries, which in 1990 totaled almost $300 billion, reduce market opportunities for farmers in developing countries. Many poor countries could sell more food to the rich countries if they were allowed to compete fairly with farmers in those countries. If the rich countries opened their markets to crops from developing countries, then the poor countries could earn the foreign exchange needed to buy seeds, fertilizer, pesticides, and farm machinery. For example, as Eastern Europe emerges from communist rule, one way for it to earn much-needed foreign exchange would be to feed prosperous Western Europe, but Western European subsidies to its own farmers frustrate this hope.

In the 1960s Western Europe was a net food importer, but by the late 1980s it had become an exporter and had begun to stockpile surpluses of cereals, fruits, vegetables, dairy products, wine, olive oil, and sugar. The European governments take about $50 billion from general tax revenues, add $85 billion paid by consumers as higher prices, and give it to their farmers. Western Europe has about 10 million farmers, compared with only 2 million in the United States. Despite such subsidies, agriculture is still only 3 percent of Western Europe's total economic output. The United States spends about the

Figure 7–15. Saudi Arabia has chosen to invest some of its oil income into irrigation of its desert landscape to achieve agricultural self-sufficiency. In 1991 wheat cost Saudi Arabia about $600 per ton to grow, and yet in that year Saudi Arabia produced so much that it sold wheat to New Zealand for $80 per ton, $45 per ton below the world market price. This pattern of world production and trade was obviously not the result of physical conditions for agriculture, but it bankrupted wheat farmers in regions more naturally suited to wheat growing than Saudi Arabia. (Courtesy of Aramco World Magazine.)

Focus on Agriculture and Food Supplies in Japan

Japan, a mountainous archipelago with only limited flat fertile land, exemplifies a rich country that subsidizes its agricultural sector. The average size of a Japanese farm is about 3 acres (1.2 ha), about one-half percent of the U.S. average, but the 9 percent of all farms that are larger than this produce 44 percent of the crop. Japan had 30 million farmers as recently as 1965, but today agriculture is the sole source of income for only about 450,000 Japanese. Japanese farmers use a great deal of machinery, but farming such small plots is still inefficient. Furthermore, about half of Japan's scarce nonmountainous land is farmed, and so the Japanese are forced to live in tiny and expensive quarters.

Japan does not have a one-person, one-vote system of representation. The Japanese courts have upheld a 3–1 disparity in population of districts represented in the Japanese parliament (the Diet), but the actual disparity is as high as 5–1 in favor of agricultural districts. The Diet imposes stiff import quotas and tariffs on food products, and as of 1992 it still refused to allow the importation of rice. This is ostensibly a national security measure. Thus, Japanese must pay 6 to 10 times for rice what they could import it for from Thailand, Australia, or California. Japan has even occasionally exported subsidized rice.

Agriculture in Japan is carried out on tiny plots, but it is productive because it is capital intensive, and the Japanese market is protected by import taxes. (Courtesy of Japan National Tourist Organization.)

The Japanese government is today investing in farming technology, including livestock raising. On the northern island of Hokkaido, the Ministry of Agriculture's experimental stock-breeding farm controls herds by remote control radio, monitoring their health, moving them to new pastures, and even fencing them in electronically. Japan has the genetics technology and the resources to revolutionize agriculture from an inefficient industry to a high-technology industry, like electronics. Today Japan is a major food importer, but food could become a new Japanese export in the 1990s.

same amount as Western Europe on farm subsidies, but consumers pay only about $28 billion more as higher prices.

A case study: sugar The world sugar market exemplifies how agricultural subsidies redistribute world production and trade. Neither the United States nor Europe is the most efficient place in the world to produce sugar, and yet both restrict imports and maintain artificially high domestic sugar prices to subsidize local producers. U.S. sugar producers receive about three-quarters of their annual income from government assistance. The United States imposes a rigid quota system on imports, and assigning a country a large quota for sugar at the inflated U.S. market price is a form of financial assistance that does not consider the country's producing capacity or economic needs. This system also forces U.S. consumers to pay between $3 and $5 billion more per year for sugar than world market prices. Sugar seems inexpensive to U.S. consumers, but the U.S. domestic price for sugar hovers at about 5 times the world market price.

Sugar from sugar beets competes with cane sugar. Europe subsidizes its sugar beet industry so extravagantly that it exports sugar, to the dismay of Australia, the Philippines, and Caribbean cane sugar producers.

The inflated price of sugar in the rich countries makes alternative sweeteners (such as corn-based or even purely chemical sweeteners, about which there are health concerns) price-competitive in soft drinks. This enables producers of artificial sweeteners to charge higher prices.

World sugar markets are not the only agricultural

markets that are distorted by national subsidies. In 1991 the European Community sold 100,000 tons of beef to Brazil at a subsidized price so low that it bankrupted Brazilian cattle ranchers, who can actually produce beef at prices lower than Europeans can. The U.S. market consumes 3.3 billion pounds of peanuts per year, but only 1.7 million pounds are allowed to be imported. This stifles the much cheaper peanut production that would be possible in Argentina, Ghana, Sierra Leone, and Senegal. Numerous additional examples of distorted world markets in foods could be cited.

The politics of subsidies How do the agricultural interests, who represent only a small minority of the U.S. population, exercise such influence over policy? Part of the answer lies in the politics of the U.S. Congress. The agriculture committees of both houses are dominated by farm interests. Urban members of Congress vote for farm bills as long as they also contain provisions for food stamps and school lunches. Farmers in the state of Iowa alone receive more subsidies each year than the World Bank and all other international agencies are able to provide to the entire African continent for combined agricultural research and food aid.

It should be noted that the subsidies are not all going to struggling family farms. In the United States in 1989, 324,000 farming concerns out of a total of 2.2 million had sales of over $100,000. This 15 percent of farming concerns received over 60 percent of all farm support payments.

In 1988 the United States called for a worldwide end to subsidies within 10 years, but the rich countries cannot agree on what constitutes a subsidy. The United States subsidizes the water supply for California farmers but denies that that is a "farm subsidy." Throughout the Central Valley, sprinklers irrigate green fields of alfalfa, cotton, and rice—crops more suited to monsoon lands than to a desert, and more than 50 percent of federally irrigated land in California is devoted to crops that are already in surplus. About 85 percent of all water used in California goes to farmers while competition for this water intensifies from California's burgeoning cities and suburbs. In 1990 California alfalfa fields alone took more water than used by the people of the San Francisco Bay area and Los Angeles combined. California farmers and urbanites are not the only contestants for the water. The federal Endangered Species Act requires water regulators to maintain high water levels in the Sacramento River to preserve the environment of the chinook salmon. In February 1992, 6 years of drought forced the first cutoffs of water to California farmers in 40 years, and the allocation of water among farmers, cities, and salmon will remain hotly contested.

The issue of farm subsidies is complicated, and in each country it is a volatile domestic political issue. The overall result, however, is to increase surpluses in the rich countries and to discourage food production in the poor countries. Land and price reform are helping to encourage farming in some countries, but these efforts are undermined by the export of Western surpluses. What quantities of food the world's farmers would be capable of producing under free market conditions and in which countries those foods would be produced is unknowable given the world's current manipulated agricultural economy.

WORLD WATER RESOURCES

Water is essential to life, and yet humankind has not shown care in managing Earth's water resources. All the fresh water in the Earth's lakes, rivers, and streams totals less than 0.3 percent of the Earth's total store of water. This supply is continually replenished by precipitation, but, unfortunately, much of that is polluted even as it falls, or it is polluted or wasted as it runs off the land to the sea.

The hydrologic cycle (discussed in Chapter 2) provides plentiful water for total human needs, but neither the water nor the population is evenly distributed. Some countries are water-rich. Iceland, for example, has been estimated to have enough excess precipitation to provide 878,000 cubic yards (672,000 m^3) of water per person per year. Other countries are water-poor. For example, the Persian Gulf nation of Bahrain has virtually no fresh water and must depend on desalination of seawater. Furthermore, citizens of different countries use different amounts of water. The average U.S. resident uses 62 times as much each year as the average resident of Ghana.

Sixty-nine percent of the water withdrawn from the Earth is used for agriculture, but this share-allocation also differs greatly from one country to another. In the United States, industry claims 46 percent, agriculture 42 percent, and domestic uses 12 percent. In China, industry takes only 7 percent, agriculture 87 percent, and domestic needs 6 percent.

Water is becoming increasingly scarce, and severe shortages are occurring. Tapping of ground water to increase supplies should be avoided unless the supplies will be replenished, but groundwater depletion is already common in parts of India, China, and the United States. Competition for water from international rivers such as the Nile, Jordan, and Euphrates is intensifying.

Shortages can be dealt with either by increasing local supplies (capturing more runoff in dams or tapping ground water) or by conserving local supplies, especially through more efficient water use. Only about 37 percent of all irrigation water in the world, for instance, is taken by crops; the rest is lost in runoff, evaporation, and leakage. The percentage lost could be reduced.

Pollution of water supplies by both organic and industrial wastes is increasing. Organic wastes are *biodegradable*—that is, they can be broken down by bacteria and other organisms—but waters can be over-

ABC NEWS ***Economics and Politics of Water in the West***

Usable water is a precious commodity. Not only do we need water for consumption, but also for agricultural and industrial activities. More than 70 percent of the Earth's surface is covered by water, but only 2 percent of all the water within the Earth's realm is considered fresh and therefore usable for human activity, and more than half of this is locked in polar ice caps and glaciers.

Securing access to adequate supplies of fresh water is a major problem in many western states. Although large supplies of fresh water exist, they are unevenly distributed. Consequently, certain individuals and groups control large water reserves and therefore also control the access of others to this water. In the West, water rights are bought and sold like any commodity. In addition, no central authority exists to regulate water use. Although this system enables some people to make substantial profits, it places many others in a highly vulnerable position. For farmers, inadquate water supplies can mean loss of valuable cropland. Cities might have to choose between limiting their growth or paying more for water.

The video titled "Economics and Politics of Water in the West" examines this issue in greater detail. If you have an opportunity to view this video, consider the following questions.

1. As the population continues to grow in the West, and with it the demand for water, what methods might be considered to ensure that water is distributed to all segments of the population?
2. Can you foresee any changes that would affect the demand for water in the West?

loaded, and their oxygen depleted. In addition, organic wastes also contain disease-causing organisms, called *pathogens*, which can be transmitted to people who use the water. Industrial wastes might not be easily degraded, and they too may be toxic.

Pollution of rivers and lakes is potentially reversible, but less is known about the quality of the Earth's groundwater reserves. Ground water is cut off from the oxygen in the atmosphere, and therefore its capacity for self-purification is low. In the United States and Europe, where ground water is a significant source of fresh water, between 5 percent and 10 percent of all wells have already been found to be contaminated. This contamination is less visible than the pollution of popular beaches, and so it provokes less public outcry, but it is just as serious. Some of the advanced countries have recently begun to battle these contaminating processes, but for much of the world's population, clean water remains precious and rare. Waste treatment is practically nonexistent in most poor countries.

The United Nations proclaimed the 1980s the International Drinking Water Supply and Sanitation Decade. It sought to provide safe drinking water and appropriate sanitation for everyone by 1990. Those goals were not met, but safe drinking water was brought to hundreds of millions more people.

In September 1990 representatives of 115 nations met in New Delhi, India, to bring the decade's efforts to a formal conclusion. The conference working paper warned that "some 80 countries, supporting 40 percent of the world's population, already suffer from serious water shortage." The World Health Organization estimated that 1.2 billion people in developing countries were still without safe water, 243 million in urban areas and 989 million in rural areas (see Figure 7–16).

WORLD MINERAL RESOURCES: FUEL AND NONFUEL

There are many types of resources—climatic, biotic, and cultural—but most people think of mineral resources first when the word is mentioned. *Organic minerals*, including coal, petroleum, and natural gas, contain carbon compounds, and these are useful fuels. Most minerals, however, are *inorganic*. About 3000 minerals are known, but fewer than 100 are common, and even fewer are of major importance to humankind. Roughly 75 percent of the chemical elements are metals. **Metals** are elements that are usually heavy, reflect light, can be hammered and drawn, and are good conductors of heat and electricity. Most metals occur naturally as minerals in combination with nonmetallic elements. The mixtures of minerals from which metals are extracted are called **ores.** Useful nonmetallic minerals include the organic minerals plus clay, salt, building stone, and gemstones.

Mineral resources are not evenly distributed throughout the world. Some areas contain substantial quantities and varieties of minerals, whereas other areas have almost none of any value. Few clear conclusions can be drawn about how this distribution relates to the distribution of human population or of national wealth, because the key to wealth is the ability to *use* resources. The maps of the distributions of mineral supplies, of mineral production, of manufacturing, and of consumption of minerals all differ.

The citizens of the world's rich countries consume or use vastly greater quantities of most minerals than do the citizens of the world's poor countries. This trend has caused some observers to worry about future supplies or even to assert that all people on Earth cannot be raised

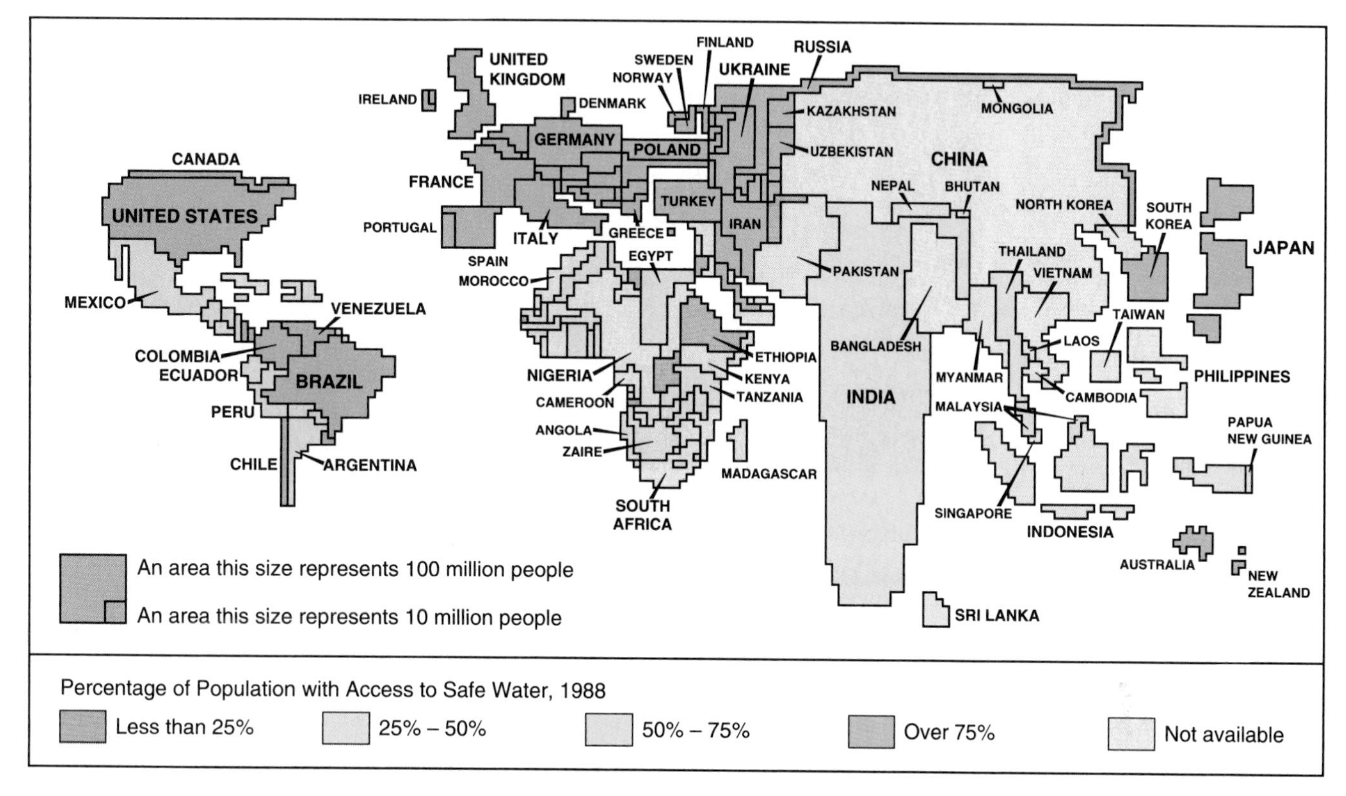

Figure 7–16. This cartogram reveals that many people still suffer limited or unsafe supplies of drinking water. (Data from World Resources Institute, *World Resources 1992–93*. New York: Oxford University Press, 1992, pp. 274–75.)

to the level of material welfare enjoyed in today's rich countries. Each person could not possibly have so many *things*. Technology, however, is astonishingly flexible, which offers reason for optimism that higher standards of living can be achieved for all.

Resources and Reserves

The **crustal limit** of a mineral is the total amount of that mineral in the earth's crust. This amount is determined by geology, and it is fixed. The amount of that crustal limit that is currently or potentially extractable is called a **resource.** Enormous quantities of minerals are known to lie on the ocean floor, but these have not yet begun to be inventoried, let alone exploited.

The **reserves** of any mineral are the amounts that can economically be recovered for use. Thus, the word *reserves* is not a geographical or geologic term, but an economic term. Many factors affect the prices of minerals and therefore the profitability of extracting them. The price of a mineral fluctuates according to the factors of supply, of demand, and of the perception of future supplies and demands. Commodity markets actually trade in what are called **futures contracts,** which are agreements to buy or sell commodities at a given price at a future date.

The price of a mineral falls and reserves increase if the cost of extracting the mineral falls. This often happens because of improved technology. Drillers can get much more oil out of each well than they used to, and new platform design allows drilling in deeper waters. Miners are everywhere exploiting mineral deposits that would have been considered waste rock just a few years ago. Nevada mines, for instance, can afford to dig up a ton of rock to get back as little as 0.025 ounce (0.71 g) of gold, worth today about $9. The mechanization of coal mining in the United States between 1975 and 1990 increased efficiency so greatly that the price of coal dropped 44 percent in real terms; reserves soared (see Figure 7–17). What are only resources of a mineral in one country could actually be counted among reserves in another country that is more technologically advanced. Thus, measures of holdings of mineral reserves in different countries can be deceptive. An increase in the capital invested in mining can multiply reserves.

Mineral Substitutes

The introduction of a cheaper substitute for any mineral reduces demand for that mineral. Some metals, such as antimony, tin, cadmium, selenium, and tellurium, are easily replaced. Tin has been losing out to glass, steel, aluminum, and plastics. Almost all plastics are made from petrochemicals. Plastics were invented in the 1860s, and they have replaced metals, wood, and fibers in many products. In 1989 the world consumed 120 million tons

Figure 7–17. In the technique of longwall mining shown here, a rotating cutting drum moves back and forth along the face of the coal seam. This efficient method of mining lowers the price of coal and increases coal reserves. (Courtesy of U.S. Bureau of Mines.)

of polyvinyl chloride, polystyrene, polypropylene, and polyethylene, plus 11 million tons of nylon, acrylics, and polyester. This is less than the 778 million tons of crude steel produced in that year, but it is significantly more than the 97 million tons of aluminum or the 8 million tons of copper produced. Stronger plastics are being introduced: Some are already stronger than steel (see Figure 7–18). In the future, more plastics may be made of agricultural raw materials.

Sometimes different metals replace others as technology advances. In the United States, for example, many industrial products substitute aluminum and other metals for steel. Magnesium, for instance, is finding more uses in manufacturing.

Strong but lightweight ceramics are replacing metals in many uses, and this technology presents one of the most challenging frontiers of scientific research today. Ceramics are basically composed of sand. In the future, manufacturers will make more and more products out of sand, a raw material that is inexpensive and can be found throughout the world. Another new chemical process offers the possibility of manufacturing polymers out of silica instead of petroleum. This will reduce the need for oil and oil products. Scientists are on the verge of synthesizing spider web fiber, which is one of the strongest substances known. Even concrete, an old invention, is being made so much better and stronger that it is replacing both metals and plastics in a wide variety of uses. Concrete may be one of the surprise "miracle products" of the future.

Figure 7–18. These plumbing fixtures are made, not out of metal, but of DuPont Delrin acetal resin. They are easier to manufacture than are parts made of metal (these are manufactured in Malaysia), and they last longer. (Courtesy of DuPont.)

Platinum is today one of the world's most valuable minerals. It is used chiefly in jewelry (38 percent of world demand in 1989) and in catalytic converters, devices that reduce air pollution from automobiles (37 percent). The price of platinum has risen with demand for catalytic converters, but its price will collapse if a catalytic converter is developed that does not require platinum. Platinum jewelry, therefore, although beautiful, is probably not a good investment. Gold makes beautiful jewelry, but its primary role is as a standard of monetary exchange. If gold ever loses this role, its price will collapse.

Scientists often find entirely new uses for minerals. Petroleum was used as a lubricant for centuries, but it has been recognized and used widely as a fuel only since the mid-nineteenth century. Other minerals, such as asbestos, go out of use because of technological or ecological obstacles to their use.

Dematerialization

New and more efficient industrial processes squeeze more output from each unit of raw material input. This is called the **dematerialization** of manufacturing. Cellular telephones, for instance, are being introduced in the poor countries that lack telephone wiring. They save raw materials by leapfrogging the technology of wiring. Even where wiring is still necessary, copper is being replaced by hair-thin optical fibers that are made, basically, of sand. As technology increases supplies and cuts demands, the world uses less raw material input per unit of total economic output. In 1990 the world used about 25 percent less copper per unit of gross product than

Figure 7–19. The U.S. aluminum industry recycled 55 billion cans in 1990. (Courtesy of Reynolds Metal Company.)

in 1970, 40 percent less iron ore, and 50 percent less tin.

Dematerialization has its limits. Some products need a minimum mass to function properly. Heavy cars, generally speaking, are safer than light ones. Dematerialization is also not necessarily environmentally beneficial. Lightweight products may be thrown away, rather than repaired. If the lightweight new product is a plastic, it may be harder to recycle than its predecessor.

Renewable and Recyclable Resources

Animal or vegetable natural resources, such as fish, livestock, or trees, naturally renew themselves, and they can be harvested at the rate at which they reproduce themselves through time. These are called **renewable resources**. Supplies of renewable resources will diminish only if they are harvested at a rate above their natural reproduction rate.

Other resources can be reused, and these are called **recyclable resources.** The word *recycling* is used to describe two distinct processes. One process is the recovery and reuse of industrial wastes. Industrial wastes pollute, but recycling them or rerouting them to serve as the raw material for another industrial process both increases the supply of raw materials and reduces pollution. Industrialized societies today worry less about using up the world's raw materials than about cleaning up the emissions and discharges that result from industrial processes. Even some renewable resources can be recycled, as when wood is reused.

Another form of recycling is the reuse of obsolete

Figure 7–20. New York City's Fresh Kills landfill is the world's largest, but it will soon be filled to capacity. As the cost of acquiring landfill sites rises, recycling of refuse becomes more cost-competitive. (Courtesy of City of New York Department of Sanitation.)

or discarded goods for their material content, in other words, "mining" our refuse dumps (see Figure 7–19). Whether this is economically feasible depends on its cost compared with the cost of using virgin materials. A rise in the price of virgin materials encourages recycling, but a fall in their price discourages recycling. World prices for lead, for instance, fluctuate, and so the percentage of lead in U.S. car batteries that was recycled in various years through the 1980s fluctuated between 90 percent and 75 percent. Recycling does, however, offer additional savings. It may reduce reliance on imported minerals, and it reduces the need for landfill space for refuse. Landfill space is becoming scarce (see Figure 7–20). Recycling can be limited by the high cost of collecting the materials. In the United States, for instance, the costs of collecting and sorting municipal waste are so high that only 10 percent to 12 percent of it is being recycled, compared with the 80 percent to 85 percent that theoretically could be. Nevertheless, the United States is making great efforts to recycle mineral raw materials. In 1989 scrap supplied 36 percent of the country's total consumption of aluminum, 43 percent of copper, 43 percent of iron and steel, and 61 percent of lead. The possibility of recycling can affect the price and thus the measured reserves of any mineral.

Because so many factors affect the supplies and prices of raw materials, the prices of these materials on international commodity markets fluctuate from moment to moment. A labor stoppage at a mine in South America can drive up the price of a mineral on the New York and London markets within minutes. Rumors of a discovery of new supplies of another mineral in Africa can lower the price of that mineral. The announcement of the invention of a successful substitute at a scientific laboratory in Asia can collapse the price of any mineral. Political upheaval can bring about sudden changes in countries' economic policies. Under the political framework of the Soviet Union, for example, Russia supplied raw materials to the other republics and to its political allies in Eastern Europe at subsidized low prices. After the fall of the Soviet Union, however, Russia sold substantial quantities of raw materials on world markets, thus driving down world raw material prices. For all these reasons, tables or maps of measurable reserves of major minerals should be dated to the very hour of their composition.

Do Mineral Reserves Ensure Wealth and Power?

Few minerals are so vital that possessing them gives the producing country real leverage in world affairs or guarantees wealth. Some minerals, however, are so important to industrial processes that they are referred to as **strategic minerals.** Chromium, for instance, is necessary in the manufacture of stainless steel. The leading producers of chromium are South Africa and Russia. The United States generally relies on supplies from South Africa and Turkey.

The United States is the world's largest importer of strategic minerals. One reason often advanced for maintaining political alliances with governments that might be considered objectionable—for example, South Africa under apartheid—is to secure supplies of strategic minerals. In fact, the costs of adjustments that would have to be made were supplies from specific countries or regions interrupted have been measured to be small. The United States also stockpiles strategic and vital minerals.

Many minerals are in global oversupply, and so possessing them does not guarantee wealth. On the contrary, if any country depends on the export of such a mineral for a high percentage of its foreign exchange earnings, then that country's economy is vulnerable to disruption by a fall in the price of that mineral. To avoid a decline in prices for minerals, producing countries have occasionally joined together in **cartels,** organizations that agree to control and limit production. Cartels have seldom succeeded, however, because of a lack of cooperation among members. Any nation can gain a short-term advantage by dumping supplies on world markets, collapsing prices. An example of an unsuccessful cartel is the Tin Council, with 22 member countries, which went broke in 1985. The Organization of Petroleum Exporting Countries (OPEC) has enjoyed unique success in maintaining high prices for its product, despite the fact that in a completely free world market petroleum would be inexpensive.

WORLD RESOURCES, PRODUCTION, AND CONSUMPTION OF MINERALS

Mineral deposits are unevenly distributed over the Earth; this distribution has favorably endowed some countries in relation to their size, their population, or both. The effort that has been expended on exploration varies greatly, and it may be that a lot more of some minerals remain to be found. Some countries, such as the United Kingdom and Japan, have been explored extensively for minerals. Others, such as Brazil and Zaire, remain largely unexplored. We can only guess about the quantities of any minerals that might still be discovered. Scientists do know that the distribution of minerals is related to processes in the formation of the Earth's crust. As geologists learn more about these processes and their distribution, better guesses can be made as to where new mineral resources might be found.

Nonfuel Minerals

The richest sources of nonfuel minerals have proved to be areas of old, worn-down rock, called *shields.* The

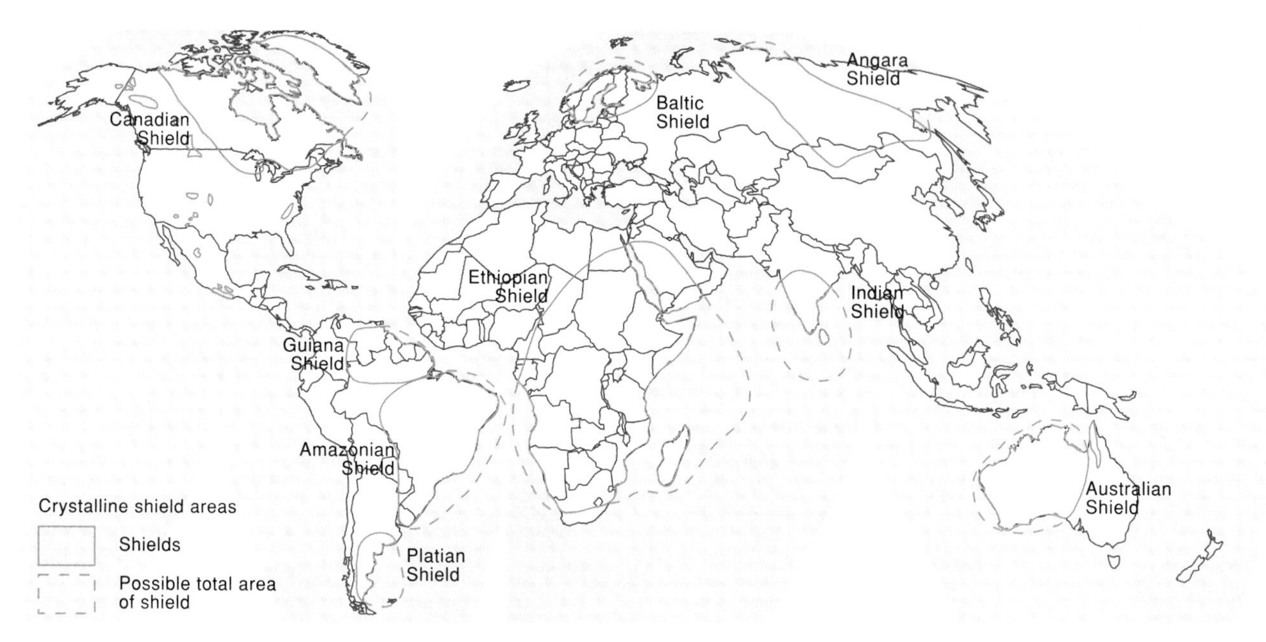

Figure 7–21. The cores of continental landmasses are ancient rock complexes called shields, which are often rich in valuable minerals.

geography of these shields has located great concentrations of the world's nonfuel mineral resources in the United States, Canada, Australia, South Africa, Brazil, and Russia (see Figure 7–21).

Several populous countries have few or virtually no reserves of nonfuel minerals. These include some rich countries, such as France and Japan, that can import nonfuel minerals, but also countries that have limited industrial development, such as Bangladesh and Nigeria.

Some countries boast huge deposits of just one or two minerals. Chile, for example, has about 25 percent of the world's copper reserves, but it has few other valuable minerals. Copper exports represent one-third to one-half of Chile's annual exports. The Chilean economy exemplifies one that is tied to world prices of one ore, which makes it vulnerable to world price shifts. If the price of copper dips, Chileans suffer.

The countries with the largest reserves of any mineral are not necessarily the ones that produce the most of it. Some countries have substantial reserves of a mineral, but they lack the technology or the capital to exploit their known deposits. They may be crippled by local civil unrest or by administrative incapacity, they may lack the necessary transportation infrastructure, or they may be the victims of international boycotts. Other countries reduce their output to conserve the resource.

Most ores need to be processed, or **smelted**—melted or fused to separate the metallic constituents—before they can be manufactured into goods. Many poor countries, however, export unprocessed ores. Rich countries, on the other hand, typically import, process, and consume many more ores and minerals than they possess. Figure 7–22 reveals specific rich countries that import and process great quantities of minerals of which they do not have their own reserves: Germany, Japan, the United Kingdom, Belgium, and others. These are world manufacturing centers.

Energy and Fuels

The world distribution of energy supplies is worth examining separately. There are several main fuels, and each has its own preferred uses and problems associated with its use. Overall, so many new reserves are being discovered that total world fuel reserves per capita are rising.

Commercial energy produced for sale rather than for direct use accounts for about 85 percent of human energy use. The traditional fuels wood and dung make up most of the noncommercial energy used, and among the poorest people in some of the world's poorest countries, deforestation has triggered a serious local energy crisis. Hydroelectric power is an important source of renewable energy, but the other renewable forms of energy production—solar, tidal, and wind power—today contribute only negligible total amounts.

The maps of the world geography of current energy supplies reveal the same conclusion as those of nonfuel minerals: The countries that have the most of any resource are not necessarily those that produce the most, and the countries that produce the most do not necessarily consume the most (see Figure 7–23). In addition, those countries that consume the most on a per capita basis do not necessarily produce or consume the most overall (see Figure 7–24).

Petroleum

Petroleum is the world's principal source of commercial energy, supplying 37 percent of world needs. Global oil

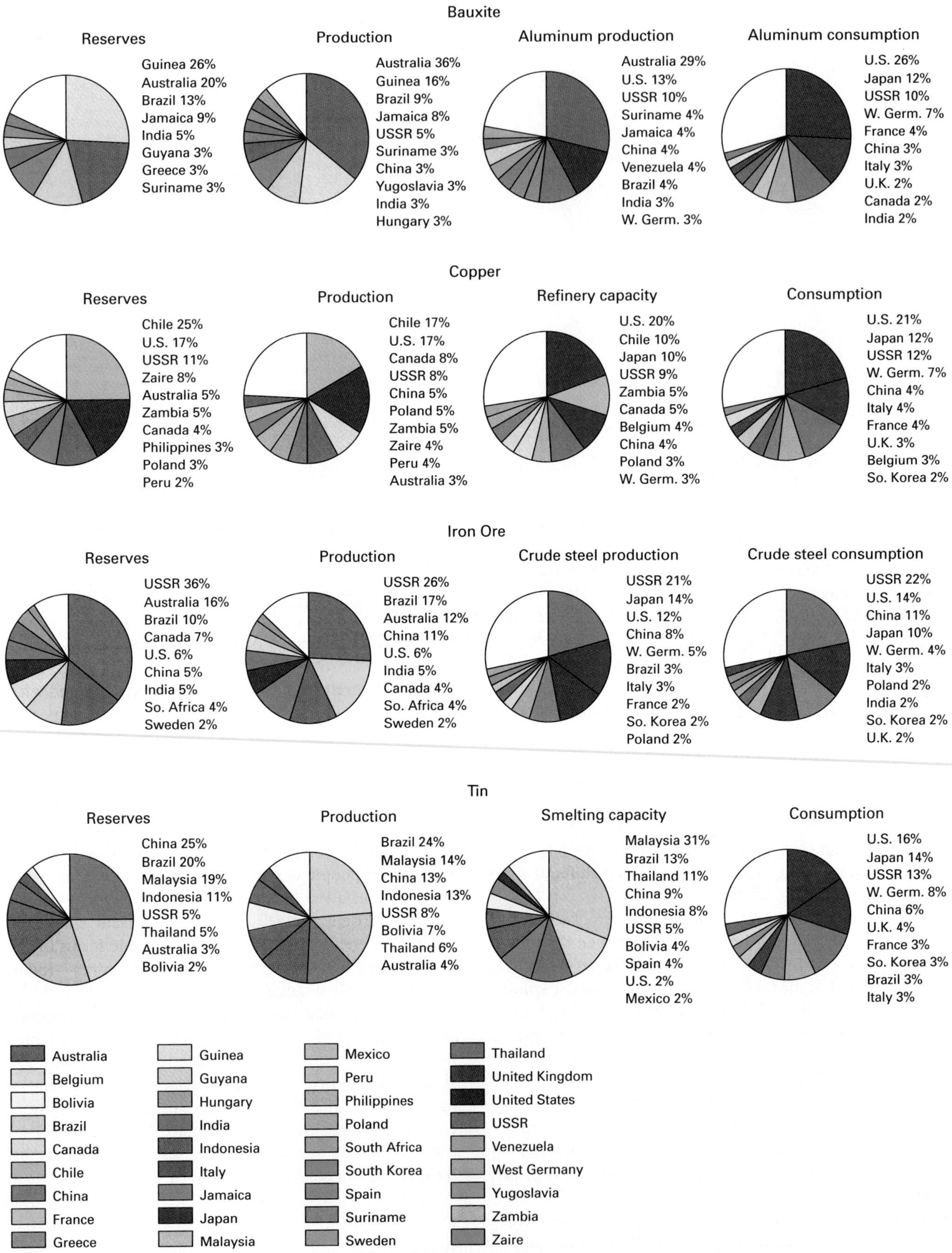

Bauxite

Reserves

Guinea 26%
Australia 20%
Brazil 13%
Jamaica 9%
India 5%
Guyana 3%
Greece 3%
Suriname 3%

Production

Australia 36%
Guinea 16%
Brazil 9%
Jamaica 8%
USSR 5%
Suriname 3%
China 3%
Yugoslavia 3%
India 3%
Hungary 3%

Aluminum production

Australia 29%
U.S. 13%
USSR 10%
Suriname 4%
Jamaica 4%
China 4%
Venezuela 4%
Brazil 4%
India 3%
W. Germ. 3%

Aluminum consumption

U.S. 26%
Japan 12%
USSR 10%
W. Germ. 7%
France 4%
China 3%
Italy 3%
U.K. 2%
Canada 2%
India 2%

Copper

Reserves

Chile 25%
U.S. 17%
USSR 11%
Zaire 8%
Australia 5%
Zambia 5%
Canada 4%
Philippines 3%
Poland 3%
Peru 2%

Production

Chile 17%
U.S. 17%
Canada 8%
USSR 8%
China 5%
Poland 5%
Zambia 5%
Zaire 4%
Peru 4%
Australia 3%

Refinery capacity

U.S. 20%
Chile 10%
Japan 10%
USSR 9%
Zambia 5%
Canada 5%
Belgium 4%
China 4%
Poland 3%
W. Germ. 3%

Consumption

U.S. 21%
Japan 12%
USSR 12%
W. Germ. 7%
China 4%
Italy 4%
France 4%
U.K. 3%
Belgium 3%
So. Korea 2%

Iron Ore

Reserves

USSR 36%
Australia 16%
Brazil 10%
Canada 7%
U.S. 6%
China 5%
India 5%
So. Africa 4%
Sweden 2%

Production

USSR 26%
Brazil 17%
Australia 12%
China 11%
U.S. 6%
India 5%
Canada 4%
So. Africa 4%
Sweden 2%

Crude steel production

USSR 21%
Japan 14%
U.S. 12%
China 8%
W. Germ. 5%
Brazil 3%
Italy 3%
France 2%
So. Korea 2%
Poland 2%

Crude steel consumption

USSR 22%
U.S. 14%
China 11%
Japan 10%
W. Germ. 4%
Italy 3%
Poland 2%
India 2%
So. Korea 2%
U.K. 2%

Tin

Reserves

China 25%
Brazil 20%
Malaysia 19%
Indonesia 11%
USSR 5%
Thailand 5%
Australia 3%
Bolivia 2%

Production

Brazil 24%
Malaysia 14%
China 13%
Indonesia 13%
USSR 8%
Bolivia 7%
Thailand 6%
Australia 4%

Smelting capacity

Malaysia 31%
Brazil 13%
Thailand 11%
China 9%
Indonesia 8%
USSR 5%
Bolivia 4%
Spain 4%
U.S. 2%
Mexico 2%

Consumption

U.S. 16%
Japan 14%
USSR 13%
W. Germ. 8%
China 6%
U.K. 4%
France 3%
So. Korea 3%
Brazil 3%
Italy 3%

Australia
Belgium
Bolivia
Brazil
Canada
Chile
China
France
Greece
Guinea
Guyana
Hungary
India
Indonesia
Italy
Jamaica
Japan
Malaysia
Mexico
Peru
Philippines
Poland
South Africa
South Korea
Spain
Suriname
Sweden
Thailand
United Kingdom
United States
USSR
Venezuela
West Germany
Yugoslavia
Zambia
Zaire

Figure 7–22. The geography of the reserves of any mineral may be very different from the geography of production, processing, use, and consumption of that mineral. These data are for 1989 from the U.S. Bureau of Mines.

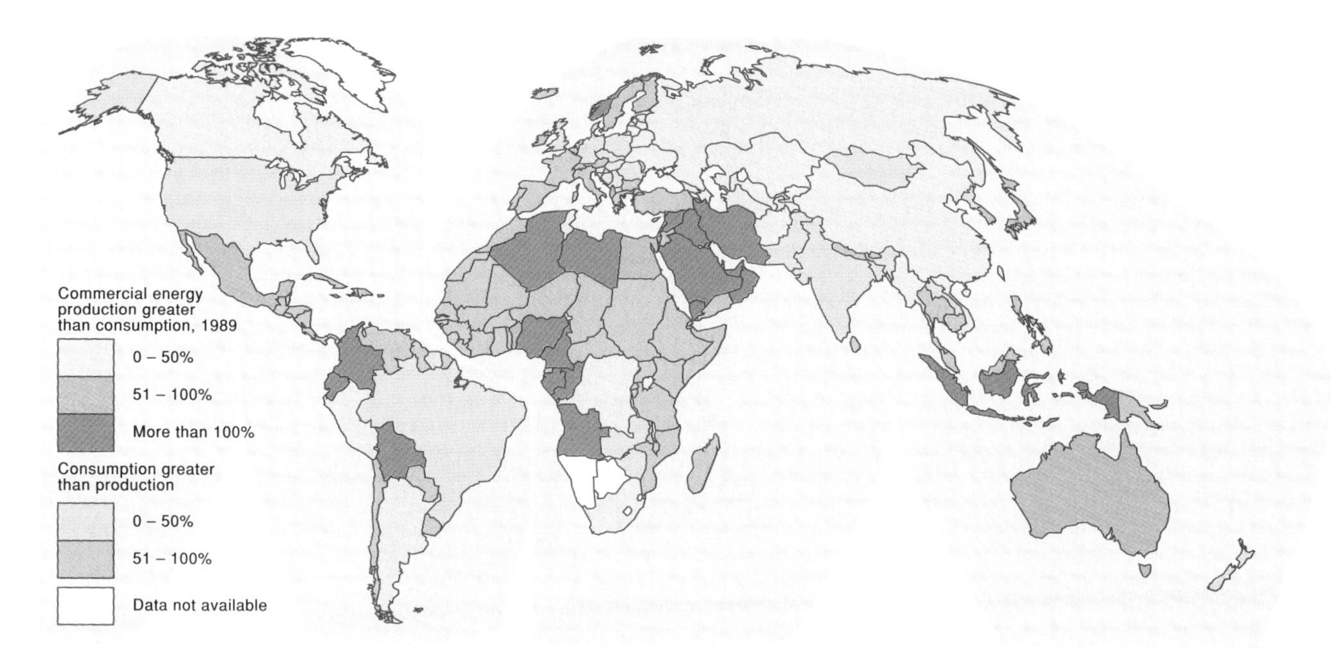

Figure 7–23. Some of the greatest energy-deficit nations on this map spend a considerable share of their national earnings on imported energy. (Data from World Resources Institute, *World Resources 1992–93*. New York: Oxford University Press, 1992, pp. 316–17.)

reserves in January 1991 were about 1 trillion barrels (see Figure 7–25 on p. 230). This figure includes only barrels of liquid petroleum. Additional trillions of barrels of oil are chemically locked into geological formations known as **oil shales** and **tar sands.** Because the costs of releasing liquid petroleum from these formations is higher than that of extracting oil from wells, estimates of quantities are inexact, and this petroleum is not counted among world reserves or resources. The United States, Canada, and Venezuela each contain over 1 trillion barrels of oil in shales or sand formations. Each time that anyone has predicted the commercialization of shales or sands over the past 100 years, however, new discoveries of liquid petroleum have dropped the price from that source, and so oil shales and sands have never been developed commercially. Liquid oil remains today the number-one commodity in world trade (see Figure 7–26 on p. 231).

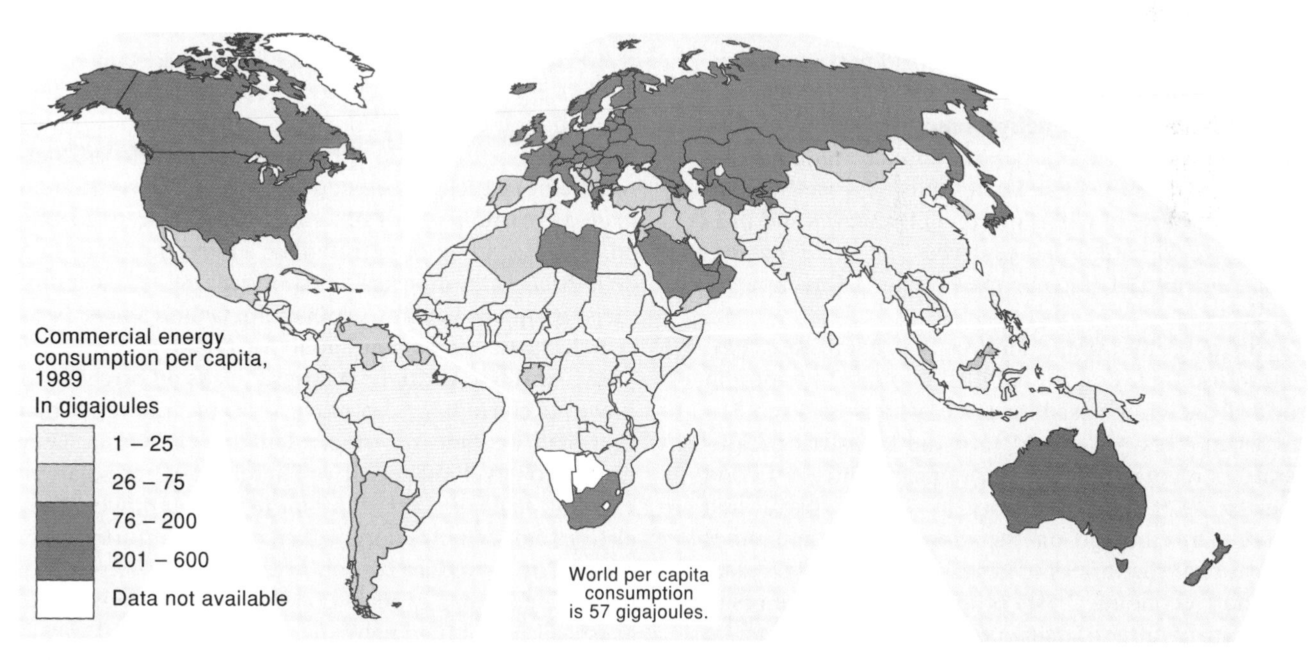

Figure 7–24. Energy consumption per capita is one of the most revealing statistical measures of people's standard of living. The nations enjoying the highest per capita consumption are not all the same as those having the greatest energy resources. (Data from World Resources Institute, *World Resources 1992–93*. New York: Oxford University Press, 1992, pp. 316–17.)

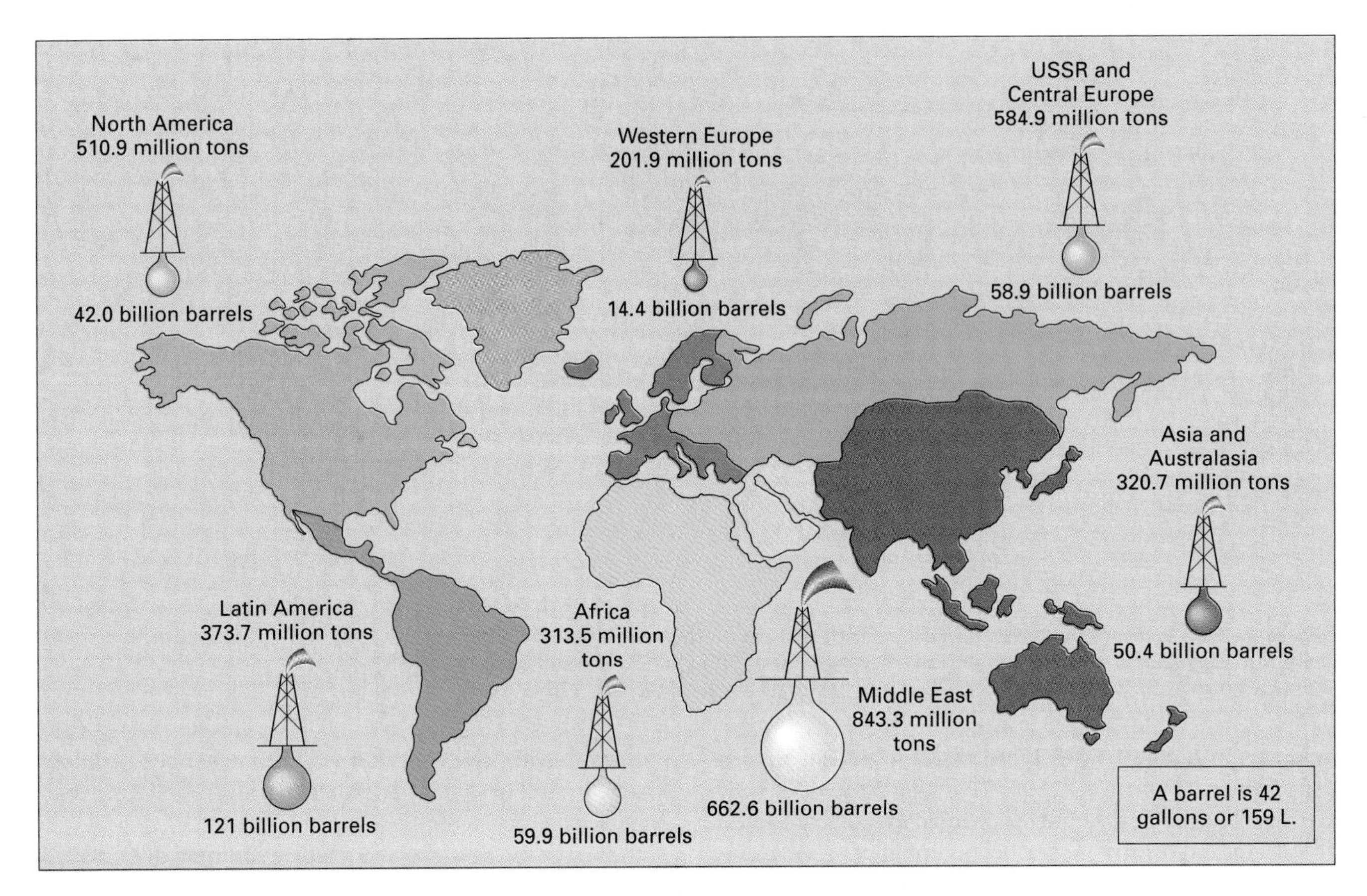

Figure 7–25. World petroleum reserves at the end of 1990 and production during 1990 . (Data courtesy of British Petroleum.)

The Organization of Petroleum Exporting Countries The Organization of Petroleum Exporting Countries (OPEC), founded in 1960, is a cartel that has enjoyed some success in limiting world petroleum supplies to maintain high prices. As of 1992 its 13 members were Iran, Iraq, Kuwait, Libya, Saudi Arabia, Venezuela, Qatar, Indonesia, the United Arab Emirates, Algeria, Nigeria, Ecuador, and Gabon. These countries together held about two-thirds of world petroleum reserves and one-third of world natural gas reserves. The cartel has exercised substantial but not total power to manipulate supplies.

In 1973 the Arab countries that dominate OPEC tried to use their control of oil as a political weapon in world affairs. They raised the price of oil from $3 to $13 per barrel. This "oil shock" cost the industrial nations about 2 percent of their gross national product. A second shock in 1979 raised the price to $40 per barrel and imposed an embargo on the United States and other international supporters of Israel. These price increases were not "economic," that is, they were not the result of scarcity of oil or of rising production costs. In fact, at that time oil from the Arab states could be produced for less than $1 per barrel.

The non-Arab OPEC countries raised prices along with the Arabs because many of them (Indonesia, for example) are dependent on oil export revenues. In international trade oil is priced in dollars, and so the countries whose currencies were rising against the dollar, most notably Japan and what was then West Germany, were cushioned against the price increases. Many of the world's poorest countries, however, suffered a crippling blow to their development hopes. The enormous profits the oil exporters made (popularly referred to as "petrodollars") were deposited in U.S. banks. The banks then loaned the money to Latin Americans—largely to buy oil. This recycling of petrodollars sowed the seeds of a debt crisis Latin America faced in the late 1980s and of the subsequent banking crisis in the United States. Nevertheless, the 1970s were good for OPEC.

OPEC and the industrial world As oil prices rose, however, the industrial countries became frightened that they had become too dependent on OPEC oil. They moved to develop alternative energies, such as solar energy and shale oils, and to promote nuclear power. They also developed new energy-efficient manufacturing techniques. In all industrial countries the consumption of energy per dollar of real economic output shrank. Japanese oil imports actually dropped between 1973 and 1979, whereas the economy grew by 20 percent. In the United States, the 1975 Energy Policy Conservation Act mandated the adoption of energy-saving technologies, and aggressive conservation and technology programs were launched to achieve energy independence.

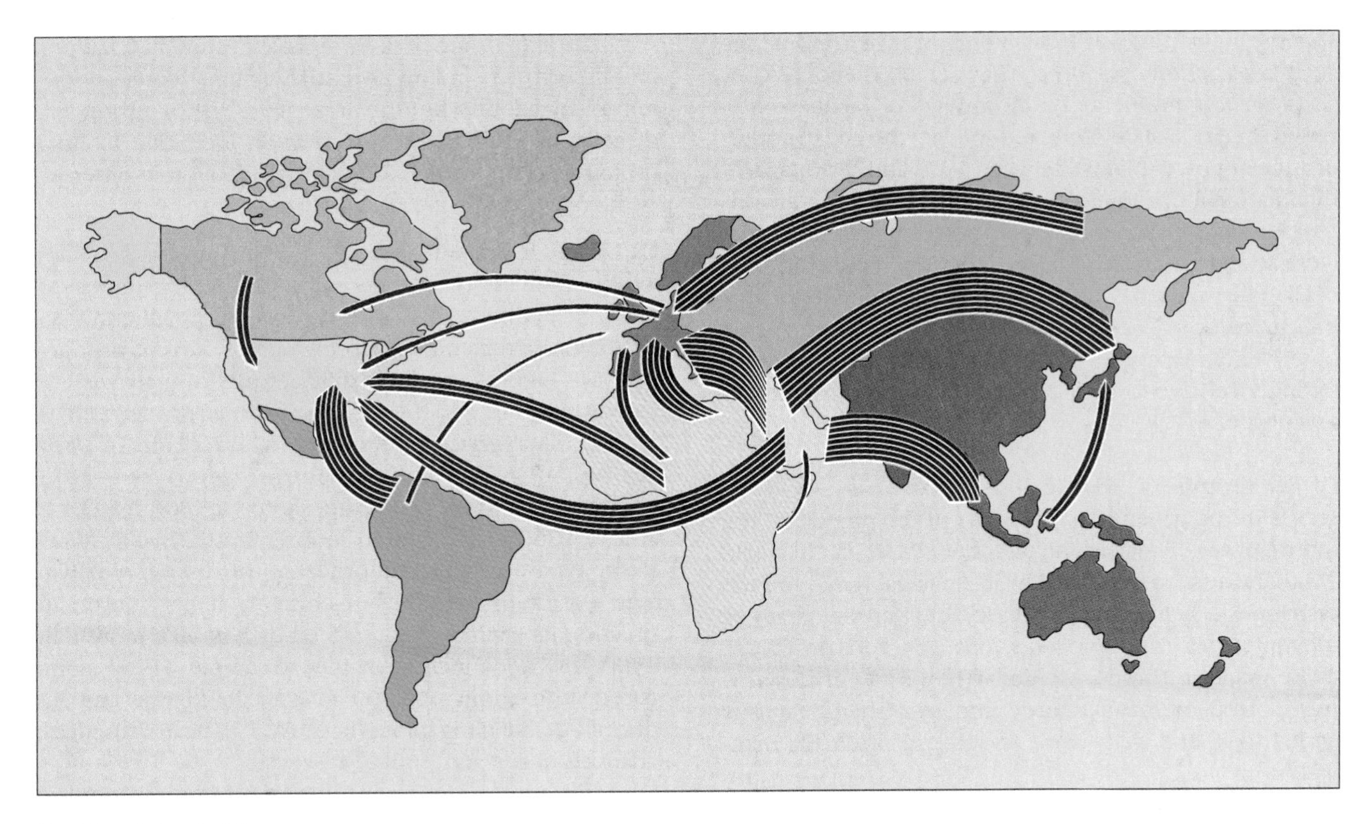

World trade in oil, 1990

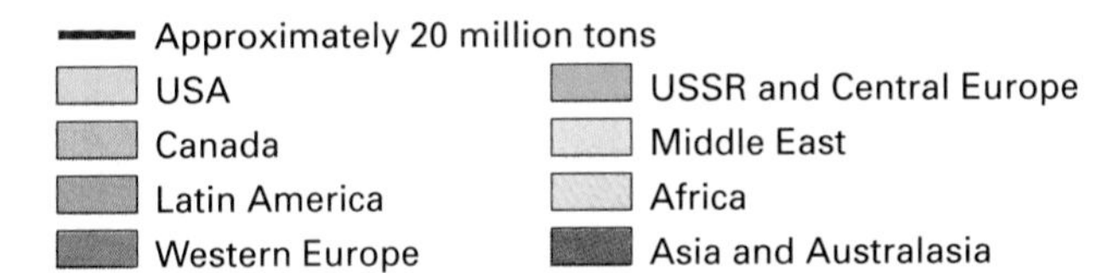

Figure 7–26. Petroleum is the number-one item in world trade. (Data courtesy of British Petroleum.)

High prices stimulated new worldwide exploration. This exploration resulted in discoveries of large reserves and in new production, albeit at high cost, in Alaska, the North Sea, Mexico, Brazil, India, and Yemen. As a result of conservation in the industrial markets and production from new sources, the 1980s saw wild price fluctuations, shrinking demand for oil worldwide, and, eventually, steep falls in OPEC revenues. By 1986 oil prices had collapsed below $10 per barrel. The world oil market had clearly turned around against OPEC, and OPEC had no choice but to lower its own prices. This move restimulated worldwide demand, which has been rising at about 3 percent per year since 1986.

Between 1973 and 1986 OPEC learned that the best way to keep the world dependent on oil is to keep oil affordable, that it is essential to avoid price wars, and that to lessen competition from alternative fuels, consumers in the industrial countries must be assured that their oil needs will always be met.

After OPEC lowered its prices, the oil-producing areas that had been developed in the late 1970s decreased their production. They never could compete with Mideastern oil on the basis of price alone. The consumer countries also slackened their efforts to find alternative sources of energy. Oil production in the United States declined, and the energy-conservation initiatives of the 1970s were allowed to wither through the 1980s. When Iraq invaded Kuwait in August 1990, U.S. domestic gasoline prices were the lowest they had been in real dollars since 1950. The nation had once more become dependent on OPEC oil. In 1982 imports had fallen to only 28 percent of U.S. oil consumption, but in 1989 imports accounted for more than 50 percent of the nation's oil needs for the first time since 1977. This trend reflected both growing demand and declining domestic production.

Japan's response to the oil crisis Japan's policies after 1973 contrast sharply with those of the United States. In 1973 oil accounted for 90 percent of Japan's energy imports and nearly the same proportion of total energy consumption. Japan responded to the escalating price of oil in ways that reduced the country's future vulnerability. Government bureaucrats devised plans to shift the economy away from energy-intensive industries such as aluminum smelting. They imposed energy conservation requirements on all aspects of Japanese life, raised the price of gasoline, and sped up the construction of nuclear power plants. Between 1973 and 1990 Japan's total national economic output doubled without any increase in national energy consumption. Japan's measures were so successful that the country now suffers the least of all industrialized countries when oil prices rise. Japan's industries are so energy-efficient that whenever the price of oil rises, Japanese manufacturers actually become stronger relative to their competitors in North America or Europe.

OPEC members' distribution networks Oil reserves in the ground are only part of the picture of the international oil market. An oil supplier's weight in world councils must be measured by its financial reach and the distribution network of refineries and gasoline stations it owns to get its oil to clients. Also, generous profits are to be made in these activities. When a raw material is transported or manufactured into something, value is added to it, and that **value added** can profit the manufacturer.

Today each OPEC country's national oil company is no longer content merely to sell crude petroleum on world markets. Rather, each company is trying to control supplies from the ground to the ultimate consumer and to capture the profit from value added at each step. The producers buy their own fleets to transport their oil, their own refineries to refine it, their own chemical companies to manufacture from it, and their own international distribution networks to retail the products. These national oil companies are replacing the largest private companies in Europe and in the United States. Between 1980 and 1990 Saudi Arabia almost quadrupled its domestic refining capacity (see Figure 7–27). It also purchased half of Texaco's refining and marketing system in the eastern United States. The National Petroleum Corporation of Nigeria is a partner in a major oil refinery in Kansas, and the state-owned Petroleos de Venezuela owns Citgo, with its Louisiana and Illinois refineries and its network of over 7000 service stations across the United States. The company also has refining ventures throughout Europe. The Kuwait Petroleum Corporation owns a refining and distribution network in Europe, and Libya owns a refinery in Germany and a distribution network in Italy.

The process is spreading throughout Asia, where petroleum consumption is growing rapidly in the industrializing countries. Despite new discoveries of petroleum in Thailand, Malaysia, Papua New Guinea, Myanmar, and Vietnam, Asia may become increasingly dependent on Mideastern oil by the year 2000, and OPEC members have begun to buy up Asian refining and distributing capacity. In 1992 Saudi Arabia contracted to build a refinery in Japan.

The expansion of oil companies from OPEC nations might seem to contain potential dangers for major oil-importing nations, but the presence of the oil producers directly in their markets also is an incentive to help promote economic stability. The economic interdependence of nations is escalating.

OPEC production quotas The biggest problem within OPEC is that of assigning production quotas to member nations. The cartel parcels out production according to a formula of percentages of current production, but OPEC's members differ greatly in their wish to use oil supplies as a weapon in world affairs, as well as in their oil reserves and their needs for income. In 1990 per capita GNPs of OPEC countries ranged from $270 in Nigeria and $560 in Indonesia to $15,860 for Qatar and $19,860 in the United Arab Emirates. Iraq's claims to the territory of Kuwait predated the discovery of Kuwait's great oil wealth, but certainly Iraq's interest in pressing its claim was intensified by Kuwait's wealth. In 1990 per capita income in Iraq was about $1950, compared with about $16,000 in Kuwait. Algeria and Indonesia demand big quotas because they have immediate financial needs and limited reserves.

The more refining and international marketing capacity that the individual producing countries accumulate, however, the less likely they will be to obey OPEC's quota system. They will be more likely to produce however much crude petroleum best suits the needs of their own integrated systems. This may ultimately destroy OPEC.

Some of the greatest potential producers have not increased production because they do not presently need the additional revenues. Kuwait needs money to rebuild after the devastation of the 1990–1991 occupation and war, but other Arab states still have huge reserves and

Figure 7–27. Refineries such as this one in Saudi Arabia add considerable value—and thus profit—to the production of crude oil. (Courtesy of Aramco World Magazine.)

few immediate needs. The richest Arab oil countries have hundreds of billions of dollars invested overseas, and that sum is growing merely by earning interest. Since the mid-1980s Kuwait has earned considerably more each year from international investments (for example, the prestigious Saks Fifth Avenue retail chain) than from exports of oil. Several Arab nations could live forever from their pools of accumulated capital without ever again tapping their pools of oil.

Saudi Arabia, which has traditionally commanded 25 percent of OPEC's total production quota, could triple its exports overnight. Were it to do so, world oil prices would plummet. In other words, potential Saudi production overhangs the world oil market. That country holds the power to dictate world oil prices.

At the time of the Iraqi invasion of Kuwait the cost to produce Mideast oil was about $2 per barrel. Saudi Arabia and the United Arab Emirates alone had proven reserves sufficient to supply the entire world's current consumption for about 30 years. Allowing the principal producing countries a profit well above their production costs and also considering that they are selling their nonrenewable resources, market economic forces would have set the world price of oil at about $10 per barrel. The world market price at that time, however, was over $20 per barrel, and it quickly rose to $40. This price defied the logic of supply and demand, but it reflected fears concerning the security of supplies. The principal oil-exporting countries reaped immense profits, and the world's poor oil-importing countries suffered another devastating blow.

Petroleum Production in the Former Soviet States Through the 1980s the Soviet Union was the world's biggest producer of petroleum. Ninety-two percent of production was in Russia, 4 percent in Kazakhstan, and 2 percent in Azerbaijan. In 1990 known oil reserves in Russia alone were twice those of the United States. Huge new resources are being discovered regularly, and there are still many unexplored but promising areas. Reserves in Russia and Kazakhstan might ultimately total more than those of Saudi Arabia. Petroleum from these sources might flood world markets and collapse world oil prices. This actually happened in the 1880s and again in 1928 when the Russian state needed money. In the early 1990s, however, Russia was still committed to selling its oil to its Commonwealth partners for 5 percent of the world market price. Selling that oil on world markets would have raised about $200 million per day.

Alternatively, political turmoil could entirely cut off these resources from world markets. In the early 1990s oil production was falling steadily, due to outdated equipment, lack of capital to buy spare parts, and political unrest. The governments of several republics invited Western oil companies to share in exploration and production of oil. The companies had the capital and the expertise to stop the decline in production, but delays in negotiating new contracts with each of the newly independent republics slowed the commitment of substantial Western capital. Oil production continued to fall in the early 1990s, but it may pick up again in the future.

ABC NEWS *Energy Needs Versus Alaskan Environment*

The Exxon oil spill in Prince William Sound focused national attention on environmental conditions in Alaska. In addition to the questions regarding who is responsible for the accident and who will clean up the area, people are asking "Can it happen again?"

The effects of the accident will not be fully realized for a number of years. It does appear, however, that given our need for petroleum—and we do have a heavy reliance on oil and oil-based products—that achieving a balance between exploration and environmental concerns will be difficult.

Alaska, in particular Prudhoe Bay, is one of the few remaining large domestic oil deposits. As the nation's dependence on petroleum increases, Alaskan oil will likely be targeted for future drilling. If not, the country risks becoming even more dependent on foreign sources, particularly OPEC.

Environmentalists, in contrast, emphasize the ecological hazards of implementing new large-scale exploration and shipping in Alaska. Not only are future accidents possible, but the area's flora and fauna could be endangered as well. Indeed, many oil companies want to drill in the Arctic National Wildlife Refuge, which is home to many species of animals.

As you watch the following video, consider these issues.

1. Outline the problems which have been highlighted in regard to the oil industry's activities at Prudhoe Bay.
2. Now that the problems at Prudhoe Bay and the incident at Prince William Sound have been brought to the attention of the public, do you think that the oil companies will respond to citizens' outrage by acting in a more environmentally responsible manner?

The Future of Petroleum Supplies If the Arab states that dominate OPEC continue to limit their production to keep prices high, then world oil prices in the 1990s will probably be determined by four factors:

1. Developments in the Commonwealth of Independent States;

2. The policy decisions of the non-Arab members of OPEC, particularly those members that need income from oil exports, as to whether to invest enough capital in oil-producing facilities to sustain their production levels into the 1990s;
3. The development of technologies to exploit oil shales and tar sands;
4. The development of alternative energy sources, either fuel or renewable, such as solar and wind resources.

Coal

Coal is the world's second largest commercial fuel source. It supplies about 30 percent of total global energy. China, which relies on coal for more than three-fourths of its total energy needs, is the world's largest consumer.

In 1991 the United States held 24 percent of world reserves, the USSR 22 percent (in Ukraine, Russia, and Kazakhstan), and China 15 percent. In general, however, coal reserves are widespread and abundant, and the use of coal is expected to rise (see Figure 7–28). The leading exporters have been Australia, the United States, South Africa, Russia, Poland, and Canada, but new suppliers to world markets may join these.

Increasing use of coal presents considerable environmental challenges. New techniques of surface mining, where the ore is scraped from the surface rather than brought out of mines, reduce the price of coal, but they threaten to scarify landscapes unless the land is carefully reclaimed. Runoff of waste from surface mining can pollute waterways. The pollutants released in burning coal contribute to the air's content of particulate matter and also to the ingredients of acid rain. These pollutants can be reduced by adding ***scrubbers*** to smokestacks, devices that remove the impurities from the smoke before releasing it. They are expensive, but technology is reducing their cost and improving their effectiveness.

Natural Gas

Finding natural gas has always been a disappointment to oil drillers, because bringing natural gas to market requires expensive pipelines or other equipment that is not available everywhere. Therefore, the world uses about five times as much oil as natural gas. Enormous volumes of natural gas have been found but left untapped, even in the world's poorest countries. Gas is nevertheless more efficient than oil in many applications, and it is cleaner. Burning gas produces virtually no sulfur dioxide or nitrogen oxides and less than half the carbon dioxide of oil or coal.

The amount of world natural gas reserves is the energy equivalent of about 665 billion barrels of oil, and

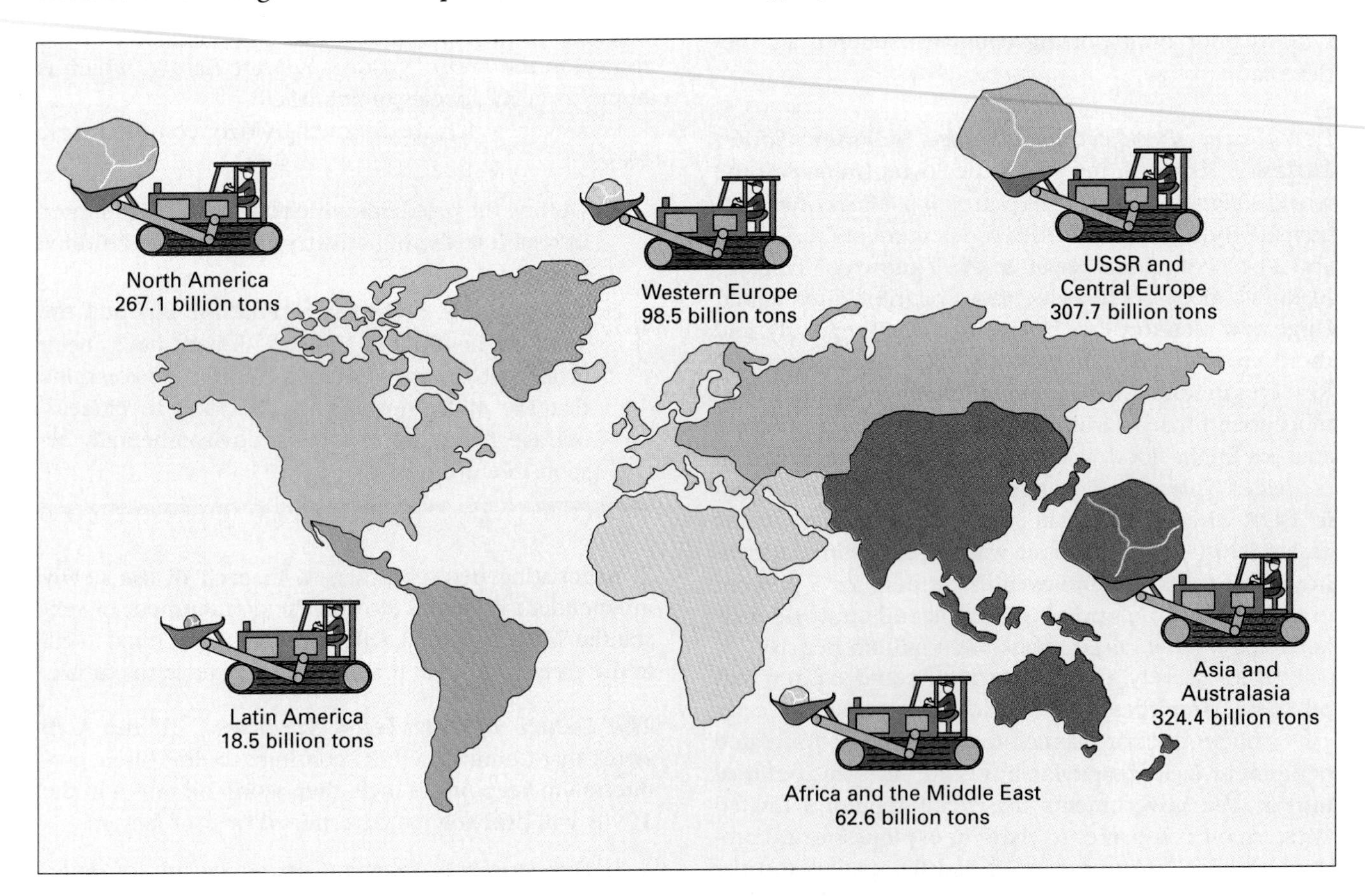

Figure 7–28. World coal reserves, 1990. (Data courtesy of British Petroleum.)

the geography of its distribution is not the same as that of oil. The USSR had 38 percent of world reserves of natural gas in 1991 (mostly in Russia). Iran, with the world's second highest reserves, had only 14 percent, and no other country held more than a 4 percent share. Many countries that import oil actually have their own substantial resources of gas. A 1989 World Bank study identified eight less-developed countries that rely on imported petroleum and yet could deliver domestic natural gas to their own population centers for less than one-quarter of the cost of the same energy in the form of oil: Bangladesh, Pakistan, Tanzania, Egypt, Tunisia, Thailand, India, and Morocco. If these countries had the capital to tap their own natural gas, they could lower their oil import bills, enjoy cheaper and cleaner domestic power, and profit from gas exports.

Even the United States might profit from greater reliance on natural gas. Demand is growing steadily, passing that for industrial oil for the first time in 1989. In that year newly discovered reserves kept pace with production, and 93 percent of U.S. consumption came from domestic production. Huge reserves remain in the Arkoma Basin in Arkansas and Oklahoma, in the Tide Formation in New Mexico, and in pockets throughout coal seams in the Rocky Mountains. If the United States were magically to transform all its electricity production and all its car engines to natural gas and methanol, the economy could run more cheaply, pollution problems would decline, independence in international affairs could be more easily achieved, emissions of greenhouse gases could be cut, and the trade deficit could fall.

Nuclear Power

By 1992 about 430 nuclear reactors had been constructed in 26 countries. Several energy-deficit countries have turned to nuclear power particularly for generating electricity. France relied on nuclear power for over 70 percent of its electricity, Belgium 66 percent, South Korea 53 percent, and Hungary and Taiwan 49 percent. In the United States nuclear generators provided only 19 percent of electricity. China, Argentina, Brazil, India, and Pakistan were the only developing countries operating commercial reactors.

The nuclear reactors currently in operation rely on **nuclear fission,** in which the nuclei of certain heavy atoms decay, giving off energy. The most common element used in reactors is uranium, which is abundant in the world, particularly in Australia and South Africa. In addition, the dismantling of thousands of nuclear weapons made during the cold war has also made available hundreds of tons of the potential fuel plutonium. Uranium-powered reactors convert uranium into plutonium as they operate. Plutonium is highly toxic, but using the world's plutonium to provide energy could make good use of obsolete weapons.

The development of nuclear fission capacity has been hampered everywhere by fears of pollution and of explosions and by the problems in disposing of the radioactive waste. Accidents at Chernobyl in Ukraine and at Three Mile Island in the United States reinforced the fears, and many of the nuclear reactors in operation in Eastern Europe, Ukraine, Belarus, and Russia were being closed down in the early 1990s because they presented unacceptable levels of pollution or danger. The world is just now learning the extent of nuclear contamination of Russia, Ukraine, and Kazakhstan by explosions and inadequate disposal of nuclear wastes under the Soviet government. In 1990 the Swiss voted to forbid the construction of any new nuclear reactors for the rest of this century, but not to abandon nuclear power altogether.

Despite these dangers, nuclear power does not impose some of the environmental problems associated with the use of other fuels: mining practices that scarify and pollute the countryside; polluting spills during transportation; the release of gases that contribute to global warming; and acid rain. Regulated nuclear power may turn out to be the environmentally preferred source of energy.

Nuclear energy could become far more attractive if scientists could harness **nuclear fusion,** the sun's energy source, as an alternative to fission reactors. Fusing hydrogen atoms into larger helium atoms releases huge bursts of energy. Fusion power would offer the attractive possibility of the direct production of electricity. Power plants would become unnecessary because electric devices could have their own individual power packs. The principal problem of fusion is that the process occurs at extremely high temperatures. Researchers are working to achieve low-temperature fusion.

In the United States the private sector has borne the burden of developing nuclear power to a degree unique among countries. Each nuclear power plant was individually designed and manufactured for a specific power company. Power plants were not standardized, and so the nuclear energy industry was not able to accumulate knowledge with experience. This, in turn, raised legitimate questions about the reliability of U.S.-made nuclear power plants. No new nuclear power plant has been ordered in the United States since 1978. Legislation to standardize construction of nuclear power plants could open a new era of development in the United States.

In France, by contrast, the government tightly controls nuclear energy production. Plants are standardized to the highest specifications, and any weakness or failure that shows up in the design or operation of any individual power plant can immediately be examined or replaced at all plants. As a result, the French safety record has been exemplary.

Japan has invested heavily in nuclear energy to reduce its reliance on oil imports. Nuclear plants today provide about one-third of Japan's electricity, and Japan plans to have nuclear power generating 40 percent of its electricity by 2000. As the only country ever to have suffered a nuclear bombing, however, Japan has understandable hesitations about nuclear power. Nevertheless, the technological sophistication of Japanese plants is so high that the level of radioactivity in the cooling water in them is 0.2 percent of that in U.S. plants, and the radiation doses to plant workers are only 7 percent to 10 percent of those in U.S. plants. In February 1991 Japan had 37 plants in operation and 16 under construction.

Although the United States obtains less of its energy from nuclear power than do other countries, because its overall consumption is so great it produces more nuclear energy than any other nation. In 1990 the United States accounted for 34 percent of the total global output of nuclear power, followed by France (13 percent), Japan (11 percent), the Soviet Union (9 percent), and Germany (7 percent).

Water Power

Water power is a particularly cheap form of energy. Thus, several tropical countries that have great amounts of rainfall have enormous untapped potential. Electricity, however, is technically difficult to transport. Few countries successfully export it, and then only to immediate neighbors. Therefore, if hydropower is to be tapped, it will have to be used locally.

Technology has multiplied the world's potential hydropower by increasing the efficiency of turbines. Less drop is required to produce electricity in what are called *low-head dams*. Hoover Dam on the Colorado River, built between 1931 and 1936, has a drop of over 1000 feet (305 m), but today efficient dams can be built with drops of less than 20 feet (6 m).

The world distribution of installed hydroelectric capacity differs greatly from the distribution of potential. The United States consumed 13 percent of the world total in 1990, Canada 12 percent, the USSR 10 percent, China 6 percent, and Norway 4 percent. Zaire, Brazil, and Indonesia, however, have estimated potential as great as either the United States or Canada.

Future Alternative Energy Supplies

The current major sources of energy will not necessarily be those of the future, and so a word must be said about what alternatives are on the horizon. Each of these may turn out to be either a turning point or just a mirage. Each is today far too costly to compete with the current sources of power, and it cannot be foretold when or where any of them might reach the price level at which it could replace power from petroleum, coal, or natural gas. Scientific breakthroughs, however, could make either of the following feasible.

First, the price of ***photovoltaic cells***, which make electricity directly from sunlight, is falling rapidly. Photovoltaic cells are now commonly found on pocket calculators, but larger-scale applications of the technology are being explored.

Second, scientists and engineers have turned their attention to the problems of exploiting the renewable energy sources of the wind and the tides. France has built the most advanced tidal facilities, which tap the power of the Atlantic Ocean tides. Several countries have built concentrations of giant windmills, called ***wind farms***. For these windmills to function effectively, the wind must be steady, not too strong or too weak. With technological improvement in the efficiency of the turbines to turn the wind energy into electrical energy, wind farms may become important in the future.

SUMMARY

Resources are things that humans can consume or put to use. Inventories of resources are expanding as humans learn to produce more of some things and also to use things previously thought useless. This justifies optimism for the future. Nevertheless, many people live in conditions of deprivation. That is humankind's tragedy.

The production and distribution of food is the result of complex interaction among environmental, technological, political, and economic factors. The result is that some countries are glutted with food, and supplies in other countries are insufficient to provide the entire population with a nutritious diet. The environmental and technological capability of the globe to produce enough food for the entire human population is secure today and in the future. Feeding all humankind adequately and at the least expense, however, would require redistributions of inputs, changes in national economic policies, and freeing world trade in food.

The hydrologic cycle provides plentiful water for total human needs, but humankind has not shown care in managing Earth's water resources. Most of the water withdrawn from the Earth is used for agriculture, but the share-allocation differs greatly from one country to another. Overall, water is becoming increasingly scarce. Shortages can be dealt with either by increasing or by conserving local supplies.

Pollution by both organic and industrial wastes is increasing. Pollution of rivers and lakes is potentially reversible, but less is known about the quality of groundwater reserves. Some of the advanced countries battle contaminating processes, but for much of the world's population, clean water remains precious and rare, and waste treatment is practically nonexistent.

The amounts of any mineral that can economically be recovered for use are called reserves. Reserves increase if the cost of extracting the mineral falls or if substitutes are found. Some animal and vegetable natural resources naturally renew themselves, and supplies will diminish only if they are harvested at a rate above their natural reproduction rate. Other resources can be recycled.

The world geography of both minerals and energy supplies reveals that the countries that have the most of any resource are not necessarily those that produce the most, and the countries that produce the most do not necessarily consume the most. The key to wealth is the ability to use resources.

KEY TERMS

resource (p. 207)
dumping (p. 207)
positive check (p. 207)
preventive check (p. 207)
green revolution (p. 209)
scientific revolution in agriculture (p. 209)
aquaculture (p. 214)
diminishing returns (p. 214)
greenbelt (p. 219)
metal (p. 222)
ore (p. 222)
crustal limit (p. 223)
reserve (p. 223)
futures contract (p. 223)
dematerialization (p. 224)
renewable resource (p. 225)
recyclable resource (p. 225)
strategic mineral (p. 226)
cartel (p. 226)
smelt (p. 227)
oil shale (p. 229)
tar sand (p. 229)
value added (p. 232)
nuclear fission (p. 235)
nuclear fusion (p. 235)

QUESTIONS FOR INVESTIGATION AND DISCUSSION

1. Which food crops are raised in your local region? What quantities are exported? Where?

2. Does your region produce any crops that are industrial raw materials, such as any fabric raw materials or industrial oils?

3. What livestock is raised in your local region? What quantities are exported? Where? Does that livestock supply any raw materials for manufacturing other than food processing?

4. Are there any significant food processing plants in your region?

5. Where did the materials in your breakfast come from?

6. Go to your local grocery store and investigate the latest storage and preservation packaging techniques. How have these allowed food to be preserved longer and to be more widely distributed?

7. What is the most common building stone in your area?

8. Is recycling practiced in your community? Does any local industry depend on recycled raw materials?

9. What fuels do local power plants use? Where do they come from?

10. Seek explanations for cases in which a given country has substantial mineral reserves, but they are not locally utilized.

11. What hampers the exploitation of the enormous concentrations of cheap hydropower potential in several of the countries listed? What political factors? Economic factors?

12. The OPEC cartel may crumble for any of several reasons. Among the most probable reasons are that its members may quarrel over the allocation of production quotas; the members may find themselves so committed to their individual refining and marketing networks that they ignore cartel rules and compete for markets; new liquid resources may be discovered and brought to market from nonmembers; or technology may drop the price of oil from sands or shales to a level competitive with liquid oil. How would any of these developments affect the world distribution of wealth? World trade? World political affairs? The production of energy in the United States or its import? The mix of energy supplies used in the United States?

ADDITIONAL READINGS

LOWRY, J. H. *World Population and Food Supply.* London: Edward Arnold, 1987.

U.S. DEPARTMENT OF AGRICULTURE. Economic Research Service. *World Agriculture: Situation and Outlook Report.* Washington, D.C., quarterly.

U.S. DEPARTMENT OF THE INTERIOR. Bureau of Mines. *Mineral Facts and Problems.* Washington, D.C., annually.

VIOLA, HERMAN, and CAROLYN MARGOLIS, eds. *Seeds of Change: A Quincentennial Commemoration.* Washington, D.C.: Smithsonian Institution Press, 1991.

WARNOCK, JOHN W. *The Politics of Hunger: The Global Food System.* New York: Methuen, 1987.

WORLD RESOURCES INSTITUTE, in collaboration with the UN Environment Program and the UN Development Program. *World Resources: A Guide to the Global Environment.* New York: Oxford University Press, 1986, 1987, 1988–1990, 1990–1991, 1992–1993.

CHAPTER 8

The Geography of Languages and Religions

Language is a key factor of culture, allowing the preservation and diffusion of knowledge and traditions. (Courtesy of Lidy Gijsen.)

Language and religion are two of the most important forces that define and bond human cultures. Their influences are so pervasive that many people take them for granted and cannot objectively observe their influences in the lives of others—or even in their own lives. Nevertheless, peoples who share either of these cultural attributes often demonstrate consistencies in other aspects of behavior. Furthermore, peoples who share either of these cultural attributes can often more easily cooperate with one another in other ways, such as in international affairs. Sharing these attributes may help people understand one another in ways that are unique and profound.

Each language and each religion originated in a distinct cultural hearth. Although the carriers of each of these cultural attributes have diffused around the world, each language and each religion still predominates within a definable realm. These realms of language and of religion are two of the most important of all types of culture regions.

THE GEOGRAPHY OF LANGUAGE

Many social scientists believe that language is the single most important cultural index. A **language** is a set of words, plus their pronunciation and methods of combining them, that is used and understood to communicate within a group of people. Each language has a unique way of dealing with facts, ideas, and concepts. Variations in languages result in variations in how people think about time and space, about things and processes. Exact translation from one language to another is virtually impossible. The language an individual speaks helps structure his or her perception and logic, and the comparative study of languages is one of the richest ways of understanding human psychology.

Any group of people who communicate exclusively with one another will soon develop "their own language," whether they are nuclear physicists or a social clique at a high school. If human beings were to appear suddenly in today's highly interconnected world, one worldwide language might develop. The great variety of languages spoken today testifies to the relative isolation of groups in the past. The distribution of any language illustrates the pattern of dispersal of its original speakers or their cultural impact on others.

The number of different languages recognized varies with the accepted definition of language. The term *language* is usually reserved for major patterns of difference in communication. Minor variations within languages are called *dialects,* but scholars do not agree on the amount of distinctiveness necessary for a pattern to be considered a language. Some scholars, for instance, accept Danish, Swedish, and Norwegian—which are mutually intelligible—as distinct languages, whereas others argue that they are dialects of one language. A **standard language** is the way any language is spoken and written according to formal rules of diction and grammar, although many regular speakers and writers of any language may not always follow all those rules. A country's **official language** is the one in which government business is normally conducted, official records are kept, signs are posted, and so forth.

Individual languages change through time, but religious classics or classics of literature can exert a powerful force for stabilization. In English, for example, the works of William Shakespeare and the 1612 King James translation of the Bible have molded the language, and yet parts of even these works may be difficult for many English speakers to read today.

Most scholars agree that there are more than 3000 distinct languages, at least 30 of which are spoken by 20 million people each. About 60 percent of all the world's people, however, speak 1 of just 14 major languages. The language with the most speakers is Mandarin Chinese (Guoyo), spoken by 850 million people. English is the primary language of about 400 million people worldwide, and it is either the only official language or one of several official languages in about 50 countries.

English serves as a second language for hundreds of millions of people, and so it is the world's **lingua franca,** that is, a second language held in common for international discourse. Other languages have served as lingua francas in the past. Latin long served Western civilization, and Swahili served throughout East Africa. Swahili developed among black peoples and in communication with Arab traders, and so it has many Arab words.

Arabic derives special transnational importance as the language of the Koran, the sacred scriptures of Islam. The Koran has been translated, but Muslims are still encouraged to study the original. Arabic is the official language in some 20 countries today.

In contrast to these widespread languages, some languages are extremely local. Linguists have discovered fully developed languages in New Guinea spoken by only a few hundred people in certain valleys. These languages have developed in total isolation over long periods of time. They are utterly incomprehensible to people just 20 miles away in the next valley.

National Languages

There is no exact correspondence between languages and the countries of the world. A few languages, such as Icelandic and Japanese, are associated almost exclusively with one country, but several languages are shared by many countries. The relationship between languages and nationalism is very complicated.

In European history, language has been interpreted as the basis of nationalism. As early as 1601, Henri IV of France seized French-speaking territories from the Duke of Savoy and declared to his new subjects, "It stands to reason that since your native tongue is French, you should be subjects of the King of France." This logic, however, never stopped the French from seizing the territory of non-French speakers whenever they could.

The notion of language as a basis for nationality eventually spread among the Germans. Martin Luther's translation of the Bible into German fixed the common language, as the Koran had fixed Arabic (see Figure 8–1). Thus, among the Germans as among the Arabs, language, religion, and nationalism became intertwined.

Language and Nation Building **Philological nationalism** is the idea that "mother tongues" have given birth to nations. This idea persists despite the fact that standard languages usually were and still are actually the product of self-conscious efforts at nation building by centralizing governments. Standardized languages cannot emerge before mass schooling and mass literacy or, alternatively, universal service in a national army. Usually these centralizing pressures transform the language of a small percentage of the population, the political or cultural elite, into the national language. In 1789, 50 percent of the French population did not speak French at all, and only 12 percent to 13 percent spoke it correctly. In 1860 only 2.5 percent of the population of the new Italian state spoke Italian for everyday purposes.

Figure 8–1. This is the title page of the last edition of Martin Luther's German translation of the full Bible printed before his death in 1546. This book launched the Protestant Reformation and fixed the German language. It can still be read by German speakers today. The page's rich decoration of Christian symbols includes the rewards of heaven and the terror of hell. (Courtesy of the Pierpont Morgan Library, New York. PML19474.)

Several national languages have appeared still more clearly as the result of deliberate political pressure. Some, such as Romanian in the nineteenth century and modern Hebrew, were virtually invented. In 1919 Hebrew was actually spoken by no more than 20,000 people, but the Israeli state has nurtured it for nationalistic reasons. Other new states have cultivated new national languages to unite diverse populations. The Indonesian government, for example, has succeeded in codifying and establishing one common language, Bahasa Indonesian. In 1990 Slovak became the official language of the Slovak part of Czechoslovakia, although hardly anyone used it. It will flourish only if the Slovaks nurture it. Not all countries succeed in reviving or creating national languages. When Ireland achieved independence in 1922, the government tried to enforce the use of Gaelic rather than English. The people themselves did not accept it, however, but continued to use English.

A minority people in a country often refuses assimilation and clings to its language as a gesture of cultural independence. The Kurdish minority in Turkey, an estimated 10 million people, clings to Kurdish, although public use of the language was outlawed from 1984 to 1991. Kurdish minorities in Iraq and Iran have also suffered political and linguistic discrimination.

Language in Postcolonial Societies European colonial powers forced their language on their colonies, regardless of how many native languages were already spoken and whether many Europeans actually settled in any given colony. The conqueror's language became the language of government, administration and law, economic development, and usually education. When the colonies gained their political independence and had to choose official languages, many kept their former ruler's language. Today's world distribution of official languages partly reflects the map of former European empires (see Figure 8–2).

Figure 8–2, however, differs markedly from a map of the distribution of the world's languages actually spoken (Figure 8–3). Across much of the world, the official language is not actually spoken by most of the population. The official language map shows six international languages (English, French, Spanish, Portuguese, Arabic, and Russian) covering most of the globe, and yet only about 20 percent of the world's population speaks any one of these six.

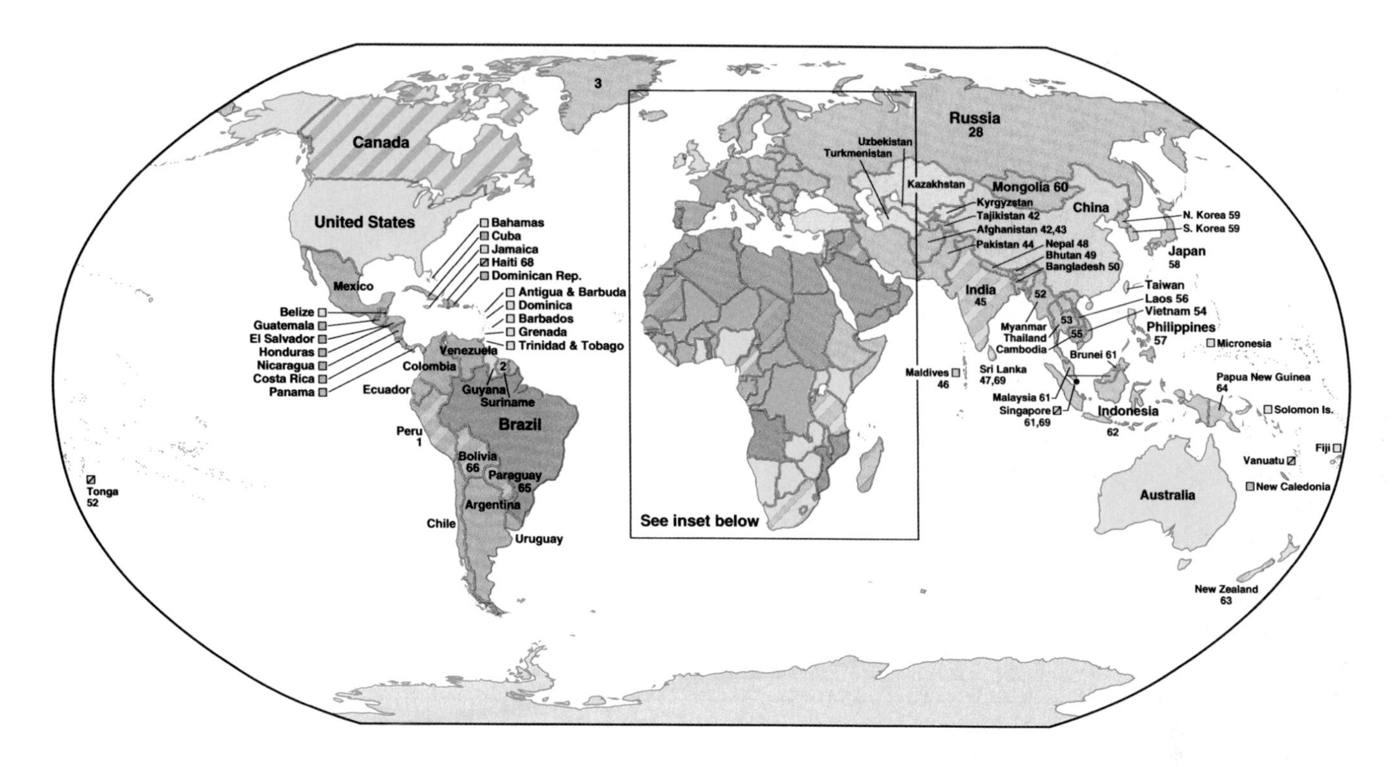

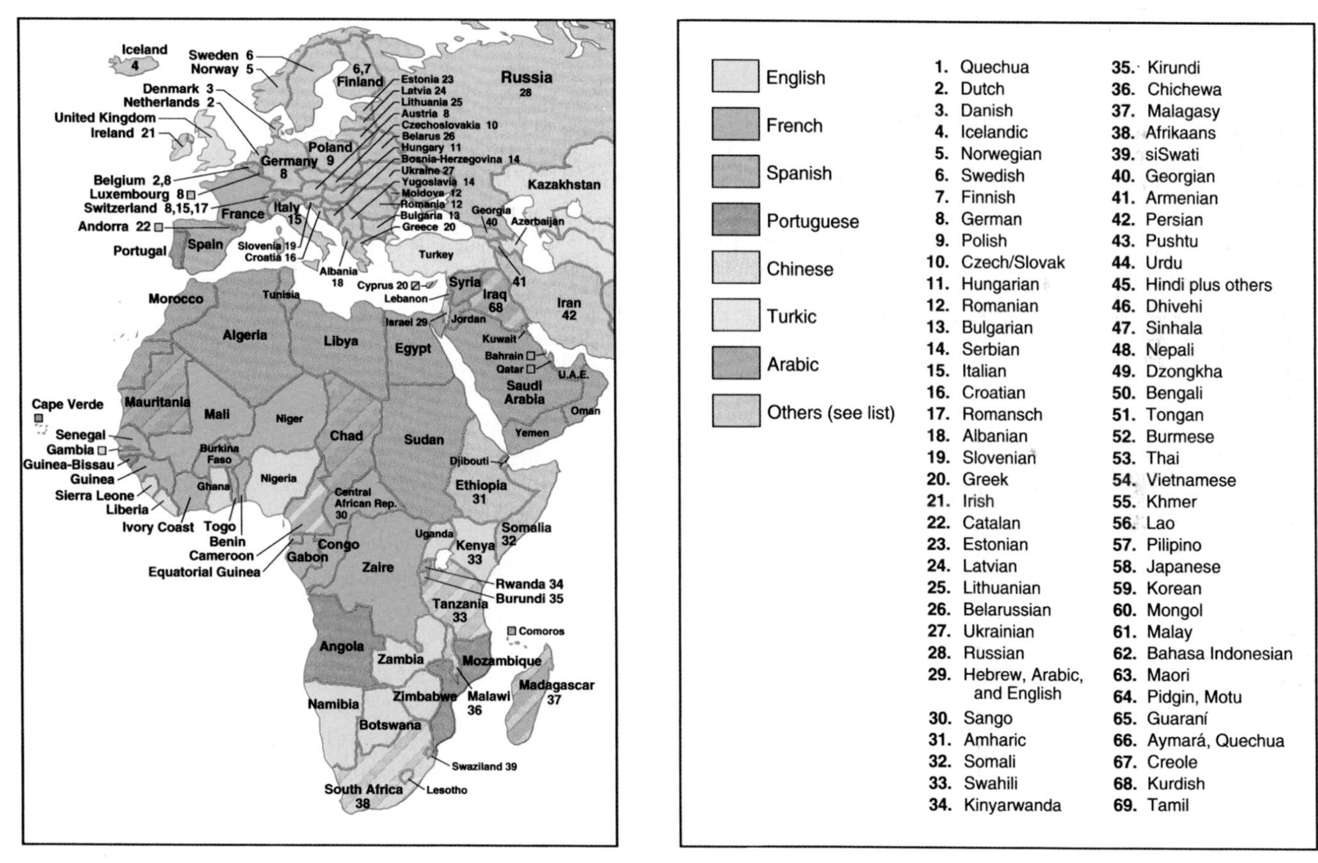

Figure 8–2. Several forces are at work in this map. One is the legacy of colonialism, which diffused European languages across much of the world. Increasing international trade and travel emphasizes the usefulness of such a linguistic legacy. In many countries that were formerly colonies, however, growing national pride is a competing force that encourages governments to recognize native languages. Therefore, a growing number of countries recognize two or more official languages—one local, and one international. In Kenya the backlash against colonialism was so strong that the country terminated the recognition of English as an official language in 1974.

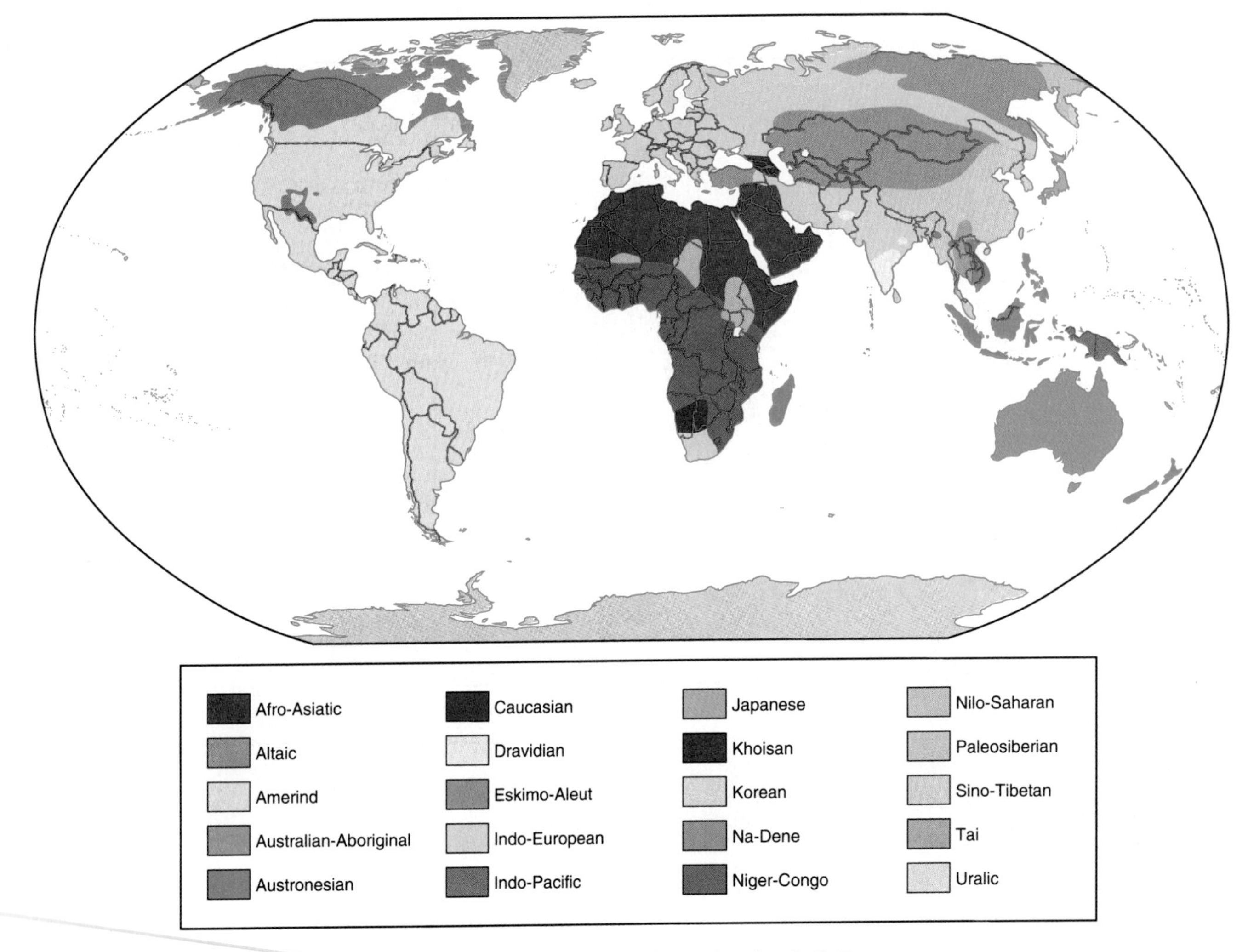

Figure 8–3. The map of the world's living language families may be said to "underlie" Figure 8–2, the map of official languages. (Western Hemisphere adapted from *Language in the Americas* by Joseph H. Greenberg with the permission of the publishers, Stanford University Press. © 1987 by the Board of Trustees of the Leland Stanford Junior University. Eastern Hemisphere adapted with permission from David Crystal, *Encyclopedia of Language*. New York: Cambridge University Press, 1987.)

In some cases the newly independent countries retained their former ruler's language to prevent arguments about which of the native languages should become the new national language. Many African countries have retained English or French for this reason, and attempts to change the official language can trigger internal strife. In Algeria the Berber minority protests violently whenever the government tries to replace French with Arabic. Over 250 languages are spoken in Nigeria, and advocates of a national language could not decide among Hausa, Yoruba, or Ibo, the three major native tongues. Therefore English was chosen, although only about 20 percent of the population speaks it well, and still fewer use it as their first language. Another reason for the continuing preferences for European languages is that the governments cannot afford to offer schooling in the many indigenous languages.

Linguistic rivalries maintain English as an official language of India, although it is spoken by a small fraction of the total population. The leaders of India's independence movement intended Hindi one day to be India's official language, and Hindi is spoken by at least 400 million Indians. Hindi speakers, however, are concentrated in northern India, and other Indians are unwilling to accept the domination of Hindi. Whenever Hindi has been pressed on India's southern states, they have threatened to secede. India recognizes English as one of 16 official languages, and English remains the preeminent language of the upper classes and the upwardly mobile.

A former colony can rise to greater wealth and influence than its former ruler, in which case its dialect sets the international standard. This has happened with the U.S. version of English. Noah Webster published his self-consciously American *Spelling Book*—a cultural declaration of independence—in 1783, and his *American Dictionary of the English Language* appeared in 1828. Today American English competes with British English for precedence worldwide, although some scholars argue that American is no longer a dialect but a language in its own right.

Portuguese boasts 180 million native speakers, and at a 1990 international conference the governments of

Brazil, Angola, Mozambique, São Tomé and Principe, Guinea-Bissau, and Cape Verde—all former Portuguese colonies—accepted the Brazilian version as the standard. Brazilian is much simpler in spelling than is the Portuguese of Portugal, but it incorporates many Native-American, African, and even American words. The Portuguese were humiliated, and they are considering what to do. Mushrooming Brazilian influence in Portugal, through investment, trade, and media products, threatens to inundate Portugal's own culture. Brazil has 15 times Portugal's inhabitants and an economy more than 8 times as large.

Some former colonial peoples find the acceptance of a major international language useful, and it often reflects these countries' continuing ties to their former rulers.

Polyglot States Many countries such as India grant legal equality to two or even more languages. These are called **polyglot states.** Switzerland, for example, recognizes German, French, Italian, and Romansch. If various languages predominate in distinct regions of a country, the country might accept each language as official in its region. Belgium, and even its capital city of Brussels, is legally divided into French and Flemish zones. In the center of Brussels, all signs are in both languages. The precedence of languages is hotly contested, and several Belgian governments have fallen when they have been seen as giving preference to one language or the other. Canada is also officially polyglot, and language is a source of friction between the Anglophonic (English-speaking) majority and the Francophonic (French-speaking) minority (see Figure 8–4).

Figure 8–4. Canada is officially a bilingual country, and so government signs must be posted in both French and English. (Courtesy of Michael Evans/Leo de Wys.)

Polyglot states usually select one language as official for the federal center and for communications among the states, and regardless of what the country's constitution or laws might say, one language will generally be the "preferred" language of the country. Those who do not use it may find their opportunity restricted or their upward mobility blocked. Russian, for instance, was long in general use throughout the Soviet Union. Native speakers of the minority languages could rise in the national power structure only if they also spoke Russian.

Different language communities within one country may be defined not only geographically, by region, but also socially, by class. Tagalog is the language of the people native to the Philippine Islands. The islands were colonized, however, first by Spain and then by the United States, and so English and Spanish are the languages of the upper classes. Today upper-class English- or Spanish-speaking Filipinos have been heard to sneer, "We speak Tagalog only to servants." Dutch remains the lingua franca of the elite throughout Indonesia, which was a Dutch colony for hundreds of years.

Throughout Central and South America, the Native-American populations and their cultures were smothered under Spanish and Portuguese rule. Vestiges of this cultural imperialism remained even after independence. For example, the original Constitution for Bolivia, written by Simon Bolívar, accepted as citizens only those who could speak Spanish. Still today in Bolivia the majority of the Native Americans speak Aymará (25 percent) or Quechua (34 percent). Peru accepts both Spanish (spoken by 68 percent of the population) and, since 1975, Quechua (27 percent). Official recognition of Quechua may reflect the political awakening of the Native-American population.

The Geography of Orthography

The geography of languages is complicated by the geography of **orthography,** which is a system of writing. Most languages are written in **alphabets,** which are systems in which letters represent sounds. There are several alphabets in use today, and the alphabet in which a language is written can reflect a historical diffusion (see Figure 8–5).

Modern Western European languages are written in the Roman alphabet because Western Europeans were converted to Christianity from Rome. In Eastern Europe, in contrast, Russian, Belarussian, Ukrainian, Serbian, Bulgarian, and a few other languages used in Russia are written in the Cyrillic alphabet. This is the Greek alphabet as it was augmented and taught by the Greek Orthodox missionary Saint Cyril (d. 869), who converted many of these peoples to Christianity. Serbian and Croatian are one spoken language, but Serbian is written in Cyrillic, and Croatian in Roman. Thus, the line between them precisely demarcates a cultural borderline across Europe (see Figure 8–6).

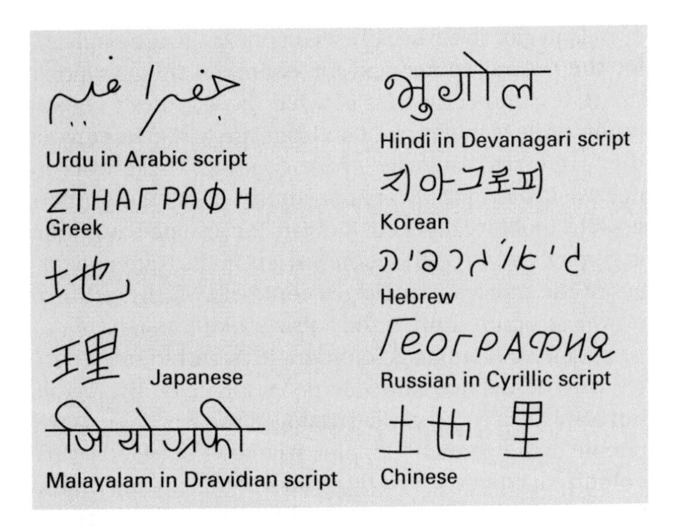

Figure 8–5. This figure shows the word *geography* written in several different orthographies.

A similar difference separates Pakistan from India. Spoken Urdu, the language of Pakistan, is the same as spoken Hindi, the language of northern India. The Muslim Pakistanis write it in Arabic script, but the Hindus of India write it in Devanagari script. Several other Indian languages are also written in Devanagari.

Sometimes a people can change the orthography in which they write their language as a self-conscious political act. Romania, for example, switched from the Cyrillic to the Latin alphabet early in the twentieth century. This change reflected a choice to be a "Western European" nation. In the same way, the languages of central Asia, called Turkish, historically came to be written in Arabic when these people converted to Islam, as discussed later in this chapter. The "distant cousins" of these central Asian peoples, however, the Turks of the country of Turkey, chose to replace Arabic with the Roman alphabet early in the twentieth century to signal a deliberate turn to Western-style modernization. In 1939 Soviet dictator Joseph Stalin forced the Turkish peoples in central Asia to replace Arabic script with Cyrillic script. His objective was to cut off these peoples from their cultural heritage and to intensify relatively minor linguistic differences among them. In 1992, however, with the collapse of the USSR, the newly independent republics of Kazakhstan, Kyrgyzstan, Tajikistan, Uzbekistan, Turkmenistan, and Azerbaijan all chose to abandon Cyrillic script. They did not choose to revert to Arabic, but to adopt Roman. Western observers hope that this reflects a commitment to join the cultural community of the West.

Nonalphabetical Languages Not all languages are written in alphabets. The major nonalphabetic forms of writing are Chinese, in which each character repre-

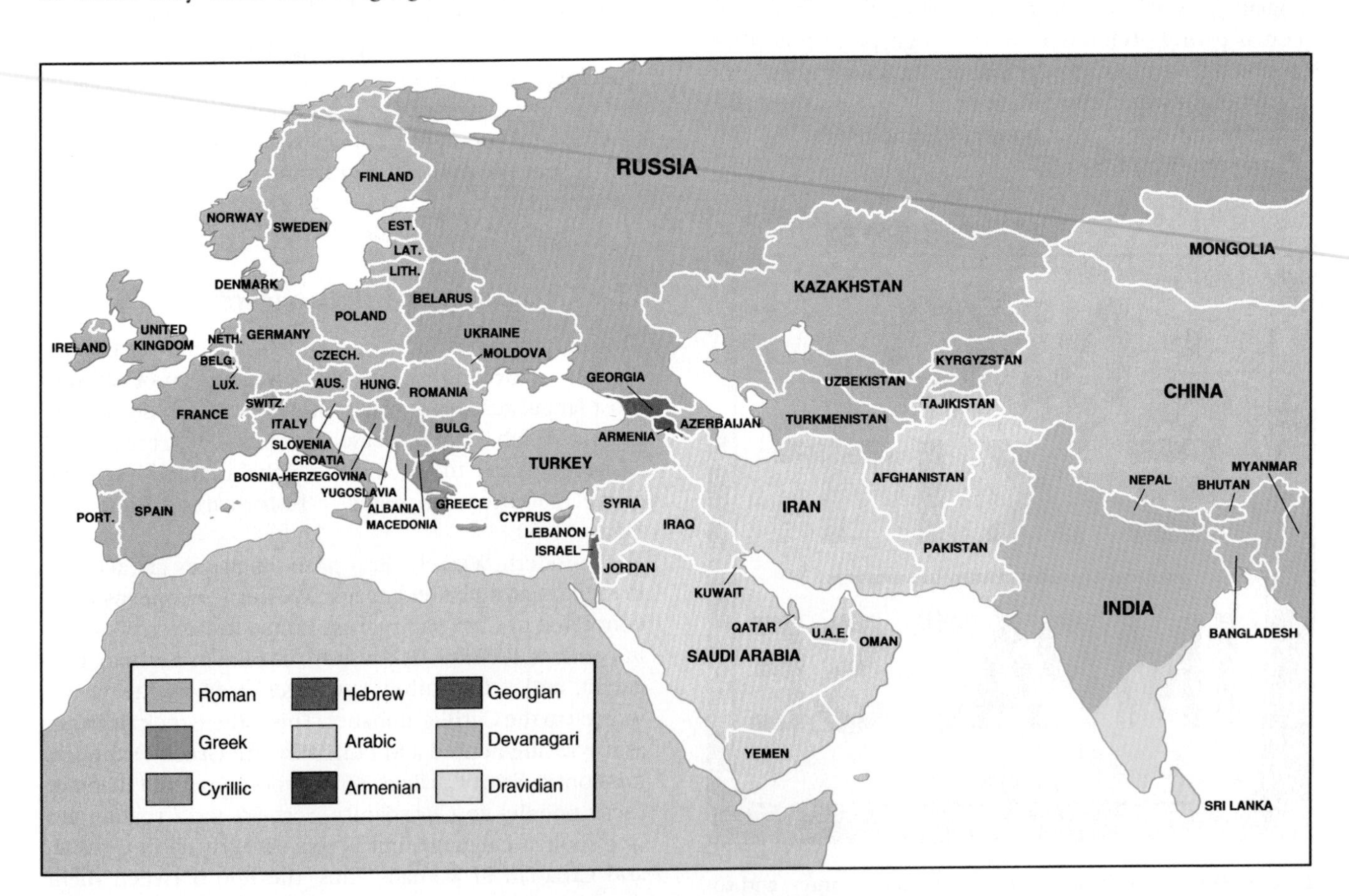

Figure 8–6. The distribution of alphabetic scripts in Eurasia reveals the diffusion of religions and other cultural markers.

sents a word or concept, and Japanese, in which each character represents a syllable. Korean writing was modeled on Chinese. The demands of contemporary international communication are forcing all these countries to convert their writing styles to the Roman alphabet. The Vietnamese changed from Chinese writing to Roman orthography in the eighteenth century when they were Christianized by Jesuit missionaries.

Many traditional Native-American, African, and Asian languages never had written forms until Christian or Islamic missionaries set out to translate the Bible or the Koran into these languages, using the missionaries' own orthographies. Although these peoples may have benefited from having their languages written, imposing *any* orthography on a people whose culture had been oral is an act of cultural imperialism. The transition from an oral to a literate culture is arguably the greatest transition in human cultural history. To purely oral people, writing is strange and magical—a conversation with no one and yet with everyone. Yet writing opens new possibilities of exact storage and communication of information.

THE DEVELOPMENT AND DIFFUSION OF LANGUAGES

The way languages develop and diffuse demonstrates the basic principles of diffusion for any cultural innovation or attribute. Any isolated group of people develops a language of its own. This language describes everything that those people see or experience together. If groups of these people break away and disperse, then each group discovers new objects and ideas, and the people have to make up new words for them. After thousands of years, the descendants of each of these breakaway groups have their own language. Each descendant language has a vocabulary of its own, but each also retains a common core of words from that earliest shared language. The ancestor that is common to any group of several of today's languages is called a **protolanguage.** The languages that are related by descent from a common proto-language make up a **language family.**

The Indo-European Language Family

In 1786 the English philosopher Sir William Jones first pronounced his theory that a great variety of languages spoken across a tremendous expanse of the Earth demonstrate similarities among themselves so numerous and precise that they cannot be attributed to chance and cannot be explained by borrowing. These languages, he concluded, must descend from a common original language. The group of languages first identified by Sir William is called the *Indo-European family* of languages, and about half of the world's peoples today speak a language from this family (see Figure 8–7). Sifting through the vocabularies of all Indo-European languages yields a common core vocabulary, which is the common ancestor of these languages, *proto-Indo-European.*

The vocabulary of proto-Indo-European tells us a surprising amount about how proto-Indo-European society was organized and how the people lived. It also hints at the language's hearth. Reconstructed proto-Indo-European has words for distinct seasons (one with snow), woody trees (including the beech and the birch), bears,

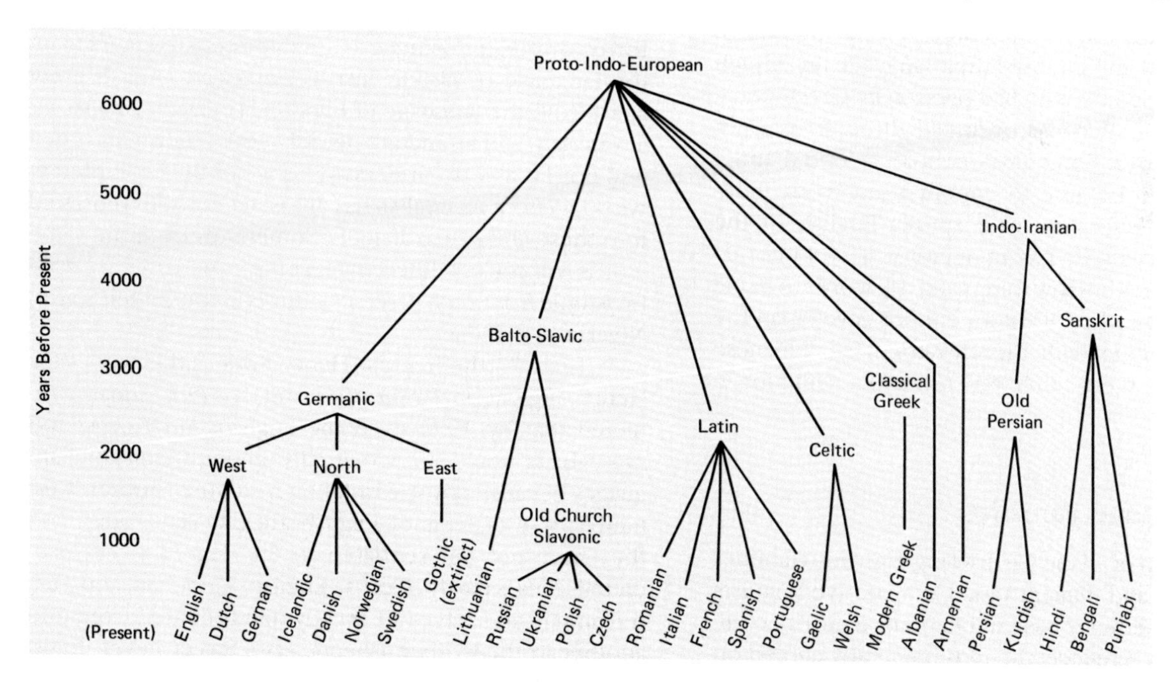

Figure 8–7. The languages in this "family tree" are all descendants of proto-Indo-European, and common word roots can be found among them. (Reprinted with permission from Carol R. Ember and Melvin Ember, *Anthropology*, 6th ed. Englewood Cliffs, NJ: Prentice Hall, 1990, p. 229.)

wolves, beavers, mice, salmon, eels, sparrows, and wasps. These things can be found together around the Black Sea, but proto-Indo-European also includes words clearly borrowed from the languages of Mesopotamia and the Near East. The word for wine, for instance, seems to descend from the non-Indo-European Semitic word *wanju.* Thus, the hearth area for Indo-European languages was probably in Anatolia, today's Turkey, some 8000 years ago. Archaeologists disagree as to whether proto-Indo-European diffused quickly, carried by warriors, or slowly, carried by individual farmers.

Hunting common Indo-European roots of words provides a fascinating study. The proto-Indo-European root *aiw*, for example, which means "life" or "the vital force," descended into Hindi as *ayua*, "life," but it also shows up as *aetas* in Latin, *aion* in Greek, *ewig* in German, and the words *ever* and *age* in English. Words that are distant cousins appear in surprising places. *Maharajah*, for example, a Hindi word for a great ruler, may seem exotic, and yet *maha* is a distant cousin of the English words *major* and *magnitude. Rajah* is a distant cousin of the English *reign* and *royal.* In this case the distant cousins clearly look or sound alike, and so they are called **cognates.** One must never assume any connection between two words from different languages, however, until it can be proved that they have a historic common root. As Chapter 5 emphasized, the possibility of cultural diffusion must always be investigated against the possibility of independent invention in two places. The study of word origins and history is called **etymology.**

Proto-Indo-European provided the basic stock for all Indo-European languages, but that does not mean that one ethnic or racial group spread out to live where all these peoples are today. Sometimes a few Indo-Europeans conquered and imposed their language on a much larger group of people who had previously developed a language of their own—as occurred throughout Latin America and Africa. Sometimes peoples adopted Indo-European just to be able to communicate with Indo-Europeans. Not everyone who speaks English in the world today necessarily has an ancestor from England.

Also, cultures borrow things and the words to name them from one another. Western culture gave Japan the word *erebata* along with the elevator itself. Japanese culture gave the West edible raw fish and a name for it: *sushi.*

Other Language Families

The classification of all the Earth's languages into families is an enormous and difficult task. Comparative linguists do not agree whether it has yet been satisfactorily completed. Figure 8–3 reflects the most generally agreed-on state of understanding.

The Indo-European family has been studied more than any other, and researchers have benefited from written sources dating back 3700 years—tablets in the extinct Indo-European language Hittite. These sources give researchers considerable confidence in including individual languages in the family and in reconstructing proto-Indo-European.

Some comparative linguists believe that after all languages have been classified into families, protolanguages can be constructed for each language family and that study of those protolanguages will eventually yield superprotolanguages. They argue that the Indo-European family, for instance, is only one of six branches of a larger group, which they call *Nostratic.* Reaching even farther back into prehistory, Nostratic itself is only one descendant of a protohuman language, or "mother tongue," that was spoken in Africa 100,000 years ago and then diffused around the globe.

The farther back into prehistory that scholars attempt to reconstruct these languages, however, the more controversial the work becomes. Some linguists believe that the time-depth of such studies is too great to reconstruct more protolanguages, let alone one prehistoric mother tongue.

The language trees worked out by the most adventurous historical linguists, however, concur to some degree with the groupings of humankind based on the genetic evidence discussed in Chapter 6 (see Figure 8–8). Geography, linguistics, anthropology, and biology all contribute to this fascinating but highly controversial research into prehistory.

Linguistic Differentiation in the Modern World

The diffusion and differentiation of languages continues today. Brazilian Portuguese has differentiated itself from the language of Lisbon just as American English grew apart from the language of London. Today an American, a Nigerian, and an Indian may all speak English, but they will not be able to understand one another completely without effort. As English has diffused it has differentiated into these and other dialects. Sometimes even the same words will convey different meanings when they are used by people from such diverse cultures as the United States, Nigeria, and India.

In 1877 the linguist Henry Sweet (the model for Henry Higgins in *Pygmalion* and *My Fair Lady*) predicted that by 1977 even the English, Americans, and Australians would speak mutually incomprehensible languages because of their isolation from one another. Certainly these dialects of English are different today, but they have not differentiated to the degree Sweet predicted. This is partly because English, American, and Australian cultures have not greatly differentiated from one another in their other aspects. Also, Sweet never could have foreseen the great increases in world communication, nor the role that nearly universal education would play in fixing language in England, America, and Australia.

Dialects usually diverge more in the way they are

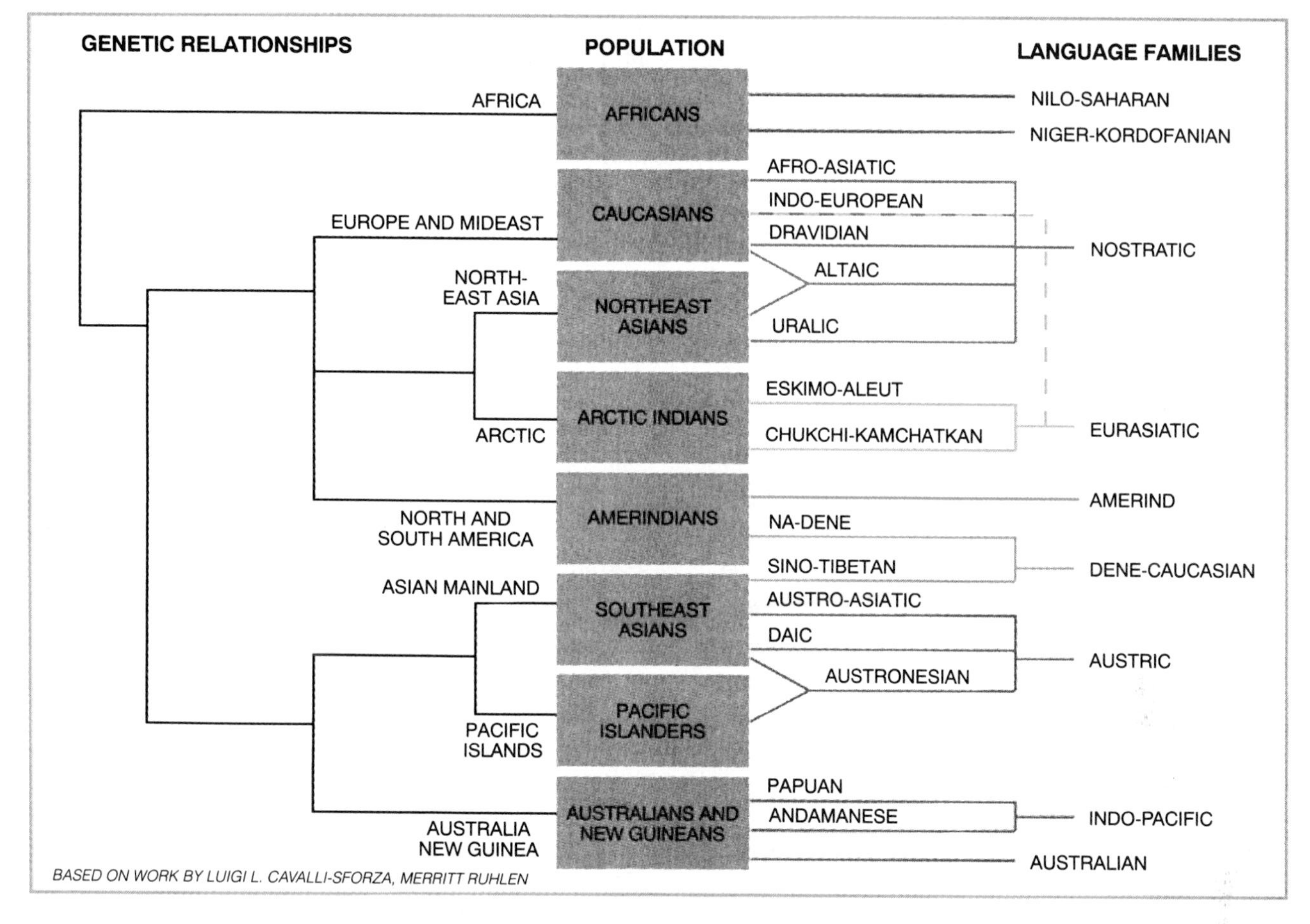

Figure 8–8. This diagram illustrates the genetic relationships among peoples as well as the relationships among language families that are currently hypothesized. The research in one field seems to reinforce the research in the other. (From Philip E. Ross, "Hard Words," *Scientific American* (April 1991): 145. Copyright © 1991 by Scientific American, Inc. All rights reserved.)

spoken than in the way they are written. This is because writings are generally more widely dispersed, and so writing tends to standardize languages. Sounds, however, are localized only among a group of people who speak together, called a **speech community.**

Other instances of language differentiation continue. Both French and Spanish are widely spoken. Both the French and the Spaniards try to "purify" their languages of linguistic borrowing but also to add new words so that their languages stay useful in the modern world. The French Academy, a scholarly institution dating back over 300 years, devises new French words for new concepts and items in international trade and discourse, and these terms become official for French. Despite the best efforts of the Academy, however, the French spoken in the 35 Francophone countries and that of Quebec are different dialects. Spanish has also experienced differentiation. The various dialects spoken and written throughout Latin America and the Philippines differ significantly from one another and from the Spanish of Spain.

When the province of Bengal was one state of the British Indian Empire, its people spoke one language, Bengali. That province was split, however, into West Bengal in India and East Bengal in Bangladesh. As a result, Bangla, the language of Bangladesh, is becoming a distinct language. This is despite shared Bengali pride in the writings of Rabindranath Tagore, the first non-European to win the Nobel Prize for Literature (1913).

Competitive Expansion and Shrinkage The map of the geography of languages is not static. The use of some languages is expanding because the speakers of those languages are diffusing around the world, are gaining greater power and influence in world affairs, or are winning new adherents to their ideas. The collapse of Soviet-dominated communist governments in the countries of Eastern Europe, for example, has practically terminated the study of Russian language and the use of Russian as a lingua franca throughout this region. Russian is being replaced by German, which indicates that historical ties with Germany are being repaired.

English is currently the predominant world lingua franca, partly because much of the world's science and business uses it. Many multinational corporations have designated English their corporate language, whatever the languages of their home countries might be. In computer terminology there is virtually no language but English. Another reason for the spread of English is the phenomenal global appeal of U.S. popular culture, especially movies and music.

Japan has recently won such tremendous wealth and economic power that Japanese might be expected to become an international language, but this has not happened. The reason is that the Japanese language developed over centuries in isolation, and it is so complicated and difficult that the Japanese actually suffer handicaps in international communication. Rather than spread the Japanese language throughout their commercial and other international links, the Japanese have been quick to adopt the languages of others, most notably English. Today Japanese power and influence actually contribute to the diffusion of English.

While some languages are growing and spreading, devising new ways of expressing new concepts and winning new speakers, others are dying out. For example, among the Celtic languages of Western Europe, Manx, the original language of the Isle of Man, and Cornish, the original language of Cornwall in western Great Britain, both have been overwhelmed by English and have disappeared. Welsh is confined to a small area and population. Countless languages or dialects throughout Africa, South America, and Asia are disappearing before they are even recorded. Because each language constructs perception and reality differently, the loss of a language forecloses an opportunity to explore the workings of the human mind.

African Languages The difference between official languages and the languages actually spoken is particularly great in Africa. The principal official languages of Africa are those of the continent's former European rulers, but there is an extraordinary richness of distinct native languages spoken, by some counts more than 1000 (see Figure 8–9). The great number reflects the minimal interaction of its many native peoples before European conquest. Most of these languages do not have a written tradition, and only 40 or so have as many as 1 million speakers. Some African governments are reviving their use or study in fresh appreciation of their peoples' pre-European heritage.

The language of Madagascar may seem curiously out of place on the world map. It is related not to the languages of nearby Africa but to those of the South Pacific. This suggests the source of early migrants to Madagascar.

Languages in the United States The population of the United States has always been composed of a great variety of peoples speaking a great variety of languages. English was the language of the principal colonial ruler and of the greatest number of European settlers, and it has always served as a lingua franca. A distinct American English nevertheless evolved. From the days of the earliest settlers, the American language adopted terms from Native-American languages. These included the names of native animals and plants unknown in Europe or Africa, products derived from them, and also place names. Each of the many immigrant groups has in turn learned American English, but each has also contributed to American English vocabulary, grammar, and diction.

Black English, an American dialect, demonstrates that the speech of parts of the African-American population retained and preserved elements of West African languages: Yoruba, Fante, Hausa, Ewe, and Wolof. Most of us speak at least one word of Wolof, the language of Senegal: *degan*, "to understand" ("Can you dig it?"). Only recently have scholars begun to study how the grammar of today's Black English diffused from West Africa to the U.S. South and from there into the cities of the North. The African elements of Brazilian Portuguese might presumably be traced not to West African languages but to the languages in those parts of Africa that were formerly Portuguese colonial sources of slaves.

Regional variations in American English persist, both in grammar and in pronunciation, despite the nationwide reach of popular media. Television homogenizes national speech patterns much less than scholars originally thought it would. The year 1991 saw publication of volume two of the *Dictionary of American Regional English*, a study of regional peculiarities begun in 1965. It got as far as the letter *H*, and it informed us, among other things, that when you put on your best clothes you are "diking up" in the South but "dickeying up" in New England.

Should American English be the official national language? The U.S. Constitution did not specify English as an official language, and many local governments throughout history and still today have found it useful to provide services and even keep official records in other languages (see Figure 8–10). During the 1970s and 1980s, however, a movement grew to declare English the nation's official language. By 1992, 16 states had statutes or constitutional clauses declaring English the sole official language, and 7 more were considering such clauses. This movement may reflect a dedication to a national language as a bonding force of a diverse population, but some people fear that it may also reflect a resentment of the changing immigration trends discussed in Chapter 6.

Language has become a civil rights issue with respect to two basic cultural traditions: education and voting. If, as is generally agreed, the inability to communicate in English is a handicap in the United States, then the learning of English becomes a key route to equal opportunity for all children. In a case involving Chinese-American children in San Francisco, the U.S. Supreme Court ruled that "students who do not understand English are effectively foreclosed from any meaningful education" (*Lau* v. *Nichols*, 1974).

This ruling triggered a national concern to identify local school districts in which English was the students' second language and to introduce bilingual educational

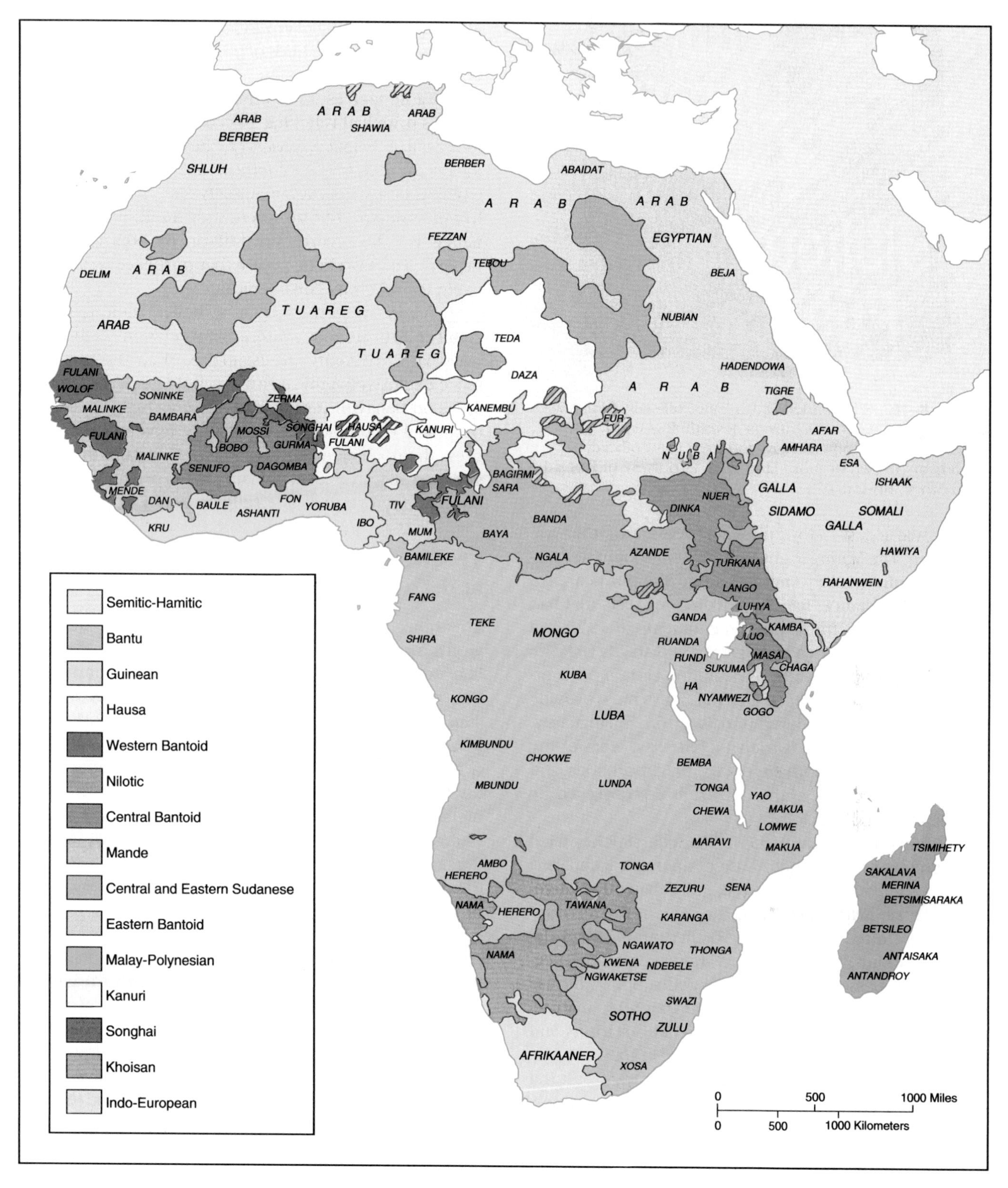

Figure 8–9. Most of the official languages in Africa are the languages of the former European rulers, but this map shows the extraordinary number of native languages, many of which in fact are spoken by a majority of the population.

programs in those districts. Arguments persist, however, over whether the only purpose of these programs should be to ease the students' transition to English or whether they also should preserve the languages and cultures that immigrant children bring to school. Some people argue that using the education system to acculturate all children to the English language denigrates the richness of their native inheritance. Their opponents contend that preservation of other languages threatens to lock the youngsters into second-class citizenship.

Figure 8–10. The United States is not officially bilingual, but governments at many levels nevertheless find it necessary to post official signs and even print official documents in two or more languages. (Courtesy of Donald Dietz/Stock, Boston.)

Equal access to the political process is another volatile issue. The Voting Rights Act of 1965 sought to draw many disenfranchised Americans into the political process by, for instance, banning the literacy tests that had been common in the South. Amendments in 1975 and 1982 mandated that voter registration districts (usually counties) whose populations contain a certain percentage of minority groups unskilled in English must provide registration and election materials in other languages. Today those electoral districts include vast areas of the country, including both inner-city neighborhoods and rural areas. Elections in these districts are observed by the Department of Justice.

From the 1880s until the 1950s, federal policy tried to discourage or eliminate Native-American languages. When the Europeans first arrived, more than 500 Native-American languages were used throughout the territory of today's United States; by 1990 that number had declined to about 200. In that year, however, federal legislation was passed to "encourage and support the use of Native-American languages as languages of instruction." Native Americans support 14 radio stations and some magazines and newspapers printed in Native-American languages. Language is such an important cultural index that many Native Americans view linguistic survival as cultural survival. Many social scientists would agree.

THE GEOGRAPHY OF RELIGION

The world distribution of religions forms a mosaic even more complex than the distribution of languages (see Figure 8–11). Each religion originated in one place and spread out from there, but today communicants of various religions mingle around the globe, and religious affiliations cut across lines of politics, race, language, and economic status.

Table 8–1 estimates the adherents of each of the world's principal religions. Because no worldwide religious census is taken, however, these figures cannot indicate the depth of anyone's belief nor the degree to which a person's religion actually affects his or her behavior. The strictest adherence to traditional religious beliefs is called **fundamentalism.** By contrast, **secularism** is a lifestyle or policy that purposely ignores or excludes religious considerations.

Several Eastern systems of belief, such as Confucianism, Taoism, and Shinto, are not religions in the same way that Christianity and Islam are. Their teachings are more exclusively ethical and psychological. That is, they focus on behavior rather than on believing a set of philosophical or theological arguments. Several do not address theological questions such as the nature of God or gods, or life after death. Many Eastern people are adherents of more than one religion.

The World's Leading Religions and Their Distributions

This book cannot examine the fine points of each religion's message. For that a text in comparative theology is needed. In studying the origins, diffusions, and distributions of religions, however, it will be necessary to summarize briefly the basic teachings of the world's major religions. That knowledge will clarify how the distribution of people who hold these beliefs affects the distributions of other things: forms of government, laws, dietary habits, civil rights, and the organization of economies. Religion is a vital force in human affairs, and understanding the messages of different religions contributes substantially to understanding many other aspects of human life.

Judaism Judaism has only about 18 million adherents, but it was the first of the great **monotheisms**—religions that preach the existence of one God—to emerge in history. It rests on a belief in a pact between God and the Jewish people that they would follow God's law as revealed in the *Pentateuch*—the first five books of the Old Testament. This covenant was granted to Abraham. Both Jews and Arabs claim descent from Abraham—Jews through his son Isaac, and Arabs through another son, Ishmael. Judaism encompasses a great variety of Orthodox (fundamentalist) and Reform *sects*, or subdivisions of a religion.

Judaism does not *proselyte*, or seek to convert others. It is actually difficult for anyone to convert to Judaism. Therefore, the diffusion of Judaism is largely the diffusion of individual Jews.

Judaism developed historically in the Near East over many centuries. Under the Roman Empire, Jewish

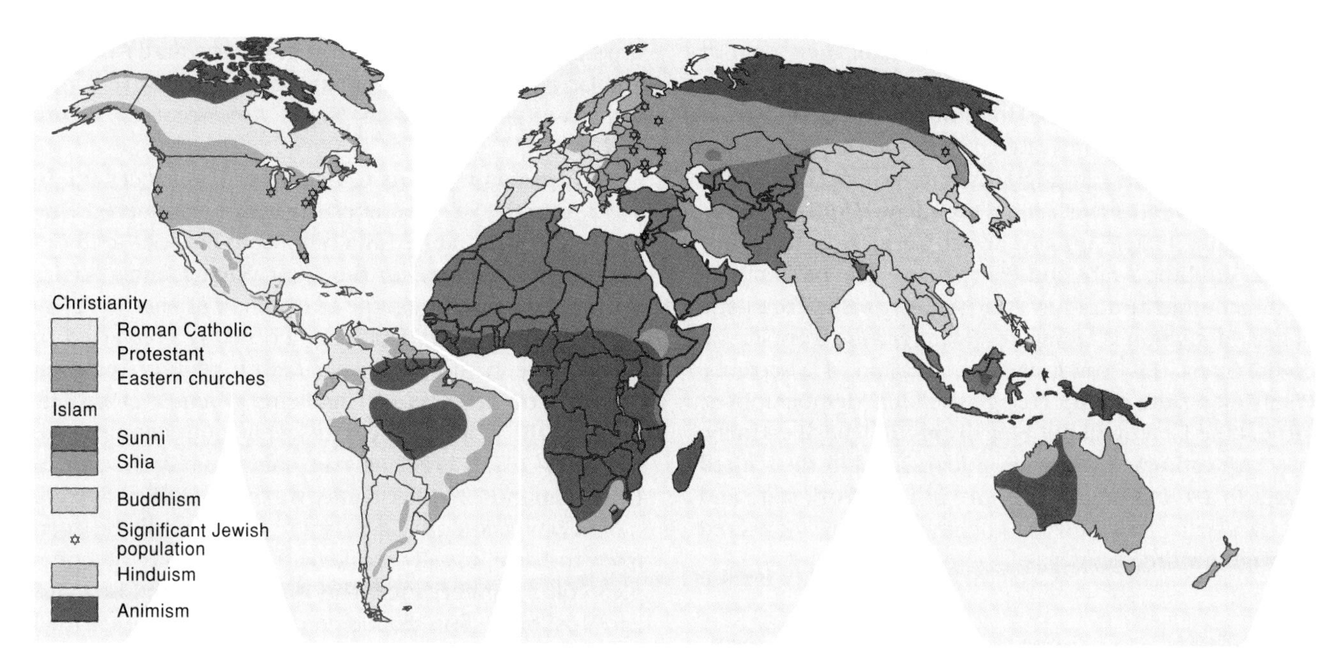

Figure 8–11. This map shows the predominant faith or faiths in each region, but it cannot show small minorities.

communities were established outside Judea. In A.D. 70, the Romans destroyed the temple in Jerusalem, and the Jews scattered in the *Diaspora.* During the Middle Ages many European Christian rulers persecuted and expelled Jews, but Jews returned to Western Europe during the Enlightenment of the seventeenth and eighteenth centuries. They were required to live in segregated communities called *ghettoes.* Legal emancipation came only in the nineteenth century. Millions of Jews came to the United States from Eastern Europe at that time, and migration from Russia and surrounding lands continues.

During the Diaspora many Jews visited Jerusalem if they could. When the nation-state idea developed in Europe, many Jews came to accept **Zionism,** the belief

TABLE 8–1 ***The Number and Distribution of Adherents of the World's Major Religions***

	Africa	Asia	Europe	Latin America	Northern America	Oceania	U.S.S.R.	World	Percent
Christians	317,453,000	257,926,000	412,790,000	427,416,000	237,261,000	22,316,000	108,498,000	1,783,660,000	33.1
Roman Catholics	119,244,000	121,311,000	262,026,000	397,810,000	96,315,000	8,095,000	5,551,000	1,010,352,000	18.8
Protestants	84,729,000	79,969,000	73,766,000	16,930,000	95,610,000	7,415,000	9,790,000	368,209,000	6.8
Orthodox	27,698,000	3,587,000	36,080,000	1,730,000	5,964,000	568,000	93,056,000	168,683,000	3.1
Anglicans	26,063,000	694,000	32,879,000	1,275,000	7,284,000	5,640,000	400	73,835,400	1.4
Other Christians	59,719,000	52,365,000	8,039,000	9,671,000	32,088,000	598,000	100,600	162,580,600	3.0
Muslims	269,959,000	625,194,000	12,545,000	1,326,000	2,642,000	101,000	38,959,000	950,726,000	17.7
Nonreligious	1,840,000	700,523,000	52,289,000	16,828,000	25,265,000	3,246,000	84,477,000	884,468,000	16.4
Hindus	1,431,000	714,652,000	703,000	867,000	1,259,000	355,000	2,000	719,269,000	13.4
Buddhists	20,000	307,323,000	271,000	530,000	554,000	25,000	404,000	309,127,000	5.7
Atheists	307,000	158,429,000	17,563,000	3,162,000	1,310,000	527,000	55,511,000	236,809,000	4.4
Chinese folk religionists	12,000	183,361,000	60,000	71,000	121,000	20,000	1,000	183,646,000	3.4
New Religionists	20,000	138,767,000	50,000	520,000	1,410,000	10,000	1,000	140,778,000	2.6
Tribal Religionists	68,484,000	24,487,000	1,000	918,000	40,000	66,000	0	93,996,000	1.7
Sikhs	26,000	17,934,000	231,000	8,000	252,000	9,000	500	18,460,500	0.3
Jews	327,000	5,484,000	1,465,000	1,071,000	6,852,000	96,000	2,220,000	17,615,000	0.3
Shamanists	1,000	10,044,000	2,000	1,000	1,000	1,000	252,000	10,302,000	0.2
Confucians	1,000	5,883,000	2,000	2,000	26,000	1,000	2,000	5,917,000	0.1
Baha'is	1,451,000	2,630,000	90,000	785,000	363,000	76,000	7,000	5,402,000	0.1
Jains	51,000	3,649,000	15,000	4,000	4,000	1,000	0	3,724,000	0.1
Shintoists	200	3,160,000	500	500	1,000	500	100	3,162,800	0.1
Other religionists	420,000	12,065,000	1,466,000	3,501,000	482,000	4,000	330,000	18,268,000	0.3
Total Population	661,803,200	3,171,511,000	499,543,500	457,010,500	277,943,000	26,854,500	290,664,600	5,385,330,300	100.0

Source: Data from 1992 Encyclopaedia Britannica Book of the Year (Chicago: Encyclopaedia Britannica, Inc., 1992, p. 269.) Used with permission.

that the Jews should have a homeland of their own in Palestine. When Israel came into existence in 1948, many Jews who had survived the Holocaust in Europe and many Mediterranean Jews migrated there. Israel has continued to welcome Jewish immigrants from around the world, although several Jewish sects repudiate Zionism.

Today Israel contains about 3.7 million Jews, but also 2.5 million Arabs and Palestinians. The birth rate is higher among the non-Jews, and so the Jewish percentage of the population is falling. Jewish migration from Russia could arrest this trend, and the creation of a separate Palestinian state would change the local demographics. Otherwise Israel will not be a distinctly Jewish country.

An estimated 4.3 million Jews lived in the United States in 1990. The majority live in New York state and in California, but Jewish communities are scattered throughout the country.

Christianity Christianity emerged from Judaism, but part of the widespread appeal of Jesus's teachings rests on his emphasis on God's love for all and on a prophetic urgency that stressed the nearness of the Kingdom of God. Jesus's life was short, and his own journeys spanned little more than 100 miles. Saint Paul and other followers however, preached his gospel throughout the cities of the Roman Empire. *Pagan*, a word that originally meant a country dweller, became a pejorative term for non-Christians. By the end of the fourth century, Christianity was the Empire's official religion. Christianity survived the disintegration of the Empire and spread throughout Europe.

Unlike Judaism, Christianity does proselyte. Many Christians believe that conversion of others is a duty, and European Christians often cited religious conversion as their purpose for exploring and conquering other lands. Christian missionaries were the chief agents of the partial Europeanization—of religion, language, social mores, and the acceptance of secular authority—that made many natives more tractable to European rule.

The rapidity and the completeness of Christianity's diffusion, especially throughout the Western Hemisphere, requires analysis. Certainly its message of all-embracing love carried strong appeal in lands in which many of the gods had been portrayed as terrifying and cruel. Furthermore, the Christian heaven after death is democratic and accessible to all classes of people. Christians also flexibly adapted some native practices. The timing of Christmas, for instance, adopts the celebration of the winter solstice.

In Latin America, some natives might have adopted Christianity because they were astonished and demoralized by the Spaniards' invulnerability to the diseases that were wiping out the Native Americans. Those diseases had been brought by the Spaniards themselves, but neither the Spaniards nor the Native Americans understood that. The Spaniards convinced the natives to acquiesce to Spanish secular authority and to seek both physical and spiritual salvation in the Spaniards' church. Similar beliefs prevailed among both the Native Americans and the whites in North America (see Figure 8–12).

Christian missions to Asia were less successful. In 1542, exactly 50 years after Christians conquered Granada—the last Arab Muslim outpost in Europe—and Columbus first voyaged to the West, Saint Francis Xavier arrived in Goa to begin the conversion of Asia. By 1549 Saint Francis was in Japan, which nearly converted to Christianity in the early seventeenth century but then expelled or persecuted Christians when it closed its doors to the world in the mid-seventeenth century (see Figure 8–13). Christians had some initial success in China, but in 1723 Christianity was banned there as well.

Conversion to Christianity, and, as will be seen later, to Islam, usually weakens the continuity of a people's cultural inheritance. This is because both Christians and Muslims believe that conversion is a genuine spiritual rebirth. Conversion is a deliberate repudiation of much of what came before, and so it can destroy a people's earlier culture more completely than can political conquest.

Christian sects and their distributions Roman Catholics make up the largest single denomination among Christians today. They believe that the only true Christian church is that headed by the pope in Vatican City, a tiny independent city-state within Rome (see Figure 8–14). The great many Protestant denominations (Lutherans, Anglicans, Baptists, Methodists, and others), were named for their protest against the Church of Rome. Many smaller Christian sects include the Georgian, Armenian, and Ethiopian churches and the Copts of Egypt. All Orthodox Christians recognize the spiritual leadership of the patriarch of Constantinople (Istanbul), but they are subdivided into national churches, such as the Serbian, Greek, and Russian. Orthodox Christians in the United States have never formed a national Orthodox church but have clung to their national affiliations.

Roman Catholicism dominates in the Mediterranean basin, throughout Latin America, and wherever else Mediterranean peoples colonized or converted, as in former French or Portuguese African colonies and the Philippines. Vietnam has the second highest Roman Catholic population in Asia, after the Philippines.

Protestant sects dominate in northern Europe and wherever northern Europeans have settled or converted: North America, Australia and New Zealand, and the parts of Africa that were either formerly English colonies (today usually Anglican) or German colonies (Lutheran).

Orthodox Christianity predominates in Eastern Europe because these peoples were converted by missionaries from Constantinople, which came to be known as "the second Rome" after the fall of Rome in A.D. 476. After Constantinople itself fell to Muslim Turks in 1453, Moscow assumed the title of "the third Rome." Thus,

Figure 8–12. John Chapman wrote that in his painting *The Baptism of Pocahontas at Jamestown, Virginia, 1613*, the Native-American princess Pocahontas "appeals to our religious as well as our patriotic sympathies, and is equally associated with the rise and progress of the Christian church, as with the political destinies of the United States." The painting strongly links Christianity and European conquest. It hangs in the U.S. Capitol. Do you think that it should? (Painting from U.S. Capitol Art Collection. Courtesy of the Architect of the Capitol.)

Figure 8–13. The Basilica of Bom Jesus in Goa, India, is the final resting place of the missionary Saint Francis Xavier. Saint Francis visited India, Japan, and China, and he is today the patron saint of Roman Catholic missionaries. (Courtesy of Father Vasco do Rego.)

Figure 8–14. This is one of the world's smallest principalities, the Vatican City. It is entirely within the city of Rome, but it is ruled by the pope according to a treaty signed between the papacy and Italy in 1929. The official language here is Latin. (Courtesy of Louis Renault/ Photo Researchers.)

Roman Catholic Western Europe came to be culturally divided from Orthodox Eastern Europe.

The Uniate church of Ukraine bridges the gap by using the Cyrillic alphabet and following the Eastern rite while recognizing some spiritual authority of the pope. Soviet dictator Joseph Stalin persecuted the Uniate church to squelch Ukrainian nationalism, but it survived underground and revived in the late 1980s.

The changing geography of Christianity Today the world map of Christianity is being redrawn. This is primarily the result of two struggles within Christianity. These struggles are on the surface theological, and yet they also affect politics, economics, and many other aspects of life in the societies in which they are being fought. The outcomes of these struggles will change world cultural geography. One struggle is the battle between Evangelical Protestantism and Roman Catholicism, and the other is the challenge of liberation theology within the Church of Rome itself.

Evangelical Protestantism Several theological points differentiate Protestantism from Roman Catholicism, but among the most important are that Roman Catholics believe that Jesus gave Saint Peter unique responsibility for founding the Christian Church, that Saint Peter became the first bishop of Rome, and that the continuing line of bishops of Rome (the popes) retains this special responsibility. The crossed keys on the Vatican flag represent Jesus's having given Saint Peter the keys to heaven; there is no salvation outside the Church of Rome. The belief that a church or priests intercede between God and humankind is called **sacerdotalism.**

Protestants deny sacerdotalism. Evangelical Protestants (from the Latin for "bringing good news") emphasize salvation by faith through personal conversion, the authority of Scripture and each individual's responsibility to read the Scriptures, and the importance of preaching as contrasted with ritual. The evangelicals emphasize the ability of individuals to change their lives (with God's help), and this message offers to many people a new sense of personal empowerment. Thus, Evangelical Protestantism frequently brings a revolutionary force into traditionally rigid or stratified societies.

Today Evangelical Protestantism is growing and spreading throughout the world. In some places it is replacing other non-Christian religions, but in many places it is replacing Roman Catholicism. It is spreading around the Pacific rim (especially into South Korea), into southern Africa, and throughout Latin America. At least one-fifth of the total Latin American population was Protestant by 1990. This percentage was even higher in Brazil, which counts from baptism the world's largest Roman Catholic population. In Brazil today full-time Protestant pastors outnumber Roman Catholic priests, and as many as 600,000 people convert to Evangelical sects each year. Evangelical Protestantism has replaced Roman Catholicism in several countries as the most widely practiced faith. (Table 8–1 counts baptized Roman Catholics.)

This changing geography of religion is in itself an important new cultural pattern, but, in addition, the spread of Evangelical Protestantism affects other aspects of cultural geography. In most Latin American states the Church of Rome has traditionally enjoyed special privileges and a role in education, and so its teachings have been enacted into law. Today, however, a rising share of elected officials throughout Latin America are Protestant, and elections are referred to outright as "holy wars."

The diffusion of Evangelical Protestantism is also at least partly responsible for changing the dynamics of population growth. Evangelical Protestants are often as opposed to abortion as Roman Catholics are, but they are not always as adamantly opposed to other forms of birth control. The loosening dominance of Roman Catholicism in Latin America and even in Spain and Italy has been accompanied by a rise in government-sponsored family planning, the distribution of birth control devices, and the introduction of sex education into school curricula.

Peru, for instance, in 1990 elected a president who was baptized a Roman Catholic but who was elected on a program dedicated to population planning. He said, "The populace knows how to distinguish between religious and social affairs. . . . A sector of the Church is restricting the freedom of citizens with its medieval opinions and recalcitrant positions." This is a powerful criticism of Church leaders in a country that is nominally 90 percent Roman Catholic and that has one of the highest annual population growth rates in Latin America.

Other possible results from the spread of Evangelical Protestantism have been hypothesized, and they bear watching through coming years. One is the spread of literacy. This follows from Protestants' individual responsibility to read and study Scripture. Literacy has risen in areas where Evangelical Protestantism has spread, but the exact cause-and-effect relationship is difficult to measure. If Evangelical Protestantism raises literacy rates wherever it diffuses, this trend may carry further ramifications—greater political participation, for example.

Still other possible consequences of the spread of Protestantism have been hypothesized in the areas of women's rights and even economics. These are discussed later in the chapter.

Liberation theology The Roman Catholic church in Latin America is itself divided over the issue of **liberation theology.** In 1968 a conference of Latin American bishops accepted the concepts expressed in the book *The Theology of Liberation* by the Peruvian Father Gustavo Gutiérrez. Liberation theology puts the problems of overcoming poverty at the heart of Christian theology. It recommends political activism, using the Church's institutional framework to organize the population in the struggle for social justice and equality. As a result, many Latin American bishops actively proposed national redistributions of land and income. The Church began to provide legal services for the poor, and some priests attacked elite power structures from their pulpits.

The diffusion of liberation theology has provoked political repercussions: spiritual disarray and political polarization in Argentina and Chile, a new sense of self-confidence in Brazil, and conflict between the "popular Church" and the Church hierarchy in Nicaragua. It has also triggered militant opposition among conservatives threatened by changes on either religious or political grounds. For example, in El Salvador during the 1980s, nuns, priests, and even Archbishop Oscar Romero were assassinated by right-wing elements of the military regime.

Pope John Paul II has tried to temper liberation theology. He has instructed Father Gutiérrez to refrain from writing or teaching, and he has redrawn dioceses and appointed new bishops who have closed Church legal-services offices, land-rights offices, and liberal seminaries.

The most conservative Latin American governments have favored Evangelicals as a counterforce to liberation theology. In Guatemala, for example, government death squads murdered politically active priests during the 1980s, and one Roman Catholic bishop was forced into exile. Meanwhile, Protestant sects, with government encouragement, converted more than 30 percent of the population. Jorge S. Elias, elected president in January 1991, was the first Protestant ever elected president of a Latin American country. The Roman Catholic bishops of neighboring Honduras accused the U.S. CIA of subsidizing Protestantism as a way of undermining the Church of Rome and pacifying the people. In Mexico, Protestants denounce Catholics as agents of the Vatican, and Catholics denounce Protestants as "Yankee" agents out to sabotage traditional Mexican culture. Examples such as these of the interplay between theology and politics may help us understand important social and political movements through coming decades.

The future of Christianity Christianity, whether Roman Catholic or Protestant, is still spreading rapidly (see Figure 8–15). Evangelical Protestantism is winning new converts worldwide, and in 1991 the pope called for renewed missionary efforts. The observance of Christianity might increase throughout Eastern Europe as political liberalization proceeds there, but overall the European core area of Christianity is of diminishing importance in relative numbers. In Africa, by contrast, the number of Christians has grown, chiefly by conversion, from fewer than 10 million in 1900 to 236 million in 1985. At this rate African Christians will number nearly 400 million by 2000 (see Figure 8–16).

In Latin America, Christians overall numbered 62 million in 1900, 392 million in 1985, and will number 571 million in 2000. In South and Southeast Asia the corresponding numbers are 19 million, 148 million, and 225 million. South Korea was by 1991 about 30 percent Christian—80 percent of them Protestant. South Korea may be the second Asian nation to have a Christian majority, after the Philippines.

Christianity is losing adherents throughout the Near East, where a resurgent Islam threatens to overwhelm it. In 1990 there were about 12 million Christians left in the Arab world, only 6 percent of the total population. The most vigorous Arab Christian community is in Egypt, where 6 million Copts comprise about 10 percent of the total population.

The movement of Christianity's center to the expanding populations outside Europe might inspire changes in Christian theology and practice. This trend exemplifies the general process of cultural diffusion by which vigor on the periphery challenges the preeminence of the historical core area, as already demonstrated with languages. Christian practices and doctrines that

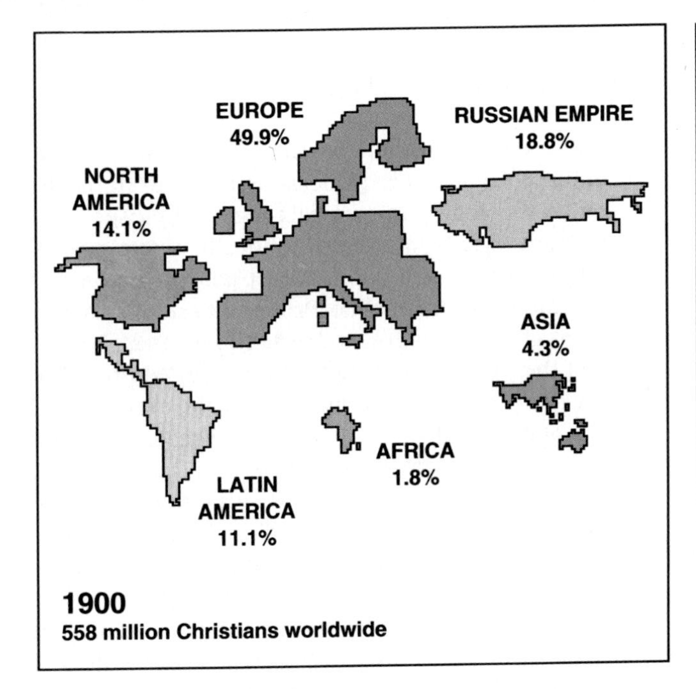

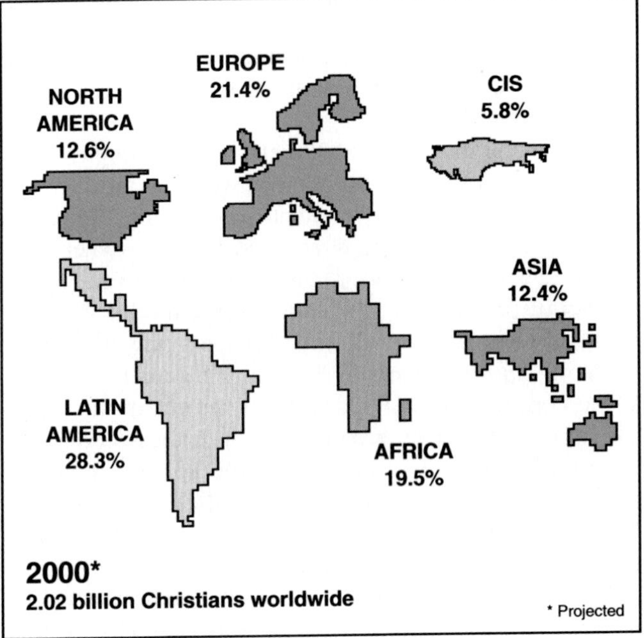

Figure 8–15. This cartogram shows that in terms of numbers of communicants, the European core area of Christianity is losing its predominant status. The rising number of converts elsewhere have challenged traditional Christian practices that are rooted in European culture. (Adapted with permission from the *Christian Encyclopedia* as printed in *The Economist*, Dec. 24, 1988, p. 62.

derive from European culture are under pressure for change, and aspects of native religions that were blanketed by Christianity are reviving. Several examples are listed below.

- Although the worldwide spread of grape and wheat crops was partly due to Christians' need for wine and bread for the mass, local churches around the world have considered adopting local foods for this sacred rite.
- At a world convocation of Roman Catholic bishops in 1990, a Brazilian bishop spoke in favor of ordaining married men in his country. The Vatican remains opposed to this change, but the bishop of Nassau, the Bahamas, emphasized that on such questions "the Church should not be tied to cultural vestiges typical of the European experience." It is revolutionary that such initiatives should come from non-European Christians.
- Efforts are being made in India to incorporate pre-Christian cultural and Hindu concepts into Roman Catholic practice. These efforts, called **interculturation,** have provoked tremendous controversy. Elsewhere **syncretic religions** that combine Christianity with traditional practices are winning converts. Candomblé and Umbanda in Brazil are examples.
- At the 1991 meeting of the World Council of Christian Churches, a Korean priest, accompanied by gongs, drums, and clap sticks, summoned "the spirit of the Amazon rainforest" and "the spirits earth, air, and water raped, tortured and exploited by human greed." Representatives of hundreds of millions of Christians threatened to abandon the World Council. A performance such as this underscores the problem of how to express Christian belief in terms meaningful to a great diversity of cultures without abandoning essential Christian distinctiveness.

Islam Monotheism was brought to the Arabs by Muhammad (570?–632), who founded the religion called *Islam*, which means "submission [to God's will]." "One who submits" is a Muslim. The Arabic word for the one God is *al-elah*, or Allah. This word is cognate to the Hebrew *eloh*, "god," and *elohim*, "the God." Muslims believe that Muhammad was the last of God's prophets, who also included Adam, Noah, Abraham, Moses, and Jesus. The five essential duties of a Muslim, called the *Five Pillars*, are belief in the one God, five daily prayers, generous giving of alms, observance of one month (called *Ramadan*) of fasting, and, if possible, a pilgrimage (*hajj*) to Mecca at least once in one's lifetime.

Despite Muhammad's flight from his native city of Mecca in 622 (the *Hegira*)—a temporary setback and the year from which Muslims date their calendar—he had converted and united most Arabs by his death. Just as Christians see their religion as building on Judaism, adding the New Testament to the Old, so Muhammad envisioned his teachings as a continued evolution of monotheism. Muhammad's holy book, the Koran, specifically instructs Muslims to protect Christians and Jews, whom Muhammad expected to be among the first to

Figure 8–16. The world's largest Christian church is the Basilica of Our Lady of Peace in Yamoussoukro, the Ivory Coast. The country's president, Félix Houphouët-Boigny, converted to Roman Catholicism as a teenager, and he built the basilica as a pilgrimage center for Africa's Catholics and as a bulwark against Islam and native religions in his country, which was about 10 percent Christian in 1990. (Courtesy of Gideon Mendel/Magnum.)

embrace Islam. He directed Muslims to face Jerusalem in prayer, as altars in Christian churches do, but he later changed the direction toward Mecca (see Figures 8–17 and 8–18).

Islamic expansion As Islam diffused, the Arabic of the Koran became the language of peoples throughout the Near East and across North Africa who were ethnically quite different. They had not previously been Arabic speakers, but as they converted, all came to be known as Arabs. To the east, Persia (today's Iran) had its own ancient culture, and although the Persians converted to Islam, they retained their own language, Farsi, an Indo-European language. Therefore, today's Iranians are not Arabs, nor are any of the peoples to the east of Iran.

Within slightly more than a century after Muhammad's death, Islam stretched from the Atlantic coast of Spain across North Africa and through Persia and southwest Asia to the borders of China. An Arab army met a Chinese army in the Talas River valley in central Asia in 751, just 19 years after another Arab army had faced a Frankish army in Tours in today's France thousands of miles to the west. Within this realm, dominance eventually passed from the conquering Arabs to the non-Arab majority, including the Persians and the several Turkish

Figure 8–17. This is the Kaaba in Mecca, where the biblical patriarch Abraham and his son Ishmael are believed to have built the first sanctuary dedicated to the one God. Through millennia the temple was given over to the worship of idols, but Muhammad threw out the idols, rededicated the Kaaba, and had its walls covered with paintings of Abraham, Mary, and Jesus. That building has since been replaced, but this is probably the longest continuously revered spot on Earth. (Courtesy of Aramco World Magazine.)

Figure 8–18. Many sites in Jerusalem are sacred to Jews, Christians, and Muslims. Jews and Christians believe that God prevented Abraham from sacrificing his son Isaac on this rock (Mount Moriah), thus ending human sacrifice. King David later built the Jews' temple here. Muslims believe that Muhammad ascended to heaven from this spot. The beautiful building standing here now, the Dome of the Rock (begun in 643), is the oldest existing monument of Muslim architecture, although its style is faithful to earlier local Byzantine traditions. Christians believe that Jesus ascended to heaven from the Mount of Olives in the background. Just in front of and below this picture is the only wall remaining of the Jewish temple that was destroyed by the Romans in A.D. 70, triggering the Jewish Diaspora. (Courtesy of Israel Ministry of Tourism.)

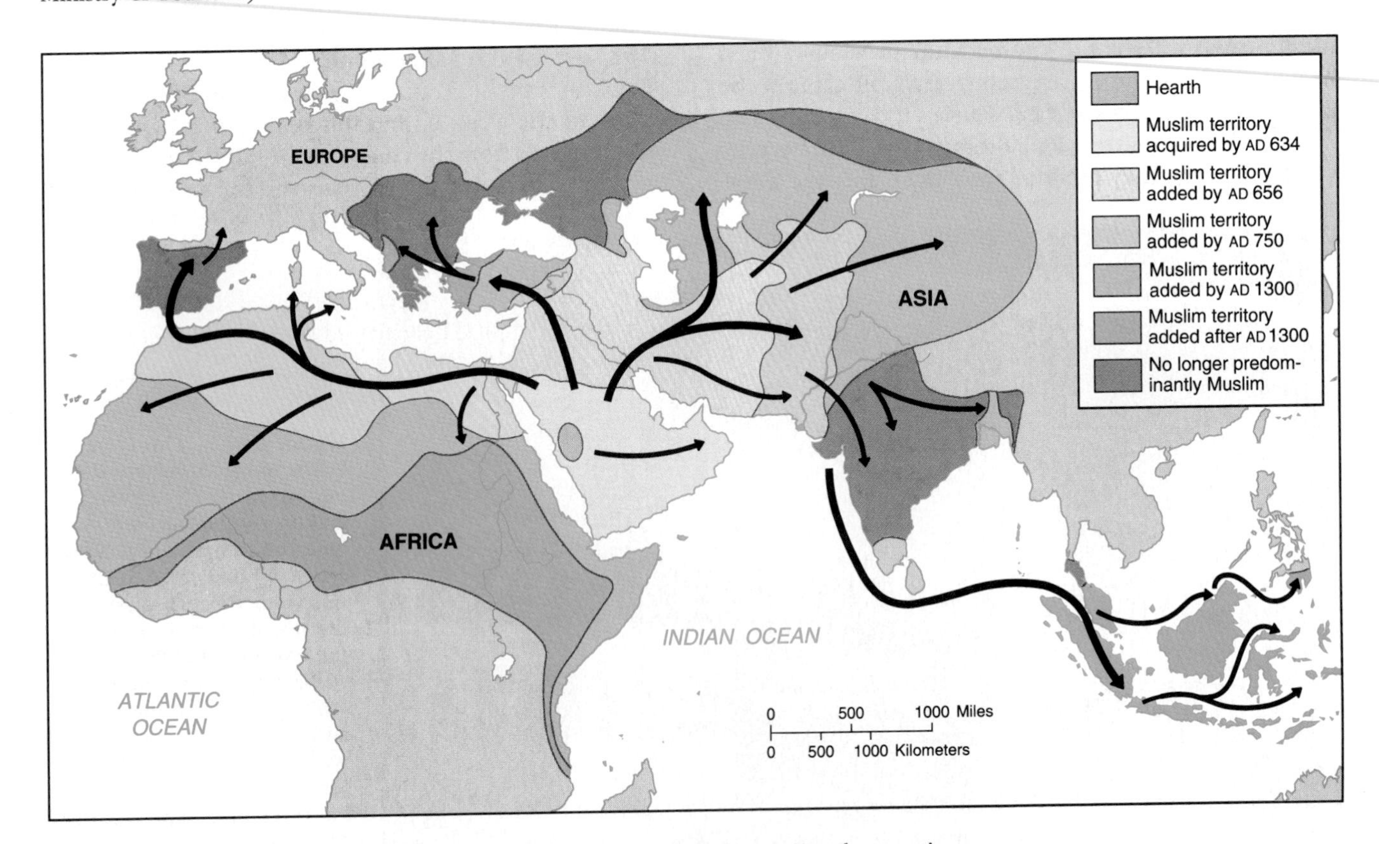

Figure 8–19. The first 9 centuries of Islam were a period of almost continual expansion, and by 1800 Arab and Indian traders had spread the faith through much of island Southeast Asia.

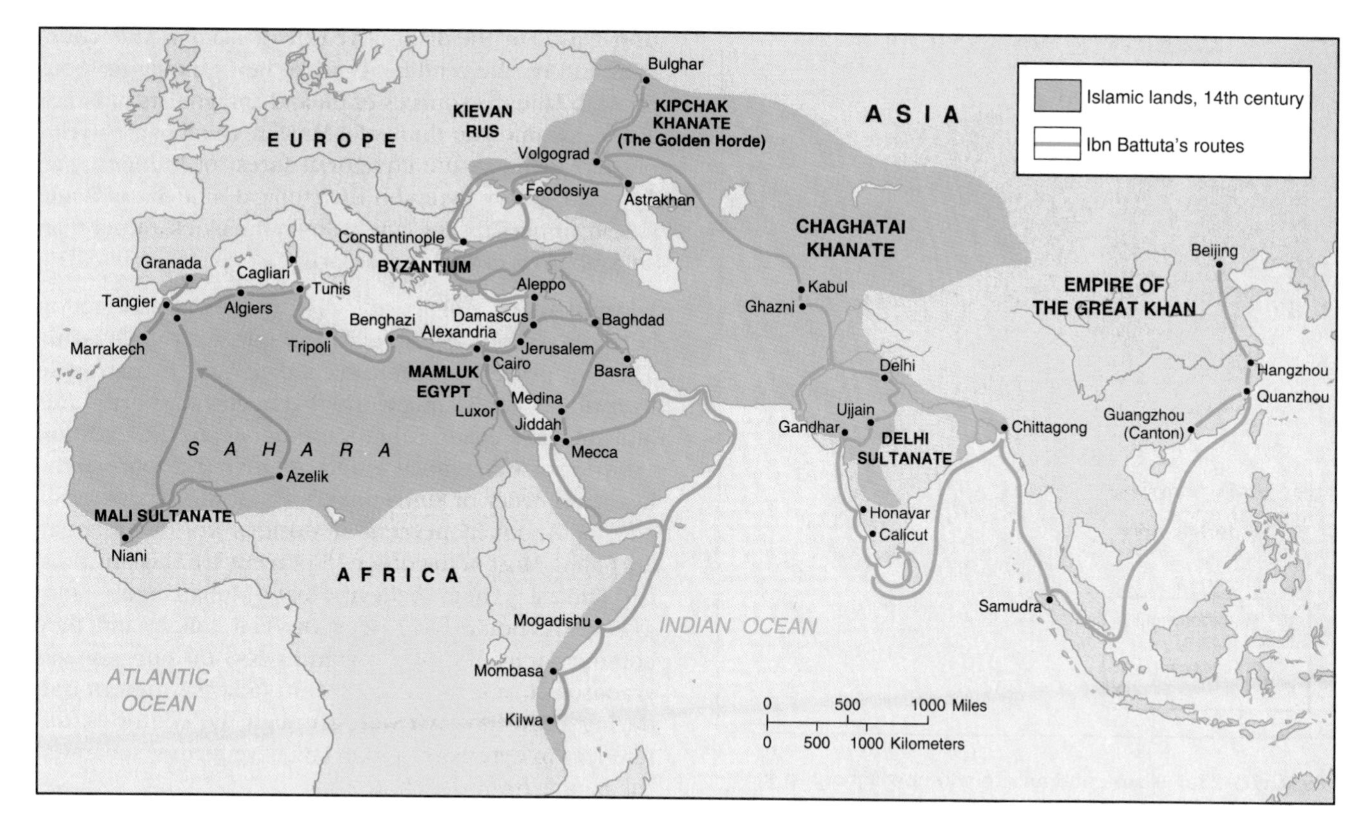

Figure 8–20. In the fourteenth century the scholar Ibn Batuta traveled from Morocco to China, and everywhere he was received at the courts of princes. Several even appointed him as a judge thousands of miles from his own home. His geographic writings are a superlative description of life across the vast Islamic realm of his day.

peoples. Islam later expanded down into South and Southeast Asia (see Figure 8–19).

This culture realm greatly exceeded the Christian culture realm in extent, power, and riches for hundreds of years. Learned travelers crossed its length and breadth, and their descriptive writings constitute some of the greatest works of historical geography (see Figure 8–20).

Islam denigrates the earlier cultures of its converts, just as Christianity can. Everything before Islam was, in Arabic, *jahiliya*, "from the age of ignorance." This leaves little room in these peoples' historical consciousness for their pre-Islamic past, and so they often lack interest in it. For example, despite Persia's brilliant antique history, for contemporary Iranians the glory began with the coming of Islam. Pakistan is a new Muslim state, but the land contains ruins of civilizations thousands of years old, and yet contemporary Pakistanis disdain them. Many people in Muslim countries view their own ancient cultural landscapes without interest. They may even discourage tourists from viewing pre-Islamic ruins. Similarly, the study of the history and art of Egypt under the pharoahs is the result of Christian European historical interest. The fact that Egyptians are interested in it today demonstrates that the newer cultural force of Egyptian nationalism competes with Islam for the primary loyalty of the Egyptian people.

As Islam filtered south across the Sahara Desert from North Africa, its message reached the black peoples of the sahel and savanna (see Figure 8–21). There it competes still today with Christianity, which entered sub-Saharan Africa from the Europeans' coastal incursions. From Senegal to the Congo the coastal areas are Christian, and inland areas are Muslim. When several of these countries first won independence, the westernized political leaders were Christian (for example, President Léopold Senghor in Senegal and President Félix Houphouët-Boigny in the Ivory Coast), but the majority of the population was Muslim. Christian politicians continue to play prominent roles, but as democracy matures in these countries, Muslims often come to power.

Within Europe, only Albania, long ruled by Turks, is predominantly Muslim. In 1944, when the Communist government forbade any religious observances, Albania was 70 percent Muslim.

The Muslim world today Today Muslims are distributed from Morocco to Indonesia, north to the frontiers of Siberia and south to Zanzibar, with outposts throughout the world. Rome boasts a splendid new mosque. The 1.5 million to 2 million Muslims in the United States are mostly immigrants from Islamic areas in Asia, Africa, and the Near East, or their descendants. About one-third are African Americans. The leading Mus-

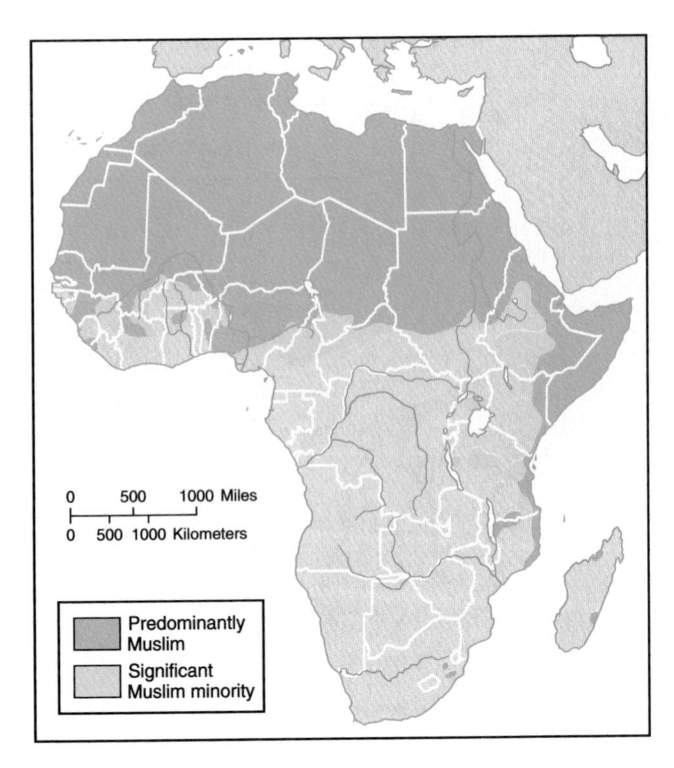

Figure 8–21. Islam continues to win new adherents in Africa today.

lim states today, in numbers, are Indonesia, Bangladesh, India, and Pakistan.

Indonesia's actual status is dubious. More than 90 percent of Indonesians put "Muslim" on their identity card, but this may reflect the fact that it is illegal to have no religion. This law is a relic of the government's intense anticommunism. Islamic fundamentalism could become a threat to civil peace in Indonesia if it becomes an anti-Chinese movement. Chinese make up a small percentage of the Indonesian population, but, as noted in Chapter 6, they dominate the economy.

Muslims believe in proselyting, as Christians do, and this necessarily injects an element of competition into the relationship between the two religious communities. Many Muslims in the Islamic realm are fundamentalist, but this fundamentalism may evolve or modify as more Muslims move outside the Islamic realm, just as Christianity has evolved as it has diffused.

The competition for converts between Islam and Christianity can spill over into the writing of laws, which is discussed later in the chapter, and it has triggered murderous street rioting in many countries. In an effort to avoid religious violence, the Constitution of Nepal guarantees freedom of conscience to people of all religions but absolutely forbids proselyting. This frustrates both Nepalese Christians and Nepalese Muslims.

Although many Americans tend to equate Islam with Arabs, Arabs actually constitute only a fraction of Muslims. There are more Muslims in Indonesia and Pakistan than in all the Arab countries combined. Americans' association of Islam with Arabs is a political preconception caused by the distinctive role of Islam in Arab countries and by the tendency of U.S. news media to focus on Arab lands as sources of oil and antagonists of Israel. The next time you think of a Muslim, think of a 17-year-old girl living in the equatorial forest of Indonesia, an 11-year-old boy living on the tropical mud-flat of Bangladesh, or perhaps the family down the block, rather than an Arab on a camel in a desert.

Islamic sects Theoretically, there is no distinction between temporal (political) and spiritual rule in Islam. The two principal Islamic sects date back to a struggle over rule of the Islamic world that occurred shortly after Muhammad's death. Sunni Muslims accept the tradition (*Sunna*) of Muhammad as authoritative and approve the historical order of Muhammad's first four successors, or *caliphs*. About 85 percent of Muslims worldwide today are Sunni. Most of the other 15 percent are Shia Muslims, or Shiites. They believe that Muhammad's son-in-law Ali should have been the first caliph, and they commemorate the martyrdom in 680 of Muhammad's grandsons in a battle at Karbala in today's southern Iraq during a Muslim civil war. Through the centuries, differences in ceremony and in law have further separated the Sunnis from the Shiites.

Shiites form the majority in Iran and in southern Iraq, and important minorities in Kuwait, Lebanon, Syria, Saudi Arabia, and Pakistan. Animosity between Sunni and Shia can be fierce. Countries that have a Shiite majority often refuse to cooperate in international affairs with countries in which the majority is Sunni. Countries that contain significant shares of both sects are split by the enactment into national laws of either Sunni or Shiite interpretations of Muslim religious teachings. Several countries, including Lebanon and Pakistan, have suffered civil disturbances between the two groups. Even the *hajj* to Mecca has regularly been disrupted by violent clashes between Sunnis and Shiites.

Hinduism and Sikhism In Asia the tradition of Judaism, Christianity, and Islam confronts an equally ancient religious tradition, that of Hinduism. The oldest Hindu sacred texts (the Vedas) date from about 1800 B.C., but the religion originated somewhere in central Asia long before that. It entered the Indian subcontinent with the coming of central Asian peoples about the time of the writing of the Vedas. Today it is confined almost exclusively to India and Nepal, where it is the official state religion (see Figure 8–22).

Hindus believe in one God, Brahma, who has evolved a triple personality: Brahma, the creator; Vishnu, the preserver; and Siva, the destroyer. These are coequal, and their functions are interchangeable. All other Hindu gods, saints, or spirits are emanations of the one true God.

Hinduism classifies people in a hierarchy of classes called **castes.** The four main castes are (1) the Brahman, or priestly caste; (2) the Kshatriya, or warrior caste; (3)

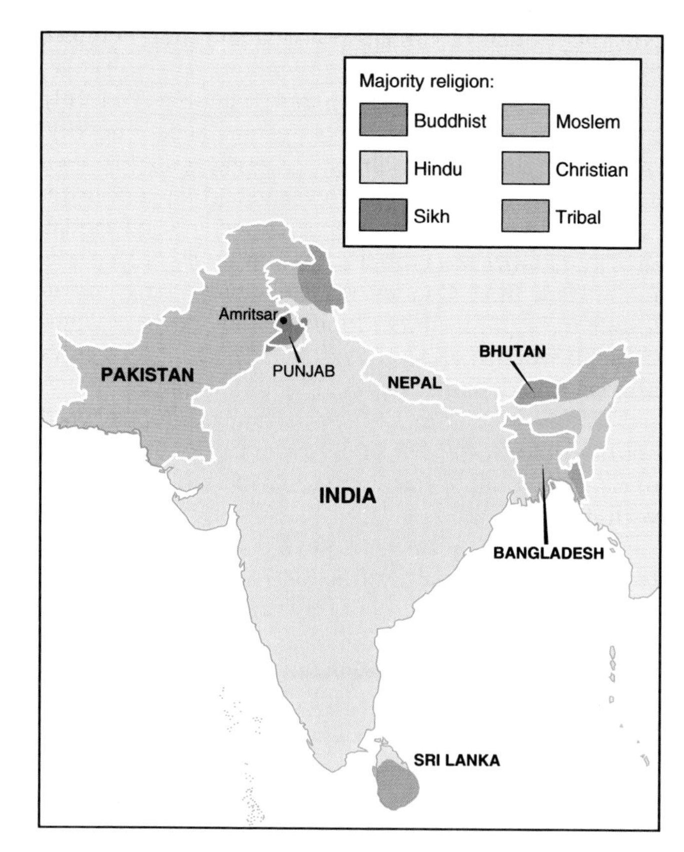

Figure 8–22. Hinduism is today restricted almost exclusively to India, Nepal, and to places where Indians have migrated, such as a few countries around the Caribbean Sea and Fiji in the Pacific Ocean. Sikhs form a majority in the Indian state of Punjab, but smaller Sikh communities can be found throughout India.

the Vaisya, or tradesman and farmer caste; (4) the Sudra, or servant and laborer caste. Each of these groups is split into hundreds of castes more narrowly defined, and each of those into dozens of subcastes. Many castes are defined by occupation. People are expected to socialize, marry, and stay within the caste into which they were born. A group of people called *untouchables* are considered so low that their status is below the formal structure of the four castes. Caste discrimination was abolished in the Indian constitution of 1950, but it still structures Indian life.

Hindus believe in *reincarnation*, that is, individual rebirth after death. The caste into which you are born is not haphazard; rather, it depends on your behavior in an earlier life. This teaching, called *karma*, discourages ambition, because only by keeping to your place in this life can you hope to enjoy a better position in your next life. The goal of Hindus is liberation from the cycle of death and rebirth.

Sikhism is an offshoot of Hinduism based on the teachings of Guru (teacher) Nanak (ca. 1469–1539). Nanak tried to reconcile Hinduism and Islam, teaching monotheism and the realization of God through religious exercises and meditation. Nanak opposed the maintenance of a priesthood and the caste system. Under a series of gurus the Sikhs had their own state in northern India, but they were eventually conquered by the British.

The Sikhs have done well economically throughout India, but many dream of the restoration of an independent Sikh state, to be called *Khalistan*. The Sikh's holy temple is in the city of Amritsar in the Indian state of Punjab (see Figure 8–23). This state is largely Sikh, and it might provide a territorial base for an independent Khalistan.

Buddhism Siddhartha Gautama (ca. 563–ca. 483 B.C.) was a northern Indian Hindu prince who, through meditation, achieved the status and title of Buddha, or Enlightened One. His teachings centered around the Four Noble Truths: (1) life involves suffering; (2) the cause of suffering is desire; (3) elimination of desire ends suffering; (4) desire can be eliminated by right thinking and behavior. This cessation of suffering is called *nirvana*, or total transcendence.

Figure 8–23. The Sikhs' 400-year-old Golden Temple in the Indian city of Amritsar stands in the middle of a sacred lake (in Sanskrit *amrita saras*, "pool of immortality"). The building was occupied by armed Sikh extremists in 1983; the following year Indian troops stormed it and drove out the occupants in a bloody encounter. Sikh fanatics retaliated by assassinating India's Prime Minister Indira Gandhi. (Courtesy of Jehangir Gazdar/Woodfin Camp.)

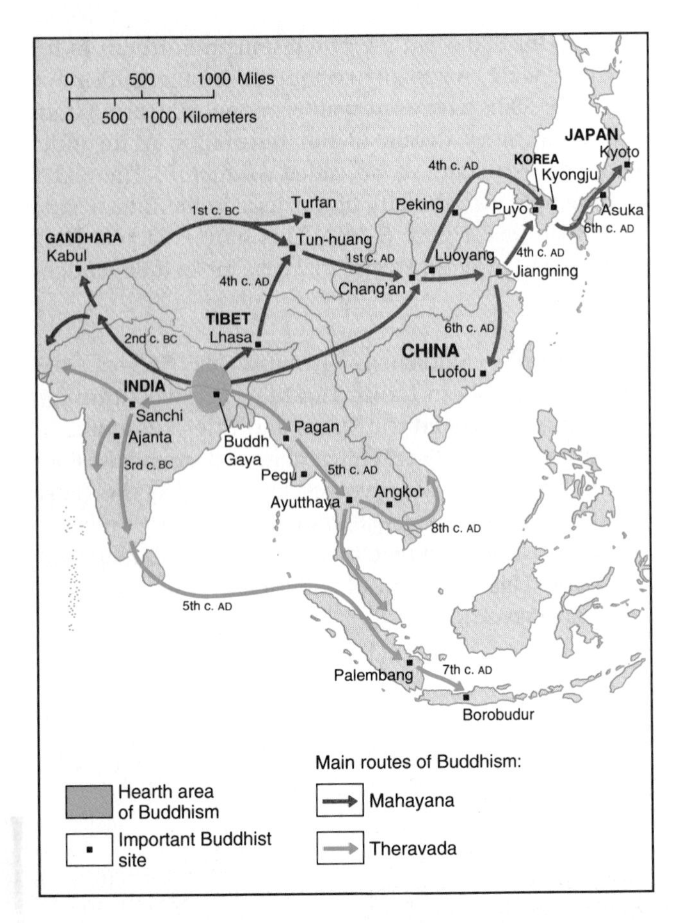

Figure 8–24. Buddhism was only a local religion of northern India for hundreds of years, but then its influence expanded throughout Asia. It has largely been abandoned in India.

As Buddhism diffused out of India, sects and schools arose. The Theravada school ("doctrine of the elders") diffused to the South, and Mahayana Buddhism ("great vehicle") diffused northward (see Figure 8–24). Mahayana idealizes the concept of the *bodhisattva*, someone who merits nirvana but postpones it until all others have achieved enlightenment. In Bhutan, Tibet, and Mongolia, Buddhism evolved a special form called *Lamaism*, which is known for its elaborate rituals and complex priestly hierarchy. Chinese Buddhists produced a new theory of spontaneous enlightenment, called *Ch'an*. This diffused into Japan as *Zen*.

Buddhism has several hundred million followers, but its adherents are hard to count because it is not an exclusive system of belief. Its practice has declined in India, but today it is the state religion in Thailand and Sri Lanka, and it may soon achieve that status in Mongolia. Buddhist philosophy has also won considerable influence in the modern Western world. Much popular "New Age" philosophy derives from Buddhism.

Other Eastern Religions Confucianism is a philosophical system based on the teachings of K'ung Fu-tzu (ca. 551–ca. 479 B.C.). He taught a system of "right living" preserved in a collection of sayings, *The Analects*, that governed much of China's political and moral culture for 2000 years.

Confucianists hold that people may attain heavenly harmony by cultivating knowledge, patience, sincerity, obedience, and the fulfillment of obligations between parents and children, subject and ruler. These moral precepts permeate life in many Eastern societies, and today the word *Confucian* is often popularly used as a synonym for these qualities. Confucianism influenced Western philosophy and political theory at the time of the Enlightenment, when it appeared as the realization of Plato's utopian dream of a state ruled by philosophers.

Taoism is a Chinese religion and philosophy based on the book *Tao-te Ching* (third century B.C.) attributed to Lao-tse. It advocates a contemplative life in accord with nature.

Shinto is the ancient religion native to Japan. It includes a set of rituals and customs involving reverence for ancestors, celebration of festivals, and pilgrimage to shrines, but there is no dogmatic system or formulated code of morals (see Figure 8–25). It is specifically nationalistic, and it traditionally recognized the emperor as divine. The Emperor Hirohito renounced this divinity in 1946, following Japan's defeat in World War II.

Animism and Shamanism **Animism** is a belief in the ubiquity of spirits or spiritual forces. Animistic religions may be basically monotheistic, but they recognize

Figure 8–25. These beautiful buildings make up part of the Toshogu Shinto shrine in Nikko, Japan, about 75 miles (120 km) north of Tokyo. This city of shrines and memorials is visited by millions of Japanese every year to learn about their history as well as to perform rituals. (Courtesy of Alon Reininger/Contact Press/Woodfin Camp.)

hierarchies of divinities who assist God and personify natural forces. Millions of Africans believe in animism, and animists are also found among Native Americans in the Amazon basin and among native peoples in the interior mountains of Borneo and New Guinea.

Animism is frequently accompanied by **shamanism.** A shaman is a medium who characteristically goes into autohypnotic trances, during which he or she is thought to be in mystical communion with the spirit world. Shamanism is practiced among the peoples of Siberia, the Inuit, some Native-American peoples, and in Southeast Asia and on East Indian islands. Almost everywhere both animism and shamanism are yielding to Muslim and Christian proselyting. Meanwhile, the world's largest pharmaceutical companies race to tap shamans' ethnobotanical knowledge.

THE IMPACT OF THE GEOGRAPHY OF RELIGION

The distribution of religions affects the distributions of many other things. Religion influences what people eat; their tolerance for others; what men think of women, and vice versa; when and how they work and when and how they celebrate; which types of behavior they encourage and discourage; and many other aspects of life. The influence of religion is so pervasive that it is difficult to isolate and identify precisely. Clearly, however, many people choose to obey God's word as revealed to them in their religion's teachings.

Religion and Politics

Almost all countries guarantee freedom of religion, and most governments observe a form of secularism. Many governments nevertheless favor one religion over others implicitly or explicitly. Therefore, the world political map partly fixes or stabilizes the map of world religions, just as it fixes the map of world languages.

In today's world there are few **theocracies,** a form of government where a church rules directly. The Vatican is one. Morocco may be a modified theocracy, because the king's legitimacy derives partly from his descent from Muhammad (see Figure 8–26). Many theocracies have existed in history: Tibet, for example, and Massachusetts Bay Colony in British North America. Utah was once a Mormon theocracy, and although it is today increasingly cosmopolitan, no Utah politician can ignore Mormon church leaders (see Figure 8–27).

Israel has grown more strictly Jewish since its founding, and more religious strictures have been enacted into national law. In 1991, for example, laws were passed banning the production or sale of pork. Laws prohibit activities on the Sabbath.

Figure 8–26. The King of Morocco alternates between modern westernized clothes and the robes traditional to his role as hereditary religious leader. (Courtesy of A. Nogues/ Sygma.)

Christianity from its beginning distinguished between religion and the political order. This distinction allowed Europe to develop secular government, secular knowledge, and a secular culture. Nevertheless, several countries are today explicitly Christian or at least support various Christian sects with public funds. These include the United Kingdom, Germany, Ireland, Sweden, Norway, Denmark, Iceland, Finland, Argentina, and Peru. Argentina's constitution includes a commitment to Christianize the indigenous populations. Roman Catholicism held official status in Italy until 1984, in Spain until 1988, and in Colombia until 1991.

Countries that are officially Islamic include Mauritania, Libya, Saudi Arabia, Yemen, Oman, Qatar, Bahrain, Iraq, Iran, Comoros, Maldives, Afghanistan, Pakistan, Bangladesh, and Malaysia. Islamic states enact Islamic teachings into law, called *Sharia*, and establish Sharia courts to determine whether secular law conforms to Islamic teaching. If the Sharia court rules against a secular law, that law is usually repealed. Islam views the principal function of government as enabling the individual Muslim to lead a good Muslim life. This is the measure of the legitimacy of the state. For Muslims to be ruled by non-Muslims, as they often have been, is a sacrilege to them.

Recent elections in several Islamic countries reflect a steady rise in influence of fundamentalist political par-

Figure 8–27. This is the citadel of Mormonism (The Church of Jesus Christ of Latter-day Saints) in Salt Lake City. In 1847 this site was surrounded by hundreds of miles of wilderness, but Mormon leader Brigham Young recognized it as an oasis that Mormon industriousness could make bloom and where they might escape the persecution they had suffered in the East. The building in the form of an overturned hull of a ship (it was in fact built that way) is the Tabernacle, and the neo-Gothic building with spires is the Temple. Today Salt Lake City is a metropolitan area of over 1 million people and the capital of Utah. (Courtesy of the Church of Jesus Christ of Latter-day Saints. Used with permission.)

ties, and many observers fear that these parties see democracy as a means to an end—the creation of an Islamic state—rather than as a system to be valued for itself. Iran under the Ayatollah Khomeini (ruled 1979–1989) was virtually an Islamic theocracy. Even before assuming power, Khomeini had written "We don't say that the government must be composed of the clergy, but that the government must be directed and organized according to the divine law, and this is only possible with the supervision of the clergy." Even Indonesia has granted official status to certain Sharia rulings on family law.

Religion and Women's Rights Religions differ in the attitudes that their teachings advocate toward women. In many cases, however, the actual practice of any religion varies from place to place. These variations reflect variations in other aspects of the cultures among places, and therefore it is difficult to determine exactly how much variation can be attributed to the predominant religion of a place. The essay on the scientific method discussed problems such as this in attributing explanatory power among variables.

For example, the Catholic Church is centralized, and it is opposed to the ordination of women. It observes that restriction worldwide. The worldwide Anglican Church, however, is less centralized. Its official head, the Archbishop of Canterbury, has approved the ordination of women, but local communities of Anglicans have different attitudes, and these reflect local variations in general social attitudes toward women's rights. In U.S. culture, for example, women's rights are protected in many areas, and the American Anglican Church (the Episcopal Church) has consecrated a female bishop (see Figure 8–28). Anglican communicants from some other world regions, however—most notably Africa, where women's rights are not always assured—have refused to recognize the U.S. female bishop.

Variations in practice can also be found across the Islamic realm. Islam originated in an Arab culture that granted women few rights, but the list of women's rights in the Koran was liberal for the time and place. Islam outlawed female infanticide, made the education of girls a sacred duty, and established a woman's right to own and inherit property. At the same time, Islam also fixed certain discriminatory practices. It teaches that a woman's testimony in court is half that of a man's and that men are entitled to four spouses, whereas women may have only one.

In practice, women's rights vary throughout the countries where Islam predominates. Women are frequently secluded, or veiled in public, in Islamic countries, but this is compulsory only in Saudi Arabia and Iran (see Figure 8–29). Also, only in the most fundamentalist Arab societies are women generally forbidden to work outside the home.

Figure 8–28. The Right Reverend Barbara Harris is the Episcopal bishop of Massachusetts. Her 1989 installation caused international controversy over the proper role of women in the worldwide Anglican church. (Courtesy of the Right Reverend Barbara C. Harris.)

Future struggle is nevertheless almost inevitable between the expansion of women's rights in Islamic societies and an Islamic fundamentalist backlash that demands the repeal of laws that had banned polygyny (having more than one wife), permitted birth control, and given women the right to divorce in some countries. The Iranian Revolution of 1979 swept away progressive legislation passed under the shahs. In Egypt the Supreme Court in 1985 struck down a 1975 law that gave a woman the right to divorce her husband should he take a second wife. Sudan's military regime, which seized power in 1989, refused to allow women the right to leave the country unaccompanied or without the permission of a father, husband, or brother. Algeria's 1984 Family Code gives a husband the right to divorce his wife and eject her for almost any reason.

It has also been hypothesized that religious teachings about women's rights may affect women's role in politics, but this is another hypothesis that is difficult to prove. Women have served as head of government in Islamic countries, as well as in countries that are officially Christian (both Catholic and Protestant), Jewish, and Buddhist. What sort of evidence for such a hypothesis could be sought? Reading scriptures alone does not prove whether those teachings are obeyed. Could we examine the percentage of national legislators who are female? The percentage of voters who are female? Chapter 10 will examine contrasts in women's rights among the nations of the world.

Indirect Religious Influences on Government In all countries the prevailing religion can dictate national ethics or morals. Religion so deeply affects what people assume to be natural or desirable human behavior that religious prejudices are taken for granted and unconsciously translated into laws. Religion should never be underestimated as an institutional, economic, and political force. Any organized religion has its own bureaucracy, sources of income, and, in the pulpit, its own channel of communication that often reaches more of the population than government information does (see Figure 8–30).

Relations between the government and the leading religious organizations constitute a major political issue in most countries. Religious leaders may lend legitimacy to rulers, or, conversely, may provide a rallying point for political opposition. A national church may sustain a suppressed nationalism, as in Ukraine and Ireland. The Polish Roman Catholic church stood against the Communist party and state as an alternative repository of Polish identity between 1945 and 1989. Attending mass was an act

Figure 8–29. These women in Tunisia are wearing the veil required in the most fundamentalist Islamic countries. (Courtesy of Daniele Pellegrini/Photo Researchers.)

Figure 8–30. Here in Mexico City, as in many other Latin American capitals, the cathedral (to the left) and the presidential palace confront one another across the main square. (Courtesy of Mexico Department of Tourism.)

of political defiance. In 1991 Poland's bishops unsuccessfully demanded repeal of Poland's constitutional separation of church and state.

In many countries a church or religion may form a political party, as the Christian Democrats have in several European countries, or as various Islamic groups have. The rise of a Hindu nationalist party in India, Bharatiya Janata, threatens India's status as a secular country.

In many countries a religion can become a big landowner or financier, and religious organizations do not always maximize the productive capacity of their property. Real property that is restricted in ownership or in the purposes for which it can be used is called *mortmain*, literally "dead hand." Church property may also enjoy preferential tax rates or escape taxation entirely. Throughout Latin America the Catholic church is a principal landowner. In the United States, church-owned properties are usually concentrated in the cities (see Figure 8–31). Because they are free from taxation, their concentration reduces any city's ability to raise property-tax income. In turn, however, the churches themselves may provide a wide range of services (such as soup kitchens and hospitals) to a public beyond their own communicants.

In some countries a community of citizens might be alienated from their national government if co-religionists in an adjacent country enjoy more complete religious freedom. For example, the Islamic vigor of Iran provided religious inspiration to the Muslim populations of the central Asian republics when they were under the rule of the Soviet Union.

States Split by Two or More Religions When the population of a country is split between two religious

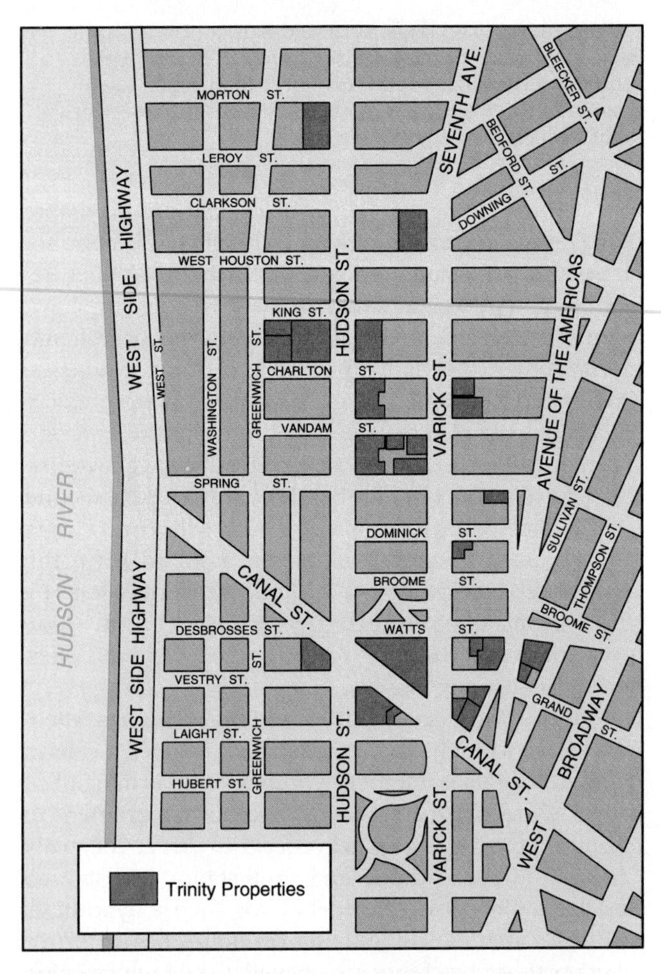

Figure 8–31. As this map of mid-Manhattan indicates, Trinity Church (Episcopal) is an active developer of its landholdings. Many U.S. cities contain great quantities of church-held property. (Courtesy of Trinity Real Estate, Parish of Trinity Church in New York City.)

communities that compete to write the laws, the country must devise a working compromise, or it can suffer internal conflict. Attempts to enforce Sharia in the Sudan, for example, have intensified conflict between the Arab Muslim-dominated north and the black Christian and animist south. In the Philippines, tensions between Christians and Muslims have nourished revolts on several islands against the central government. In Nigeria Muslim-Christian disputes over legislation have sparked recurrent riots and massacres.

When Lebanon received its independence from France after World War II, that land suffered many religious tensions. A compromise for sharing governmental offices among the religious groups lasted until 1990.

Sometimes states will enforce two separate systems of law, depending on the religion of the people involved in a case. Senegal and several other African states will allow Muslim men, but only Muslim men, more than one wife. In India Muslims are allowed to follow Muslim law in many matters of education, marriage, divorce, and property.

In some countries struggles that are ostensibly religious actually conceal other social divisions. Fighting in Northern Ireland is, on the surface, a conflict between a Protestant majority and a Roman Catholic minority. That split, however, involves much more than theology. It is a competition for jobs and opportunity, and it is complicated by the fact that both sides draw on outside support.

Religious tensions on the Indian subcontinent
Religious divisiveness can cause problems both within countries and among countries. The Indian subcontinent provides an example (see Figure 8–32). The British ruled the subcontinent (today's India, Pakistan, and Bangladesh) as one colony, "India," and they intended to grant it independence as one secular state. Many Muslims, however, demanded that the colony be divided so that they could create a Muslim state. When independence came in 1947, the Muslim population established the country of Pakistan. It originally included a western territory (today's Pakistan) and a separate eastern territory. In the months before the partition was realized, millions of Muslims fled toward areas that were designated to become Pakistan, and millions of Hindus moved to regions scheduled to become parts of India. Tens of thousands died in terrible acts of violence. Muslim-Hindu tensions were intensified by the fact that for hundreds of years Muslim minorities had in many areas ruled Hindu majorities (see Figure 8–33).

In 1971 eastern Pakistan broke off from Pakistan to form a separate Islamic country, today's Bangladesh. Meanwhile, religio-political turmoil continues in what is left of Pakistan. Islam has not proved a sufficient bond to unite Sunnis (about 70 percent) and Shiites (about 20 percent), and tribal loyalties and identifications subdivide the population still further. About one-tenth of the population is not Muslim, and under a system of separate electorates non-Muslims control just 10 of the 307 seats in Parliament. Their candidates do not represent geographical districts, and so they have to campaign across the entire country.

India itself emerged as a secular state with a majority Hindu population. Between 10 percent and 13 percent of India's population, however, is Muslim, and the conversion of many untouchables to Islam or Buddhism is upsetting India's internal religious balance. Muslims and Hindus regularly battle on the streets of India's cities, and India's Muslims were persecuted during India's wars with Pakistan in 1965 and again in 1971.

Conflict continues on the Indian-Pakistani border. In 1947 India and Pakistan fought over the border state of Jammu and Kashmir. Its population was mostly Muslim, but the state was eventually split, with India taking about two-thirds. The population of Indian Kashmir today is about 60 percent Muslim, but India refuses to allow a plebiscite. It is important to India's image as a secular country to have a state with a Muslim majority.

The Geography of Religion in the United States
Several of the American colonies were founded as theocracies, and most of them retained established churches until the War for Independence. In forging the new United States, however, James Madison included in the Bill of Rights an amendment prohibiting the federal government from establishing any religion—the world's first such clause. This was intended to protect the churches from government interference.

Even in the United States, however, the shifting borderland of morals, ethics, and religion leaves lawmaking open to religious pressures. The Supreme Court has upheld the writing into law of "Judeo-Christian moral and ethical standards" (*Bowers* v. *Hardwick*, 1986).

The United States is arguably the most religious nation in the developed world. Ninety percent of Americans say that they have never doubted the existence of God, and 40 percent attend religious services in a typical week. The percentage who say that religion is important in their life, however, is steadily falling from a high of over 80 percent in the 1950s to about 60 percent in 1990. The Census Bureau has avoided studying the distribution of religions within the United States, but Figure 8–34 on page 270 reproduces one map of the variations of religious affiliation across the country.

During the period 1989–1990 the City University of New York conducted a nationwide poll on religious affiliation. The poll did not ask about baptism, church contributions, or regular worship. It asked only about religious self-identification. A full 86.4 percent of Americans identified themselves as Christian. This is a degree of religious commitment, and of religious homogeneity, almost unique in the world. Only 3.7 percent of Americans claimed any religion other than Christianity. This included about 4.3 million Jews, 1 million Buddhists, and

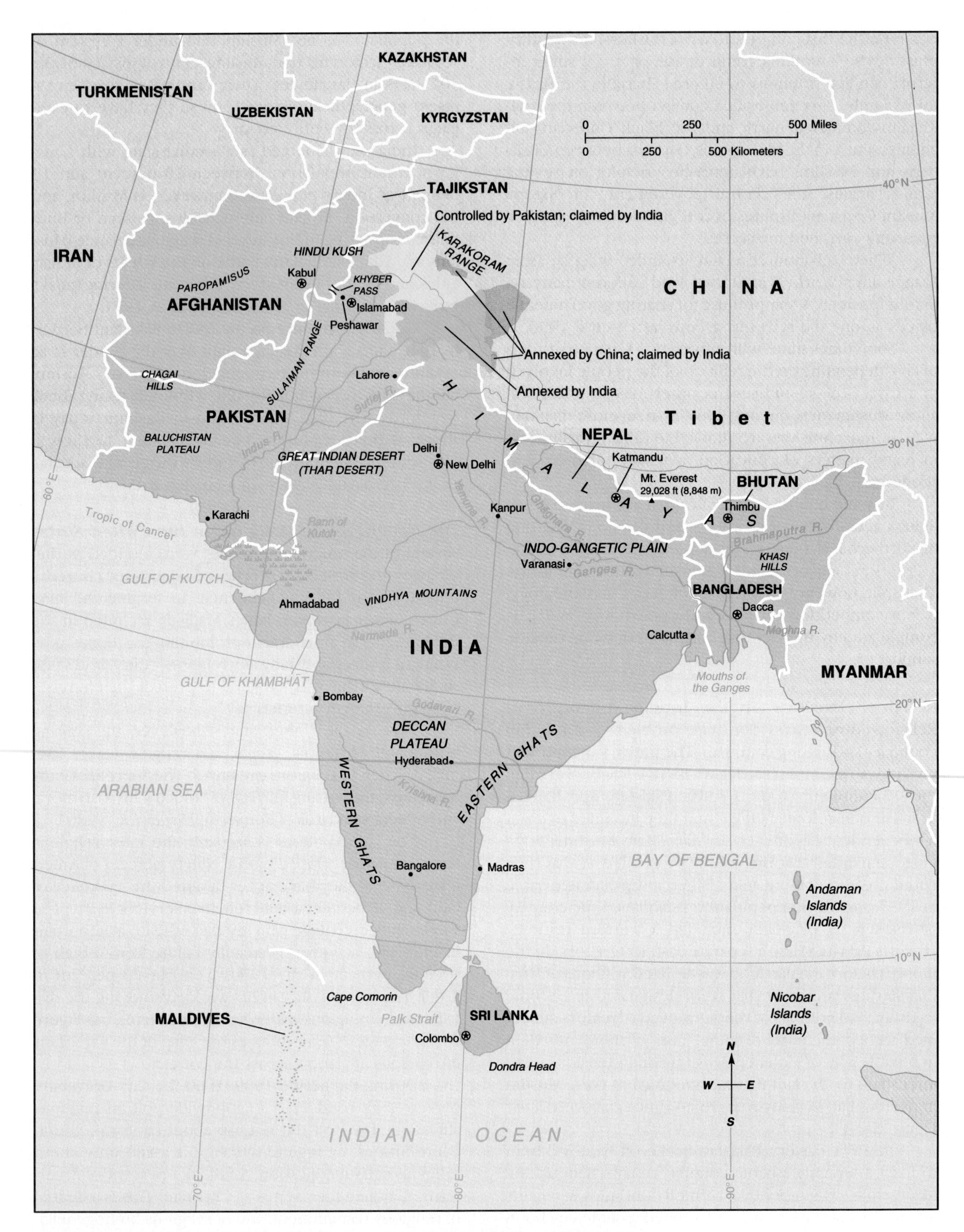

Figure 8–32. Antagonism and border disputes persist between India and Pakistan and between India and China. These three nuclear powers comprise about 40 percent of the human population.

Figure 8–33. The Taj Mahal in Agra, India, completed in 1648, is the mausoleum a Muslim ruler built for his beloved wife. India's Muslim rulers were so rich and powerful that their name, Mughals, has come down to us in English as a word for a business executive, a *mogul*. (Courtesy of the Agency for International Development.)

500,000 Hindus. About 0.5 percent of the population, or 1.4 million people, claimed to be Muslim, a number much lower than pollsters had anticipated. Some 40 percent of these were African Americans, although they made up less than 2 percent of the African-American population. Only 7.5 percent of the U.S. population claimed to have no religion, and 2.3 percent refused to answer. Some regional variations in religious affiliation revealed by this poll are shown in Table 8–2.

The U.S. separation of church and state has an especially important ramification in central-city African-American neighborhoods. In central cities, African-American churches often serve as cornerstones of community life to a greater degree than in any other communities. They provide services ranging from financial credit to remedial reading lessons. More than three-quarters of African Americans belong to a church, and nearly half go to church each week. The churches are potential building blocks of government community-assistance programs, and yet if the government were to subsidize this social infrastructure or allocate public resources through it, the action might provoke a constitutional battle.

Religion and Dietary Habits

Different religions espouse different beliefs concerning the sacredness of plant and animal life, and this affects the farming and dietary practices of their adherents. Buddhists are generally vegetarians. Hindus believe that cattle are sacred; therefore, they will not eat beef, and they use cattle only as draft animals and as a source of milk.

TABLE 8–2 ***Percentage of U.S. Population Claiming Certain Religious Affiliations and the Five States Within Which the Largest Percentage Claimed That Affiliation***

Catholic	26%	Baptist	19%	Methodist	8%	Lutheran	5%	Jewish	2%	No religion	8%
R.I.	62	Miss.	55	Del.	27	N.D.	37	N.Y.	7	Ore.	17
Mass.	54	Ala.	51	Iowa	16	Minn.	34	N.J.	4	Wash.	14
Conn.	50	Ga.	51	S.C.	16	S.D.	30	Fla.	4	Wyo.	14
La.	47	N.C.	47	W.Va.	15	Wis.	26	Mass.	4	Calif.	13
N.J.	46	D.C.	47	Kan.	15	Neb.	16	Md.	3	Ariz.	12

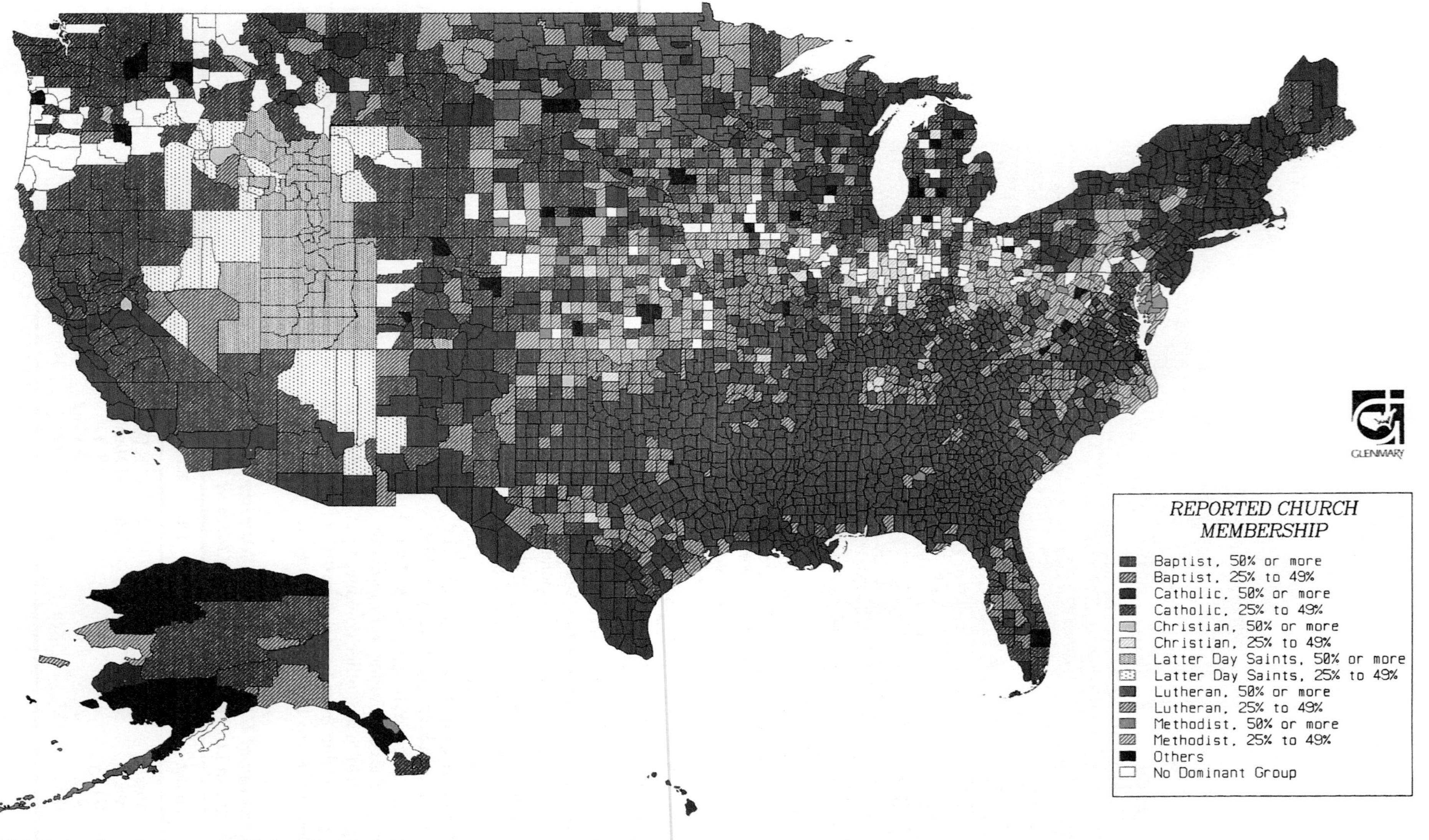

Figure 8–34. This map of U.S. church affiliations reveals the predominance of Baptist membership in the Southeast and of Mormons in Utah and adjacent regions. The predominance of Lutherans in the northern Midwest reflects the Scandinavian and North German backgrounds of that region's pioneer settlers, and the distribution of Roman Catholics reflects, among other factors, the migration to the United States of Hispanics and Southern Europeans. (Reproduced with permission from Martin B. Bradley, Norman M. Green, Jr., Dale E. Jones, Mac Lynn, and Lou McNeil, *Churches and Church Membership in the United States: 1990*. Atlanta, GA: Glenmary Research Center, 1992.)

Figure 8–35. In India cows are sacred and cannot be molested, no matter how much of a nuisance they become. (Courtesy of Helen Marcus/Photo Researchers.)

Outside observers have tried to measure in economic terms the advantages and disadvantages of preserving these cattle. Aging and unproductive cattle may overgraze the land or may compete with the human population for limited food supplies. Sacred cows roam the streets of Indian cities unmolested, befouling them and congesting them (see Figure 8–35). On the other hand, the cattle provide dung as a fuel and manure, and those that die of natural causes are eaten by non-Hindus. Whatever the balance of these economic arguments might be, to a religious Hindu this is a religious matter not open to economic debate.

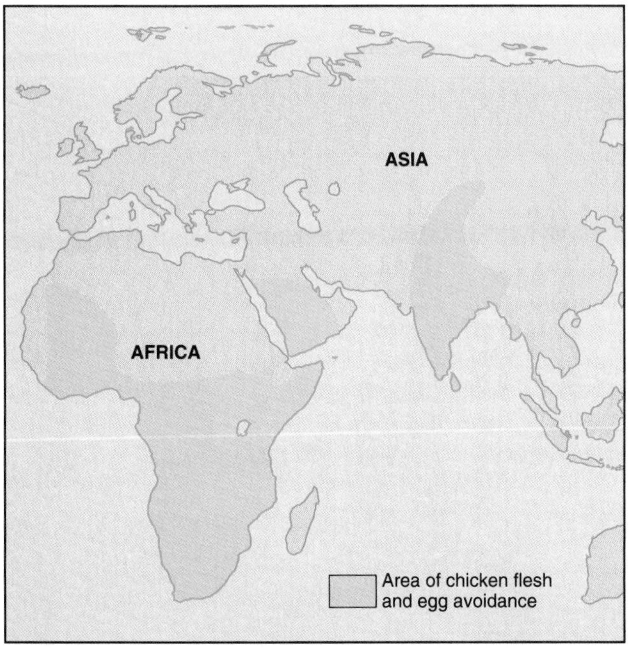

Other religions prohibit certain foods. Jews and Muslims both refuse pork, and Muslims are prohibited from drinking alcohol. Altogether, a significant share of humankind limits its diet because of religious beliefs (see Figure 8–36). This affects even world trade in food. Table 7–2 shows that several of the greatest importers of pulses are countries that limit their consumption of meat for religious reasons. Pulses are an alternative source of protein. Annual per capita consumption of meat in India, for example, is only 4.4 pounds (2 kg), and India is a major importer of pulses.

ASIA

AFRICA

Area of chicken flesh and egg avoidance

Figure 8–36. The worldwide distribution of chicken flesh and egg avoidance.

Religion and Economics

Observers have long suggested that certain connections might exist between religion and economic behavior. Many of these suggestions are conjectural and must be taken only as hypotheses.

One hypothesis is that religions can affect the way people view moneymaking. If that is true, then religions can influence the accumulation of money and capital, and ultimately the economic development of whole societies. As one example, the scriptures of most religions bar the charging of interest. They view it as unfairly taking advantage of another person. Christian teachings, however, have evolved to differentiate between two reasons for borrowing money: for needs, and for investment. It may be sinful to charge interest on money borrowed to buy food, but it may not be sinful to charge interest if the borrower invests the money for his or her own profit—buys a truck to increase the profit in his or her own business, for example.

The religious leaders in some societies have not evolved or do not accept such a differentiation. Islamic fundamentalists, for example, still condemn any notion of interest. Pakistan's Federal Sharia Court ruled in 1992 that interest is un-Islamic and therefore illegal. This ruling threatened Pakistan's entire modern financial system, as well as the country's international trade and financial links. Compromises have to be defined in each Islamic country.

Weber and the Protestant Ethic German sociologist Max Weber linked Protestantism and capitalism in *The Protestant Ethic and the Spirit of Capitalism* (1904). According to Weber, Protestantism encourages individualism. Therefore, with the rise of Protestantism in Western Europe, acquisitiveness, as the result of actively exercising individual ability in the marketplace, became a recognized virtue. Today this ethic seems to characterize adherents of almost any religion, and so it is usually referred to as the *work ethic*.

If, however, there is any true advantage in the Protestant Christian attitude toward moneymaking, then it might be hypothesized that the worldwide expansion of Protestantism might encourage individualistic capitalism. This explains why the spread of Protestantism is widely reported and happily received in the U.S. business press, which clearly accepts the hypothesis of a link. *The Wall Street Journal*, *Business Week*, *Forbes*, and other publications all devote extensive coverage to the spread of Protestantism. They report it as economic news, not just religious news.

If the corporate managers who read these publications believe that Protestantism is hospitable to capitalism, then they will direct corporate investment to the areas undergoing conversion. Thus, a belief can launch its own fulfilment, whether or not it can objectively be shown to be true. This demonstrates once again the idea of cognitive behavioralism.

The Catholic Church and Capitalism In 1991 Pope John Paul II issued an encyclical on economic matters, *One Hundred Years*. This document asserted a superiority of capitalism over socialism. "On the level of individual nations and of international relations," wrote the pope, "the free market is the most efficient instrument for utilizing resources and effectively responding to needs." At the same time, however, the pope insisted that capitalism must be tempered with concern for social justice and human dignity.

Religion and Economics in Asia The Confucian tradition in East Asian societies may exemplify another religious influence on economic development. The Confucian view of leadership by an intelligent elite with the moral obligation to guide the people enhances the status of jobs in the government bureaucracies. Ambitious young graduates compete for positions within the Japanese Ministry of Finance or the Korean Economic Planning Board the way ambitious young Americans compete for jobs at investment banks. The ability to attract talent gives these Asian governments more legitimacy and more competence in dealing with businesses than the U.S. government can bring to bear, which probably has helped to build successful economies.

Also, diligence, obedience, and high savings rates characterize the peoples of neo-Confucian Japan, South Korea, Taiwan, Hong Kong, and Singapore, and of every Chinese settlement in the world. It is risky to hypothesize a cause-and-effect relationship, but the coincidence must be noted. The government of Taiwan emphasizes the coincidence in full-page ads in U.S. business publications calling attention to the island's Confucian tradition as good for business. The Taiwanese may just be pandering to what they believe are Western preconceptions.

Figure 8–37. The Cathedral of Santiago, in the northwest corner of Spain, is believed to hold the relics of Saint James. It was begun in 1078, although later additions, such as this baroque facade, have covered many original features. Pilgrims from throughout Europe and the world have sought it out. In medieval times the traditional routes that they took and the staging points at which they rested played important roles in trade and in the diffusion of ideas. (Courtesy of Spanish Tourism Office.)

Several East Asian leaders have praised their own societies' traditional communitarianism and decried Western Christian individualism. Zealous Evangelical Protestant missionaries have even been arrested in Singapore, accused of upsetting that nation's Confucianism. The missionaries' intolerance caused them to be branded, ironically, as Marxist.

Impacts of Religion on the Landscape

The geography of the various religions affects so many aspects of human geography that a few can receive only a brief mention. Different religions recognize different holidays and pace human activities through the year differently, as is often reflected in national laws. Most Christian countries use the solar calendar. Jews use a calendar of 12 months plus an additional month 7 times in 19 years. Muslims use the lunar calendar. Muslims celebrate the Sabbath on Friday, Jews on Saturday, and Christians on Sunday. Muslims recognize an entire Holy Month and accordingly curtail their activities.

Environmental degradation today afflicts areas of all religious persuasions. The Worldwide Fund for Nature is translating all the world's major religious texts into English specifically to allow scholars to compare what the various scriptures say about conservation. Muslim, Christian, Buddhist, Hindu, Muslim, Jewish, Sikh, Shinto, and Taoist organizations are all cooperating.

Religions designate holy places and places of pilgrimage. These places attract visitors and play a major role in cultural diffusion (see Figure 8–37). Fairs and markets grow up at sites of pilgrimage. They still attract millions of visitors, many of whom, admittedly, come only as tourists. Every Muslim is expected to visit Mecca at least once, and the Saudi royal family derives prestige from its role as protector of the holy places. Muslims' pilgrimages to Mecca bring together people from around the world.

We stated earlier that the diffusion of religions can affect the diffusion of languages. Muslims everywhere are encouraged to read the Koran in its original Arabic, and Martin Luther's German translation of the Bible defined the German language and, arguably, the German nation.

Religions mark toponymy. Islamabad ("the place of Islam") is the capital of Pakistan; St. Paul is the capital of Minnesota. Each religion also marks the landscape with easily identifiable religious structures, burial grounds, and specifically religious settlements.

At the local level, religious administrative patterns, such as Roman Catholic dioceses, organize and focus community spirit and, in many cases, education. Many Roman Catholic New Yorkers are more aware of what parish they live in than can name their city council district, and it is through the parish that they make friends and meet potential spouses.

Religion means "rebinding," of people with their gods and of people with one another. The study of such binding will always be a key consideration of human geography.

SUMMARY

Language and religion define and bond human cultures. Peoples who share either of these cultural attributes often demonstrate consistencies in other aspects of behavior and often cooperate with one another in other ways.

The distribution of any language illustrates the pattern of dispersal of its original speakers or their cultural impact on others. The great variety of languages spoken today testifies to the relative isolation of groups in the past. There are more than 3000 distinct languages, but about 60 percent of all people speak 1 of just 14 languages. The language with the most speakers is Mandarin Chinese. English is the world's lingua franca. In contrast with these widespread languages, some languages are extremely local.

A few languages are associated almost exclusively with one country, but several languages are shared by many countries, and some polyglot countries officially recognize several languages. The idea that languages gave birth to nations persists despite the fact that standard languages usually were the product of self-conscious efforts at nation building by centralizing governments.

Languages can be categorized into families on the basis of common ancestry, and they continue to diffuse and differentiate. The use of some languages is expanding; other languages are dying out.

The population of the United States has always been composed of a great variety of peoples speaking a great variety of languages. Regional differences persist, and some people think English should be the official national language.

Religion is a vital force in human affairs, and understanding the messages of different religions contributes substantially to understanding many other aspects of human life.

Judaism, Christianity, and Islam form one ancient religious tradition, originating in the Near East and diffusing from there. Christianity came to predominate in Europe, and it spread worldwide with European expansionism. Today two struggles divide Christianity: one between Protestants and Roman Catholics, and another within the Church of Rome itself. At the same time as Christianity continues to win new converts, practices and doctrines that derive from European culture are under pressure for change, and aspects of native religions that were blanketed by Christianity are reviving. Islam today strongly influences the politics of the countries where it predominates.

East Asia is dominated by Hinduism and its offshoots, Buddhism and Sikhism. Other Eastern religions include Confucianism, Taoism, and, in Japan, Shinto. Several of these are not religions in the same way that Christianity and Islam are. Their teachings are more exclusively ethical and psychological, and they do not address basic theological questions. Many people are adherents of more than one of these. Animism and shamanism have many followers in the less-developed parts of the world.

The teachings of various religions differ regarding politics, women's rights, government, diet, economics, and environmental attitudes. Thus, the geography of religious communities influences the geography of these other aspects of human culture.

KEY TERMS

language (p. 239)
standard language (p. 239)
official language (p. 239)
lingua franca (p. 239)
philological nationalism (p. 240)
polyglot state (p. 243)
orthography (p. 243)
alphabet (p. 243)
protolanguage (p. 245)
language family (p. 245)
cognate (p. 246)
etymology (p. 246)
speech community (p. 247)
fundamentalism (p. 250)
secularism (p. 250)
monotheism (p. 250)
Zionism (p. 251)
sacerdotalism (p. 254)
liberation theology (p. 255)
interculturation (p. 256)
syncretic religion (p. 256)
caste (p. 260)
animism (p. 262)
shamanism (p. 263)
theocracy (p. 263)

QUESTIONS FOR INVESTIGATION AND DISCUSSION

1. How many languages are spoken in your community? Does the local school system have bilingual or English as a Second Language (ESL) programs? Where do your community's non-English speakers come from?

2. How many religions have temples, churches, or synagogues in your community? What, roughly, is the statistical breakdown of religious affiliations of the people of your community? Does this reflect the source areas of settlers to your region? Does each place of worship anchor a residential neighborhood of that religion's communicants?

3. How many religiously oriented parochial schools exist in your community? With which religions are they associated?

ADDITIONAL READINGS

CHADWICK, HENRY, and G. R. EVANS, eds. *Atlas of the Christian Church.* New York: Facts on File, 1987.

COOPER, ROBERT L., ed. *Language Spread: Studies in Diffusion and Social Change.* Bloomington: Indiana University Press, 1982.

CRYSTAL, DAVID. *The Cambridge Encyclopedia of Language.* Cambridge, England: Cambridge University Press, 1987.

AL-FARUQI, ISMA'IL R., and DAVID E. SOPHER, eds. *Historical Atlas of the Religions of the World.* New York: Macmillan, 1974.

MARTIN, DAVID. *Tongues of Fire: The Explosion of Protestantism in Latin America.* London: Basil Blackwell, 1990.

MARTY, MARTIN E., and R. SCOTT APPLEBY, eds. *Fundamentalisms Observed.* Chicago: University of Chicago Press, 1991.

RUHLEN, MERRITT. *Guide to the World's Languages.* Vol. 1, *Classification.* Palo Alto, Calif.: Stanford University Press, 1987.

SOPHER, DAVID. *The Geography of Religions.* Englewood Cliffs, N.J.: Prentice-Hall, Inc., 1967.

CHAPTER 9

Cities and Urbanization

Chicago and its suburbs spread out for miles from these skyscrapers along the Lake Michigan lakefront. Chicago incorporated in 1837 with just a few hundred residents, but already by 1900 its population had risen to more than 1.5 million. (Courtesy of Mark Segal/Tony Stone Worldwide.)

Every settled society builds cities because some essential functions of society are most conveniently performed at a central location for a surrounding countryside. A **city** is a concentrated nonagricultural human settlement. Cities may provide a variety of services, including government, religious services, education, trade, manufacturing, wholesaling and retailing, transportation and communication, entertainment, business services, and defense. The region to which any city provides services and upon which it draws for its needs is called its **hinterland.**

In many cultures, subsistence farmers build their dwellings together in one place to which they return after the day's work in the fields, but no independent activities are carried on at that place. Some of these strictly residential settlements are home to thousands of people, but they are not real cities.

In the earliest writing that can be read, hieroglyphics, the ideogram meaning "city" consists of a cross enclosed in a circle. The cross represents the convergence of roads that bring in and redistribute people, goods, and ideas. The circle denotes a moat or a wall. Few modern cities have walls, but cities do have legal boundaries, and within those boundaries they usually exercise a degree of self-government. The process of defining a city territory and establishing a city government is called **incorporation.**

Sometimes several cities grow and merge together into vast urban areas called **conurbations.** In the northeastern United States, for instance, one great conurbation stretches all the way from Boston to Washington, D.C. This conurbation has been called *Megalopolis*, which is Greek for "city of millions." The world's largest urban areas are generally called **metropolises,** Greek for "mother cities."

In several countries one large city concentrates a high degree of the entire national population or of national political, intellectual, or economic life. These cities are called **primate cities.** Paris, for example, is primate within France, and Bangkok is primate within Thailand. Not all countries, however, have a primate city. The United States does not. Whether a country has a primate city depends on national history and social and economic organization.

Today in all countries urban populations are growing faster than rural populations. This process of concentrating people in cities is called **urbanization.** The United Nations predicts that about one-half of the world's population will live in urban areas by the year 2000, compared with 30 percent in 1950. The rate of urbanization, however, is not the same in all countries (see Figure 9–1). One of the purposes of this chapter is to explain why this is so.

The geographic study of cities, **urban geography,** includes three topics:

1. The study of the functions of cities and their role in organizing territory;
2. The comparative study of urbanization as it occurred in the past and as it is continuing in different countries today;
3. The study of the internal geography of cities, that

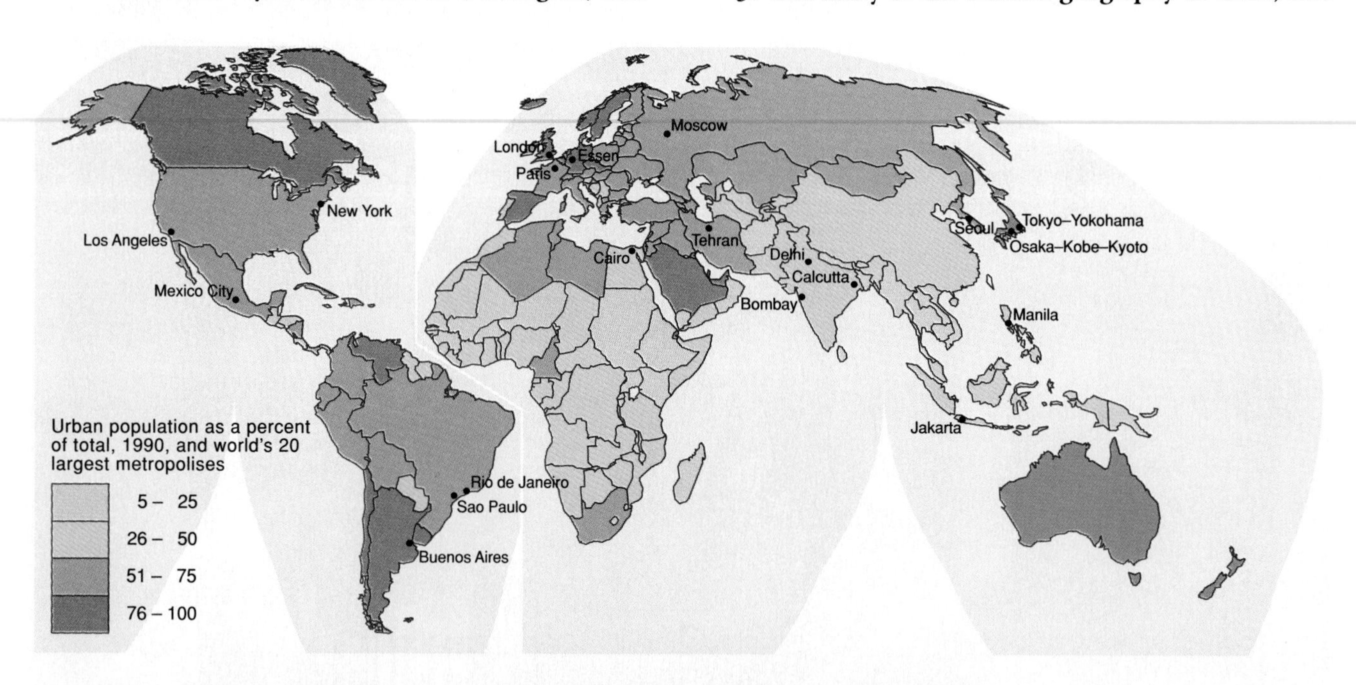

Figure 9–1. This map shows the varying degrees of urbanizaton of different countries and the world's largest metropolises. Different countries use different definitions of "urban area," ranging from settlements as small as 200 people to as large as 30,000, but most countries use a minimum of 2000 to 5000 people. The U.S. Bureau of the Census accepts a minimum of 2500 people. (Data from World Resources Institute, *World Resources 1992–93*. New York: Oxford University Press, 1992, pp. 264–65.)

is, the internal distribution of housing, industry, commerce, and other aspects of urban life.

This chapter reviews each of these topics and then examines the growth and internal geography of U.S. metropolitan areas. This examination illustrates many of the principles of urbanization in situations that may be familiar. The chapter ends with a brief examination of key ideas of urban and regional planning.

URBAN FUNCTIONS

In many traditional societies, the ritualistic and political importance of cities far outweighed their economic functions. Cities were the seats of government officials, or even of the gods themselves, and so they were appropriately embellished (see Figure 9–2). Sometimes their layouts were coordinated with the cardinal points of the compass or with celestial bodies because the people believed that human activities ought to conform to certain "forces" in the universe. This system of designing a city is called **geomancy.**

As societies develop, the economic role becomes paramount in most cities. Cities bring people and activities together in one place for greater convenience. This process is called **agglomeration.** Agglomeration facilitates the convenient **division of labor,** which is the separation of work into distinct processes and the apportionment of these processes among different individuals to increase productive efficiency. Cities also promote and administer the regional specialization of production throughout their hinterlands. This regional specialization may be thought of as a territorial division of labor. With industrialization, cities become important centers of production as well.

The Three Sectors of an Economy

Economic activities are generally divided into sectors: the primary, secondary, and tertiary sectors. This terminology indicates the degree to which each sector is removed from direct involvement with the earth's physical resources.

Workers in the **primary sector** extract resources directly from the Earth. Activities in this sector include agriculture, fishing, forestry, and mining. Agriculture usually provides the bulk of primary-sector employment. Workers in the **secondary sector** receive the commodities and raw materials produced by the primary sector and transform them into manufactured goods. Construction is included in this sector. All other jobs in an economy are within the **tertiary sector,** sometimes called the **service sector.** The tertiary sector includes a great range of occupations, from a store clerk to a surgeon, from a movie ticket seller to a nuclear physicist, from a dancer to a religious leader. Because this sector includes so many occupations, some scholars have tried to split the tertiary sector into a tertiary service sector and a *quaternary sector*, called an *information sector*. There is no consensus as to what this sector includes, however, and no agencies of the U.S. government or of the United Nations recognize this sector in statistics. Therefore, it is not used in this book.

As economies grow, the balance of employment and output tends to shift from the primary sector toward the secondary and tertiary sectors. These sectors require urban settings, and therefore, as an economy's secondary and tertiary sectors grow, its cities also grow. This growth of the secondary and tertiary sectors is called *sectoral evolution*, and it will be examined in detail in Chapter 11.

Western culture often contrasts the countryside with the city—the former "natural," the latter human-

Figure 9–2. This is the world's largest wooden structure, the Todaiji Temple in Nara, Japan. It houses a colossal statue of Buddha 53 feet (16 m) high. Both the temple and the statue were originally made in the eighth century, when Nara was Japan's capital city. Today Nara is not a large city or important in Japanese economic life, but much of it has been preserved as a treasure of Japanese history and culture. (Courtesy of Japanese National Tourist Office.)

made (artificial) and "unnatural." In fact, fields and pastures are as human-made as are factories or city streets. Natural landscapes of forests or grasslands are transformed into both types of cultural landscapes, and economically the two types depend on each other.

The Economic Bases of Cities

Cities must import, at the very least, their food. The cities must, in turn, provide services or "export" something to the outside. Many cities produce and export manufactured goods, but the exports of a city are not necessarily things that leave it. They may be things or services that people come to the city to buy. If people go to Houston for heart surgery, for example, then heart surgery is counted as an export of Houston. Vacations are an export of Miami Beach; gambling is an export of Las Vegas. In this economic sense, capital cities export government.

Some of the work in any city produces the city's exports, but other work serves the needs of the city's own residents. The part of a city's economy that is producing exports is called the **basic sector,** and that part of its economy serving the needs of the city itself is called the **nonbasic sector.** A city's basic and nonbasic sectors may be in the primary, secondary, or tertiary sectors of the economy as a whole.

Jobs in the basic sector of a city create jobs in the city's nonbasic sector. For example, if a local factory makes a product that is sold around the world, the factory workers will spend their earnings by shopping locally, getting their hair cut locally, and purchasing other local goods and services. Earnings from exports circulate and recirculate through the local economy. Therefore, each job in the basic sector actually supports several nonbasic sector jobs. When a factory worker buys a shirt, the store clerk can get a haircut; the barber, in turn, might eat at a local restaurant, and so on. Thus, jobs in the basic sector have a **multiplier effect** on jobs in the nonbasic sector.

Cities can be classified economically by examining each city's basic and nonbasic sectors and by comparing these sectors among different cities. New York, Detroit, and Hollywood, for instance, each contain a number of dry cleaners and doctors. These people, for the most part, work within the nonbasic sectors. In terms of exports, however, workers in New York provide specialized financial services, Detroit exports cars, and Hollywood makes movies.

There are two ways of measuring economic specialization. It is possible to analyze either (1) a city's employment structure or (2) the flow of money through the city's economy. The former approach assumes that any city that has an unusual concentration of workers in a specific job category must be exporting that product or service. A city that has a concentration of auto workers is presumably an auto-assembly site, and so on (see Figure 9–3). Methods of analyzing the flow of money through economies are called **input-output models.** Advanced urban geography courses study both methods in detail.

The Locations of Cities

The location of any city depends on a balance of the site and situation factors discussed in Chapter 5. A mining town is obviously the result of its site. Some mining towns virtually sit on top of valuable ore deposits but are otherwise isolated (see Figure 9–4). Cities often develop at waterfalls to exploit hydropower for industry. The United States's first planned industrial city—Paterson, New Jersey—was built at the Great Falls of the Passaic River in 1791.

Other cities are more the result of their geographic situations. Cities frequently grow up at places where two different physical areas meet, such as land and water, forest and prairie, and mountains and plains, or along the border between two cultures. Tombouctoo was located on a key trade route, on the border between two physical environments and also two cultures. Many cities spring up as transportation hubs or at bottlenecks, such as at a bridge across a river, or where two political jurisdictions funnel trade through border checkpoints. Some cities grow at sites where the method of transportation necessarily changes. These are called *break-of-bulk points.* A seaport is an example. Another is the head of navigation of a river. If there is a waterfall on a navigable river, cargo has to be offloaded and reloaded beyond the waterfall up or down river, or else the cargo might be shifted to rail or truck.

Louisville, Kentucky, for example, was laid out in 1773 at the falls of the Ohio River. The river is navigable both up and down river from Louisville. The opening of the Portland Canal in 1830 allowed ships to pass around the falls, but by then the city was well developed. It provided many services to its rich developing hinterland and to westward-moving pioneers. The bridge across the Ohio River focused north-south traffic, and later the railroad lines also focused on the city.

If a situation is favorable, a great city may arise on an unfavorable site. Many of Asia's coastal cities were built by European merchants or conquerors at sites that provided access to the sea and that may have been defensible, but they sit on deltas or the swampy foreshores of tidal rivers. Karachi, Pakistan; Madras and Calcutta, India; Colombo, Sri Lanka; Yangon, Myanmar; Bangkok, Thailand; Ho Chi Minh City, Vietnam; and Guangzhou, Shanghai, and Tianjin, China, all face to this day formidable problems of drainage, water supply, construction, and health.

One of human geography's great paradoxes is that one of the world's largest cities, Mexico City, is located on one of the world's most unfavorable places to build a city: a drained lake bed in an earthquake zone in a basin of interior drainage at a high elevation in a dry

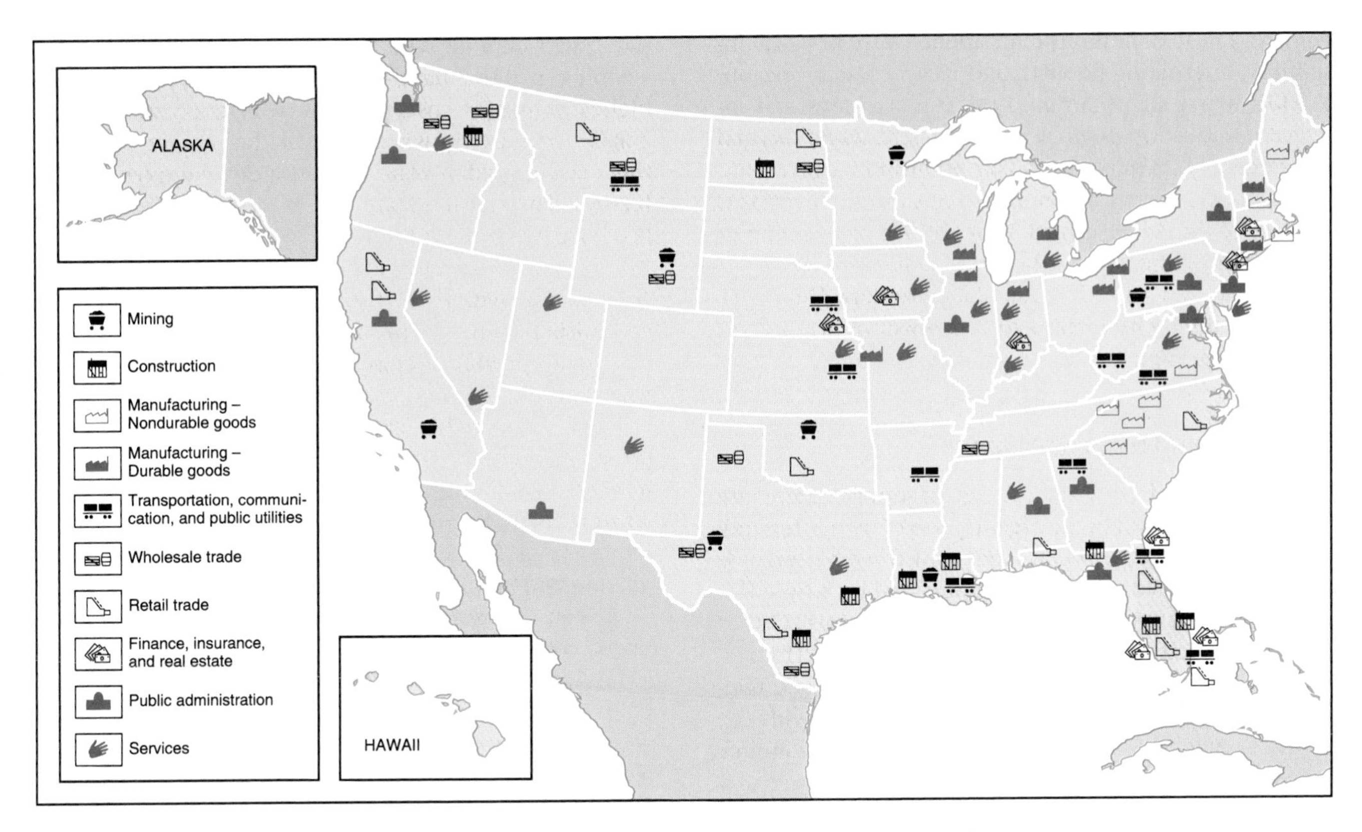

Figure 9–3. This map shows cities that have specialized economies; it does not indicate the cities' individual ranking in each of the activities. Los Angeles, for instance, is the country's greatest manufacturing center, but it does not appear on this map because its economy is diverse. (Adapted from J. Clark Archer and Ellen R. White, "A Service Classification of American Metropolitan Areas," *Urban Geography* 6(1985):122–51.)

Figure 9–4. Butte, Montana, is an example of a city whose location was determined by site factors, in this case its proximity to a rich copper mine. (Courtesy of Lawrence B. Dodge, Butte, MT.)

climate. These conditions combine to cause physical instability, alternating flooding and lack of water, and the world's worst air pollution. Both human lungs and internal combustion engines are inefficient when oxygen is reduced by 23 percent, and air pollution is aggravated by local windstorms. Today's Mexico City, however, was the site of the Aztec capital Tenochtitlán at the time of the arrival of the Spaniards. The Spaniards maintained the site as their capital, and so have modern Mexicans. Thus, this great city testifies to the power of history and geographical inertia.

How Cities Organize Space

The existence of a city and the task of supplying any city with its needs result in the organization of the hinterland. The relationships between cities and their hinterlands have inspired two models for understanding why cities are distributed across territory the way they are. One is the **isolated city model** of Johann Heinrich von Thünen, and the other is Walter Christaller's **central place theory.** Both von Thünen and Christaller started with the simplest imaginary landscape, a homogeneous flat surface across which population density and resources are uniformly distributed. Such an imaginary surface is called an **isotropic plain.** Transportation cost is determined according to straight-line distance.

Von Thünen asked the question "If there were a city on such an isotropic plain, how would the territory around the city be organized?" Christaller asked, "How would cities be distributed on such a plain?" These two questions are complementary.

Von Thünen's "Isolated City" Johann Heinrich von Thünen (1783–1850) was a Prussian aristocrat who wondered: What product can I grow on my estates and market most profitably? Thinking about this question led him to question how various crops become distributed across any countryside. What is the influence of the city on the territory around it? The conclusions he drew about the general distribution of crops represent one of the first examples of deductive geography.

Von Thünen noted that the different uses to which parcels of land are put result from the different values placed on the land. These values are called the *rent value* of land. On an isotropic plain, transport costs are the only variable, and so differences in rent reflect the transport costs from each farm to the market. The greater the cost, the lower the rent that can be paid if the crop produced is to be competitive in the market. In addition, perishable products such as milk and fresh vegetables need to be produced near the market, whereas less perishable crops such as grain can be produced farther away.

Transport costs on an isotropic plain are uniform in all directions, and so a pattern of concentric zones of land use will form (see Figure 9–5). The intensity of cultivation, that is, the amount of labor and capital applied, will decline with distance from the market. Perishable crops that have the highest market price and the highest transport costs per unit of distance, such as vegetables, will be grown closest to the market. A hint of this market garden zone around a city survives in New Jersey, "The Garden State." New Jersey market gardens historically supplied the cities of Philadelphia and New York. Today New Jersey is the most heavily industrialized state in the nation, and yet it also ranks as a leading agricultural state. It still specializes in greenhouse products, dairy products, eggs, and tomatoes, which are typical market garden commodities.

Less perishable crops with lower transport costs per unit of distance will be grown farther away from the market. Concentric rings of mixed farming and grain farming form around the inner ring of market gardens. In the rings farthest outward from the city, livestock grazing and similar extensive land uses predominate.

Von Thünen elaborated his model by adding a navigable river flowing through the town (see Figure 9–6). He noted that the areas along the river's banks would enjoy greater accessibility to the city market. In other words, the cost-distance to the city market was lower along the riverbank, and so each of the zones would be expanded out along the river.

Some scholars have tried to apply von Thünen's model to entire countries. They have plotted the use of cropland in the United States, for instance, as if Megalopolis or Chicago were a central market. Trying to apply

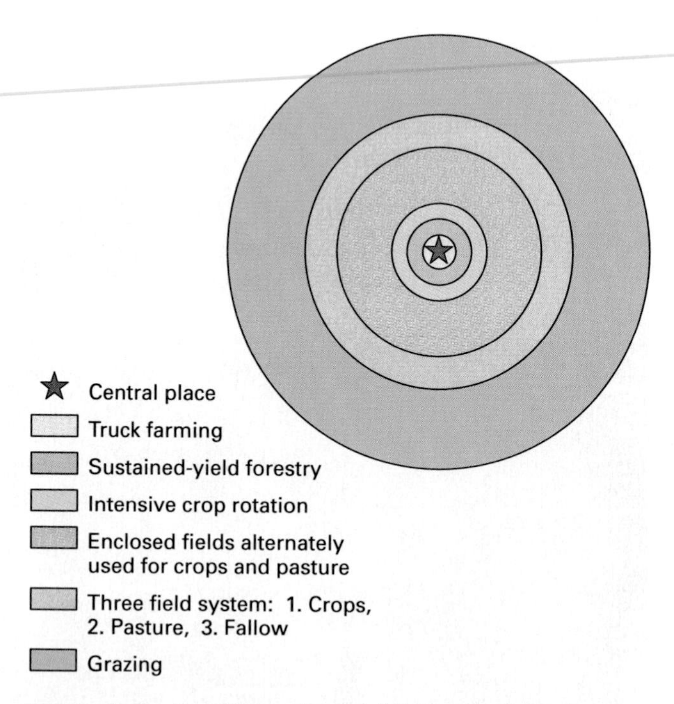

Figure 9–5. Von Thünen's model shows rings of decreasingly intensive land use concentrically outward from the city. The ring of sustained-yield forestry may be surprising, but forestry was then an intensive land use. The forest provided grazing for livestock, fuel, and raw materials for building and many other purposes.

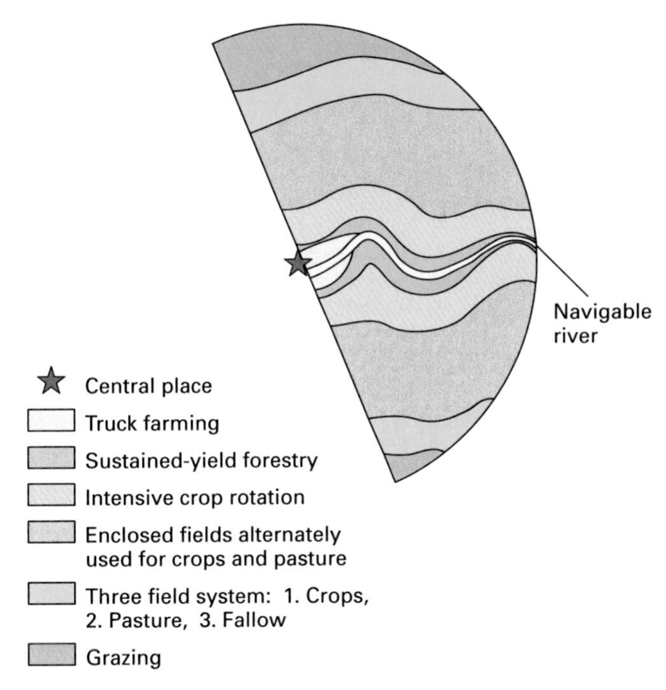

Figure 9–6. Water transportation is cheaper than transportation overland. Thus, if a navigable river flows through the town, the cost-distance to the city is lower along the river. Therefore, the activity of each ring would extend outward along the river.

von Thünen's model to such vast territories under contemporary conditions, however, strains the model's usefulness. The model focuses on the role of distance from cities as markets, but in any study of the geography of agriculture proximity to the market is only one consideration. Physical environmental conditions, governmental regulations, the economic system, the pattern and the regulation of the transport system, and still other factors must be included as well.

Christaller's "Central Places" Walter Christaller (1873–1969) asked a question complementary to the one asked by von Thünen: If cities are to be distributed to serve as convenient centers for exchange and other services across an isotropic plain, how will cities be distributed? What will be the pattern of towns and their hinterlands?

To answer these questions, Christaller developed his central place theory. This model has three requirements: (1) The hinterlands must divide the space completely, so that every point is inside the hinterland of some market; (2) all markets' hinterlands must be of uniform shape and size; and (3) within each market region, the distance from the central place to the farthest peripheral location must be minimal.

The only pattern that fills these three requirements is a pattern of hexagons—six-sided figures (see Figure 9–7). Therefore, market towns will be distributed across an isotropic landscape as foci of a grid of hexagons. The distributions of towns in many real landscapes have been shown to approximate this model (see Figure 9–8).

Hexagons appear repeatedly in both nature and in industrial design. Honeycombs are hexagonal, and so probably is the pencil in your drawer. A hexagonal shape allows the greatest possible number of pencils to be cut out of a block of wood.

Urban Hierarchies Chapter 5 discussed diffusion up or down a hierarchy of cities. There are many small towns, fewer and more widely spaced medium-sized cities, and still fewer big cities. A hierarchy of cities exists because the more specialized a service or product is, the larger the number of potential customers that is needed for that product or service to be offered.

No product or service can be offered without a minimum number of customers. This minimum demand is called the **threshold** for that product or service. A dry cleaner, for example, needs a minimum number of customers to earn a living. Most people periodically need a dry cleaner, and so the threshold of demand for dry cleaning is low. Each neighborhood or small town can support one. Similarly, many people frequently buy fresh bread and milk. Therefore, small groceries can be found in each neighborhood.

Few people, however, take tuba lessons or buy diamond bracelets. The threshold of demand sufficient to support tuba teachers or expensive jewelers is high, and so only larger towns will be able to support these enterprises. On an isotropic plain on which people and buying power are equally spaced, dry cleaners and gro-

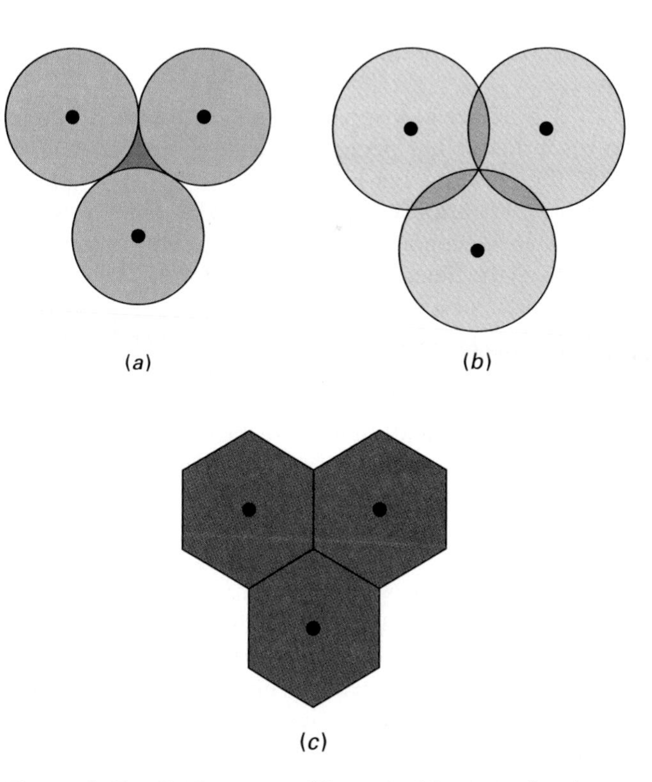

Figure 9–7. Circles cannot fill space without overlapping one another, but circles collapse into hexagons that neatly fill space.

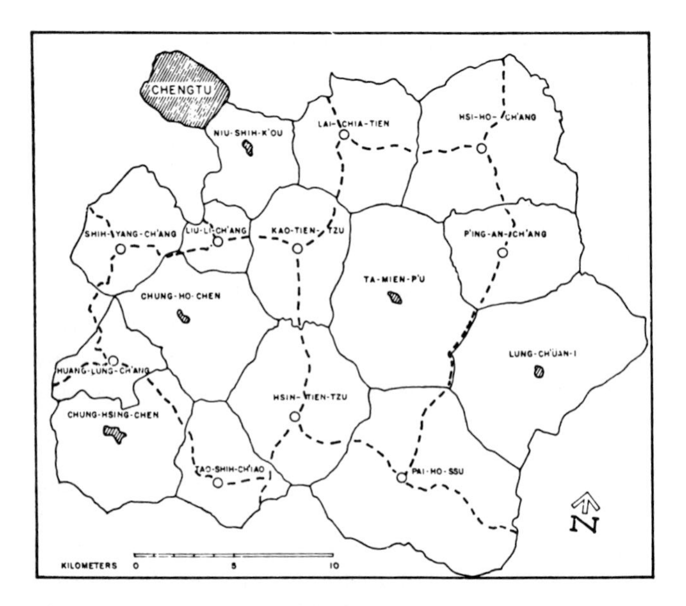

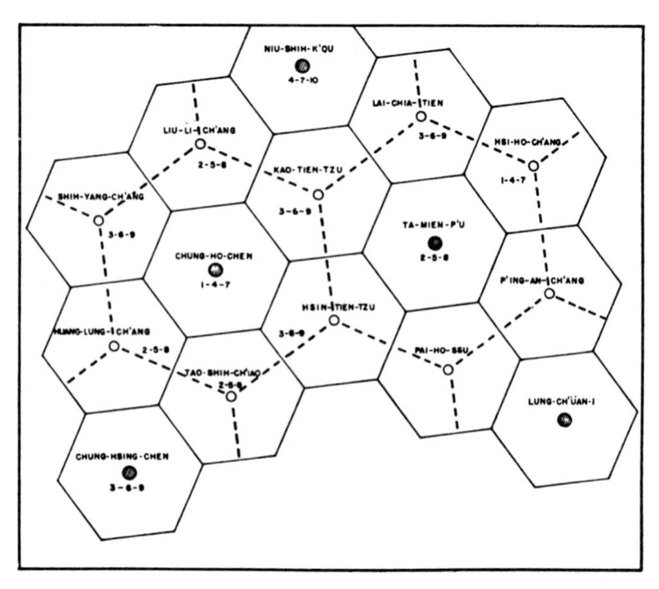

Figure 9–8. The Chinese landscape is not an isotropic plain, but the distribution of market towns nevertheless reflects the geometric ideas of efficiency proposed by Christaller's model. (Adapted with permission from G. William Skinner, "Marketing and Social Structure in Rural China," *Journal of Asian Studies*, Vol. 24, No. 1, Nov. 1964.)

cers will be closely spaced; tuba teachers and jewelers will be widely spaced.

In real landscapes, factors in addition to distance must be considered. For example, if one area has a concentration of rich people, then the density of jewelers will be greater there. Quantity of purchasing power substitutes for numbers of people. If another area has a high population density, then grocery stores will be denser there.

A provider of a service can do one of two things to reach his or her necessary threshold of customers. First, the provider can be *itinerant* (travel from place to place). Traditional societies that are not much above a subsistence economic level may not be able to support markets all the time. Markets may be periodic. Therefore, the provider of a good must travel to town *X* when it holds its market each Wednesday, town *Y* each Thursday, and so forth. Alternatively, the provider can set up shop at one convenient place and wait for people to come. Convenience and accessibility are the principal purposes of cities. Agglomeration of services in cities saves travel costs, and it allows the thresholds to be reached for more specialized goods and services.

The range of goods and services in a city offers small businesses the opportunity to rent pieces of equipment and to hire services or temporary employees only when they need them. This saves the businesses the cost of investing in equipment or hiring people full-time. These available services or goods are called **external economies,** and their availability lowers initial costs for new business ventures. Thus, cities are called incubators of new businesses. If a company grows large enough to justify buying its own equipment or hiring full-time employees for a service, then the company has achieved an **internal economy.**

The patterns of urban hierarchies Christaller's model of evenly spaced market towns with hexagonal hinterlands can be elaborated to represent a hierarchy of cities. In the isotropic plain the larger cities that in the urban hierarchy are "above" the small market towns must be more widely spaced than the small market towns. The larger cities must also be evenly spaced, however, and their hinterlands must also be hexagonal. Therefore, the distribution of the larger cities is represented by a grid of larger hexagons superimposed on the grid of small hexagons (see Figure 9–9).

These models have been applied in planning new cities in countries with unsettled lands. The land that the Dutch have claimed from the sea behind new dikes is similar to an isotropic plain. There Dutch geographers have planned new market towns on hexagonal grids. Brazil has adapted the model to settle territories in Amazonia. The government built what it calls an "Agrovila" every 6 miles, with a school, a health-care center, and a post office. Every 25 miles the government placed an "Agropolis;" these offer the services of an Agrovila, plus sawmills, stores, warehouses, banks, and other commercial services. Each 85 miles Brazil built a "Ruropolis," a city, and established light industry there. The ratio of 6, 25, and 85 miles cannot be reconciled with the proportions of a "pure" geometric grid, but it has been found useful by the Brazilian planners. The planned towns serve as growth poles.

When improvements in transportation allow people to travel farther to obtain services or goods, the

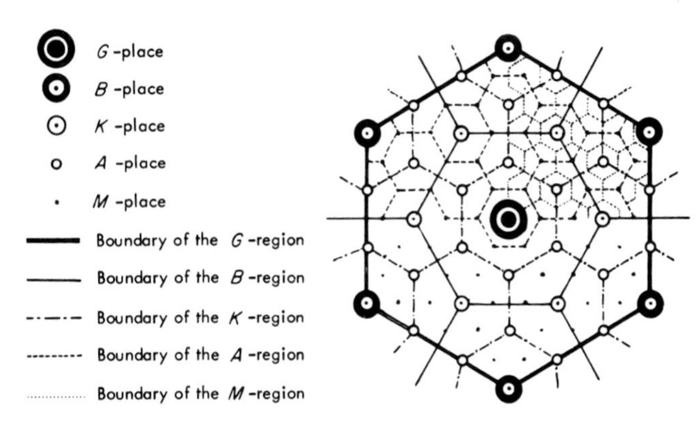

Figure 9–9. The distribution of cities on an isotropic plain could be represented by a grid of large hexagons superimposed on the grid of the small hexagons that represent the hinterlands of the small market towns. Geometry defines at least three ways of doing this. The pattern illustrated is called the K3 model because in it the hinterland of each city includes a hinterland that the city serves as a market town plus one-third of the hinterlands of each of the six surrounding contiguous market towns. When a resident of a market town needs a higher-order service available only in a city, he or she finds each of the three cities equally convenient. Therefore, that person might visit each city occasionally, and so news diffuses across a landscape organized in this way. (Reprinted with permission from Walter Christallen, *Central Places in Southern Germany*, trans. Carlisle Baskin. Englewood Cliffs, N.J.: Prentice Hall, 1966, p. 61.)

smallest central places may lose their reason for existence and disappear. This is happening in the United States. Across the farm states of Kansas, Nebraska, Iowa, and the Dakotas most small towns developed as commercial centers for farmers. They offered grocery stores, banks, hardware stores, farm-implement dealers, automobile dealers, and feed stores. Technological advances, however, starting with the tractor, as well as advances in transportation and communication and other economic forces have increased the size of farms and cut the number of farmers. As a consequence, there is less need for the small-scale central places. Farm states are today punctuated by wholly abandoned towns, and the number of rural towns is expected to continue to decline. Only small towns within a reasonable distance of a metropolitan area can survive, or even prosper. These become **exurbs,** the name given to settlements that make up the outermost ring of the ever-expanding metropolitan areas.

WORLD URBANIZATION

Urban geographers compare urbanization as it occurred in the past and as it is continuing today. The rapidity of worldwide urbanization presents many nations with both challenges and opportunities for the welfare of their populations.

Early European Experiences of Urbanization

In the seventeenth century, Holland became the first modern urban country, with over one-half of its population living in towns and cities. The principal economic functions of the cities were tertiary-sector functions. The cities arose as administrative and commercial foci of Holland's global shipping, banking, and trading activity, and the urban populations could be supported by Holland's highly productive agriculture. This urbanization occurred before the Industrial Revolution.

In Britain, a larger country with a more varied economy, urbanization occurred with industrialization. Over one-half of its people lived in cities and towns by about 1900. Several developments over the previous 200 years had resulted in the concentration of Britain's population.

1. Improvements in agricultural technology—part of the Agricultural Revolution—reduced the number of agricultural workers necessary. Landowners found it profitable to release employees and to evict tenants. The resulting rural depopulation is described in Oliver Goldsmith's poem "The Deserted Village" (1770), which laments that "rural mirth and manners are no more."
2. These displaced people migrated to the cities. There, many were absorbed by the concurrent **labor-intensive** stage of the Industrial Revolution. An activity is labor-intensive if it employs a high ratio of workers to the amount of capital invested in machinery. Other newcomers to the city were absorbed by the tertiary sector. The largest class of employed workers in cities was actually domestic servants—a tertiary-sector job. In 1910 domestic servants formed more than one-third of the total British labor force. The United States and other rich Western countries were not far behind.

   Rural-urban dislocation caused appalling hardship. The history of urban life in nineteenth-century England records overcrowding in dreadful slums, malnutrition and starvation, crime, attempted assassinations of political figures, and death (see Figure 9–10). The descriptions of the miseries of England's slum populations in Charles Dickens's novels still haunt our imaginations. Cities have never easily absorbed all those who have flocked to them.
3. Population pressures were somewhat relieved by emigration, or else by *transportation*, that is, the forcible exportation of criminals and debtors throughout the British Empire. Australia and the colony of Georgia absorbed many of these transported people.

This British experience provided the world's first model for modern urbanization, and most of today's de-

Figure 9–10. Gustave Doré's illustrations of London in the 1860s won attention as art, but they did not achieve his purpose of provoking solutions to the squalor and depression he depicted. (Courtesy of Bettman/Hulton Archive.)

veloped countries experienced similar histories. In 1800 the 21 European cities with populations of 100,000 or more held about 4.5 million people, or one thirty-fifth of the total European population. By 1900 there were 147 such places with a total population of 40 million, or one-tenth of the total population.

The word *model* as used here does not mean that the British experience of urbanization was perfect or that it should be imitated. It imposed terrible hardship on millions of people. Britain provides a model only in the sense that Britain experienced these forces of urbanization *first*, and so that experience can be compared with the current situation.

In the British story the push of rural displacement and the pull of urban job opportunity were not coordinated. Governments have since learned how better to manipulate both forces to ease the process of urbanization, and yet some aspects of contemporary urbanization are causing as much hardship as the experience did in Britain.

Urbanization Today

Today urbanization is occurring in many places without concomitant economic development, especially in the world's poor countries. Burgeoning populations swamp the cities' ability to absorb the people and to put them to work. The populations overload the cities' infrastructures of housing, water supply, sewerage, internal transportation, and education. Reliable water, electricity, and telephone services are becoming increasingly rare. From the tops of new skyscrapers in many modern cities, the view presents vast shantytowns of the desperately poor, called *bidonvilles* in Francophonic West Africa, *favelas* in Brazil, *ishish* in the Mideast, and *kampungs* in Indonesia (see Figure 9–11). Living conditions in these pullulating cities are no better, for the majority, than those that existed in Europe in the nineteenth century.

Rapid urbanization is a problem, but problems can also be opportunities. New urban populations present concentrated labor forces that might be put to productive work in ways that could raise the living standards of all people. To do so, however, presents a tremendous challenge.

The rapidity of urbanization often reflects deteriorating conditions in the countryside. To that degree urbanization is a rural problem, but in many of the poor countries it is intensified by the sharp cultural differences between urban and the rural populations. Chapter 5 noted that many cities in today's poor countries did not evolve out of local needs and economic development, but rather were outposts of international commerce grafted onto the local societies. Chapter 6 noted that particularly throughout Latin America and Africa cultural differentiation is partially a remnant of the westernization and urban prejudices of the national elites. These conditions can make it harder to deal with rapid urbanization.

At least eight other circumstances differentiate urban growth today from the historical British experience.

1. *The commercialization and mechanization of agriculture has accelerated, and modern agricultural technology requires capital investment.* The new technology may be available only to those who are already rich, affect only crops that are grown for export, or be applied only to the production

Figure 9–11. The immigration of Bolivian peasants into La Paz has inflated the city's population to more than 1 million and surrounded the modern downtown skyscrapers with rings of slums. Sights like this are typical of the growing cities in the poor countries. (Courtesy of Agency for International Development, John Metelsk.)

of industrial crops rather than to food crops. In almost all poor countries, agricultural innovation widens the income gap between rich and poor and accelerates the displacement of rural workers. At the same time, rising rural populations multiply pressures on the land. National radio and television penetrate the countryside, spreading an image of the city as a place of opportunity.

Rapidly growing urban populations may threaten the stability of national governments. Urban food riots are common in Africa, and even in relatively well-off Venezuela, urban riots over the price of food left over 250 dead in March 1989. Therefore, many governments subsidize urban food supplies by underpaying the farmers. As we saw in Chapter 7, this only discourages the farmers from producing more food. The effect of such policies in accelerating urban growth is clear in the contrast between most African countries, on the one hand, and, on the other, Zimbabwe and Malawi. These two countries have paid their farmers good prices, and urbanization has been slower, even though they have worse-than-average shortages of farmland among African countries.

2. *Today many countries are in the second stage of the demographic transition, and the introduction of medical care and hygiene into the cities triggers natural population increases.* In the past, cities everywhere suffered natural population decreases. That is, urban death rates were higher than urban birth rates. "Cities," wrote Jean Jacques Rousseau 200 years ago, "are the burial pit of the human species." High death rates resulted from the lack of sanitation, the spread of disease, and starvation among the poor. Without a steady influx of new residents from the countryside, the town populations would have decreased.

   Today, however, modern medicine and hygiene lower urban death rates. Natural population increases replace natural population decreases and compound the population growth that occurs from in-migration from the countryside. Modern medicine and hygiene later diffuse throughout the countryside from the cities.

3. *Today's cities are less able to employ the displaced rural population than were cities in the past because urban economies have changed.* Historically, urban economies offered entry-level job opportunities for the unskilled—jobs available for anyone with a strong back and a willingness to work. Domestic servitude absorbed many workers, and manufacturing and construction absorbed others. Today, however, domestic servitude has declined as an opportunity in all countries. There may also be fewer jobs in manufacturing and construction, and many of the jobs that exist in these industries demand skills. One hundred years ago a worker's transition from a farm to a factory or construction site was easier than it is today. Manufacturing and construction are less labor-intensive than they used to be; they are increasingly **capital-intensive.** Machinery has replaced workers. The petroleum industry, for example, is so capital-intensive that several OPEC countries have high per capita incomes, but they have difficulty devising policies to spread that wealth among their people. In Venezuela the petroleum industry provides 80 percent of the nation's exports but only 2 percent of national employment. Urban unemployment there triggered the food riots in 1989.

   The gap between rural skills and the skills demanded for manufacturing jobs is great, but the gap is even greater between rural skills and urban service-sector skills. Few societies have succeeded in catapulting rural labor forces into urban tertiary-sector jobs that require skills in one generation, but that challenge faces many societies today.

4. *For most societies, the safety valve of emigration has been stopped.* Some countries still welcome skilled professionals, but, as we noted in Chapter 6, migration opportunities for the unskilled and for rural workers have decreased.
5. *The interdependence of the world economy reduces the control local governments exert over local economic activities.* This situation restricts the ability of local governments to regulate urban economic opportunity. Factory openings or closings in Bangkok are determined in Tokyo. Worker opportunity in Caracas, Venezuela, is regulated from New York. International trade and financial links affect both the countryside and the cities. The poor countries today cannot pass through the stage of labor-intensive industrialization because their goods must compete in world markets. National life is increasingly conditioned by international forces.
6. *As urban life has become more technologically complicated, it has become more difficult to accommodate the migrants from the countryside.* Enormous special challenges and costs confront any modern city if a large proportion of its residents are first-generation urban residents. First-generation urbanites do not always know the importance of hygiene in crowded urban conditions, when a trip to a hospital emergency room is really necessary, how to save water, and how and when garbage is collected. Even New York City elementary schools have offered experimental courses on such rules and regulations of urban life. Conditions in the slums of poor countries are unimaginable.
7. *Residents of the cities in many countries are cosmopolitan mixes of ethnic groups that may not even be native to the region. Ethnic animosities can intensify economic problems.* Throughout Latin America, for instance, the cities are predominantly white or mestizo. Any measures intended to restrict urban migration are interpreted, often correctly, as racist discrimination against Native Americans.

   Where the native majority dominates national politics, they may discriminate against the urban populations. They see the cities' ethnic pluralism as an affront to nationalism. Chapter 6 noted this in Uganda, Kenya, and other countries in East Africa, and in Malaysia.
8. *Many governments favor urban projects over rural projects for reasons of status, even though these urban projects may not really be best for the country.* This enhances the perceived opportunity in the cities, but it may not provide real opportunity. In 1990 Doctor Edouard Saouma, the Director of the UN Food and Agriculture Organization, sharply criticized many poor countries for spending too much time and energy on "showcase" urban projects while neglecting food production and the needs of the rural poor. Money was spent on urban industry and services, he argued, at the expense of rural development and agriculture. "Development too often takes no account of rural people, and very few nonagricultural job opportunities have been created in the countryside. Little or nothing has been done to remedy the discomfort and unhealthiness of living conditions." Young farm people seek to escape, "drawn by the mirage of possible jobs and a better life. In fact they are merely exchanging poverty for destitution." They wind up in overcrowded shantytowns, straining the limited resources of city governments, and depriving the farms of people.

Government Policies to Reduce Urban Migration

People migrate from the countrysides to the cities because of the balance of push and pull factors. Governments might regulate this migration, therefore, either by reducing the attractiveness of the cities or else by raising the quality of rural life.

Reducing the Pull of Urban Life Forceful measures to limit urbanization are not new in human history. The Russians have been required to have internal passports since the days of Peter the Great over 200 years ago. Recently the world has even witnessed brutal instances of compulsory "ruralization." When North Vietnam absorbed South Vietnam in 1976, the new rulers forced millions of South Vietnamese urbanites into the countryside and put them to work raising food using labor-intensive methods. Thousands, if not hundreds of thousands, died. The Vietnamese government argued that more would have perished had they remained in the cities, which had bloated on U.S. financial assistance. The government of Kampuchea (Cambodia) similarly relocated urban populations to the countryside in the late 1970s. In 1990 almost 1 million urban Myanmarese were forcibly relocated to rural settlements, and Cuba has periodically sent urbanites to work in the countryside.

Many cities try to discourage newcomers in less brutal ways. They restrict housing and economic opportunity. Building codes that ban substandard housing, for instance, are often ineffective in stopping the growth of slums, but they make squatters' settlements illegal. Therefore, their residents do not get city services and will not risk investing to improve the property. The city, in turn, cannot collect property taxes. These "illegal" communities hold 30 percent to 60 percent of the total populations—70 percent to 95 percent of newcomers—in the cities of the poor countries. Cities also try to restrict small businesses in residential areas, but backyard workshops may thrive anyway. Cities may discriminate against new urbanites in education, housing permits, business licenses, or job opportunities. Hawkers are chased from city centers.

In some countries, frustrated city authorities have even bulldozed squatters' settlements with only 1 or 2 days of warning. This has happened in major cities throughout Latin America, Africa, and Asia, leaving tens of thousands of people absolutely homeless.

In the extreme, the urban governments may round up squatters and drive them out. Tanzania's Human Resources Deployment Act declares that anyone who cannot show proof of employment in Dar es Salaam can be deported to a state farm; thousands have been moved.

China enforces a system of household registration. Chinese are not supposed to move without getting their registration changed, but people move anyway. In 1990 China counted 80 million "surplus" rural workers and an additional 50 million migrant laborers, called *floaters*. In major cities they make up more than 10 percent of the population, freely coming and going. Urban street committees watch over the behavior of local residents, but they cannot control people who do not belong there in the first place. Many urban residents find the floaters useful. They bring produce to the city or provide services not provided by central planning.

Rural Investment Instead of trying to drive people out of the cities, governments might invest in rural development to raise rural standards of living, to keep people in the countryside, and also to raise national food production. National budgets could be redirected toward investments in agricultural productivity, rural health care and education, and housing, roads, and other infrastructure. Some governments use television to discourage rural migration to the city. Ghana, for instance, broadcasts to the villages films suggesting that life is better there.

Governments throughout history have organized work on labor-intensive construction projects even with little investment (see Figure 9–12). Agricultural terraces, roads, and dams can be built by hundreds of thousands of workers equipped with shovels and wicker baskets just as well as by relatively few workers equipped with expensive earth-moving machinery (see Figure 9–13).

Figure 9–12. The U.S. government sponsored voluntary labor-intensive projects during the Depression. The Reforestation Unemployment Act of 1933 created the Civilian Conservation Corps to alleviate unemployment and at the same time initiate national reforestation. Workers also cleaned streams and beaches, constructed small reservoirs and dams, and refurbished roads, bridges, and trails in national parks. Other programs such as the Civil Works Administration, the Public Works Administration, and the Works Progress Administration also initiated labor-intensive infrastructure projects. ("The Spirit of 1938" by Harry L. Rossoll; courtesy of the National Archives, Records of the Forest Service.)

The Economic Vitality of Cities

Most of the previous paragraphs about urbanization in the developing countries may have seemed pessimistic. Urban growth has been viewed as a problem for which various possible solutions were listed. In the discussion of efforts to restrict entry, however, hints of urban vitality appeared in backyard shops, in the small-scale retailing of street hawkers, and in the lives of the Chinese floaters. These activities provide optimistic indications that urban migration, balanced between management and liberty, can provide a reservoir of economic vitality that can be harnessed for national growth.

The book *The Other Path* (1986) is a study of urbanization in Peru by the economist Hernando De Soto. It became a worldwide bestseller. De Soto noted that in many cities a substantial share of the population is engaged in productive activities that do not appear in official accounts. The people may not have licenses to do what they are doing, they may be avoiding taxes, or for some other reason their activity escapes official notice. These activities make up the **informal** or **underground sector** of an economy. Every city in the world has such a sector, but studies suggest that in the cities of the poor countries this sector may be particularly important.

De Soto estimated, for example, that in Lima, Peru, the informal economy employs fully 60 percent of the population and produces 40 percent of all goods and services. The informal economy is open, self-regulated,

Figure 9–13. Construction of even massive infrastructure projects such as this dam in India can begin with women carrying baskets of dirt on their heads. In the rich countries, labor-intensive construction methods have yielded to capital-intensive methods. (Courtesy of the Agency for International Development.)

and democratic. In Lima, De Soto estimated, the poor own and control 95 percent of the public transportation network of private taxis and vans. This represents capital of at least $1 billion. De Soto estimated that the urban land and housing occupied by the poor is worth an additional $17 billion. None of this is legal, however, and so it cannot be used as collateral. If the government simply recognized this activity as legal, the business operators could more easily borrow additional investment capital, and their businesses could grow. Furthermore, the government could tax the activity. Therefore, the government's refusal to recognize what is happening handicaps the country's economic growth and vitality.

De Soto's book illustrates how the preconception that urban migration is necessarily a problem blinds governments to the recognition that urban immigrants' industriousness could be an asset for economic growth. Millions of people survive, even thrive, in the burgeoning cities. Something terrifically dynamic is going on, and geographers, economists, and government officials at all levels are challenged to understand and measure it.

THE INTERNAL GEOGRAPHY OF CITIES

Urban geographers study not only the distribution of cities in the landscape but also the distribution of activities within cities. This distribution may be caused by economic forces, by social factors, or by deliberate actions of the government.

The Role of Economic Factors

The distribution of people and activities is, for the most part, determined by economic factors. Certain locations are more desirable than others, either for business or for living purposes, and these can demand a higher rent.

Many factors influence desirability. Some areas may be considered more prestigious than others. Mental maps of the prestige value of different parts of a city powerfully influence rents. As a general rule, residential neighborhoods can be classified by the income level of their residents.

The accessibility of a place is usually a principal determinant of its rent. The success of a department store, for instance, will depend partly on whether customers can reach the store easily. A city's most convenient and busiest intersections are the most valuable for commerce.

Models of Urban Geography Models of the internal geography of cities based on accessibility alone echo von Thünen's agricultural zone model. Rent decreases concentrically out from the center. Before industrialization and the development of mass transport systems, city growth was limited by the time that it took to walk across the cities. Few cities exceeded a radius of 4 miles (7 km). As populations increased, the cities extended their boundaries, the way a tree grows rings. If a city had defensive walls, new walls periodically had to be built. Paris, for example, built five concentric rings of outer walls between 1180 and 1846. The last of these defines the boundaries of Paris today.

Concentric zone model Figure 9–14 is a concentric zone model of urban growth and land use. The core of the city, called the **central business district (CBD),** concentrates office buildings and retail shops. Landowners usually maximize the density of use on this valuable land by building up, and so a traditional CBD is identifiable by tall buildings as well as crowded streets. Even within the CBD, clusters of functions appear. Lawyers' offices, for instance, cluster near courts or near the offices of their client firms. Retail stores of one type, such as jewelry stores, may cluster so that consumers can comparison-shop. Merchants find that business attracts additional business.

The CBD is surrounded by less intensive business uses such as wholesaling, warehousing, and even light industry—that is, nonpolluting industries that require relatively small quantities of raw materials. Residential land use surrounds this urban core.

This concentric zone model was devised by sociologists to describe the density and quality of housing patterns observed in U.S. cities early in the twentieth century. The model proposed that the inner zones of the city were continuously expanding. They would "invade" the contiguous outer zones and eventually replace those land uses, or "succeed" them, by pushing those land uses still farther out from the expanding CBD. Cycles of invasion and succession described the replacement of populations of one income level by another, and also the observed replacement of one ethnic group by another among the various urban immigrant communities.

The concentric zone model can be refined by considering the effect of transportation routes. Accessibility is determined not by centrality alone but by transportation systems. Since the nineteenth century, changes in transportation systems have been the single most important factor in determining the spread of built-up areas. New means of transport—first canals, railroads, and tramways—spread out radially from the heart of the city, although their paths were modified by topography. Industrial and residential growth took place in ribbons or fingers along these radial routes. Wedges of open land were usually left between these routes.

Sector model A second classic sociological model of urban housing, the *sector model* (see Figure 9–15), assumes that high-rent residential areas expand outward from the city center along the new transportation routes. Middle-income housing clusters around high-rent housing, and low-income housing lies adjacent to the wholesale and light-manufacturing areas.

Multiple nuclei model Figure 9–16 recognizes the development of several nodes of growth within an expanding city area. These *multiple nuclei* may each concentrate a different function. This model best describes the expansion of U.S. metropolitan areas, as will be discussed in detail below. Development stretched out along the highways, and eventually suburban highway interchanges provided new transportation foci for shopping malls and offices. Individualized automobile transportation gradually filled in the wedges of open land between the original radial routes in and out of the central city.

Non-North American models The foregoing models were all devised to describe the North American experience, but other models have been sketched to describe developments characteristic of other cultures. Figure 9–17, for example, describes Latin American cit-

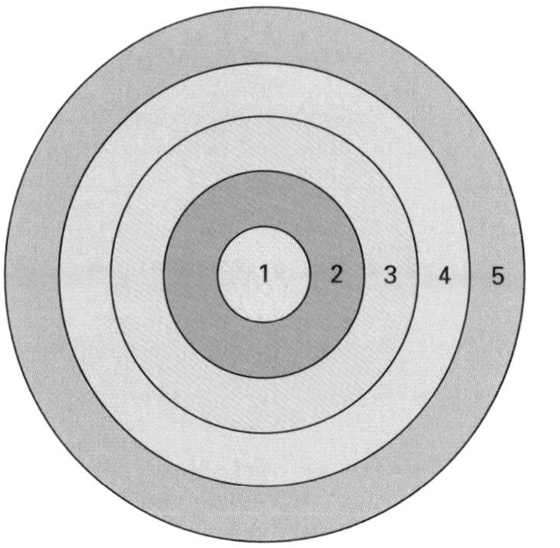

1 2 3 4 5

1. Central business district
2. Zone of transition
3. Zone of independent workers' homes
4. Zone of better residences
5. Commuters' zone

Figure 9–14. In a concentric zone urban model, cities form as a series of rings around the central business district.

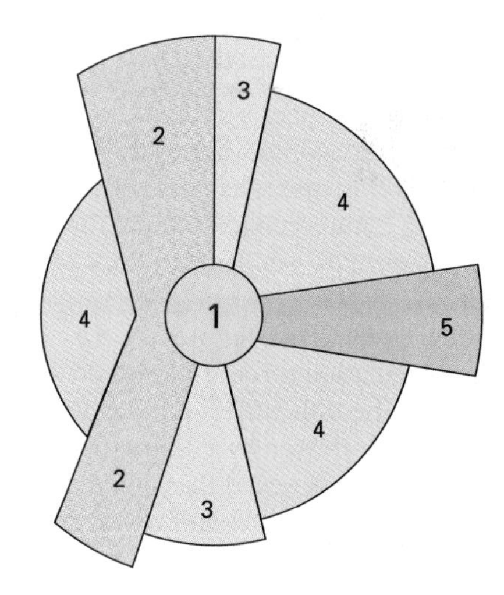

1. Central business district
2. Transportation and industry
3. Low-class residential
4. Middle-class residential
5. High-class residential

Figure 9–15. In a sector model of urban form, a city grows out from the central business district in wedges or corridors of various land uses.

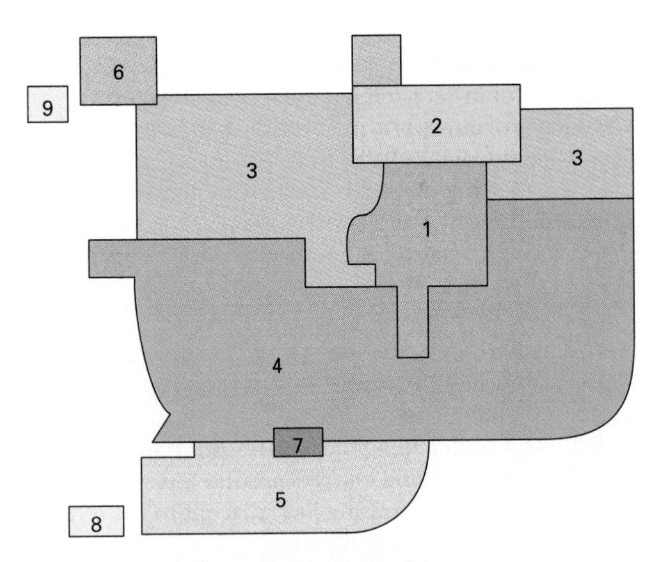

1. Central business district
2. Wholesale, light manufacturing
3. Low-class residential
4. Middle-class residential
5. High-class residential
6. Heavy manufacturing
7. Outlying business district
8. Residential suburb
9. Industrial suburb

Figure 9–16. In a multiple nuclei urban model, activities cluster around nodes of various activities.

ies. There CBDs thrive partly as a result of continuing reliance on public transit and partly because high-income populations choose to live close to the CBD. A commercial spine such as a boulevard extends out from the CBD, and amenities such as opera houses, chic stores, and elegant parks follow along this spine. Zones of more modest housing and value surround this elite zone, and the periphery is dominated by squatter slums.

Social Factors in Residential Clustering

Social considerations play a role in residential segregation in cities. Some people want to live with people like themselves, and this causes clustering. Ethnic groups or immigrants of a common background, for example, may want certain services, such as grocery stores offering their traditional foods.

In other cases, however, people may live together because discrimination forces them to do so. In any specific case it may be difficult to determine the degree to which the people cluster by choice or because of discrimination. Chapter 8 noted that in the past Jews were legally segregated in ghettoes, but today the word *ghetto* means any residential concentration of any one kind of people. The factors causing residential segregation of any group—of Chinese Americans in "Chinatowns," Italian Americans in "Little Italys," Hispanic Americans in "barrios," or of African Americans discussed later in this chapter—must be evaluated carefully in each case. Sometimes the words placed in quotation marks above are considered insulting, but they usually are not meant to be.

Religion is another social consideration that frequently causes clustering. People sharing a religious faith may cluster around their house of worship, and in-migrants of that faith will seek that neighborhood. Lan-

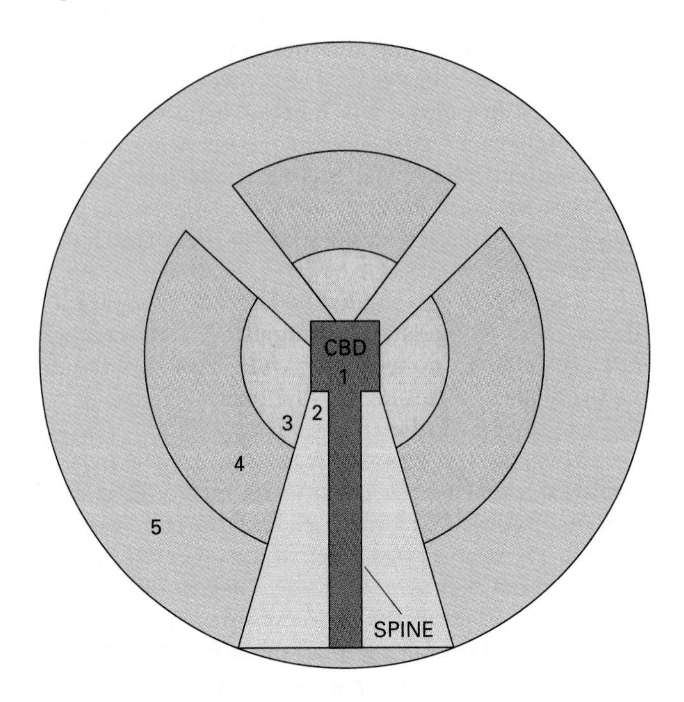

Commercial/industrial areas
CBD = Central Business District, the original colonial city
SPINE = High quality expansion of the CBD, catering to the wealthy

Elite residential sector

Zone of maturity
Gradually improved, upgraded, self-built housing

Zone of accretion
Transitional between zones 3 and 5, modest housing, improvements in progress

Zone of peripheral squatter settlements slum housing

Figure 9–17. In contrast with the three other models, which draw on North American experience, the Latin American model of urban growth shows an elite residential sector following a spine of high-value land use stretching out from the central business district. (Adapted with permission from Ernst Griffin and Larry Ford, "A Model of Latin American City Structure," *Geographical Review*, 70(1980), 406.)

guage communities frequently form, as do communities of the elderly and communities of gays and lesbians. Almost any factor of social bonding can encourage the creation of an identifiable residential neighborhood.

The Role of Government

Land use may be determined by government. Restricting the use to which various parcels of land may be put is called **zoning.** Governments can zone to distribute land uses according to plans. Industrial and commercial districts, for example, are usually kept away from residential neighborhoods. Land can even be set aside for different types of industries, and residential areas can restrict types of housing—single-family homes or apartment buildings.

Some urban planners, however, believe that the separation of land uses may be emphasized to an unnecessary and counterproductive degree. Many zoning laws, for instance, are remnants of the days when all industry was noisy, polluting, or smelly, and so the separation of industry from housing was desirable or even necessary for public health. Today, however, separating homes from jobs might lead to excessive commuting. Perhaps light industry or commercial activities and housing can be integrated. Zoning might integrate homes with workplaces and amenities so that they are easily accessible by public transit or even walking.

Across vast conurbations, the distributions of activities can be determined by the fact that each of the individual incorporated jurisdictions may exercise local taxing and zoning powers. This is typical of U.S. metropolitan areas, as we shall see.

CITIES AND SUBURBS IN THE UNITED STATES

The distinguishing evolution of metropolitan form in the United States has been the explosive growth of cities out across the countryside. Growing cities have spilled over their legal boundaries into areas called **suburbs.** Some suburbs are entirely residential, but others develop their own service centers for the surrounding residential population. Some suburbs may be older cities that were once distinct but have been engulfed by the growth of a larger neighbor, but others are new settlements that are incorporated in their own right. What defines an area as a suburb is its economic and social integration with a larger population nucleus nearby. The terms *town*, *village*, and *hamlet* are inexact, but they generally designate settlements smaller than cities, although they may be incorporated.

Many of the developments in the following discussion are now taking hold elsewhere around the world, but they occurred first in the United States—largely because of the nation's prosperity.

Early Suburbs

Most large U.S. cities included large manufacturing districts by the late nineteenth century. They were noisy and dirty, and they often attracted a working class, largely immigrant, whom many long-established residents found to be unpleasantly "different." These biases pushed those who could afford to leave the city to do so.

At the same time another factor pulled many people out of the city: a cultural preference for rural or small-town life to life in big cities. This preference for a return to nature, to the land, or to open spaces—even if only a suburban yard—is a manifestation of the broad cultural movement known as Romanticism. Many Americans fell in love with the idea of "the country." Therefore, when the railroads brought older rural communities on the outskirts of the city within commuting distance, many people who had the time and the money to live outside the city and to commute to work did so. In other cases, the well-off built entirely new towns (see Figure 9–18). "Streetcar suburbs" sprang up when that new means of transportation appeared. Some of these planned suburbs were beautiful, but often they became completely built up, merged into other settlements, and lost their identities altogether.

The automobile, however, ultimately opened the nation's landscape to suburban growth. For those who disliked urban life, the suburb was the solution. "We shall solve the city problem," wrote Henry Ford in 1922, "by leaving the city."

Government Policies and Suburban Growth

The migration from the cities to the suburbs was slowed by the Great Depression in the 1930s and by World War II in the early 1940s. Following the war, however, government policies established a new balance of push and pull forces that encouraged the movement of investment, residents, and jobs out of the central cities into the suburbs.

Government Loans After the war the Federal Housing Administration (FHA), established during the 1930s, was given the assignment of solving a housing shortage and also of forestalling a return of the Depression. Housing construction often serves as an economic pump in the United States because U.S. construction methods are labor-intensive and because new houses create demand for furnishings and other goods. Before this time, mortgages had been limited to one-half or one-third of the value of a house. With FHA-guaranteed loans, however, down payments shrank to less than 10 percent of the price. Interest rates fell, and suddenly thousands of families could afford new houses.

The FHA loans were not available for just any house. The agency favored new single-family housing over the

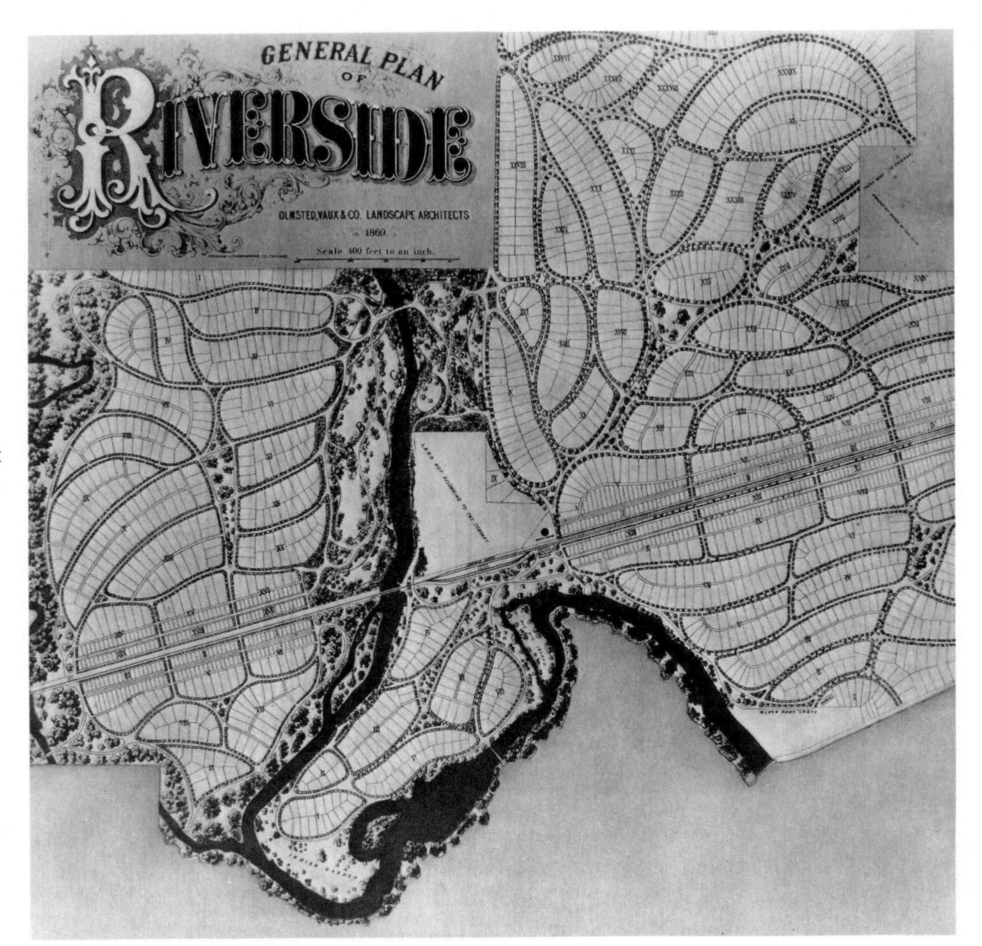

Figure 9–18. Riverside, Illinois, laid out 9 miles (14 km) from the center of Chicago in 1869, was the first U.S. application of landscape architectural design to a real estate subdivision. Designers Frederick Law Olmsted and Clarence Vaux allowed for two straight business streets paralleling the railway into the city, but all residential streets were curved to slow traffic. Open spaces contribute to the sense of breadth and calm enjoyed by "the more fortunate classes" for whom Riverside was designed. (Courtesy of National Park Service, Frederick Law Olmsted National Historic Site.)

rehabilitation of older houses or apartment buildings. Also, whole classes of houses, such as Baltimore's traditional 16-foot-wide (5-m) houses, were ineligible because they failed to meet minimum lot sizes. These policies encouraged the abandonment of the housing stock of the central cities and the relocation of housing investment to the suburbs.

These benefits did not apply equally to every neighborhood or for everyone. The FHA was against what it termed "inharmonious racial or nationality groups." In some places the presence of one nonwhite family on a block was enough to label the entire block "Negro" and to cut it off from FHA loans. Government policy helped segregate the suburbs.

In addition, property owners in many suburbs set *restrictive covenants* on the land. These were legal agreements that the land would never be sold to people of a designated race or religious group. Such covenants are no longer legal in the United States, but they were common as late as the 1970s.

Tax Policies Tax and financial incentives also directed U.S. capital into new housing. Tax benefits to homeowners included the deductibility of both mortgage interest payments and of local property taxes from gross taxable income. These two benefits alone often made buying a new house cheaper than renting. By 1990 the annual tax loss to the U.S. Treasury of these two benefits totaled over $70 billion. This is, in effect, a subsidy to homeowners that is 10 times greater than government expenditure on subsidies for public housing.

Other tax benefits protected homeowners' rising equity from capital gains taxation and even allowed older people to sell their houses without paying tax on a portion of any capital gains enjoyed. Also, savings and loan institutions were allowed to pay higher interest on savers' money than commercial banks were, on the condition that the money in savings and loans was directed into housing. (This last condition was true until 1980.)

In 1944 Congress created the Veterans Administration Housing Program, called "Homes for Heroes," and soon billions of dollars of mortgage insurance were being pumped into the FHA and the Veterans Administration's plans. By 1947 the Levitt Company was completing 30 single-family homes in a new Levittown in a Long Island potato field each day, and similar developments were springing up on the outskirts of every major U.S. city (see Figure 9–19). Nationwide housing starts jumped from 114,000 in 1944 to 1,696,000 by 1950.

The expansion of the suburbs through these programs brought fulfillment of the dream of home ownership to an increasing share of U.S. families. The percentage of U.S. housing that was owner-occupied leaped from 44 percent in 1940 to 62 percent in 1960.

Figure 9–19. In building their first Levittown, 25 miles (40 km) east of Manhattan, the Levitt family changed U.S. homebuilding techniques. The land was bulldozed and the trees removed, and then trucks dropped building materials at precise 60-foot (19-m) intervals. Construction was divided into 27 distinct steps. At the peak of production more than 30 houses were completed each day. Through the years owners have personalized them so much that few visitors today can see that the houses were originally identical. (Courtesy of UPI/ Bettman.)

Homeownership signaled middle-class status for a rising share of the population. Expanding homeownership increased the number of citizens who profited from the many homeowner subsidies. Therefore, it has reduced the political possibility of rescinding any of these subsidies, which are still in place.

New suburbs incorporated as the population spread across the landscape, and the Census Bureau devised a term to measure population growth in these sprawling conurbations: **metropolitan statistical area (MSA;** originally standard metropolitan statistical area, *SMSA*). The bureau defined a metropolitan area as "an integrated economic and social unit with a recognized large population nucleus." Thus, MSAs are the principal central cities and their suburban counties (except in New England, where the definitions are in terms of cities and towns). Two or more contiguous MSAs form a **consolidated metropolitan statistical area (CMSA).** MSAs and CMSAs are regions defined for statistical measurement. They are not governmental jurisdictions.

By 1990 the nation's 284 MSAs contained 77.5 percent of the total population, 31.3 percent in the central cities and 46.2 percent in their suburbs (see Figure 9–20). These MSAs covered only 17 percent of the country's land surface.

The Suburban Infrastructure

Unplanned sprawl of single-family homes is a uniquely expensive way of housing a population. It requires, first, roads. Individualized transportation also demands energy, and so does heating and cooling individual homes, no matter how well insulated they are. Dispersed housing also requires enormous investment in sewerage, water pipelines, telephone lines, and electric wiring. The cost of providing electric wiring for 100 people in an apartment building is much less than the cost of wiring for the same 100 people spread out in 20 single-family homes over 4 or 5 acres (1.5–2 ha).

Infrastructure costs are further inflated by the fact that suburbs do not expand contiguously outward from the city, like the waves from a stone tossed into a pond. Each developer wants to buy land as cheaply as possible, and so the developer does not buy land at the edge of growth, but beyond the edge of growth. This is called *leapfrogging*. The infrastructure network cannot be advanced regularly. This increases costs.

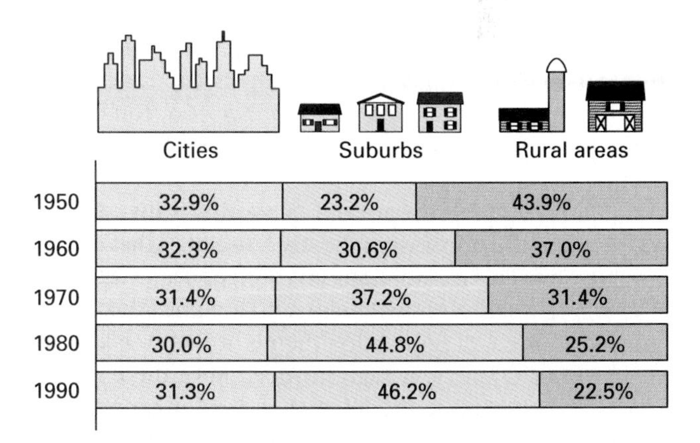

Figure 9–20. By 1970, a higher percentage of the U.S. population (37 percent) lived in the suburban portions of the MSAs than lived in the central cities or in rural areas. This trend has accelerated in the ensuing decades, as indicated by the statistics for 1990.

Filling in the leapfrogged areas always came later, and then some infrastructure had to be rebuilt. Many suburbs originally relied on individual home wells for water, for instance, and their sewage was treated in individual home septic tanks. As the inner suburbs matured and density increased, water supplies became polluted. Homeowners had to pay for wholly new public water mains and sewerage.

Suburbs are spacious, and U.S. urban areas spread at the rate of about 1 million acres (405,000 ha) per year. This required that a good share of the country's most productive and valuable farmland be paved over, such as the Long Island potato fields covered by Levittown. The reason for the original location of many cities had been the agricultural productivity of their immediate hinterlands, but that agricultural productivity disappeared under concrete and crabgrass. A rising percentage of the nation's food now has to be grown in less-than-optimal conditions. This requires expensive fertilizer or irrigation. In addition, the food has to be transported farther, consuming still more fuel, perhaps requiring refrigeration or special handling, and further boosting food prices. By 1990 a food item consumed in the United States had traveled an average of 1200 miles (1920 km).

Suburban real estate developers offered farmers tempting prices for their land, but even if the farmers did not want to sell, taxes forced them to. Property taxes are calculated on land's potential value, not its current-use value. A farmer who was making a small profit by farming could not afford to pay property taxes calculated on the land's potential value as housing. Even if a farmer could somehow meet the annual property tax bill, then the farmer's heirs would eventually have to sell the farm to pay inheritance taxes. These too are based on land's value for potential use, not its actual use. Only recently have these tax laws been changed to preserve greenbelts around some cities.

In the 1950s and 1960s, U.S. citizens did not worry about costs. Between 1950 and 1973 median family income doubled in real terms. Demands for housing mushroomed with the baby boom and the splintering of families into separate households, partly as the result of a rising divorce rate. The average number of occupants of a U.S. household shrank from 3.67 in 1940 to 2.63 in 1990.

Today a greater share of U.S. wealth is invested in housing and the necessary infrastructure than that of any other nation. This tremendous investment has succeeded in bringing a sense of well-being and a higher quality of housing to the U.S. population than is arguably available in any other country. A high proportion of the nation's housing stock is substantially new, and a high proportion of Americans enjoy private ownership of spacious, free-standing, well-equipped homes.

Social Costs The suburban lifestyle has imposed social costs on those who enjoy it. Americans generally sort themselves out residentially so distinctly that sociologists and mass marketers can confidently predict an astonishing amount of information about people on the basis of their address alone. The selection of junk in mailboxes in each zip code is fine-tuned. In low-density suburbs, it has been argued, local racial and social homogeneity may have caused conservatism and conformity.

Since the 1950s the number of women employed outside the home has been rising. More children are being left home alone in both cities and suburbs, but this presents acute problems in sprawling suburbs, because suburban children can seldom rely on public transportation to get home or to activities. Today property developers and suburban employers must invest in day-care (or "educare," educational day care) centers.

The Movement of Jobs to the Suburbs

The suburbs first expanded as bedroom communities for the middle-class workers of the central city. Thousands of workers left their suburban families each morning to go to work in the city. This flow of commuting diminished not with the suburbanites' return to the city but with the dispersal of more urban activities across the suburbs.

The completion of the Interstate Highway System and related metropolitan expressways reduced the geographical advantage of the central-city CBD. Many locations on expressways enjoy accessibility, and the best locations are at the intersections of two expressways. Retailers soon took advantage of these new crossroads, and by the mid-1960s giant shopping malls sprouted to tap the suburban market. Soon developers put up office buildings near the malls, and corporations joined them in building their own spacious office parks.

Manufacturing establishments abandoned the central city too, for several reasons. New light-industrial facilities proliferated in the suburbs because new technology favors horizontal rather than the vertical building styles characteristic of the central cities. In addition, light industries relocated to escape central-city congestion, and because energy costs are higher in the cities, as are taxes, wages, and rent. Warehousing also relocated out from the inner-city railroad yards to the suburban highway interchanges. Airports grew to provide tens of thousands of jobs in both freight and passenger services; few U.S. airports are within the legal boundaries of central cities.

In the 1950s suburban growth was fed by young married couples who wanted to raise children away from the cities. By the early 1970s, however, the proliferation of jobs in the suburbs became the driving force for new housing. The suburbs surpassed the central cities in employment, and job opportunity became a pull factor for continuing suburbanization (see Figure 9–21).

The demand for suburban-style housing, corporate space, and pleasant surroundings continues to throw ex-

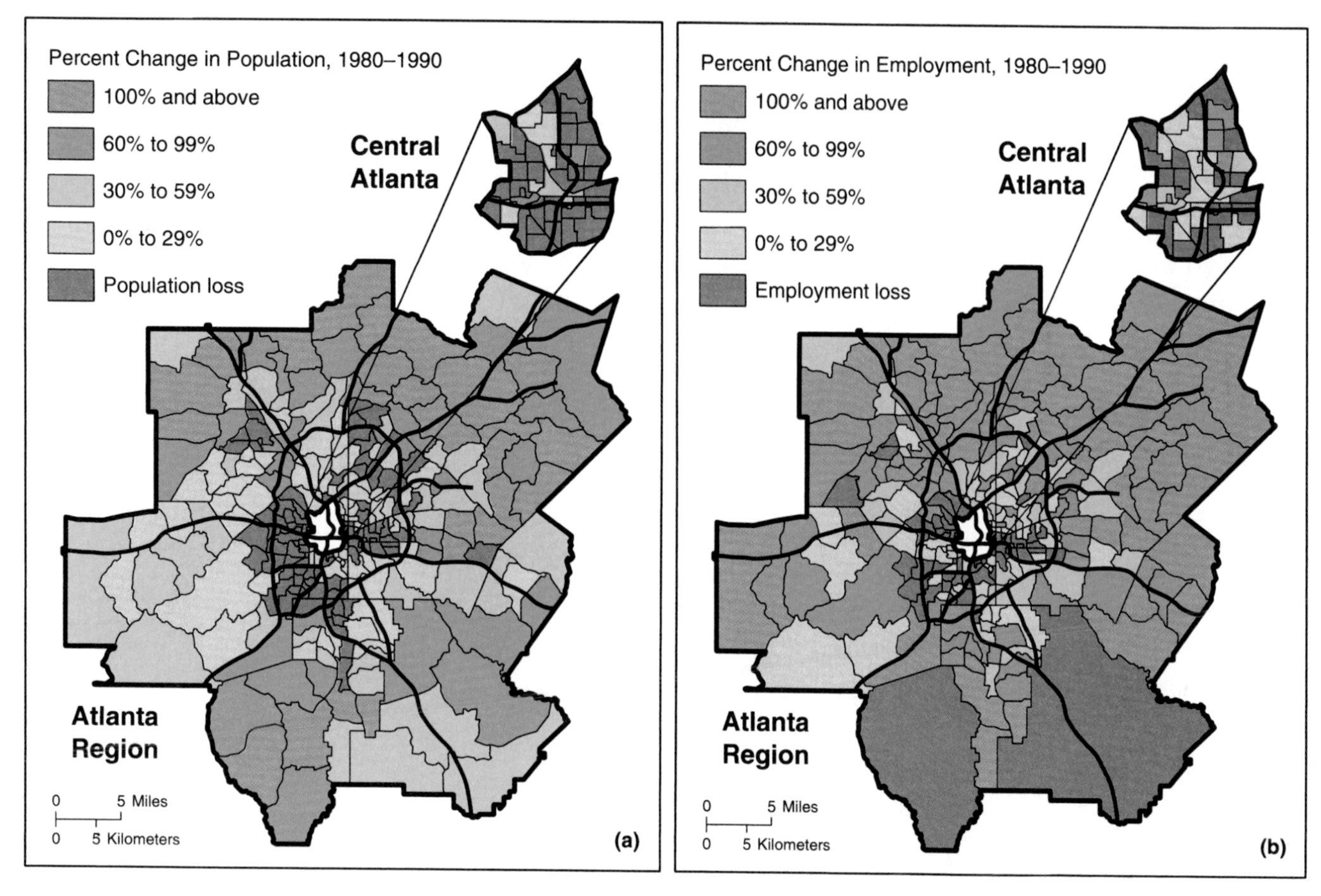

Figure 9–21. In Atlanta, as in most U.S. metropolitan areas, the greatest population growth (a) and job growth, (b) are occurring in the regions located the farthest from the central city. (Adapted with permission from *Atlanta Region Outlook May 1991*, Atlanta Regional Commission.)

urbs farther and farther out from the central city. People willingly spend hours commuting each day to enjoy more spacious and affordable housing. The exurbs' skilled labor pools and cheap land draw the next wave of development: light-industrial and back-office jobs such as assembly operations and regional processing centers for insurance companies and banks.

Satellite Cities The suburbs have generated their own foci of activity, often called **satellite cities** or **edge cities.** Some satellite cities are older towns that have been engulfed by the growth of their larger neighbors, but others are entirely new. Many today boast greater retail sales and contain more office space than the old central cities. They even offer amenities formerly found exclusively in the central cities: art galleries, theaters, sports teams, and fine restaurants. Some of these suburban centers are well designed, and the newest balance local housing and employment. Residents can walk to work or ride local public transit. They form self-contained villages (see Figure 9–22). New satellite developments may even be planned to allow mass-transit service into the central city (see Figure 9–23). Thus, the advantages of the original planned suburbs such as Riverside (see Figure 9–18) could be reproduced.

The question of private property One significant difference, however, between traditional downtowns on the one hand and most new villages and suburban malls on the other is that the latter are entirely private property. This difference has raised important questions in property law.

People cannot be banned from a traditional downtown, and petitioning on a public sidewalk is constitutionally protected. People may, however, be banned from private property, and constitutional rights, such as political pamphleteering, may be restricted. In many communities malls are the only public gathering places. If the mall owners are to be allowed to decide who may speak in them, then mall owners can determine the public's access to ideas. Candidates for political office have been banned from busy malls owned by their opponents, and so have labor union organizers. The regulation of access to private property has become an issue of free speech and association.

U.S. law recognizes that private properties can perform functions traditionally associated with government. This is called the *public function doctrine*, but the U.S. Supreme Court has ruled that the Constitution does not protect citizens' access to shopping centers against the wishes of the owners (*Pruneyard* v. *Robins*, 1980).

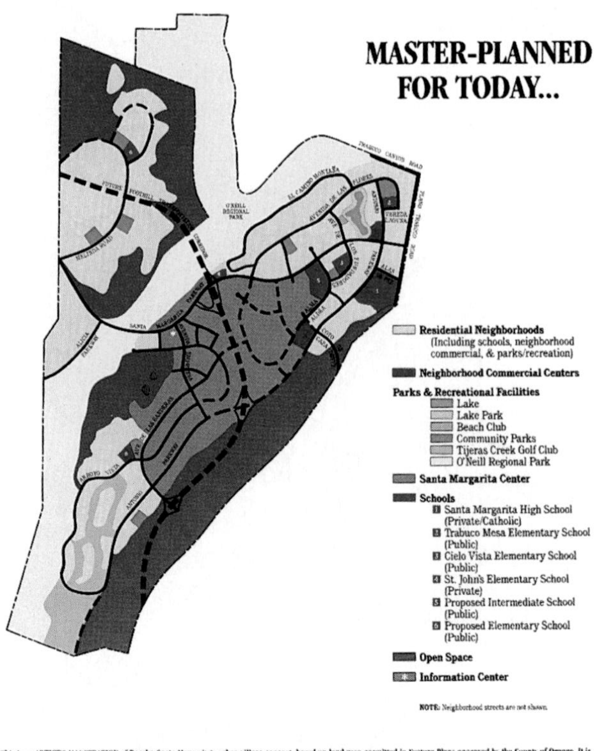

Figure 9–22. Rancho Santa Margarita, a new community 50 miles (80 km) southeast of Los Angeles, has won considerable national attention. It will not be fully developed until at least the year 2000, but by 1991 it already had some 5000 homes. About 75 percent of the people who work in the town either live in Rancho Santa Margarita or commute from within 10 miles (16 km). Many of these new suburban communities around the country are getting some people out of their cars and back on their feet and reintegrating home, work, and recreation. (Courtesy of Santa Margarita Company.)

The Court left it to the individual states, however, to decide whether their individual state constitutions protected access. As of 1992 only five states had upheld public rights, and ten others had favored property owners. Each case is different, and each state has balanced public and private rights slightly differently. The question of protecting free speech in places where people gather remains.

Today increasing numbers of rich and even middle-class people live, work, and shop in private environments, send their children to private schools, enjoy private recreation facilities, and travel by private means. This has been called "the secession of the successful." These private environments are patrolled by private security officers, one of the country's fastest-growing occupations. (They constitute almost 2 percent of the nation's total labor force and considerably outnumber public police officers.) The privatization of space is a new form of economic and social segregation. This geographic trend carries profound influence throughout U.S. political and social life.

Changing Commuting Patterns When the suburbs were dormitories for the central-city workers, those workers commuted in and out of the city, and radial mass-transit systems focused on the CBD could serve the needs tolerably well. Today that pattern has changed. Most commuting is no longer from the suburbs into the central city, but from one suburb to another (see Figure 9–24). It depends almost entirely on individualized transportation, clogging metropolitan highways. Mass transit cannot serve populations that are spread out at low densities.

As more family members have gone to work, the number of cars on suburban highways has increased faster than the local populations. This is true even in the older, closer-in suburbs that are losing total population to exurbs. Nassau County, a suburb contiguous to New York City, reported a 2.6 percent population decline between 1980 and 1990 but a 15 percent rise in passenger-car registrations. Suffolk County just beyond Nassau experienced a 3 percent population increase but a 29 percent increase in passenger cars. Everywhere rush hour starts earlier and eases later each day, and even Saturdays now generate traffic jams. It is the day many suburban working families choose to shop.

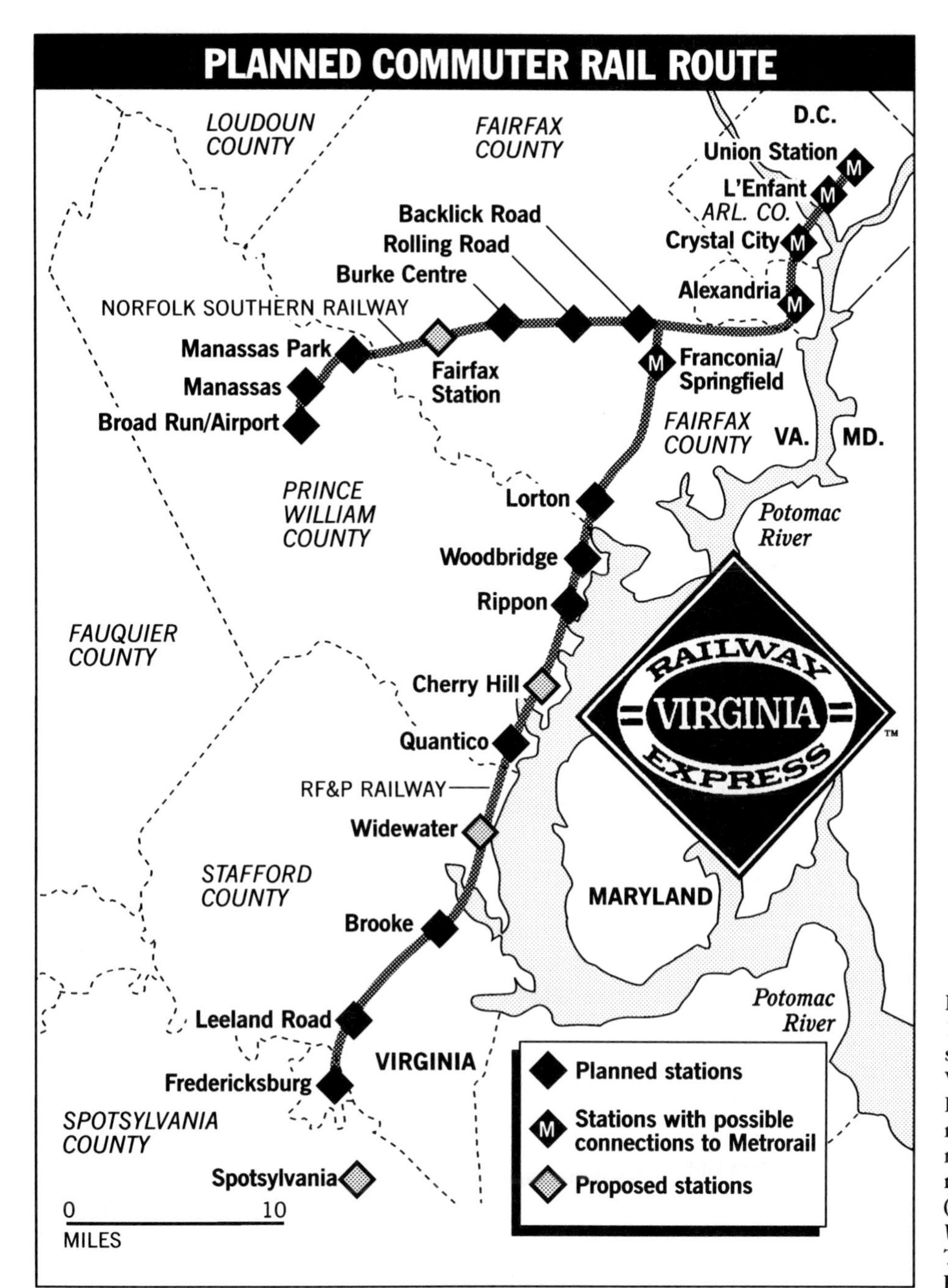

Figure 9–23. Through the 1990s the opening of a new suburban rail system in Washington, D.C., the Virginia Railway Express System, will redistribute development and might stimulate new planned railroad communities there. (Map by Dave Cook of *The Washington Post.* © 1991 The Washington Post. Reprinted with permission.)

The cars generate poisonous air pollution, especially when they are idling. In 1987, almost 70 percent of urban interstate highways were officially congested during peak hours, an increase from 40 percent in 1975.

Telecommuting It is easier to move information than to move people, and so more people are working at home at computer terminals. This is called **telecommuting.** Much clerical work or other work that does not require face-to-face contact might abandon cities within the next few decades. In 1992, between 3 million and 5 million employees of private companies in the United States were already working from home on computers, and the state governments of California and Washington had telecommuting programs. California has even established suburban satellite telecommuting centers.

Telecommuting eases the pressure on transport facilities, saves fuel, reduces air pollution, reduces the demand for office space, reduces absenteeism, and has been shown to increase workers' productivity. It also allows employers to accommodate employees who want more flexible work arrangements, thus opening employment opportunities to more people. We do not know what it will mean for the future geography of large cities if the number of telecommuters continues to rise.

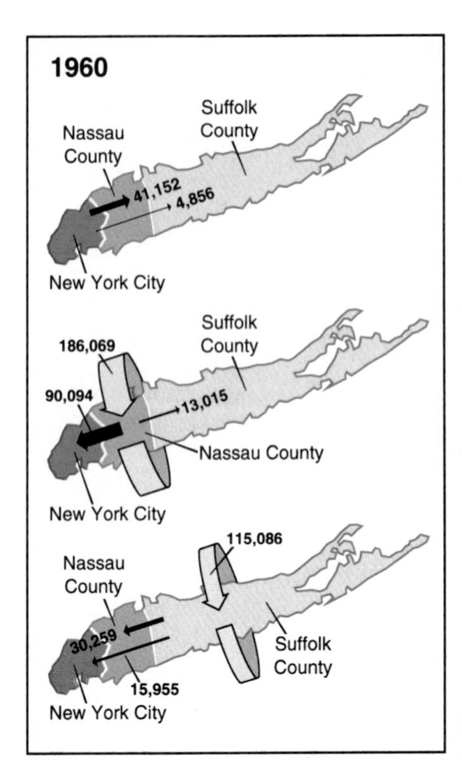

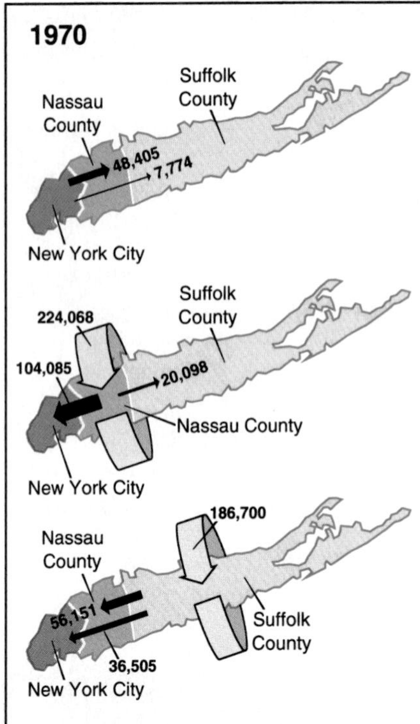

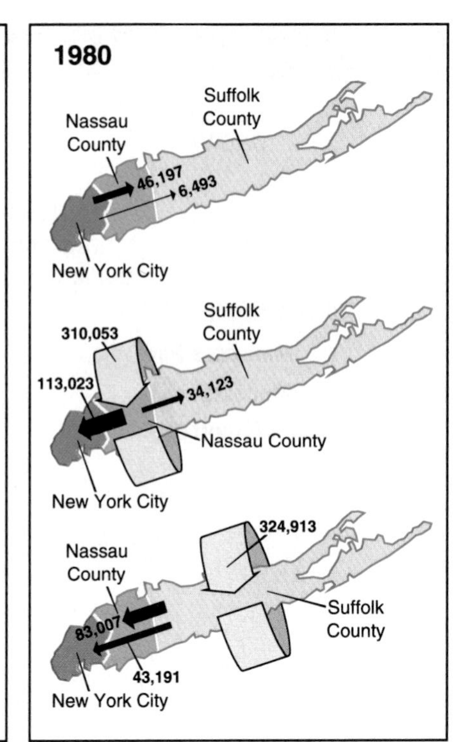

Figure 9–24. On New York's suburban Long Island, as in suburban areas around the country, commutation into the central city has yielded to commutation within and among suburbs. Individual automobile transportation necessarily replaces mass transit.

Developments in the Central City

The exodus of people and jobs from near the CBD to the suburbs drained many central cities of economic vitality. One reason for this was the failure of federal policies based on an idea called *filtering*. Policymakers thought that if new suburban houses were available to middle-class people from the cities, the urban poor could move into the older city dwellings. This would solve the housing problem for lower-income families.

That policy proved mistaken. The pent-up demand for housing that had developed during the 1930s and early 1940s was essentially satisfied by the end of the 1950s, but financial mechanisms in place such as FHA loans and tax incentives still made it profitable to give up a house in the city and to move to the suburbs. Much of the central-city population remaining behind was financially incapable of maintaining the inherited housing stock. Consequently, many central-city neighborhoods deteriorated. The term "inner-city neighborhood" became a euphemism for slum.

Economic Decline Central cities can thrive as long as (1) their economies offer a complete range of job opportunities from entry-level jobs for the unskilled up to the most specialized, and (2) family stability, education, and other social systems help people ascend the socioeconomic ladder. In other words, cities do not have to retain their middle classes, but their economies have to offer the lower classes opportunity to become middle class. Unfortunately, the departure of the middle classes to the suburbs occurred at the same time as two other developments.

First, urban economies were transformed by the out-migration of jobs. Traditionally, as has been seen, urban entry-level job opportunities could be found in the secondary and tertiary sectors. Workers could then climb the ladder of job skills and income as their careers advanced, or at least their children could enjoy the opportunity for better careers. Manufacturing and warehousing jobs, however, migrated to the suburbs, and in the central city even construction no longer offered entry-level opportunities for the unskilled. Workers on a construction site today are highly skilled. The tertiary sector could not offer sufficient entry-level opportunity, either. Positions of domestic servitude, the great employer of 100 years ago, evaporated, and not enough jobs opened in other unskilled service occupations. The loss of these jobs broke the rungs of the ladder of upward mobility for many people.

Second, new waves of unskilled workers poured into the central cities from rural regions. The stream of rural migrants to the cities included African Americans from the rural South, Puerto Ricans, and poor immigrants.

This influx continued long after the city economies could provide entry-level job opportunities and upward mobility for these new residents.

The loss of jobs and economic opportunity from the central city and the concomitant development of the surrounding suburbs have hollowed out many major U.S. metropolitan areas. They have been called "economic doughnuts." Many CBDs have lost their purpose. Commercial, professional, and financial offices have relocated to the suburbs, and middle-class retailing has followed. Many U.S. downtowns consist of only a government center, a convention center, and a few hotels; the streets are deserted after 6 P.M.

The Service Economies Some downtowns have enjoyed spectacular growth in financial, information, and specialized technical services. In these areas skyscrapers have replaced rusty factories (see Figure 9–25). New York's leading export, for instance, which for decades was garments, by the late 1980s was legal services. Between 1950 and 1987 the number of jobs in manufacturing fell from over 1 million, representing 30 percent of all jobs, to 383,000, or 11 percent of all jobs. Jobs in services jumped from 507,000 (14 percent) to 1.1 million (31 percent), and in finance, insurance, and real estate from 336,000 (10 percent) to 547,000 (15 percent). This shift in job opportunity is often called a switch

Figure 9–25. No U.S. city illustrates a transformed economy better than downtown Pittsburgh, which was transformed from a dirty and smoky industrial area (top) into a new park and gleaming service center, a "Golden Triangle" (bottom). This is where the Allegheny River (upper left) meets the Monongahela (at right) to form the Ohio. Three Rivers Stadium is off to the left. [(top) From The Pittsburgh Gazette-Times, *Story of Pittsburgh*, 1908. Courtesy of Professor Arthur G. Smith, University of Pittsburgh. (bottom) Courtesy of Professor Arthur G. Smith, from *Pittsburgh Then and Now*. Pittsburgh: University of Pittsburgh Press, 1990.]

from *blue-collar jobs*, performed in rough clothing and usually involving manual labor, to *white-collar jobs*, salaried or professional jobs in which work does not involve manual labor.

The people who held the best new jobs, notably the young, urban professionals ("yuppies"), revitalized older residential neighborhoods in a process known as **gentrification.** In some cities (including Seattle and San Francisco) the demand for new housing for these white-collar workers spurred the governments to limit construction of new office towers or to require builders of new commercial space to construct residences as well.

In some cases immigrants have contributed to the rejuvenation of central cities. Some immigrants bring capital, and for the most entrepreneurial of these newcomers, the city's traditional advantages of agglomeration and external economies revive the city's function as incubator of new businesses.

The shortcomings of gentrification These upscale urbanites, however, are only a small fraction of the total inner-city population. The new white-collar jobs being created do not equal the number of older blue-collar jobs being lost, and the workers who lose their blue-collar jobs often require retraining or education before they can capture one of the new openings. Despite the publicity that the new urbanites and their lifestyle receive, the percentage of jobs in most central cities held by commuters is rising—especially the percentage of the best jobs.

At the same time the **labor force participation rates** of central-city populations—that is, the percentage of the population that is currently employed or even looking for a job—are falling. The gap between average city and suburban incomes continues to widen. In 1960 median family income within New York City was 93 percent of that of suburban families. By 1985 that percentage had fallen to 55 percent. In the same period the incomes of Detroiters fell from 85 percent to 50 percent of suburban incomes; those of Baltimore declined from 90 percent to 50 percent. In many U.S. metropolitan areas, inner-city incomes are less than one-half of suburban incomes.

The successful people in the central city construct restricted areas that are as highly privatized as the suburban developments, but they cannot improve cities as a whole. The same may be said of the much-publicized remodeling of urban cores into recreation areas or tourist attractions as theme parks (see Figure 9–26). The gap between rich and poor, or successful and unsuccessful, in the central cities widens.

The Urbanization of African Americans In the years following World War II, a large percentage of the rural migrants into U.S. cities were African Americans. The movement of southern blacks to northern cities had actually begun during World War I, when job opportunities in war-related industries had pulled rural blacks north at the same time as a boll weevil infestation in southern cotton fields had thrown many black laborers out of work. These pulls and pushes accounted for most of the 1.5 million black migrants to the North between the First World War and the Depression.

After the 1940s the main push factor was the revolution in southern agriculture. The mechanical cotton picker, introduced in 1944, made many agricultural workers obsolete. As the blacks were pushed off the farms, they continued to be pulled into the cities, and particularly the cities of the North, by the promise of economic opportunity and enhanced civil liberties. Be-

Figure 9–26. Baltimore's Inner Harbor area offers new attractions and a convention center at the edge of the CBD. The *U.S.S.* Constellation tied at dockside is the oldest U.S. warship afloat (launched 1797). The tall building is a World Trade Center, and the modern building at the right is a new aquarium. The low buildings house restaurants and shops. (Courtesy of Baltimore Area Convention and Visitors Association.)

Focus on Changing Employment Patterns in Milwaukee, Wisconsin

Metropolitan Milwaukee exemplifies several trends affecting major U.S. cities. An evolving economy can create winners and losers along geographical and racial lines.

From the 1940s to the 1970s Milwaukee won fame for its factories, foundries, and breweries. One in three jobs was in manufacturing. A person with a high school diploma could get a job, buy a house, and live well. Between 1970 and 1990, however, the number of manufacturing jobs shrank from 220,000 to 173,400, whereas the number of nonmanufacturing jobs grew from 461,200 to 591,000. The metropolis's black population, concentrated in the central city, could not capture those new jobs. Black unemployment rose from 17 percent to 20.1 percent, whereas white unemployment shrank from 5.3 percent to 3.8 percent. The number of black individuals receiving welfare rose from 64,317 to 71,113, whereas the number of whites fell from 27,595 to 19,493. By 1991 more than one-half of all Milwaukeeans were on some form of public assistance.

In the late 1980s Milwaukee was able to recast its manufacturing sector and slightly increase the number of manufacturing jobs. This was largely because the falling dollar increased the amount of manufactured goods that the city's factories sent abroad. Most of the black manufacturing workers, however, have the least seniority, and they have not been able to capture even the new manufacturing jobs. Total manufacturing jobs are down to one-quarter of Milwaukee's jobs. Black men stand idle on street corners just blocks from the breweries and factories that used to employ them while white tertiary-sector employees work in those buildings, now converted to offices.

tween 1945 and 1970, 5 million more blacks followed the 1.5 million who had migrated north since 1910. This migration of southern rural blacks to the cities of the North was one of the greatest migrations in human history. It was more than 10 times the number brought to the United States forcibly through more than 2 centuries of slavery.

These newcomers arrived in the central cities just as the central cities were losing their ability to provide entry-level job opportunities, and they often faced discrimination in employment and segregation in housing. Deteriorating conditions of life in new African-American ghettoes eventually triggered civil unrest. In August 1965 a riot raged for a week in the Watts section of Los Angeles. It left 34 people dead and more than 1000 injured, and it required military occupation of 46 square miles (119 km^2) to halt the violence. The Watts riot was followed by 150 major riots and hundreds of minor ones that summer and the next three summers. In Detroit in July 1967 43 people were killed, 14 square miles (36 km^2) of the city were gutted by fire, and 15,000 troops were brought in to restore order.

Since the 1960s the nation has made great strides in civil rights, and a great many of the black urban in-migrants have achieved success. Others, however, have been left behind. If the first urban generation failed to find employment, skills, and upward mobility, the second and third generations may have, also. Many families have been trapped in poverty.

Numerous aspects of deprivation have concentrated in the central cities: poverty, substandard housing, inadequate education and nutrition, and violent crime, with the deterioration of the nuclear family structure and the formation of gangs. Drugs have made everything worse. Murder rates have soared, and life expectancy among black males has tumbled. The deterioration of conditions in the central cities has itself become a push factor driving businesses and middle-class residents out. Los Angeles erupted again in April 1992 when four white police officers were acquitted of beating a black motorist. Over 13,000 police and troops restored order after three days of riots, fires, and looting left over 50 people dead and 4000 buildings burned.

Only in pockets of the central cities are conditions at their worst, and even the worst slums of New York City, Chicago, Detroit, and Los Angeles scarcely compare with the conditions of life for many people in Mexico City, Lagos, or Calcutta. The inner-city second or third generation living in deprivation, sometimes called the *underclass*, numbers under 3 million people, less than 1 percent of the national population. Still, the conditions of deprivation and the lack of opportunity contrast starkly with the national self-image.

Racial segregation in the inner cities The 1990 census revealed some progress in reducing racial segregation in U.S. cities. ***Racial isolation*** is defined as the percentage of African Americans living in census blocks of 300 to 400 people that were at least 90 percent African American. In the 50 largest metropolitan areas, that percentage dropped from 43.7 in 1980 to 37 percent in 1990. Seven of the 50 areas—Buffalo, Cincinnati, Cleveland, Detroit, New York, Philadelphia, and Seattle—experienced increased segregation, but many metropolises seem integrated, and several others seem to have made substantial steps toward integration.

In some of these cases, however, the reduction in black geographic isolation may be due to in-migration of Hispanic and Asian minorities rather than to black integration with whites. It may also reflect neighborhoods in transition to minority dominance. In May 1991 Hispanics rioted in Washington, D.C., a city with an African-American mayor and council president. This may foreshadow a new round of civil rights struggles in which Hispanic-black antagonism replaces black-white confrontation.

Efforts to Redistribute Jobs and Housing

Several programs have been proposed to provide job opportunities for America's inner-city unskilled. One is to rebuild the blue-collar economies of the central cities by encouraging industry to locate there in **urban enterprise zones,** where manufacturers receive government subsidies. The federal government has been slow to act, but several states and cities have designated such zones in their poorest communities (see Figure 9–27). Legislation must be carefully drafted to encourage labor-intensive job opportunity. Some of these zones have enjoyed success, but the problems that drive industries from the central cities are difficult to overcome.

One irony and tragedy of urban sprawl is simultaneous labor shortages in the suburbs and unemployment in the central cities. There is little low-cost housing in the suburban job growth areas, and so it has been suggested that the central-city poor be moved out into new subsidized suburban housing. The federal government actually built suburbs for low-income people in the 1930s: Greendale outside Milwaukee, Greenhills outside Cincinnati, and Greenbelt outside Washington. Today, however, the political impetus for such experiments is gone.

Existing suburbs often zone to exclude subsidized housing or even private apartments with young families. These types of housing impose net burdens on community finances. Communities zone exclusively for expensive single-family homes because only these pay property taxes sufficient to cover the costs of the services that they and their residents require. Owners of undeveloped property, however, usually want to maximize development on their land. They battle such *exclusionary zoning.*

A clause in the Fifth Amendment to the U.S. Constitution called the *takings clause* protects private property against seizure by the government without just compensation. Property owners argue that zoning can so

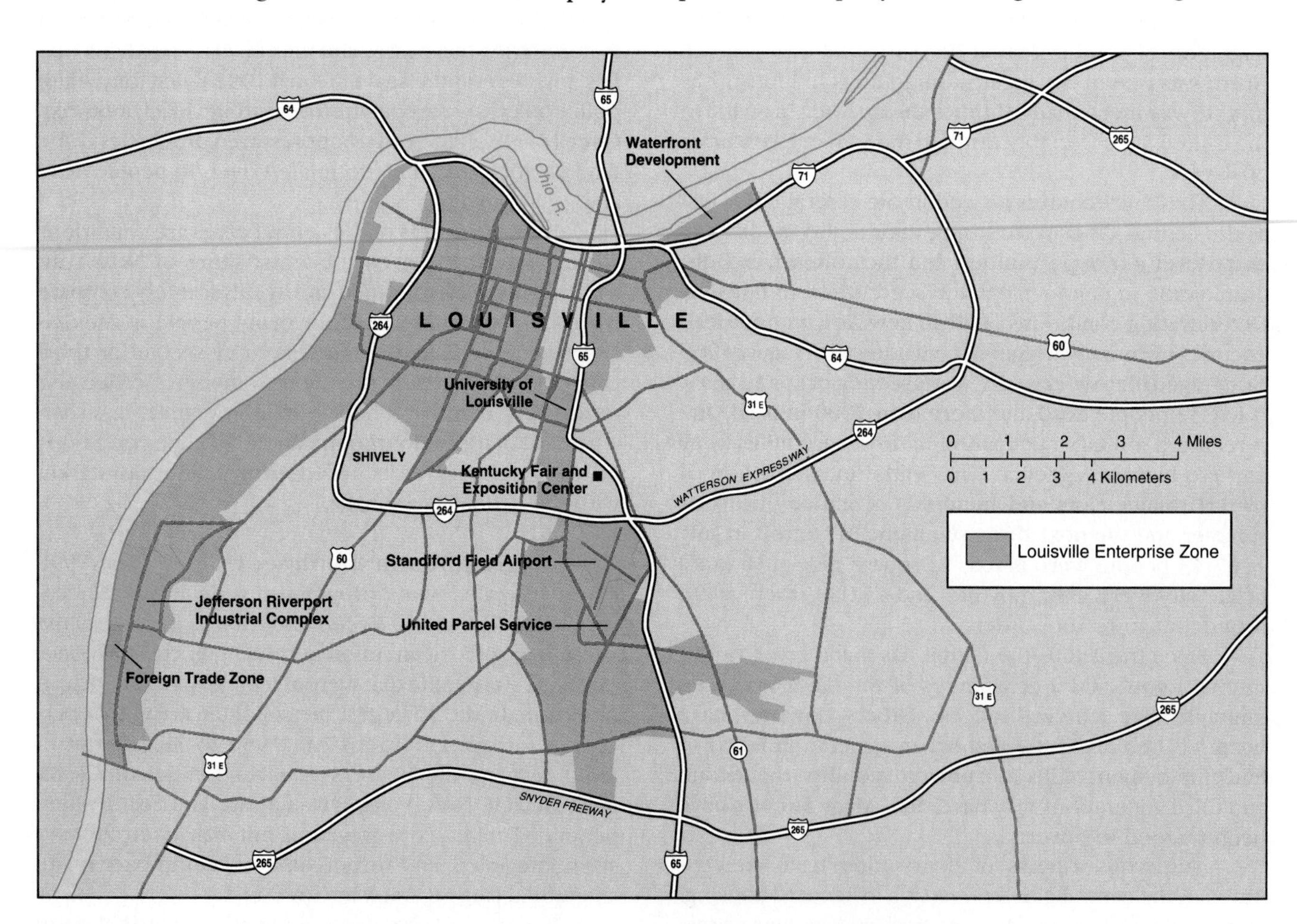

Figure 9–27. Louisville's urban enterprise zone claims to have captured almost $2 billion in new investment and to have created almost 15,000 new jobs. (Courtesy of the Louisville-Jefferson County, KY, Office for Economic Development.)

severely restrict the uses to which land can be put that the zoning virtually confiscates the land. This is a fine point in the interpretation of law, but in recent years several state supreme courts have banned exclusionary zoning. Several have actually required local communities to zone for all sorts of housing: high-density apartments and low-density single-family homes, as well as housing affordable to people of all income levels.

Alternatively, inner-city poor people might be provided with mobility to suburban job opportunities. Several city governments, private employment agencies, and even suburban employers have instituted bus services for this purpose.

Another strategy is to educate and train the central-city population to capture the tertiary-sector jobs that are opening in the central city. The quality of education in the central cities, however, is generally discouraging. Rich suburban schools are equipped with modern equipment, whereas inner-city schools often lack basic texts.

Governing Metropolitan Regions

The branch of geography that deals with the boundaries and subdivisions of political units is called **political geography.** Governing conurbations presents special problems of particular interest to political geographers.

The legal boundaries of most U.S. cities were originally drawn to include some surrounding land for future growth, and when the city outgrew its boundaries, it annexed suburban areas. By the 1920s, however, the suburban populations had begun to incorporate themselves to avoid annexation. This was thought to be in the U.S. tradition of local self-government, and suburbanites felt that by incorporating their own communities they were escaping the problems (and people) of the old city (see Figure 9–28).

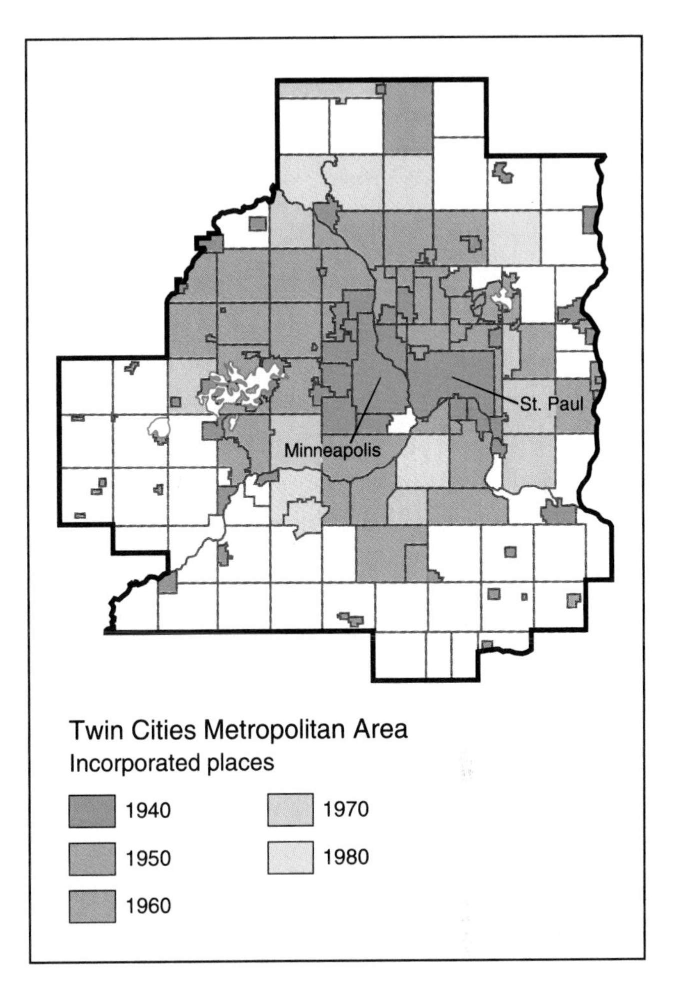

Figure 9–28. The pattern of suburban incorporation in metropolitan Minneapolis/St. Paul typifies U.S. metropolitan expansion. The Twin Cities were surrounded by tiers of independently incorporated municipalities.

The Proliferation of Governments Today autonomous municipal units form a legal retaining wall around almost every large city in the United States. Metropolitan areas cover a great number of municipalities and also a myriad of *special-district governments*, which are incorporated to deal with specific problems. For example, Nassau and Suffolk counties in suburban New York include 2 cities, 13 towns, 95 villages, 127 school districts, and more than 500 special districts, most of which exercise taxing powers for services from garbage collection to hydrant rental (see Figure 9–29). The five-county Los Angeles metropolitan region contains 160 separate governments. Los Angeles County alone has 82, and even entirely within the boundaries of the city of Los Angeles there are 7 independent city governments.

The boundaries among the many jurisdictions are not obvious to everyone, but each government has its own agenda and marks the landscape, and they jostle for authority and tax dollars. Property taxes soar, but many metropolitan area residents have no idea which government is responsible for which service. The system can discourage local participation.

The proliferation of special-district governments is not democratic. Their governing bodies are seldom elected in elections in which each citizen exercises an equal vote. Instead, they are either appointed or else chosen in weighted elections. As a result, the percentage of public funds spent by officials directly responsible to the voters shrinks. Furthermore, the jurisdictional boundaries of special districts may not conform to those of general-purpose governments but overlap them. Such overlapping or *noncoterminous* boundaries multiply the difficulties in coordinating the provision of services.

Decisions regarding metropolitan land use and the location of industries, transport facilities, and new housing cannot be made in the best interests of the entire metropolitan population. They are made on the basis of competition among the local governments. Each wants to enhance its own property tax base by attracting commercial developments that pay high property taxes but demand little in the way of local services. For example,

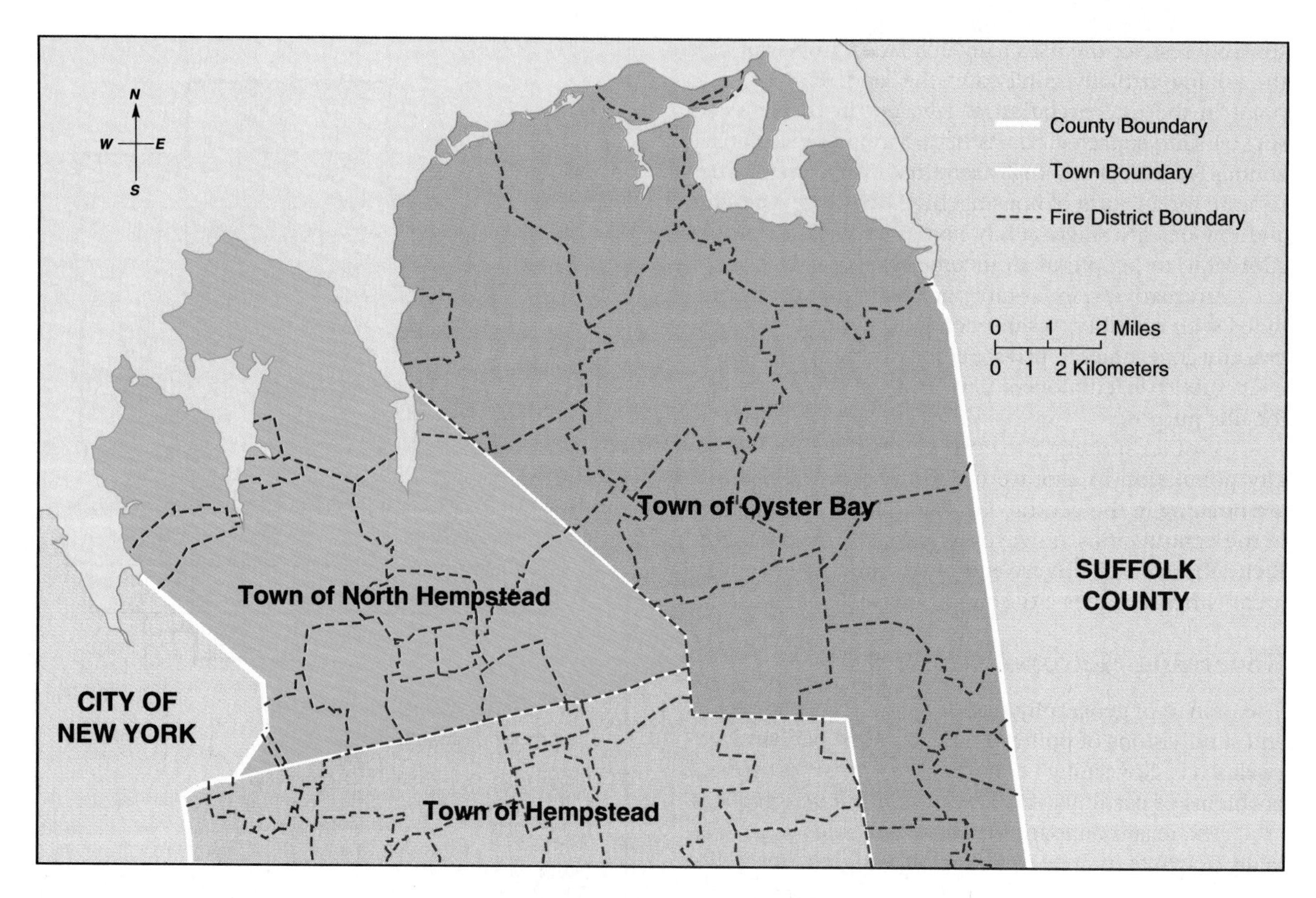

Figure 9–29. Fire districts are among the special-purpose governments in suburban Nassau County, New York. The confusion and inefficiency of this system results in higher property taxes for property owners as well as higher fire insurance premiums.

Santa Clara County, California, has zoned for 250,000 new jobs but only 70,000 new homes. Each community hopes to let surrounding towns cope with the additional costs of schooling, pollution, and congestion.

Many areas have created *councils of governments* (*COGs*), committees of officials representing each of the local governments in the region. The powers of COGs, however, are limited. Each local government holds veto power over any proposed areawide action.

Regional Land-Use Planning and Governing Regional governments can coordinate local governments to address areawide problems. In addition, state governments increasingly are assuming responsibility for governing their metropolitan areas. Several states have drawn up statewide land-use plans. California is one of several that requires local governments to draft and periodically to update plans. California has considered creating a tier of 7 regional governments above the state's 58 counties. Florida's 1985 Growth Management Act requires each of the state's counties and cities to file comprehensive development plans, including standards for water, roads, and other infrastructure. The state government reviews these plans and will not fund infrastructure until it approves a plan.

Many Americans object to planning as an infringement of liberty, and yet the lack of planning has allowed suburban sprawl, traffic jams, and air pollution. Today southern California is of necessity regulating aspects of life in ways that would have been unthinkable 20 years ago. Transportation, industrial processes, dry cleaning, and even backyard barbecuing are coming under government supervision. The regulations are made not even by any of the elected local governments but by two state agencies: the Metropolitan Water District and the South Coast Air Quality Management Board.

Race and Political Geography Sometimes the geography of race hinders governmental reorganization of a metropolitan region. In many metropolitan areas African Americans concentrated in the central city have won political control of the central-city government. A shift of some governmental responsibilities up to a regional government might improve overall governmental administration, but it would dilute the power of the central-city government, and, thus, of the African-American population. The U.S. Department of Justice has occasionally stepped in to prevent such reorganization.

In other cases, the Justice Department has prevented white-ruled central cities from annexing white-occupied suburban areas because the annexation would dilute the growing proportional voting strength of African Americans in the central city. Memphis, for example, has been prevented from annexing suburbs for that reason.

THE TRADITION OF URBAN AND REGIONAL PLANNING

Urban and regional planning are local-scale applications of many principles of urban geography. The concept of designing an "ideal city" has challenged the best minds for centuries. The founder of Western city planning was probably Hippodamus of Miletus. He laid out that city in today's Turkey according to a grid plan as early as 450 B.C.

The first modern attempt to formulate the needs of the city as a whole was the work of the British visionary Sir Ebenezer Howard. In *Garden Cities of Tomorrow* (1898), Howard outlined a plan to stop the unbounded growth of the industrial city and to restore it to human scale. He wanted to relocate excess population into new medium-sized "garden cities" in the outlying countryside. These regional cities would be ringed by greenbelts of farmlands and parks. All land would be communally owned, and each town and its surrounding region would be planned as an interlocking whole (see Figure 9–30).

Howard built two garden cities just north of London: Letchworth (1904) and Welwyn Garden City (1919). These inspired the Regional Planning Association

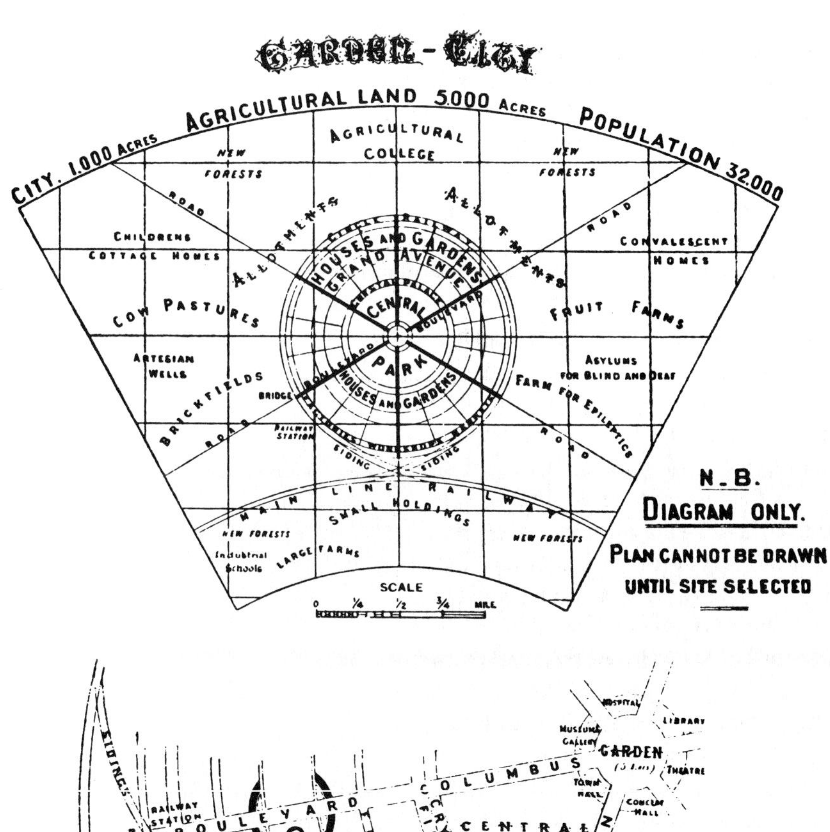

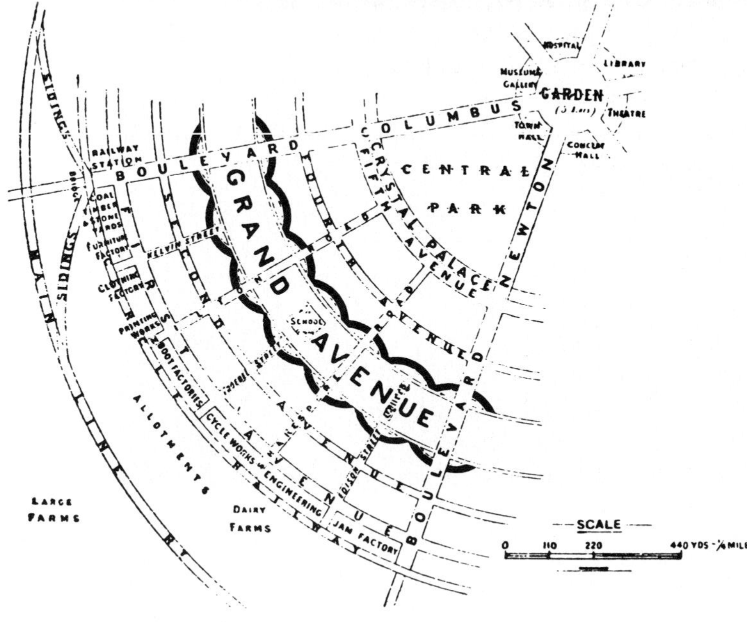

Figure 9–30. Sir Ebenezer Howard envisioned planned "garden cities" that would disperse the concentrations of population in nineteenth-century cities. (Reprinted with permission from Ebenezer Howard, *Garden Cities of Tomorrow*. London: Faber & Faber, 1946.)

of America, a private nonprofit organization, to construct two planned communities in the New York City area: Sunnyside Gardens, Queens (1924); and Radburn, New Jersey (1928). Neither is a complete garden city, but both are harmoniously designed communities that have greatly influenced urban planning in the United States and Europe. Some of the new exurban villages reproduce features of Howard's garden cities.

Probably the most important city planner of the twentieth century was the Swiss architect Charles Édouard Jeanneret-Gris (1887–1965), better known by his professional name, Le Corbusier. In a celebrated plan of 1922, he proposed to raze the crowded, run-down historical core of Paris, preserving only the central monuments. In its place he wanted to build a "Radiant City" of tall office buildings and apartments, spaced so far apart that each glass tower would be surrounded by green space and have a fine and wide view. The concentration of facilities within high-rise slabs would liberate the city from its environment. It could be placed anywhere. Le Corbusier brought together two concepts: the machine-made environment, standardized, technically perfect to the last degree; and, to offset this, the natural environment, treated as open space, providing sunlight, pure air, greenery, and views.

Le Corbusier planned Chandigarh, a new capital for the Punjab State in India, in 1950, but the world's supreme Radiant City is Brasília, the capital of Brazil, designed by Lucio Costa in 1957 (see Figure 9–31). Unfortunately, the gigantic scale of the city demands a completely motorized population. That is the problem with excessive openness. The city in a park can become a city in a parking lot. The Australian capital at Canberra, planned by Walter Burley Griffin of Chicago, has less openness, and it is generally considered superior (see Figure 9–32).

Le Corbusier's 1930 plan for the little town of Nemours in Algeria, with a geometric grouping of domino structures, set the international fashion for high-rise slabs for the next 50 years. These ideas were disseminated worldwide by the 1933 Athens Charter of the International Congress of Modern Architecture (*CIAM*). In the charter's codification, largely by Le Corbusier, the functions of the city—housing, work, recreation, and transport—provided the city's framework. The charter called for separation of high-rise development, industrial zones, parks and sports fields, and a hierarchical street system for traffic at different speeds. These ideas diffused to dominate urban planning around the world.

Many people believe today that the widespread adoption of Le Corbusier's ideas produced a half-century of monotony—not merely of detail and of style, but of insensitivity to place, to the essential difference between one place and another. The high-rise slabs ringing every big city in the world from Mexico City to Singapore look much alike (see Figure 9–33).

Figure 9–31. Brasília, Brazil's capital since 1960, arose on a largely unpopulated open plateau 603 miles (970 km) northwest of Rio de Janeiro. As this picture reveals, it sometimes lacks sufficient water to keep the grass green or to operate the elaborate fountains. (Courtesy of Professor Barbara Weightman.)

Figure 9–32. Canberra became home to Australia's Parliament in 1927, but foreign missions and federal government departments moved here only in the 1950s. (Courtesy of Australian Overseas Information Service.)

Figure 9–33. Does anything about this typical modern skyscraper city tell you where it is? Look carefully. Could it be in South America? Canada? Indonesia? This is, in fact, a district of Tokyo. (© 1987 Japan National Tourist Organization.)

Figure 9–34. The Pruitt-Igoe Housing Project in St. Louis, Missouri, was demolished when it was determined that the design of the buildings may have contributed to the social problems among the resident families. Other high-rise low-income public housing projects have since been demolished across the country, and low-rise housing is widely preferred for public projects today. (Courtesy of UPI/Bettman.)

In 1961 a group of younger architects broke away from CIAM and proclaimed that architecture was more than the art of building; it was the art of transforming people's entire habitat. The School of Architecture at the University of California at Berkeley was reconstituted and renamed the School of Environmental Design.

Most recently, many urban planners worldwide have come to criticize the concept of high-rise living. Low-rise dwellings can achieve the same density of habitation as Le Corbusier's "towers in a park" can, and many people feel more content living in low-rise dwellings. High-rise public housing projects, it turned out, can breed a sense of alienation and helplessness, and many have been razed across the United States (see Figure 9–34). Today we still work to design more humane urban environments.

SUMMARY

Every settled society builds cities, because some essential functions of society are most conveniently performed at a central location for a surrounding countryside. The region to which any city provides services and upon which it draws for its needs is called its hinterland. Today all countries are experiencing urbanization. The United Nations predicts that about one-half of the world's population will live in urban areas by the year 2000.

In many traditional societies, the cities' ritualistic and political importance far outweighed their economic functions. As societies develop, however, cities' economic role usually becomes paramount. Cities offer the convenience of agglomeration for the division of labor, and they promote and administer the regional specialization of production. With industrialization, cities become important centers of production as well. As economies grow, the balance of employment and output in them shifts from the primary sector toward the secondary and tertiary sectors.

Some of the workers in any city produce the city's exports. This is called the basic sector. Other workers in the nonbasic sector serve the needs of the city's own residents. Jobs in the basic sector have a multiplier effect on jobs in the nonbasic sector. Cities can be classified economically by examining each city's basic and nonbasic sectors and by comparing these sectors among different cities. Economic specialization can be analyzed by examining either a city's employment structure or else the flow of money through the city's economy.

The relationships between cities and their hinterlands have inspired two models of understanding why cities are distributed across territory the way they are. One is von Thünen's isolated city model, and the other is Christaller's central place theory.

England provided the world's first model for urbanization, but urbanization in England imposed terrible hardship on millions of people. Today urbanization is occurring in many places without concomitant economic development, especially in the world's poor countries. People migrate from the countrysides to the cities because of the balance of push and pull factors, and governments try to regulate this migration both by reducing the attractiveness of the cities and by raising the quality of rural life. The preconception that urban migration is necessarily a problem blinds governments to the recognition that urban immigrants' industriousness could be an asset for economic growth.

Urban geographers also study the distribution of activities within cities. This distribution may be caused by economic forces, by social factors, or by deliberate actions of the government.

The distinguishing evolution of metropolitan form in the United States has been the explosive growth of cities out across the countryside. Growing cities have spilled over their legal boundaries into suburbs. This dispersion has resulted from a complex of government policies, cultural choices, and other economic and social forces. Both housing and jobs have dispersed to the suburbs, and commuter traffic has been redirected. The central cities have suffered economic decline and residential segregation. Efforts are being made to attract jobs back into cities and to redistribute opportunity throughout the metropolitan areas. Governing these vast conurbations presents enormous challenges.

KEY TERMS

city (p. 276)
hinterland (p. 276)
incorporation (p. 276)
conurbation (p. 276)
metropolis (p. 276)
primate city (p. 276)
urbanization (p. 276)
urban geography (p. 276)
geomancy (p. 277)
agglomeration (p. 277)
division of labor (p. 277)
primary sector (p. 277)
secondary sector (p. 277)
tertiary or service sector (p. 277)
basic sector (p. 278)
nonbasic sector (p. 278)
multiplier effect (p. 278)
input-output models (p. 278)
isolated city model (p. 280)
central place theory (p. 280)
isotropic plain (p. 280)
threshold (p. 281)
external economy (p. 282)
internal economy (p. 282)
exurb (p. 283)
labor-intensive (p. 283)
capital-intensive (p. 285)
informal or underground sector (p. 287)
central business district (CBD) (p. 289)
zoning (p. 291)
suburb (p. 291)
metropolitan statistical area (MSA) (p. 293)
consolidated metropolitan statistical area (CMSA) (p. 293)
satellite city (p. 295)
edge city (p. 295)
telecommuting (p. 297)
gentrification (p. 300)
labor force participation rate (p. 300)
urban enterprise zones (p. 302)
political geography (p. 303)

QUESTIONS FOR INVESTIGATION AND DISCUSSION

1. What are the site characteristics of your city? Why was the site selected?

2. What are the situational relationships of your city? What principle transport routes converge on your city?

3. Drive from the center of your city to the countryside, and observe the changing land uses. Does the density of residential population decrease?

4. What range of services is available in your town's Yellow Pages?

5. What are the biggest employers in your town? Which enterprises do the biggest dollar volume of business? What are your town's basic economic activities?

6. Are any planned suburbs or exurbs located around your town? Which forms of transportation do their residents depend on?

7. In 1992 the Census Bureau announced the new boundaries of the nation's MSAs. Are you in one? Did its boundaries change?

8. How many local general-purpose governments and special-purpose governments are in your metropolitan region, or the metropolitan region nearest to you? Study a map of all the forms of independent government in the area. How are special-district government ruling boards chosen? How big are their budgets, and how do these budgets compare with the budgets of general-purpose governments? What kind of regional planning body or regional government does the region have? What powers and responsibilities does it have?

ADDITIONAL READINGS

ABU-LUGHOD, J., and R. HAY, eds. *Third World Urbanization.* New York: Methuen, 1977.

CRONON, WILLIAM. *Nature's Metropolis: Chicago and the Great West.* New York: W. W. Norton & Co., 1991.

DRAKAKIS-SMITH, DAVID, ed. *Economic Growth and Urbanization in Developing Areas.* New York: Routledge, 1990.

FOX, KENNETH. *Metropolitan America: Urban Life and Urban Policy in the United States 1940–1980.* New Brunswick, N.J.: Rutgers University Press, 1990.

GUGLER, JOSEPH, ed. *The Urbanization of the Third World.* Oxford: Oxford University Press, 1988.

HART, JOHN FRASER, ed. *Our Changing Cities.* Baltimore: Johns Hopkins University Press, 1991.

JACKSON, KENNETH T. *Crabgrass Frontier: The Suburbanization of the United States.* New York: Oxford University Press, 1985.

LEMANN, NICHOLAS. *The Promised Land: The Great Black Migration and How It Changed America.* New York: Knopf, 1990.

MUMFORD, LEWIS. *The City in History.* New York: Harcourt, Brace & World, 1961.

CHAPTER 10

A World of Nation-States

The partition of the Earth into countries and the development of loyalties to those units affects almost all human activities. (Courtesy of Sandy Macys/Gamma Liaison.)

The world political map is probably the most familiar of all maps, because the Earth's division into countries, or states, is the most important territorial organizing principle of human activities. **States** are independent political units that claim exclusive jurisdiction over defined territories and over all the people and activities within them. The governments are not always able to exercise this jurisdiction completely, but, as we noted in Chapter 8, states can encourage or even force patterns of other human activities such as language and religion to conform to the political map. *Patriotism*, or strong emotional attachment to one's country, is itself a powerful cultural attribute. A country is in many ways a fixed culture realm. Sometimes states even claim jurisdiction over their citizens living outside the state, as when the United States taxes its citizens living abroad.

The idea that the whole world should be divided up into countries seems natural to us, but it is relatively new in human history (see Figure 10–1). This chapter

Figure 10–1. The dates of independence of the world's newest states emphasize that this political partitioning of the world is very recent.

explains how the idea originated in Europe and how it was diffused worldwide with European conquest. The neatness of the units on today's world political map nevertheless exaggerates the degree to which all people accept the current pattern of countries. The map suggests that all the borders are clearly demarcated and that they divide the activities on their two sides. In fact, some activities overlap the borders. The map also suggests that the areas within those borders are politically homogeneous; this is also false. Governments are only more or less successful in organizing their territory, and no territory can be sealed off. Civil wars within countries and border wars among countries continue around the world.

This chapter also reviews the political geography of states, analyzing how they demarcate their borders and subdivide their territories for government and administration. In addition, it tries to measure and map the degree to which various governments either place restrictions on their people's freedom or encourage full development of their peoples' potential. The countries

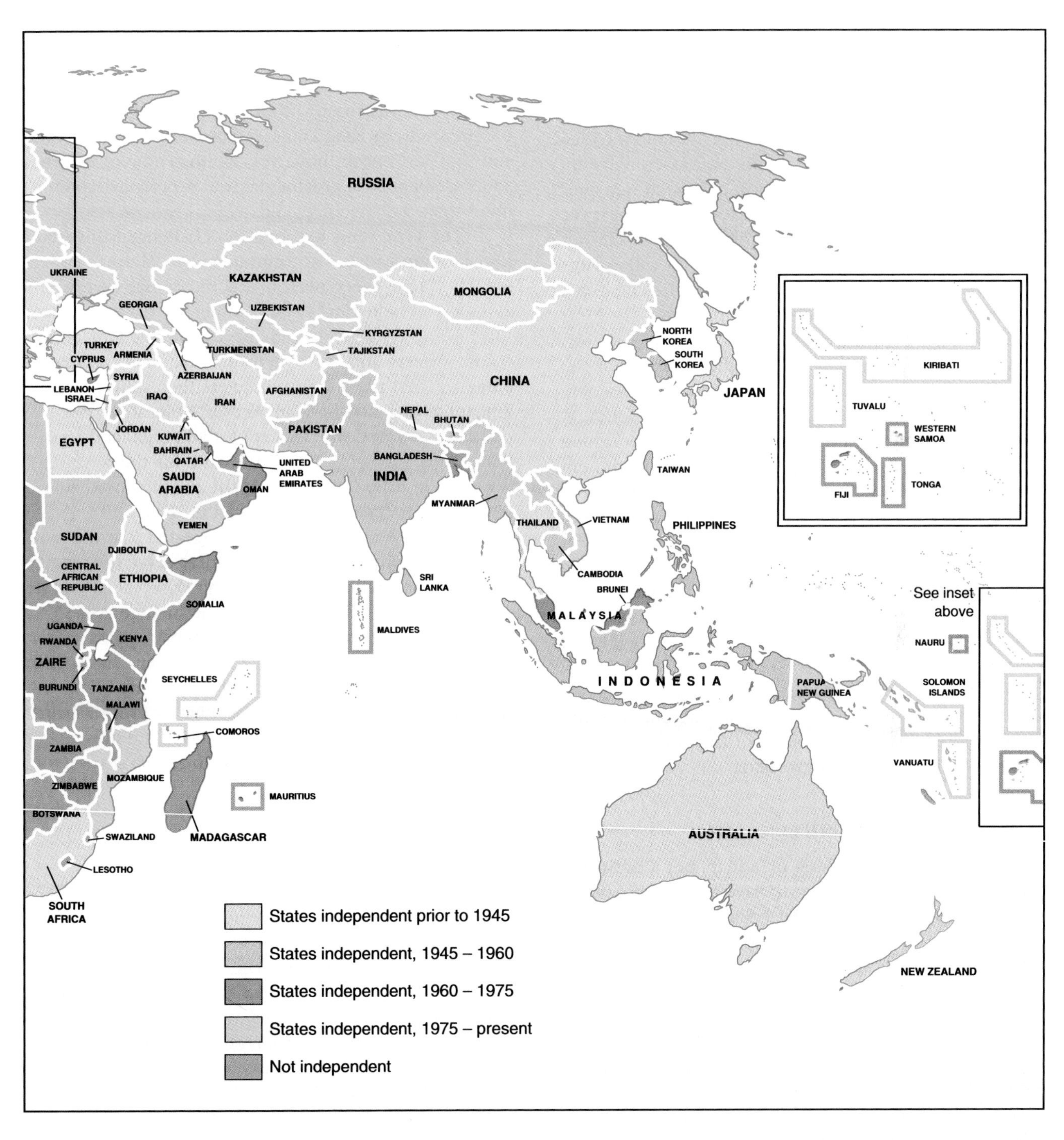

also try to organize their territories economically, that is, to integrate and build national economies. That is the subject of Chapter 11.

THE DEVELOPMENT OF NATION-STATES IN EUROPE

The idea that a state claims exclusive sovereignty over a demarcated space and all the people and resources within it originated in medieval Europe. Under the Roman Empire the Roman Catholic church was geographically organized into dioceses. When the empire fell, the Church and its system survived. As the Church converted people to Christianity, it also converted them to the idea of territorial political organization. Before conversion, the kings of the peoples in Europe had ruled not over fixed territories but over groups of followers, wherever they wandered. This system of government over a group of people rather than a defined territory is called **regnum.** The Church taught the principle of rule over a defined territory, which is called **dominium.** The Merovingian kings (fifth through eighth centuries), for example, called themselves "kings of the Franks," but the later Capetians (tenth through fourteenth centuries) settled down and called themselves "kings of France."

In the nineteenth century, anthropologists argued that conversion from the regnum form of government to dominium was an evolutionary step in human society. This argument, however, might be interpreted as an excuse for nineteenth-century imperialism. The argument that native forms of government everywhere were "backward" compared with European forms offered a justification for European conquest. At least it served as a rationalization after the conquests. British sociologist Herbert Spencer (1820–1903) extended Darwin's theory of evolution to insist that "Nature's law" called for "the survival of the fittest," even among cultures and whole peoples. This phrase was not Darwin's. Spencer's theory is called **social Darwinism.**

The Idea of the Nation

A state is a territory, a thing on a map, but a **nation** is a group of people who want to have their own government and rule themselves. The feeling of nationality may be based in a common religion or language, but it does not have to be, as the Swiss and many other nations demonstrate. A group sharing a sense of nationalism may or may not share any other attributes. Nationalism is a cultural concept in its own right.

Nationalism is one expression of **political community,** which is a willingness to join together and form a government to solve common problems. Today nationalism is in most places the most powerful level of political community. (Chapter 13, however, examines whether worldwide concern over environmental pollution may nurture global political community.)

The evolution of nationalism is part of the story of the evolution of **legitimacy,** the question of who has the right to rule any group. Upon what sort of consent of the governed, if any, is rule based? Even the most totalitarian states today claim to represent the people.

Chapter 8 noted that Christianity explicitly distinguishes religion from secular life. European emperors and kings signified the state; they ruled over earthly matters with the sanction of the Church—by ***divine right***—but they required only obedience, not loyalty or personal identification with the state. That is why kings could often assign and reassign thrones among themselves and redraw the political map without significant protest by the people. France's King Louis XIV insisted flatly, "I am the state." The Church, however, as protector of people's souls, demanded a greater degree of personal commitment than the sovereign did.

The Protestant Reformation challenged this traditional church-state accommodation. Martin Luther preached that every person was his or her own priest and carried individual responsibility for his or her own soul. This belief sabotaged the divine right of consecrated priests and, by extension, that of consecrated kings. In 1581 the Dutch, who were then subjects of the king of Spain, adopted an Act of Abjuration that renounced (abjured) the theory of divine right and argued that a king had an obligation to rule for the welfare of the people. If he did not, then the people could abjure his rule over them. This declaration turned the notion of divine right upside down. It suggested that the king, or any government, could rule only to serve the people.

Thomas Jefferson drew on the Dutch act when he composed the U.S. Declaration of Independence of 1776, which is, among other things, a bill of particulars against King George III. Prior to that time, even the English themselves had risen up against their King Charles I, defeated his forces in battle, tried him for crimes against "his" people, and executed him in 1649. The subsequent English Bill of Rights (1689) recognized that sovereignty lay not in the king but in the people.

The Nation-State

The Swiss philosopher Jean Jacques Rousseau (1712–1778) laid the foundation for allegiance to the state *as* the people. Rousseau believed that in nature people were merely physical beings, but when they united in a *social contract* they were capable of perfectibility. For Rousseau, politics was a means to moral redemption. Rousseau's ideas swayed France, and the French Revolution gave birth to the French nation. The 1789 Declaration of the Rights of Man stated: "The principle of all sovereignty resides essentially in the nation; no body nor individual may exercise any authority that does not proceed directly from the nation."

The nation demanded personal dedication and allegiance from its citizens. Therefore the perfect state was a **nation-state,** a state ruling over a territory containing all the people of a nation. This theory assumed that the nation would develop first and would then achieve a territorial state of its own. Some scholars have argued that several of the nation-states had historical **core areas** that were the historical homelands of certain nations. In many other cases, however, core concentrations of settlement or activity developed only after the state had come into existence.

Within any state, the general set of rules and regulations for governing is called the **regime,** and that is usually formalized in a constitution. The word **government** refers to the people who are actually in power at any time.

Constitutions and laws never fully explain how any government works because each political community has a unique political culture. A **political culture** is the "unwritten rules," that is, all the unwritten ways in which, in any culture, written rules are interpreted and actually enforced. Political communities differ widely, for example, in granting deference to wise elders, rich people, or religious leaders; in their tolerance of bribery; and in the rigidity or laxity with which they enforce laws. Political culture reflects other aspects of a people's culture, such as their religion. This is true everywhere and at every level of government from city council districts to the UN General Assembly.

The European Nation-States

The Napoleonic Wars that followed the French Revolution carried the idea of nationalism across Europe. Armies, which formerly had been comprised of hired professionals, became nations in arms. The French Revolution introduced military service for the entire male population for years at a time. The national army was called "the school of the nation." Mass conscription would emerge almost everywhere as a building block of nationalism. Napoleon was defeated in 1815, but the Napoleonic Law Code that he had imposed left a widespread legacy. It swept away aristocratic privileges and strengthened the middle class (see Figure 10–2).

The idea of nationalism matured in Europe during the nineteenth century, and new nation-states struggled to emerge from old empires and feudal states. In some cases, this struggle produced a competitive nationalism, and this competitiveness has caused many wars through the nineteenth and twentieth centuries. For example, several countries, including Serbia and Bulgaria, claimed the maximum extent of territory over which their people had ever wandered or ruled. This lead to overlapping territorial claims.

After World War I, U.S. President Woodrow Wilson advanced the ideal of the nation-state, which he called **national self-determination.** The victors in the war redrew the map of Europe to break up the defeated German, Austro-Hungarian, and Ottoman Turkish empires (but not their own empires) and to grant self-determination to several European nations in new nation-states, including Poland and Czechoslovakia (see Figure 10–3). Vladimir Lenin, the new Communist ruler of the Russian Empire, had criticized the empire as "a prison-house of nations," but when he reorganized the empire under a new totalitarian government, he disguised it as a union of nations. That action will be examined in Chapter 12.

Figure 10–2. This 1812 painting by Jacques-Louis David was Napoleon's favorite portrait of himself. He is shown neither in imperial robes nor astride a horse in battle but as a law giver. The pen and scattered documents, the hour on the clock, and the dying candles reveal that the emperor has worked all night on the Law Code. The code remains today the basis of law in 30 countries in Europe and beyond, including Quebec and the U.S. state of Louisiana. (Courtesy of National Gallery of Art, Washington, Samuel H. Kress Collection.)

Therefore, several European nations arguably existed before they achieved their own independent territories and governments, that is, their own states. Even in Europe, however, the national governments consci-

Figure 10–3. Several new states appeared on the European map after World War I, as the German, Austro-Hungarian, Russian, and Ottoman empires were dismembered. Several of the new states, such as Yugoslavia and Czechoslovakia, did not actually represent nations, but were composed of diverse populations. The new Poland was unsatisfied with the eastern border it was originally awarded, and it seized more territory in a war against Russia.

entiously inculcated patriotism in their citizens, and few states have ever achieved a clean match between people and territory. The European map was redrawn again after World War II (see Figure 10–4), but several states still today pursue *irredentist* claims (from the Italian for "unredeemed") on their neighbors' territory. Hungary, for example, claims the Romanian province of Transylvania, where many Hungarians live. Hungary was forced to cede Transylvania to Romania in 1920, but it forcibly retook it during World War II and had to surrender it again after that war. In 1989, Hungarian demonstrations in Transylvania triggered the fall of the Romanian Communist regime, and ethnic rioting continued after that.

THE FORMATION OF STATES OUTSIDE EUROPE

At the same time as the idea of nationalism was maturing in Europe, the Europeans were actually enlarging their empires. They were not willing to recognize that their colonial subjects had national rights. They argued that non-Europeans were inferior or "not yet ready" for political independence. After World War I the Europeans did not offer national self-determination to their subject peoples outside Europe. The winners just took the losers' colonies while retaining their own colonies.

In fact, the individual colonies seldom represented self-conscious political or cultural communities. A few cultural entities might have been recognized as historical nations if they had been located in Europe: perhaps China, Japan, Korea, Vietnam, Iran, and Egypt. The political patterns of the colonies, however, were the conquerors' creations, and they were often only a few decades old. The European imperialists carved up the world for their own convenience. No matter what level or form of political organization existed among the native peoples, the Europeans created **superimposed boundaries.** Some colonial boundaries split native political communities. Others combined two or more into one colony (see Figure 10–5).

The Europeans often used the native rulers as intermediaries between themselves and the people, especially if a colony included several groups. This form of government was called **indirect rule.** Direct rule might have united the several peoples of any colony in resentment against the foreigner. Later, when these colonies were granted independence, indignation against

Figure 10–4. After World War II the Soviet Union expanded considerably. It retook what Poland had won in the earlier war and later gave Poland German territories, thereby effectively shifting Poland about 150 miles (240 km) to the west. It gave half of German East Prussia to Poland, and it swallowed up the three Baltic states. They had been part of the old Russian Empire but had enjoyed independence since 1919. The USSR also detached Ruthenia from Czechoslovakia and incorporated it into Ukraine, and Bessarabia from defeated Romania. Bessarabia was combined with a part of Ukraine to form a new Republic of Moldavia—today's Moldova. With the collapse of the Soviet Union, Moldova may rejoin Romania.

Yugoslavia seized territory at the head of the Adriatic Sea from Italy, but the port of Trieste, which Yugoslavia wanted, remained Italian.

Figure 10–5. Traditional African political-territorial organization was usually small-scale. The Europeans superimposed colonial boundaries on this map for their own convenience at a conference in Berlin in 1884–85.

the colonial power often provided the only basis for a new nationalism.

The colonies were administered by bureaucracies made up in some cases of Europeans, in other cases of natives trained in European ways, and in still others of foreign peoples imported by the Europeans. When the colonies received their political independence, these bureaucracies had a vested interest in maintaining the existing units and borders. Therefore, today's world political map is not a map of nations that have achieved statehood. It is a vestige of colonialism.

The Threshold Principle

As late as the 1920s, international diplomats upheld a belief that a nation had to have some minimal population and territory to merit self-determination. This principle, called the **threshold principle,** was responsible for yoking together, for instance, the Czechs and the Slovaks, who had never formed a political community.

The threshold principle was abandoned after World War II, and so today the world map reveals many tiny, independent nation-states (see Table 10–1). Some, such

TABLE 10–1 ***A Few of the World's Smallest Independent States***

	Area (km^2)	Area (m^2)	Population (in thousands)	GDP (in U.S. dollars)
Bahrain	620	239	521	3500 million
Barbados	430	166	263	1300 million
Grenada	340	131	85	130 million
Maldives	300	116	218	136 million
Monaco	1.9	0.733	29	NA
Nauru	21	8.1	9	90 million
San Marino	60	23.16	23	NA
Tuvalu	26	10	9	4.6 million
Vatican City	0.438	0.169	0.774	NA

as Singapore, thrive economically, but others rely on foreign subsidies. Few are militarily defensible, and the real independence of some of these microstates is dubious. Four of the smallest, the tiny Caribbean states of Grenada, Saint Lucia, Saint Vincent and the Grenadines, and Dominica, may merge into one larger state, but even that new state would still have only 500,000 people, and its economy would precariously depend on only two activities: tourism and the export of bananas.

The concept of the territorial nation-state received a new twist in 1988. The Palestine Liberation Organization (PLO) claimed to represent the Palestinian nation and declared the existence of a state of Palestine, with Jerusalem as its capital. The state has no territorial reality, but several countries have extended symbolic diplomatic recognition.

Cultural Subnationalism

When the entire population of a state is not bound by a shared sense of nationalism but is split by several local primary allegiances, then that state is said to suffer **cultural subnationalism.** In some states, many people grant their primary allegiance to traditional groups or nations that are smaller than the population of the whole new state. These traditional identifications may be strong enough to trigger civil war. Two or more groups within one country's borders may threaten to break the state apart. Other peoples' bonds of affinity may extend beyond the state's borders, and this may inspire international disputes. Subnationalism is among the forces called **centrifugal forces,** which tend to pull states apart. A strong sense of nationalism shared throughout the whole state is among the competing **centripetal forces,** which bind a state together. Multinational empires have been held together by force in history, but politics in a multinational state can be divisive.

Subnationalism in Africa Subnationalism has particularly plagued African states. Somalia, for example, contains only a fraction of the Somali people. They overlap into Ethiopia, Kenya, and Djibouti, and so Somalia claims territories from its neighbors. The Ewe people of West Africa are split between Ghana and Togo. Along the southern borders of the Sahara in West Africa, the Taureg people ruled the Mande, Fulani, Songhai, Zarma, and Hausa peoples for centuries before the coming of the European imperialists. Today the Tauregs remain a self-conscious nation, but they are a minority in each of the new countries formed among their former subjects: Algeria, Libya, Mali, and Niger. The Tauregs disrupt all these countries and even encourage fighting among them.

Civil wars in Africa Many countries that have gained independence since World War II have suffered almost endless civil strife. In several colonies groups actually expressed their fear of independence if it required them to be bound in a political unit with a hostile group. Foreign rule had blanketed and frozen hostilities that were released again with independence. In some cases government collapsed in a war of each against all, as occurred in the early 1960s in the country that was then called the Republic of the Congo. (That country is today Zaire, not today's People's Republic of Congo.) In other cases one or more regions have attempted to break away from the new state. The Ibo people of Nigeria, for example, declared their region to be the independent state of Biafra in 1967, and the central government fought for 3 years to restore its authority. Casualties were estimated at more than 1 million. In Ethiopia a military junta overthrew the emperor in 1974 and ruled until it was overthrown in 1991 by separatist armies representing the two provinces of Eritrea and Tigre, which may achieve independent statehood in the 1990s. Meanwhile, Ethiopia absorbed hundreds of thousands of refugees from the Sudanese civil war to Ethiopia's west and several hundred thousand more from the civil war among clans in Somalia to the south. In both Burundi and Rwanda the majority Hutu and the minority Tutsi have slaughtered each other. Angola and Mozambique dissolved in civil wars after gaining independence in 1975. Peace was signed in Angola only in 1991. In some cases outsiders prolonged wars for economic or ideological reasons or just to keep potentially unfriendly neighbors busy at home. South Africa, for example, interfered in Angola and Mozambique.

Civil wars and threats of civil wars help explain the number of authoritarian governments. The armies are frequently the only truly national organizations, and they still earn the epithet "the school of the nation." As of January 1, 1990, 38 of the 43 sub-Saharan African countries had single-party or no-party systems, and more than half were led by soldiers. The only three that governed themselves by civilian majority rule—Senegal, Gambia, and Botswana—were notably ethnically homogeneous. The authoritarian rulers argued that iron rule was the only alternative to having their countries torn apart by tribalism, but this argument was at least partly a rationalization to enable them to hold onto power.

In the early 1990s authoritarian government began to fall in Africa, as it had in Eastern Europe and the Soviet Union. In March 1991 General Kérékou of Benin became the first African leader ever to be voted out of office. This was after 18 years of military rule. The 1990s will therefore test whether authoritarian rule is necessary to hold the African countries together.

Subnationalism in Asia Subnationalism has also characterized the Asian countries that are remnants of colonialism. Slaughter continues in Sri Lanka between the 75 percent of the population that are Sinhalese and Buddhist and the 18 percent that are Tamil and Hindu. In Malaysia a 41-year civil war ended only in 1989. Myanmar too has suffered, as have the Philippines and all of Southeast Asia.

Many countries on the world map are not nation-states (see Figure 10–6). At the same time there are many groups that are politically self-conscious and are, there-

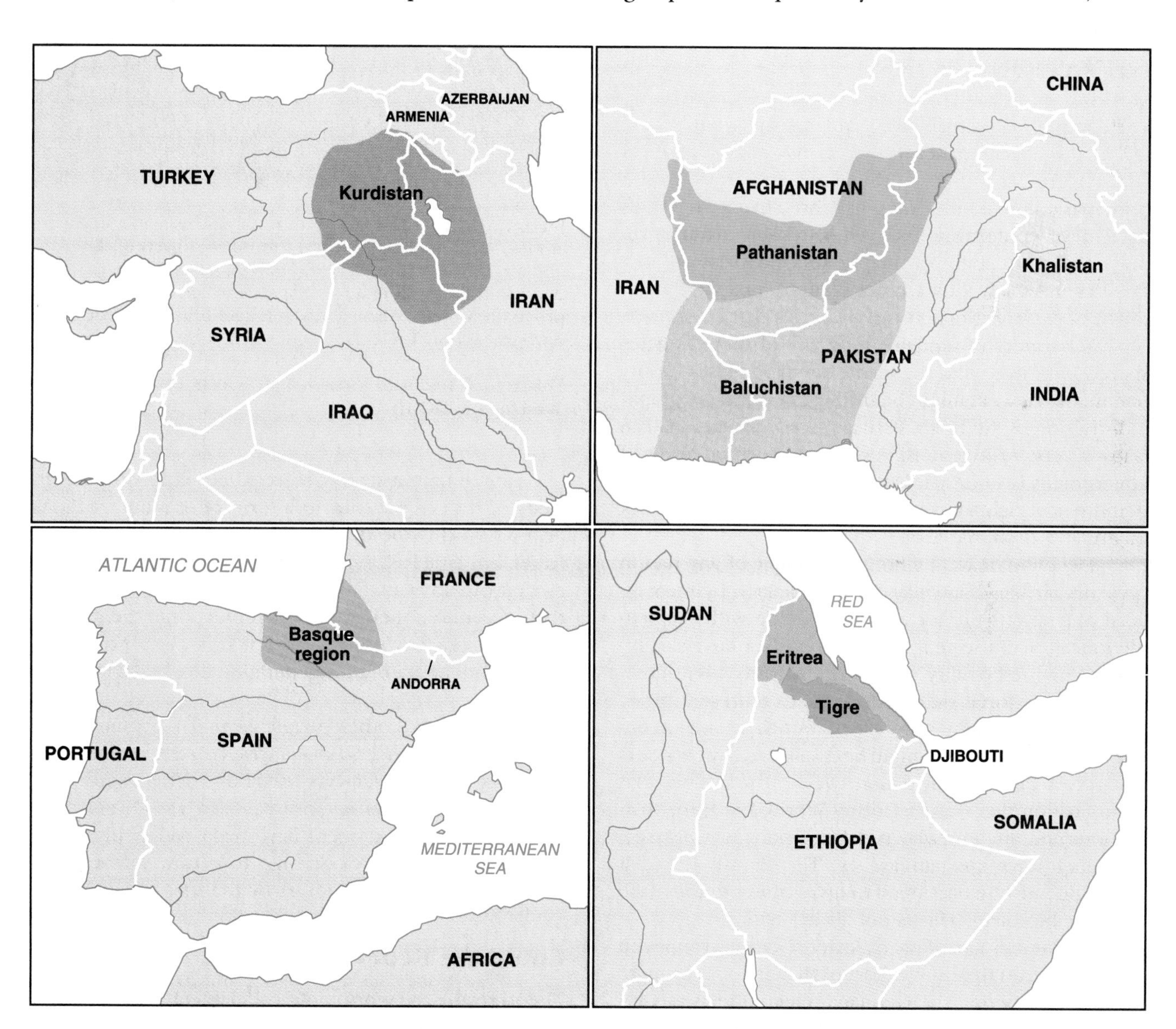

Figure 10–6. These are a few of the nations that do not have states of their own on today's world political map but may achieve statehood in the future. The power of the present states, however, may succeed in dissolving these nations.

fore, arguably nations but are politically submerged and do not show up on that map. Hundreds of African and Native-American groups might claim national self-determination. In Asia the Kurds were promised a national homeland at the end of World War I, but they remain split and submerged. Pathanistan, a nation composed of related peoples in Pakistan and Afghanistan, also claims national self-determination in Asia. Even certain European groups, such as the Basques, lack their own states.

Submerged nationalities often present threats of local violence, as events in Southeast Asia, the Mideast, Africa, Yugoslavia, and throughout the former USSR have continued to demonstrate. Such local violence may become more frequent in the 1990s. The idea of national self-determination retains its powerful attraction.

EFFORTS TO ACHIEVE A WORLD MAP OF NATION-STATES

The complete division of the world into stable and peaceful territorial nation-states may or may not be an ideal goal. At present, however, the nation-state remains the principal political unit. States use three major strategies to achieve conformity between states and nations: redraw the world political map; expel people from any country in which they are not content, or exterminate them; or forge nations in the countries that exist now.

The rigors of implementing any of these policies should remind us that the territorial nation-state is only one theory of what might be the best political-geographical division of the Earth. Perhaps territorial nationalism is not a very positive ideal. A return to regnum forms of government is improbable, but the universal guarantee of individual rights in any political-territorial framework might be a desirable goal.

Deciding exactly who is a member of any nation presents problems anyway. Nations defined by their official champions do not always coincide with the self-identification of the people concerned. Nor do the members of any nationality necessarily identify themselves with any territorial state that claims to represent them. Nations are generally self-defined, but there are exceptions. A group of people might be considered by outsiders to be part of a certain nation even if they reject that identification themselves. During World War II the British interned anyone who had been born in Germany, including Jews and antifascists. The British felt that everyone must be presumed to owe loyalty to the place of his or her birth. This is not always true.

Sometimes a national majority rejects a group who had considered themselves part of that nation. Throughout the Nazi period the Germans rejected Jews as Germans. During World War II the United States interned Japanese Americans in concentration camps in the West (see Figure 10–7).

Figure 10–7. These U.S. citizens of Japanese ancestry were among those confined to camps in isolated parts of the country during World War II. The U.S. government distrusted them, although no Japanese American ever proved traitorous. Many actually fought bravely on the European front. The United States is now paying reparations to the families who suffered. (Courtesy of the National Museum of American History, Smithsonian Institution, Washington, DC.)

Some nation-states project their nationality to citizens of other countries in a form of regnum. Millions of Jews worldwide deny any allegiance to the state of Israel, but Israel's Law of Return guarantees them Israeli citizenship anyway. Other people repudiate allegiances that the lands of their ancestors project onto them. Some countries, such as Yugoslavia, draft unlucky tourists who are second- or even third-generation descendants of emigrants.

In contrast, other people exploit a presumed nationality. Some descendants of Irish citizens, for instance, whatever their citizenship, travel on Irish passports because in some places an Irish passport wins friends. Argentines of Italian descent have been taking advantage of presumed Italian citizenship to return to the prosperous land of their forebears by the tens of thousands.

Could We Redraw the Map?

Theoretically the world political map could be redrawn until everybody was content being in the state he or she was in. Unfortunately, this would open endless disputes and provoke new wars. Some countries would split apart, and numerous territories would have to be transferred from one country to another. No satisfactory solution could exist for the states in which different national or ethnic groups intermingle or in which the cities are populated by people of one group and the countryside by people of another group. Disputes would also arise over the distribution of natural resources. Each group would claim the most generously endowed territory as its own, or at least a "fair share" of what had been the entire state's endowment.

The governments of most existing states oppose redrawing international borders. At the 1964 Cairo Conference of the Organization of African Unity, the African states pledged themselves to respect the existing international borders, even though they resent them as a colonial legacy and find it difficult to govern within them. Figure 10–8 maps the few border changes in Africa and Asia since decolonization.

Mass Expulsions and Genocide

A second strategy to unify a population into a nation is to expel or exterminate peoples who are not accepted as members of the nation. The first of these policies is tragic, and the second is abominable. Both, however, actually have been implemented when groups have attempted to carry the logic of territorial nationalism to its illogical extreme. That is the reason why these actions

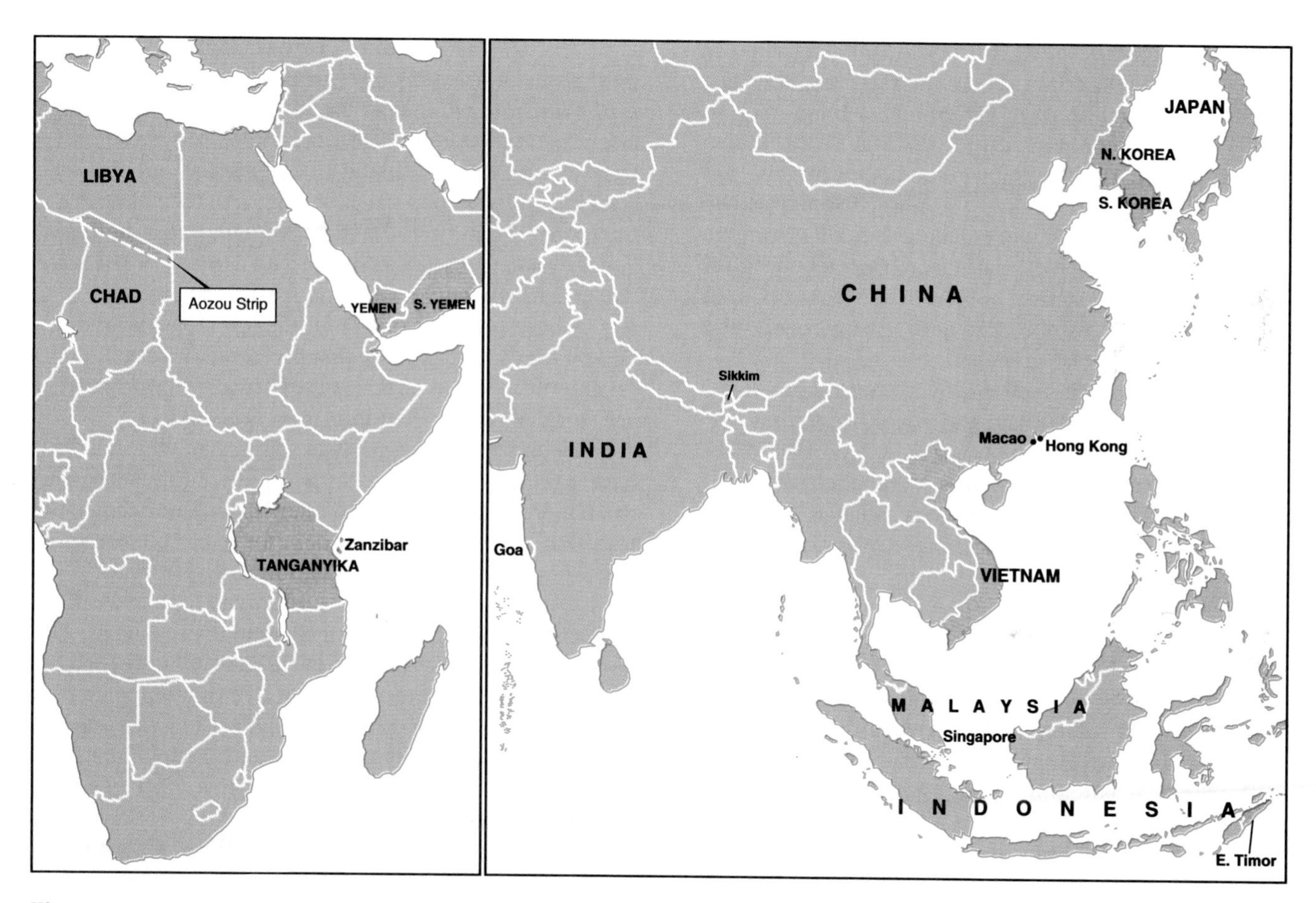

Figure 10–8. The only significant border change in Africa since decolonization has been the merger of Tanganyika and Zanzibar to form Tanzania in 1964. A stretch of territory across northern Chad about 100 miles (80 km) wide known as the Aozou Strip is claimed and currently occupied by Libya.

A few colonial borders have been redrawn in Asia. In 1965 Singapore separated from Malaysia. In 1976 North and South Vietnam joined into one Socialist Republic of Vietnam. Korea, which had been annexed by Japan in 1910, was divided by victorious Soviet and American troops in 1945, and it remains split into South and North Korea. North Korea tried to conquer South Korea, but South Korea rebuffed it with UN assistance (1950–1953). Yemen, which had received independence from Turkey in 1918, and South Yemen, which had received independence from the United Kingdom in 1967, merged in 1990. India annexed the Portuguese colony of Goa in 1961 and the formerly independent country of Sikkim in 1975. In 1975 Indonesia seized the eastern half of Timor from Portugal, but the United Nations refuses to legitimize this action. China will absorb Hong Kong from the United Kingdom in 1997 and Macao from Portugal in 1999.

must be listed among "theoretical" means to achieve a map of nation-states, but it must also be reemphasized that a world of territorial nation-states may not be a worthy ideal. The Native-American populations of the United States and of Argentina, the natives of Australia and New Zealand, and the Jews of Germany have come the closest to being exterminated.

Mass expulsions and genocide occurred in southern Europe during and after World War I. The Turks massacred Armenians in 1915, and after the Greco-Turkish War of 1922, they expelled about 1.5 million Greeks from Asia Minor. Greece responded by expelling 400,000 Turks. Later, Germany attempted to bring into the country many Germans who lived in other countries and to eliminate Jews within Germany. After World War II, Germans were expelled from Poland and Czechoslovakia. Mass expulsions continue. In 1989 Bulgaria expelled an estimated 100,000 ethnic Turks. In 1991 as many as 1 million Kurds fled or were driven out of Iraq.

Tragedy accompanied the partition of the Indian subcontinent in 1947, and it is to be hoped that this did not provide a foretaste of what might happen as the Commonwealth of Independent States emerges from the collapsed Soviet Union. The future of that new political unit is uncertain. Censuses of the late 1980s counted some 60 million "displaced Soviets" living outside the regions of their ethnic identity.

Probably the most ethnically homogeneous states today are Iceland, Japan and the two Koreas; they are 99 percent homogeneous. China is 92 percent Han. These countries exist, more or less, within their historical frontiers. Japan, an island state that isolated itself for hundreds of years, may be the closest realization of the concept of a nation-state.

Forging a Sense of Nationality

Today the existing countries are struggling to weld their populations into nations. This reverses the theoretical order in which the formation of the nation precedes the achievement of statehood.

The significance of this transition for world cultural geography can scarcely be exaggerated. It means that the world political map is struggling to make the maps of other human activities conform to it. Chapter 8 noted that governments can affect the geography of religion and language, but, in addition, governments can affect the geography of political community, of law, of land use, and, as we shall see in Chapter 11, of economic activity. Countries can encourage the circulation of people, goods, and ideas within their territories, and they can restrict or discourage circulation across their borders. Countries can never control these activities entirely, but modern governments exercise incomparably greater power over human activities than did governments in previous eras. Mass conscription, mass education, and the mass media are tools that even the most totalitarian governments lacked in the past.

Chapter 5 emphasized that maps of human activities reflect a balance of the forces for change and the forces for stability. The maps of any activity at any time are only temporary, and the units on today's world political map are relatively new. More than half of today's countries containing about half of the Earth's population and covering a great percentage of the Earth's surface—including virtually all of Africa and South and Southeast Asia—are less than 50 years old. As recently as 1991 the Soviet Union covered one-sixth of the Earth's land surface, and yet today that area is divided among 15 new political units. No one can predict how long today's world political map will remain fixed, but the existing governments are trying to stabilize it.

U.S. history recounts many struggles to weld one nation out of many diverse groups. The national motto, *E pluribus unum,* means "out of many, one." In Europe a few nation builders faced their task self-consciously. Italy was politically unified in the 1860s, and then the statesman Massimo D'Azeglio observed: "We have made Italy; now we must make Italians." The Polish hero and later president Joseph Pilsudski said flatly, "It is the state that makes the nation, not the nation the state."

These observations from Europe's past were echoed by Julius Nyerere, the first president of Tanzania. He commented on the African states: "These new countries are artificial units, geographical expressions carved on a map by European imperialists. These are the units we have tried to turn into nations." Many Asian states face the same difficulty, and even many Latin American countries that have been independent for over 100 years still have not welded their populations into nations.

Many states have to undermine cultural subnationalisms before they can forge new nationalisms. This is what the European imperialists carefully avoided doing. It can be profoundly destructive psychologically to at least the first generation. Typical is Kenya's campaign to abolish the ancient traditions and distinctiveness of the Masai people, once one of Africa's most powerful nations, and to integrate them into the modern state. The Masai must go to school and conform to a new style of life dictated by the government. They must surrender their age-old lifestyle of nomadic cattle herding, settle down, and take up farming (see Figure 10–9).

Some countries have absorbed the leaders of the traditional subnational groups into the new political structure (see Figure 10–10). Cameroon, for example, still allows the 17 traditional kings within its borders powers of judgment over minor crimes, and it defers to their ceremonial importance. The several kings in Nigeria and the Ashanti kings in Ghana have counseled their peoples to accept and adapt to modern state structures. In Malaysia a paramount ruler is elected every 5 years from among the 9 traditional Malay sultans. In some cases

Figure 10–9. These Masai warriors visiting downtown Nairobi are objects of bewilderment (and, it seems, hostility) to their modern westernized compatriots. Kenya is forcibly acculturating the Masai to modern norms. (Courtesy of Robert Caputo.)

traditional elites may include religious functionaries, particularly in Islamic countries.

India is the world's largest experiment in bringing diverse groups of people together under common democratic political structures. It is an empire of nations. Some of the centrifugal forces that afflict India are based on religion, some on ethnicity, and others on caste or language, and several of these distinctions are regional. This might encourage separatism. Changes of government seem to be a sign of confusion, but they also prove that new coalitions are forming, preventing control by one set and allowing grievances to be ventilated. Furthermore, the Indian civil service, a legacy of British rule, remains a strong centripetal force. Political stability will

Figure 10–10. Many of Africa's traditional kings retain symbolic privileges and esteem. (Courtesy of Brian Brake/Photo Researchers.)

be secure as long as politicians avoid fanning the resentments of the various groups, but the manipulation of religious groups or castes as voting blocs could tear India apart.

How States Build Nations

Nationalism can be built, and as today's countries forge their populations into nations, the world political map becomes increasingly important as a map of cultural units. Different countries rely on different instruments of nation building, and the choice reflects differences in aspects of their cultures.

Religion Chapter 8 noted countries in which a national church served as a building block of nationalism: Ireland, Ukraine, and Poland. The Orthodox churches are traditionally national churches, and some Protestant denominations are also rooted in specific nations: Lutheranism among the Germans, Presbyterianism among the Scots, and Anglicanism in England. The Roman Catholic church is transnational, but the 1965 Vatican Council II encouraged countries to form national councils of bishops. These councils have already begun to reflect cultural differences among countries that are significant enough to surprise and upset the Vatican.

Figure 10–11. Why isn't this boy in school? He is a soldier in the army of Laos. Many countries draft high percentages of their people to keep them under control. (Courtesy of Nancy and Douglas Paley.)

Armed Forces In many countries the armed forces are still "the school of the nation," and a high proportion of the people may be serving (see Figure 10–11). The armed forces may consume a high percentage of the national budget, and in many countries they provide a disciplined labor force for building infrastructure. Defense against external enemies is not always an important military function. The forces' primary purpose is often to hold the state together, either brutally by force or just by training and socializing young people. Even the U.S. armed forces have been assigned nation-building tasks. For example, in 1947 President Harry Truman ordered the racial integration of the armed forces decades before the society at large was ready to integrate.

Education National school systems are another tool with which states create nations. Rousseau wrote: "Education must give souls a national formation, and direct their opinions and tastes in such a way that they will be patriotic by inclination, by passion, by necessity." Most countries have a national curriculum and national textbooks. In France, it is said, the minister of education in Paris can look at his watch at any time of day and say precisely what all French schoolchildren are studying.

Schools inculcate youngsters with the society's values and traditions, its political and social culture. This is called the process of **enculturation, or socialization.** Schools propagate national culture, literature, music, and artistic traditions. In schools, national history and geography lessons leave the rest of the world outside that focus of concern (see Figure 10–12). Classroom maps often claim irredenta.

President Ronald Reagan's farewell address emphasized the teaching of history as the inculcation of patriotism:

> An informed patriotism is what we want. And are we doing a good enough job teaching our children what America is and what she represents? We've got to teach history based not on what's in fashion but what's important: Why the Pilgrims came here, who Jimmy Doolittle was, and what those 30 seconds over Tokyo meant.

The United States relies strongly on the schools to forge a nation, and yet it allows a unique diversity of programs and standards across the country. Education is not mentioned in the Constitution, and so responsibility for it remains with the individual states. The states in turn have generally delegated the task down to local school districts. Figure 10–13 illustrates the complex pattern of local school districts in one metropolitan area—Minneapolis–St. Paul, Minnesota.

Figure 10–12. This cartoon illustrates that the way people are taught their national history can engender either friendly or antagonistic feelings toward their neighbors. Today European countries are harmonizing their schoolbooks to encourage the formation of a European political community. (Reproduced from *European Community*, 117, October, 1968, p. 8.)

International comparisons indicate that U.S. elementary and secondary schools are relatively weak. These comparisons have spurred demands for nationwide standardization of curricula and testing. A 1989 conference of all 50 state governors concluded: "We believe that the time has come . . . to establish clear national performance goals, goals that will make us internationally competitive." Local districts could still provide laboratories for experimentation in designing school programs, but national standardization could end one of the nation's oldest and most revered traditions of local government.

0 5 10 15 Miles
0 5 10 15 Kilometers
St. Paul
Minneapolis
Twin Cities Metropolitan Area
School districts
District boundary
County boundary

Figure 10–13. Local school districts in the Minneapolis–St. Paul metropolitan area.

Manipulating Symbols Every country has an **iconography,** a set of national symbols, including a national anthem and flag, and these are emphasized to schoolchildren (see Figure 10–14). In the United States the flag is almost sacred. Americans pledge allegiance to it (see Figure 10–15). China forbids desecration of its flag or its use in advertising.

The map is another icon of great power. Children can be taught to respect, even to cherish, the size, shape, topography, resources, and variety of their land. The U.S. weather map has demonstrated appeal on television, and the editors of *USA Today* were surprised to discover that the weather map is one of the paper's most popular features. Few people actually read the map or even care about the weather at the other end of the country, but the map seems to pull all Americans together. The power lies simply in the image.

Figure 10–14. In the center of the flag of the Republic of Korea, Yang over Um symbolizes the dualism of the cosmos. The four outer signs read, clockwise from top right, heaven, water, earth, and fire. The flag was designed to suggest the richness of Korean traditional philosophy and culture.

Figure 10–15. In 1984 these protesters outside the Republican National Convention in Houston, Texas, tore down and burned a U.S. flag. The national flag is probably the most important national icon, and a protester was convicted of violating a Texas law against defiling the flag. The U.S. Supreme Court, however, accepted the action as a form of speech protected by the Fifth Amendment, and it overturned the Texas law. (Courtesy of John Keating.)

Media National media can also be used to build a nation's cultural integration. The media may reach and sway more of a national population than any other shared experience, even the army or the school system. Two-thirds of India's people, for example, are illiterate, but 80 percent to 90 percent of them can be reached by state-controlled radio and television.

Government media monopolies may inform the population, educate it, win its allegiance, or command it. The United States is the only country in the world in which commercial television broadcasting preceded public programming, and today only the United States, Sri Lanka, and Norway do not require television stations to donate air time to political candidates.

Many countries protect their national airwaves against foreign influence as a way of safeguarding their national culture. American trade negotiators refuse to acknowledge the cultural importance of these barriers, and so the United States protests against them as strictly economic protectionism. The media, however, are probably the principal definers of U.S. culture as well. In the nineteenth century the distribution of popular magazines helped create the U.S. nation, and still today the media play a key role. What is an American? A person who knows who Jimmy Doolittle is? A person who knows who played in last year's Super Bowl? Or a person who can complete the lines to a popular television advertising jingle? The international spillover of U.S. media has been accused of subverting Canadian nationalism, and one reason why East Germany was never able to create an East German nation, despite 40 years of trying, was that East Germans could not be prevented from watching West German television.

Political Parties In some countries political parties are important building blocks of nationalism. They politicize and mobilize populations, recruit people, and give them a sense of participation. They can broaden the government's base of support, but they can also be very divisive.

National variations in the role of political parties demonstrate how local political cultures affect government. Most new states adopted one of two government forms that grew out of European political experience, but the forms do not actually work the same everywhere. One model, originated by the United States, has a strong president elected independently of the legislature. This model was copied, for the most part, throughout Latin America and later in France's former African colonies. In the alternative democratic form, the executive is elected from among the members of the legislature. This form was left behind in Britain's former colonies. Both forms assume the existence of several competing political parties. Multiparty democracy may be appropriate in some cultures, but in others one-party government does not necessarily mean totalitarianism. Traditional African political systems, for instance, are based on consensus building through dialogue. This system is called *existential democracy,* as distinct from the *adversarial democracy* that Westerners practice. Existential government conforms with traditional African legal systems and religion, which are more concerned with the maintenance of social harmony than with retribution.

An example of existential democracy took place in the People's Republic of Congo in 1991, when 1500 delegates met in Brazzaville for 3 months to revise the country's constitution. Revolutionary changes were introduced, and yet few formal votes were taken. The conference reached agreement through debate and consensus. When the conference had concluded, all delegates dipped their hands into the reflecting pool in the conference hall ceremonially to wash away any ill feelings that might have accumulated.

Asian politics too are notably existential, in conformance with traditional Confucian values. The Japanese constitution, for instance, was imposed during the U.S. occupation of Japan, and yet the Japanese govern themselves under that constitution in a distinctly Japanese way. Consensus on most issues is reached before any vote is taken, and so formal votes are seldom contests or measures of power.

Labor Unions National labor unions may serve as still another building block of nationalism. "Labor" or "workers' " political parties can even govern. In totalitarian states the government may sponsor official labor unions and curtail the formation of independent unions. In Poland the independent Solidarity labor movement overwhelmed the Communist party candidates for parliament and deprived the communists of any pretense of legitimacy. In the former Soviet Union, independent unions also played a key role in bringing down the communist government.

Each of these six institutions—a country's religious institutions, armed forces, schools, media, political parties, and labor unions—may be either a centrifugal force or a centripetal force. Geographers study their presence or absence in any country, the mix or balance among them that is unique to each country, and their interaction with other aspects of each country's culture. Each institution has a geography within each country. One or more institutions may be equally influential across the entire territory, concentrated in only one region, or spill over the international borders. Both regionalism and internationalism are centrifugal forces.

HOW DO COUNTRIES ORGANIZE THEIR TERRITORY POLITICALLY?

Each state demarcates its territory's borders, and it subdivides its territory for political representation or administration. This chapter examines the methods used and some of the results. Maritime boundaries and the law of the sea will be examined as international affairs in Chapter 13.

International Borders

Many people think of rivers or mountain ranges as "natural" borders, perhaps because these features often resemble lines on a map. This simple assumption, however, is false, and attempts to enforce it have triggered numerous wars throughout history. Rivers bind the peoples on their two banks as much as rivers divide them, and mountain borderlines can be drawn either from high peak to high peak or else up and down the mountain valleys that reach between the peaks. Mountain valleys can actually provide foci for political organization. The early Swiss united the valleys on either side of mountain passes. The passes did not divide peoples but gave them a common interest.

The borders among the states of the United States illustrate various types that can be drawn. Some are rivers (Washington-Oregon and the Mississippi Valley), and some follow mountains (Idaho-Montana, Virginia–West Virginia, North Carolina–South Carolina–Tennessee). Most, however, are geometric lines imposed across the landscape.

Some international borders are defended with minefields and watchtowers, but others are relatively open and free. The U.S.-Canadian border is the world's longest undefended border. International borders can be clear to anyone flying overhead even if there are no border structures. Borders may be evident by contrasting land uses on the two sides, by different types of landholdings, or by discontinuity in the transportation networks (see Figures 10–16 and 10–17).

An international border across a sparsely populated area may not be marked or supervised at all, and even its exact location is disputed only if valuable resources are discovered in the region. This has happened along the inland borders of South American countries. In 1941 Peru attacked Ecuador and annexed 55 percent of Ecuador's territory. The discovery of oil in the former Ecuadorian territory in what is today northern Peru revives the possibility of renewed war between these two countries. The jungle border between Brazil and Venezuela is also disputed. Similar disputes have occurred in Mid-

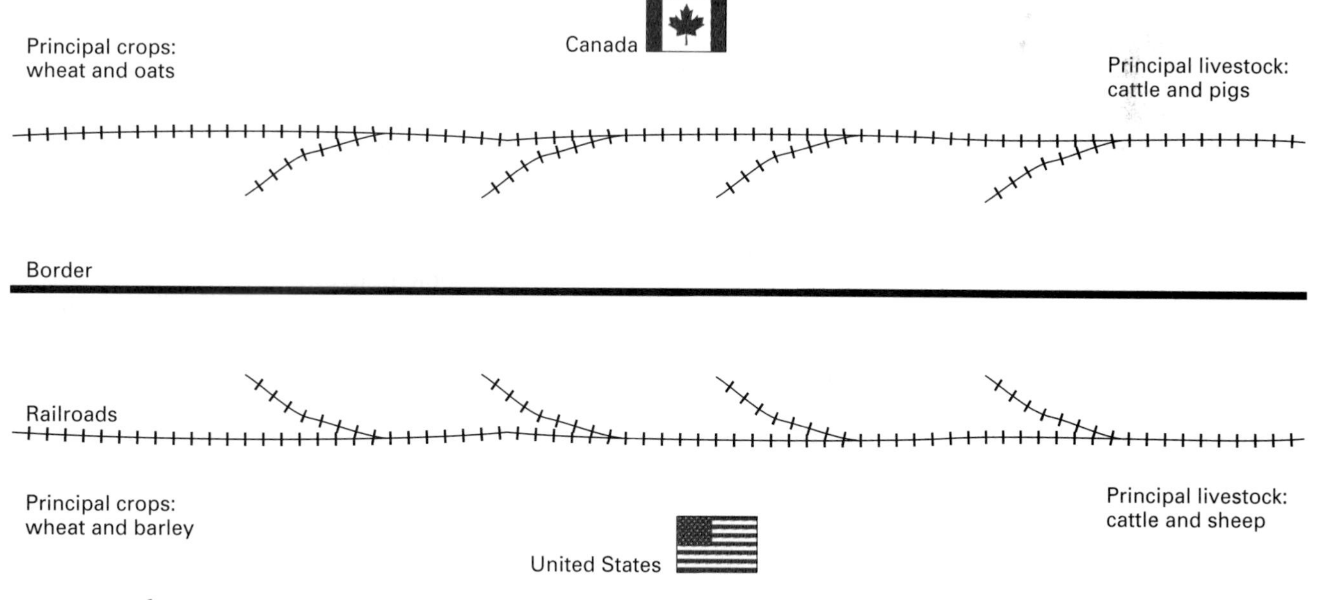

Figure 10–16. This schematic drawing illustrates two effects of policies governing the U.S.-Canadian border. The predominance of different crops and livestock on the two sides is a result, not of different environmental or market conditions, but of the two countries' agricultural support policies. The breaks in the railroad network illustrate how countries often try to restrict circulation and trade across national borders.

again, presumably, because of Mexicans' deposits. The border zone has also developed complementary manufacturing, which is discussed in Chapter 12.

States often want to control the passage of information as well as of people and goods over their borders. Today, however, direct people-to-people links via telephone lines or satellites have proliferated. Telephone connections have been used to break into computers in foreign lands and to read, steal, or tamper with privileged information. This practice challenges international law. In which country was the crime committed?

Governments can regulate only from within the country's borders, so they cannot regulate broadcasts into a country if those broadcasts originate outside its borders. Electronic broadcasts into a country can be jammed, but efforts to do so are expensive and are not always successful. Globally, radio is still more important than television, and U.S. broadcasts of Radio Free Europe and Voice of America blanket the world. The United States broadcasts Radio and Television Martí into Cuba, and Cuba retaliates by televising propaganda to the United States.

Some countries welcome international transmissions. In 1990 Pakistan began to allow transmission of U.S. Cable News Network (CNN) into Pakistani homes. Pakistanis have been astonished to see the range of debate allowed in the United States on every issue. Such direct international dissemination of the U.S. political style might represent a triumph for democracy and sow the seeds for democratic awakenings. CNN has also contributed to the growing worldwide predominance of American English over British English.

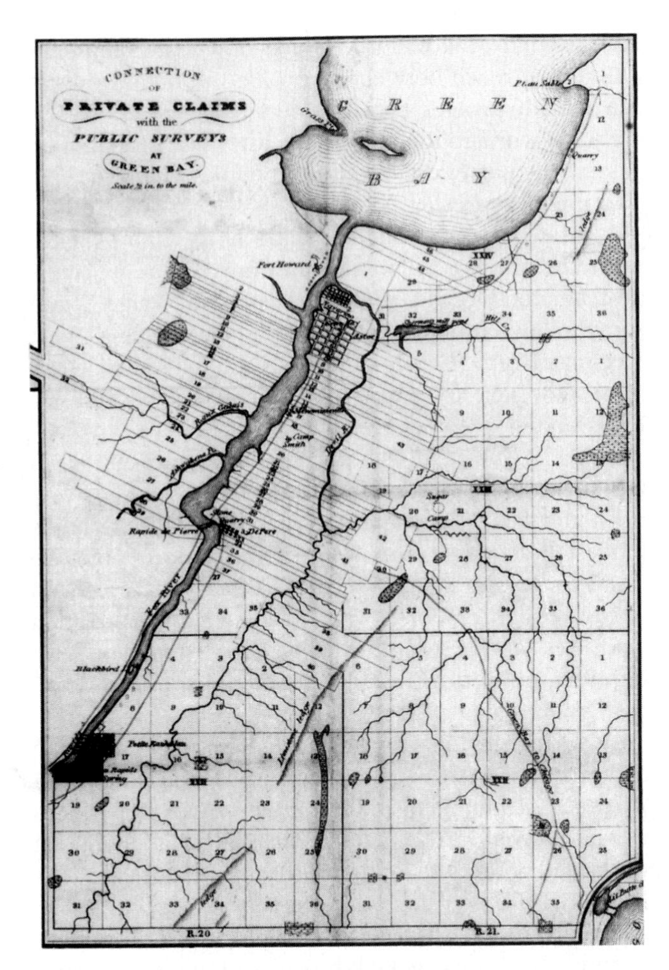

Figure 10–17. This early map of the region of Green Bay, Wisconsin, illustrates two different systems of landholding that are usually found in different countries. The French who first settled here defined landholdings according to their traditional *long-lot system*, which extends holdings back from the riverbank. Later, U.S. surveyors recognized the preexisting long lot claims but subdivided all land not already claimed in a pattern of regular squares, called the *township-and-range system*. This system was chosen for use in the United States by the Ordinance of 1785. (Map by S. Morrison, E. Swelle, and J. Hathaway. Courtesy of The State Historical Society of Wisconsin.)

eastern deserts. Under a 1966 agreement, Saudi Arabia and Kuwait shared jurisdiction and oil revenues from a 5700-square-mile (14,763-km^2) "neutral zone" between them. From 1983 to 1988 the states donated revenues from the zone to Iraq for its war against Iran.

Borders may attract complementary economic activities on the two sides. In the 25 countries along the U.S.-Mexican border in the states of California, New Mexico, Arizona, and Texas, the population rose 30 percent between 1980 and 1990, almost 3 times the national figure. Those counties' populations are notably younger, more Hispanic, and poorer than the national norm, and yet retail spending per household in those counties is extraordinarily high. Mexicans presumably come over the border to shop. The banks in those 25 counties hold assets far above what might normally be expected—

The Shapes of Countries

The shape of a country may affect its ability to consolidate its territory and control circulation across its borders. A circle would be the most efficient shape on an isotropic plain because a circular state would have the shortest possible border in relation to its territory, and that shape would allow all places to be reached from the center with the least travel. States with shapes the closest to this model are sometimes called *compact states*. Poland and Zimbabwe are examples. *Prorupt states* are nearly compact, but they have at least one narrow extension of territory. Thailand is an example. If these extensions reach out to navigable waterways, the extensions are called *corridors*. The special importance of corridors to landlocked states is discussed in Chapter 13. *Elongated states* are long and thin, such as Chile or Norway, and *fragmented states* consist of several isolated bits of territory. Pakistan was a fragmented state before one fragment broke off to create independent Bangladesh. *Archipelago states* such as Japan or the Philippines, which are made up of strings of islands, are fragmented states. Still other states, called *perforated states*, are interrupted by the territory of another state enclosed entirely within

CRITICAL THINKING: *The Drama in China*

China still tries to limit the information available to its people, but events in 1989 demonstrated how permeable to information national borders have become. During May and June of 1989 the Chinese government granted U.S. television networks direct satellite transmission facilities to report a visit by Soviet Premier Gorbachev. That visit, however, triggered student demonstrations in Beijing's Tiananmen Square demanding political liberalization. The Chinese government demanded an end to television transmissions, and the demand itself was broadcast live worldwide. This did not stop videofilming. Tapes of continuing demonstrations were smuggled out of China.

On June 4, columns of army tanks attacked the protesters in the square, killing untold numbers—perhaps thousands. Tapes of this slaughter were broadcast around the world.

This papier mâché copy of the Statue of Liberty was carried through Beijing's Tiananmen Square during political demonstrations in June 1989. Seldom in history has a more stirring symbolic sight been presented to U.S. viewers. The building in the background is the tomb of China's communist leader Mao Zedong, and it stands in what the Chinese have for thousands of years viewed as the symbolic center of the universe. (Courtesy of Alon Reininger/Contact Press.)

The severity of this repression split the army itself into factions, and this factionalism opened the possibility of civil war. This danger made it essential for the Chinese government to monopolize information about what was happening so that various army commanders could not coordinate their activities against the government. On June 8, however, the Voice of America initiated television broadcasts into China that were bounced off satellites over the Indian and Pacific oceans. The broadcasters had no way of knowing who watched the broadcasts or what they did with the information, but they knew that most satellite dishes in China were operated by the army.

For the United States to broadcast information directly to the Chinese army was international interference in a volatile domestic political situation. The director of the Voice of America commented only: "It makes available to the Chinese military information about what is going on in China—information that may not otherwise be available to the Chinese military." The Voice was at the same time broadcasting into China 11 hours each day of Chinese-language radio programming to an estimated 60 million to 100 million people.

Meanwhile the 40,000 Chinese students studying in the United States faxed daily news summaries and pictures from U.S. news sources to every fax machine in China for which they had a number: universities, government offices, hospitals, and businesses. These senders never knew who was receiving the information or what they were doing with it, but the information got there. People in China had no other way of learning what was going on in their country. When the Chinese government later announced a telephone number for informers to call to report the names of "criminal enemies of the people," Chinese students in the United States flooded these lines naming the leaders of the Chinese government themselves.

The Chinese government succeeded in repressing the student movement, but millions of Chinese for the first time had access to information about events in their own country. This might influence future internal affairs.

It should be noted that the United States also has laws and governmental directives that are intended to keep out of the country the presentation of opinions that the government considers "subversive." Many foreign artists, writers, and political figures have been prohibited from entering the United States personally, even though they are interviewed abroad and featured on U.S. television and newsmagazine covers.

Questions

What methods do governments use to prevent the influx of information and foreign ideas?

Can these methods be entirely successful, given modern technology?

Do governments ever have the right to restrict the flow of information? Why or why not?

them. South Africa, for example, is perforated by Lesotho, and Italy is perforated by Vatican City and by San Marino.

The shape of a state's territory can influence the government's ability to organize that territory, but this is not always true. A topographic barrier such as a mountain chain may effectively divide even a compact state. Bolivia and Switzerland, for example, are compact in shape, but mountain chains disrupt their interiors. For some of their regions, trade across borders is easier than trade with other regions of their own country. The people throughout an archipelago state, by contrast, may be successfully bound by shipping. History records many cases of successful political organization of the shores surrounding a body of water, such as the Roman Empire around the Mediterranean Sea.

Before drawing any conclusions from the shape of a state alone, we must consider the distribution of topographic features and of the state's population and resources, as well as the presence of any centrifugal forces such as economic or national ties that straddle the state's borders.

Territorial Subdivision and Systems of Representation

All countries subdivide their territory, and the subunits share responsibility for government with the national capital city. Each country has a unique balance of powers between its local governments and its national government, but we generally call those countries in which the balance of power lies at the center **unitary governments,** and those in which the balance of power lies with the subunits **federal governments.** Today most national governments are unitary. The newly independent and poorer countries have not widely adopted the federal form, perhaps out of fear that centrifugal forces, once recognized, could pull the country apart.

The definitions of unitary states and of federal states represent models, but in fact countries with federal constitutions may be highly centralized through government by a one-party political system or a military regime. India's federal constitution allows the prime minister to impose direct rule from New Delhi when law and order have broken down in a state. All governmental systems actually leave much open to improvization and continual redefinition.

Unitary Government In unitary states the central government theoretically has the power to redraw the boundaries of the subunits. This flexibility offers constant opportunity to accommodate geographical shifts in national population distribution or economic growth. It might actually be a good idea if every country periodically redrew the borders of its internal subdivisions. Interests become vested in any given pattern, however, and so unitary governments seldom redraw internal boundaries.

The United States is a federal government, but each of the 50 states is internally a unitary government. Each state government has the power to change its internal organization and redraw its counties or towns and municipalities. The states seldom change their internal organization, however, largely because political parties are organized at the local level, and they resist change. Political parties are not recognized in the U.S. Constitution, and yet they are a part of U.S. political culture that conditions every aspect of political life.

China is constitutionally "a unitary multinational state," and five regions of minority (non-Han) peoples are granted constitutional status (see Figure 10–18). Inner Mongolia, Xizang, Xinjiang, Ningxia, and Guangxi form an arc around Han China. They constitute 42 percent of China's territory but hold only about 8 percent of the republic's population (see Figure 10–19). Other smaller subdivisions provide local autonomy for about 50 national groups.

Federal Government A federal government assumes that diverse regions ought to retain some local autonomy and separate voices in the central government. The protection of diversity, however, may perpetuate economic or social inequality.

Today the individual states of the United States are only dubiously culture regions or even genuine political communities. Each controls its own education system, which usually inculcates state history and geography, just as education systems do in sovereign nation-states. Each state boasts an extensive iconography: state flag, bird, flower, and so forth. Texas and Hawaii have histories as independent countries. States have personalities, as revealed by the different types of legislation they pass, but if Puerto Rico becomes a state, it will be the most distinct state culture region.

Many observers have suggested redrawing the map of the states. A new pattern could improve the states' conformity with physiographic or economic regions, making each state more compact or equal in population or in economic strength (see Figure 10–20). The U.S. Constitution, however, protects the states against restructuring against their will (Article IV, Section 3).

The United States may be merging into one political arena anyway. The media are national, and even local media get most of their national and international news from nationwide services. Citizens are associating more with nationally famous politicians than with their own representatives in Washington. Two facts provide evidence. First, voter turnout is much higher for presidential elections than it is for House or Senate contests, or even elections for state and local officials. This suggests a weak sense of identification with local representatives. Second, members of Congress are raising increasing percentages of their campaign funds outside their own constituencies. More than half of the U.S. senators, for example, raise more than 50 percent of their funds outside their

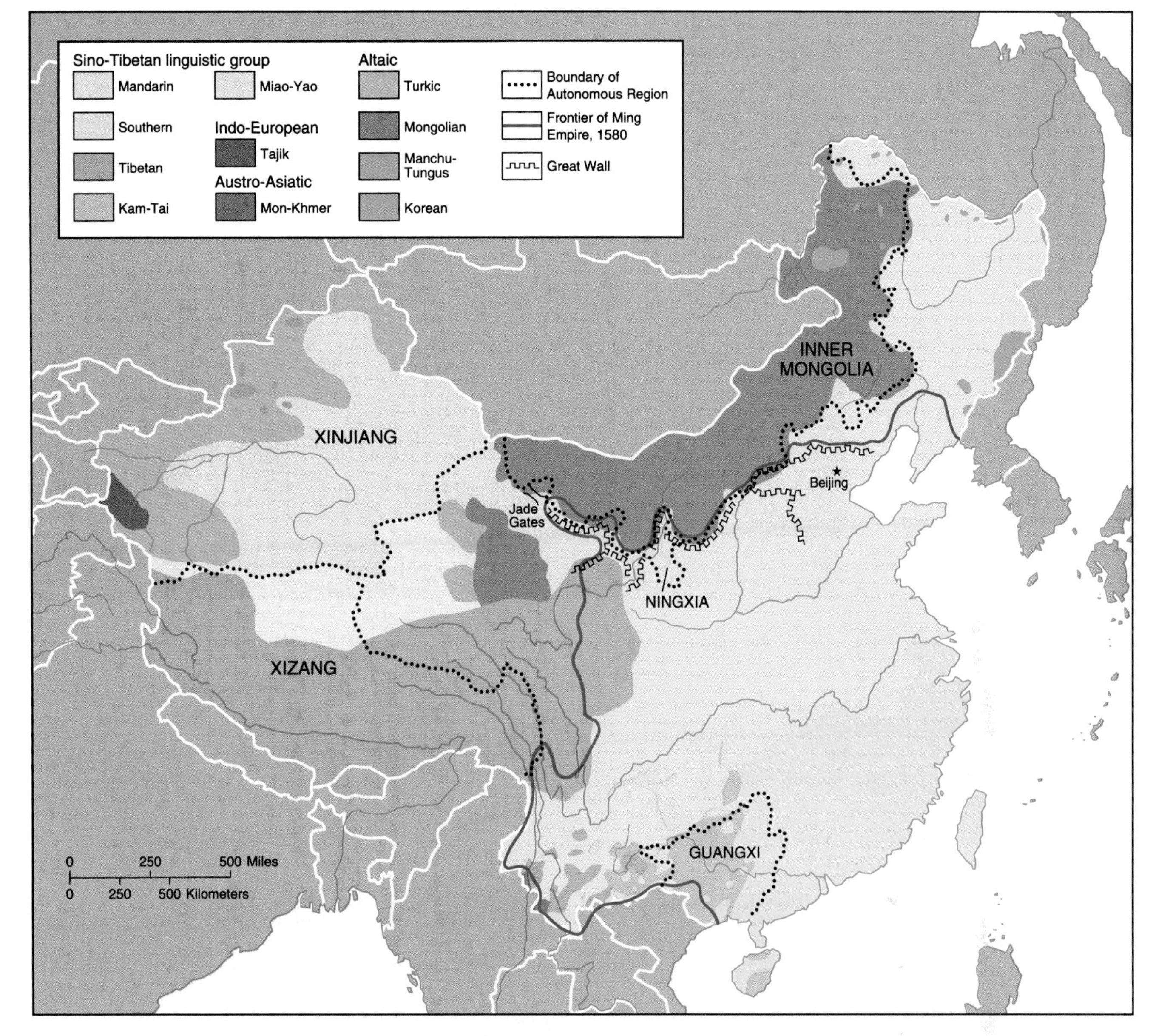

Figure 10–18. The Jade Gate to central Asia and the Great Wall against the Mongols marked for centuries the outermost limits of cultural China, but Chinese rule has intermittently expanded out over non-Chinese peoples and then shrunk back to the limits of ethnic (Han) China. Today China's international borders are in the middle of the two extremes. China's constitution guarantees a degree of political and cultural autonomy to the peripheral minority peoples, but the Turkish Muslims of Xinjiang must be drawn to their ethnic relations in newly independent Uzbekistan, Kyrgyzstan, Tajikstan, and Kazakhstan. This is a new centrifugal force for China.

home states. When Senators Albert Gore of Tennessee and Larry Pressler of South Dakota raise at least 95 percent of their funds outside their home states, whom, or which constituency, do these senators represent? This system is legal, but it might corrode the federal character of the government.

The federal patterns of some countries reflect polyethnicity (for example, Russia, Myanmar, India, Czechoslovakia). In other countries federalism protects local rights or allows each area to serve as a laboratory for legislation that can be adopted elsewhere if it works. Wisconsin, for instance, might devise a successful school program that other states might copy, New Mexico a highway program, or Maine environmental legislation.

Sometimes the units of a federal country are older than the federal framework. They came together only on the condition that they would retain certain powers (the original 13 United States, Germany, Yugoslavia, Canada). Federalism may also serve a country that is relaxing central control. Spain, for example, is moving toward federalism. Spain's culturally distinctive peripheral provinces have never been welded into one nation, and today

Figure 10–19. The Great Wall of China was built to separate the ethnic Chinese people from the Mongols. It still does, but at times in history the Chinese have been able to extend their political rule beyond the wall over the Mongols. The wall is entirely within China today. (Courtesy of Paul Lau/Gamma-Liaison International.)

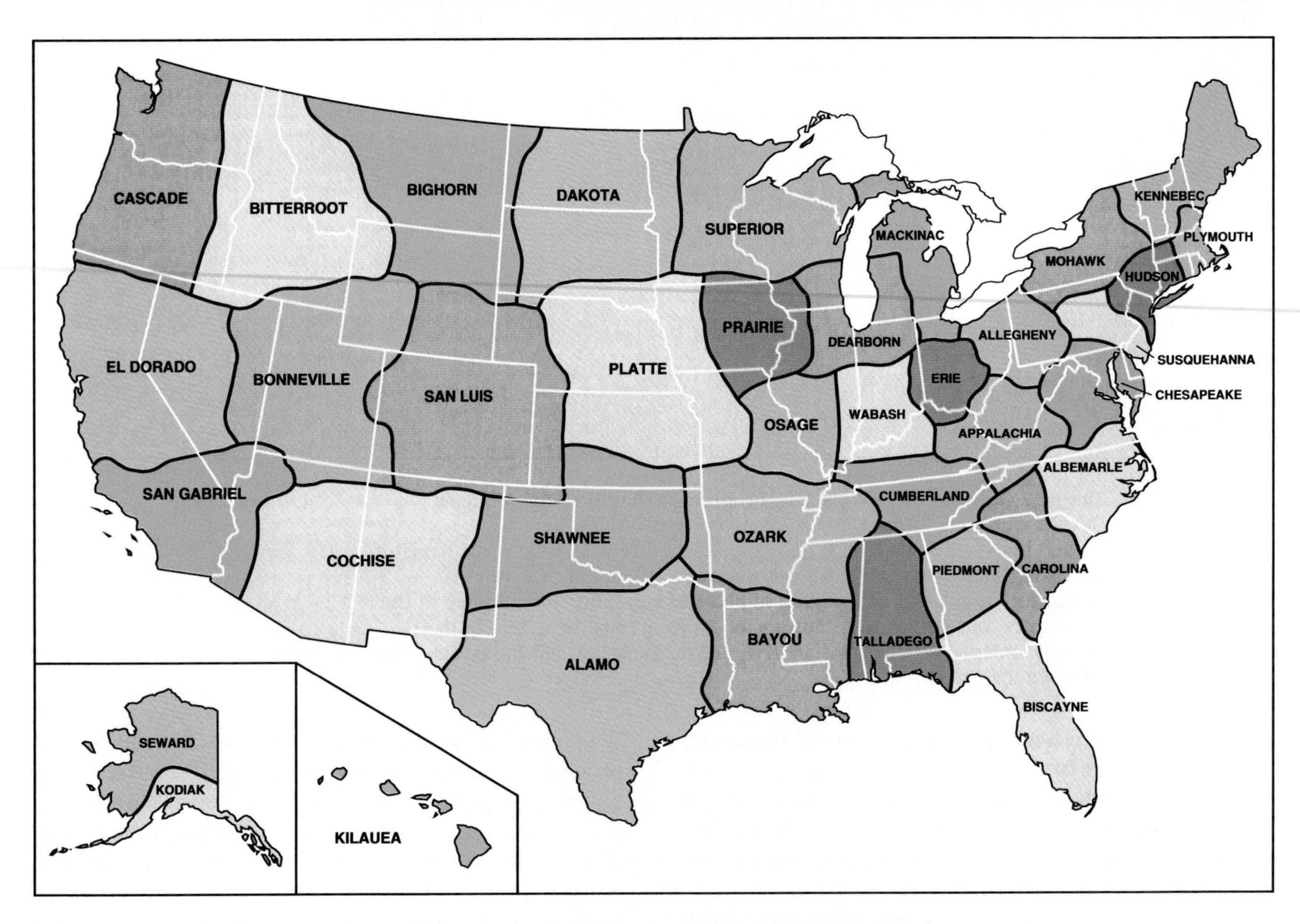

Figure 10–20. This suggested redivision of the United States into 38 states was drawn as an intellectual exercise by G. Etzel Pearcy, a distinguished geographer who long served as Geographer to the United States—a government title dating from 1781. These territorial units may be more equivalent in population or product, or they may better reflect true economic or cultural regions than do the present states. (Adapted with permission from G. Etzel Pearcy, *A Thirty-Eight State U.S.A.*, Monograph No. 2. Redondo Beach, CA: Plycon Press, 1973.)

Focus on The U.S. Civil War

The 13 original U.S. states existed before the federal government did. They formed a federal government in the Constitution, and they gave it only specific powers. The states retained all others. Through the early years of U.S. history, the people identified primarily with their states, and they feared the aggrandizement of federal power. Defenders of states' rights argued that the states protected the people from possible federal tyranny. According to John C. Calhoun of South Carolina, the state governments were "intermediary powers" between the people and the federal government. The doctrine had much to be said for it, despite the fact that it was invoked by the Southern states to defend an institution as monstrous as slavery.

The relative balance of power between the states and the central government was ultimately resolved in a civil war. In conducting this war, President Abraham Lincoln achieved two ends. First, he upheld the central government's jurisdiction over the entire territory of the United States. States would not be allowed to secede. In addition, he exercised federal power directly over each and every citizen. The states could no longer claim a role as "intermediary powers" between their own citizens and federal power. Several federal laws enacted during the war breached the traditional powers of all state governments, North and South. These included a federal military draft, new direct federal taxes, and banking legislation.

The powers of the individual U.S. state governments have increased, but their power relative to that of the federal government has been reduced, and there is evidence that the prestige of the state governments has continued to sink. In a 1988 poll, 51 percent of Americans did not realize that the individual state governments have their own constitutions. Despite the expansion of federal power, however, the allocation of responsibility between the state governments and the federal government is still open to negotiation in areas as diverse as road building, product liability laws, and welfare.

they have demanded and received considerable autonomy from Madrid. The dissident Basque people in the north enjoy local autonomy, but some Basques demand total independence. The situation is complicated because their homeland overlaps into France.

If loosening the federal ties proves insufficient, a federal state may dissolve in civil war. Post–World War II Yugoslavia was governed as a federation of six republics, one of which contained two autonomous provinces. Yugoslavia's peoples' religions, histories, cultures, languages, economies, and national feelings were so disparate that Yugoslavia was held together only by totalitarian central government. Civil war broke out in 1991, and in 1992 the republics of Croatia, Slovenia, Bosnia-Herzegovina, and Macedonia (which may change its name) won recognition as independent states. Only Serbia and Montenegro remained as Yugoslavia.

Canada's federal system Canada was formed as much to preserve British institutions in North America as to respond to any indigenous nationalism. The British North America Act of 1867 united the provinces of Upper Canada (Ontario), Nova Scotia, New Brunswick, and Lower Canada (Quebec). It invested executive authority in the sovereign (to be carried on by a governor-general) and legislative authority in a bicameral legislature. Other provinces later joined the confederation, and the 1931 Statute of Westminster removed all legal limitations on Canadian legislative autonomy (see Figure 10–21).

Canada's federal government is distinctively different from that of the United States. Canada's provinces retain more powers than do the individual states of the United States, but they have no significant institutionalized recognition at the level of the national government. The Canadian Senate is only appointive, and it exercises little power or influence. The office of governor-general, a potential centripetal force, also has little power or influence.

This arrangement has allowed the two most populous provinces, Ontario and Quebec, to dominate the national agenda. Canadian economic protectionism has benefited manufacturers in these two provinces and disproportionately disadvantaged the peripheral Atlantic and western provinces by denying them cheaper imports. The federal government's 1980 National Energy Program, for instance, effectively expropriated some $50 billion (Canadian) of windfall profits from Canadian oil producers in the West. As the West's population has increased, however, it has gained political power. Many Canadians are demanding that the members of their Senate be elected by the various provinces and more clearly represent their interests, as in the United States. The form of government in Canada may change in the 1990s. We will refer to this issue again in Chapter 12.

Special-Purpose Territorial Subdivisions The power of any local government can be undermined by the creation of special districts. If the special-purpose regions do not conform to the pattern of the political subunits, problems may arise in jurisdiction. This was demonstrated already in the governing of conurbations in Chapter 9, and the same analytical principles apply at the level of the study of states.

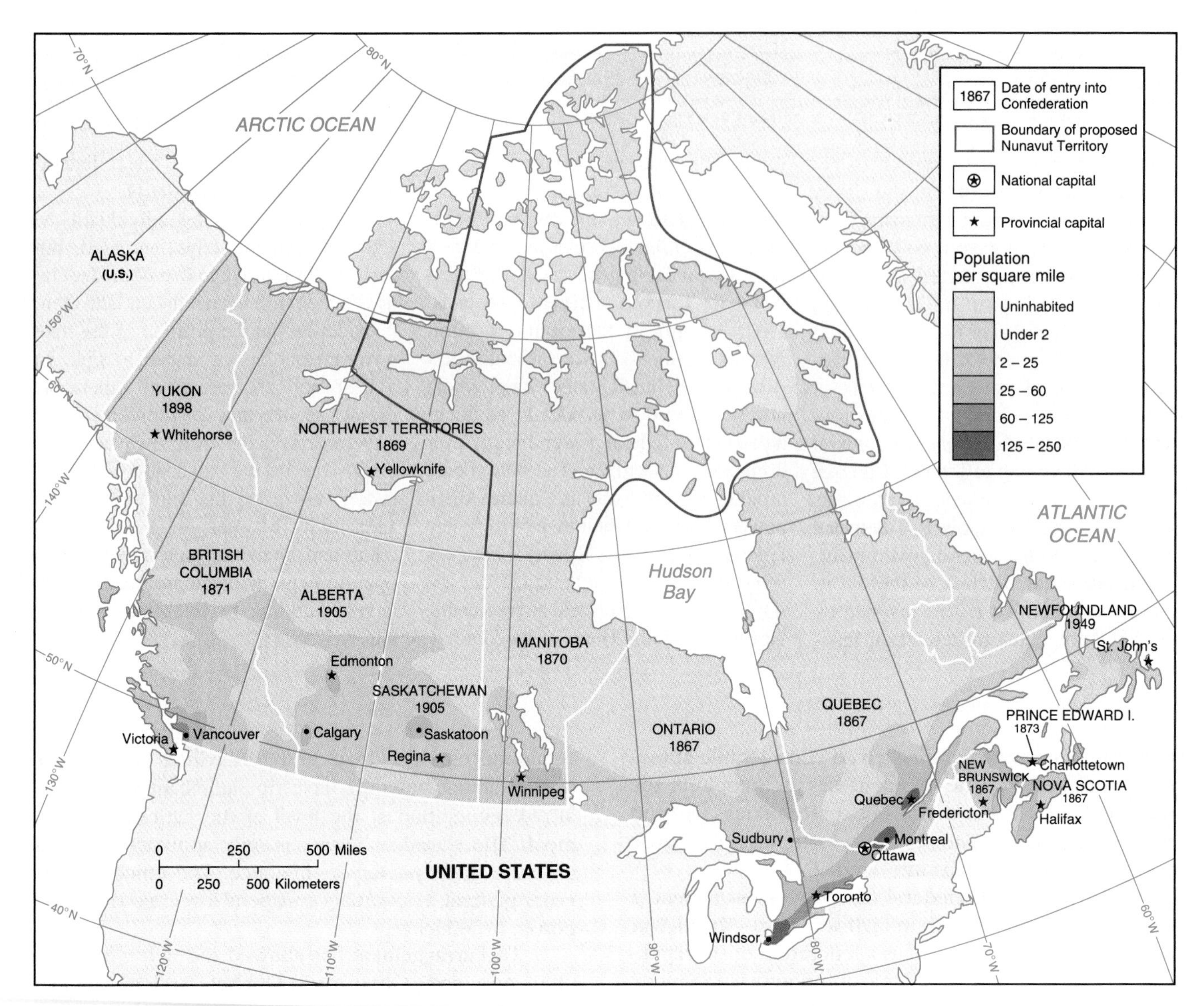

Figure 10–21. The units of federal Canada came together only slowly into the country's present configuration. The 1867 British North America Act united the provinces of Upper Canada (Ontario), Nova Scotia, New Brunswick, and Lower Canada (Quebec). In 1869 the Northwest Territories were purchased from the Hudson's Bay Company; Manitoba was carved from this territory and admitted into the confederation as a fifth province in 1870. British Columbia joined in 1871, and Prince Edward Island in 1873. Alberta and Saskatchewan were formed out of previously provisional districts and admitted in 1905. Newfoundland, previously an independent country, joined only in 1949.

Several countries divide their territory into economic planning regions, and these regions can gather almost as much power as the legal constituent subunits. The former Soviet Union did this, and France has, too. In the United States the Federal Power Commission, the National Labor Relations Board, the Bureau of Reclamation, the Federal Trade Commission, the Federal Communications Commission, and the Federal Reserve System have each subdivided the country into a different pattern of districts (see Figure 10–22). None of these patterns conforms to the pattern of the states, and this has sabotaged the states' power to govern their territories.

The United States is also subdivided by federal district courts and, above them, 11 federal courts of appeal (see Figure 10–23). No one would argue that these judicial districts represent cultural or political communities, and yet it is possible for federal law to be interpreted differently in each of them until a case reaches the Supreme Court. That usually takes years. People a mile apart may be living under different interpretations of federal laws.

Legal efforts to protect local differences may have unexpected consequences. Two of today's most controversial topics, abortion and pornography, illustrate this. In 1973 the Supreme Court allowed the states consid-

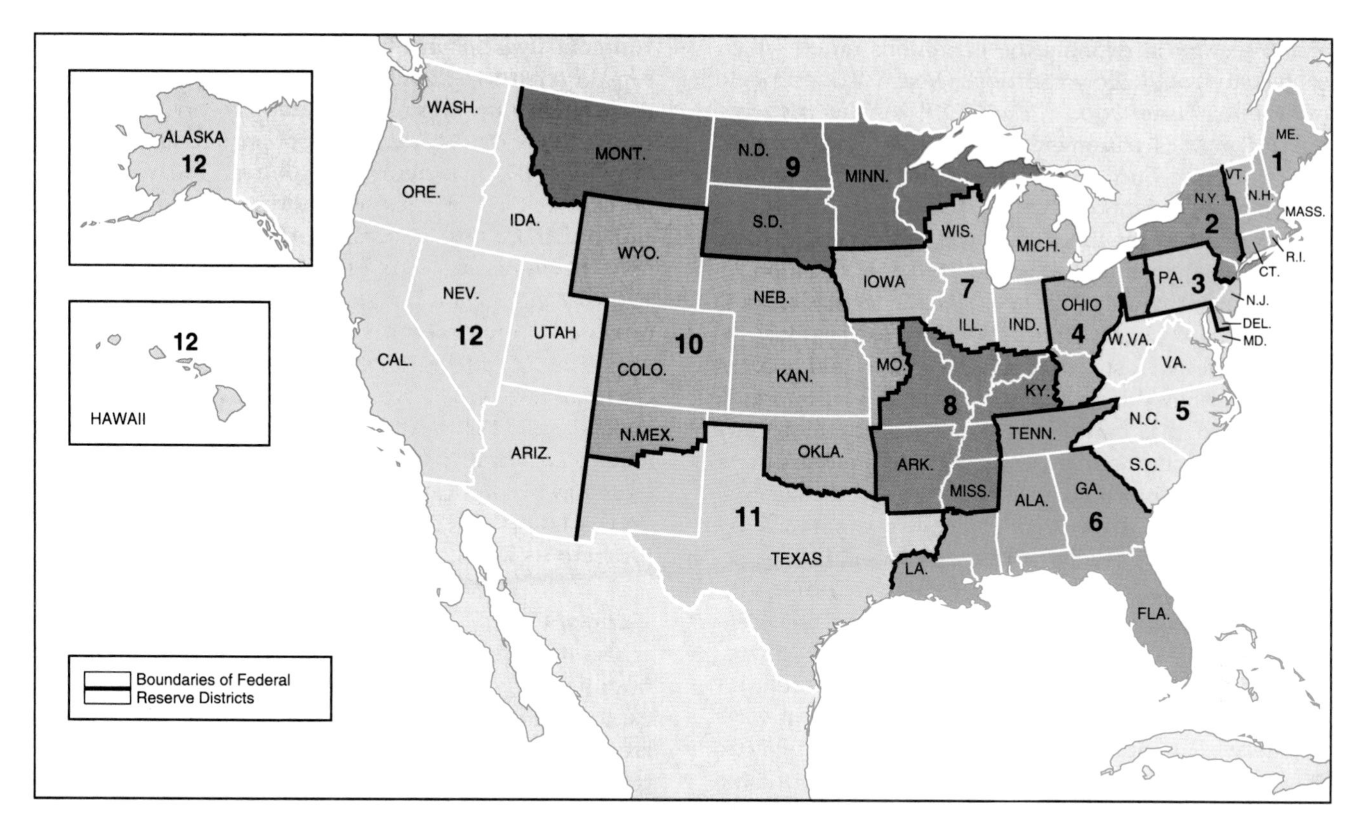

Figure 10–22. U.S. Federal Reserve Districts do not conform to the states. When two different districts enforce different policies in two parts of a state, the state government's taxing, bank, and business regulations can be confused.

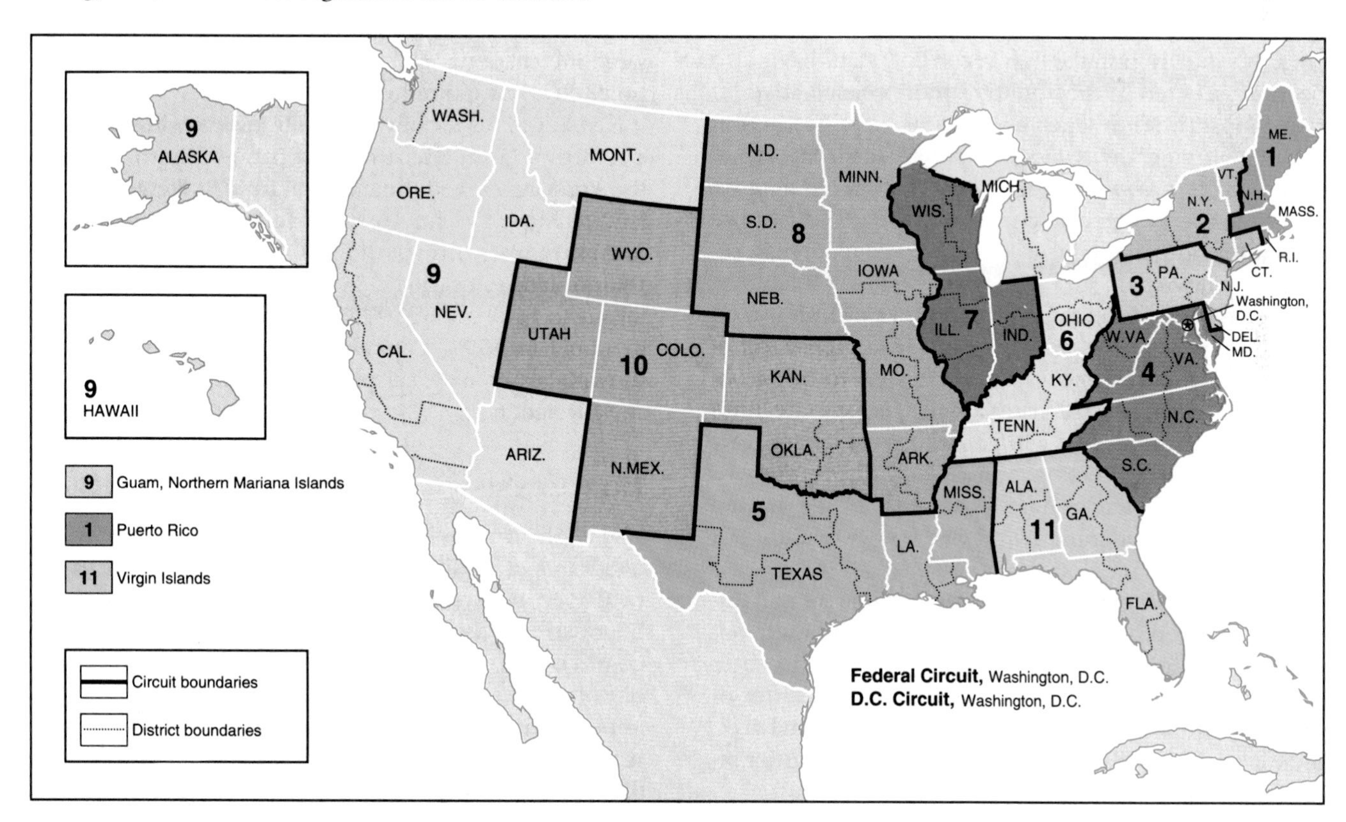

Figure 10–23. Federal law is not uniformly applied throughout the country as long as contradictory interpretations by any two of these courts remain unresolved by the Supreme Court.

erable leeway in defining the conditions under which each state would allow abortions (*Roe* v. *Wade*; upheld in *Planned Parenthood* v. *Casey*, 1992). This triggered political battles nationwide. Many Americans crossed state lines to protest abortions in states other than their own. Moreover, whereas women with money could cross state borders to get abortions, poor women had to obey the laws of their home state. The abortion question remains on the national agenda. Federal legislation or a new Supreme Court ruling may define national law.

Legislative restriction of pornography and obscenity also presents difficult questions. In 1973 the Supreme Court ruled that courts can regulate expression only "when the average person, applying contemporary community standards, would be likely to find that the material appealed mostly to prurient interest, was patently offensive and was entirely lacking in serious literary, artistic, political, or scientific value" (*Miller* v. *California*). This ruling might work in a country of isolated communities, but not in a country where materials are distributed nationally. The most conservative communities can prosecute what they view as pornography in ways so punitive as to deny the material to other communities. The New York–based Home Dish Satellite Network, for instance, which beamed X-rated films to 30,000 subscribers around the country, was driven into bankruptcy after a district attorney in Montgomery County, Alabama, (where 30 households received the service) brought criminal charges for violating Alabama's antiobscenity laws. Nationally popular rap recording stars have also been prosecuted in the country's most conservative jurisdictions. The Fifth Amendment to the U.S. Constitution prohibits trying the same person twice for the same crime (*double jeopardy*), and yet each local jurisdiction can bring charges against nationally distributed materials no matter how many juries elsewhere have already refused to convict.

Today the federal government has even assumed an active prosecutive role. The Anti–Drug Abuse Act of 1988 grants federal prosecutors the right to bring obscenity charges against materials that cross state lines even if local authorities will not or cannot do so. How can freedom of expression for nationally distributed materials be protected against local suits? Is the United States one cultural community? Can it be? Should it be?

Federal Territories Most federal states contain territories that are not included in any of the subunits but that are administered directly from the federal center with, perhaps, some local government. These territories may constitute a significant share of a state's total area, including capital districts, colonies, strategic frontier areas, and federal territories.

Territories usually have only limited local government until some presumed future time when they will be ready for full statehood. If the territories are environmentally inhospitable or without resources, they may remain territories indefinitely. Most of the land area of the United States was once federal territory, but the territories were quickly organized, populated, and admitted into the Union as states. By contrast, today some 40 percent of Canada remains in territorial status, but the situation in Canada is changing. Mexico's two territories may become states upon reaching populations of 120,000 and demonstrating "the resources necessary to provide for their political existence."

The U.S. government still owns about one-third of the total national land area (see Figure 10–24). The 1976 Federal Land Policy and Management Act requires the federal government to receive full value for any lands traded away, and Congress still wrestles with the question of how the federal government should exploit or preserve its lands.

Capital Cities In some countries the federal capital is also the capital of a component state (Bern: capital of both its own canton and Switzerland). Austria's federal capital, Vienna, is a federal state. Some federal capitals are governed directly by the federal government (Australia's capital territory, Mexico's Distrito Federal). Residents of Washington, D.C., complain that they have no voice in the federal government; if the city were a separate state it would rank 44th in population.

Countries may move their capitals. Several have done so because they believed that their old capitals were international cities not truly representative of national life, but grafted onto national life. A new capital, it is believed, especially one built inland, will symbolize a rebirth and rededication of a national spirit. In 1918 the Russians moved their capital from St. Petersburg inland to Moscow; the Turks moved theirs from Istanbul to Ankara after a revolution in 1923; Brazilians moved theirs inland to Brasília in 1960; Tanzanians moved theirs inland to Dodoma in 1975; and Nigerians moved theirs to Abuja in 1991. Capitals are frequently designed as showplaces of national pride to impress both citizens and foreigners.

How to Design Representative Districts

The subfield of political geography that studies voting districts and voting patterns is called **electoral geography.** In any system of representative government, the boundaries of the voting districts can determine the outcomes of the ensuing elections. This can occur in either one of two ways. First, if the electoral districts are unequal in population, then the ballots cast by some voters outweigh those cast by others. Second, district lines can be drawn in ways that include or exclude specific groups of voters, so that one group gains an unfair advantage. This division is called **gerrymandering** (see Figure 10–25 on page 338).

Focus on Washington, D.C.

No city better illustrates the pride a new nation takes in its capital than Washington, D.C., the world's first city built exclusively as a seat of government.

The decision to locate the new capital in the South was part of a political compromise. In exchange for a southern location for the capital, the southern states agreed to allow the federal government to assume the states' debts, most of which were owed by northern states. George Washington selected the spot for "Federal City," and for plans he turned to Pierre L'Enfant (1754–1825). Construction began on the president's house in 1792 and on the Capitol the following year. In 1800 Congress moved from Philadelphia and held its first session in the city, which was named for President Washington just after his death (December 14, 1799).

L'Enfant wrote that the "mode of taking possession of, and improving, the whole district at first must leave to posterity a grand idea of the patriotic interest that promoted it." L'Enfant had been born in Versailles, a city in France in which every avenue and vista focuses on the king's palace. The purpose of that design was to emphasize that power radiates from the king. L'Enfant adopted this plan for the capital of the new republic. For Washington, however, L'Enfant focused all streets on the Capitol. This urban design was intended to symbolize that in the new republic all power derives from "the people in Congress assembled."

L'Enfant's plan was beautiful, but it was too ambitious. It was designed to accommodate many more residents than the city would grow to house for at least 75 years. Meanwhile, the grandeur of the plan hindered its modest beginnings, and the city began to develop haphazardly.

During the Civil War, Confederate troops threatened the city several times, but President Lincoln insisted that work continue on the new dome for the Capitol to symbolize the nation's commitment to pressing on. After the Civil War the city retained its largely rural, unkempt aspect, but in 1889 L'Enfant's plans were reexamined, and in 1901 work was begun to resurrect his original design. Washington began to assume the gracious aspect it offers today.

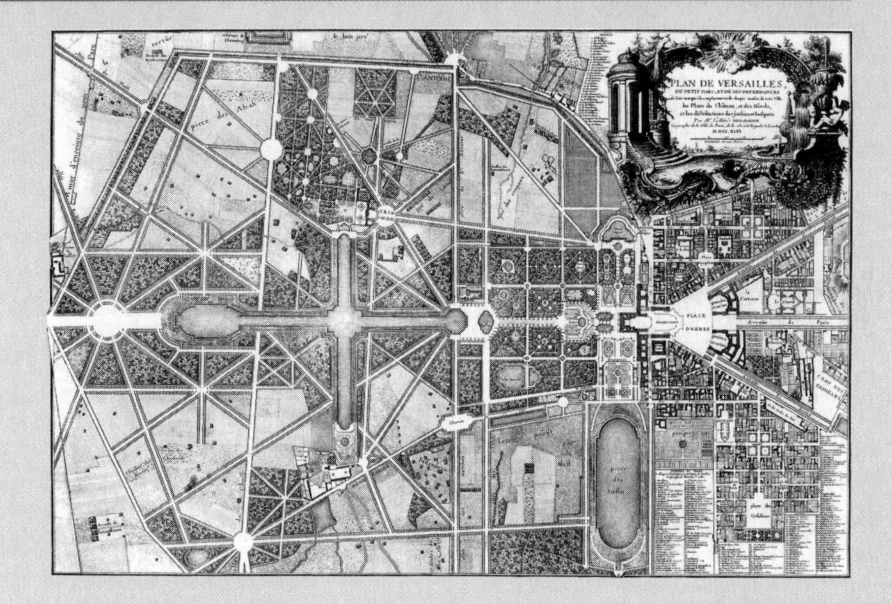

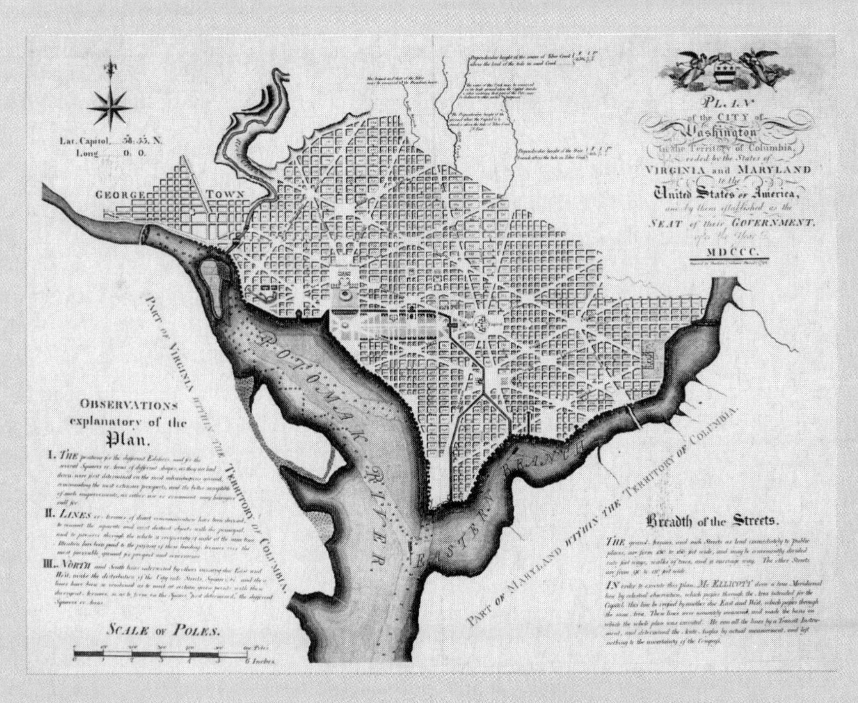

Washington, D.C., reflects the design of the French palace and gardens of Versailles, but the symbolism is different. (Reproduced with permission from the Collections of the Library of Congress.)

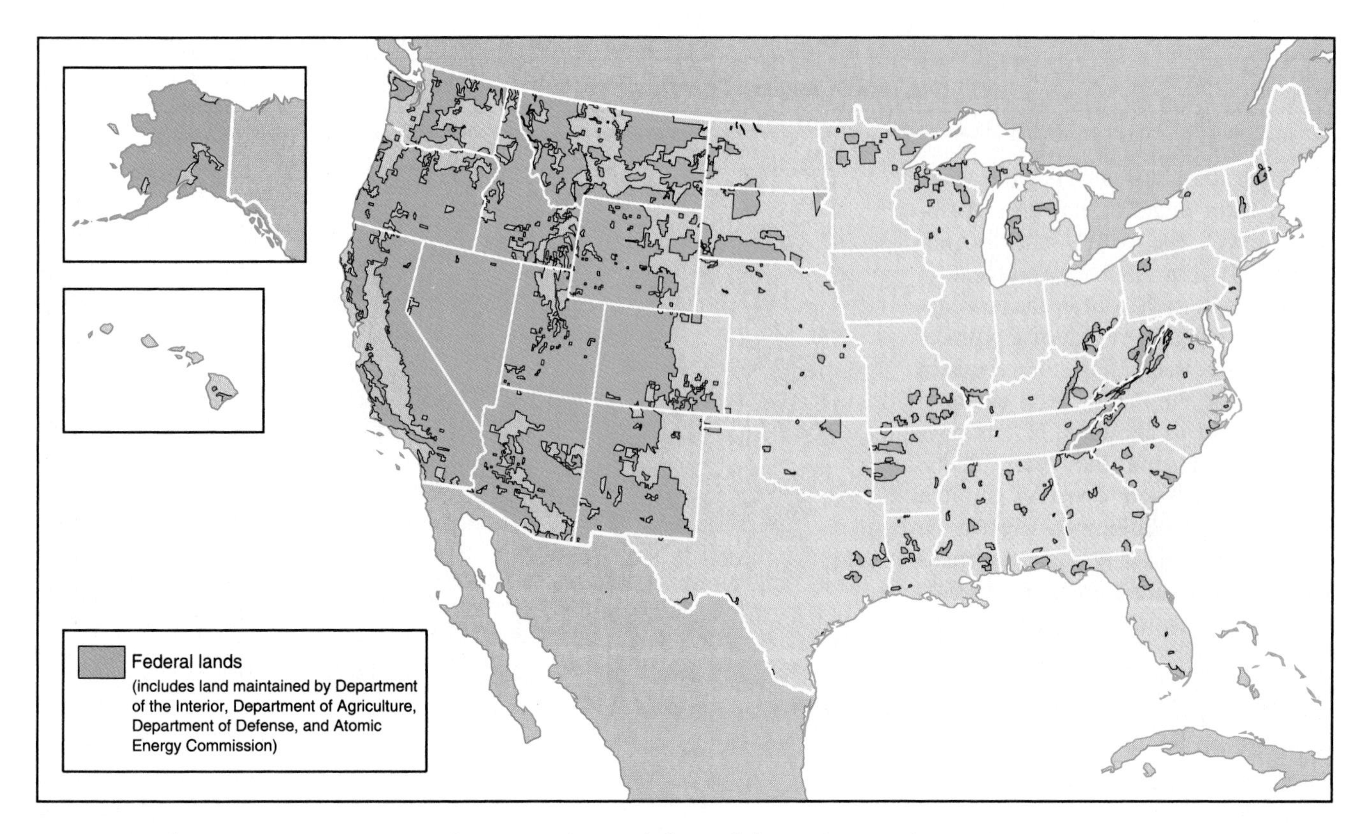

Figure 10–24. The U.S. government still owns a substantial share of the total national territory.

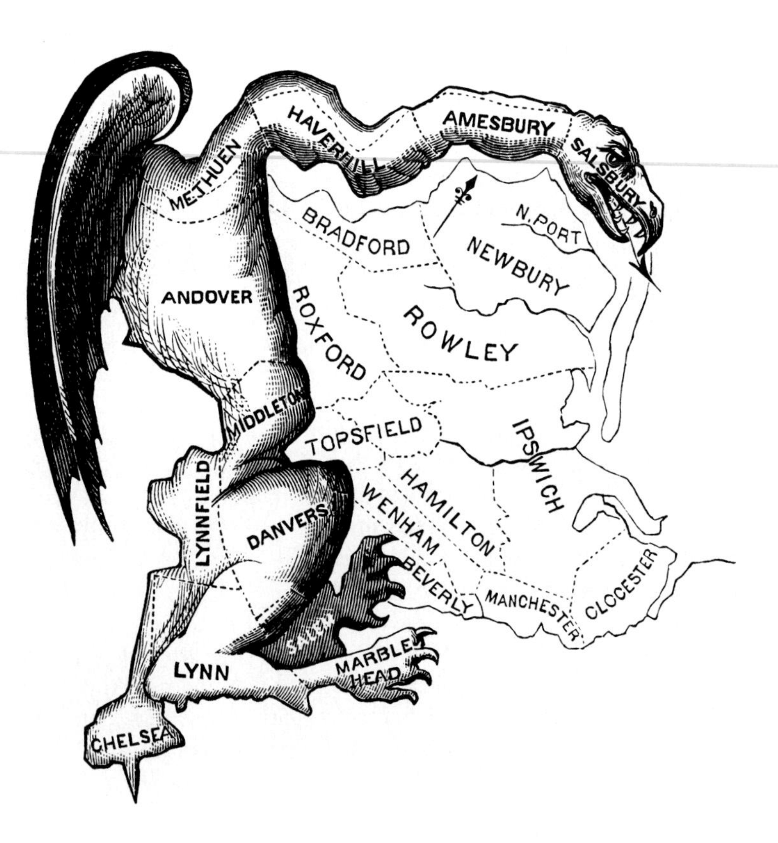

Figure 10–25. The original gerrymander was a district created in Massachusetts in 1812 that was designed to concentrate the Federalist Party vote and thus to restrict the number of Federalists elected to the state senate. The district configuration was at first likened to a salamander, but later it picked up the name of Governor Eldridge Gerry, the Anti-Federalist who signed the districting law. ("It looks like a salamander." "No, by golly; it's a Gerrymander!") (Reproduced from James Parton, *Caricature and Other Comic Art*. New York: Harper Brothers, 1877, p. 316.)

The electoral systems in many countries tolerate inequalities in numerical representation. Chapter 7 noted that this is true in Japan today. It was true throughout much of U.S. history, but in 1962 the Supreme Court ruled that for state and local general-purpose governments the number of inhabitants per legislator in each district must be "substantially equal" (*Baker* v. *Carr*). In 1962 Tennessee was still electing state legislators on the basis of a 1901 apportionment. The population had urbanized, so that by 1962 1 rural vote was equal to as many as 19 urban votes. Many other state legislatures contained similarly disproportionate representation, but redistricting now follows each decennial census. After each census the state legislatures also redistrict for federal representation for the new decade. The Supreme Court has extended the one-person, one-vote rule to apply to all levels of political representation in the United States except the governing boards of the special-district governments in metropolitan areas.

Representation in the U.S. Federal Government
The U.S. federal government itself is exempt from the one-person, one-vote principle. Neither the House of Representatives nor the Senate nor the presidency is based on population count.

The Constitution says "Representatives shall be apportioned among the several States according to their respective numbers, counting the whole number of persons in each State, excluding Indians not taxed." (AMENDMENT 14; SECTION 2). This means that no district may straddle two states, and courts have ruled that "whole number of persons" means both resident citizens and aliens, even illegal aliens.

The Constitution also specifies that "each State shall have at least one Representative." (ARTICLE I; SECTION 3). Several states have so few residents that without this constitutional protection, they would have to share a representative. There are 435 seats in the House of Representatives. Therefore, assigning one representative to each state leaves 385 seats to be allocated among the states by population. The Constitution's mandates cannot be reconciled to a one-person, one-vote ideal. In 1990 the 500,000 people of Wyoming had a representative, but each representative from California represented a district of about 620,000 people. Each state is also represented by two senators, no matter how few residents it has. Each of the senators from Wyoming therefore represented about 250,000 people, whereas each senator from California represented about 15 million (see Figure 10–26).

The president, in turn, is elected indirectly in the electoral college, an assembly also based on the states. Each state has as many electors as it has total seats in Congress, and the candidate who wins a majority of a state's popular vote wins all that state's electoral votes. In the 1988 presidential election Democratic candidate

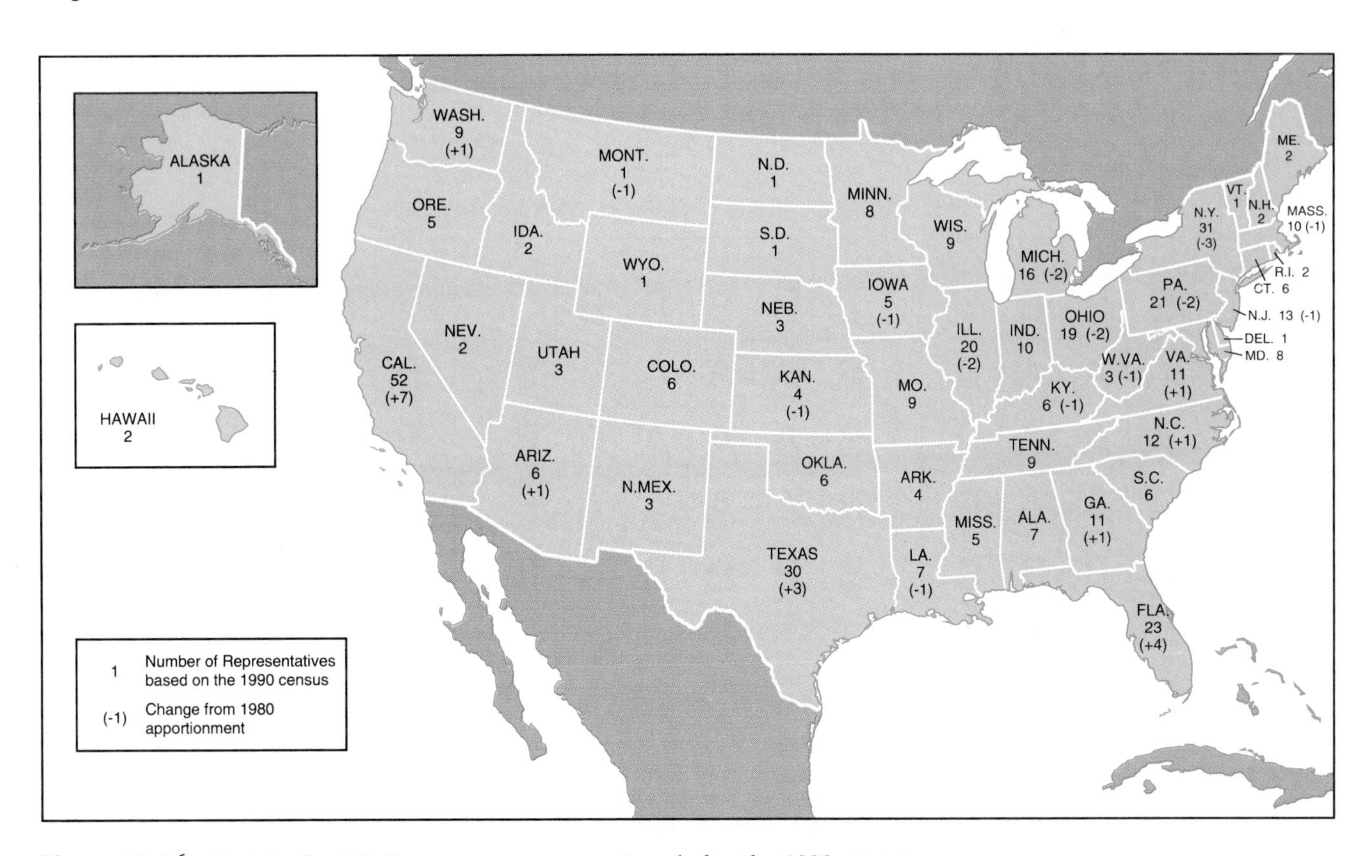

Figure 10–26. Seats in the U.S. Congress were reapportioned after the 1990 census.

Michael Dukakis won 46.1 percent of the national popular vote but only 20.8 percent of the electoral vote. In 1888 Grover Cleveland won the popular vote over Benjamin Harrison, but Harrison won the electoral vote and became president.

Today some states are rapidly increasing in population, and others are holding steady or losing population. Therefore, a population imbalance is widening among the states. Congress is becoming less representative, and a repeat of the unbalanced presidential election of 1888 is becoming more possible.

Gerrymandering The following discussion of the techniques of gerrymandering will draw mostly on U.S. examples because they will be most familiar to most readers. These techniques, however, apply wherever government is established with representation by territory. Figure 10–27 illustrates the problem of drawing district boundaries in an area of heterogeneous population. It is easier to define the problems of districting than it is to define equitable solutions. Substantial equality of population count in each district is legally required, and contiguity and compactness of districts is desirable because it eases communication within districts. The question still arises, however, whether homogeneity or heterogeneity of population within a district is preferable. Some people feel that districts should have a common social or economic characteristic, whereas others feel that balanced or integrated districts are preferable.

In 1972 a federal court asked a university geography professor to oversee redistricting in the state of Washington, but today computers allow any legislator to call up any district on a screen, shift a boundary, and get an instantaneous readout of what the voting behavior, racial composition, and other population characteristics would be in the newly drawn district. Technology has made it easier to gerrymander (see Figure 10–28).

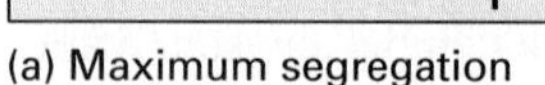

(a) Maximum segregation (b) Maximum integration

Figure 10–27. If we want to draw three districts of roughly equal populations, and if we have three population bases (x, y, z) defined by race, income, or any other characteristic, we can draw the boundaries in many different ways: (a) illustrates maximum segregation; (b) demonstrates maximum integration.

Voting At-Large A dominant majority can retain power without bothering to gerrymander by holding all elections in **at-large systems of voting.** In at-large elections there are no districts. All members of a government are chosen by the entire electorate. At-large systems al-

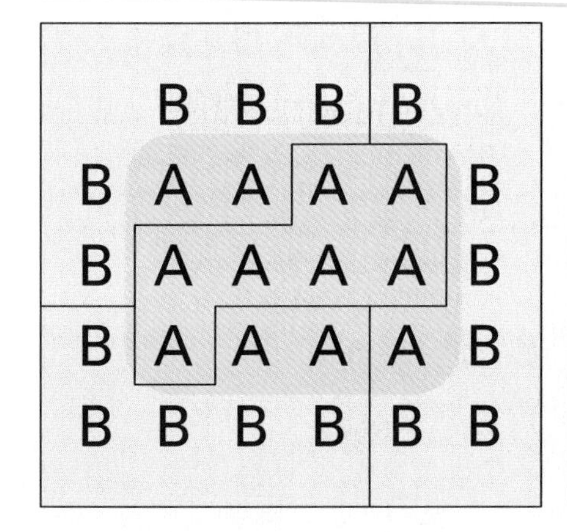

(a) Opponent concentration: A's control one district

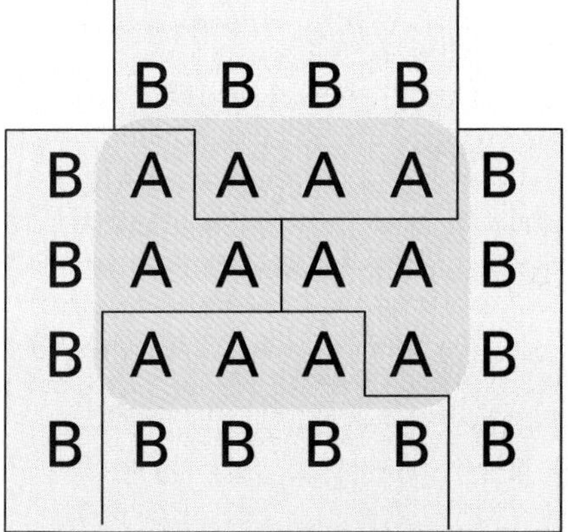

(b) Opponent dispersion: A's control no district

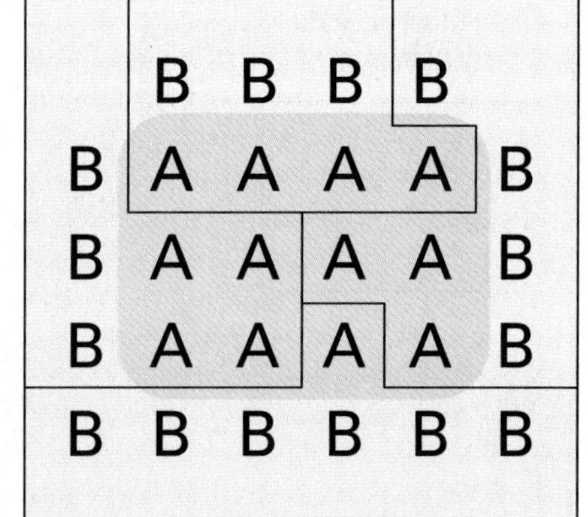

(c) Even division: A's and B's each control two districts

Figure 10–28. This figure illustrates types of gerrymandering. In each case the B's have an absolute majority, and so they have the power to draw the district lines. Drawing (a) illustrates containment of the opponents (the A's) so that the A candidate wins that one district with an unnecessarily large majority, but A's cannot win anywhere else. Drawing (b) splits up the A's so that they do not form a majority anywhere and cannot elect even one representative.

CRITICAL THINKING: *Garza* v. *County of Los Angeles*

In 1990, federal courts ruled that the Los Angeles Board of County Supervisors had used a combination of unconstitutional gerrymandering methods in drawing their voting districts to exclude Hispanic people from representation. Hispanics made up 35 percent of the county's total population, but none had ever sat on the five-member board.

The court recognized that the lack of a Hispanic on the board did not prove that unconstitutional discrimination had occurred. Alternative hypotheses can partly explain the lack of Hispanic representation. Hispanics may be underrepresented because:

- A high percentage are aliens.
- A high percentage fail to register.
- A high percentage are children too young to vote.
- A high percentage simply are not interested in politics.

Although each of these hypotheses has some validity, the federal court ruled that their combined explanatory power could not account for the total exclusion of Hispanics unless intentional discriminatory gerrymandering had also taken place.

Board district lines were redrawn, and a district was created in which Hispanics made up 71 percent of the population, although only 51 percent of registered voters. Voter turnout in the subsequent 1991 special election was only 21 percent, but a Hispanic won the seat on the board.

Questions

Does the fact that a particular group is underrepresented or not represented in a governmental body mean that discrimination has occurred?

How can discrimination be proved?

How much power should the courts or other institutions have to remedy such situations?

low a slight majority of one group to elect a government that is entirely of their group. The makeup of the community will not be reflected in a balance. Courts across the United States have reviewed cases in which at-large elections resulted in severe underrepresentation of distinct interest groups, and they have required many local governments to switch from at-large to district systems of representation.

MEASURING AND MAPPING INDIVIDUAL WELFARE

Countries differ greatly in the restrictions they place on their people's freedom and in the degree to which they encourage and achieve full development of their peoples' potential. Predominant attitudes toward some issues, such as women's rights, may be determined by cultures, and the regions of those cultures may cover more than one country. International Arab culture, for example, as noted in Chapter 8, seems relatively hostile to women's rights. Nevertheless, countries are the units within which laws promote or restrict individual advancement and opportunity, and so it is possible meaningfully to analyze and compare the policies of different countries.

A society in which the most capable people can rise to the top on merit alone is called a **meritocracy.** Most countries claim to be meritocracies. Rigid social stratification can be as unhealthy as arteriosclerosis is in a human body. Hurdles to individual advancement may include stratification of the population by caste, by race, or by employment restrictions. The modern labor movement has won important victories for working people, but in some countries labor unions may protect their members' jobs at the expense of opportunity for other individuals. In the United States, for example, some labor unions have been bastions of racism.

Many countries of diverse populations have *affirmative action* programs designed to lift to national standards of achievement those segments of the population that may be lagging because of historical lack of opportunity. Laws in both Malaysia and in Indonesia, for example, favor ethnic Malays over the more entrepreneurially successful Chinese. Nigerian law recognizes efforts of school and job-development programs to "reflect the federal character." This is a code for federally mandated ethnic quotas. Affirmative action programs have been crafted in the United States in both the public and the private sectors.

The major form of discrimination in the world is sexism, and the most powerful boost to individual achievement is education. Individual freedom is a value sought by all individuals. The variations in these social characteristics from country to country are examined next.

Sexism

Almost everywhere women are worse off than men are. They have less power, less autonomy, less money, more work, and more responsibility. National variations in sexism can be approached through a number of quantitative measures, beginning with the question of life itself. The variations in the sex ratios in the populations of the various countries provide clues to the geography of discrimination against women.

National Sex Ratios It is biologically natural for about 105 or 106 male children to be born for every 100 females. This ratio occurs everywhere in the world except in countries such as China, India, Korea, and Bangladesh where amniocentesis and abortion are available and where there is a strong preference for male children. The demographic impact is dramatic. In South Korea, for example, where fetal testing to determine sex is common, male births exceed female births by 14 percent. It must be assumed that female fetuses are aborted at a higher rate than male fetuses.

It is natural for more males to be born than females, but, given similar care, women tend to live longer than men. Therefore, the natural ratio of females to males in a human population should be 1 to 1, or slightly higher. The United Nations estimated in 1990, however, that, worldwide, men outnumbered women. Figure 10–29 shows that the national ratios of women to men vary around the world.

What determines national sex ratios? Some scholars who have looked at Figure 10–29 have noticed that the ratios of females to males are high in the rich countries of North America, Western Europe, and Japan. They have hypothesized that sex ratios vary with national wealth. Almost all countries with deficits of women are more or less poor, and few rich countries have such a deficit. Rising national wealth might be expected to reduce female mortality for two reasons, both of which follow from the fact that female mortality at childbirth is potentially high everywhere. First, birth rates are generally lower in richer countries, thus reducing maternal mortality. Second, improved medical care is generally available in richer countries during that critical period.

This hypothesis fails, however, when we examine the ratios in the poor countries. These countries do not show a consistent pattern. Some poor countries do not have deficits of women. Tropical Africa, for example, ravaged by poverty, hunger, and disease, has a substantial excess rather than a deficit of women.

In fact, variations in sex ratios can be explained only by a combination of national economic and cultural factors. In the states of North America, Europe, and Japan, women may suffer many kinds of discrimination, but they do not seem to suffer in access to medical care. Moreover, in these areas social and environmental differences tend to increase mortality among men. In most of Asia and Africa, however, women do not get the medical care, food, or social services that men get. As a result, fewer women survive than would be the case if they had equal care. In India, for example, the death rate is consistently higher for women at almost all ages except immediately after birth. A 1990 UNICEF study found that Indian girls are not fed as well as boys and get less medical attention.

Great variations exist within each world region. In Asia, Japan is at European status, and in some countries of Southeast Asia women outnumber men. In South Asia, however, the ratios of women to men are low. India and Pakistan exhibit two of the lowest ratios in the world, except for those Middle Eastern countries with large numbers of immigrant single males.

Sharp contrasts exist even within individual countries. In the northern Indian states of Punjab and Haryana,

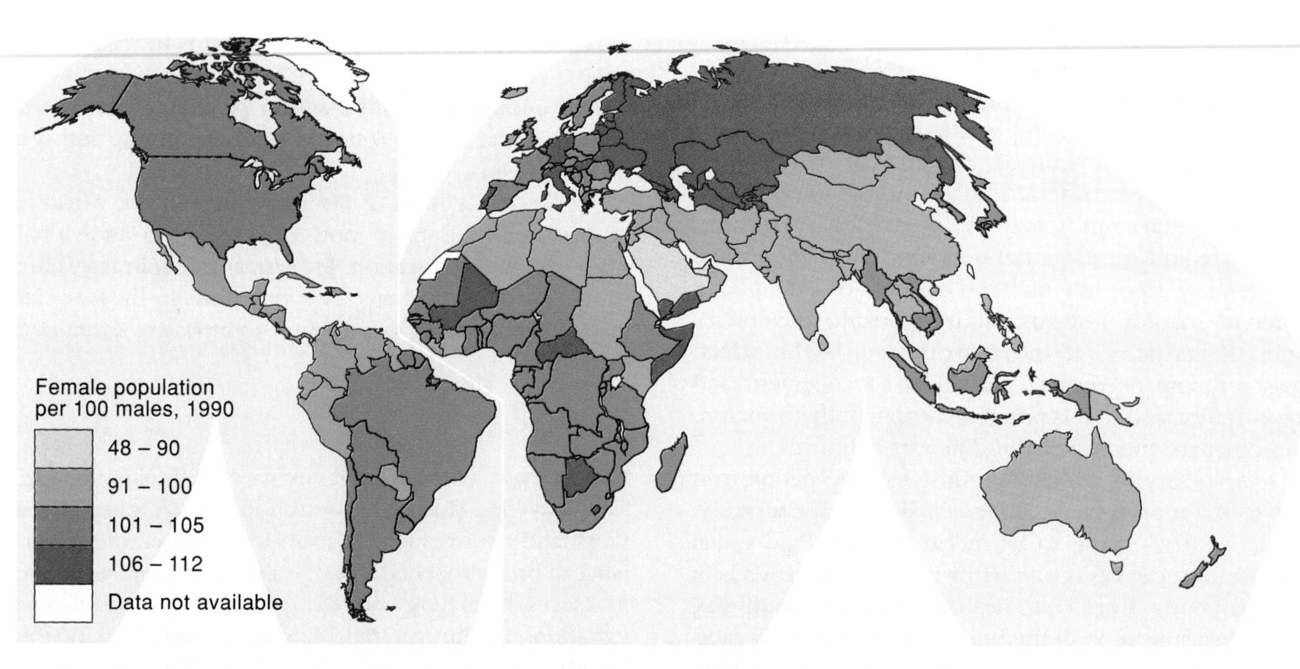

Figure 10–29. The ratios of females to males vary greatly around the world. (Data from United Nations, *The World's Women: Trends and Statistics*. New York's United Nations Publications, 1991, pp. 22–23.)

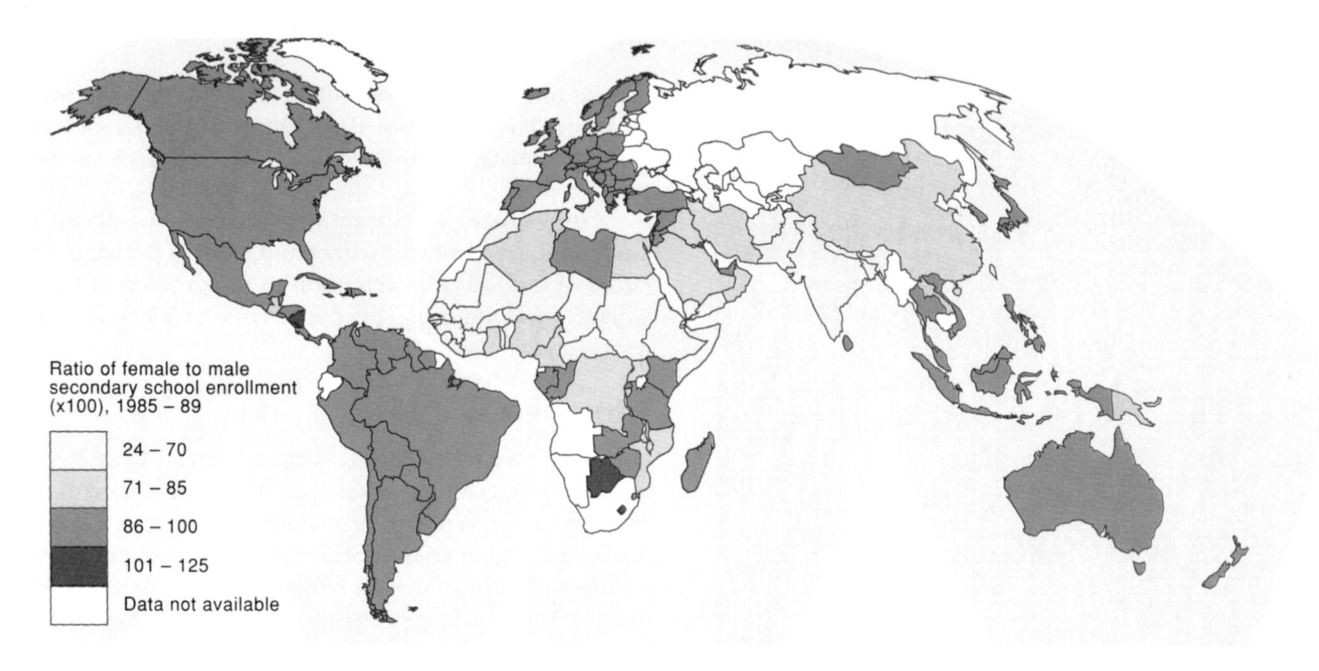

Figure 10–30. Varying national ratios of males to females in school may indicate sexism. (Data from United Nations, *The World's Women: Trends and Statistics*. New York: United Nations Publications, 1991, pp. 50–51.)

for instance, which are among the country's richest, the ratio of women to men is a remarkably low 0.86, whereas the state of Kerala in southwestern India has a ratio higher than 1.03, similar to that in Europe, North America, and Japan.

In China, the status of women seems to have risen since the Communists came to power in 1949, but most recently it seems to be declining. This may be the result of the birth control drive and the one-child policy. The preference for boys may be unfavorably affecting the survival prospects of female children, possibly explaining a rise in the female infant mortality rate. China's 1990 census revealed a female/male ratio of 0.938, but for children under 1 year of age the ratio was 0.898.

It seems almost universally true that women's positions are generally more favorable if they can earn an income outside the home. That reduces boys' advantage as potential supporters in the parents' old age, and females suffer less relative deprivation from birth. Later in life working women can rely on their own resources. A lot more needs to be learned about the attitudes of different cultures toward female employment outside the home.

Some demographers hypothesize that the societies that limit women's rights do not count all women in the census. If this undercounting of females actually occurs, then the resulting statistics would distort the sex ratio imbalances or even report gaps where they do not in fact exist. Certainly women are better off in countries that acknowledge female deprivation and seek to remedy it. Efforts to do this can be influenced by education and suffrage.

Comparisons of Women's and Men's Welfare

The UN Decade for Women (1975–1985) measured and mapped striking differences in women's welfare around the world. It should not be surprising to learn that women in rich countries are generally better off than women in poor countries.

Fewer data, however, focus on the contrasts between the welfare of women and that of men in individual countries, and those are the data that must be isolated to compare sexism in one country with sexism in another. The differences between the percentages of males and females in school in individual countries might reveal sexism (see Figures 10–30 and 10–31). A country's failure to educate young girls matters not just in itself but also because improving female literacy is one of the most effective ways to control population growth and to reduce infant mortality.

Some issues are almost exclusively women's issues, such as the female genital mutilation that is common across Africa and in the Near East and maternal death rates. Other issues affect both sexes but principally affect women: abortion rights, birth rates, equal pay, rape, domestic violence, maternity care or maternity leave, availability of contraception, child welfare and rearing practices, and even marriage and divorce rights.

One index of women's rights might be the percentage of the labor force that is female. The problem with this statistic, however, is that modern statistical measurements of the labor force measure only paid occupations, and in many countries women are not paid for their labor—particularly homemaking, farming, or even small-scale domestic manufacturing. Statistics report that in 1990 women made up 50 percent of the labor force in the Soviet Union, but everywhere else their participation rate was below that of men. It was about 40 percent in North America, slightly less in Japan, Europe, Australia, and New Zealand, and only 20 percent

Figure 10–31. Males and females attend this school in Kenya together, but generally worldwide, females have less chance to attend school. (Courtesy of Agency for International Development.)

or even less in South Asia and the Mideast. These figures would suggest that the women of Africa are not working. Observations of African life, however, reveal that the women are indeed working, they just are not getting paid. Better statistical measures of women at work are needed.

Women's participation in politics might serve as another index of sexism. In most democratic countries men got the vote before women. By 1992, however, Bhutan and Kuwait were the only countries where men could vote but women could not. Several countries reserve legislative seats for women. In Uganda, for example, each of the 34 parliamentary districts reserves an at-large seat for a woman.

The collapse of communist government in Eastern Europe has paradoxically restricted women's rights. The communist regimes rigorously upheld job security, maternity benefits, rights of divorce, and other social guarantees as a matter of Communist party doctrine. Women lost many of these rights in the transition to democracy and free-market economies. Layoffs in state industries struck women hardest because men's jobs were considered more important. Newly elected parliaments in every country had smaller percentages of women than had been the case under communism. Abortion rights in many Eastern European countries might be curtailed.

Regardless of national constitutions and laws, the relative income and welfare of women and men can be a product of their traditional roles in national culture. Many countries provide formally for sexual equality in the law, but few protect specific job, inheritance, property, or marriage rights or protect women from domestic violence. To learn more about the world geography of sexism, statistical and analytical tools will have to be improved.

In February 1992 the United Nations sponsored a conference in Geneva, Switzerland, of over 50 wives of heads of state to draw attention to the problems of the world's rural women. This conference may become an annual event.

Education

The more advanced a society is, the larger the investment it requires in its human resources, that is, education. Any country's greatest potential resource is its people. Ireland is poor in natural resources, and yet it advertises in U.S. business magazines that "A Single Investment Has Made Ireland The Wealthiest Nation In Europe: Ireland . . . invest[s] in education.. . . Ireland is rich in the resource that matters most . . . skilled people." Ireland has been so successful in educating its people that it must now attract jobs for them or else they emigrate.

A country's wealth and the education levels of its citizenry usually rise together. In the rich countries of Western Europe and Asia and in the United States, adult literacy is almost universal. The percentages drop only slightly in Latin America but markedly in Africa. Few African countries boast literacy rates, even for the preferred males, of much over 50 percent (Figure 10–32). Secondary-school enrollment figures tell the same story. The children in the world's rich countries are in school; the world's poor are not.

The World Geography of Freedom

Some people in the world live in freedom; others do not, but it is difficult to quantify freedom. Surprises appear when comparing statistics. Many civil libertarians ask why the United States tops all countries in the percentage of its population that is incarcerated. In 1990, 426 Americans per 100,000 of population were incarcerated. South Africa ranked second among countries with 333/100,000 overall, and the Soviet Union, on the verge of collapse, was third overall at 268/100,000. Western European countries generally range between 35 and 120 per 100,000, and Asian countries rank between 21 and 140. Does this mean that the United States is the world's most totalitarian country? Or does it mean only that if you break the laws, there is a higher probability than anywhere else that you will get caught and go to jail? Many observers might respond that just living in many countries is the equivalent of being in jail. How are such statistics to be interpreted?

Freedom House is a research institute that monitors political and civil freedom around the world. In the measurement of political rights, the institute assigns the highest scores where elections are fair and free and where the people who are elected, rule. The institute defines

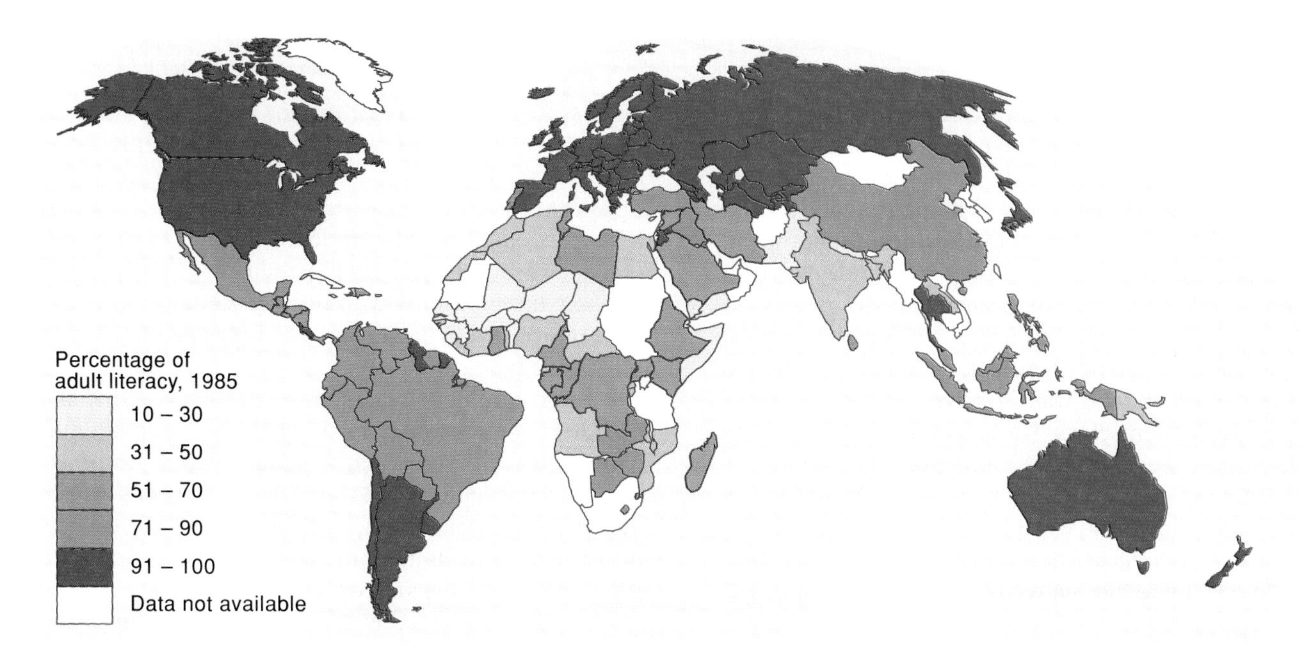

Figure 10–32. This map of national literacy rates reveals considerable variations. (Data from World Resources Institute, *World Resources 1990–91*. New York: Oxford University Press, 1990, pp. 262–63.)

civil liberties as including freedom of expression, assembly and demonstration, religion, and association. It rates highest those states that protect individuals from political violence and from harm inflicted by courts and security forces. These states have free economic activity and strive for equality of opportunity. Freedom House assigns each country a score of 1 to 7 in each of these two categories and then sums the scores in an overall rating of "Free," "Not Free," or "Partly Free."

Freedom House findings, as of December 1991, are reproduced in Figure 10–33. These rankings change rapidly as governments rise and fall, but the attempt to rate countries reveals wide disparities in freedom.

In 1991 the United Nations ranked freedom in each country according to a scale of 40 criteria reported in 1985. Each criterion was given equal weight, and they ranged from "the right to teach ideas and receive information" to the right to "interracial, interreligious or civil marriage." Sweden and Denmark scored 38; Iraq 0. The United States and Australia tied for sixth place, guaranteeing 33 of the 40 freedoms, although 12 countries ranked above them.

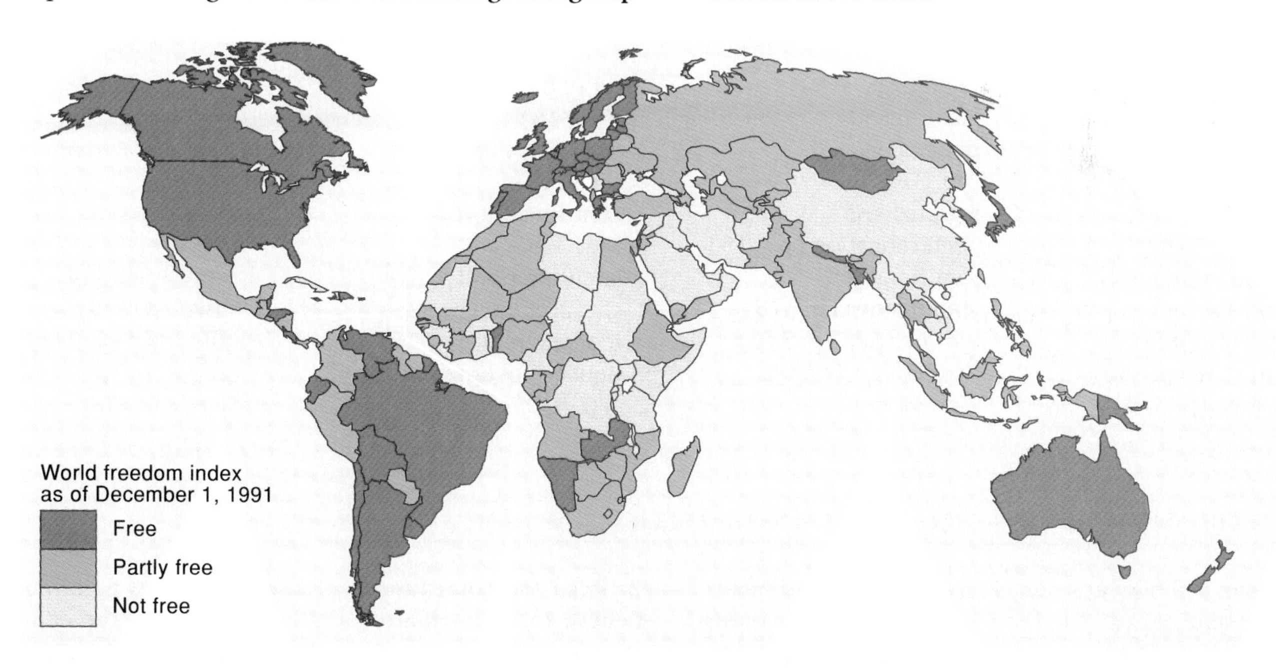

Figure 10–33. This map attempts to quantify the degree of freedom enjoyed by the population in each country.

SUMMARY

The earth's division into countries, or states, is the most important territorial organizing principle of human activities. The idea that the whole world should be divided into countries is relatively new in human history. It originated in Europe and diffused worldwide with European conquest. The neatness of the units on today's world political map nevertheless exaggerates the degree to which all people accept the current pattern of countries.

A state is a territory, a thing on a map. A nation is a group of people who want to have their own government and rule themselves. The ideal state is theoretically a nation-state. Several European nations existed before they achieved their own states. The political patterns of the colonies, however, were the conquerors' creations. The European imperialists carved up the world for their own convenience, no matter what level or form of political organization existed among the native peoples. Later these units gained independence. At first only large units were granted independence, but later this threshold principle was abandoned. Thus, today's map reveals many tiny, independent states.

Many states suffer from cultural subnationalism. These and other states employ three major strategies to meld diverse populations into a nation: redraw the world political map; expel people from any country in which they are not content, or exterminate them; or forge nations in the countries that exist now. The rigors of implementing any of these policies should tell us that territorial nationalism may not be a very positive ideal.

The existing countries struggle to weld their populations into nations. Six institutions—a country's church or churches, armed forces, schools, media, political parties, and labor unions—may be either centrifugal forces or centripetal forces. Geographers study the presence or absence of these institutions in any country, the unique mix or balance among them in each country, and their interaction with other aspects of each country's culture. Each institution has a geography within each country.

Each state demarcates its territory's borders, and it subdivides its territory for political representation or administration. Governmental subunits share responsibility for government with the national capital city. Each country has a unique balance of powers between its local governments and its national government. Federal and unitary governments are two forms. The power of local governments can be undermined by the creation of special districts, and most federal states contain territories that are administered directly from the federal center. Capitals are particularly important. Electoral geographers note that in any system of representative government, the boundaries of the voting districts can determine the outcomes of the ensuing elections.

Countries vary widely in the degree to which they liberate and train their populations. The major form of discrimination in the world is sexism, and the most powerful boost to individual achievement is education. Individual freedom is a value sought by all individuals. The incidence of each of these three social characteristics varies from country to country.

KEY TERMS

state (p. 312)
regnum (p. 314)
dominium (p. 314)
social Darwinism (p. 314)
nation (p. 314)
political community (p. 314)
legitimacy (p. 314)
nation-state (p. 315)
core area (p. 315)
regime (p. 315)
government (p. 315)
political culture (p. 315)
national self-determination (p. 315)
superimposed boundaries (p. 316)
indirect rule (p. 316)
threshold principle (p. 317)
cultural subnationalism (p. 318)
centrifugal forces (p. 318)
centripetal forces (p. 318)
enculturation (p. 324)
socialization (p. 324)
iconography (p. 325)
unitary government (p. 330)
federal government (p. 330)
electoral geography (p. 336)
gerrymandering (p. 336)
at-large system of voting (p. 340)
meritocracy (p. 341)

QUESTIONS FOR INVESTIGATION AND DISCUSSION

1. When and how was the pattern of local governments in your state drawn? What responsibilities do the various local governments have? Could you improve the pattern, reassign responsibilities to new districts or to different levels of government?

2. Read the constitutions of a few countries that you believe to be totalitarian and centralized. Do the constitutions reveal that? Does the U.S. Constitution accurately reveal the division of power between state and federal governments today?

3. Who represents you in the U.S. House of Representatives? In the Senate? Where do those individuals' campaign funds come from?

4. How is your city divided into council districts? Your state into counties or federal representative districts? Are these gerrymandered?

5. Has your state ever initiated any legislation that was later copied nationwide?

6. Which Federal District Court and Federal Appeals Court jurisdictions do you live in? Can you find an issue on which the interpretation of federal law differs in contiguous jurisdictions?

7. Is your state overrepresented or underrepresented in the electoral college?

8. Although many countries prohibit flag burning, no country has laws forbidding people from tearing or burning a national map. Why do you think this is so? Should governments have the power to outlaw either activity? Why or why not?

9. In 1991 the Yugoslav republics of Croatia and Slovenia declared themselves independent. The Yugoslav army, which was, in fact, largely Serb, invaded these republics and tried to suppress them, but in 1992 the major European countries recognized their independence. Shortly thereafter, UN forces arrived to patrol and keep the peace. Do you think regions of a country ought to be able to secede? If you favor a vote, do you think that majority approval should be required throughout the entire country or only in the areas seeking to secede? Should foreign countries interfere in the interest of protecting civil rights?

ADDITIONAL READINGS

Barone, Michael, and Grant Ujifusa. *The Almanac of American Politics.* Washington, D.C.: The National Journal, annually.

Hobsbawm, E. J. *Nations and Nationalism since 1780.* New York: Cambridge University Press, 1990.

Human Development Report. New York: Oxford University Press, published annually for the UN Development Program.

Johnston, R. J., David Knight, and Eleonore Kofman, eds. *Nationalism, Self-determination, and Political Geography.* London: Croom Helm, 1988.

Morrill, Richard L. *Political Redistricting and Geographic Theory.* Washington, D.C.: Association of American Geographers, 1981.

Olorunsola, Victor O., ed. *The Politics of Cultural Subnationalism in Africa.* New York: Doubleday & Co., Anchor Books, 1972.

Paxton, John, ed. (since 1969). *The Statesman's Yearbook.* New York: St. Martin's Press, annually since 1864.

Taylor, Peter J. *Political Geography: World-Economy, Nation-State, and Locality.* London: Longman, 1989.

The World's Women 1970–1990: Trends and Statistics. New York: United Nations, 1991.

CHAPTER 11

National Paths to Economic Growth

This scene from Lima, Peru, illustrates that although some countries have achieved high incomes and standards of living for their people, others struggle to provide basic services and goods. Each country devises a program to build and to regulate its national economy. (Courtesy of Agency for International Development.)

Most people today agree that one of the principal tasks of any government is to promote the welfare of its people, including their economic welfare. Each country devises a program to build and to regulate its national economy and to ensure that economic well-being is distributed across the national territory. Therefore, the world political map is also a map of local economies and economic policies. Many states need economic growth not only to raise standards of living but also to hold the nation together. Economic growth may convince the population that the state works for them, at least economically.

Some countries have achieved high incomes and standards of living for their people; others have not. Throughout this text we have referred to the *rich countries* and the *poor countries.* The terms rich and poor are fairly straightforward and descriptive. Other terms that are often used include references to the *developed countries* and the *undeveloped countries* or *underdeveloped countries.* These words, however, have sometimes been used to suggest that development is a process, a direction in which all countries are headed, that some have gone farther in this direction than others, and even that there may be something "wrong" with the countries "lagging behind."

This chapter emphasizes that levels of wealth and standards of living of different countries are actually difficult to measure and compare. There are many ways to measure economic status and welfare, and these measures do not always rank countries in the same order. The chapter is divided into three sections. The first examines several ways of measuring countries' wealth, and then it maps the rich and the poor countries. In that section we investigate the meaning of the popular term *The Third World.* The second section of the chapter examines the mechanisms that cause wealth to be created in some places and not in others. How do we explain the world distribution of wealth? In Chapter 7 we introduced the paradox that the countries that have the most resources are not necessarily the richest. It is not the possession of resources that makes countries rich, but the ability to use resources. The third section of this chapter investigates the variety of national economic geographies. Each country organizes its economy differently, manages the distribution of economic activities within its territory, and regulates participation in world trade.

ANALYZING AND COMPARING COUNTRIES' ECONOMIES

The study of statistics was first developed in the nineteenth century as a branch of political science, designed to collect and analyze numerical facts about states. British prime minister Benjamin Disraeli doubted the value of statistics. He knew that countries count things differently, that countries might lie to exaggerate their power or achievement or to conceal relative backwardness, and that even with the best of intentions, numbers often are not reliable and are open to misinterpretation. Disraeli concluded, "There are three kinds of lies: lies, damn lies, and statistics."

How Can We Identify Rich and Poor Countries?

Consider the following facts:

In 1989 the leading Japanese economic newspaper, *Nihon Keizai Shimbun*, trumpeted that as of 1987 Japan had become the world's richest country. It had $43.7 trillion worth of land, factories, stocks, and other wealth,

ABC NEWS *The Japanese*

Although Japan does not possess abundant natural resources, it has become one of the world's preeminent economic powers. The extent to which this economic expansion has been translated into a higher standard of living for the Japanese people, however, remains uncertain. Many people wonder whether Japanese consumer felt needs are the same as those of Americans.

The video titled "The Japanese" explores the realities of life in Japan. It explains ways in which Japanese culture differs from U.S. culture. For example, Japanese cultural traditions place greater emphasis on group cooperation and less on individual accomplishment. The video also maintains that some popular portrayals of the Japanese have some basis in reality. For example, Japanese employees put in long hours, and Japanese schoolchildren consistently outperform all competitors on international math and science exams.

At the same time, however, individual Japanese pursue many of the same goals as Americans, such as owning a home and being able to afford some of the many consumer goods produced in their country. Moreover, although Japan is a wealthy country, the high cost of living prevents most Japanese from enjoying a wealthy lifestyle. If you have an opportunity to watch "The Japanese," consider the following questions.

1. What has been the price of economic success to the Japanese?
2. How different from U.S. culture does Japanese culture really appear?

Figure 11–1. The city of Tokyo was "worth more" than the entire United States of America in 1987, due to high Japanese land values and the exchange rate between the U.S. dollar and the Japanese yen. (Courtesy of Japan National Tourist Organization.)

whereas the United States in that year had, according to the U.S. Federal Reserve Bank, $36.2 trillion worth of assets. Japan's population was a little less than one-half that of the United States, and so Japan's wealth per capita was more than double that of the United States. This wealth was due partly to all-time highs on the Japanese stock market and also to the explosion in Japanese land prices (see Figure 11–1).

This statistic may not seem to make sense. Rising land prices do not make a nation richer. They simply represent a transfer of wealth to landowners from those who wish to own land, in this case, overwhelmingly, other Japanese. Also, within a few months of this announcement, the Japanese stock market fell 39 percent. Trillions of dollars of Japanese assets evaporated. The process whereby paper wealth alternately appears and disappears is an ongoing one.

Most people think that statistics describe facts, but actually statistics are matters of choice and value. The vocabulary used to measure national wealth is based on theories devised by the British economist John Maynard Keynes (1883–1946).

Gross Product and Its Limitations

The two most commonly used measures of a country's wealth are its gross domestic product and its gross national product. **Gross domestic product (GDP)** is the total value of all goods and services produced within a country. **Gross national product (GNP)** is the total income of a country's residents, no matter where it comes from. Profits that a U.S. company receives from overseas, for instance, are included in GNP but are not included in GDP. Conversely, profits generated within a country by a foreign-owned firm are included in GDP, but if they are taken out of the country, they are not included in GNP. Most countries' net income from abroad is small compared with their domestic economy, and so there is little difference between the two statistics. Some countries, however, either receive large payments from abroad or send money abroad, and so for them the difference is great. Kuwait's GNP, for example, is 135 percent of its GDP, because it enjoys the income from its investments abroad, as discussed in Chapter 7. The U.S. GNP is about 101 percent of its GDP. A country's GNP can be smaller than its GDP, if profits are removed from the country each year. Canada's GNP is 5 percent smaller than its GDP, Ireland's 13 percent smaller, and Brazil's about 14 percent smaller. GNP is generally more useful in comparing standards of living in different countries, but GDP provides a better guide to analyzing countries' internal economies.

Both GNP and GDP are deceptive. Both begin with the idea that most activities and goods are of no value or of little value if they cannot be exchanged—they have no value in their simply being used. Therefore, these measuring techniques underestimate the activities of hundreds of millions of people who provide entirely for their own needs or who exchange very little except through barter. Farmers' incomes, for example, are probably undercounted in both the rich and the poor countries. In the poor countries, however, farmers make up greater percentages of the population. Furthermore, in the poor countries many farmers are producing food for subsistence, and so this production is not counted, whereas in the rich countries most farmers are business executives producing foods for sale. Therefore, undercounting farmers' production reduces and distorts the total measured national products of the poor countries more than it reduces that statistic for the rich countries. Measures of gross product underestimate the material welfare of the people in the world's poor countries, although by noting this we do not mean to exaggerate the quality of life those people may have or to suggest that they live well.

Because the statistics undercount subsistence areas, they exaggerate the degree to which modern areas, especially cities, dominate national economies. In extreme cases, a single city can provide most of a country's measurable output. Abidjan, with 15 percent of the Ivory

Coast's population, accounts for 70 percent of all economic and commercial transactions in that country. São Paulo, with 10 percent of Brazil's population, contributes 25 percent of that country's measurable economic activity.

Statistics particularly overlook the work of women (see Figure 11–2). Unpaid homemaking has no counted value. If homemakers cooked for one another instead of for themselves, measured national product would soar.

Statistics cannot measure activity that is illegal and therefore clandestine. Colombia's profitable drug exports do not seem to exist. Neither do illegal drugs in the United States, and yet the federal government admits that the drug trade is one of the nation's largest industries. Nor does gross product include the activities in the informal sector discussed in Chapter 9.

Figure 11–2. The grain carried in the bucket on this woman's head is not counted in measures of Niger's national product, nor is her labor. If the grain were carried in a truck, however, it would increase Niger's gross product. (Courtesy of Agency for International Development, Carl Purcell.)

The Geography of Exchange Rates Comparisons of economies are further complicated by the lack of common measures of value. Gross products are measured in local currencies, but the exchange rates among them are changeable and may be manipulated. This is of special concern to geographers not only because it hinders comparison of one place with another but also because variations in exchange rates trigger flows of goods, investment, and people. Between 1985 and 1990, for example, the U.S. government deliberately engineered a 50 percent decline in the value of the U.S. dollar relative to the currencies of Japan and the Western European countries. This devaluation had extensive geographic results. Exports of U.S. manufactured goods rose 80 percent from 1985 to 1990; new exports boosted the output of the nation's manufacturing belt (the industrial region from Pittsburgh to Green Bay; see Figure 9–3), as exemplified by the story of Milwaukee told in Chapter 9. Billions of dollars of foreign investment poured into U.S. industry, and millions of foreign tourists came to visit. In 1989, for the first time ever, foreign visitors spent more money in the United States than U.S. citizens spent abroad. As the dollar fell, goods flowed out, but tourists flowed in.

These new flows of people and goods diffused culture. Tourists took away impressions of American life, and Americans received new impressions of foreigners. Many more Europeans and Japanese could afford U.S. clothes, cassettes and compact disks, videos, and other cultural exports when those goods were half-price. Millions more saw U.S. movies and television shows.

Reciprocally, the exports to the United States from U.S. trading partners sagged, their economies suffered, and prosperity shifted across their national landscapes. Billions of dollars that U.S. manufacturers had planned to invest in factories abroad were invested in the United States instead, and communities from the Philippines to Brazil went without new vocational training programs, roads, and schools. Places dependent on U.S. tourists suffered recessions.

Geographers watch changing exchange rates to understand changing patterns of local prosperity. When the dollar rises again—and it surely will—the flows will reverse.

Gross Product and the Environment Measures of gross product fail to assess environmental damage. In standard techniques of bookkeeping, machines and buildings are counted as capital assets. As they age, their declining value is deducted from income. When a country's natural resources are exploited, however, annual depreciation is not deducted. Many countries sell off their timber and minerals, destroy their fisheries, mine their soils, and deplete their aquifers, and national accounting treats the proceeds as current income. The loss of the natural resources is not deducted as a depreciation of national assets.

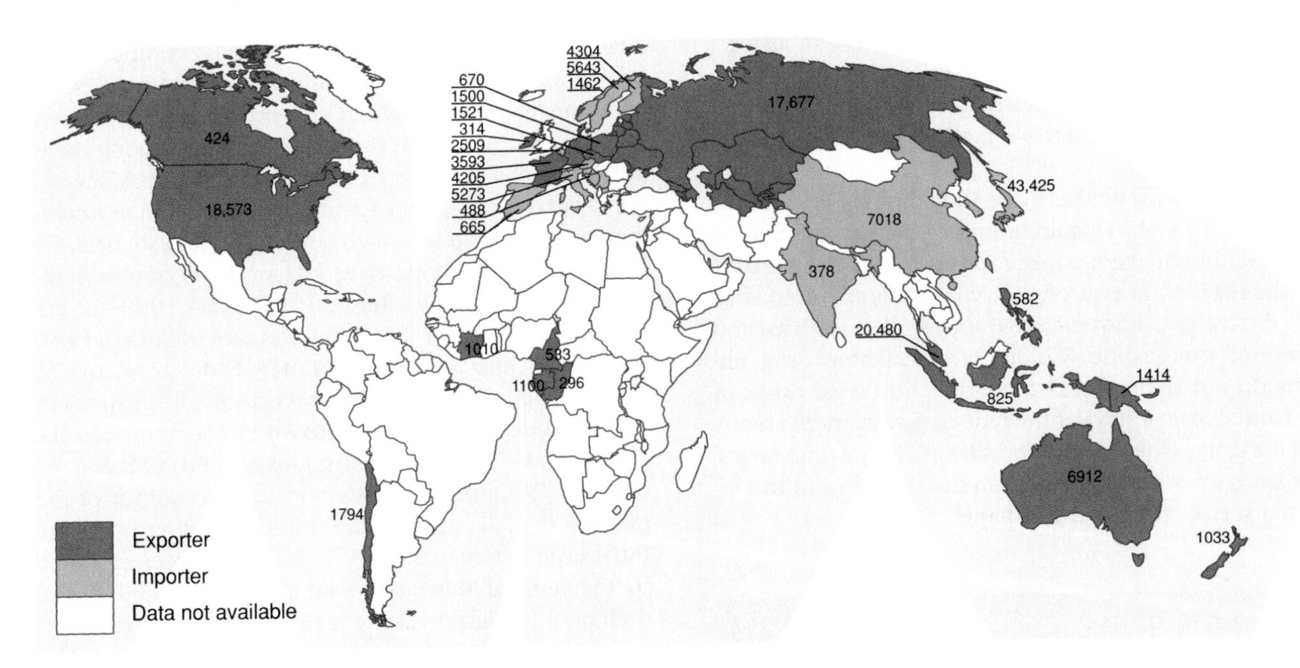

Figure 11–3. World wood trade. Top exporters and importers of roundwood (average annual exports and imports, 1985–87, in thousands of cubic meters).

Renewable assets such as forests, for example, can be harvested perpetually at the rate at which they renew themselves. Overharvesting, however, depletes them. In 1989 Indonesia and Malaysia accounted for 80 percent of the total world sales of tropical hardwoods, but their forests are disappearing. In West Africa and Southeast Asia, the hardwood trade is already necessarily drawing to an end. Gambia, Senegal, Togo, and Benin already have been virtually denuded, and the Philippines cut over 90 percent of its old-growth hardwood forests between 1960 and 1990. Ghana, Nigeria, and the Ivory Coast are struggling to save forest remnants. In 1989 the World Bank estimated that fewer than 10 of the 33 countries then exporting tropical forest products would still be able to do so by the year 2000 (see Figure 11–3).

If natural resources were treated as assets, then statistics would demonstrate that protecting the environment is in a country's best interest. It would be clear that there is no competition between environmental protection and economic growth, but that these two goals are related. A new statistic, **gross sustainable product (GSP)**, subtracts from gross product the value of destroyed or depleted natural resources. Costa Rica's GNP, for example, grew on average by 4.6 percent per year between 1970 and 1989. More than a quarter of this apparent growth disappeared, however, when a calculation of GSP adjusted for depreciation of Costa Rica's natural resources.

No matter how gross product is calculated, whether GNP, GDP, or GSP, data that present product per capita are doubly dubious. The second variable, the population, is also uncertain. Few countries have truly accurate population counts.

The Gross National Product and the Quality of Life

Figure 11–4 is a cartogram of the countries of the world with each country drawn in proportion to its GNP. This cartogram is strikingly different from a territorial map of the world. Some of the smallest countries—Belgium, for instance—have the biggest GNPs. Figure 11–4 also differs from the population cartogram pictured in Figure 6–2. Some countries with small populations have enormous economies.

Figure 11–5 establishes categories of GNP per capita and maps these on a regular world map. Figure 11–6 then transfers that information onto the population cartogram. Of all these figures, Figure 11–5 is perhaps the least useful. The territorial size of the countries does not explain wealth.

Noneconomic Measures of National Welfare

Per capita gross product is only one measure of national welfare. A picture of world standards of living and welfare can be improved by combining it with other statistics. We have already examined world death rates (Figure 6–7), food supplies (Figure 7–12), education (Figure 10–32), freedom (Figure 10–33), the availability of safe water supplies (Figure 7–16), and per capita energy consumption (Figure 7–24). You should not be surprised to discover that there is a great deal of coincidence among these statistics. The countries with the lowest death rates generally have the best food supplies, education, and safe drinking water, and their citizens consume more energy. These are all measures of the standard

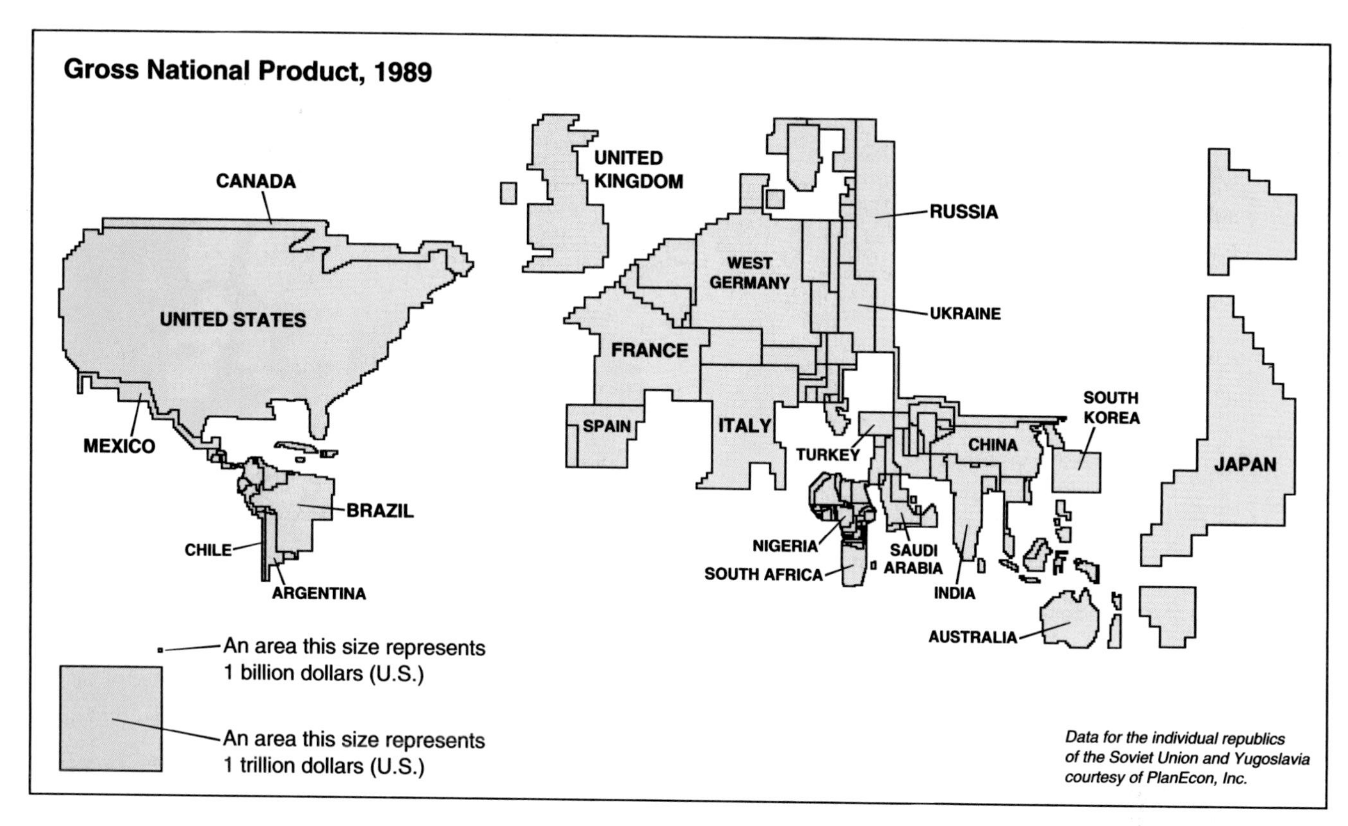

Figure 11–4. The size of the countries on this world map corresponds to their gross national products. The United States, Europe, and Japan grow considerably beyond their territorial relative size. In contrast, Africa shrinks. (Data from World Resources Institute, *World Resources 1992–93.* New York: Oxford University Press, 1992, pp. 236-37.)

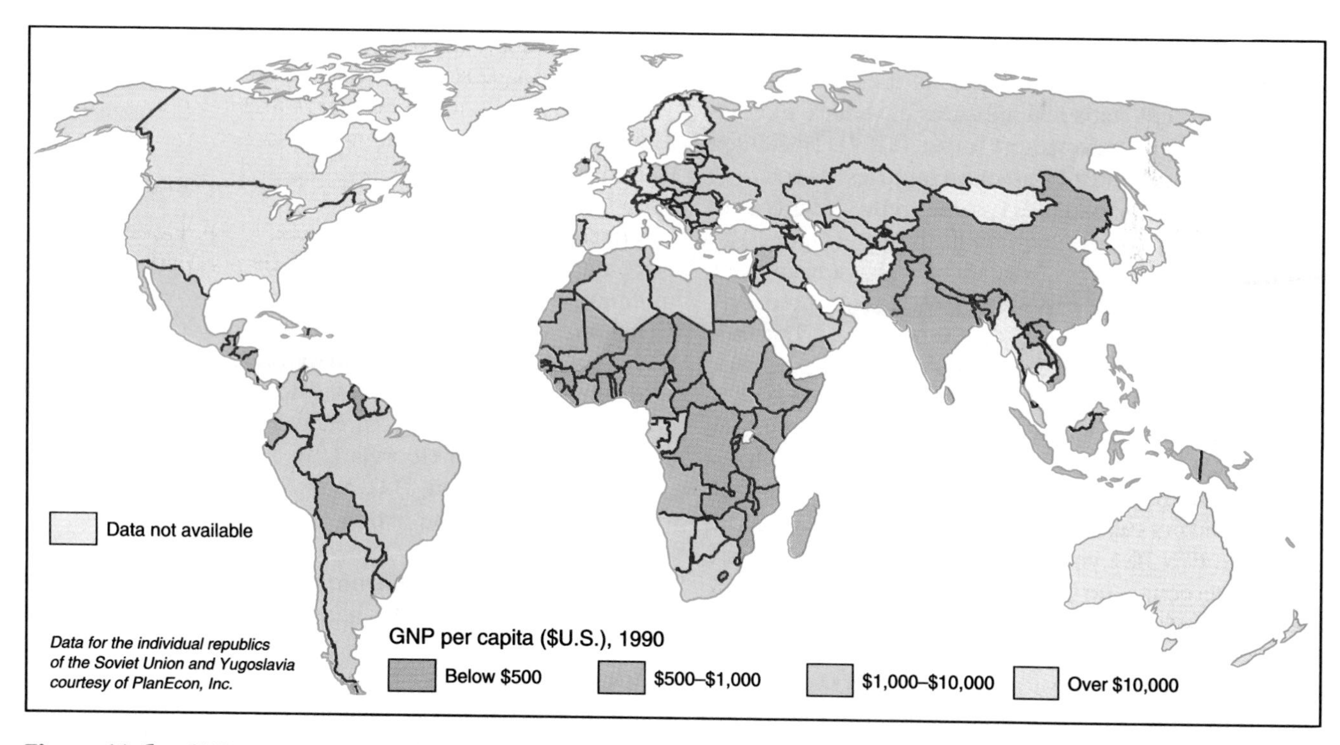

Figure 11–5. GNP per capita, 1990. (Data from World Resources Institute, *World Resources 1992–93.* New York: Oxford University Press, 1992, pp. 236-37.)

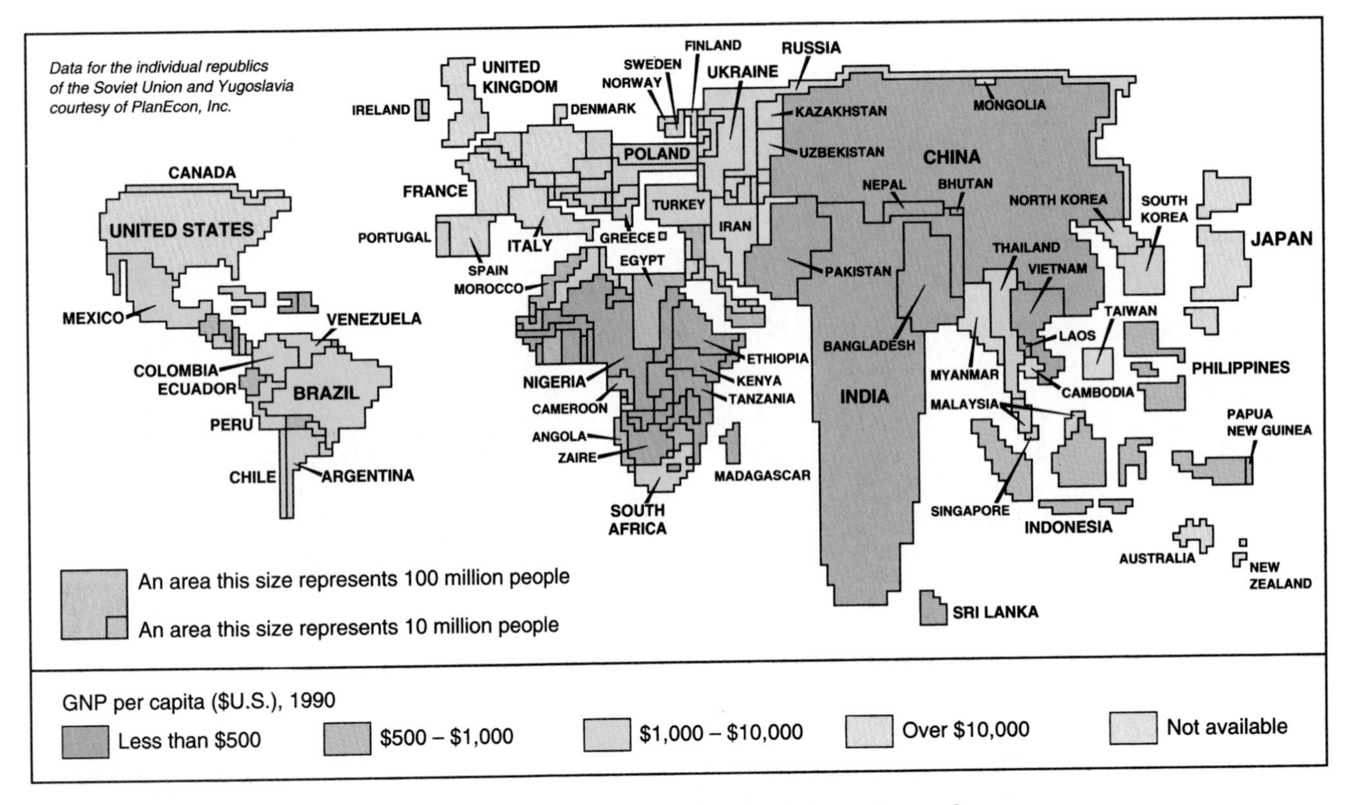

Figure 11–6. This figure presents visually what share of the earth's population enjoys each category of GNP per capita. Over half the world's population lives in countries in which per capita GNP is less than $500. (Data from World Resources Institute, *World Resources 1992–93*. New York: Oxford University Press, 1992, pp. 236-37.)

of living, and the highest-ranking countries are much the same as those that have the highest per capita GNPs.

These coincidences between per capita product and the other measures of welfare, however, are not exact. Therefore, the United Nations tries each year to balance different statistical measures of welfare to compile a **Human Development Index (HDI)**. This index compares three statistics among countries: purchasing power (a figure that adjusts income to the local cost of living), life expectancy, and adult literacy (Figure 11–7). HDI rankings differ from those based on any one criterion alone. In the report for 1991, for example, Uruguayans reported a GNP of $2470 per capita, but they enjoyed purchasing power of $4891 per capita because goods are cheap there. Their life expectancy was 72.2 years, and 95.3 percent of adults were literate. That gave Uruguay an HDI rank of 32. Citizens of the Republic of Korea (South Korea) reported a per capita GNP of $3600 but enjoyed per capita purchasing power of $4887, could expect to live 70.1 years, and had an adult literacy rate of 94.7 percent. The Republic had an HDI rank of 35. Saudi Arabia (per capita GNP $6200) scored per capita purchasing power of $4963, life expectancy of 64.5 years, and 57.9 percent literacy for an HDI rank of 69. The HDI ranking of these three countries was the reverse of their ranking by per capita GNP. The United States in 1991 ranked sixth in per capita GNP but seventh in HDI.

Any system of rankings that balances several criteria is problematic because the results depend on the variables chosen. Statistical reference volumes in the library, such as the several statistical yearbooks published by the United Nations, the World Bank, and the U.S. government, facilitate comparisons among the quantified descriptions of life in each country. General attributes of wealth and of "the good life" are found in some countries but not in others.

Variations within Countries Variations within countries can be even greater than variations among countries. These internal variations may reveal uneven national economies or inequity or social injustice. In the United States the 1988 infant mortality rate was 10 infant deaths per 1000 live births. The figures for the various states ranged from 7.2 in Hawaii (the same as in Denmark) to 12.6 in Georgia (the same as in Cuba). The rate in Washington, D.C., was 23.2, the same as on the Indian Ocean island-nation of Mauritius.

The Direction of Change Studying a statistic that changes over time helps geographers to assess whether the situation it describes is getting better or worse. Statistics examined per capita through time reveal a relationship between population growth and changes in the statistic measured. A country's GNP can grow, for example, but if the population grows faster than the GNP, GNP per capita will fall. For the countries of tropical

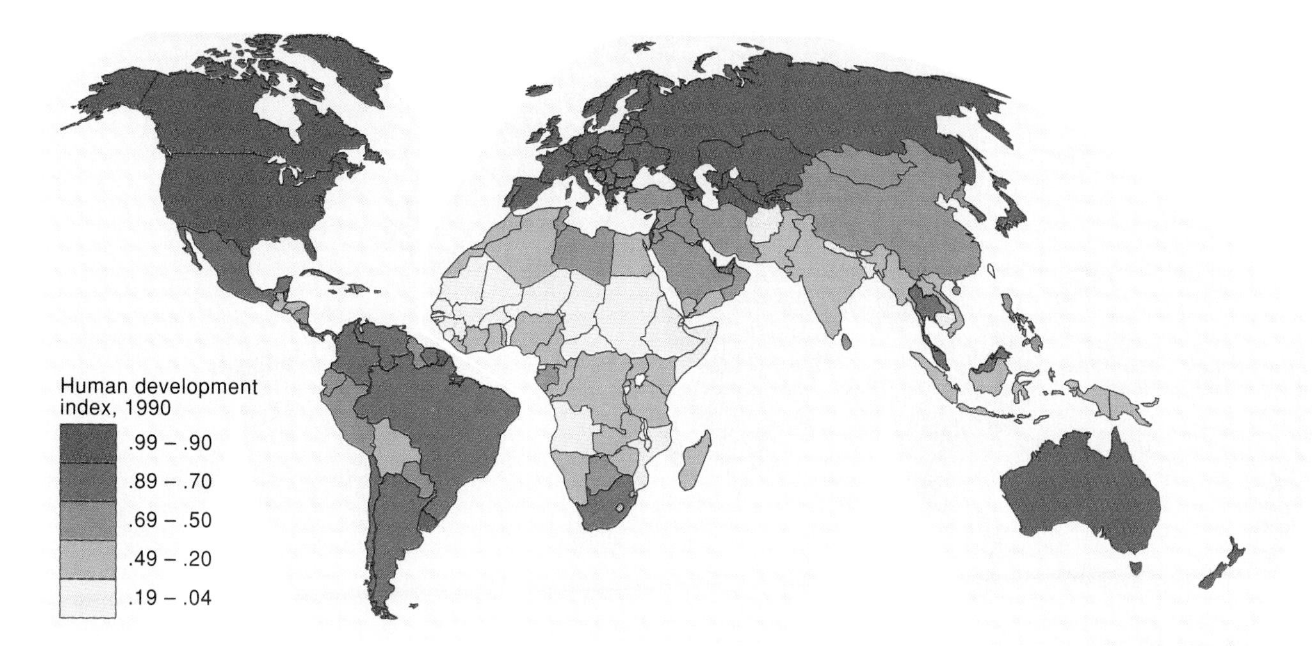

Figure 11–7. This map indicates where life is relatively good, as measured by the UN's 1991 Human Development Index.

Africa, per capita incomes rose in the 1960s, flattened out in the 1970s, and fell in the 1980s. This was due partly to the deterioration of African economies, but it was also due partly to rapid population growth.

Comparing Countries' Sectoral Evolution

Geographers are interested in more than the size of each country's economy. They also want to know what the people are doing for a living and which activities are producing the wealth in each country.

Employment Shifts among Sectors Over a long period most countries that are rich today have experienced shifts in the distribution of jobs among economic sectors. This shift was defined in Chapter 9 as **sectoral evolution**. Before industrialization, most of a country's labor force is occupied in the primary sector (see Figure 11–8). Societies with the bulk of their employment in this sector are called **preindustrial societies** (see Figure 11–9).

With industrialization, the *proportion* of the labor force employed in the primary sector declines. This is true despite the fact that both the *total number* of people in the sector as well as the value of the sector's output may rise. The primary and extractive activities remain crucial to many economies, but they provide a diminishing share of jobs.

Many jobs lost in agriculture are initially replaced by new opportunities in industry, and the proportion of workers in the secondary sector increases until, at no precisely defined point, certain societies come to be called **industrial societies**. In the United States, for instance, manufacturing employment overtook farm employment during World War I, and by 1925 blue-collar workers in manufacturing industries had become the largest single occupational group. Many of the richest countries today are industrial societies.

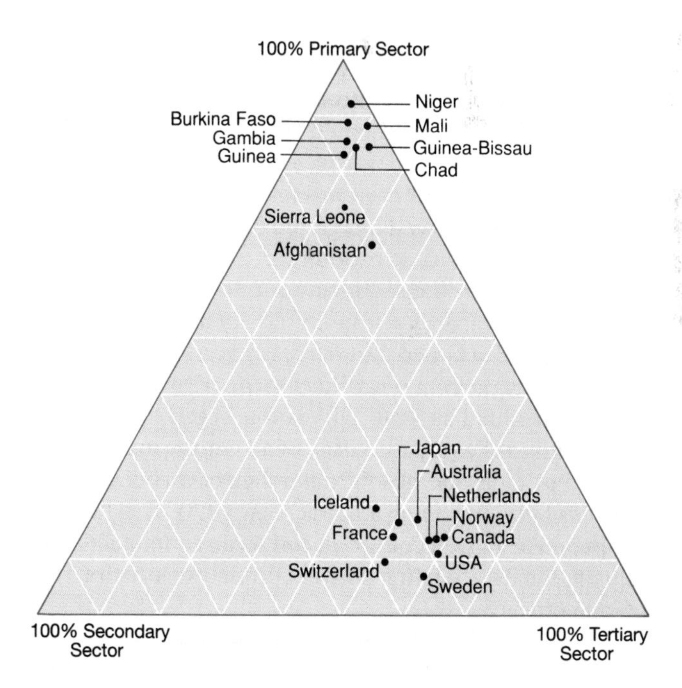

Figure 11–8. Sectoral shares of the national labor force. This figure shows the proportions of the labor force in each country that work in each of the three sectors of those country's economies. The 20 countries shown are the top 10 and the bottom 10 ranked by HDI. The workers in the poorer countries are concentrated in the primary sectors, whereas more of those in the richer countries are in the secondary and tertiary sectors. (Data from UNCTAD.)

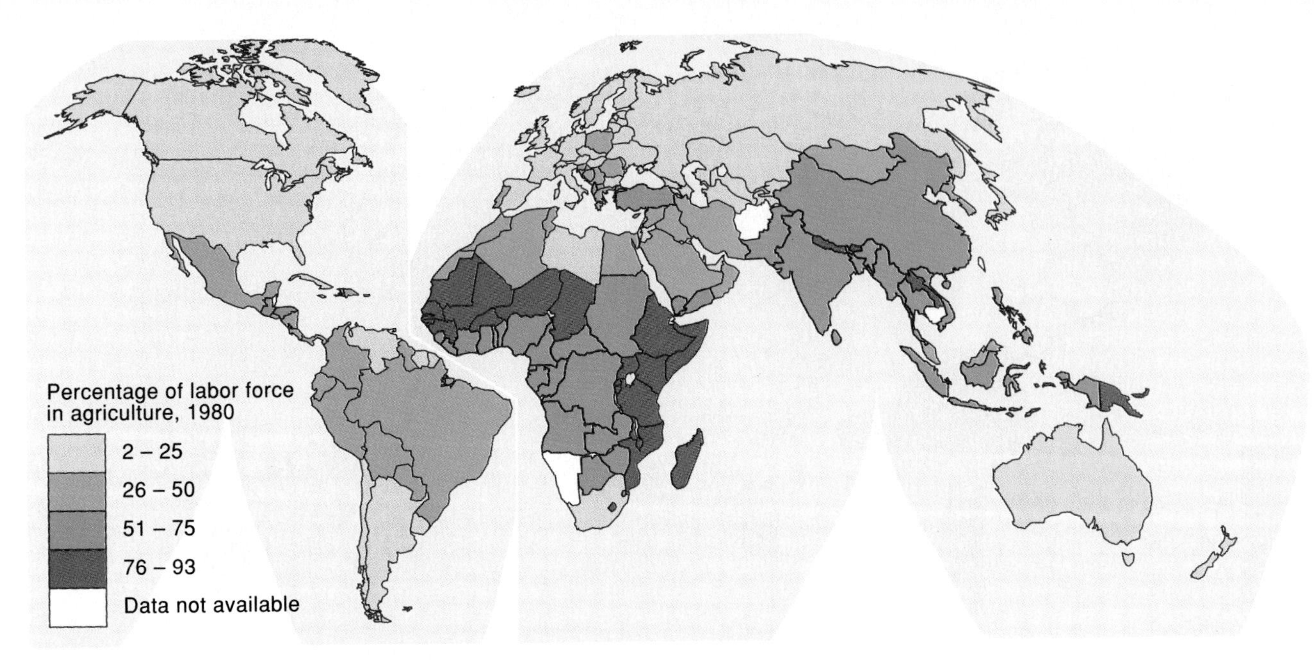

Figure 11–9. Percentage of the labor force in agriculture. Many of these countries with high percentages of their labor force in agriculture are barely able to feed themselves, whereas several countries with only small percentages of their labor force in agriculture enjoy food surpluses. This pattern reveals the importance of capital investment and technology in agricultural production. (Data from World Resources Institute, *World Resources 1992–93.* New York: Oxford University Press, 1992, pp. 264-65.)

Continuing evolution of some countries' economies has drawn a higher percentage of workers into the tertiary sector. Services accounted for 28 percent of all jobs in the United States in 1869 and almost 50 percent by 1929. Sometime in the 1940s the proportion first exceeded 50 percent, and the United States was hailed as the world's first **postindustrial society**. By 1990 tertiary employment represented about 75 percent of the nation's jobs. This trend has been observed in most other advanced nations and even in a few poor nations whose tertiary sectors are well integrated into the international economy—tourist destinations, for example.

At first both the secondary and the tertiary sectors increased their shares of total employment as the primary sector's share dropped, but in recent years in the United States the proportion of workers in the secondary sector has begun to fall as well. The nation began losing manufacturing jobs as a percentage of all employment about 1960, and even the absolute number of manufacturing jobs has been declining since at least 1970. This has not happened in all rich countries, and some economists fear that it signals a decline in the competitiveness of U.S. manufacturing.

Shifts in the sectoral distribution of a country's labor force—from farming to factory work to data processing—affect urbanization, social life, personal values, individual social and geographical mobility, and every other aspect of a nation's culture.

The Sectors' Contributions to Gross Product
The share that each sector of an economy contributes to total national product is not necessarily the same as the share of the national labor force that sector employs (see Figure 11–10). Some countries have high proportions of their workers in the primary sector, but those

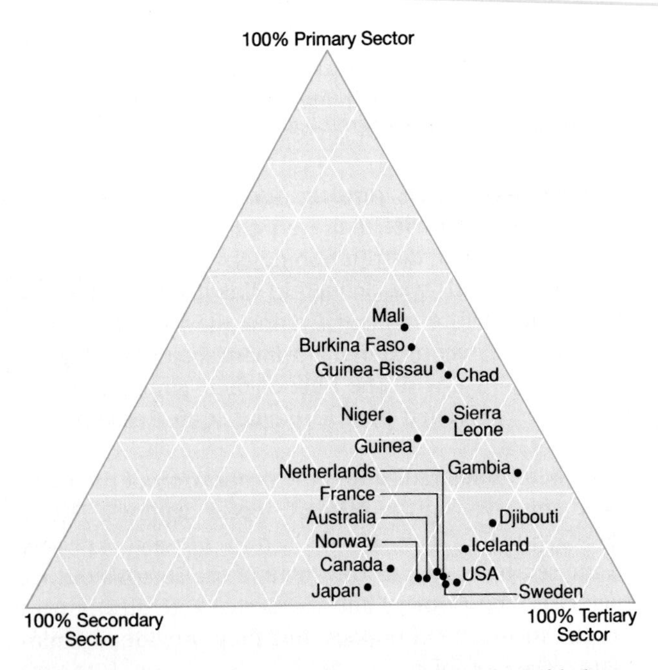

Figure 11–10. Origins of GDP, by sector. Value added per worker is usually highest in the secondary sector, and so the percentage of a country's GDP that is produced in the secondary sector almost always exceeds the percentage of the country's labor force in that sector. The large percentages of the labor force in the poor countries that are working in the primary sectors are producing a relatively small share of GDP. (Data from UNCTAD.)

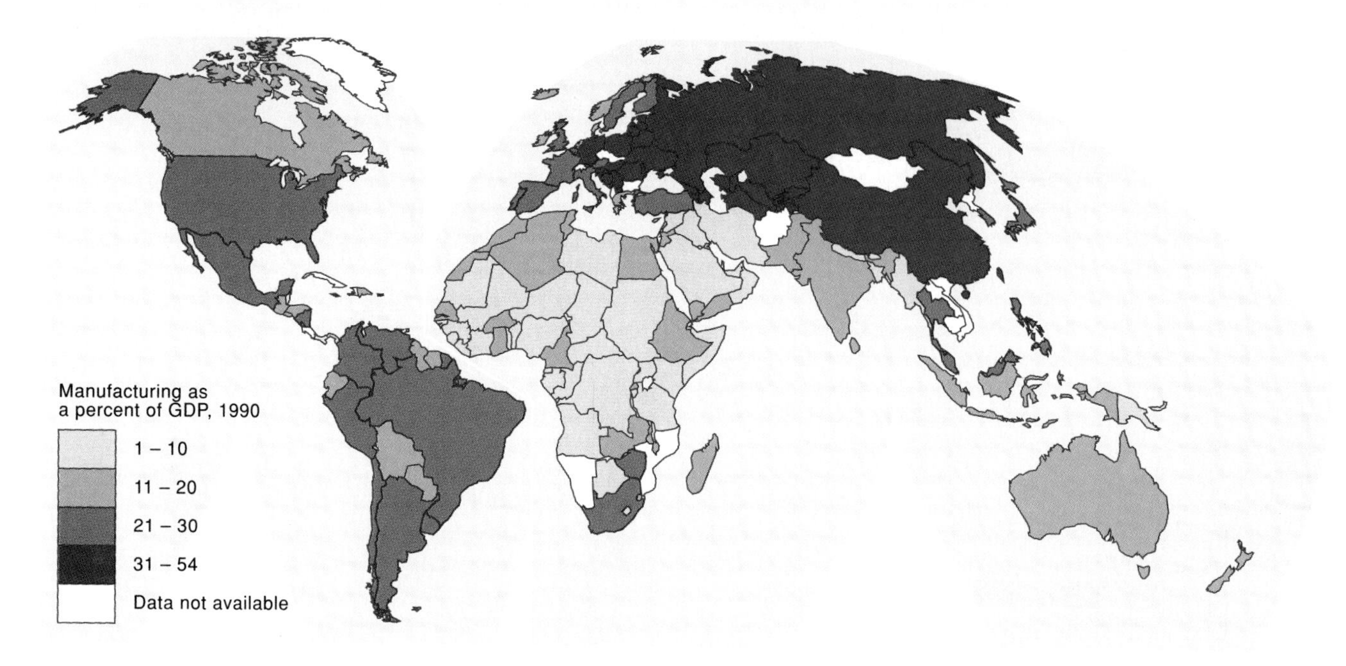

Figure 11–11. Secondary sector output as a percentage of GDP. The term *industrial society* has no exact formal definition, but this map reveals the countries in which industry contributes the highest share of GDP. (Data from UNCTAD.)

workers are producing relatively little of the measurable national output.

As a rule, the value added per worker in manufacturing is usually highest, and so the secondary sector contributes a larger share of total product than it employs of the labor force (see Figure 11–11). The share of the U.S. labor force in manufacturing, for instance, is falling, and even the total number of jobs in manufacturing is falling, but that sector's contribution to GNP has held fairly steady since World War II—between 20 percent and 23 percent annually.

The ability to add value to a raw material and to capture some of that value added as profit helps explain why the miller usually gets richer than the farmer, the manufacturer richer than the miner—the miller and the manufacturer are in the secondary sector. Later we shall see how this helps explain why some *places* get rich while others stay poor.

Where Is the Third World?

The term *Third World*, introduced by the French demographer Alfred Sauvy in 1952, reflected attitudes during the cold war following the end of World War II. An ideological line was drawn between the capitalist countries led by the United States and the communist-ruled countries headed by the Soviet Union. At the height of the diplomatic pulling and tugging, it was expected that all countries should line up on one side or the other. Many did so, but a few, led by President Tito of Yugoslavia, Nasser of Egypt, Nehru of India, Nkrumah of Ghana, Makarios of Cyprus, Sukarno of Indonesia, and Emperor Haile Selassie of Ethiopia, clung to a precarious neutrality. These maverick states came to be known, collectively, as the **Third World**, to distinguish them from the **First World** of the Western bloc and the **Second World** of the Soviet bloc.

Over the years the term *Third World* gained economic and sociological connotations. The First World was interpreted as a haven of science and rational decision making: progressive, technological, efficient, democratic, and free. The First World included as allies in the cold war all reasonably well-to-do, noncommunist countries, even though some of them were politically repressive, such as several Latin American states, South Africa, South Korea, and Taiwan. The Second World was also defined as modern, powerful, and technologically sophisticated, but it was dominated by totalitarian communist governments.

The countries of the Second World never achieved economic growth equivalent to that in the First World, and the collapse of communist regimes in 1989 exposed their true poverty. The Soviet Union achieved brilliant scientific advances, maintained an immense military establishment, and supported revolutionary strife in other countries from Cuba to Ethiopia, but its military efforts ultimately contributed to its economic collapse.

In contrast with the First World and the Second World, the Third World was seen as a world of underdevelopment, overpopulation, irrationality, religion, and political chaos. The term also carried a racial, or even racist, aspect: Most Third-World countries were nonwhite. Many Third-World countries were still colonies in 1939, but they "emerged" after World War II. According to the theory, they were poor but struggling to develop. Third-World societies were interpreted as a zone of competition between the First- and Second-World models of modernity. Some Third-World coun-

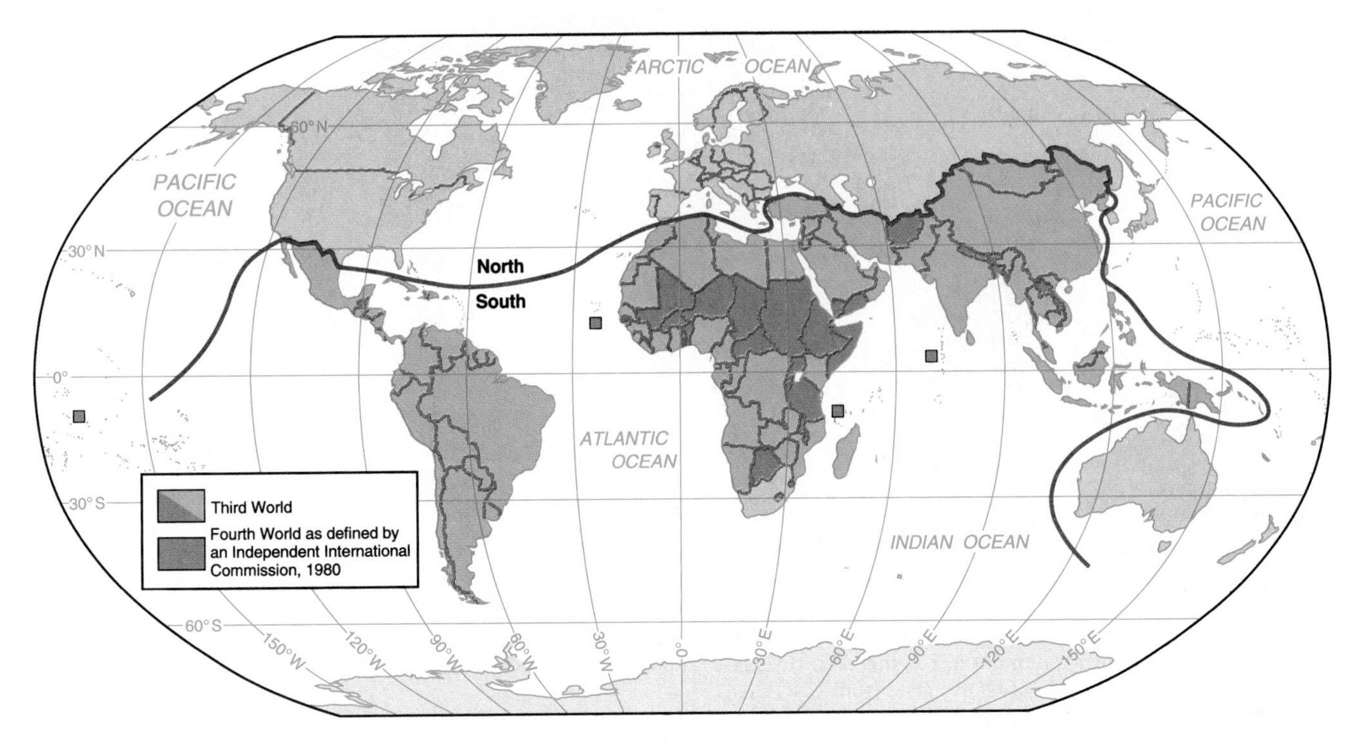

Figure 11–12. Because the term *Third World* confused cold war political alliance with economic status, efforts were made to "refine" it. Some scholars subdivided the Third World into a poorest Fourth World. Others tried to differentiate a rich "North" of the world from a poor "South."

tries won financial aid from both the First and Second worlds by playing them off against each other.

In 1980 an Independent Commission on International Development Issues introduced a new distinction between the world's rich "North" and the poor "South." The South was subdivided into a poor Third World and a still poorer Fourth World (see Figure 11–12).

Today the terminology of First, Second, and Third worlds is clearly out of date. As we saw in the discussion of regions in Chapter 2, a region by definition must be homogeneous in certain aspects. Historically, the unifying element in the Second World was the existence of communist regimes. With the collapse of these regimes in Eastern Europe and the former Soviet states, however, the Second World as a political bloc simply ceased to exist.

Similarly, the term Third World lumped together for political reasons peoples with extremely diverse histories and cultures who inhabit a broad range of physical environments that possess a great variety of natural and cultural resources. Each of these societies is richly unique. If we consider just the countries whose leaders made up the original political Third World—Yugoslavia, Egypt, India, Ghana, Cyprus, Indonesia, and Ethiopia—they cannot be homogenized economically, culturally, or sociologically. It is condescending and ultimately insulting to refuse to recognize and to appreciate their differences. Indiscriminate categorizing not only prevents understanding but also prevents helping. If we want to understand and to help the world's poor countries, we must recognize and appreciate their differences.

In the 1990s the term Third World will disappear. It will be remembered as a relic of a political cold war mentality that was indifferent to all other distinctions among the world's nations and cultures.

WHICH MECHANISMS DETERMINE THE GLOBAL DISTRIBUTION OF WEALTH?

A comparison of the map of per capita GNP (Figure 11–5) with the map of world soils (Figure 3–27) and the several maps of resources in Chapter 7 reveals that the countries most richly endowed with fertile soils and minerals do not necessarily have the highest per capita products. Conversely, some of the world's richest peoples live in environments with meager endowments.

The theory of environmental determinism discussed in the essay following Chapter 4 would find this pattern inexplicable, but the study of geography can actually help explain it. As early as 1957, just on the eve of the wave of decolonization that brought independence to over one-half of today's countries, Norton Ginsburg's presidential address to the Association of American Geographers emphasized that in the coming years having resources would not be as important as being able to put them to work. The geography of resources alone does not explain the geography of wealth.

Downstream Activities

Possessing raw materials and selling those raw materials in world markets is generally less profitable than capturing the value added by manufacturing and transporting goods, or even advertising, insuring, marketing, and financing these goods. In economic discussions it is common to use the image that all products "flow" from their sources as raw materials to their ultimate consumers, and so the economic activities closer to the consumers are called **downstream activities.**

The value added to raw materials downstream in the secondary and tertiary sectors surpasses the value of the original raw materials, whether the raw materials are mineral or agricultural. The value added by refining copper ore and manufacturing things out of it, for example, quickly surpasses the value of unrefined copper. The value added by grinding wheat into flour is greater than the value of the wheat. This principle holds true no matter how valuable the original raw material may be. Diamonds are valuable, but the value added to them by diamond cutters and polishers is greater than the value of the original uncut diamonds. The greater the value added, the greater the potential for profit. Chapter 7 noted that the OPEC members are eager to increase the value they add to their crude oil by refining and marketing it. They can capture a percentage of this added value as profit.

The places that provide the downstream activities capture the greatest value added, and they prosper. The places that export raw materials do not enjoy the same growth. They may even have to buy goods manufactured out of raw materials that they originally exported themselves. As long as Chile exports copper to the United States, for example, and buys back goods manufactured out of copper, the United States will grow richer than Chile. As long as Botswana exports uncut diamonds to Israel and buys back engagement rings, the Israelis will grow richer than the Botswanans.

The term value added has even been extended to **conceptual value added**, as in advertising. If a $10 million advertising campaign convinces people to spend $100 million more for brand *X* than for brand *Y* when those brands are, in fact, nearly identical, then $90 million of value has been added "conceptually." That $90 million can provide a lot of jobs. The conceptual value added to products by the advertising industry in New York City today is far higher than the value added by manufacturing in many of the world's greatest manufacturing cities.

If consumers consistently buy a particular brand that they have come to trust, they will try different goods sold under that brand name ("I like Brand X plum jam, and so I will try Brand X strawberry jam.") The right to use that brand name, called the ***trademark***, is itself a valuable commodity—more valuable than any good to which it may be attached. The geography of adding value to goods downstream seems abstract, but the geography of these activities explains in the 1990s what places are like and what people do for their living.

The Japanese Economy Japan is the most astounding and paradoxical success story in economic geography today. The Japanese islands have few natural resources, but on this meager natural geographical endowment the Japanese have developed a great economy. How?

The answer lies in Japan's cultural resources and in its trading patterns. The key cultural resources include the people's education and labor skills, technology, and the country's style and degree of cooperative organization. All these are carefully managed by the government. Japan's participation in international exchange is also carefully regulated. The Japanese import raw materials, transform them into manufactured goods, and export these finished goods plus an array of valuable services around the world (see Figure 11–13). The Japanese also invest their profits and savings around the world. Much else about Japanese culture is still distinct. It developed among the people over a long period of relative isolation, and they value and protect it.

Japan's economic success despite its lack of raw materials demonstrates that a place need not have its own raw materials to accumulate wealth. It can prosper by importing raw materials and adding value in manufacturing, or by providing other downstream services. A

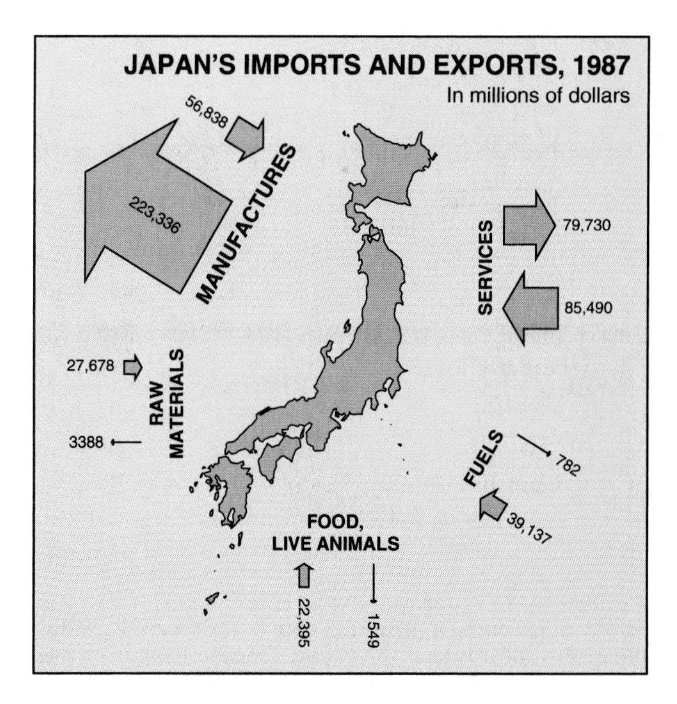

Figure 11–13. Japan's imports and exports. Japan imports 8 tons of fuel, food, wood, and other raw materials for each ton of goods that it exports.

TABLE 11–1 ***Value Added in Manufacturing Products***

Product	Value added ($/pound) Rough 1990 estimates
Satellite	20,000
Jet fighter	2,500
Supercomputer	1,700
Airplane engine	900
Jumbo jet	350
Video camera	280
Mainframe computer	160
Semiconductor	100
Submarine	45
Color television	16
Numerically controlled machine tool	11
Luxury car	10
Standard car	5
Cargo ship	1

country can achieve economic growth by adding higher and higher value to raw materials.

Table 11–1 gives the approximate value added in manufacturing certain products. In the 1950s Japan was a relatively poor country, but it began to manufacture the products at the bottom of the scale. It has prospered by steadily progressing up the scale in its exports, and it has surrendered the manufacture of the items that are low in the value-added scale. Japan invests in factories to manufacture these products in less developed countries dominated economically by Japan: Malaysia, Thailand, and Indonesia.

The Problems of Today's Poor Countries Today's poor countries have not been able to capture the value added by downstream activities. Most of them export just unprocessed raw materials (see Table 11–2), and they may be dependent on the export of just one or two raw materials (see Figure 11–14). The welfare of their people rises and falls with world prices over which they have no control, and world prices of commodities have slowly declined. In 1989 the world market prices of the 33 leading nonfuel commodities in world trade were only 70 percent of what they had been in 1975.

Exporters of some raw materials have tried to form cartels, such as OPEC, but few have succeeded. If production of the commodity can easily be increased, then huge national stockpiles hang over the international market. Any nation can gain a short-term advantage by dumping stockpiles on world markets, collapsing prices. The 74-member International Coffee Organization broke down in 1989, and world coffee prices halved in 4 months. In the countries dependent on coffee exports, suddenly there was no money for schoolbooks; national roads went unpaved; hospitals ran out of medicines. Erratic swings in the prices of commodities can be as harmful to national economies as steadily falling prices.

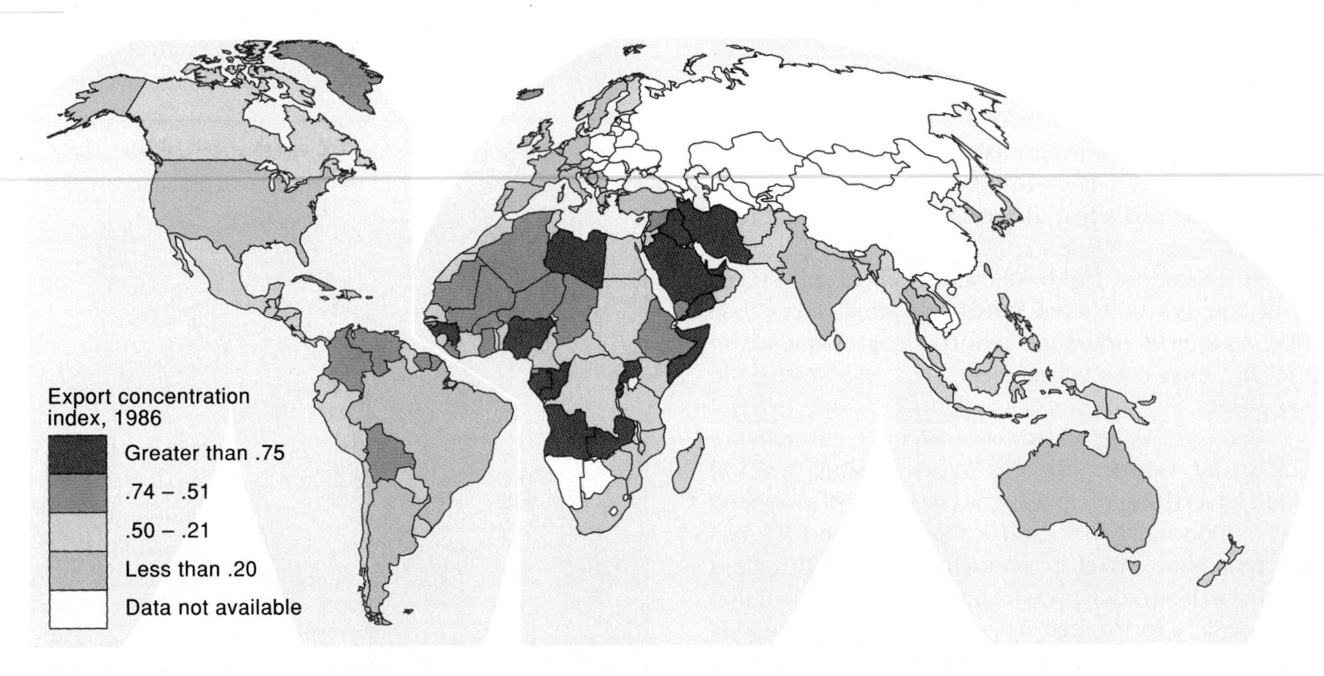

Figure 11–14. Countries' export concentration index. If a country is able to export a variety of products, it is more economically secure if production of any one product falls or the market for it is lost. The export concentration index mapped here ranges from 0 to 1, with 1 representing extreme concentration. As a rule, the poorest countries have the highest indices. Their economies are most vulnerable to production falls or world market shifts. The rich cartelized oil exporters deviate from the rule. (Data from UNCTAD.)

TABLE 11–2 ***Leading Visible Exports of Selected Countries***

Country	Export	Percentage of total exports
	South and Central America	
Bolivia	Natural gas	54
	Non-ferrous metals	16
	Tin	15
Chile	Copper	32
	Fruit and nuts	11
	Concentrated metal ore	10
	Animal feed	8
El Salvador	Coffee	57
Venezuela	Crude petroleum	38
	Petroleum products	27
	Africa	
Liberia	Iron ore	62
	Crude rubber	18
Nigeria	Crude petroleum	95
Tanzania	Coffee	50
	Spices	11
	Cotton	10
	Copper	5
Zaire	Copper	50
	Coffee	24
	Crude petroleum	16
	Diamonds	8
	Asia	
Bangladesh	Clothing	21
	Textiles	30
	Fresh and salted fish	13
Fiji	Sugar	70
Indonesia	Crude petroleum	38
	Natural gas	19
	Plywood	6
Iran	Crude petroleum	87
	The United States (For comparison)	
Road motor vehicles		9
Office machines		7
Aircraft		7
Misc. electric machinery		6
Misc. nonelectric machines		5
Non-electric-powered machinery		4
Organic chemicals		3

Source: The 1989 UN *Handbook of International Trade and Development Statistics*.

Alfred Weber's Theories of the Location of Manufacturing

Most countries strive to industrialize to capture the high value added by manufacturing. One of the earliest scholars to identify the locational determinants of manufacturing was Alfred Weber (1868–1958). Weber's analysis focused on the role of transport costs, and he devised models that differentiate **material-oriented manufacturing** from **market-oriented manufacturing.**

Material-oriented manufacturing locates close to the source of the raw material for one of two reasons. In the first case, the raw material is heavy or bulky, and the manufacturing reduces that weight or bulk. The steel industry is an example. The value added in manufacturing steel is low relative to the cost and difficulty in transporting the raw materials (iron ore, coal, and water), and so steel mills generally locate where the raw materials can be found or cheaply assembled. The second type of material-oriented manufacturing locates near the raw material because the raw material is perishable and needs immediate processing. Canning, freezing, and the manufacture of cheese from milk are examples. Switzerland provides a classic case study. The Swiss prospered by adding value to milk, a perishable raw material that is high in bulk but low in value, by making it into cheese, a less perishable product that is low in bulk but high in value.

Market-oriented manufacturing, Weber's second category, locates close to the market either if the processing increases the perishability of the product (baking bread, for instance) or if the processing adds bulk or weight to the product. A can of soft drink, for example, consists of a tiny amount of syrup plus water in a can. The syrup can be transported easily and cheaply, but canned soda is expensive to move. Therefore, the water is added and the beverage canned close to the market.

Weber elaborated his models of the location of manufacturing by adding additional considerations, such as the availability of a labor force. In that three-factor model, manufacturers situate factories to minimize the cost-distance from three points: the location of the raw materials, the labor force, and the market. The optimum location for any specific manufacturer depends on the balance of these costs in his or her business.

The Global Distribution of Manufacturing Today

Since Weber's original studies transportation costs have steadily fallen (look back to Table 5–1), and the value added in manufacturing has increased; many manufactured products are more and more sophisticated. Also, Weber's studies focused on the role of locational determinants for manufacturing within countries. Today, as world regions increasingly become interconnected through trade, the question of locational determinants may be addressed on a global, rather than a national scale.

The percentage of the final value of many manufactured goods that is attributable either to transport costs or to the value of the raw material is falling. These two ***inputs*** to manufacturing represent shrinking percentages of products' final value. What percentage of the price of a television-video unit is the value of the iron,

CRITICAL THINKING: *The United Fruit Company*

Chiquita Brands International's modern refrigerator ship *Edythe L.* carries Central American fruits to Northern Europe in containers that can be taken off the ship directly onto railroad cars or truck chassis. This containerized shipping offers great convenience. (Courtesy of Chiquita Brands International.)

The story of the United Fruit Company exemplifies the geography of value added. In the nineteenth century the company established banana plantations in Central America. Several countries came to depend economically on the export of bananas, and many aspects of their local cultures were consequently transformed. Their politics, for example, were manipulated internally by banana interests and externally by relations with the United States; their social structures reflected peoples' positions in the dominant export-oriented cash-crop economy; the peoples' diet came to depend on imported staples. The United Fruit Company owned the banana plantations and the ships that transported the bananas, and it marketed the bananas.

Eventually many Central Americans began to resent the foreign control of the plantations, and they accused the company of "Yankee imperialism." The company responded by abruptly surrendering many of its plantations to local governments.

This might sound like a financial sacrifice, but what really happened? The company was no longer tied to one supply (its own plantations), and so it could bargain with several potential suppliers. Competition among potential suppliers reduced the price of bananas. The producers increased production to raise their incomes, but they increased production faster than demand could be raised. This only lowered prices still further. The new plantation owners got poorer. Transporting bananas and marketing them—adding conceptual value through name-brand advertising—is considerably more profitable than growing them, and it supposedly causes far fewer headaches for the company's management.

Eventually the company even changed its name to Chiquita Brands International, and it is now branching out and applying its valuable trademark to marketing other fruits from a great variety of sources worldwide. By focusing on downstream activities, the company transformed itself from a supplier of raw materials into a marketing company. Its profits fattened, while the countries that supply the raw material got poorer.

A similar tale could be told about almost any raw material or food supplier in the world today and its relationship with multinational corporations that control every aspect of handling and processing goods from the raw material to the final consumers. Compare the story of the United Fruit Company with the story of OPEC's petroleum marketing networks told in Chapter 7. In that case the suppliers of raw materials are themselves seizing control of downstream activities.

Questions

Why can oil producers capture downstream activities, but not food producers?

Does this reflect a difference between oil and food?

How many alternative sources of oil are there?

How many alternative sources of food?

copper, and other raw materials? A minuscule percentage. Lowering transport costs and increasing value added means that high-value-added manufacturing is largely released from locational constraints; it is increasingly footloose. This idea was introduced in Chapter 5.

Even though manufacturing is increasingly footloose, it is still necessary to investigate why manufacturing frequently does not locate in the countries that have raw materials. Local processing would add value and generate local employment and wealth, but many resource-rich countries have never been able to develop local industries. We must identify other factors, other locational determinants for manufacturing, that are more important than the location of the raw materials. What are they? What locational determinants have made it possible for some countries to grow rich from industrialization while others, equally or even better endowed with raw materials, remain poor?

Alfred Weber's models considered the locations of the raw materials, the labor force, and the market. At least five other considerations, however, must be added to these. One is the availability of *capital*, because manufacturing is increasingly capital-intensive. A second is the availability of *technology*. Third, footloose industries seek places with *hospitable governmental regulations*; that usually means low taxes and little environmental regulation. *Political stability* is a fourth significant consideration in situating industries. Manufacturers hesitate to invest capital in volatile political environments. *Inertia* is still another consideration in the geography of manufacturing, as it is in most other geographical patterns. Because factories are major investments and because networks of suppliers and trained labor forces develop around them, they are not quickly abandoned.

Weber was aware of the importance of these factors, but he focused on the role of cost-distance. Because this factor has shrunk in importance since Weber's time, the balance among locational determinants for many industrial processes has tipped away from those Weber considered. The balance among the locational determinants most relevant to each manufacturing process is different, but today's highest-value-added manufacturing processes rely increasingly on inputs of skilled labor, capital, and technology. An environment of political stability is essential.

Europe today concentrates much of the world's manufacturing, for reasons that stem from historical consideration of the locational determinants. Europeans brought about the Industrial Revolution first, giving Europe a technological lead, and then Europe politically and economically overwhelmed most of the rest of the world. The manufacturers, who were also the colonial rulers, situated the manufacturing in their own country to retain the value added by manufacturing. It was not their intention to enrich their colonies; in some cases they actually intended to retard the development of manufacturing in the colonies. For example, from 1651 until 1776, at the very beginning of the Industrial Revolution, Britain's Navigation Acts required that American raw materials be shipped to Britain so that Britain could capture the value added by manufacturing from them. Iron from New Jersey had to be sent to Britain, where nails were manufactured to be shipped back to New Jersey. The colonists' resentment of this exploitation helped trigger the War of Independence. Later, as advancing technology caused value added in manufacturing to soar, many colonies in Africa and Asia suffered the same deliberate restraint on their industrialization.

When manufacturers situated factories in countries other than their own, they chose other rich countries; there they could find markets and labor forces. Today the advanced countries still have the technological lead and the best-trained labor forces. They usually are politically stable, and they provide the eventual market for most goods. This geographical system has prevailed for so long that it has a tremendous inertia.

The industrializing efforts of today's poor countries suffer from a number of problems that reinforce one another. The countries are poor, and so they are not as important as markets as the rich countries are. In addition, cheap goods manufactured in the rich countries can flood their domestic markets. Their poverty increases their political instability. They need capital, and their people need education. The only locational determinants many poor countries can offer are raw materials, inexpensive labor (usually only unskilled labor), and hospitable regulatory environments.

Locational Determinants Migrate The geography of locational determinants is not fixed forever. The locational considerations for manufacturing are increasingly footloose, and so manufacturing can migrate around the world if the mix of attributes of any place changes and attracts it.

There are, in fact, reasons why the geography of determinants will constantly evolve and change. One is that the mix among relevant determinants changes for new products, and so new locations can offer the right mix of determinants to manufacture new goods. A second reason for constant change is that at each place the local balance among the relevant factors continuously evolves. For example, if a place attracts manufacturing because it offers a low-wage labor force, its local standard of living and local wages will rise. This will drive away the industries that developed there originally to take advantage of the low wages. The simplest industrial jobs that earn the lowest wages will migrate away from that place to still poorer places where wages are lower.

The manufacture of men's dress shirts for the U.S. market offers an example. Shirt making is labor-intensive, and so manufacturers are constantly searching for cheaper labor. In the 1950s, U.S. shirt manufacturers located in Japan to use the low-paid workers there. When the costs of labor and real estate rose there, the companies moved to Hong Kong. As Hong Kong's factories gave way to offices, the shirt makers moved again, first to Taiwan and Korea and then, in the 1970s and 1980s, to China, Thailand, Singapore, Indonesia, Malaysia, and Bangladesh. Toward the end of the 1980s and into the early 1990s, Costa Rica, the Dominican Republic, Guatemala, Honduras, and Puerto Rico became shirt-manufacturing centers.

These Caribbean basin countries offer cheap labor, and U.S. manufacturers who locate in them enjoy U.S. tax advantages granted under a U.S. government program, the Caribbean Basin Initiative, that was designed to help these countries industrialize. These countries also offer proximity to the United States. Relative proximity means that oversight is easier and that shipments can be faster. Speed is important for three reasons. First, fashion demands faster introduction of new styles—a consideration once restricted to women's clothing but now relevant to men's clothing. Second, renting a ship for an additional week is expensive. Third, capital tied up as shirts in the hold of a ship in transit is not earning any profit. One shipload of shirts may be worth $10 million. What is the interest on that money each *day* that the ship is on its way to the United States? How much does that add to the retail cost of the shirts?

Interest Rates and Economic Growth

A good deal of human geography is about diffusion, exchange, and trade. Interest rates, or what economists call "the cost of money," are of special concern to geographers for three reasons.

One reason is that variations in interest rates from place to place cause money to migrate. High returns attract capital investment, and capital investment may bring development.

Geographers also study the costs of overcoming distance—the friction of distance. It costs money to move anything. The costs of the expensive machinery and of the fuel and labor are apparent, but there is another cost in addition to these. Whenever anything is in transit, the value that it represents is idle. The owner is therefore losing an opportunity to invest that value in some other activity that might pay a return every minute. This loss is called an **opportunity cost**, the amount of return sacrificed by leaving capital invested in one form or activity rather than another. The opportunity cost of capital is the reason why people want things to move *fast*. Goods in transit are costing their owner both moving costs and also opportunity costs. If an entire society or country has poor transportation, then much of whatever capital it has is tied up in unproductive raw materials or in materials in transit. Fast and efficient transportation releases capital for productive investment. That is one reason why improving transportation boosts economic growth.

A third reason why interest rates concern geographers is that geographers study what Chapter 5 defined as the inertia of an infrastructure. Societies make huge investments in infrastructure. Transportation facilities—roads, bridges, railroads, airports—are among the greatest of these. Money is usually borrowed to finance the construction. The interest rates on that debt can be high, and they can determine what gets built and how it is operated.

A Case Study: U.S. Railroads The history of railroads in the United States provides an instructive example of how the interest rates affect transportation and the growth of cities. Railroads are of special importance because they were the first relatively cheap means of overland transportation invented, and they remain the cheapest and most efficient means of transporting goods overland over long distances. A steel wheel on a steel rail wastes very little energy in friction. Large sums of money were borrowed to build the railroads. Thus, the railroads had to pay their operating expenses (fuel, maintenance, employee salaries), and they also had to pay back the debt plus the interest on the debt. All those costs had to be paid before the railroads could make any profit.

All but one of the railroads stretching inland to the Midwest from the cities of the eastern seaboard had to forge their way through the mountains. The exception was the New York Central Railroad. It followed the Hudson River valley north from New York City up to Albany, after which it turned west to take advantage of the relatively flat Mohawk River valley and the southern shore of Lake Erie. The competing Pennsylvania Line had to push up and over the mountains of central Pennsylvania (see Figure 11–15).

By using a relatively level route, the New York Central saved both operating expenses and interest expenses. The New York Central was cheaper to operate because its engines did not have to burn as much fuel chugging up and over mountains. In addition, the New York Central had been cheaper to build, and so it had lower interest payments to make on its original construction debt. These interest costs were high. In the nineteenth century, interest payments amounted to one-quarter of all expenses for U.S. railroads. The New York Central had lower interest expenses, and so a higher percentage of its income was profit. New York City benefited by having this relatively low-cost route into the Midwest hinterland.

Still today the interest rates on capital that must be borrowed to build infrastructure can represent a high continuing cost, and thus interest rates can determine whether a project will get built and how it will be operated and maintained.

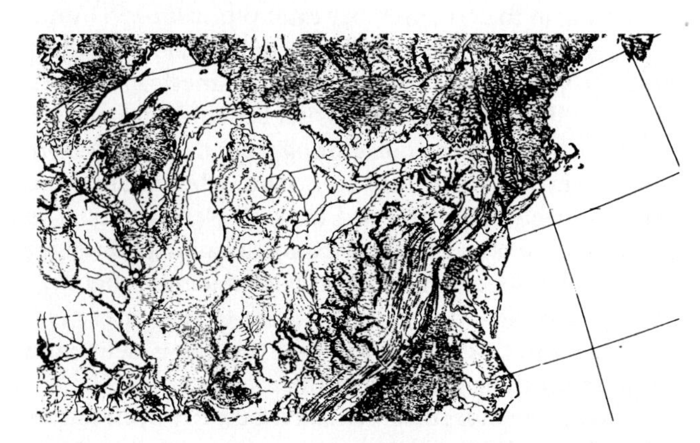

Figure 11–15. The Hudson-Mohawk Corridor offers New York City a relatively lowland route into the Midwest. (Map by A. K. Lobeck; reprinted by permission of Hammond, Inc.)

If you have capital, interest can compound in your favor. If you pay interest on borrowed money, however, you can become a prisoner on a treadmill of interest charges. Most poor countries are today prisoners on such a treadmill. High percentages of the value of their exports go to pay interest charges on their national debts. Many U.S. consumers who pay interest charges on their credit cards or revolving charge accounts are in a comparable position.

Technology and the Future Geography of Manufacturing

The balance among the inputs to manufacturing is continuously changing. The value of the labor input is generally a shrinking percentage of the value of manufactured goods—no more than 5 percent to 10 percent in the consumer electronics industry.

Technology is of increasing importance. When Japan was industrializing, it bought technology. By one estimate, the Japanese paid foreign manufacturers a total of only $10 billion for patents and licenses between 1950 and 1980. That must rank as the shrewdest investment any nation ever made. Today Japan builds factories abroad to carry on low-technology manufacturing, but it retains the highest technology at home. The world geography of technology today can be suggested by a list of holdings of key patents (see Table 11–3). This list indicates that scientific research is concentrated in a few countries. The world geography of manufacturing *tomorrow* may be guessed by mapping investment in research. It is difficult to compare this worldwide because of the variability of exchange rates and because some research is carried on in the private sector and more by various levels of government. Nevertheless, many observers feel that the United States may be falling behind both Germany and Japan in investment in research, measured both as a percentage of GNP and in actual money terms.

TABLE 11–3 ***The 15 Leading Holders of Internationally Influential Patents, 1991***

Country	Number of patents
United States	104,501
Japan	76,984
Germany	17,643
Great Britain	8,795
France	7,672
Netherlands	5,737
Switzerland	5,002
Canada	1,156
Sweden	1,124
Italy	1,106
Taiwan	1,000
South Korea	400
USSR	400
Belgium	330
Panama	301

Based on an index devised by CHI Research, Inc.

Technology and capital contribute the most rapidly increasing shares of the final value of most manufactured goods, but these inputs are the two that any poor country is least able to develop on its own. These inputs are, however, exceptionally footloose. In the manufacture of computer disk drives, for example, the raw materials are an insignificant fraction of the disks' final value, and so computer disk drives can be easily manufactured and distributed worldwide from almost anywhere. Why then are more than one-half of all of them manufactured in Singapore? They were not invented there. The decisive locational determinants for this high-value product are the availability of skilled inexpensive labor, technology, capital, and political stability in Singapore. Each of these factors, however, is footloose. How long will Singapore continue to dominate world manufacture of disk drives? What could change this situation?

The development of computer software is one of the world's fastest growing industries, and it is even more footloose than is the manufacture of computer hardware. All it requires is a personal computer, an electric outlet, and a skilled programmer. This multi-billion-dollar industry can migrate on an airplane. In 1990 the United States commanded 57 percent of the $110 billion global market for software and related services. This is greater than the total production of almost any of the world's raw materials. Of the world software industry, Japan held 13 percent; France, 8 percent; Germany, 7 percent; Great Britain, 6 percent; Canada, 3 percent; and all other countries of the world, only 6 percent. How long will this distribution last?

Air freight costs are falling, and because more manufactured products are becoming smaller and of higher value, they can be moved by air. Fort Worth, Texas, has opened the world's first airport at which planes can taxi directly from the runway into an industrial park. Sponsors hope that it will attract manufacturers who rely on far-flung foreign and domestic suppliers and who sell their finished goods to buyers around the globe. Could such an airport–industrial park just as well be in Alaska or on an island in the South Pacific?

Many tertiary activities are footloose. Multinational corporations can relocate their offices across a country or around the world wherever a mail service or a few electronic cables can reach. New Yorkers' bank credit card charges are processed in South Dakota. Why South Dakota? Why not in Paraguay? The Boeing and Bechtel multinational corporations have located software development facilities providing many jobs in Ireland. Why not in Costa Rica, the Sudan, or Madagascar? Chapter 13 investigates these questions in the context of examining the global economy.

THE VARIETY OF NATIONAL ECONOMIC-GEOGRAPHIC POLICIES

Now that we have mapped the rich countries and the poor countries and have analyzed some reasons for this world distribution of wealth, we turn our attention to the diverse ways that countries organize their domestic economies. Each country manages the distribution of economic activities within its territory and regulates its participation in world trade and investment.

Political Economy

Each country defines a set of principles to organize its economic life. The study of these principles is **political economy**. Direct government participation ranges theoretically from a communist system, in which the government itself owns both the natural resources and the productive enterprises, to a capitalist system, in which the state defers to private enterprise. The economy of each country lies between these two extremes. Most countries have *mixed economies*.

Even among those countries generally considered capitalist, there is a great range of government involvement in the economy. The political economy a country chooses usually reflects other characteristics of that country's culture. The United States, for example, generally favors a system that minimizes the government's role. This is called **laissez-faire capitalism**, from the French for "leave us alone," supposedly said long ago by a French businessman to a government bureaucrat. The U.S. government has had an enormous impact on the national economy through its regulations, its funding of research and development of new products, and, particularly in the past 50 years, through the defense budget. Nevertheless, the government has never developed an explicit industrial policy.

Most Latin American countries, by contrast, have traditions of considerable government ownership of the nation's assets and industries. In the late 1980s and early 1990s, however, these governments began selling the assets to private investors, or **privatizing** them. The same is true in Western Europe, where, as of 1991, nationalized companies still accounted for about 25 percent of GNP in France, 11 percent in Germany, and 3.5 percent in Great Britain, which began the privatizing trend in the early 1980s. Eastern European countries are rapidly privatizing as they abandon communism.

In several Asian countries the government does not own many companies outright, but it plans and regulates the economy. This system is called **state-directed capitalism,** and it conforms with these countries' Confucian bureaucratic cultural traditions discussed in Chapter 8. South Korea, for example, is considered a capitalist, free-market economy, and yet the government proposes national economic plans and subsidizes new industries. Industry is dominated by combinations of corporations that are affiliated by interlocking directorships and ownership. Known as *chaebol*, they would be broken up as illegal trusts in the United States. The economies of Japan and Taiwan are similarly organized.

The economies of communist countries have often been referred to as *planned economies*. This term, however, does not recognize the significant amount of government planning in the state-directed capitalist economies, whose recent success has provoked the suggestion that the United States adopt an explicit industrial policy. Many Americans, however, distrust government planning as both an intrusion on liberty and as economically inefficient.

In all systems, the government usually provides those services that are unprofitable to the provider but that diffuse benefits throughout the economy. These services include education, transportation, and other infrastructure. The widespread benefits of these services are called **positive externalities. Negative externalities**, by contrast, are costs that one person's activity imposes upon someone else, as when a corporation pollutes the environment. The cleanup cost is borne not by the corporation alone but by the entire society.

Each state also manipulates its national budget and expenditures, interest and exchange rates, and money supply to promote growth and high employment. The tools for manipulating these are imperfect, but national governments' economic planning and management, the sheer weight of their revenue and expenditure, and their role in redistributing income have made national economic policies more central to most peoples' lives than ever before.

Domestic Economic Geography

Countries try to boost their economic growth by organizing their populations, territory, and resources for production. A strong sense of nationalism can encourage the population to work together, but the population also needs education and training. The territory must be at peace, the population should feel safe within it, and the country should constitute a single market for raw materials, goods, and labor.

Internal Population Mobility Internal population mobility assists each individual in realizing his or her own potential, and it also generally detaches people from their traditional identifications and encourages their development into national citizens. Employers can draw on the potential of the entire national population. No society realizes the ideal of total fluidity, but freedom of internal migration is a step in that direction.

Workers can normally be expected to move from areas of high unemployment to areas of low unemployment, and so variations in regional unemployment can provide a measure of labor mobility and national cultural homogeneity. Great variations suggest that something is

Focus on The Economy of Singapore

Singapore's exports equal roughly 200 percent of its GDP. How can a country export twice what it produces? The answer is that Singapore does two things: First, it imports raw materials and semifinished products and reexports them after adding value by further processing. Second, it serves as a transshipment point for other countries' exports. A transshipment point is called an *entrepôt*.

During its first 25 years of independence (1965–1990), Singapore lured world manufacturers with an inexpensive but highly trained and disciplined labor force and government-supported factories. Singapore achieved a per capita income comparable to those of Ireland and Spain. Today, however, Singapore wants to increase the value it adds. Its economy must mature from dependence on factory workers to researchers, software engineers, and biotechnology specialists. The research on new computer screens, for example, the advancing edge of technology, is increasingly locating in Singapore.

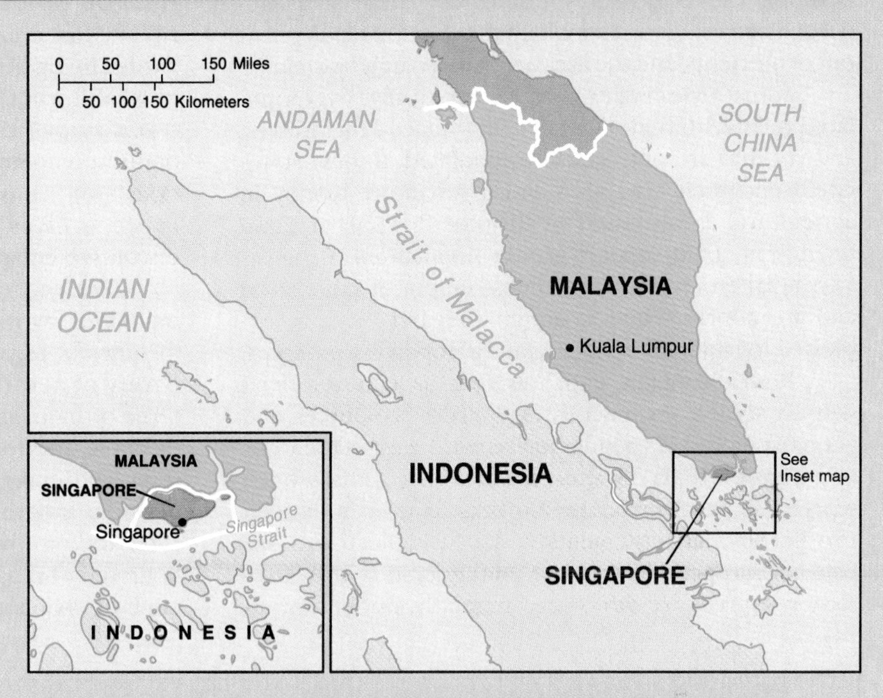

Singapore, at the narrow Straits of Malacca, was not founded by the peoples native to the region. The situation's strategic value was most immediately recognizable to a globally seafaring people, and so it was the British who founded the city as a naval base early in the nineteenth century.

Singapore is also trying to capture air traffic as it has captured shipping by building Asia's largest airport, called Airtropolis. Airtropolis's director has said, "Since its founding Singapore has been an entrepôt for goods and trade. As with cargo, so with air passengers—we import and export, and we hope to provide a bit of value added on the way." Singapore will profit handsomely if each traveler passing through Airtropolis buys just a cup of coffee and a magazine.

Singapore's stability and its transport services enhance its role as an Asian service center. In 1991 the Disney Company chose Singapore as its Southeast Asian distribution center. The countries of the region will be flooded by over 16,000 Disney products. Will the children of Southeast Asia be able to resist Goofy and Donald Duck T-shirts? If not, will they also remember their culture's traditional fables and mythical characters? Today financial services provide one-third of Singapore's gross product, and the city-state hopes to become a regional medical center.

discouraging people from moving. Factors preventing mobility can include racial or ethnic animosity among different localized groups; lack of nationwide availability of certain cultural products and services (religious services, foods, and other consumer goods); regional variations in unionization and difficulty in joining unions; variations in wage scales; variations in workers' benefits and their potential loss due to relocating; and nationwide acceptance of degrees or certification for professionals.

Italy exemplifies a country in which unemployment rates in one region, the south, are usually two or more times as high as in the north. This pattern reflects continuing cultural differences between the regions. Unemployment rates vary across the United States, but on the whole Americans are a mobile population. Every year almost 20 percent move to a different house and almost 3 percent to a different state. This movement illustrates the relative homogeneity of the national culture. When multinational economic unions form, such as the European Community, barriers to migration throughout the union are difficult to eliminate.

States with Frontiers Not all states occupy and utilize their full territory. The world population map in Figure 6–1 reveals that many countries have core areas of dense population and development, called **ecumenes,** but they also have internal **frontier** areas, undeveloped regions that may offer potential for settlement.

There may be environmental limitations on exploiting these frontier areas. Brazil, for example, contains vast, sparsely settled areas into which the Brazilian government has encouraged expansion in recent decades, but the results have often been economically disappointing and ecologically disastrous. These areas undergo deforestation, after which the thin soil rapidly runs out of nutrients, and the area becomes an ugly wasteland.

Some African states are, paradoxically, overpopulated relative to their current ability to feed themselves, and yet they are only sparsely populated. If these states could end their civil wars and invest more money in agriculture, as discussed in Chapter 7, some of these territories could support greater populations. Canada's vast northern regions are almost empty of population and are poorly suited to agriculture, but they are exploited for mineral, timber, and hydropower resources.

Russia contains expanses that have never been densely settled or even fully explored for resources, and yet most of Russia's population remains west of the Urals (see Figure 11–16). Scattered mines have actually been worked as penal colonies. Workers have been enticed to other Siberian settlements by salaries higher than those paid in European Russia, but immigration is still slow. New resources are discovered regularly, but enormous investment will be required to develop them at the site or else to bring them out for manufacturing elsewhere.

How Do Governments Distribute Economic Activities? Most countries try to maintain a fairly equal standard of living throughout their territory. Great disparities of wealth from region to region are a centrifugal force that may pull the state apart (see Figure 11–17). Economic competition among regions may dominate national politics as politicians weigh the domestic regional impact of every program and devise new ones to reduce disruptive imbalances.

The geography of any country's resource endowment will favor some regions over others. The relative fortunes of regions may change, depending on the discovery of new resources, shifting trade patterns, or patterns of innovation. In the area of today's Belgium, for example, the lowland coastal Flemings grew rich from trade and commerce in the medieval and Renaissance periods, and they dominated the highland Walloons. Later the discovery of coal and the industrialization of Walloonia brought Walloons prosperity. Today the decline of coal and steel manufacturing and the develop-

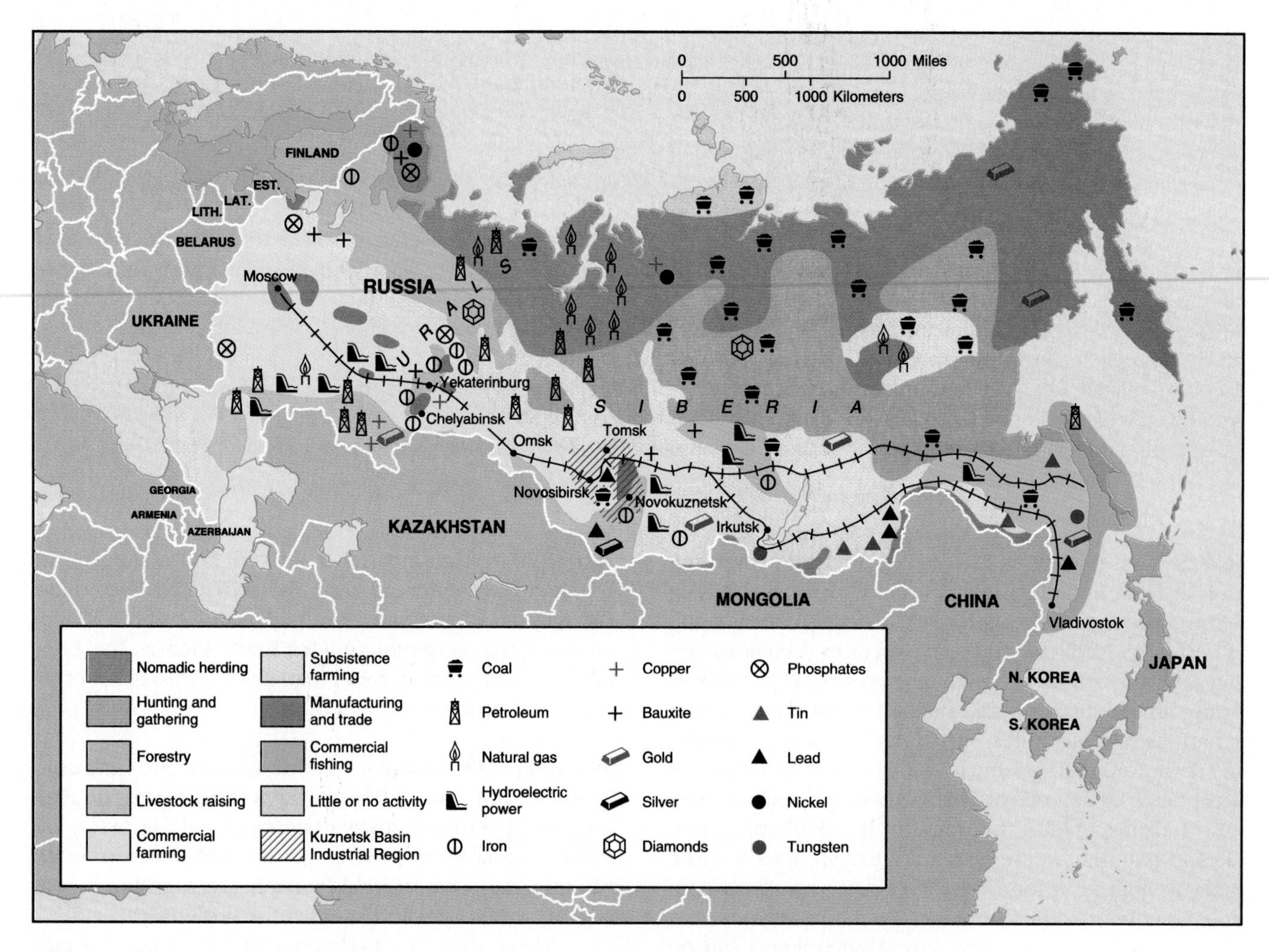

Figure 11–16. Siberia remains almost uninhabited, and although it contains vast quantities of raw materials, enormous amounts of capital will be needed to exploit them.

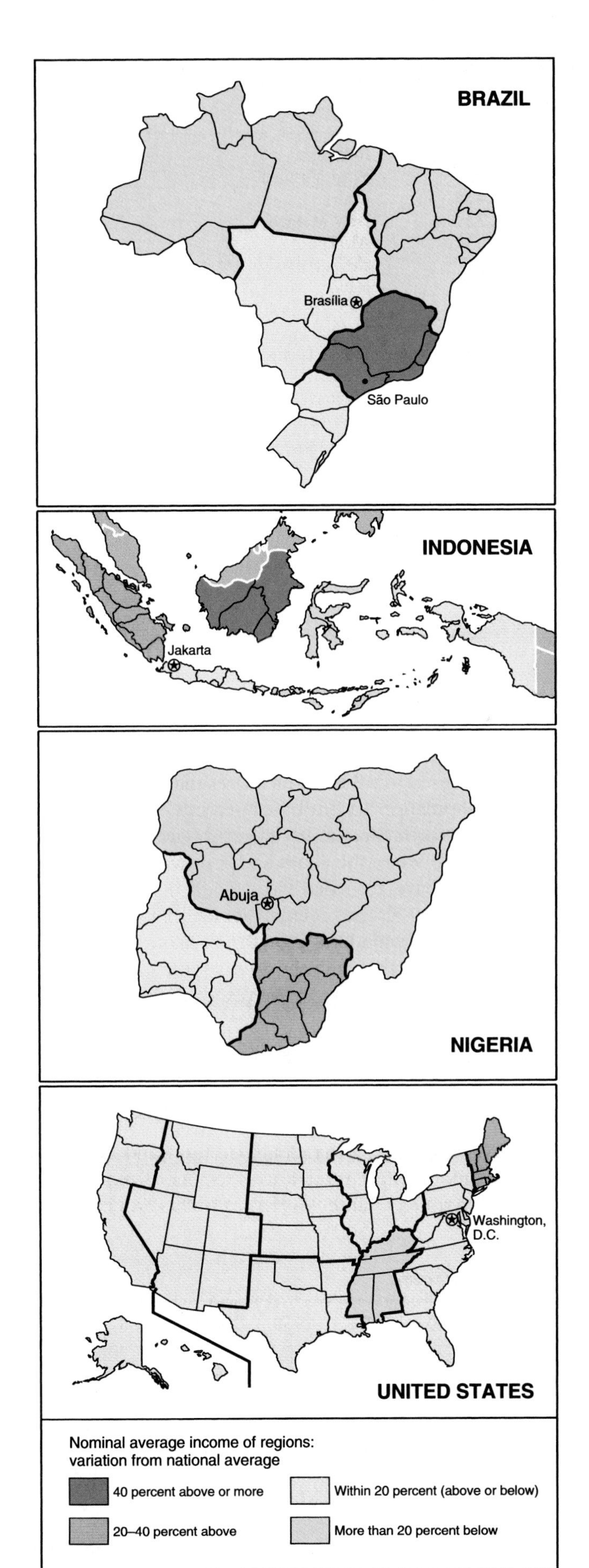

Figure 11–17. Most countries try to maintain a relatively uniform level of income across the national territory. Of these four countries, Indonesia suffers the greatest variation; the United States shows the least. (Adapted with permission from The World Bank, *World Development Report 1991*, p. 41. *Sources:* Indonesia, Bire Pusat Statistik 1989; Nigeria, World Bank; Brazil, IBGE 1987; United States, U.S. Department of Commerce, Bureau of the Census, 1990.)

ment of a trade and service economy has swung the pendulum of prosperity back to the Flemings.

The national distribution of prosperity can also change as the sectors of the national economy evolve. As employment shifts from sector to sector, job opportunity shifts from place to place. The expansion of the secondary and tertiary sectors, for example, brings urbanization. These shifts might occur faster than workers can be retrained or relocated, and so certain regions of a country may suffer while others thrive.

Countries often devise special development programs for poor regions. For example, the United States created the Tennessee Valley Authority (TVA) in 1933, an Appalachian Regional Commission in 1964, and a Lower Mississippi Delta Development Commission in 1988 (see Figure 11–18). These agencies devised pol-

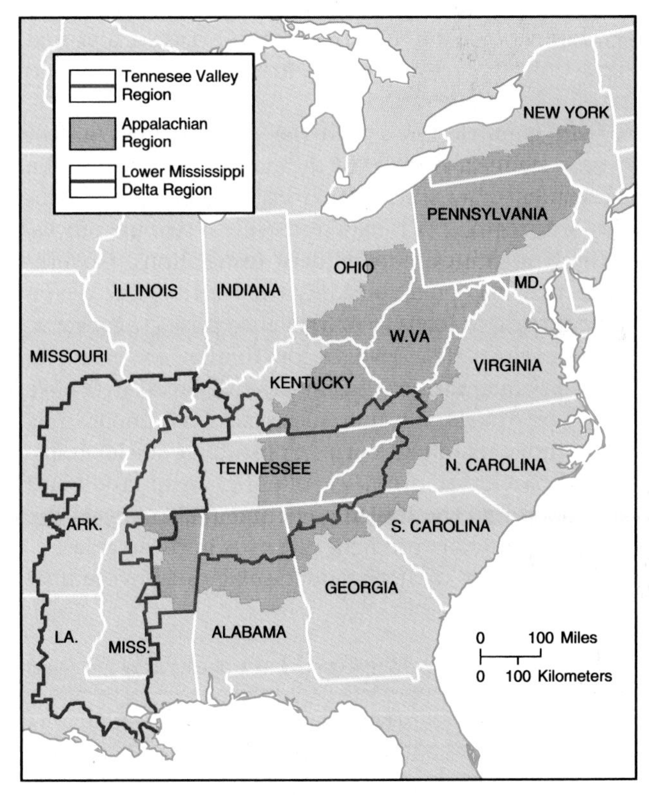

Figure 11–18. The three regions on this map—the Tennessee Valley, Appalachia, and the Lower Mississippi Valley—have been designated as significantly below national levels of income and welfare, and federal programs have attempted to boost their economies. The Tennessee Valley Authority has been in existence since 1933; the Lower Mississippi Valley Commission studied its focus region, reported on conditions, and then dissolved.

TABLE 11–4 ***Gross Products of U.S. States, 1986 (in millions of dollars)***

State	Gross product	State	Gross product	State	Gross product
California	533,816	Maryland	76,504	Nebraska	26,521
New York	362,736	Minnesota	75,626	West Virginia	24,096
Texas	303,510	Louisiana	74,426	Utah	24,008
Illinois	209,666	Tennessee	72,328	New Mexico	23,603
Pennsylvania	183,559	Connecticut	70,639	Alaska	19,575
Florida	177,729	Colorado	59,177	Nevada	19,426
Ohio	176,102	Alabama	55,007	Hawaii	19,320
New Jersey	154,765	Arizona	53,253	New Hampshire	18,518
Michigan	153,240	Kentucky	53,135	Maine	17,326
Massachusetts	115,526	Oklahoma	49,814	Rhode Island	15,205
Virginia	104,155	South Carolina	44,727	Idaho	13,170
Georgia	102,922	Iowa	43,836	Montana	12,163
North Carolina	100,961	Kansas	42,472	Delaware	11,706
Indiana	84,922	Oregon	41,728	Wyoming	11,673
Missouri	83,534	Mississippi	31,830	North Dakota	10,733
Washington	77,683	Arkansas	31,633	South Dakota	9,802
Wisconsin	76,922	District of Columbia	28,791	Vermont	8,636

Source: Statistical Abstract 1991.

icies to boost the regional economies with government assistance. The TVA built dams, and the two other commissions encouraged increased federal highway construction and other programs in their focus regions. Several federal programs today encourage the economic development of Puerto Rico, but the states claim that these only drain jobs from other parts of the United States.

In countries in which the national government owns a significant share of the national economy, the government may situate factories or other enterprises in poorer regions or it may purposely distribute offices, research institutes, and military installations. Governments may also offer subsidies, tax waivers, free development sites, or loans to private enterprises to locate in poorer regions. The Spanish government, for example, subsidizes international businesses that invest in Spain's poorer provinces, and Italy subsidizes investment in its South. India encourages firms to locate in officially designated "backward areas" by offering cheap credit and tax reductions. The Indian government also lowers the freight rates on government railroads in those areas.

No matter what economic policies a government devises, some regions may dominate a national economy (see Table 11–4). The top four states account for one-third of the U.S. GNP. Other states suffer population losses. The central region of the United States has been losing population since World War II. Kansas alone has over 2000 officially registered ghost towns (see Figure 11–19). The Census Bureau estimates that Iowa will suffer a net population decline of 9.4 percent between 1990 and 2000; North Dakota, 9.3 percent; and Nebraska, 4 percent. Meanwhile the South and Southwest, known as the Sunbelt, have absorbed virtually all U.S. population growth since 1970, and regional incomes have risen with population and economic growth. As Table 11–5 indicates, regional incomes in the South, which historically have lagged behind those in the rest of the country, are now catching up.

National Transportation Infrastructures

We have emphasized that transportation and communication allow different territories to specialize production and to trade. Chapter 9 referred to this process with

TABLE 11–5 ***U.S. Regional Incomes as a Percent of National Income***

	1840	1860	1880	1900	1920	1930	1940	1950	1960	1970	1980	1989
United States	100	100	100	100	100	100	100	100	100	100	100	100
Northeast	135	139	141	137	132	138	124	115	114	106	105	115
North Central[a]	68	68	98	103	100	101	103	106	101	100	100	96
South	76	72	51	51	62	55	65	72	80	85	86	90
West	—	—	190	163	122	115	125	114	105	101	103	104
Mean deviation (%)				37			28			12	9	

[a] North Central is equivalent to what was the 'West' before the Civil War.

Source: US Bureau of the Census.

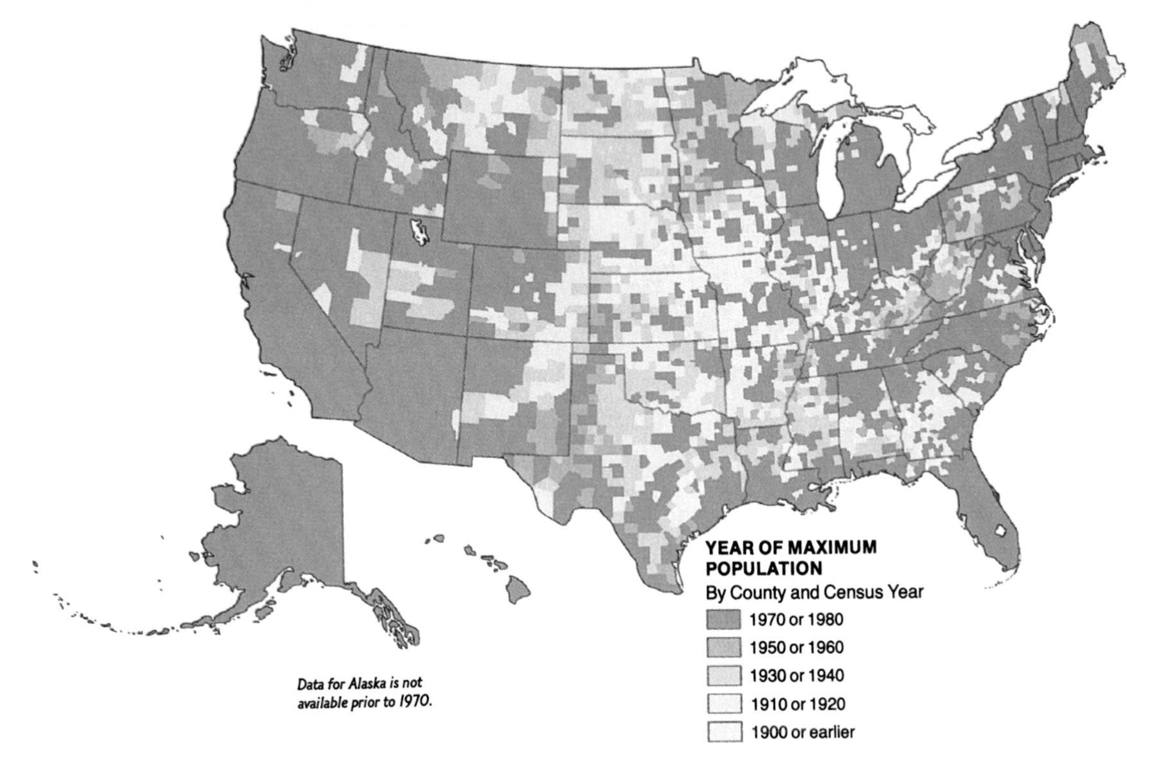

Figure 11–19. Year of maximum populations. Some counties in the central U.S. farming region and some old cotton farming centers in the South counted their maximum populations before 1900. The populations of many other areas across the Great Plains and in Appalachia peaked by 1940, when the prosperity that had come from farming and mining was past. The national population today seems to be concentrating at the margins of the country. (Reprinted with permission from National Geographic Society, *Historical Atlas of the United States.* Washington, D.C.: National Geographic Society, 1988.)

the image of "the territorial division of labor" and compared it with the division of labor among workers. In any country, the design and the regulation of the national transport infrastructure are two of the most important considerations of internal geography. In some countries transportation is slow and difficult. Some regions may be virtually inaccessible (see Figure 11–20). Other countries have fine transportation networks, which make vir-

Figure 11–20. Malaysia's road-building programs cut through the forests and open national frontier regions to development. (Courtesy of Caterpillar Inc.)

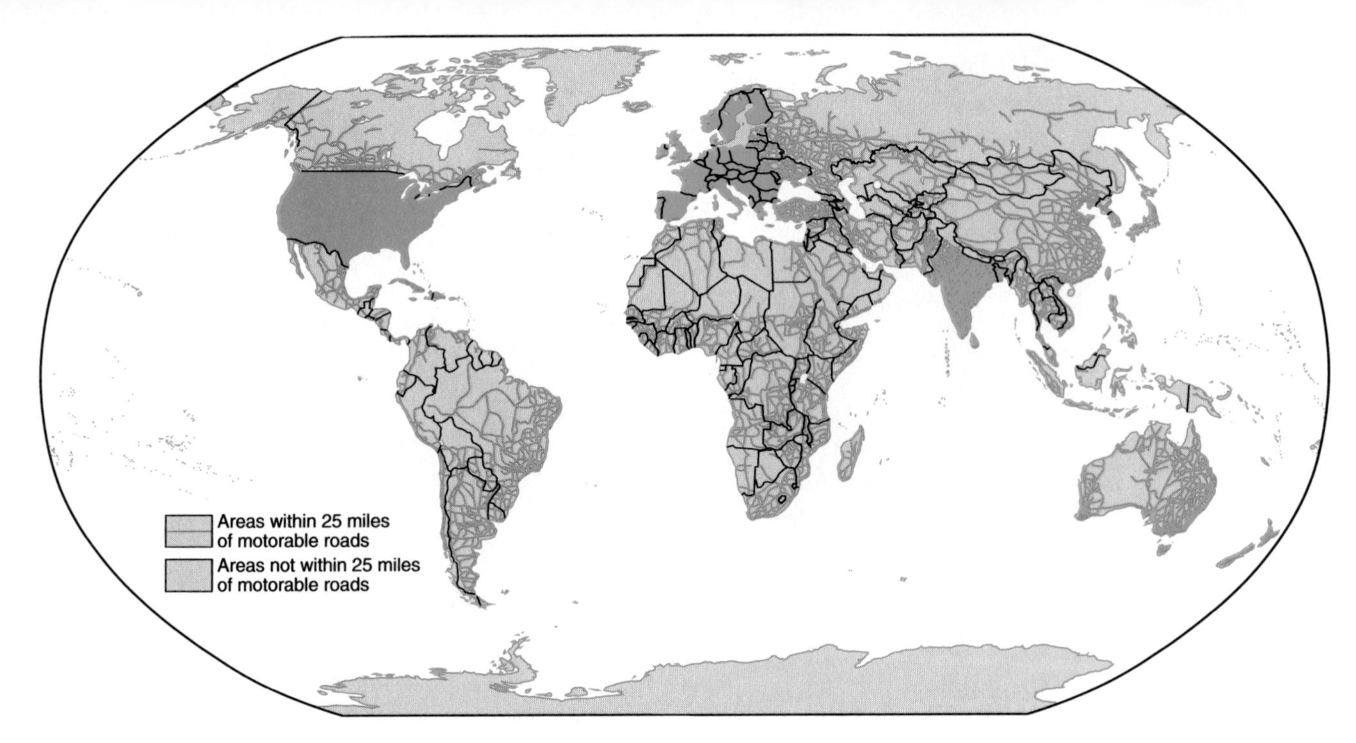

Figure 11–21. Compare this map of world roads and highways with the world population distribution shown in Figure 6–1. Many densely settled regions, such as eastern China, would benefit from improved networks. (Data adapted from Rand McNally, *Goode's World Atlas*, 18th edition. 1990 by Rand McNally R. L. 92-5-76.)

tually every place easily and cheaply accessible from every other place (see Figure 11–21).

Transport and Political Control Some countries' transportation networks reflect considerations of political control. In much of Southeast Asia and in parts of South America and Africa, central control has never extended throughout the country's entire territory. Some regions may be sparsely occupied and developed; others may be populated but in open rebellion. The central governments invest in roads specifically to "occupy" the territory.

Some national transport networks reflect concentration of power in the national capital. In the ancient Roman Empire "all roads led to Rome," and still today many countries route traffic moving from one peripheral region to another through the capital. Center-periphery connections are superior to periphery-periphery links. Such a network may be economically inefficient, but it is politically desirable.

Transport and Economic Growth The most important reason for national investment in transportation, however, is usually economic growth. We have already seen that fast and efficient transportation releases capital for productive investment. Improved transportation allows the exploitation of natural resources, regional specialization of production, and domestic trade among regions. India exemplifies a country whose economic growth is held back by the inadequacy of the road system. Traffic in passengers and goods on India's roads has increased 30-fold since independence, but the total road mileage in India has only quintupled. Roads throughout India are overcrowded, and thousands of villages lack an all-weather road. The government hopes to connect every village of more than 500 people to an all-weather road by the year 2000. Accessibility would allow regional specialization of production and cash cropping, especially of higher-value-added crops such as fruits that spoil quickly. This would also improve the Indian diet.

In many countries the pattern of the transport network presents a special problem: It is not integrated but consists of lines penetrating into the country from its ports. These roads or railroads are called *tap routes*. Just as some trees have tap roots that reach deep down into the soil to draw up water, so these tap routes stretch into the interiors from the seaports to bring out the exploitable wealth. Tap routes make it easier to get into or out of a country than to get around inside it.

A case study: African railroads Today Africa illustrates tap-route transportation geography. Africa is relatively impenetrable by water. Most of its coastline is unindented, and it offers few natural harbors. The continental interior is a high plateau, or tableland, and its great rivers drop thousands of feet through rapids and waterfalls only a few miles inland from the coast (check Figure 6–4). This is true even if the rivers are navigable up on the plateau in the interior. These rapids and waterfalls offer potential for hydroelectric power, as discussed in Chapter 7, but they hamper African foreign trade.

The railroad first offered Europeans the possibility of penetrating the African interior, and they used it to explore and conquer. Each colonial power built railway lines to haul agricultural products and minerals out of its African empire. This pattern of tap routes affects African trade and politics to this day (see Figure 11–22). Many African railway lines provide the only freight service in areas where there are virtually no roads at all. In some cases the legacy of the colonial map left successor states landlocked and dependent on lines across foreign territory to the sea (see Figure 11–23).

Today many of Africa's railroads are in bad shape. Deterioration of the rail network hikes transport costs and lowers income for many African states. Railroads are potentially more economical than trucks for carrying heavy freight over Africa's vast distances, but governments have not invested in their railroads adequately to maintain them, and civil wars have in many cases de-

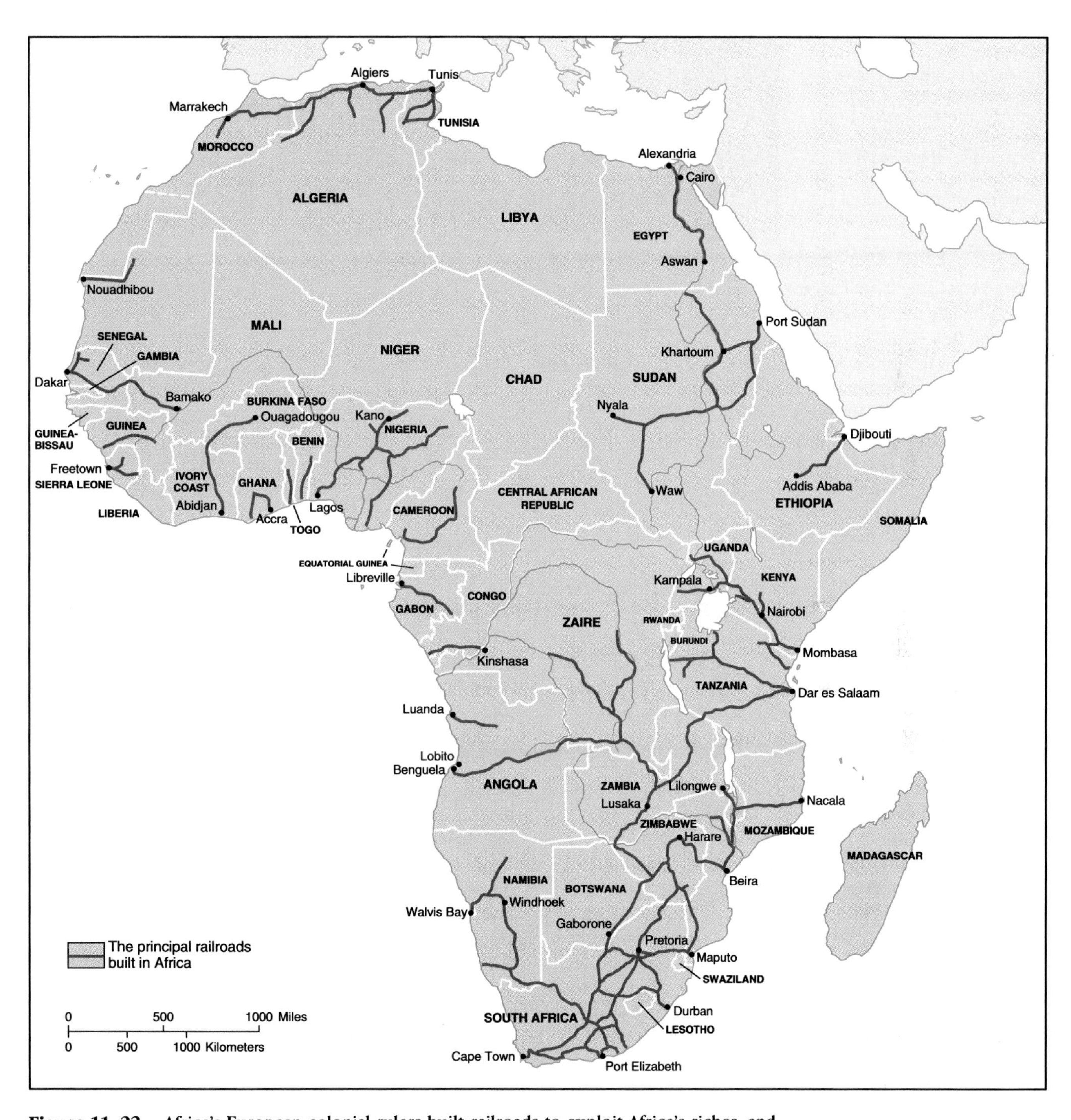

Figure 11–22. Africa's European colonial rulers built railroads to exploit Africa's riches, and these lines rarely provide internal circulation for today's political units. Many lines shown on this map only intermittently offer service because they are disrupted by civil strife or else they are inadequately maintained.

Figure 11–23. These citizens of landlocked Burkino Faso are boarding the country's most important rail link—a line to Abidjan in the Ivory Coast, where many Burkinans find work. (Courtesy of Agency for International Development, Carl Purcell.)

stroyed trackage. If African governments invested in railroads, the financial return in economic development could be generous.

The condition of Africa's railroads is paralleled by the condition of its roads. In one example, the Belgian colonialists built about 88,000 miles (142,000 km) of usable roads in what was the Belgian Congo, but by 1990 that network had shrunk to about 12,000 miles (19,000 km), of which only about 1400 miles (2300 km) was paved. Much of that was rutted and riddled with potholes.

Regulating Transportation The responsibility for providing transport facilities and services may lie either in a country's public sector (the government) or with the private sector (private businesses). Private-sector services must be able to make a profit, or else no entrepreneur will provide them. Governments, however, will subsidize facilities that provide positive externalities. Therefore, most countries blend public and private services.

Mapping a network is only a first step toward understanding what moves and where it moves. Movement can be manipulated by conditions and charges that are "hidden" behind a map. Many countries manipulate freight rates on government-owned railroads, for example, to distribute industry. This is true despite the fact that transportation costs are of decreasing importance as a locational determinant for industry. We noted earlier that the Indian government reduces freight rates on railroads in and out of poor regions to subsidize economic development in those regions. Sweden fixes rates to distribute industry evenly around the country.

In the United States today even the distribution of farm products is manipulated because of a transport-cost factor in the price-support system. It works as follows: The Department of Agriculture sets the minimum price for milk, for example, at the farm in regions called *marketing orders* across the country. Then it adds a bonus depending on the distance of each producer from Eau Claire, Wisconsin, which is considered the center of national milk production. This affects the geography of U.S. milk production by subsidizing production in marketing orders located far from Eau Claire. These policies promote the production of roughly twice as much milk as Americans drink and stimulate inefficient production even in inappropriate climates such as Texas and Florida.

In all countries the various modes of transport—trucks, railroads, airlines—compete for traffic. This competition can extend into the political arena as a competition for subsidies, and these subsidies can in turn affect domestic economic geography. In Kenya, for example, the trucking industry battles to prevent national investment in railroads, with considerable success. In most European countries, in contrast, the railroads are owned by the government, and they receive generous subsidies. In the United States the trucking industry, the car-driving public, the oil industry, and the highway-construction industry make up a lobby so powerful that it muffles the voice of the railroads. As a result, trucks pay only a fraction of their real fuel costs or of their highway-use costs, and they do not pay for the many negative externalities that they impose, such as air pollution, traffic accidents, and costs due to congestion. This constitutes an enormous subsidy to the trucking industry and enables it to capture much long-haul traffic that is actually best suited to railroads. Thus, political competition distorts economic competition, and inefficiency is built into the U.S. economy. The nation still has about 254,000 miles (409,000 km) of railroad track, and the railroads still handle much bulk traffic (see Figure 11–24). The nation has shifted freight to trucks, however, and railroad trackage has been abandoned at a rate of about 3000 miles (4830 km) per year (see Figure 11–25).

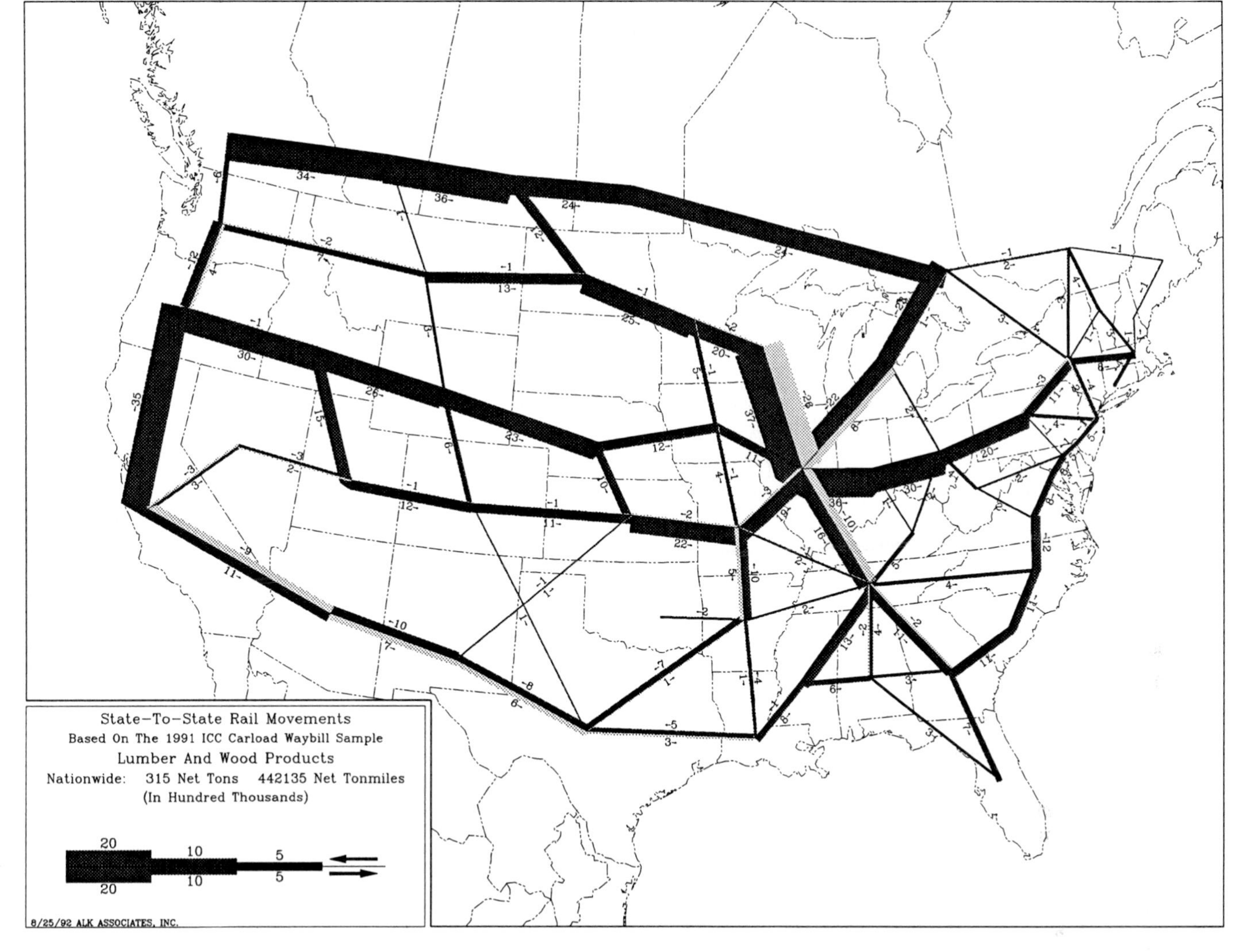

Figure 11–24. Many bulk commodities still rely on the nation's railroad network. This map indicates the flow of lumber and wood products. (Courtesy of ALK Associates Inc.)

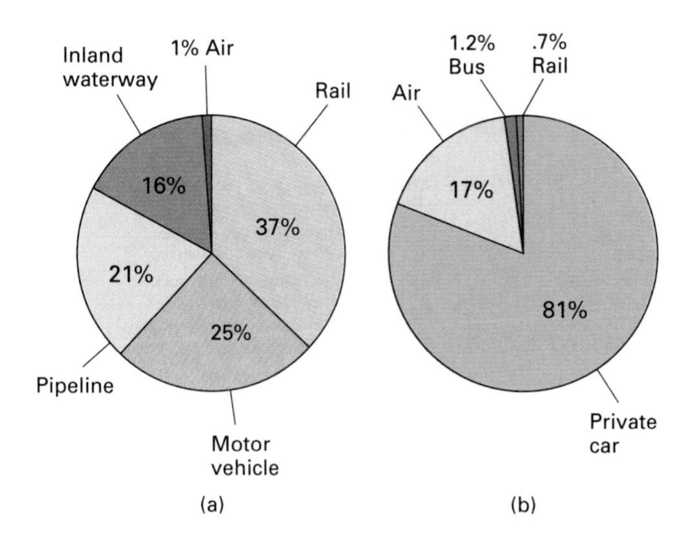

Figure 11–25. These pie graphs illustrate the percentage shares of movement of ton-miles of freight (a) and passenger miles (b) in the United States in 1989. In the United States today, freight moves by rail and truck; people move by private car and by air. Rail passenger services, which once dominated intercity traffic and still do in some European countries, have shrunk.

Countries may also manipulate domestic freight rates to direct traffic to favored seaports, and the port facilities themselves may be subsidized to attract business and create jobs. The Netherlands, for example, organizes its national transport network and fixes prices on it to capture European commerce for its port of Rotterdam.

These examples provide only a hint of the tremendously complicated manipulation of freight and delivery charges built into each country's transportation infrastructure. Geographers often learn that the answer to their question Why there? lies hidden in these policies. They must be considered as additional factors modifying models of the location of economic activity such as those of Heinrich von Thünen or Alfred Weber.

National Communications Infrastructures

The dissemination of information can be as important as the dissemination of goods and materials. Maintaining diverse sources of information reflects and furthers both

Focus on The U.S. Transport Infrastructure

The United States today has one of the world's finest transport infrastructures. Its development allowed the exploitation of the nation's raw materials, the creation of one mass market, and regional specialization of production across the country. Today's regional specialization of land uses reflects environmental endowments, but each environment's productive capabilities could not be realized and specialized without transportation and exchange.

Today, however, complaints are rising about the deterioration of the U.S. transport infrastructure. The Federal Highway Administration reported that in 1990, 57 percent of U.S. primary and secondary roads needed repair, and 39 percent of the nation's bridges were structurally deficient or obsolete. In that year Americans spent over 2 billion person-hours tied up in traffic jams, and airport congestion caused periodic gridlock from coast to coast. Delayed deliveries of goods cost the national economy an estimated $35 billion. Estimates of what the United States must spend to repair what it has and to build what more it needs are as high as $3.3 trillion. Raising and allocating investment in infrastructure might be a major national issue in the 1990s.

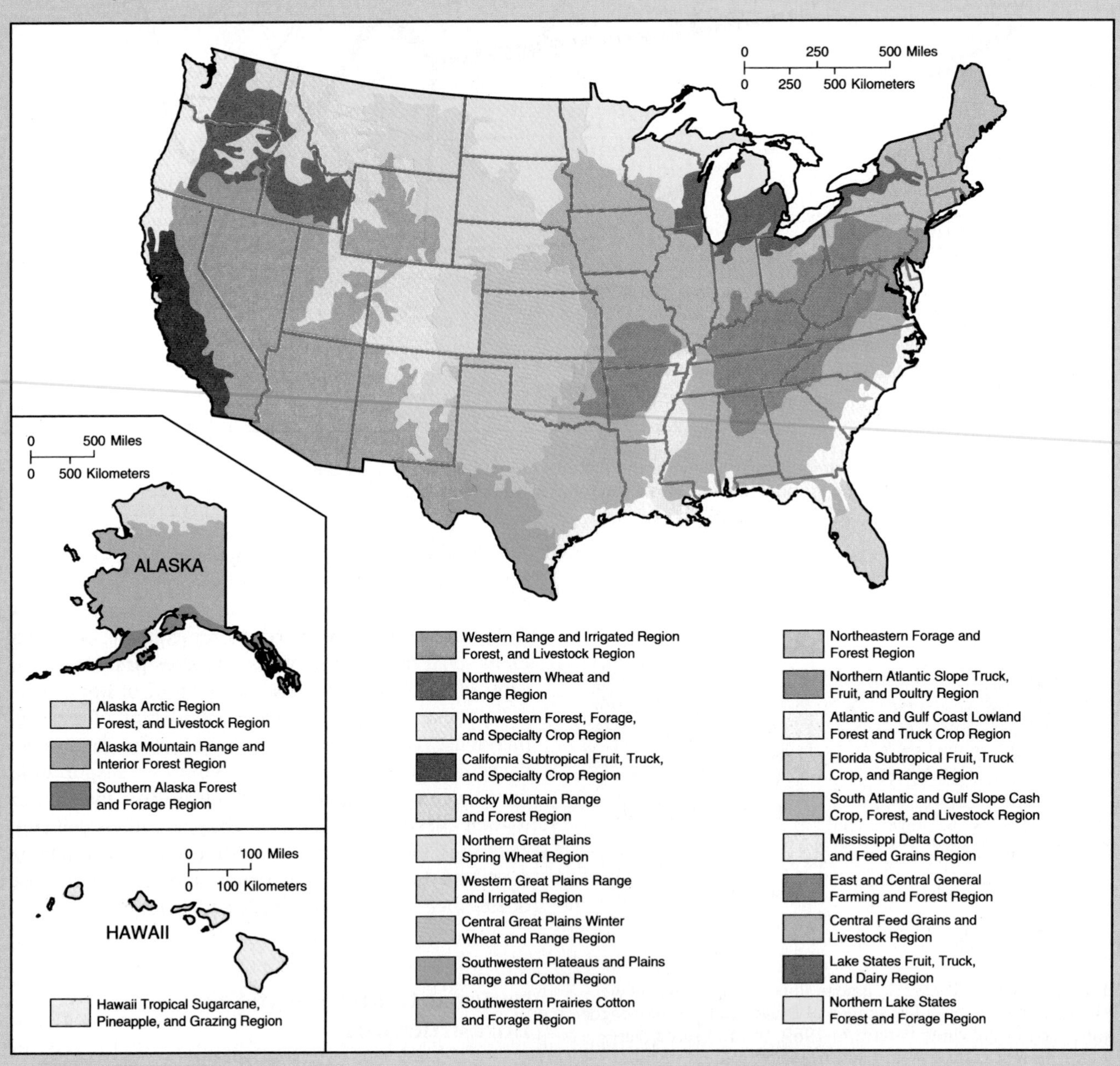

U.S. land resources. Resources are only potential resources until transportation allows regional specialization for maximum output and exchange. (After *The National Atlas of the United States.*)

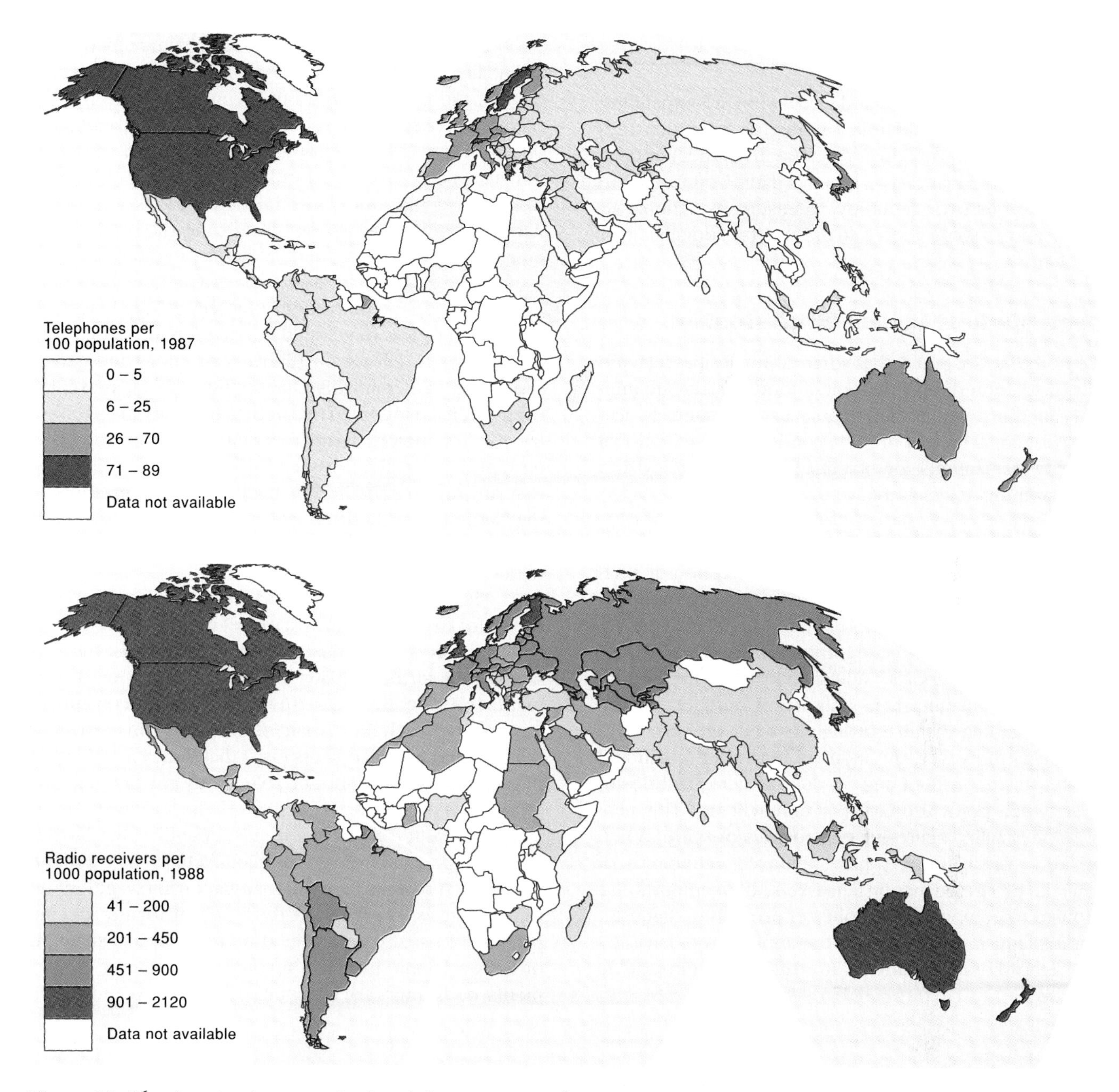

Figure 11–26. A national communications infrastructure can be as important as a national transportation infrastructure. Communications both reflects and boosts economic development, and it may be argued that communications are essential for political liberty as well.

economic development and political liberty. There are several ways of distributing information, and the availability of each type can be measured (see Figure 11–26).

Home mail delivery, for example, is guaranteed only in the world's richest countries. In 1988 Americans mailed 41 percent of all cards and letters mailed worldwide; the Japanese, in second place, mailed only 8 percent of the world total. The availability of telephones is another indicator of connectivity. The United States and Canada have almost 40 percent of the world's telephones; Africa, about 1 percent. The fax machine uses telephone wires, and it is rapidly replacing postal service.

Communications can substitute for a great deal of transportation, at greater convenience and lower cost, as we saw with telecommuting in Chapter 9. The U.S. government is considering building a nationwide fiber-optic network to deliver video, voice, and data into U.S. homes. A 1991 engineering study estimated that a network could replace 10 percent to 20 percent of all travel miles now logged by commuters, shoppers, and travelers. This would cut pollutants, reduce fuel consumption, and free people's time. France is considerably in advance in bringing computer and video services into individual homes, and Japan is at work on a national network to be finished by 2000.

TRADE POLICIES

World networks of trade and circulation are expanding, but countries still differ in the degree to which they participate in international trade and investment or open their domestic markets to imports. Each country wants to develop its own economy, to ensure its national security, and, to some degree, to protect its national culture or identity.

Open Markets Versus Protectionism

We noted in Chapter 7 that some countries protect and subsidize domestic production of food or strategic minerals for purposes of national security. Countries also protect industries that provide goods critical to national security (armaments, for example) or culture (television programming, for example).

The Import-Substitution Method of Growth In some cases a government may view protection of a national industry from imported competition as only a temporary measure. This is done in the belief that newly developing industries, called **infant industries,** cannot compete with imports. When industries are getting started, they manufacture only small numbers of the product, and so manufacturing costs per unit are high. Eventually, however, the industry may build a national market for its product and produce larger quantities of it, at which point the cost per unit will fall. This progression is called achieving **economies of scale.** When domestic manufacturers have achieved economies of scale in protected national markets, then imports are allowed to compete. This policy of protecting domestic markets and favoring domestic manufacturers is called the **import-substitution method of economic growth**.

This form of economic protectionism helped some countries industrialize in the past, including, to some degree, the United States, but it runs economic risks. If a country raises tariff walls to protect a particular industry, it is making all national consumers subsidize that industry. Some protected industries may *never* be able to survive competition with imported goods, and so the national public may go on subsidizing them. This may be an inefficient investment of national resources.

Some economists argue that as a national economy evolves, it ought to surrender low-technology industrial specialties to less developed countries and concentrate its productive efforts on higher-technology industries. We noted earlier in this chapter that the Japanese, for example, have followed this policy. In an advanced country, low-technology industries that are being allowed to disappear are called **sunset industries.** One problem that an import-substitution national policy presents is that without competition it is impossible to distinguish between infant industries and sunset industries in any country at any time.

The manufacture of garments in the United States today, for example, may be a sunset industry. The United States protects this industry behind tariff walls, thereby saving jobs in the industry but raising the price of garments to U.S. consumers. How much should consumers pay to save each U.S. garment worker's job? Garment workers and manufacturers insist that if they are allowed to go broke, they will not be able to buy any of the goods or services other U.S. workers produce, and so all Americans will suffer. Some economists believe that worldwide competition from goods manufactured in low-wage countries threatens to lower manufacturing wages in the rich countries. The height of the tariff wall for any specific product is usually determined by the political strength of the economic interests involved. In the case of garments, consumers are less well organized and combative

CRITICAL THINKING: *The Jones Act*

The 1920 Jones Act requires that all maritime trade between U.S. ports be carried in ships owned, built, and staffed by Americans. The act's original justification was that the nation needs a merchant fleet in times of war, and so the industry must be protected in times of peace.

The law has not worked as intended. Today, U.S.-owned ships make up less than 3 percent of the world's fleet—not enough to play a significant role in case of war. The act costs U.S. citizens about $1 billion per year in additional transportation costs, and it incidentally benefits domestic rail, trucking, and air freight companies. These companies are not required to use U.S.-built vehicles, and so they get a competitive advantage over coastal shipping. If the Jones Act were repealed, cost-distances among U.S. ports would fall, and a significant share of the freight that is now moved from one port city to another by railroad or truck could move more cheaply by ship.

This protective law passed for national defense does not achieve its purpose, costs each U.S. resident $4 per year, and distorts the national transport system and cost-distances among U.S. cities.

Questions

Are there reasons for protecting the U.S. merchant fleet?
Is there a better way of doing it?

than are the garment workers and the manufacturers because this one issue is less central among their concerns.

Is the garment industry a strategic industry that should be saved because it is vital to U.S. national interests? Are there any such vital industries? How about the automobile industry, the computer industry, or the camera industry?

Export-Led Economic Growth The alternative to the import-substitution method of economic growth is the **export-led method.** Countries that adopt this program welcome foreign investment to build factories to manufacture goods for international markets. In that way they can achieve economies of scale immediately. The countries that have grown the fastest in recent decades have generally followed export-led programs.

The success of these policies demonstrates again that what happens *at* places increasingly is the result of what happens *among* places. We introduced this point in Chapter 5. Therefore, geography studies not only the characteristics of individual places but how places relate to each other. Export-led policies could not succeed without global capital markets to facilitate international capital investment and global marketing networks to distribute the manufactured goods. Because export-led industrialization is a result of international developments, we will hold further examination of these developments for Chapter 13.

Growth Policies Versus Distributive Policies

We noted earlier in this chapter that governments try to maintain an equal standard of living throughout the country because regional disparities in welfare can be a centrifugal force. International trade makes this goal more difficult to achieve whether the country tries import-substitution policies or export-led growth policies.

If a country adopts protectionism, all regions of the country may not profit equally, and that can intensify regionalism. The United States in the nineteenth century, for example, used tariffs to protect infant industries in the East and invested the money in internal improvements for the West. This infuriated Southerners because the South exported raw materials and had to pay inflated prices for manufactured goods. Southern resentment of economic protectionism helped bring about the Civil War. We noted in Chapter 10 that today economic protectionism intensifies regionalism in Canada. The regions that produce raw materials do not benefit from the protection extended to the manufacturing regions.

The alternative export-led growth policies can also disrupt national regional investment plans. This is because the investment comes from outside, and different regions of the country can profit to different degrees. China best illustrates how disruptive this can be.

China's Domestic Political-Economic Geography China exemplifies on a colossal scale how regional imbalances in economic growth can be intensified by foreign investment and how these regional differences can become a centrifugal force threatening to tear a country apart.

China has always had an economic and cultural dichotomy, and an antagonism, between an outwardly focused southeast coastal region and an inwardly focused central region. Chinese capitals were in the north or center of China, in the heartland of Chinese civilization. None was ever in the south. The southern coastal cities were intent on trade, but the central government never valued commerce and cosmopolitanism. On the contrary, the ruling bureaucrats strictly limited the merchants' contacts with outsiders.

In the nineteenth century, European powers forcibly "opened" China's southeastern ports as trading and developmental centers. Chinese merchants, protected by foreign powers, escaped the restrictions imposed by the national rulers. Industrial development, inspired by international trade, clustered along the coast. The growth of the cities of Xiamen, Fuzhou, Tianjin, Shanghai, and Hong Kong resulted from these political conditions over which the Chinese government, ridden by strife and split by civil war, had little control. The government was forced to concede to the foreigners rights on Chinese soil, and this humiliation is still keenly felt.

The communist government that Mao Zedong brought into power in 1949 deliberately set out to shatter the lopsided geography of industry. In Mao's words: "This irrational situation is a product of history." China's first 5-year plan distributed industry inland, favoring such cities as Wuhan, Chongqing, Taiyuan, Guiyang, and Lanzhou. It was hoped that industry would boost the inland regions out of their relative economic backwardness into "socialist modernity." The policy of favoring the interior also intended to protect new industry from foreign invasion. From 1949 to 1979 foreign trade plummeted, China's coastal areas stagnated, and the cities wilted. The coastal merchants were even persecuted. In 1952 Shanghai merchants' assets were seized, and many merchants were exiled to China's northwestern frontiers.

Economic growth since Mao Mao died in 1976, and in 1979 the Chinese government reversed its policies. It admitted that Mao's regional egalitarianism was naive not only politically but economically as well. Pouring scarce resources into backward inland areas only deprived those regions with more potential for growth of needed capital. The government launched a new experiment that reversed the previous 30 years of economic policy. It designated five "special economic zones" and 14 "open cities" along the coastline (see Figure 11–27). These areas were allowed freer capitalistic practices than were permitted elsewhere in China. In 1987 Communist party general secretary Zhao Ziyang said: "The coastal areas should have their enterprises edge into the

Figure 11–27. In 1979 China designated special zones and cities where capitalist-style economic growth has boomed.

world market and further take part in international exchange and competition in a well-guided, planned, systematic manner, and make great efforts to develop an export-oriented economy. We must regard it as a strategic issue."

The designated coastal areas leaped to create new wealth. The province of Guangdong, adjacent to Hong Kong, experienced an industrial growth rate of 20 percent per year. Local factories sold shares and borrowed from farmer cooperatives. The state sector's percentage of total industrial output fell from almost 100 percent in 1978 to as low as 14 percent in the coastal provinces. This is lower than it is in several European capitalist countries. Hong Kong manufacturers built factories in China to take advantage of the cheaper land and labor, and within 10 years they employed over 3 million mainlanders (see Figure 11–28). A host of international corporations built factories in the special zones.

Shanghai reclaimed its role as a trading city. The central government allowed the city to keep a higher percentage of local revenues to reinvest in local infrastructure. This could entice new manufacturing, and the city even issued bonds on world financial markets. By 1989 Shanghai was responsible for 10 percent of China's industrial output, 12 percent of China's exports, and 14 percent of total national income. A quarter of China's foreign trade passed through the port.

China had engaged in very little foreign trade in 1979, but by 1988 foreign trade represented one-third of the national income. This is an extraordinarily high proportion for such a big country. From 1978 to 1988, total Chinese exports rose from $10 billion to $47 billion, and imports from $11 billion to $55 billion.

As China's seaboard provinces grew in wealth, the inland provinces slipped behind noticeably. The historical contrast between the interior and the periphery reemerged. Party leader Zhao noted this, but he hoped to convince the inland areas that they eventually would benefit from these policies. During a visit to two poorer inland provinces, Shanxi and Shaanxi, in June 1988, he said, "Development in the coastal areas will definitely contribute to the development in the interior part of the country. After the coastal areas have developed, it is impossible for the interior part of the country to remain undeveloped."

Economic liberalization triggered demands for political liberalization, but those demands were crushed. Political repression diverted some foreign investment to Indonesia, Malaysia, Thailand, and other Asian countries, but large sums continue to pour into coastal China. The income gap between the coastal provinces and inland areas is still growing. In 1991 for the first time since 1949 the percentage of total national industrial output that was from the state sector fell below 50 percent.

Partly as a result of the surge in China's food production that occurred with privatization noted in Chapter 7, partly as a result of attracting foreign investment capital, and partly as a result of the slowing rate of China's population growth, the country has achieved a rate of economic growth estimated at 6 percent or 7 percent per year. This has come about with repressive government, but it means that the population of China, a full 23 percent of the human species long living in grinding poverty, might achieve its goal of middle-class material comfort within 50 years. This is one of the most important "news items" of the late twentieth century.

Nevertheless, China's leaders delicately balance trade against autarky, and the inland regions against the coast politically and economically. When they fear that control is slipping out of their hands, they are willing to repress the population and shock and offend world opinion, as they did in 1989. The residents of Hong Kong, which is to be turned over to China in 1997, have been promised that they will be allowed to continue living under a capitalist system, but they must remember that in 1952 the Chinese government was willing to squash successful capitalists in Shanghai to maintain political control, and the central Chinese government may choose to suppress capitalism again.

Taiwan Taiwan is a principal source of investment funds for China. This island-nation boasts the world's 25th largest economy, and it is the world's 13th largest trading entity. Taiwan's 20 million people enjoy a per capita income of almost $8000 per year, in contrast to China's $300. These contrasts underline a bitter political division.

Figure 11–28. Traditional Chinese fish, rice, and tea have yielded to hamburgers, fries, and soft drinks at this McDonald's in Guangdong, China, a special economic zone in which foreign investment has created a thriving industrial concentration adjacent to Hong Kong. The setting of prices in Hong Kong dollars reflects the integration of the local economy with that of Hong Kong. (Courtesy of Francis Li/Gamma Liaison.)

In the 1930s a civil war raged in China between the Communists led by Mao, and the Kuomintang, a political party led by General Chiang Kai-shek (1887–1975). In 1949 the Communists drove the Kuomintang offshore to Taiwan, where it established a separate government. The ethnic Chinese form a minority among the native Taiwanese.

Both the government in Beijing and that on Taiwan claim that there is only one China and that they represent it, but at present China and Taiwan are in fact two separate countries. The UN General Assembly ousted the Taiwan government and seated the People's Republic in its place in 1971, and the United States recognized the People's Republic as the sole government of China in 1978.

In Taiwan state-directed capitalism has achieved phenomenal economic growth. A government economic council guided the creation of an industrial infrastructure that has successfully evolved electronics, computers, and other high-technology goods. Today Taiwan has been transferring low-wage, low-technology industries such as shoes and plastics across the strait to Fujian province, and Taiwanese are investing throughout China.

Some political accommodation will probably be reached in the 1990s between the government on Taiwan and the government in Beijing. The two governments might agree on joint economic ventures and joint representation in foreign affairs, for instance, without completely merging their domestic governments. The Beijing government, however, has threatened to attack Taiwan if Taiwan declares itself an independent state. The economic integration of Hong Kong and Taiwan with mainland China, and even together with the considerable economic vitality of the Overseas Chinese scattered from Singapore to Indonesia, is proceeding rapidly.

SUMMARY

Every country devises a program to build and to regulate its national economy and to distribute economic well-being across the national territory, and so the world political map is also a map of regionalized economies. Gross national product, gross domestic product, and gross sustainable product are three ways of measuring wealth, although each undervalues subsistence production and unpaid labor. Maps of these values nevertheless reveal the great differences in incomes around the world. Other measures of welfare, such as infant mortality rates and consumption of energy, may be combined to reveal standards of living. The United Nations publishes a Human Development Index. Any figures of welfare may vary greatly within countries. When statistics are examined per capita and through time, they reveal a relationship between population growth and economic growth.

Analysis of sectoral evolution of various countries reveals that some countries are preindustrial, some industrial, and others postindustrial. The contribution that each sector makes to national product is not the same as the share it employs of the labor force. Value added is high in the secondary sector, and so the secondary sector usually contributes a share of national wealth that is larger than its share of national employment.

The term Third World combines economic measures with cold war political values. When the First and Second worlds competed, the term had political validity, but the collapse of the Second World must reawaken our attention to the great differences among poor countries.

The world distribution of wealth is not the same as the world distribution of raw material resources. Wealth depends on the ability to put resources to work. Downstream secondary and tertiary sector activities produce great wealth by adding high value. Models of the locational determinants of manufacturing focus on transport costs, but in the world today these are of decreasing importance. Today's map of global manufacturing is a result of historical developments modified by new balances among relevant locational determinants. Inputs of capital and technology, for example, are increasingly important. These are footloose, and so manufacturing will presumably be redistributed almost continuously in the future, always according to the local mixes of the most important inputs.

Each nation organizes its own political economy, and this organization usually reflects other aspects of local culture. Nations try to knit and develop the national territory and to maintain an equal standard of living throughout it. Transportation networks allow regional specialization and trade, and their improvement releases national resources for production. Communication infrastructures both reflect and boost economic development as well as, potentially, political liberty.

Countries try to regulate their participation in international trade and investment. Import-substitute policies protect national growth behind tariff walls. Export-led economic growth policies encourage participation in international trade both as a source of capital and as markets. Both methods entail economic risks, as when a country protects inefficient national producers, and also political risks, because either policy may benefit some regions more than others.

KEY TERMS

gross domestic product (GDP) (p. 350)
gross national product (GNP) (p. 350)
gross sustainable product (GSP) (p. 352)
Human Development Index (HDI) (p. 354)
sectoral evolution (p. 355)
preindustrial society (p. 355)
industrial society (p. 355)
postindustrial society (p. 356)
Third World (p. 357)
First World (p. 357)
Second World (p. 357)
downstream activity (p. 359)
conceptual value added (p. 359)
material-oriented manufacturing (p. 361)
market-oriented manufacturing (p. 361)
opportunity cost (p. 364)
political economy (p. 366)
laissez-faire capitalism (p. 366)
privatize (p. 366)
state-directed capitalism (p. 366)
positive externalities (p. 366)
negative externalities (p. 366)
ecumene (p. 367)
frontier (p. 367)
infant industries (p. 378)
economies of scale (p. 378)
import-substitution method of growth (p. 378)
sunset industries (p. 378)
export-led method of growth (p. 379)

QUESTIONS FOR INVESTIGATION AND DISCUSSION

1. Do the rich countries today have greater power to restrict development in the poor countries than they did in the past? Do they actually do so? How? Is this deliberate, or is it just the result of "economic forces"? If the latter, which forces are involved?

2. What international webs of interdependence (banking, shipping, loan structures, and so on) ensnare today's poor countries? How?

3. At the opening of international currency markets on May 4, 1992, the following values were quoted against the U.S. dollar:

Australian dollar	0.7545
Brazilian cruzeiro	0.00043
British pound	1.7775
German mark	0.6068
Indian rupee	0.0348
Japanese yen	0.00751
South African rand	0.3487

Check these values in today's newspaper. Calculate the percentage each has risen or fallen against the dollar or against one another. What flows of goods or of tourists (remember, they move in opposite directions) should these fluctuations have triggered?

4. From the *United Nations Demographic Yearbook*, pick your own list of three or four variables to compare the quality of life in various countries.

5. Can you hypothesize an explanation for the variations in infant mortality recorded across the United States?

ADDITIONAL READINGS

CENTRAL INTELLIGENCE AGENCY. *Handbook of Economic Statistics.* Washington, D.C., annually.

CHISHOLM, PETER. *Modern World Development: A Geographical Perspective.* Totowa, N.J.: Barnes & Noble, 1982.

THE ECONOMIST INTELLIGENCE UNIT. *Country Reports.* New York and London. Quarterly. Ninety-two separate reports cover the economies of 165 countries.

GINSBURG, NORTON. "Natural resources and economic development." *Annals of the Association of American Geographers* 47(3): 197–212.

INDEPENDENT COMMISSION ON INTERNATIONAL DEVELOPMENT ISSUES. *North-South: A Programme for Survival.* Cambridge, MA: MIT Press, 1980.

THE INTERNATIONAL BANK FOR RECONSTRUCTION AND DEVELOPMENT. *World Development Report.* New York: Oxford University Press, annually for the World Bank.

LARDY, NICHOLAS. *Foreign Trade and Economic Reform in China 1978–1990.* Cambridge, England: Cambridge University Press, 1992.

OWEN, WILFRED. *Transportation and World Development.* Baltimore: Johns Hopkins University Press, 1987.

PEET, RICHARD, and NIGEL THRIFT, eds. *New Models in Geography: The Political Economy Perspective.* London: Unwin Hyman, 1989, 2 vols.

REPETTO, ROBERT, et al. *Accounts Overdue: Natural Resource Depletion in Costa Rica.* Washington, D.C.: World Resources Institute, 1992.

———. *Wasting Assets: Natural Resources in the National Income Accounts.* Washington, D.C.: World Resources Institute, 1989.

WADE, ROBERT. *Governing the Market: Economic Theory and the Role of Government in East Asian Industrialization.* Princeton: Princeton University Press, 1991.

CHAPTER 12

Regional International Organizations: Knitting or Unraveling?

This load of cars rumbles uninterrupted across the border between France and Luxembourg, which is marked only by a sign of the flag of the European Community: 12 gold stars in a field of blue. Across 12 countries of Europe the territorial scale of organization of all activities is expanding as the countries unite their territories in an economic union that is evolving into a complete political union. (Courtesy of the Delegation of the Commission of the European Community.)

Individual countries remain the highest level of absolute territorial sovereignty. A few organizations, however, have tried to regulate activities across the territories of several countries. **International organizations** coordinate activities among nation-states, whereas **supranational organizations** exercise powers over the member nation-states. Nation-states surrender some powers to supranational organizations, but they retain their ability to make war, thereby limiting the control these organizations exercise over them. Therefore, the distinction between international and supranational describes allocations of power that are in fact flexible and open to constant renegotiation.

International and supranational organizations are not new in history. Alliances among independent governments have always been made, and these alliances have often included efforts to coordinate governmental policies across a variety of issues. Today some international organizations are united strictly for purposes of military defense against common enemies. Today's most active international and supranational organizations, however, are working for economic or broader cultural goals. They are trying to achieve across several countries many of the same things that sovereign states try to achieve within their individual borders. These goals include creating unified markets for materials, goods, and services; knitting areawide transport networks; assuring free travel throughout the territory; and guaranteeing common standards of political or civil rights. A group of countries that presents a common policy when dealing with other countries is usually referred to as a **bloc.** The goals may be achieved through cooperation among governments, but often an independent governmental body is created.

This chapter reviews three instances of international organization. These three are particularly significant because of the economic and military power of the participants and because of the degree of autonomy surrendered by the participants to a central body or at least to a common policy. Two are examples of activities and territory being knit; one involves a formerly united country that is unraveling. The first case, the European Community (EC), is an international bloc that is coordinating activities across the territory of 12 Western European nation-states. Economics is the driving force for unification, but cooperation is based on a degree of preexisting cultural unity, and the EC explicitly seeks eventual political union. The EC has created the framework of a genuine supranational government.

The second case is the Commonwealth of Independent States (CIS), an international organization that is evolving to replace the Union of Soviet Socialist Republics (USSR). The Soviet Union was a Russian-dominated multinational empire that centralized political and economic organization across a full one-sixth of the Earth's land surface. In the late 1980s, however, the empire's subject peoples demanded and won greater autonomy, and in 1991 the Soviet Union formally dissolved. Some constituent republics broke off completely, but others chose to reorganize into the new Commonwealth. Mechanisms to coordinate or govern the Commonwealth will evolve through the 1990s, but it is certain that the degree of centralized organization within the Commonwealth territory will not be as great as it was in the Soviet Union.

In the third case, the United States and Canada are committed to achieving free trade by the year 2000, and Mexico may join this pact. This agreement is explicitly only economic, but, as Chapter 5 emphasized, achieving free and fair trade requires that many aspects of the people's lives become more similar. Free trade in architectural services, for example, might require equalizing standards of professional certification for architects; this in turn might require equalizing systems of educating architects and engineers. Other policies requiring standardization include conditions of labor, banking and tax policies, pollution control efforts, and business regulation. The North American countries are not creating any supranational governmental structures; the national governments will negotiate among themselves.

EUROPE: TERRITORIAL ORGANIZATION BEING KNIT

World War II ended in Europe in 1945 with the surrender of the Axis powers: Germany, Italy, Romania, Bulgaria, and Hungary. Japan had been an Asian Axis ally. The principal victorious Allied powers were the United States, Great Britain, Canada, France, the USSR, and China. In Europe the Allies had squeezed the Axis powers from west and east, and their victorious armies had met in the middle of the continent.

The Formation of Blocs

The Allies themselves, however, were already divided by mistrust, and Europe soon split into two competitive blocs. British prime minister Winston Churchill said, "From Stettin in the Baltic to Trieste in the Adriatic, an Iron Curtain has descended across the continent." To the east, Soviet armies installed subservient satellite regimes in Poland, Hungary, Romania, Czechoslovakia, and Bulgaria. Local communist leaders seized power in Yugoslavia and Albania without Russian help. The constitutions of all these countries were amended to guarantee Communist party dominance in national politics. Agriculture and industry within these countries were generally forcibly collectivized. Only Poland retained a private agricultural economy, and Yugoslavia some private industry and farms. These Communist party governments stated that the achievement of true communism was their goal, but none ever claimed to have achieved that goal; they called themselves socialist governments.

Germany itself was split into four zones of military occupation. Already by 1949, however, the U.S., French, and British zones were united functionally, and a new Federal Republic of Germany (West Germany), with Bonn as its capital, gained sovereignty in 1955. The Soviet Union reacted by granting its occupation zone independence as the German Democratic Republic (East Germany). Germany's capital city of Berlin had also been jointly occupied, and so it too was split between West and East Germany, even though it lay about 125 miles (201 km) inside East Germany. Its western sector was recognized as a part of West Germany, and East Berlin became the capital of East Germany (see Figure 12–1).

Austria had been annexed by Germany in 1938. After World War II it too was split into four occupation zones, which were reunited in a newly independent state in 1955. Austria's constitution stipulated that the country would forever remain politically neutral.

The Iron Curtain across Europe followed fairly closely the far older cultural division that Chapter 8 demarcated in religion and in orthography: The West was defined by Roman Catholicism or Protestantism and the Roman script, and the East by Orthodox Christianity and the use of the Cyrillic script. The Iron Curtain deviated from this line, however, in that the Baltic republics (Estonia, Latvia, Lithuania), Poland, East Germany, Czechoslovakia, Hungary, Croatia, and Slovenia had never been "Eastern European" culturally, and yet they were east of the Iron Curtain.

In Greece and Turkey, communist guerrilla fighters supplied by the Soviet Union were trying to overthrow the governments. In response, in March 1947, U.S. president Harry Truman announced his Truman Doctrine, which asserted that "it must be the policy of the United States to support free peoples who are resisting attempted subjugation by armed minorities or by outside pressures" (see Figure 12–2). Under the Truman Doctrine, the United States provided military assistance to Greece and Turkey, which helped them to defeat the guerrillas.

Territorial Changes During and after World War II the victorious Soviet Union seized a considerable amount of territory in Eastern Europe, as can be seen in Figure 10–4. Some territory was seized from enemies and some from weaker allies. In 1975, 35 countries—mostly European, but including the United States and other World War II participants—met in Helsinki, Finland, and agreed to accept European borders as inviolable. The conference did not, however, rule out revision by peaceful agreement. In 1990 the Federal Republic of Germany and the German Democratic Republic united

Figure 12–1. Already in the 1950s, West Berlin showcased Western prosperity in the middle of East Germany. As a result, so many East Berliners fled into West Berlin that in August 1961, East Germany built a wall across the city to seal in its citizens. This infamous Berlin Wall fell on November 9, 1989. (Courtesy of R. Bossu/Sygma.)

and renounced claims on territory to the east that had once been part of Germany. The unification treaty recognized Berlin as the capital of the united country.

Military Pacts The political division of Europe was accompanied by its division into two military alliances. In the West, the **North Atlantic Treaty Organization (NATO),** established in 1949, united Belgium, Denmark, France, Great Britain, Iceland, Italy, Luxembourg, the Netherlands, Norway, and Portugal with Canada and the United States. Greece and Turkey joined in 1951, West Germany in 1955, and Spain in 1982.

In 1955 the **Warsaw Pact** linked the USSR, Albania (which withdrew in 1968), Bulgaria, Czechoslovakia, East Germany, Hungary, Poland, and Romania. The USSR used this treaty to justify the continued occupation of Eastern European countries by hundreds of thousands of Soviet troops, at the expense of the occupied countries. These troops put down periodic anti-Communist upheavals, most notably in Hungary in 1956. The Communist parties of Yugoslavia and Albania refused to follow the lead of the USSR in all policies. Yugoslavia denounced

Figure 12–2. The Truman Doctrine committed the United States to a new policy that differed from the historical U.S. position in international affairs, which had been to avoid participation in foreign political quarrels (remain "neutral") unless the United States itself was attacked. The Truman Doctrine, however, led to military pacts and to military involvement around the world, including in Korea and Vietnam. (Courtesy of AP/Wide World Photos.)

the USSR in 1948, and it is historically ironic that the only time Russian forces ever marched into Yugoslavia was in March 1992, as part of a UN peacekeeping force to end the Yugoslav civil war. Albania broke with the USSR in 1961. For several years Albania considered an alliance with China, but it remained for 45 years after World War II Europe's poorest and most economically and politically isolated country.

Leonid Brezhnev assumed power in the Soviet Union in 1964. He enunciated the Brezhnev Doctrine, which asserted that if the Communist party government in any state were threatened, the other socialist states had the right to intervene. In 1968 five of the Warsaw Pact states invaded Czechoslovakia to overthrow a reform communist government. Only Romania refused to participate. The USSR renounced the Brezhnev Doctrine in 1989.

The Western and Eastern blocs thus faced each other in a protracted **cold war.** Although the two blocs never engaged directly in combat, they competed for economic growth and for influence among the new countries that gained their independence after World War II.

The pattern of alliances in Europe dissolved in 1989. The USSR directed its attention to its own internal problems and released its grip on the countries of Eastern Europe. These countries, in turn, repudiated the privileged role of the Communist party in their own countries, and within a few months they scheduled free elections. The Warsaw Pact formally dissolved in 1991, and the complete withdrawal of Soviet troops was delayed only by the Soviet economy's inability to absorb, house, and employ so many people. Germany subsidized the withdrawal.

NATO recognized the end of the cold war at a London Conference on July 6, 1990. It declared: "The Atlantic Community must reach out to the countries of the East that were our adversaries in the cold war and extend to them the hand of friendship." This does not guarantee that any military threat is entirely over. NATO will survive as a mutual defense and political structure.

Economic Blocs In 1947, just a few weeks after the proclamation of the Truman Doctrine, the United States announced a plan of financial aid to war-shattered Europe. This plan eventually became known as the Marshall Plan in honor of Secretary of State George C. Marshall (1880–1959). Between 1948 and 1952 the United States offered financial subsidies to the European countries to help them rebuild their economies. Aid was offered to the Soviet Union and to the countries of Eastern Europe, but Soviet dictator Joseph Stalin (1879–1953) rejected it. The Marshall Plan went on to become a program for the economic rehabilitation of Western Europe.

COMECON The Soviet Union and its Eastern European satellites responded to the Marshall Plan by forming a competing economic pact, the **Council for Mutual Economic Assistance (COMECON).** This arrangement brought about a partial integration of their national economies. The Soviet Union supplied raw materials and fuel to its COMECON partners and in return took manufactured goods from them (see Figure 12–3). The COMECON countries also subsidized other international communist ventures; they provided economic assistance, for example, to Cuba. In 1991 COMECON formally disbanded.

In the long term, the dissolution of COMECON will probably enhance the economic growth of the Eastern European countries, but in the short run it was disastrous for them. Their inferior manufactured goods could not compete in Western markets, and the Soviet Union, which was itself declining economically, could not afford them. At the same time, however, the Soviet Union began to demand payment for its fuel and raw materials in currencies that are freely traded on international currency exchanges, the so-called **hard currencies.** This demand imposed a tremendous economic burden on the former satellites.

Some of the Eastern European economies grew under their communist regimes, enough to surpass the Soviet Union itself in per capita measures. The imposition of Communist party government, however, prevented them from keeping up with Western Europe. Czechoslovakia, for example, had been a rich country before World War II, the equal of Germany in technology and

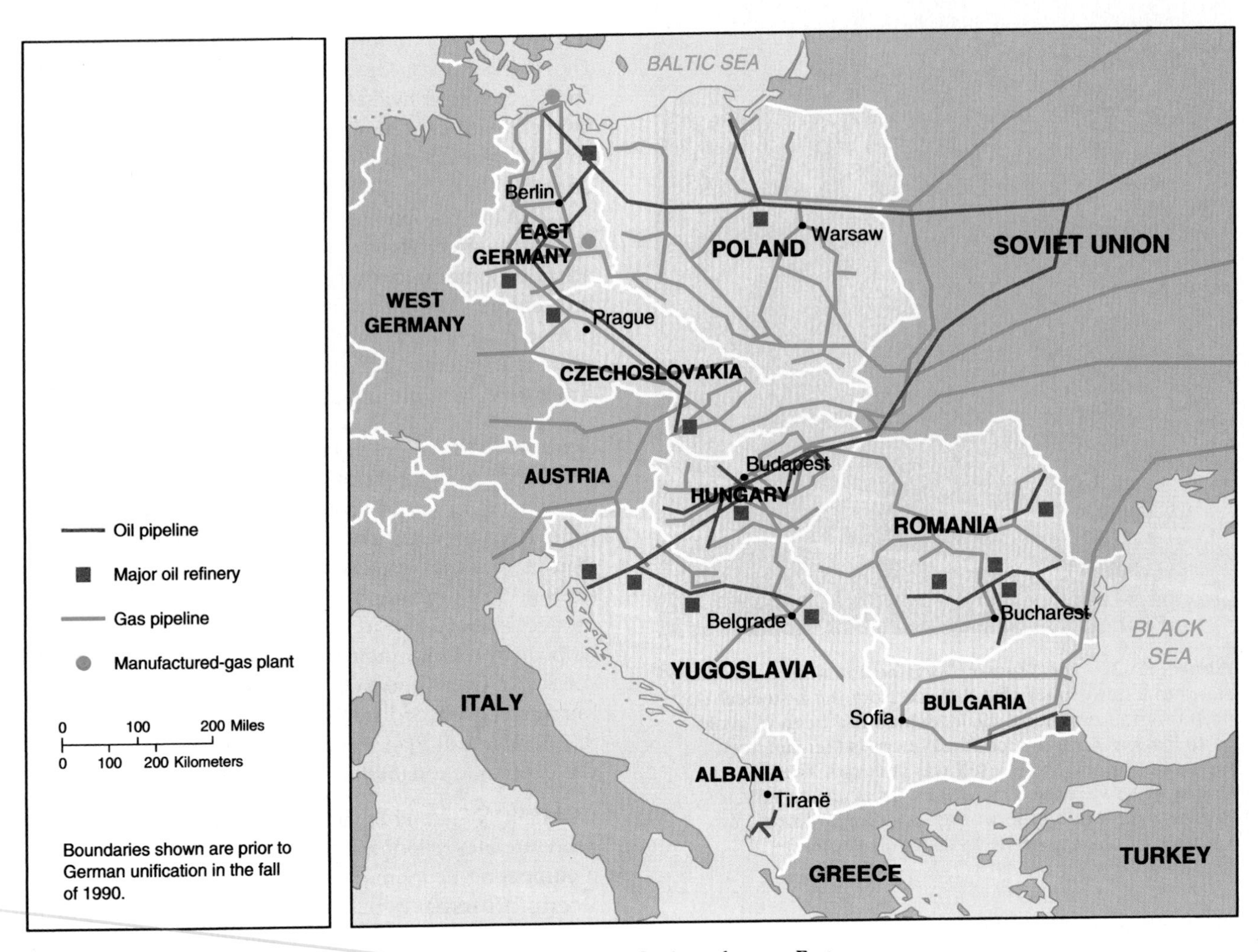

Figure 12–3. Eastern European energy supplies, 1990. Russian Soviet rule over Eastern Europe was the only case in history in which the ruling power supplied raw materials, including, as seen here, energy supplies, to its satellites.

productivity per worker, and ahead of France. During World War II its industries were not damaged, and yet by 1990 productivity per worker in Czechoslovakia, although higher than that in the Soviet Union, was less than one-half that of West Germany or France.

The Reprivatization of Eastern European Economies The bulk of Eastern Europe's socialized industries are being returned to private ownership, but this presents tremendous challenges in countries that do not have stock or bond markets, private banks, or accounting systems. Western Europeans knew a lot more about starting both political parties and private companies in 1945 than Eastern Europeans knew about either in 1990. Eastern European industries were inefficient and hugely overstaffed. Restructuring to make them competitive triggered soaring unemployment rates and other economic dislocations. It is hoped that these problems will be only temporary.

In communist Hungary, for example, all economic activity was run by state-owned enterprises or cooperatives with only a few exceptions for small businesses. An Enterprise Council controlled 75 percent of the businesses, and state authorities directly controlled the rest. In 1990 about 2000 state-owned enterprises were eligible for privatization, and the government began selling them off. The Hungarian stock market, which had been closed in 1948, reopened in 1990.

During the 1980s East Germany was estimated to be the world's eighth most powerful industrial economy. When it unified with West Germany, the German government put all East German state properties (8000 corporations employing 6 million people and with assets worth a theoretical $400 billion) into a gigantic holding company. The government intended to sell these assets and rebuild the economy on capitalist principles. These assets were, however, so dilapidated, polluting, and lacking in modern infrastructure that their sale realized just a fraction of the original estimate of their value.

In 1990 many Eastern bloc citizens and even people living in the West demanded the return of properties that had been taken from them during the collectivization drives. Questions of ownership and compensation will not be fully sorted out for years.

The service sector might offer individual opportunity in the newly privatizing economies. It could absorb many displaced workers. Eastern Europeans have long gone without a selection of services from housepainters to beauty parlor operators.

Reprivatization and pollution Reprivatization will also require that the new governments and the new private sectors agree on sharing responsibility for cleaning up polluted Eastern European environments (see Figure 12–4). The industrial policies of the various communist governments had compelled managers to meet production targets regardless of the cost or hazards to people's health or to the environment. Unbridled industrial spewing and spilling carried disastrous consequences. In East Germany, environmental conditions were so bad that the government insisted that West Germany assist in cleanup activities even before political unification. The task includes purging East Germany's rivers, tackling poisoned soil, rehabilitating sick and dying forests, and bringing fresh air back to the suffocating industrial towns. Plans require dismantling scores of obsolete, polluting factories and reviewing some 15,000 dump sites for dangerous toxic waste. To provide East Germany with clean energy will require dismantling all the large power stations that had been driven by highly polluting brown coal and replacing those and hundreds of other furnaces and plants. Similar cleanups will be required in each of the other COMECON countries. It will take years and enormous sums of money to clean up Eastern Europe's contaminated environment and to restore its people to health.

Western assistance In 1991 the United States, Japan, and the Western European countries formed a European Bank for Reconstruction and Development to help Eastern Europe's new democracies make the transition to market-oriented economies. If the rich countries of the West want to help Eastern Europe, they must either give the Easterners financial help outright or else open Western markets to Eastern goods so that the Easterners can earn a rising standard of living. Food is a potential Eastern European export, but Western European protectionism may prevent food imports.

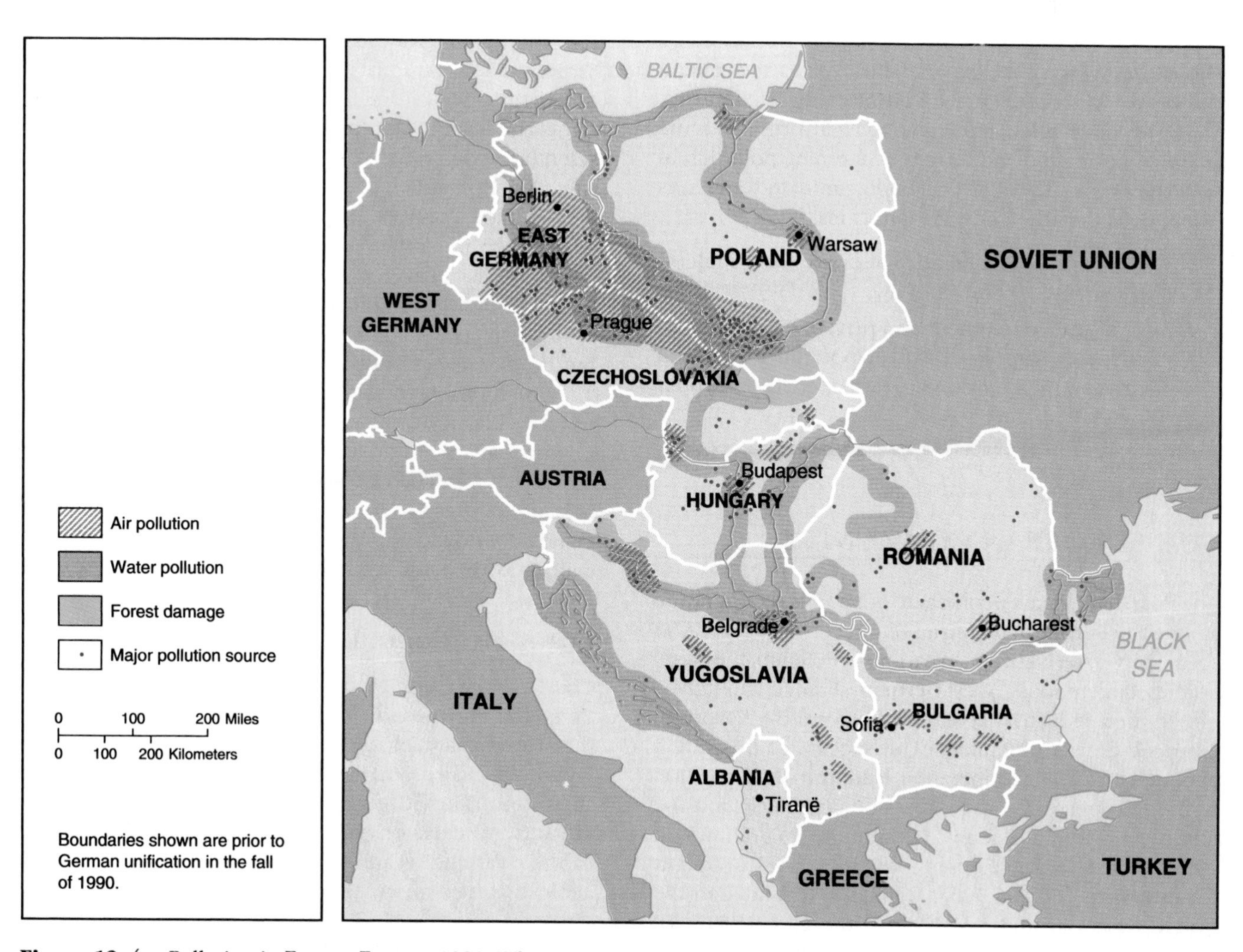

Figure 12–4. Pollution in Eastern Europe, 1990. When communist governments in Eastern Europe collapsed, and Western observers were allowed to examine industrial facilities, they found a degree of pollution that they could scarcely believe.

ABC NEWS ***Welcome to Hell***

With the collapse of the communist regimes imposed on Eastern Europe by the USSR, information concerning political oppression and economic stagnation in these countries has become more widely available. Some of the most startling information, however, concerns the state of the physical environment.

Simply put, Eastern Europe might be the most polluted region in the world. Pollutants—often toxic—from factories and chemical plants have devastated the region's water, air, and physical landscape. In the former East Germany, for example, 70 percent of all energy was produced by the burning of high-sulfur coal. A byproduct of this burning is sulfur dioxide, which hurts the eyes and lungs and helps to create acid rain. Because the government controlled the media in this region, many people remained ignorant of the extent of this problem. Those who were aware could do little, given the general lack of freedom of speech and action in these countries.

The video titled "Welcome to Hell" presents some graphic information and images concerning pollution in Eastern Europe. If you have an opportunity to watch this video, consider the following questions.

1. What steps can be taken to alleviate this problem? Who should pay for it? Why?
2. Would the government and private companies allow something like this to happen in the United States and then cover it up, as happened in Europe? Why or why not?

THE EUROPEAN COMMUNITY

The Marshall Plan encouraged economic cooperation among the European countries that received aid. In 1952 six of the Western European nations—Belgium, the Netherlands, Luxembourg, West Germany, France, and Italy—united their industrial economies to form the European Coal and Steel Community. The success of this union encouraged further cooperation, leading to the formation of the European Economic Community (EEC) and the European Atomic Energy Community (Euratom) in 1957. All three organizations merged into one **European Community (EC)** in 1967. Denmark, Ireland, and the United Kingdom joined in 1973; Greece, in 1981; and Spain and Portugal, in 1986. By 1990 the 12 member countries had a combined population of 325 million (see Figure 12–5).

Economic groupings come in three major forms. A **free trade area** has no internal tariffs, but its members are free to set their own tariffs on trade with the rest of the world. A **customs union** has a common external tariff and no internal customs. A **common market** is a customs union, but it also has a common system of laws creating similar conditions of economic production within all the member countries. The EC is forming a full common market.

The EC also declares political merger to be an eventual goal. It seeks to become a genuine United States of Europe, just as the 13 British colonies merged into a United States of America. One crucial difference, however, between the U.S. and European experiences is that U.S. unity was the creation of essentially political forces. The colonies had a common foreign policy for years before they had an internal economic policy. European integration, by contrast, has been driven by economic logic, and the states so far jealously guard their sovereign power to make foreign policy. For this reason Europe may find true federalism more difficult to achieve.

Within Europe, individual national cultures remain distinct, but European states share many cultural assumptions and traditions. Several broad cultural patterns cross national boundaries. Common traditions include democratic ideals and parliamentary institutions; civil rights and legal codes; Judeo-Christian ethics; respect for scientific inquiry; artistic traditions of Classicism, Realism, and Romanticism; and humanism and individualism. In fact, cooperation within the EC has bolstered democracy in certain countries—Portugal, Spain, and Greece—that previously were only dubiously committed to it. It may be difficult to extend EC membership to some states, such as Turkey, that are pressing for entry but do not share these European traditions.

In addition to economic integration, another reason for European unification is the urge to play a role in world affairs. In an age in which economic strength has largely replaced colonialism or military might as a measure of prestige, no European nation alone can command prominence on the world stage. Combined, however, Europe could assume a leading position.

European Community Government

The European Community is governed by four institutions: the European Commission, the Council of Ministers, the Parliament, and the European Court of Justice. Figure 12–6 on p. 392 illustrates the relationships among these four. The EC government has the power to raise funds from levies on the member countries, according to their varying wealth, and it dispenses these funds to achieve agreed-on community goals.

The European Commission The European Commission proposes legislation and is responsible for administration. Its 17 commissioners, appointed by the

Figure 12–5. The European Community and the European Free Trade Association.

governments of the member nations, are committed to act in the EC's interests, independent of their national governments. The commission sits in Brussels, Belgium, which has enjoyed considerable growth and prosperity as de facto capital of the EC.

The Council of Ministers The Council of Ministers, comprised of one minister per member nation who represents his or her government, is the final legislative body. Ministers change according to the subject under discussion. The presidency of the council rotates among member states every 6 months. Votes on most subjects required unanimity in the past, but the 1985 Single Europe Act expanded the range of issues on which council votes are decided by majority vote. This considerably reduced the sovereignty of the individual members.

The European Parliament The European Parliament sits in Strasbourg, France. It has 518 members. Parliament members have been elected directly by the citizens of the member countries since 1979, and in the parliament caucuses of international political parties ex-

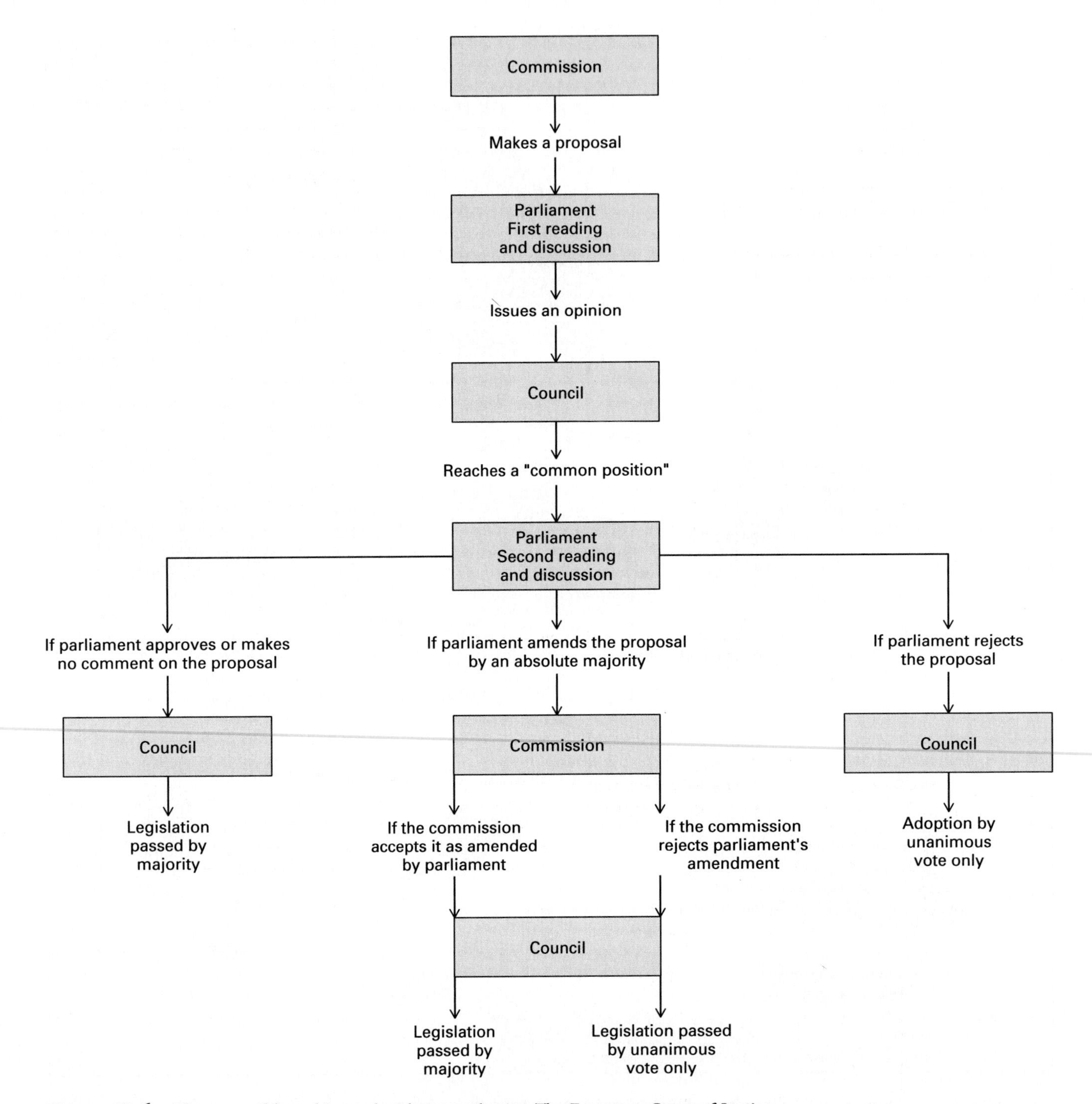

Figure 12–6. The council has ultimate legislative authority. The European Court of Justice, not shown on the chart, adjudicates EC laws and ensures that national laws conform to EC laws.

tend across national political groupings. Therefore, the parliament, and to a degree the commission, are genuinely supranational, rather than international.

The European Parliament is not, however, equivalent to a national legislature. It cannot initiate legislation, select the executive, or levy taxes. It can scrutinize proposed EC laws and act as a public forum. It can offer amendments, vote against directives, and even veto the actions of any branch of the community. It can dismiss the European Commission with a vote of "no confidence."

Parliament will almost surely gain power: It is the most democratic of EC institutions, and many Europeans are eager for the democrats to take power from the bureaucrats. Parliament's political influence and assertiveness is already outgrowing its legal authority because more influential individual politicians are seeking election to it. Former French president Giscard d'Estaing, for instance, who sits in the parliament, has said, "This is where the questions of the future will be defined. In national parliaments we discuss national problems. Here we talk about Europe."

The Court of Justice European law is adjudicated by the 13-judge European Court of Justice. The Court was originally established with the European Coal and Steel Community, and it has enlarged its jurisdiction to include all European Community laws. In cases of conflict between community law and national laws, the supremacy of community law is being steadily affirmed. Community institutions have become so important to the member states that, for example, the German government estimated that 50 percent of all West German legislation passed in 1989 was based on EC directives.

European Community Integration

The Single Europe Act of 1985 called for the creation of one economic community by the end of 1992. Integration of the EC territory as "one place," economically at least, can be seen in five aspects of its evolution: (1) creation of one market for goods, (2) creation of one market for capital, (3) transportation policies, (4) social policies, and (5) regional policies.

Creating One Market for Goods One of the EC's earliest acts was to establish a common agricultural policy. This policy has achieved several of Europe's agricultural goals discussed in Chapter 7, but it has protected inefficient farmers at great cost. The EC members with the most inefficient farmers have profited handsomely from the policy, particularly Greece, Ireland, Italy, Portugal, and Spain. Overall, however, agriculture accounts for less than 4 percent of total community gross product.

European manufacturers would like to achieve the economies of scale that U.S. manufacturers have long enjoyed. Many European firms have merged across national borders, and EC bureaucrats have standardized international business regulations.

Business integration will not homogenize the consuming habits of Europeans overnight; cultural obstacles will still exist. Companies are nevertheless creating "Eurobrands," giving products a single name throughout Europe. Many companies are supporting these with advertisements that are similar in message and execution except for the language. This presumes and supports the eventual creation of a "Euroculture."

Three trends support the supposition that a single continental consumer culture market will emerge.

1. Changes in European media should make it easier to transmit a pan-European message across national borders. These changes include the privatization of television stations (which formerly were all government-owned and -operated) and direct satellite broadcasting into homes. One-third of European homes were already receiving satellite channels by 1992. The number of television channels in Western Europe is expected to rise from 69 in 1987 to 91 by 1993, and the amount of programing will jump from 282,755 hours to 784,020.
2. Patterns of European prosperity and consumption no longer conform to national borders, but straddle them, defining new international regions. Europe's wealthiest people, for instance, are concentrated within a 250-mile (400-km) ring around Cologne. This area includes parts of Germany, France, Luxembourg, Belgium, and the Netherlands. Manufacturers and marketers are defining other specific international marketing regions.
3. Younger Europeans purchase consumer goods from many countries, and each year more are traveling and studying in other EC countries. This indicates a strong continuation of Europewide acculturation. As noted in Chapter 5, the young and the rich are usually the first groups to adopt new products and lifestyles across cultures.

The evolution of one continental market should boost the EC's economic growth and lower prices to consumers. The sacrifice demanded, however, is the individuality of the national cultures. Both U.S. tourists and Europeans have already observed that the distinctiveness of the various European national cultures is diminishing.

As the EC's common tariff wall against imports from foreign countries has risen, the community has begun to discriminate against imports. To penetrate this wall, manufacturers from many outside countries have bought or built factories in EC member countries. Many Japanese and U.S. corporations have for years managed their European operations as if Europe were "one place." This experience might make them better prepared to take advantage of new opportunities than the Europeans are themselves.

Creating One Market for Capital In 1978 the European Monetary System introduced its own monetary unit, the **European Currency Unit (ECU).** This was a theoretical unit of account and did not exist in banknotes. In 1991, however, the 12 governments committed themselves to the creation of a single currency and a regional bank by the year 1999. A single currency will end the confusion every time a traveler crosses a border. In addition, trade will be greatly assisted by the elimination of fluctuations among exchange rates and the simplification of accounting procedures. Acceptance of a single currency and central bank is another important surrender of national sovereignty, because a nation's control over its own currency affects control over its national budget, expenditures, and domestic interest rates.

Each national stock exchange has traditionally dealt in the stocks of its own national corporations, but no national exchange can raise enough money for the enormous corporations such as British Petroleum or Nestlé that operate in many countries. Therefore, one dominant international stock market will probably develop. That development will, in turn, require new rules regarding what financial information businesses must disclose to protect investors.

London is currently Europe's largest financial center, but Frankfurt is the financial capital of Germany, which represents about one-quarter of the total EC gross product. The leading European financial centers will compete for dominance, and eventually one will emerge as Europe's economic capital. If none of the European centers proves quick enough to offer a wide enough range of services efficiently, the economic capital of Europe might well become New York. Telecommunications makes this possible.

Transport System Europe's transport system will be redesigned from 12 national webs into a single coordinated web. Much of the transportation infrastructure will require modification. All railroads, for example, must be on the same gauge and electricity systems, and all highways, bridges, and tunnels must have the same construction specifications. Europeans are planning an integrated network of high-speed trains to be superimposed on the existing national networks (see Figure 12–7). This network is to be completed by 2025, and it will redistribute relative accessibility. The city of Lille, for instance, anticipates growth because of its position in the center of the Paris-London-Brussels triangle. European air transport also will need coordination by a new supranational equivalent to the U.S. Federal Aviation Administration (see Figure 12–8).

The unification of Germany required investment to bring the infrastructure of Eastern Germany up to Western European standards and to unite two networks that were formerly disconnected. Before 1990 each national network ran primarily north-south, and there were few links between them. Now the government is building new east-west roads, railroads, waterways, and telecommunications.

Germany's central European situation will be emphasized with the opening of a barge canal connecting

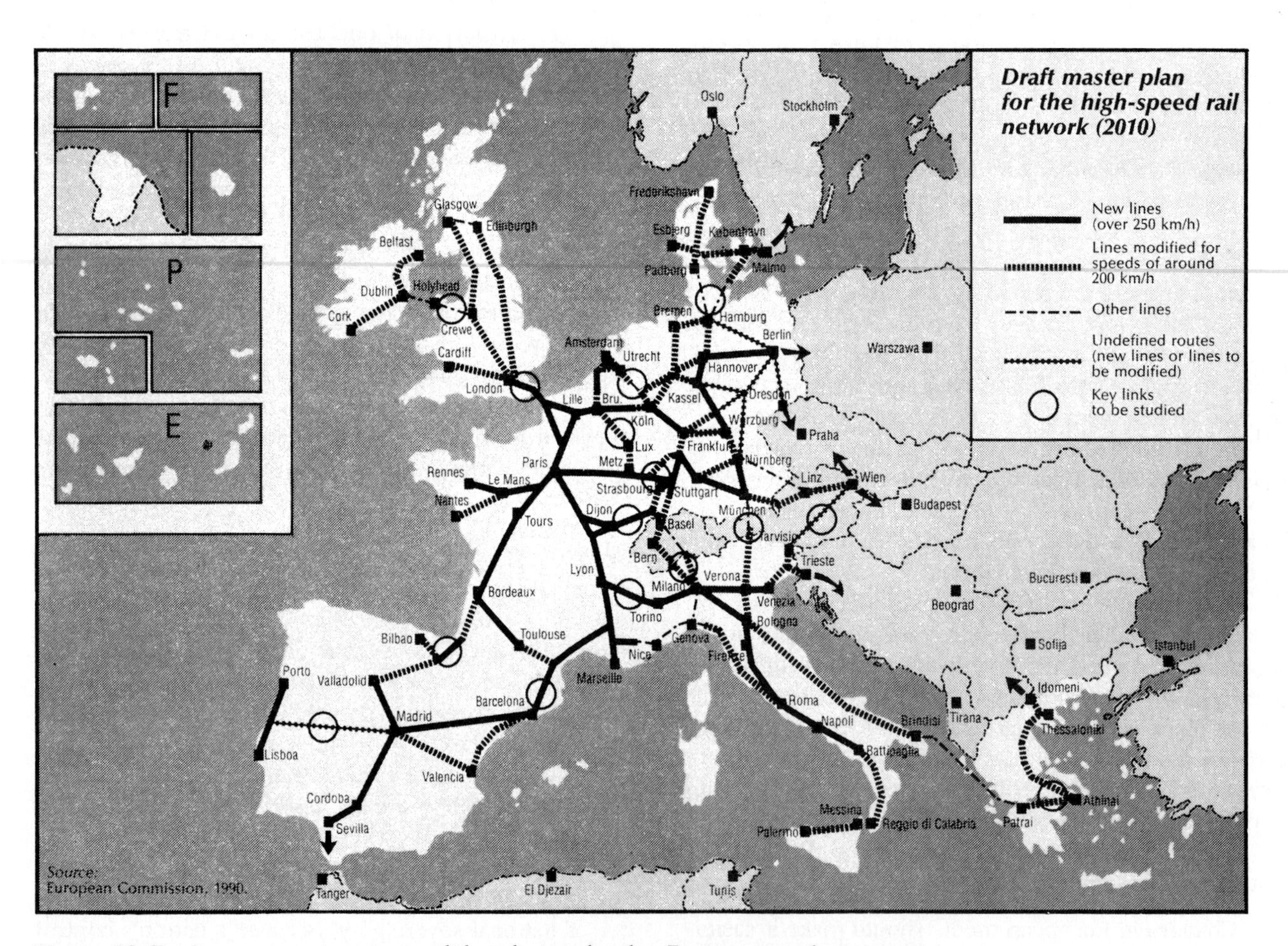

Figure 12–7. In an area as compact and densely populated as Europe, central city-to-city travel by train is a sensible alternative to air travel at little or no additional time-cost. The new system will require laying 4600 miles (7400 km) of new track and improving 12,000 (19,300 km) more. (Data from European Commission, 1990.)

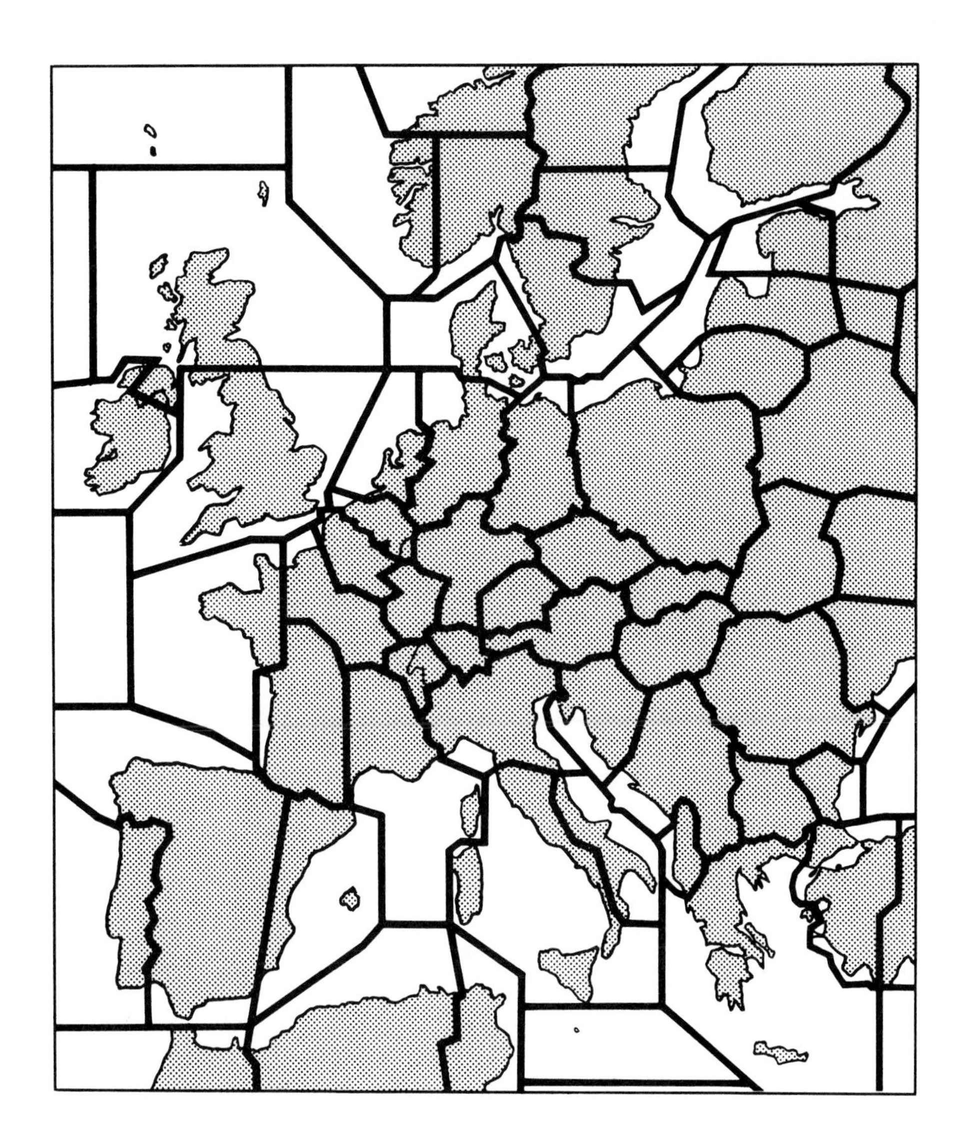

Figure 12–8. Western Europe has only one-half the density of flights of the northeastern United States, but the inefficient system of 22 national air-traffic control systems operating from 42 air-traffic control centers in 1990 caused flight delays. (Data from Association of European Airlines, 1989.)

the Rhine River with the Danube (see Figure 12–9). This new route will provide an internal trans-European highway from the North Sea to the Black Sea. Because it is opening at a time of political and economic liberalization in Eastern Europe, this canal may increase East-West trade and bring Eastern Europe more under the influence of the economies of the Western countries.

Despite international unification of the transport network, each nation will still manipulate the system to favor its own economy. The Dutch, for instance, will do what they can to maintain Rotterdam's status as the world's largest port against its rivals Antwerp, Hamburg, and Le Havre (see Figure 12–10). The Chunnel will shift European transport patterns southward, but the Dutch offer subsidies and tax breaks to attract European corporate headquarters and distribution operations to Rotterdam.

Regional Policy The Single Europe Act commits the countries "to strengthen economic and cultural cohesion." This has been interpreted to mean strengthening regional development policies. Efforts have been initiated to equalize living standards and quality of life throughout the community area, just as nations strive to equalize incomes throughout their territories (see Figure 12–11). The community furnishes funds for regional development or for basic infrastructure, such as transport and telephone service.

Community planning can transcend national borders and designate regions that straddle them. The commission is planning for the Spain-Portugal border area, for example, where it will build roads and clean polluted rivers. Private companies will then be encouraged to increase industrial investment in the north of the region and to develop tourist facilities in the south.

Social Policy The EC members have tried to strengthen cultural ties. International committees are rewriting school history books, for example, to soften nationalistic antagonisms. University admissions policies, academic standards, and definitions of degrees are being harmonized. Community institutions are also encouraging and subsidizing international scientific research projects. *Esprit*, for example, is a scientific effort to devise

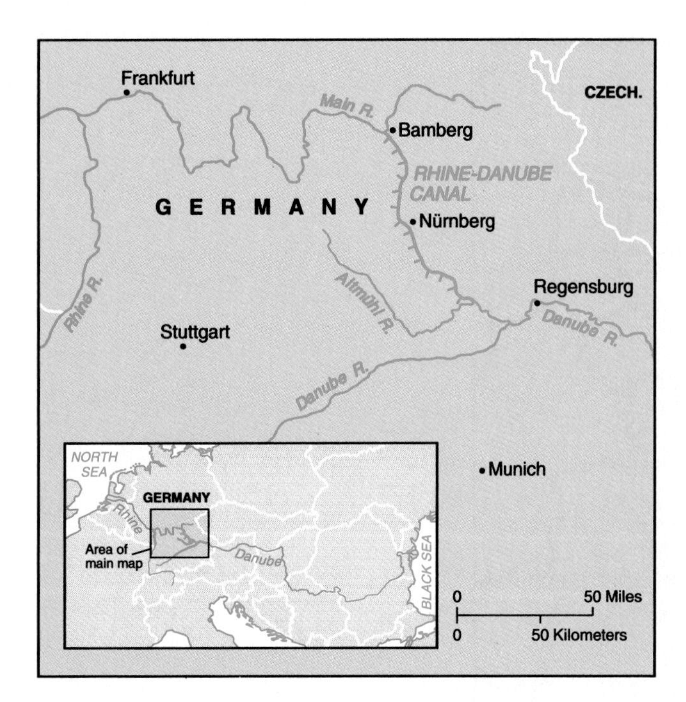

Figure 12–9. The Rhine-Danube Canal, one of the greatest public-works projects in history, will benefit countries from Holland to Romania. The Rhine already carries more traffic annually than the sea lanes between the United States and Europe, but the Danube has carried only about 40 percent as much as the Rhine.

new information technologies. Numerous pan-European cultural festivals are celebrated, including the naming each year of a European Cultural Capital, a city chosen for its pan-European cultural-historical significance.

Most specific EC social actions, however, deal with questions of worker health and safety standards, the international transfer of pension rights, the guarantee of rights to migrant workers, and mutual recognition of technical qualifications. The European Commission has written a Workers' Charter, and regulations dictate the workweek, overtime pay, and holidays. Future international rules might standardize annual paid vacations, set a minimum working age, establish maternity leave, and call for worker participation in corporate decisions. Workers' rights will be equivalent throughout the European Community. In addition, the European Court of Justice has enforced standard civil rights and antidiscrimination legislation.

Through the years, statistical measurements of many aspects of life in the EC member countries have become more similar. These measurements include statistics as diverse as median educational attainment of the population, degree of urbanization, percentage of females in the labor force, crude birth rates, and total fertility rates. Life in the different member countries is growing measurably less different and more alike.

Figure 12–10. Rotterdam is the world's greatest seaport, measured either in tonnage or by value of freight. Port facilities stretch over 30 miles (48 km) along the mouth of the Maas River, a tributary of the Rhine. (Courtesy of Havengezicht/Netherlands Board of Tourism.)

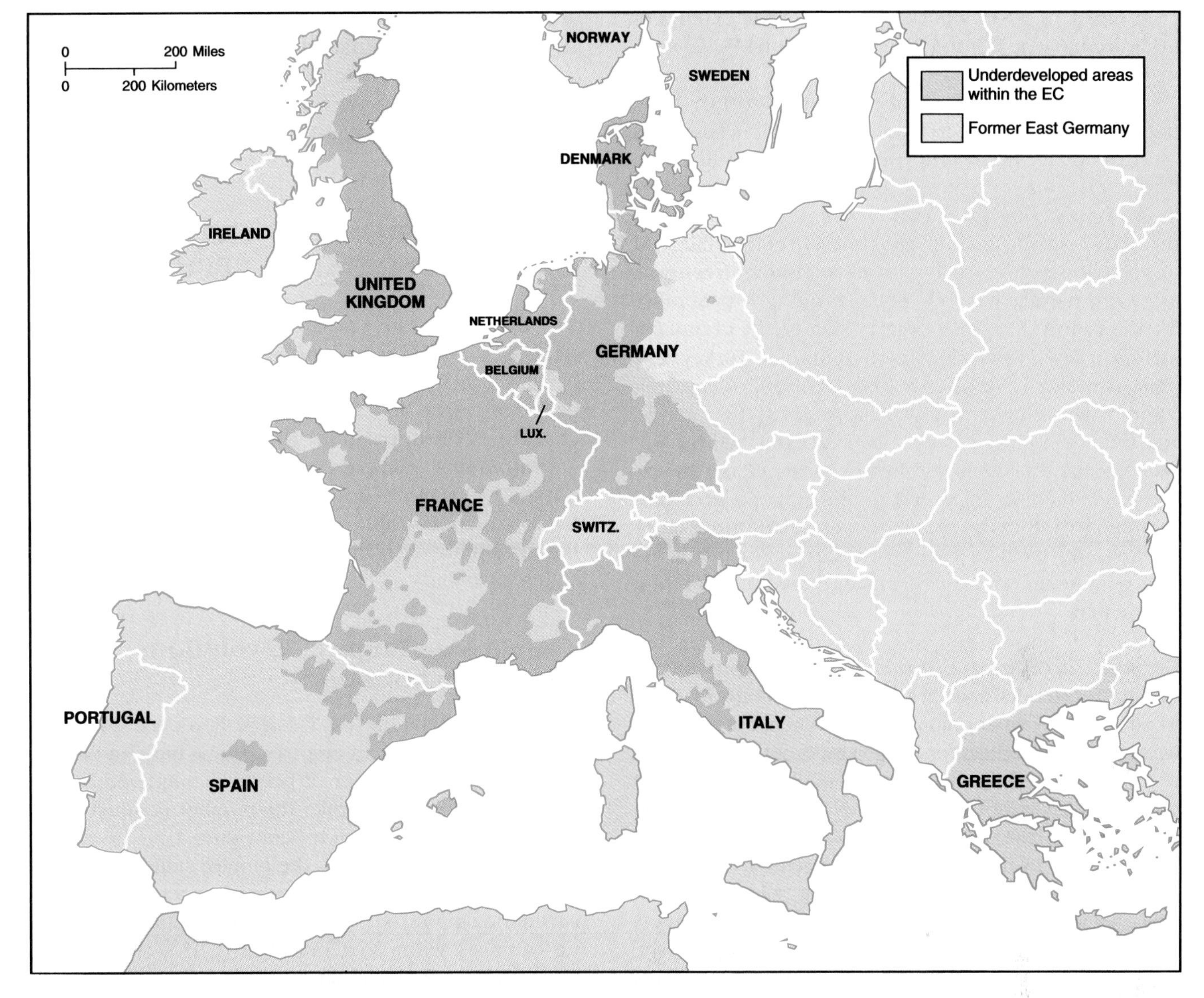

Figure 12–11. Incomes are not everywhere equal throughout the EC territory, and the Community devises special programs to assist the less-well-off areas. The areas shaded green on the map are largely areas of relatively low income, and the EC is devising special assistance programs to develop them. The area shaded yellow-green, the former East Germany, is also eligible to receive special EC funding.

The Circulation of Non-EC Citizens Community citizens may travel and settle anywhere within the EC, and in 1985 a European passport was introduced in all member countries except Germany, Greece, and the United Kingdom. Non-EC citizens, however, face common immigration and visa requirements. Unless common regulations are in force, the country with the most relaxed rules will become the gateway to the entire community. Already in mid-1991 there were an estimated 8.2 million legal and 3 million illegal aliens in community member nations.

Free circulation creates a number of problems. Each country grants political asylum to refugees on the basis of its own international relations. Therefore, creating a single set of conditions of asylum will affect each nation's international affairs. Also, most governments want to ensure that the elimination of internal borders does not give migrants who have legal residence in one country the right to live and work in others. France, for example, does not wish to extend workers' rights to the Turks currently living in Germany. The EC will list nationalities that will need visas.

Expansion of the Community

When the EC first formed, a few Western European countries chose not to join. Seven of them united in a **European Free Trade Association (EFTA):** Austria, Liechtenstein, Finland, Iceland, Norway, Sweden, and Switzerland. The EC and EFTA trade extensively with each other. In 1990 about 66 percent of the exports of EFTA went to the EC, and 27 percent of total EC exports went to EFTA members.

In 1991 the EFTA and the EC agreed that by 1993 they would create one free trade region to be known as the **European Economic Area (EEA).** The EEA will not be a customs union or a common market, nor will its members participate in all the types of integration that are occurring among the EC members. The EEA will nevertheless be the world's largest consumer free trade bloc, and merger into one political unit is a possibility for the future. This new organization might eventually include a few Eastern European countries. Turkey, Cyprus, and Malta are already "EC Associates," enjoying some privileged access to EC markets for their goods, and the community has even been approached by Morocco. Some diplomats within the EC encourage the expansion of membership; these diplomats are often called "wideners." Others believe that consolidation and more integration must take place among current members before new members are welcomed; these are called the "deepeners."

Former European Colonies Community members have extended special access to community markets to 66 of their former colonies. The French, for example, want privileged access for the products of their overseas departments, and the British want similar privileges for their former Caribbean colonies. Ninety percent of the population in the 66 countries is African, but several countries are in the Caribbean and Pacific basins.

These special privileges are being called into question, because the beneficiaries' agricultural products compete with products from southern Europe. Italy, Greece, and Portugal want to protect European markets for their own produce. The United States opposes any favoritisms as discriminatory against U.S. products.

Common Foreign Policy

The EC members increasingly consult among themselves in foreign affairs, but they lack a common foreign policy. Their inability to speak with one voice reveals that the community countries remain far from united. The foreign policies of the members did not agree during two international crises in 1991: the war against Iraq and the civil war in Yugoslavia. Several EC members have suggested increasing the role of another defense organization, the Western European Union (WEU), which includes all 12 except Denmark, Greece, and Ireland, but negotiations on developing a common foreign policy will probably continue through the century.

The collapse of Communist party governments in Eastern Europe presents the EC countries with tremendous opportunities but also tremendous challenges. We already noted in Chapter 6 the possibility of mass migration from east to west in Europe. If, however, the EC countries can consolidate the gains they have achieved and also reach out to help the countries of Eastern Europe achieve economic growth, a new age of European affluence and prestige may just be opening.

THE UNRAVELING OF THE UNION OF SOVIET SOCIALIST REPUBLICS

Throughout the late 1980s, the Union of Soviet Socialist Republics, the largest country on Earth, unraveled, and in 1991 a new, looser association, the Commonwealth of Independent States, re-formed among 11 of its 15 constituent territorial units. The exact allocation of powers within this new framework will be defined throughout the 1990s. With luck, the transformation from country to commonwealth will be achieved peacefully, but the possibility remains that the commonwealth may splinter violently.

The Russian Empire, Revolution, and Reorganization

At the same time as the Western European countries were building overseas empires, Russia built an empire across Asia. The number of peoples conquered by the Russians was no fewer than the number conquered by any other European nation (see Figure 12–12).

During World War I, the empire suffered internal rebellion as well as external attack. Subject peoples criticized the empire as a "prison house of nations," and the Communists, led by Vladimir Lenin (1870–1924), allied themselves with these nationalist minorities as long as the alliance was directed against the czarist government. The Communists pledged to reorganize and reform not only Russia but also Russia's relationship with the subject nationalities.

The Communists seized control in 1917, and despite their promises of national self-determination, they struggled to hold together all the empire's peoples and territory. They replaced the anti-Russian propaganda that had helped them gain power with a new propagandistic ideal of comradely federation. Theoretically, each subject nation of the old empire would experience an individual revolution of liberation from czarist and capitalist oppression, and then the nationalities would reunite in a new supranational union. This union, the Union of Soviet Socialist Republics, came into being in 1922.

Not all the subject nations were willing to reunite. Several fought for real independence and had to be reconquered. Other peoples actually gained freedom. Poland, Finland, Estonia, Latvia, and Lithuania all emerged as independent countries after World War I, and the province of Bessarabia broke away to join Romania. The

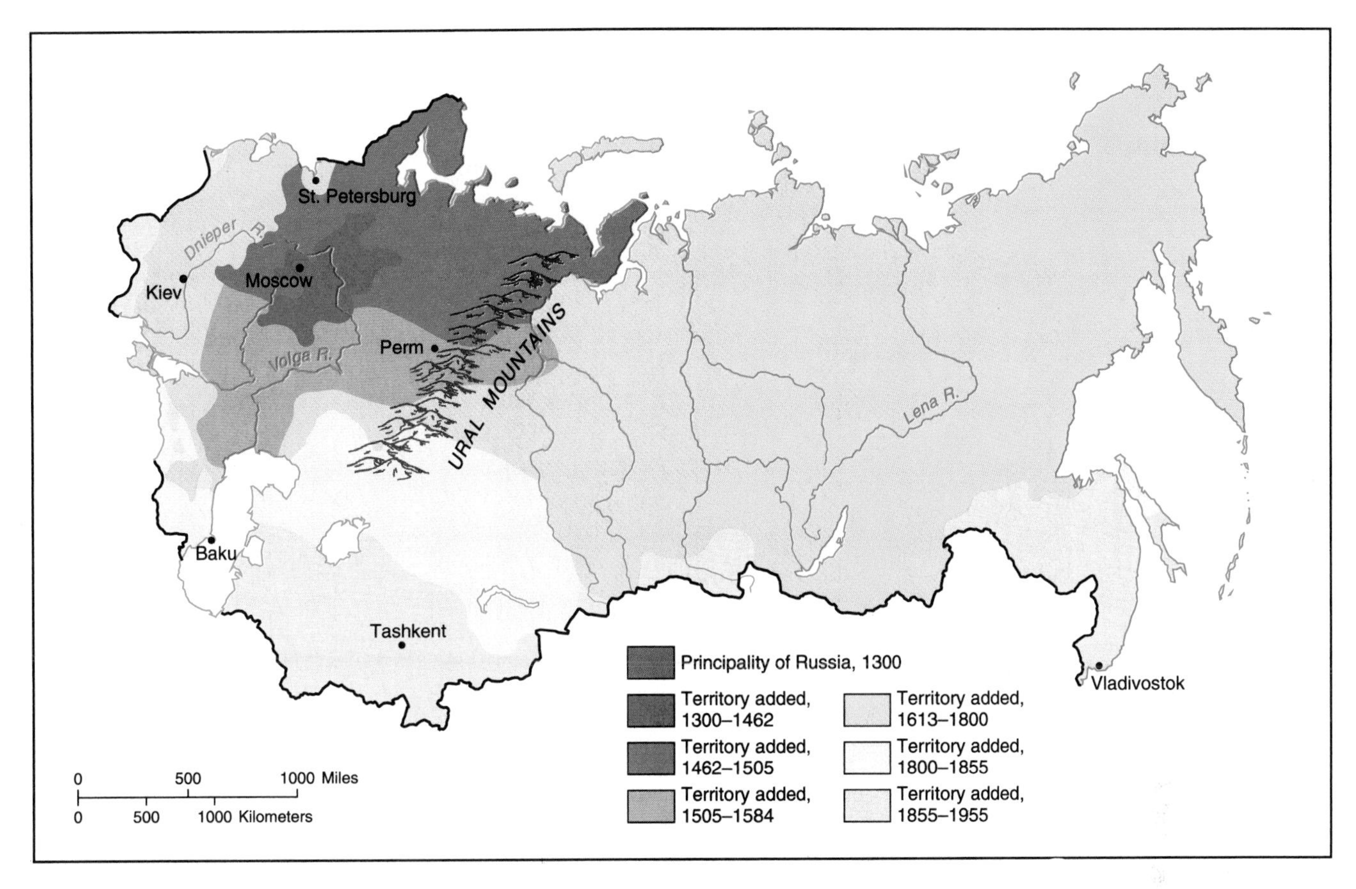

Figure 12–12. Russians' expansion from their homeland around Moscow began in the mid-sixteenth century, and by 1647 Russian cavalrymen had reached the Pacific. During the seventeenth and eighteenth centuries, Russians pushed south to the Black Sea, and Russian explorers crossed over the Bering Strait to incorporate Alaska (which they sold to the United States in 1867) and to spot settlements along the west coast of North America as far south as California. In the nineteenth and early twentieth centuries, czarist power advanced still farther into Eastern Europe and central Asia.

Soviet Union regained much of this lost territory during and after World War II. In 1990 the Soviet Union covered 8.6 million square miles (22.4 million km^2), and its population of about 290 million made it the Earth's third most populous country, after China and India.

The Political Geography of the USSR and the Commonwealth

The Soviet Union was subdivided into fifteen *union republics*, each of which was theoretically the homeland of a national group, endowed with territorial autonomy and the right of secession (see Figure 12–13). The ability to secede was ostensibly ensured by the fact that each republic had an international border or a border on an international body of water. The republics were equally represented in the Soviet of Nationalities, one of two chambers of the Supreme Soviet. Russia was the largest republic, occupying fully 76 percent of the total territory of the USSR. Moscow was the capital of both the Russian Republic and the USSR.

Several republics contained within them the homelands of less numerous ethnic groups. These groups were awarded lesser territorial autonomy. Within Russia, for instance, were 16 *autonomous republics*, 5 *autonomous regions*, and 10 *autonomous areas*.

The communist rulers redrew internal borders and even transferred whole populations. Between 1921 and 1991 there were 90 changes of borders among the republics, and when the union dissolved, a few historical border disputes were rekindled. The Crimea, for instance, was taken from Russia and given to Ukraine in 1954, and Russia demanded its return. A protocol to the founding document of the commonwealth recognizes all current borders, but these borders may not be permanent.

Russian Imperialism

The individual Soviet republics ostensibly enjoyed considerable autonomy, but in fact all power was centralized in Moscow. Russian Communists dominated the other

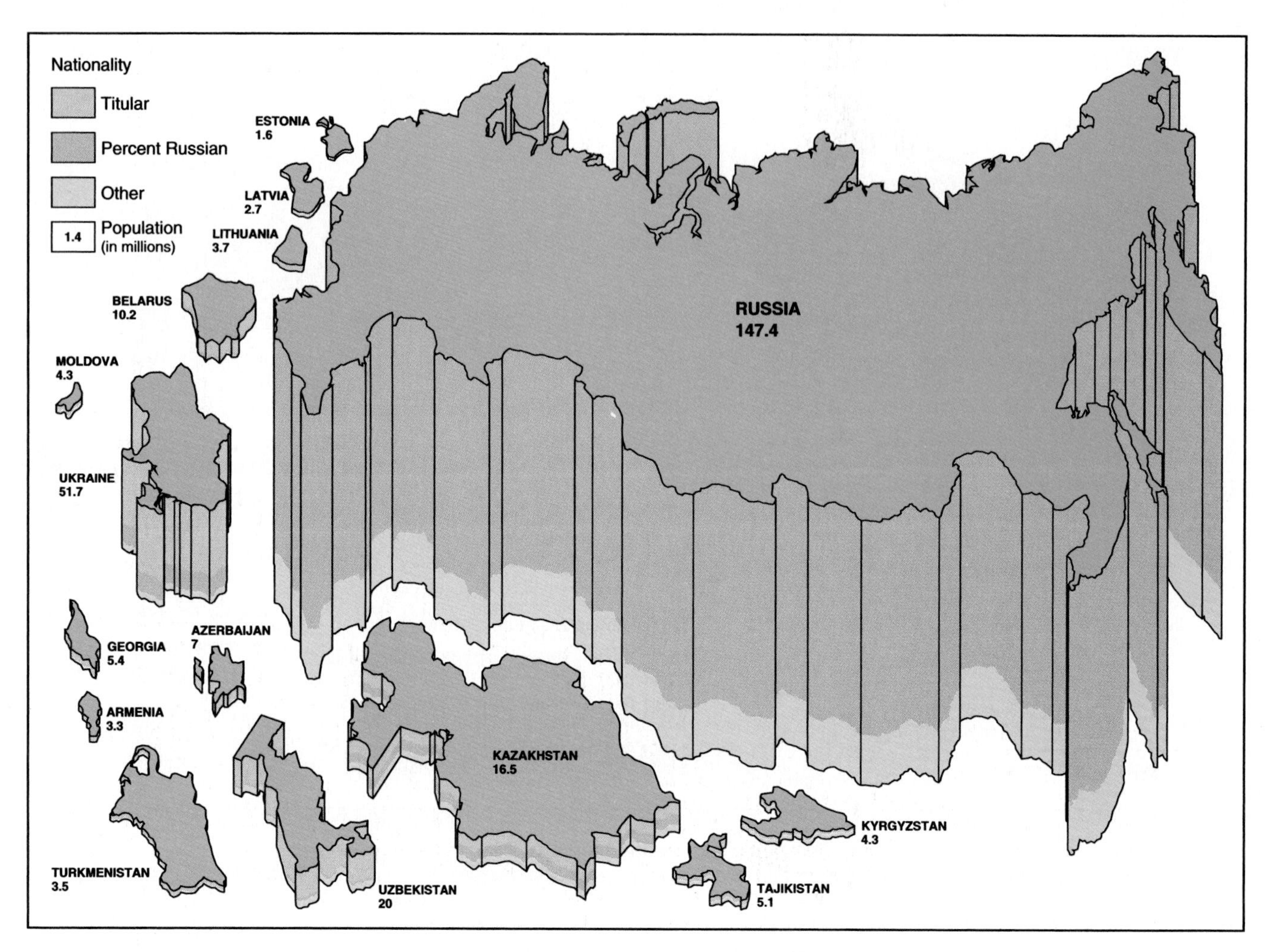

Figure 12–13. The political geography and population geography of the USSR in 1990.

nationalities in several ways. Real power never lay in the organs of government, but in the Communist party, and the leading positions in society were reserved for Party members taken from a list called the *nomenklatura*. Party membership was restricted, and Russians always dominated the Party. In 1990 Russians made up 52 percent of the population of the Soviet Union, and yet all members of the *Politburo* (the Party governing body) were Russian except one. Even in the Party hierarchies of the non-Russian republics, Russians held controlling positions.

Russification Russians also dominated the other peoples through acculturating them to a Russian norm. This process was called *Russification*. The Russian language, for example, was preferred for central government affairs, and it was one of two official languages in all non-Russian republics. Russians who resided or officiated in the minority republics wanted Russian to be the sole official language everywhere, and they rarely bothered to learn the local languages.

Under communism, religions were persecuted. Some peoples had historical national churches, and in those republics the churches provided a repository of nationalism that defied Russification. The Estonians and Latvians continued to practice their Lutheranism; and the Lithuanians, their Roman Catholicism. The Armenians have their own ancient Christian church, as do the Georgians. The Ukrainian Catholic church was banned in 1946, but it stayed alive underground.

The position of the 55 million Muslims in the Soviet Union was especially important and volatile. Five republics in central Asia (Kazakhstan, Turkmenistan, Uzbekistan, Tajikistan, and Kyrgyzstan) and Azerbaijan in the Caucasus region gave the Soviet Union one of the world's largest Islamic populations (see Figure 12–14). Muslims made up the second largest segment of the total population after the ethnic Russians themselves, and because of high birth rates among them, their percentage of the total population was rising fast. Political dissatisfaction among them, triggered partly by the revival of Islamic culture throughout the Islamic cultural realm, threatened political stability. As recently as 1986 the central government attacked Islam as "the enemy of progress and socialism." As the central government lost control, however, it allowed new mosques and religious schools to be built and fresh copies of the Koran to be distributed.

Figure 12–14. This beautiful sixteenth-century mosque in Bukhara, Uzbekistan, illustrates the city's history as a center of Islamic learning. Islam may play a prominent role in the cultural life and politics of the newly independent central Asian republics. (Courtesy of Sovfoto/Eastfoto.)

Immigration of Russians Russians left their homeland in large numbers and moved into the borderlands in search of individual opportunity and, perhaps, easier climatic conditions. In 1990 some 25 million Russians lived in the 14 other republics, more than half of them in Ukraine or Kazakhstan.

In Kazakhstan, the Russians opened new lands to agriculture. In most areas of central Asia, however, as well as in Latvia, Estonia, eastern Ukraine, and several other republics, the Russians brought industrialization, and therefore they settled in cities. The ethnic contrast between the cities and the countryside provoked antagonism.

Russians were not the only people to migrate within the Soviet Union. Under the protection of Russian hegemony, as many as one-quarter of Soviet citizens may have been living outside their national ethnic homelands by the late 1980s.

Developments from 1985 to 1991: The Gorbachev Years

The 6 years from 1985 to 1991 brought momentous change. In 1985, when Mikhail Gorbachev assumed power as union president, the Soviet economy was stagnant. To revive the economy, Gorbachev cut loose the satellites of Eastern Europe and launched two liberalizing initiatives in the Soviet Union itself. One was *glasnost*, a loosening of restraints against freedom of speech and the press. The second was *perestroika*, a restructuring of the economy and of politics. In 1990 he even proposed repeal of the Communist party's constitutional monopoly on political power. New, noncommunist labor unions, cultural associations, and opposition political parties sprang to life. In each of the non-Russian republics, nationalist movements arose to offer an alternative to the Communists. Even within the Russian Republic itself, President Boris Yeltsin renounced communism and then won reelection to his office. This proved Yeltsin's popularity and gave him a unique political legitimacy within the largest and most powerful of the republics.

Gorbachev gambled that his policies could unleash individual initiative and productivity while retaining the centralized control of the Party and of the central union government. Gorbachev lost. The tensions unleashed by *glasnost* and *perestroika* brought down the old governmental system before Gorbachev could bring a new one into place.

Ethnic frustrations rose to the surface. Nationalist sentiments had been intensified by the spread of mass education, echoes of decolonization abroad, and resentment of Russian hegemony. The minority nationalities were more conscious of their identities and aspirations in 1990 than they had been before the 1917 revolution.

Among the first victims of political unrest were the ethnic Russians living in the non-Russian republics. Several republics contemplated the introduction of residency requirements, which would effectively disenfranchise their Russian populations. Between 1985 and 1990, some 500,000 Russians migrated from the other republics into European Russia, which lacked housing and employment for them. Many moved on into Russian Siberia. In some of the republics these Russians had been the educated elite, and so their leaving may create a brain drain problem for the newly independent countries.

Disputes arose not only between Moscow and the republics but also among the various minority peoples. The Armenians and Azerbaijanis, for example, fought over territory, but their hostility was rooted in religious antagonism between the Armenian Christians and the Azerbaijani Shiite Muslims. Nearby, the minority Muslim Abhazians rose against the Georgians, and in the Turkmen and Uzbek republics outbreaks of violence occurred between Kyrgyz and Uzbeks. Several non-Russian minorities suffered persecution and even massacres. Soviet military forces with great difficulty quelled several regional civil wars.

Ethnic tension threatened the stability of the army itself. Soldiers had always served outside their home regions in ethnically mixed units, but with rising ethnic tensions some nationals could no longer be put into units with others, and some refused to serve outside their home republics. At the same time, the army was becoming increasingly essential to the economy. Military transportation and distribution facilities became crucial to prop up the crumbling civilian infrastructure. Soviet troops assumed a larger share of construction work, and military trucks were required to haul food.

The nationalities problem was complicated by tensions between the rich republics and the poor republics (see Table 12–1). The central Asian republics complained that they received too little investment, but the relatively wealthy Baltic republics disdained subsidizing the empire's poorest peoples. Russians were convinced that the borderlands benefited disproportionately from investments by the central government in industry and defense. The non-Russian republics, however, considered themselves exploited as underpaid workers and as suppliers of raw materials at artificially low prices to industry located in Russian areas. Variations in the quality of life among the republics can be seen in such basic data as the birthrates (look back at Figure 6–9), death rates (Figure 6–7), and rates of infant mortality (Figure 6–8).

To quell the widespread unrest, the central government proposed a new treaty to redefine the relationship between the center and the constituent republics. The day before the treaty was to be signed, however, August 19, 1991, a coup organized by reactionary elements of the Communist party, the army, and the KGB (state police) arrested Gorbachev. The governments of the individual republics had already gained too much power through the recent years, however, and the coup failed. The government of Russia itself held out against the coup, thereby destroying the Soviet government, and with it the Soviet Union. Within days, many of the republics outlawed the Communist party, and within weeks most of the republics had declared their independence.

The Formation of the Commonwealth In the months following the unsuccessful coup, Russian president Yeltsin moved quickly to establish the hegemony of the government of the Russian Republic and to assume responsibility in international affairs as the legal successor of the Soviet Union. Russia nationalized its natural resources, assumed responsibility for Soviet government expenses, and even offered to assume responsibility for all Soviet foreign debt and for the Soviet nuclear arsenal as well.

The death blow to the Soviet Union was a plebiscite in favor of independence in Ukraine, the second largest and second richest republic. On December 9, 1991, representatives of Belarus, Russia, and Ukraine met in the Belarussian town of Brest and declared "the USSR, as a subject of international law and geopolitical reality, is ceasing its existence." The three leaders created a new Commonwealth of Independent States that was declared to be "neither a state nor a superstate structure," but only a set of "coordinating bodies" with its "seat" (not capital) in Minsk, the capital of Belarus. Two weeks later, 11 more republics joined as "equal cofounders" of the Commonwealth. The only holdout was Georgia, then in the midst of a civil war. On December 25, Gorbachev resigned, and the next day the Soviet Parliament formally voted the country out of existence.

TABLE 12–1 ***The Economies of the Soviet Republics at the Time of the Dissolution of the Soviet Union***

Republic	Per capita GNP* (dollars)	Economic profile**
Latvia	6740	Advanced industry producing high-quality electronics and consumer goods; imports energy and raw materials
Estonia	6240	Most industrialized Baltic republic; largely self-sufficient in agriculture
Belarus	5960	Diverse highly productive industry, although dependent upon imported raw materials; important producer of fertilizers and synthetic fibers; net exporter of mixed agricultural products
Lithuania	5880	Diverse manufacturing base; imports raw materials, energy, and grain
Russia	5810	Produced more than 60 percent of total Union income; diverse exports; imports include specialty crops
Armenia	4710	Important machine-building industry; agricultural exports include specialty crops, fruits, and preserves
Ukraine	4700	Second largest economy in the Union; major heavy industry and also the Union's "breadbasket," producing about 25 percent of total Soviet agricultural output
Georgia	4410	Highly diverse economy, but exports principally specialty crops, tea, citrus fruits, wine
Moldova	3830	Intensive agriculture and food processing, with a small agricultural machinery industry
Azerbaijan	3750	Industry largely based on Caspian oil; agricultural sector mostly fruits and vegetables
Kazakhstan	3720	Major supplier of power, fuel, chemicals, metals and agricultural products
Turkmenistan	3370	Natural gas and irrigated land; important fishing industry in the Caspian Sea
Kyrgyzstan	3030	Poor in resources, but exports include sugar, fruits, vegetables, and wool
Uzbekistan	2750	Largest central Asian economy; indigenous chemical industry based on natural gas; produced two-thirds of Union's cotton
Tajikistan	2340	Exports hydroelectric power and minerals; also fruits, vegetables, and processed foods
USSR	5000	

* Courtesy of PlanEcon, Inc. Washington, D.C.

** Adapted from "The Soviet Republics: A Political and Economic Overview," Office of the Geographer, U.S. Dept. of State, February 1991.

Focus on The Baltic Republics

The three Baltic republics of Estonia, Latvia, and Lithuania seceded from the Soviet Union and were internationally recognized as independent countries in 1991 before the general dissolution of the Soviet Union. Their departure reduced the Soviet Union's territory only 0.8 percent and its population only 2.9 percent, but they were three of the richest republics.

The Baltics enjoyed 2 decades of independent democratic liberties after World War I, but they had been forcibly reincorporated into the Soviet Union in 1939. In local plebiscites even the ethnic Russians living in these republics supported independence, swayed, perhaps, by the relative wealth and ambience of freedom in their adopted homelands.

The Baltic republics are of strategic importance to Russia because they lie between it and the Baltic Sea. They even isolate bits of Russian territory. The capitals of Estonia (Tallinn) and of Latvia (Riga) are seaports. Vilnius, however, the capital of Lithuania, is inland. Lithuania's seaport is Klaipėda, which historically was a German city, Memel. Russia took this city from Germany at the end of World War II, and Russia insists that it remains Russian property.

The Baltic republics regained their independence in 1991, and today they are struggling to rebuild their cultural and economic links with Western Europe.

Russian access to the port of Kaliningrad (historically German Königsberg) is even more important. That port, also taken from Germany, is Russia's only ice-free port, and so it was home to the Soviet Baltic fleet. Kaliningrad has now been declared a free economic zone where international investors have been offered economic privileges.

The Baltic republics were the ones that could most quickly shift their trading patterns outside the Soviet Union. They formed a joint Baltic Council for economic cooperation, and they are trying to attract foreign investment to develop economies dedicated to agriculture and light industry, rather like the Finns, to whom the Estonians are ethnically related. They will probably continue to depend on Russia for raw materials and fuel.

If the Commonwealth framework is to survive, terms of the allocation of powers between the newly independent republics and the Commonwealth will have to be defined. Joint agreements are to be negotiated on foreign affairs, transportation and communication, environmental matters, and economic affairs. Is the Commonwealth, for instance, to have one currency and one army, or several of each? Soon after establishment of the Commonwealth, several republics insisted on their right to create their own currency and army; these demands justify pessimism about the future of the Commonwealth. The possible dismantling of the Common currency and military forces contrasts sharply with the EC, which is now building these institutions in common.

In response to international concern, Commonwealth leaders quickly agreed that the ex-Soviet stock of nuclear weapons would remain under a single Russian command and that all the weapons themselves would be moved onto Russian territory. Russia was granted the Soviet seat on the UN Security Council, and Russia seized all Soviet embassies abroad. The Commonwealth will have to negotiate ownership of other former Soviet assets.

Continuing Fragmentation within the Republics

The dissolution of the Soviet Union has necessitated the renegotiation of the relationships among the republics and the former autonomous regions within them. These regions all declared the precedence of locally passed laws over laws passed by the governments of the republics.

Because the threshold principle for the size of states has been abandoned in international affairs (see Chapter 10), several small regions may claim the right to national self-determination and full independence. The separatist regions within Russia form three groups. The first of these is on the edges of Russia, bordering other states. These are the Caucasian regions, Tuva, and Buryatia. They might claim full independence, but their total population is scarcely 6.5 million, and their independence would not threaten Russia. A second group consists of Muslim areas along the Volga River, such as Tatarstan, an important manufacturing center, and Bashkiria, an oil-refining center. These are economically significant areas, but they are surrounded by Russian territory, and independence would be only nominal. The third group consists of areas like Yakutia, which is Russia's richest concentration of raw materials. Large Russian minorities live there, however, or even majorities. Referendums on secession would probably fail.

In 1991 the Russian government agreed to establish a new internal Volga Republic region to provide a homeland for ethnic Germans just north of Volgograd. Such a republic existed from 1924 until 1941, but it was abolished when Germany invaded during World War II. Germany has agreed to subsidize development of this new unit, partly, perhaps, as a way of preventing these peoples' migration to Germany and partly as a way of establishing an ethnic base to serve as an eventual economic base within Russia.

The Economic Geography of the USSR and the Commonwealth

One of the reasons for the political unraveling of the USSR was its economic failure. The political unraveling in turn threatens worse economic circumstances for its peoples. This is because of the way the territory was

Focus on Ukraine

Ukraine was the second most powerful of the Union republics. It accounted for about 21 percent of total union population and product, including higher percentages of agricultural products. Holding onto Ukraine had preoccupied all Soviet leaders since Lenin, who said, "For us to lose the Ukraine would be the same as losing our head."

Ukrainians fought the Russians for their independence after World War I and again after World War II. After World War II the Soviet government enlarged Ukraine by annexing Polish territory. Many people of the western Ukraine continued to feel some allegiance to Poland. In the 1980s they watched Polish television and came to admire Solidarity, the labor movement that became a political force and overthrew Polish Communism.

The eastern Ukraine contains a great concentration of industry. The Donets coal basin (called the *Donbas*) has been the empire's main coal producer for more than a century, although costs of mining are rising. The industrial Ukraine was heavily Russified, but the Ukrainian nationalists won considerable support there too.

Ukrainian demands for freedom were aided, ironically, by an act of Stalin's. In 1945 Stalin insisted that the Soviet Union receive three seats at the United Nations. These were assigned to the USSR, Belarus, and Ukraine. Therefore, there was a precedent for accepting Belarus and Ukraine as independent actors on the world stage.

At the time of the creation of the new commonwealth, Ukraine most staunchly rejected any infringement upon its own newly won sovereignty. It demanded the right to have its own army and its own currency. These issues will be open to continuing negotiation.

Ukrainians demonstrating in Kiev to claim their independence from the Soviet Union carried their nation's historic blue and yellow flag. (Courtesy of Vladimir Sichov/SIPA-Press.)

organized economically. The two principal characteristics of Soviet economic development were *autarky*—which is discussed in Chapter 5— and *centralized state control*.

Autarky was not adopted entirely by choice. The United States and most European countries long ostracized the communist regime. As late as 1988 international trade was only 5 percent to 10 percent of Soviet gross product, and one third of that total was with the four satellites of East Germany, Poland, Czechoslovakia, and Bulgaria.

Communist Economic Organization The economic organization imposed on the Russian communist empire was dictated by theory. No communist theoretician would have chosen Russia to be the first communist experiment. Marx had taught that communism would be possible only in a country that had achieved superabundant productive capability—probably first in Germany.

Having gained control in Russia, however, the Russian Communists set out to build communism. Some private property and a few private sector services were allowed, but no one was allowed to own "means of production" in such a way as to profit from the labor of others. Natural resources became the property of the state.

Agriculture was brutally collectivized. Despite the opening of new lands to agriculture, notably the steppes of northern Kazakhstan, Soviet agricultural output was slow to increase. For one thing, despite the Soviet Union's immensity, conditions across much of the territory are unfavorable for agriculture (see Figure 12–15). Agriculture also suffered from mismanagement, lack of worker incentive, lack of investment, and government policies that kept farm prices artificially low. These factors kept productivity low, and the poor internal transportation system compounded the difficulties. The country lacked both trucks to get the food to market and roads to drive them on. As much as 20 percent of the grain harvested was lost before it got to processing plants, and the number of processing plants was insufficient. Estimates of losses for fruits and vegetables were as high as 70 percent.

Industry was owned and organized by the state, guided by planning bureaucracies. The Party and government stressed industrialization beginning with the first Five Year Plan in 1928, and heavy industry made immense strides. By the 1980s the Soviet Union had become the world's largest producer of many heavy industrial goods, but it lagged behind the West in light industry, in the production of consumer goods, and, despite superlative institutes of higher education and scientific research, in high-technology products.

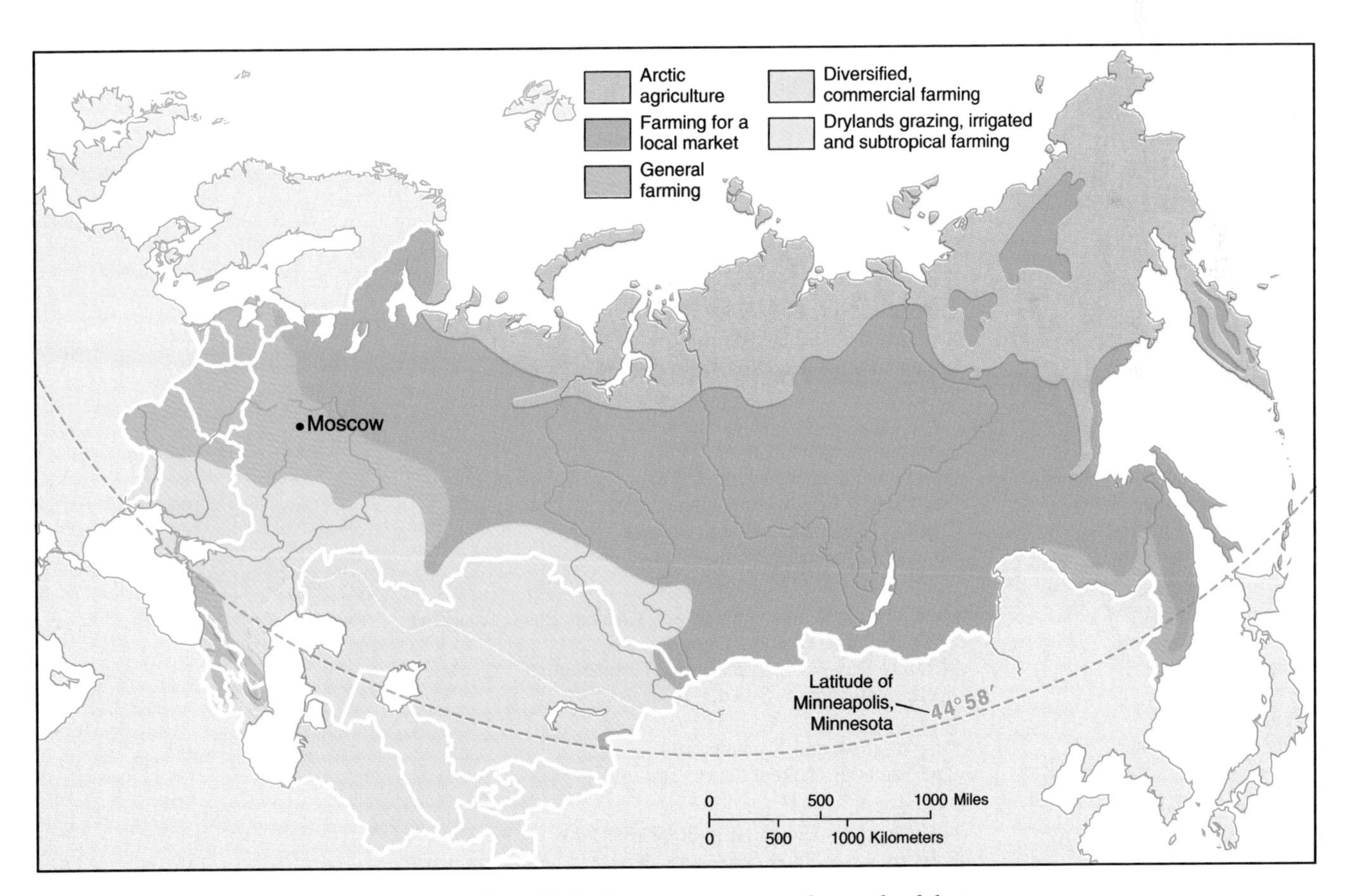

Figure 12–15. Despite the immense size of the USSR, the territory was to the north of that of the United States, and only a fraction of it was agriculturally productive.

Distribution of Industrial Production We do not have enough space to provide a detailed economic geography of the Soviet Union or the successor Commonwealth, nor an investigation of how and why those distributions came about. Because the individual republics dissolved the union, our discussion will focus on the differences among them.

Russia was overwhelmingly the dominant republic in the Soviet Union. In 1990, with just over half the Soviet population, Russia produced 91 percent of the USSR's oil, 75 percent of its natural gas, 55 percent of its coal, 58 percent of its steel, 81 percent of its timber, and 48 percent of its wheat. Russia's treasury subsidized inefficient industries in other republics. Russian Siberia supplied raw materials to other republics, most of which were shipped at below-market prices.

Under the planned economy, both production and internal trading were run by the state, and this system was so highly centralized that it will be extremely difficult to deconstruct. Each factory or farm in the Soviet Union was called an "enterprise," and in 1990 there were only 700,000 of them, compared with about 17 million in the United States. The relatively small number of enterprises in the USSR reflected economic inflexibility. In a typical capitalist system, by contrast, new enterprises may freely enter markets and compete, and freely succeed or fail.

Soviet factories received their instructions from the central government. They sent goods to other factories and shops according to the dictates of a series of Five-Year Plans. There were few direct contacts between one enterprise and another, and fewer still between any enterprise and consumers. Both production and distribution were rigid monopolies. Soviet enterprises did not have a range of suppliers and customers or any alternatives to those assigned by the state planning bureaucracy. Enterprises specialized in a few products, and their only customer was the state. There was little or no competition, and enterprises could not change from one product to another, from one supplier to another, or from one customer to another, as demand changed.

Another important geographical characteristic of Soviet industrialization was that industries were scat-

TABLE 12–2 ***Concentration of Production of Selected Items in the Soviet Union in 1990***

Product	Producer	Percent of total production
Consumer goods		
Sewing machines	Shveinaya Association, Podolsk	100
Automatic washing machines	Elektrobytpribor Factory, Kirov	90
Transport		
Trolley buses	Uritsky Factory, Engels	97
Forklift trucks	Autopogruzhchik Association, Kharkov	87
Diesel locomotives	Industrial Association, Voroshilovgrad	95
Electric locomotives and trains	Electric Locomotive Plant, Novocherkassk	70
Tram rails	Integrated Steel Works, Kuznetsk	100
Metals		
Reinforced steel	Krivoy-Rog-Stal, Krivoy Rog	55
Construction equipment		
Concrete mixers	Integrated Mill, Tuva Works	93
Road-building cranes	Sverdlovsk Plant, Sverdlovsk	75
Locomotive cranes	Engineering Plant, Kirov	100
Oil, chemicals and chemical engineering		
Polypropylene	Neftkhimichesky Combine, Perm	73
Deep-oil-well sucker rods	Ochesk Engineering Plant, Ochesk	87
Sucker-rod pumps	Dzerzhinsky Engineering Plant, Baku	100
Hoists for coal mines	City Coal Machinery Plant, Donetsk	100
Coking equipment	Kopeisk Engineering Plant, Chelyabinsk	100
Power engineering		
Hydraulic turbines	Metallurgical Works, Leningrad; Turbines Plant, Kharkov; Pipe Building Factory, Syzran	100
Steam turbines	Metallurgical Works, Leningrad Turbines Plant, Kharkov; Turbomotor Plant, Sverdlovsk	95
Metals		
Electrolytic tin plate	Magnitogorsk and Karaganda	100
Rolled stainless-steel pipes	Pipe Factories, Nikopol and Pervouralsk	96
Consumer goods		
Color-photography paper	Positive Film, Leningrad and Positive Film, Pereslavl	100
Freezers	Freezers Association and Plants, Kishinev and Krasnoyarsk	100

Source: Soviet State Committee on Statistics.

tered all over the huge area, but the production of any given item was highly concentrated in individual factories (see Table 12–2). Many of the most basic goods in any economy, such as polypropylene, stainless-steel pipes, and concrete mixers, were produced entirely or almost entirely by a single factory. Between 30 percent and 40 percent of the total value of Soviet manufactured goods was produced at single sites.

The centralization, rigidity, and geographical particularization of production in the Soviet Union meant that the republics came to depend on one another for trade. As Table 12–3 indicates, in 1988 trade among republics accounted for 40 percent or more of the income of 13 of the 15 republics. All trade within the Union was conducted at state-dictated prices. This system lacked a method of fixing market prices, and therefore each republic was able to insist that it had been exploited. This intensified antagonisms among the republics.

Economic Collapse and the New Economies By the late 1980s, all across the Soviet Union, local political authorities began to seize control of local economic activities, in defiance of Moscow. The Union's economic system collapsed. The local regions sought to achieve individual self-sufficiency by (1) seizing control of local resources and by dictating terms and prices for their sale, (2) imposing direct restraints on trade, and (3) raising prices for goods produced locally. The various republics tried to reach bilateral trade agreements, bypassing Moscow, and they even began bartering resources and goods among themselves directly. Efforts to reach and implement these agreements led to a decline in trade among the republics.

Several of the republics began privatizing their economies, but at different rates, and this discrepancy further confused trade among republics. Privatization continued after the end of the Union and the birth of the Commonwealth. In many cases these privatizations have been employee stock ownership plans, in which the workers assume ownership (which communist propaganda had insisted was theirs all along), or private sales to domestic or even foreign investors. In other cases, privatization has meant seizure of the assets by the members of the old nomenklatura.

TABLE 12–3 ***Interrepublic Trade, USSR, 1988***

Republic	Percent of net material product imported from other republics
Russia	18
Estonia	65
Latvia	62
Lithuania	64
Ukraine	37
Moldova	62
Belarus	64
Armenia	79
Georgia	52
Azerbaijan	47
Uzbekistan	45
Turkmenistan	52
Kazakhstan	42
Tajikistan	53
Kyrgyzstan	50

Source: Organization for Economic Cooperation and Development.

The individual republics continue to institute other economic reforms and to seek new outside international trade and investment partners, but the legacy of communism will be difficult to eradicate. The leaders of the Commonwealth are trying to maintain free trade among the republics without infringing on their new independence. Trade wars that cut off long-established flows of raw materials, foods, or industrial parts could be economically disastrous.

Russia and the other republics may turn to capitalism, but, as we discussed in Chapter 11, there are many forms of capitalism. State-directed capitalism such as that in South Korea, Taiwan, and Japan worked in Russia in the decades before 1914, when Russia had the fastest growth rate in Europe. Despite political and economic restructuring, the Russian Republic today still boasts one of the world's largest economies, and its schools produce one-third of the world's doctorates in science and engineering. In the year 2000, Russia, even if standing alone, will almost certainly still be the largest and one of the most powerful nations on Earth.

Environmental Pollution The constitution of the USSR said "citizens of the USSR are obliged to protect nature and conserve its riches." They failed to do so. As early as 1920 the Soviet regime passed environmental protection measures, but with the inauguration of the Five-Year Industrial Plans and the drive for all-out industrialization, these measures were ignored (see Figure 12–16).

The government tried to hide the damage rather than commit resources to undo neglect. Few environmental laws incorporated penalties, and because most industries were attached to ministries of the central government, they were not subject to any local legislation. Agricultural chemicals were often used without regard to their poisonous consequences, and industries were technologically backward and grossly inefficient in their use of raw materials. In 1990 Alexei Yablokov, director of the Institute of Biology at the Academy of Sciences, estimated that 20 percent of Soviet citizens lived in "ecological disaster zones," and 35 percent to 40 percent more in "ecologically unfavorable conditions."

In 1913 European Russia was at least on a par with the United States in health, infant mortality, and longevity. By 1990 it was well below most poor countries in every single health category, at least partly as a result of environmental pollution.

New Centripetal and Centrifugal Forces The Soviet Union had been held together by the raw power exercised by the Communist party. Today we must ask

Figure 12–16. The Aral Sea was once the world's fourth largest lake, but it has shrunk to 60 percent of its former size because the rivers that flowed into it were diverted to irrigate central Asian cotton fields. (Courtesy of Novosti/Lehtikuva/SABA.)

which centripetal forces remain to hold the republics together in any political framework. The republics were never united by any common political idea, nor do they share a desire for common defense. Some may seek Russian protection, but others feel Russia to be the greatest threat to their security. Shared economic interests might hold several of them together because the years of central planning made them highly dependent on trade with one another. These centripetal forces, however, will be balanced by powerful centrifugal forces.

Centrifugal forces tending to pull the Commonwealth apart include not only the hatred that many minority nationalities feel toward their Russian former rulers but also external attractions. In the western part of the Commonwealth, many Moldovans want to join Romania because they are ethnically related. Ukraine and Belarus are attracted to the economic power of Europe, and the European Community is already reaching out to the Eastern European former satellites.

To the south, the Islamic republics of central Asia may be linked to Russia economically, and yet Iran and Turkey are competing for influence among them. Kazakhs, Kyrgyz, Uzbeks, and Turkmen are Turkish peoples and Sunni Muslims, as the Turks of Turkey are. Turkey, which is democratic and secular, a NATO member and an EC associate member, may be a model for these peoples. Within months of the dissolution of the Soviet Union, Turkey opened embassies in each of the new republics; the president of Turkey visited each capital; Turkish airlines instituted new flights to them; Turkish universities offered thousands of scholarships; Turkish banks took on central Asian apprentices and explored new investments; and Turkish radio and television beamed new programming by satellite into the republics—in Turkish with Latin alphabet subtitles.

The Iranians seek to counter Turkish influence. The Tajiks are related to the Iranians in language and culture, although they are Sunni Muslims. The Azerbaijani of the Caucasus region are Shia Muslims, as the Iranians are. Iran has paid to build mosques and schools, and it too has initiated joint economic ventures. The central Asian republics may turn south to Pakistan and India for markets for their cotton or to use those countries' seaports. Saudi Arabia has also opened banks, schools, and mosques, and it has extended offers of liberal financial assistance. China is investigating opportunities, and even South Korea has invested in Kazakhstan and Kyrgyzstan.

The dissolution of the Union of Soviet Socialist Republics has created new units on the world political map and released political, religious, cultural, and economic forces that will re-form systems of alliances throughout Europe and Asia.

THE FORMATION OF A NORTH AMERICAN TRADE BLOC

The United States has begun organizing a free trade region consisting of Canada, the United States, and Mexico. The impetus for this move is partly defensive, out of fear of being closed out of the European market. It is also, however, partly an effort to extend economic aid to Mexico, an underdeveloped country.

The goals of this developing trade pact do not include the formation of a full customs union or common market. It has no international or supranational governing body, nor do the North American partners look toward political merger. If the zone is fully established, however, it will considerably outstrip the EC in area [8.2 million square miles (21 million km^2) to less than 1 million (2.6 million km^2)], in population (365 million to 325 million), and in total gross product ($6 trillion to $4.8 trillion).

The Canada-U.S. Free Trade Agreement

The Canadian embassy in Washington, D.C., is not near the other foreign embassies, clustered along Massachusetts Avenue in the city's northwest corner. It stands alone on Pennsylvania Avenue just 4 blocks from the Capitol. This proximity reflects Canada's longstanding special relationship with the United States.

Now, by the terms of a 1988 Free Trade Act, all barriers to trade in goods and services will be eliminated between the partners over a 10-year period that started on January 1, 1989. In addition to phasing out tariffs, the treaty removes or reduces the hurdles to cross-border investment, government procurement, agricultural sales, and movements of employees.

Economic Links In 1988 Canada and the United States were already by far the world's two leading trading partners (see Figure 12–17). The two countries had also invested in each other. Of about $100 billion (Canadian dollars) total foreign investment in Canada, 72 percent was U.S. money. This represented one-third of all U.S. investment abroad. Canada was the fourth largest investor in the United States, and since the 1988 pact more Canadian businesses have established new operations or bought U.S. businesses. This shift of Canadian businesses south may threaten Canada's traditionally large trade surplus with the United States. U.S. corporations were generally strong supporters of the free trade pact, because Canadian tariffs were two to three times higher than U.S. tariffs.

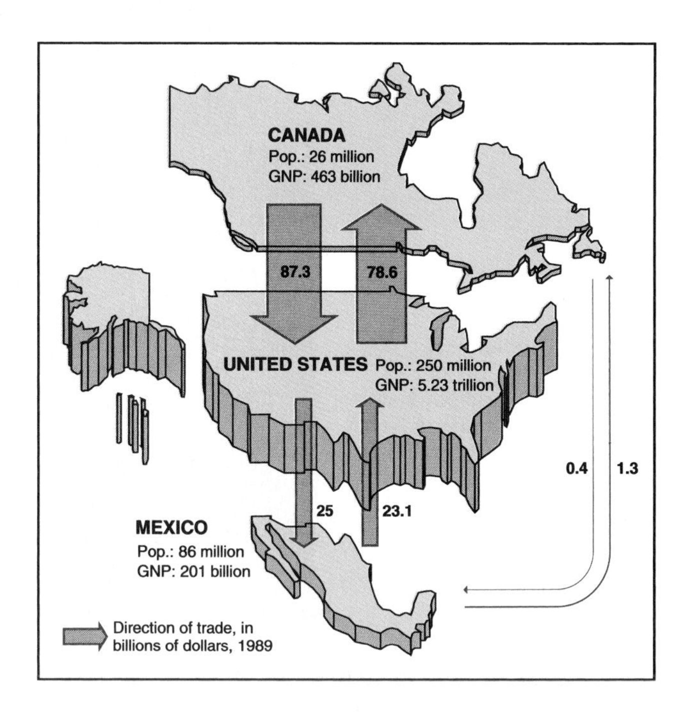

Figure 12–17. Canada takes about 20 percent of U.S. exports each year, more than twice Japan's share, and the United States buys three-quarters of Canada's exports. Trade between Canada and Mexico, however, is limited.

Canada's labor unions fought the pact. They pointed out that social benefits in Canada are generally much more generous than they are in the United States, and the unions feared a drop to U.S. standards. In Canada, labor unions represent 40 percent of the labor force, whereas in the United States unions represent only 19 percent. Moreover, in 1989 hourly compensation for production workers in manufacturing in Canada was 103 percent of that in the United States.

Canada's farmers also fought the pact. Canadian agriculture has been dominated by relatively high-cost small farms, and they will undoubtedly suffer from competition. Canadian agriculture is at a competitive disadvantage from the harsh climate, vast distances, and small scale of regional markets. Canadian small producers will also suffer competition from the big food-processing firms that can buy more cheaply in the United States.

Cultural and Nationalistic Considerations Canadians approved the pact only in a special national election. Many U.S. citizens, by contrast, remain unaware of its existence. This difference in national attitudes reflects the fact that the United States is so much more populous and economically powerful that relations between the two countries are more important to Canada than they are to the United States.

Canadians have longed feared the economic and cultural attraction of the United States as a centrifugal force disrupting Canadian nationalism, and they have tried to define and to defend their nationalism as distinctively different from that of the United States. If you look back at Figure 10–21, you will notice a thin line of Canadian settlement along the border and an almost empty empire stretching away to the north. Some 75 percent of Canadians live within 150 miles (240 km) of the U.S. border, where they absorb U.S. news and culture. In contrast, only 12 percent of Americans live within 150 miles of Canada. Canadian means of transportation and communication have always been designed to foster east-west linkages, as illustrated in Figure 10–16. One Canadian leader warned against the pact that it would "throw us into the north-south pull of the United States. And that will reduce us, I'm sure, to an economic colony of the United States, because when the economic levers go, the political independence is sure to follow."

Trade among the Canadian provinces may have served Canada as a national centripetal force, but Canadian internal trade has always been hampered by hundreds of trade barriers among the provinces. The international border to the south, by contrast, is punctured by more than 75 million border crossings per year, by common telephone and pipelines, computer and power lines, by satellites, contracts and cross-ownership, and even by common professional sports allegiances.

Economic union with the United States affects Canada's political cohesion in still another way. The people of the province of Quebec continue to debate possible withdrawal from the Canadian federation. As Canada achieves free trade with the United States, this internal political debate assumes an international aspect. Quebec's withdrawal might jeopardize Quebec's important economic ties with the United States. Conversely, secession and independence might allow Quebec to negotiate still more favorable trade ties with the United

Figure 12–18. HydroQuébec has been developing the province's great hydroelectric potential, but in 1992, the state of New York cancelled a $17-billion contract, casting doubt on future development, as well as on Québec's economic viability as an independent country. (Courtesy of J. A. Kraulis/Masterfile.)

States (see Figure 12–18). Nobody is sure what the result of the secession of Quebec would be. Thus, U.S. attitudes toward Quebec separatism may play a role in encouraging or discouraging it. In a 1991 poll 18 percent of Quebecois said that they would actually like their province to join the United States. Politicians in several other Canadian provinces have publicly discussed the possibility of requesting to join the United States as states in the event of dissolution of the Canadian federation.

Canadian cultural nationalists argue that economic union with the United States threatens the existence of distinctly Canadian broadcasting, publishing, and related cultural industries, but the pact provides for protection of these industries. It may be difficult even today, however, to distinguish Canadian cultural products from those of the United States. The majority of Canadians neither watch Canadian television, read Canadian books, listen to Canadian music, nor go to Canadian movies and plays, whereas many U.S. citizens are probably unaware of the Canadian nationality of authors Robertson Davies, Margaret Atwood, and Farley Mowat, journalists Peter Jennings and Morley Safer, and entertainers Rich Little and Michael J. Fox.

Mexico

Mexico opened negotiations on a free trade pact with the United States in 1990, and Canada soon joined the talks. Mexico is the third largest market for U.S. goods, after Canada and Japan, and 88 percent of Mexico's exports come to the United States. Canadian-Mexican trade is limited.

The integration of Mexico into a North American free trade pact might be more difficult than the integration of the U.S. and Canadian economies, and it may not be achieved for a few more years. The United States and Canada are both rich, advanced postindustrial societies with populations that are growing slowly. Mexico is a less-developed nation in the second stage of the demographic transition. Its population grew from 25 million in 1950 to 85 million in 1985, and it is expected to mushroom to over 100 million by 2000. To provide work for this rapidly rising labor force, Mexico needs economic growth and hopes to attract foreign investment. Economic growth in Mexico might reduce the push for Mexicans to migrate, legally or illegally, to the United States. Illegal Mexican migration into the United States numbers about 250,000 per year. Mexico's Hispanic culture and language contrast with the British traditions and English language held in common in the two countries to the north (except in Quebec), but that cultural contrast should not be as troublesome in building a free trade pact as it might be if the three North American countries were to try to build a full common market.

A continental free trade pact would redraw the map of continental economic geography. Grain farms in the United States and Canada could continue to fill North America's bread basket and cereal bowls, but Mexico could specialize in fruits and vegetables. This would challenge the agricultural economies of California, Texas, and Florida. If the United States dropped its high tariff on oranges, for example, Mexico could provide that product. Both the United States and Mexico now impose food tariffs of roughly 15 percent, and behind that barrier, California provided one-half of Americans' fruits and vegetables in 1990. Chapter 7 discussed current pressures to change Mexico's ejido system, but Mexicans have suggested that U.S. food processing corporations could supply the ejidos with the capital investment they need.

Mexican Industrial Policies The secondary sector of the Mexican economy has been state-directed and centralized through most of the twentieth century. In 1982 some 70 percent of the nation's economy was in the hands of the state. Mexico took up a new direction, however, in 1986. The government sought foreign investment by privatizing much of its state sector and by liberalizing its rules for foreign investment. The government sold off state-owned factories, shopping centers, copper mines, and public utilities. Mexico's transition from import-substitute industrialization with strong state participation in the economy to open markets and trade succeeded in attracting foreign investment in new factories (see Figure 12–19). This investment offset the collapse of many industries that had been protected. Many new stocks came onto the Mexican stock market, and in the late 1980s the Mexican stock market rose faster than any other in the world.

Mexicans remain reluctant to privatize PEMEX, the state oil monopoly created when Mexico's oil industry was nationalized in 1938. It is viewed as politically untouchable, and in 1989 it accounted for 6 percent of Mexico's GDP and 23 percent of export revenue. PEMEX is, however, so notoriously inefficient that the country must import gasoline, despite having some of the world's greatest proven oil reserves. Private ownership might release this huge corporation from the political considerations that affect its business decisions—its high number of employees, for example—and boost the corporation's productivity.

Mexico has built new infrastructure through cooperation between the public and private sectors. The government has granted concessions to build private toll roads that will eventually become government property after the original private builder has made a profit. This system produced almost 3000 miles (4800 km) of new roads between 1986 and 1992, and it has aroused interest as a system that may be imitated by both the U.S. federal government and states.

Mexico has been ruled by one political party without interruption since 1929, and its political apparatus makes little distinction between the party and the state. Accusations of corruption and repression abound, and

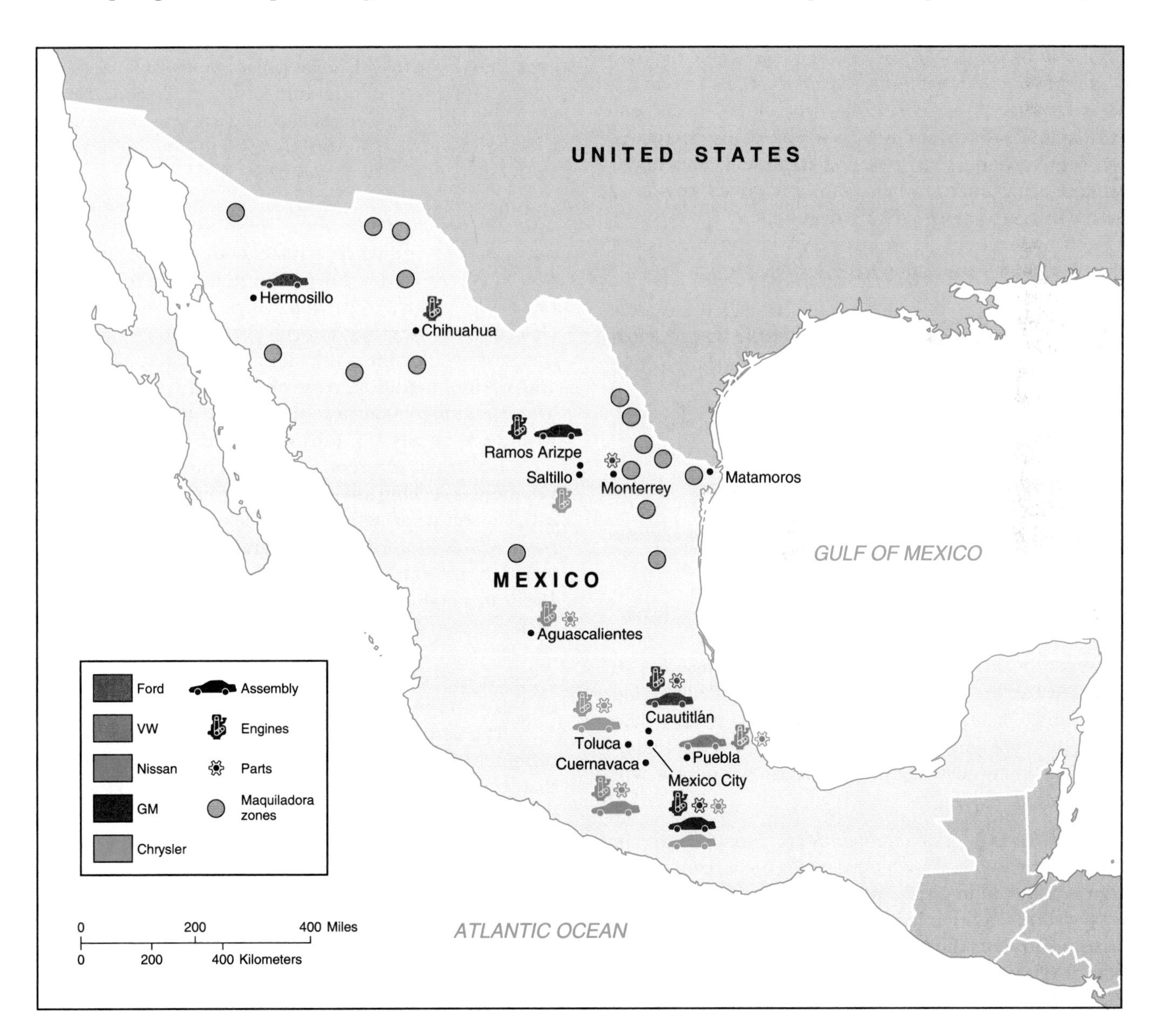

Figure 12–19. Several foreign corporations have built automobile factories in Mexico.

ABC NEWS ***Debate over Bush's Free Trade Powers with Mexico***

In a move that parallels somewhat the development of the European Community, the United States has reached a free trade agreement with Canada and is attempting to forge one with Mexico. This agreement has sparked a great deal of debate, particularly with the inclusion of Mexico into the trade pact. Perhaps the key issue in this debate is the effect of the agreement on jobs and wages in the United States.

Many U.S. companies today buy products from firms located in Mexico, called *maquiladoras*. These products are less expensive than similar ones manufactured in the United States because Mexican labor is cheaper than U.S. labor. For this reason, many critics contend that the free trade agreement will shift jobs from the United States to Mexico, where labor costs are lower and safety and environmental regulations are less stringent. Supporters of the pact respond that lowering tariffs will encourage Mexican consumers to buy more U.S. goods, which in turn will create more jobs in this country.

If you have an opportunity to watch the video listed above, consider the following questions.

1. These reports concentrate on the negative aspects of a free trade agreement between the United States and Mexico. What would be some of the positive aspects of such an agreement?
2. Could the United States make agreements with Mexican officials that would address the criticisms of the free trade pact? What would be the terms of these agreements? Is this approach feasible?

the possibility of violence looms unless democratic participation is expanded. A rebirth of democracy could persuade both foreigners and Mexicans themselves to invest in Mexico. The amount of capital that fled Mexico between 1970 and 1990 has been estimated at more than twice the total amount of foreign investment in Mexico in 1990. No country can prosper if its own people will not invest in it.

Maquiladoras U.S. companies are already active in the Mexican economy, accounting for 63 percent of foreign investment in Mexican industry in 1989. Many U.S. corporations have moved labor-intensive operations into factories just on the Mexican side of the international border. These are called *maquiladoras* (see Figure 12–20). In these factories almost 500,000 Mexican workers assemble duty-free imported components, which then are reexported to the United States. Tariffs are charged only on the value added, which is low because Mexican wages are low. The Ciudad Juarez/El Paso area has the most jobs, but Tijuana has grown into a city of over 1 million people, "the world capital of television manufacturing." Between just 1985 and 1990, the value of Mexico's manufactured exports doubled, and those exports were of increasing sophistication.

An example of future economic linkage could be the contract already signed in 1991 between the U.S. Corning Corporation and Mexico's largest glass-producing company, Vitro. The contract transfers to Vitro in Mexico responsibility for producing Corning's relatively low value consumer products. Corning's U.S. facilities will concentrate on high-margin technologies: laboratory services, optical fibers, and specialty materials.

The fact that the maquiladoras are virtually on the international border means that little improvement of the international transport network between the United States and Mexico has been required. Recent years, however, have witnessed an astonishing multiplication of international telecommunications links. These links reflect the increasing numbers of U.S. citizens traveling or retiring in Mexico, as well as the multiplying financial links between the two countries.

The Free Trade Debate There are many arguments both for and against free trade with Mexico. In 1989 hourly compensation for production workers in manufacturing in Mexico was just 16 percent of that in the United States. Mexican factories also have longer working hours, more flexible work rules, laxer safety standards, and poorer pollution controls than do U.S. factories. These arguments suggest that a free trade agreement would send many U.S. jobs south over the border.

U.S. manufacturers, however, counter that U.S. investment is seeking low-cost labor around the world, and if U.S. manufacturing is going to leave the United States anyway, it is better for it to migrate to Mexico than, for instance, to Southeast Asia. This is because when the Mexican economy grows, an estimated 15 percent of each additional dollar of income is spent on U.S. goods and services. This produces new U.S. exports. Growth in Southeast Asia does not have the same reciprocal effect on U.S. exports.

The result of a pact with Mexico will also be determined by the ultimate destination of any products that are manufactured in Mexico by U.S. corporations. If these products stay in Mexico or are exported out to the rest of the world, that will be good for both Mexico and the United States. If these products are imported into the United States, however, that will lower costs for U.S. consumers, but it will increase the U.S. trade deficit and eliminate manufacturing jobs in the U.S.

Large corporations from many other nations have invested heavily in Mexico, and the purpose of much of

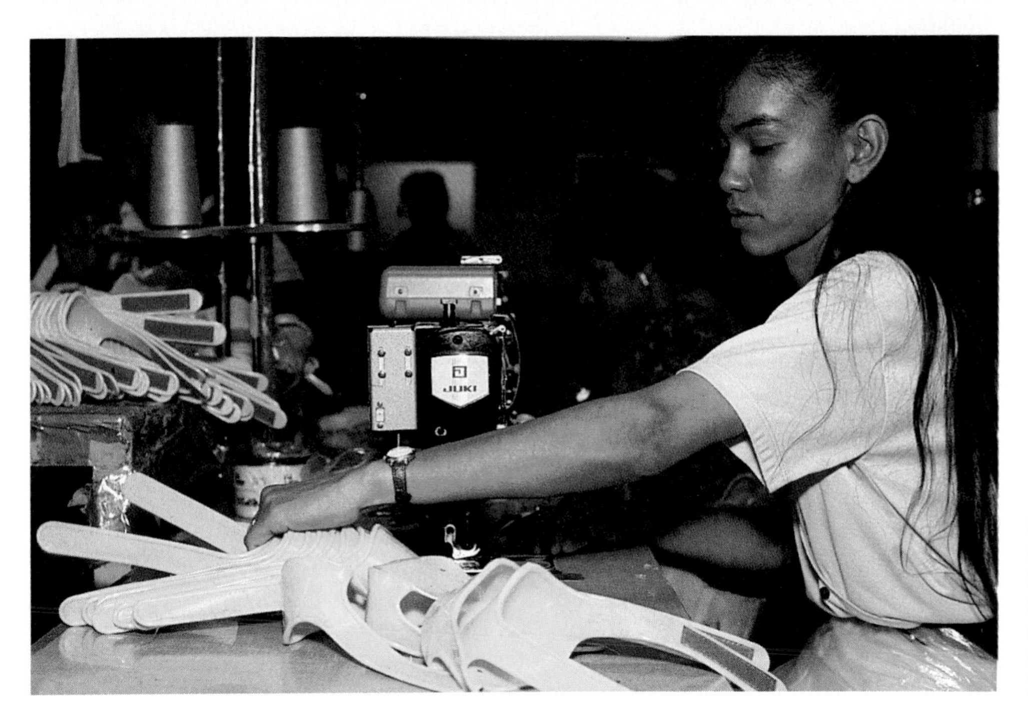

Figure 12–20. The toy manufacturer Fisher Price has invested and greatly expanded its production facilities in maquiladoras along the U.S.-Mexican border. (Courtesy of Keith Dannemiller/SABA.)

this investment may be to use Mexico as a point from which to reexport goods into the United States. To protect the U.S. economy, the trade pact must state clearly what percentage of the total value of a good entering the United States must have been added in Mexico for that product to qualify as a Mexican product. This is called a *local content requirement*. In the pact between the United States and Canada, disagreements have already arisen about whether goods assembled in Canada out of components made in other countries should enjoy free access to the U.S. market. Arguments over local content requirements typically arise whenever two countries negotiate free trade pacts that are not customs unions.

As a complement to U.S. investment in labor-intensive industries in Mexico, Mexico's own global corporations have shifted capital-intensive investments north. The Mexican cement company Cemex, for example, North America's largest, purchased several large U.S. producers in the 1980s. "The economic border between the United States and Mexico has disappeared," said the company's president; "we are North American companies." This statement may be premature in fact, but it exemplifies the ambition of many Mexican and U.S. industrialists.

The U.S.-Canada agreement and the possible extension of this agreement to Mexico is leading to the redrawing of the economic, political, and cultural geography of North America. The changing redistributions of activities will continue to be fascinating to watch.

Creating a Western Hemisphere Free Trade Region

U.S. president George Bush proclaimed a goal of forging a free trade region of the entire Western Hemisphere, "Enterprise for the Americas." In the summer of 1991 in Guadalajara, Mexico, the heads of state of every Spanish- or Portuguese-speaking country in Latin America and Europe met for the first time in history, and a number of Western Hemisphere free trade pacts formed in the early 1990s. The United States extended in 1990 the special economic relationship called the Caribbean Basin Initiative, discussed in Chapter 11, that was launched in 1984. In 1991, Venezuela, Colombia, Peru, Bolivia, and Ecuador contracted to achieve free trade among themselves (the Andean Pact), as did Argentina, Brazil, Paraguay and Uruguay (MERCOSUR). Mexico signed free trade agreements with Chile, Colombia, and Venezuela. El Salvador, Honduras, Guatemala, Nicaragua, and Costa Rica tried unsuccessfully to form a common market in the 1960s, but in 1991 they pledged to revive the effort and to achieve free trade by 1994. In addition, the English-speaking countries of the Caribbean united in a Caribbean Community and Common Market (CARICOM) and agreed to achieve free trade by the mid-1990s.

Only 15 percent of total Latin American trade in 1990 was with other Latin American countries. This is a small amount compared with the level of trade typical between industrialized neighbors, and it suggests that the various Latin American states do not produce items they can exchange to mutual profit. The signing of the new trade pacts nevertheless might encourage specialization of production, increases in productivity, and intraregional trade and growth.

Other Regional International Groups

A number of other regional international organizations exist (see Figure 12–21). Some of these are primarily military-defensive; others, economic. With the end of the cold war, however, even those that were originally military may turn into general economic and cultural as-

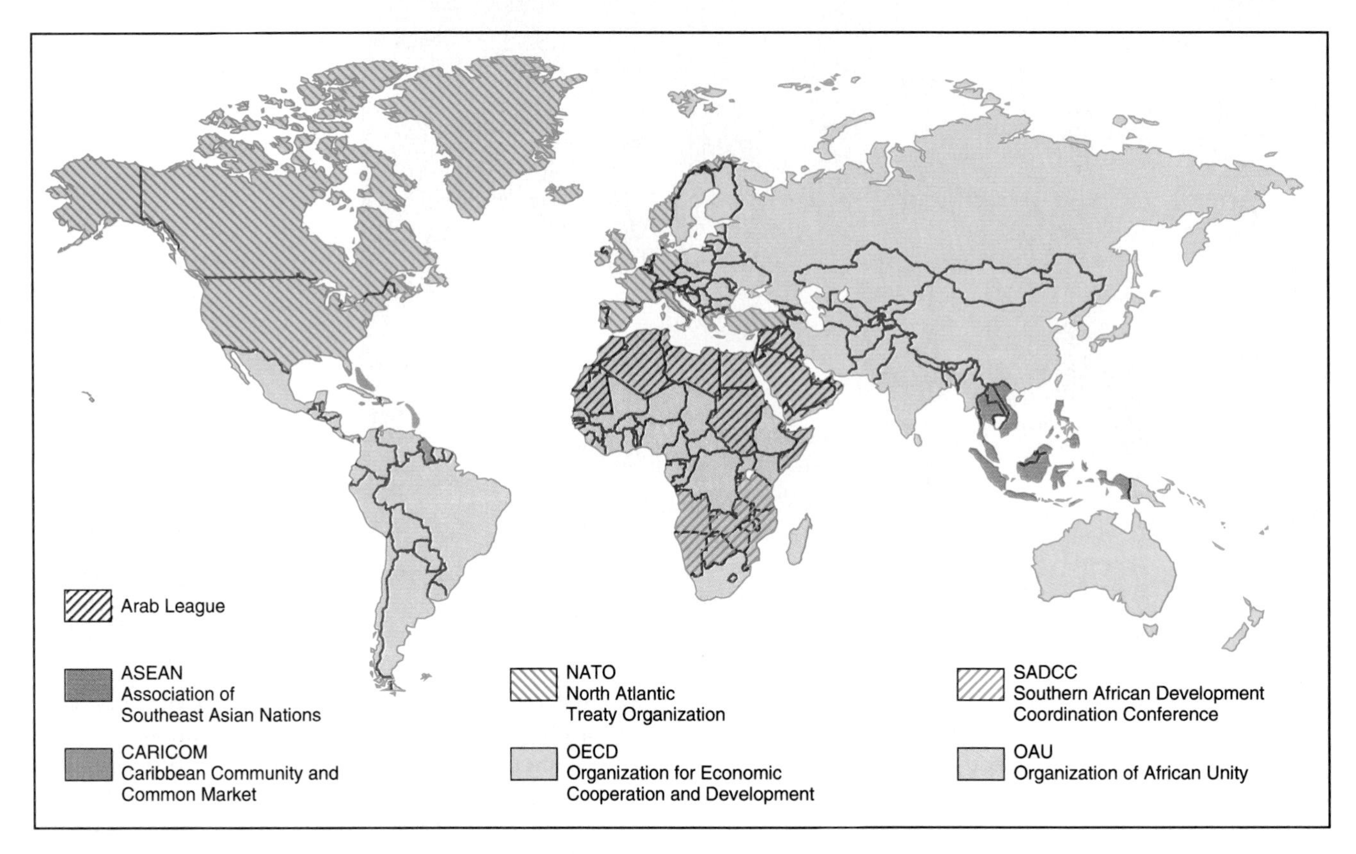

Figure 12–21. None of these organizations exercises enough power over its members to be called supranational; all are international. Some are rent by dissension among members; others have no active purpose.

sociations. For example, the Association of Southeast Asian Nations (ASEAN) began as a military alliance, but in 1992 it became a free trade organization, and Vietnam and Laos became "associates." Each of these organizations provides a framework for consultation to reach agreement whenever possible in areas of mutual concern.

With the end of U.S.-Soviet rivalry, the Earth may be organizing into three major economic blocs—Europe, North America, and East Asia. If this division rigidifies, it may splinter the global economy and impede global trade and political agreements. These topics are discussed in the next chapter.

SUMMARY

International organizations coordinate activities among nation-states, whereas supranational organizations exercise powers over the member nation-states. Today some organizations are united for military defense, but the most active are working for economic or broader cultural goals.

After World War II, the Allies split into two competitive blocs divided by an 'Iron Curtain.' Germany was split into two new states. The Western nations formed the North Atlantic Treaty Organization (NATO), and the communist nations formed the Warsaw Pact. These faced one another in a protracted cold war.

The pattern of alliances dissolved in 1989. The two Germanies united in 1990. The USSR released its grip on the countries of Eastern Europe, which held free elections. The Warsaw Pact dissolved. NATO will survive as a mutual defense and political structure.

Twelve Western European countries meanwhile merged into the European Community (EC). The EC seeks eventual political union and has created the framework of a supranational government. Integration of the EC territory can be seen in the creation of one market for goods and capital, and in transportation policies, social policies, and regional policies. The EC members consult in foreign affairs but lack a common foreign policy.

Seven other European countries formed a European Free Trade Association (EFTA). EFTA and the EC will merge into one free trade region to be known as the European Economic Area (EEA), and this may extend membership to Eastern European countries.

The Commonwealth of Independent States is evolving to replace the Union of Soviet Socialist Republics, a Russian-dominated multinational empire that centralized political and economic organization across one-sixth of the Earth's land surface. Mechanisms to govern the Commonwealth will evolve through the 1990s, but the degree of centralized organization will not be as great as it was in the Soviet Union. The Soviet Union had been held together by raw power exercised by the Communist party. The republics were never united by any common political idea, nor do they share a desire for common defense. Shared economic interests might hold several of them together because the years of central planning made them highly dependent on trade with one another. Centrifugal forces tending to pull the Commonwealth apart include animosity toward the former Russian rulers, and also external attractions.

The United States has begun organizing a free trade region consisting of Canada, the United States, and Mexico. The goals do not include formation of a customs union or common market. The pact creates no governing body, nor do the North American partners look toward political merger. All barriers to trade in goods and services between the United States and Canada will be eliminated by the year 2000. Mexico opened negotiations on a free trade pact with the United States in 1990. A continental free trade pact would redraw the map of economic and even cultural and political geography.

KEY TERMS

international organizations (p. 385)
supranational organizations (p. 385)
bloc (p. 385)
North Atlantic Treaty Organization (NATO) (p. 386)
Warsaw Pact (p. 386)
cold war (p. 387)
Council for Mutual Economic Assistance (COMECON) (p. 387)
hard currency (p. 387)
European Community (EC) (p. 390)
free trade area (p. 390)
customs union (p. 390)
common market (p. 390)
European Currency Unit (ECU) (p. 393)
European Free Trade Association (EFTA) (p. 397)
European Economic Area (EEA) (p. 398)

QUESTIONS FOR INVESTIGATION AND DISCUSSION

1. Has the European Community achieved full free trade?

2. Describe and analyze the assignment of seats in the European Parliament. Does it represent population? Economic power? Have seats recently been reapportioned?

3. The *Statistical Abstract* reports flows of trade and investment between the United States and Canada. What trends are apparent? Does one country seem to be gaining more than the other?

4. Since the collapse of the Soviet Union and the emergence of a new Russia, have many industries or services been privatized? What market incentives have gone into practice?

5. Is private property allowed in Russia? Are private farms? Is there a free market for agricultural produce? Are factories owned by stockholders or are they cooperative? Is there a free market for manufactured goods? Services?

ADDITIONAL READINGS

Collins, S., and D. Rodrick. *Eastern Europe and the Soviet Union in the World Economy.* Washington, D.C.: Institute for International Economics, 1991.

Iberry, B. W. *Western Europe: A Systematic Human Geography.* New York: Oxford University Press, 1986.

Institute for International Economics. *From Soviet disUnion to Eastern Economic Community?* Policy Analysis No. 35. Washington, D.C., 1992.

McKnight, Tom L. *Regional Geography of the United States and Canada.* Englewood Cliffs, N.J.: Prentice Hall, Inc., 1992.

Nahaylo, Bohdan, and Victor Swoboda. *Soviet Disunion: A History of the Nationalities Problem in the USSR.* New York: The Free Press, 1991.

Paterson, J. H. *North America: A Geography of the United States and Canada.* New York: Oxford University Press, 1989.

CHAPTER 13

The Globalization of Economics and Politics

The number of sovereign states has steadily increased, and yet the states face important political, economic, and ecological challenges that require cooperation. (Courtesy of Chris Jones/The Stock Market.)

In 1522 the Spanish explorer Sebastian del Cano (1476–1526) completed the circumnavigation of the globe that had begun under the command of Ferdinand Magellan, who had been killed in the Philippines. This voyage proved that the world was one spherical stage for human activity. The many local dramas being played on that stage, however, were widely separated, little related to one another, or even totally oblivious of one another. Political and economic events remained focused at the local level.

The European voyages of exploration connected all peoples, and the accelerating improvements in transportation and communication seemed to shrink the world. The British geographer Sir Halford Mackinder identified the period from 1492 until his time (1904) as "the Columbian Era," during which the world was connected. During the next age of human affairs, he wrote, the "post-Columbian era," the Earth would become "a closed political system . . . of worldwide scope. Every explosion of social forces . . . will be sharply re-echoed from the far side of the globe."

This post-Columbian era has so far witnessed two contradictory trends. One trend has been the shattering of once-vast European empires into a great number of new sovereign states. As discussed in Chapter 10, more than half of today's countries containing about half of the Earth's population and covering about half of the Earth's surface are less than 50 years old.

At the same time, however, the states' genuine independence of action has been undermined by international economic links and other transactions and travels. An advertisement for one U.S. bank shows a globe drifting in space. The headline reads: "We do business in only one place." Many activities have expanded their scale of organization to cover the whole globe, a process called **globalization.** Economic globalization has far outpaced cultural or political integration, and tensions among these economic, cultural, and political organizing activities generate much of today's international news.

This chapter first examines economic globalization. The study of world trade, of various countries' production and exports, is a time-honored specialty of geography. Today, however, world trade has accelerated so fast and multiplied to such a degree that international management consultant Peter Drucker has written, "The world economy has become a reality, and one largely separate from national economies. The world economy strongly affects national economies; in extreme circumstances it controls them." Chapter 7 explained that greater quantities of foods are moving in world trade, but the creation of a completely free global market for foods has been thwarted by national protectionism. A global market for manufactured goods is evolving, and today the states are negotiating to create a global market for services.

The codification of rules to regulate economic globalization is especially important—and especially difficult—because the number of states has risen so dramatically. After examining globalization, therefore, the chapter briefly reviews the collapse of a few of the European empires and focuses on a few international problems that linger from that process.

The chapter then takes up the instruments and institutions through which international law may be evolving toward global government. At present no effective global government exists, but at the least systems of global agreement are needed to regulate conduct among nations. International agreements also regulate activities in the areas of the Earth that all nations hold in common, such as the seas. Humankind is also beginning to realize how important it is that all peoples agree on rules to protect the global environment.

THE FORMATION OF THE GLOBAL ECONOMY

International trade has been carried on for millennia, but in the late nineteenth century new means of transportation and communication began to bind the world together to form a global economy. Since then international trade has almost without interruption multiplied faster than total world production. This trend reflects increasing regional specialization and exchange. In 1989 international trade in goods and services totaled almost $4 trillion—over $700 per capita (see Figure 13–1).

Capital and Commodities

Two developments in economic globalization accelerated in the late nineteenth century. One was the flow of capital investments from one country into another. International investments were at first limited to shares of stock and bonds representing minority holdings in foreign companies; these are called **portfolio holdings.** The owners of portfolio holdings do not take an active role in running the company. International foreign direct investment, however, soon followed. **Foreign direct investment (FDI)** means investment by foreigners in wholly owned factories that are operated by the foreign owner.

The second development was the evolution of a world market for certain primary products, foods, and minerals. Transportation and communication allow the expansion or widening of the territory that forms one market region within which producers must compete. This causes prices to become equalized across greater areas. The consolidation of an international wheat market

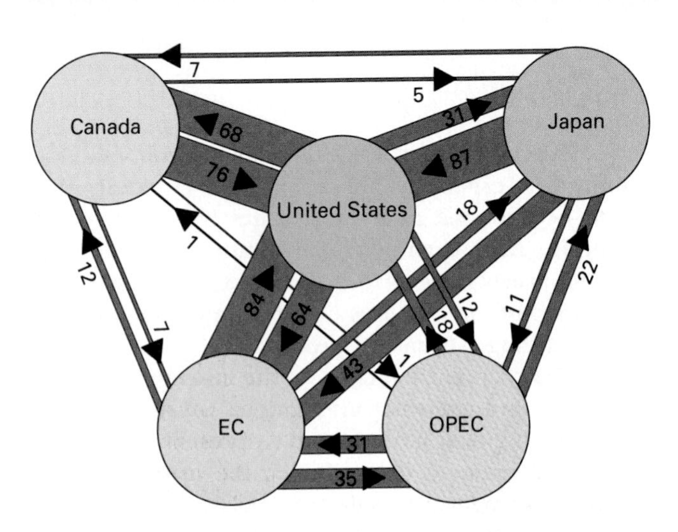

Figure 13–1. International trade among selected countries or blocs, in billions of dollars, 1989. (Data from *UNCTAD Handbook of International Trade and Development Statistics.*)

and its impact on one California ranch was captured by U.S. novelist Frank Norris:

> The most important thing in the [California ranch] office was the ticker. . . . The office . . . was thus connected by wire with San Francisco, and through that city with Minneapolis, Duluth, Chicago, New York, and at last, and most important of all, with Liverpool. Fluctuations in the price of the world's crop during and after the harvest thrilled straight to the office. . . . During a flurry in the Chicago wheat pits in August of that year . . . Harran and Magnus had sat up nearly half of one night watching the strip of white tape jerking unsteadily from the reel. At such moments they no longer felt their individuality. The ranch became merely the part of an enormous whole, a unit in the vast agglomeration of wheat land the whole world round, feeling the effects of causes thousands of miles distant—a drought on the prairies of Dakota, a rain on the plains of India, a frost on the Russian steppes, a hot wind on the llanos of the Argentine.
>
> *The Octopus* 1901

For commercial wheat farmers, the globe had clearly become "one place." Today virtually instantaneous global transmission of information is common, and instantaneity is called *real time.*

U.S. entrepreneurs tapped the country's natural resources for export. John D. Rockefeller's Standard Oil Company set the pace. When these entrepreneurs met competition from producers of natural resources elsewhere, they first bought up worldwide distribution outlets, and then they sought new foreign supplies to sell through their distribution networks. The copper industry, aluminum industry, and others followed the pattern of the oil industry. Because local U.S. entrepreneurs controlled the process and captured the profit, the United States accumulated wealth. In much of Latin America, as well as in the colonial areas of Africa and Asia, by contrast, the exploitation of local resources was undertaken by and for foreigners.

Enterprises with producing and marketing facilities in several countries are called **multinational** or **transnational corporations.** Their evolution followed four stages. These stages were later repeated in the evolution of international manufacturing, and they are now being repeated again in the tertiary sector. In the first stage, demand abroad was met by export of the item from the United States. In the second stage, the corporation established production facilities abroad to supply specific markets abroad; exports from the home country dropped. Third, the foreign production facilities began to supply foreign markets other than their own local market. Finally, the foreign production facility began to export back into the United States.

Multinational Manufacturing

Modern transnational manufacturing was launched in 1865, when the German chemical magnate Alfred Bayer built a factory in Albany, New York. Two years later the Singer Corporation of the United States located a sewing machine factory in Glasgow, Scotland, and U.S. industrialists quickly took the lead in economic internationalism. Part of the reason for U.S. leadership was that U.S. manufacturers had an immense domestic market that tariff walls protected from imported competition. U.S. manufacturers could achieve economies of scale in production for domestic markets and then win international markets. By the 1890s the United States boasted the world's greatest manufacturing economy, and a 1901 British book warned of *The Americanization of the World.* U.S. manufacturers still enjoy scale economies in a mass market, but Europeans hope to achieve a similar advantage.

After much of European industry was destroyed in World War II, U.S. corporations and products threatened to overwhelm Europe so completely that a 1960s European bestseller again warned of *The American Challenge.* U.S. companies continued to build factories abroad, and by 1988 about 17 percent of the total assets of U.S. nonfinancial corporations—that is, excluding banks, insurance companies, stock brokerages, and other companies whose "product" is money itself—were located overseas. This percentage is greater than that of any other country, and it is still rising.

Today technology allows the global integration of manufacturing processes almost in real time. For example, in the spring of 1990 a model paraded a prototype of a new fashion on a runway in Paris. Observers working for The Limited, America's largest retail garment chain, sketched the garment and faxed the sketch from Paris to a factory in Foshan, China. There an inexpensive copy of the garment was mass-produced, and this was then distributed to Limited stores throughout the United States within 1000 hours. This was weeks before the original designer had actually produced the garment.

Manufacturers have been predicting for years that the world would merge into one global market, but cultures determine markets, and cultures are stubborn. As discussed in Chapter 12, the European Community is an example of a regional international market. The countries have enough cultural and consumer similarities for corporations to attempt single advertising and sales strategies for their products.

Today, however, multinationals are attempting truly global sales strategies. In 1989 Gillette launched its "Sensor" razor with a single global advertising campaign—the world's first. This was fairly easy with razor blades, because what all customers in the world want is easily defined and highlighted. The number of global brands, however, that mean the same to everyone everywhere is small. Coca-Cola, Marlboro, and Levi's are among the most successful. They sell visions of U.S. style. This brand recognition exemplifies the world's acculturation to certain aspects of U.S. culture.

The Geography of Global Foreign Direct Investment

The amount of foreign direct investment being made today is many times the amount of foreign aid being given, and foreign direct investment has been much more successful in triggering economic growth. The following discussion is restricted to direct investment partly because no international agency knows how many cumulative trillions of dollars are held worldwide in foreign portfolio investments, although it is estimated that international portfolio transactions totaled almost $1 trillion in 1990.

In the 1980s global FDI grew three times faster than world trade and four times faster than total world output, to a total of about $2 trillion. This flow of FDI in the 1980s greatly increased and significantly redistributed the world's productive capacity.

In the 1980s companies from an increasing number of countries became active investors in other countries (see Figure 13–2). In 1980 about 50 percent of FDI was from the United States, but the European Community countries and Japan came to play increasing roles. Thus, by 1989 the United States was the source of only 30 percent of global FDI. The United States, the combined EC countries, and Japan accounted for 81 percent of FDI in 1989. The United Nations calls these three sources of FDI the Triad.

Although FDI from the United States is a declining percentage of world FDI, U.S. firms have steadily increased their ownership of the world's economy. The geographical distribution of this investment is changing. In 1950 about 40 percent of U.S. capital invested abroad was in Latin America; by 1990 only 11 percent was. In 1960, 20 percent was in Europe; by 1989, 40 percent was. Another 16 percent was in Canada, and another 5 percent in Japan.

The United States is not only the world's largest source of FDI in other nations, it is also the world's largest recipient of FDI from other countries. As the U.S. dollar fell in the late 1980s, foreigners were able to buy U.S. assets at favorable rates of exchange, and so FDI in the United States accelerated. In 1989 about 29 percent of total global FDI was invested in the United States. Over half the total was from EC countries; another 21 percent was from Japan; 7 percent was from Canada.

Figure 13–2. These silos in China are part of a joint business venture between the Chinese government and the multinational agricultural and industrial conglomerate Chaeron Pokphand, which is based in Thailand. Chaeron Pokphand's worldwide interests range from feed grains and frying chickens to ships and motorcycles. (Courtesy of C. P. Group/Chia Tai Group.)

Figure 13–3. Japan's Yaohan department store chain is rapidly expanding across the United States, offering goods from around the world. International retailing is a tertiary-sector investment activity. This store is in New Jersey. (Courtesy of Reginald Wickham.)

Forty percent of all FDI in the United States is in manufacturing. Foreign holdings of U.S. factories represented only 6 percent of the total value of U.S. factories in 1977, but by 1989 that figure had risen to 17 percent. The list of foreign firms that manufacture in the United States lengthens daily. These firms are creating more new jobs than U.S. manufacturers are, and they are among the leading exporters of goods from the United States. They are also introducing new technology and management techniques and are changing Americans' daily lives in the workplace as well as Americans' consumption habits (see Figure 13–3).

FDI in the Developing Countries Three trends characterize FDI in the developing countries. First, the share of FDI that the Triad is allocating to the poor countries is declining. In the 1970s the Triad sent one-third of all FDI monies to the poor countries; by the end of the 1980s that percentage had slumped to less than one-fifth. The rich countries are increasingly investing in one another.

Second, FDI has become increasingly geographically selective. The countries that attracted most investment were countries that chose the export-led method of economic growth discussed in Chapter 11. Their economies boomed. Four Asian nations attracted FDI and grew so fast during the 1980s that they earned the nickname "The Four Tigers": Singapore, South Korea, Taiwan, and Hong Kong (technically not a country—see below—but considered one in international trade statistics). These areas reached European levels of prosperity within a decade, and they have already become sources of capital and technology for the next tier of Asian developing lands: Thailand, Indonesia, Malaysia, and China. Vietnam and other nations now court investment. In the early 1990s even India, which had been committed to import-substitute growth since 1947, began to welcome foreign investment. The collapse of the Soviet Union, India's main economic, military, and ideological benefactor since independence, partly prompted India's change of course.

A third characteristic of FDI in the developing countries is that each Triad member has a majority share of the FDI in a distinct cluster of countries that have become economic satellites of that member of the Triad. These clusters help explain world patterns of trade. One example of this pattern involves trade between the United States and Japan. Many U.S. economists became alarmed in the mid-1980s when a large U.S. trade deficit opened with Japan in consumer electronics. In the late 1980s and early 1990s, however, this trade deficit shrank, but a new trade deficit in these goods emerged with Thailand. This caused less alarm because Thailand is, after all, a developing country and a long-time U.S. ally. As Table 13–1 indicates, however, Thailand falls within the Japanese cluster; the electronics factories in Thailand are owned by Japanese. Thus, the consumer electronics that the United States imported from Thailand profited the Japanese corporations as well as the Thai economy. In a similar fashion, many of Europe's imports from Latin America are the products of U.S. corporations.

TABLE 13–1 ***Triad Foreign Investment Clusters***

United States	European Community	Japan
Argentina	Brazil	Hong Kong
Bolivia	CIS states	South
Chile	Croatia	Thailand
Colombia	Czechoslovakia	
Mexico	Hungary	
Panama	Poland	
Philippines	Slovenia	
Saudi Arabia	Yugoslavia	
Venezuela		

Note: Clusters identified by the U.N. Centre for the Study of Transnational Corporations, 1991.

In coming years the developing countries, including Eastern Europe and the Commonwealth of Independent States will have to compete to attract investment. Even those rich countries that fail to maintain roads, education, and other national infrastructure might lose their manufacturing base and fail to attract new investment.

In 1990 the UN General Assembly unanimously adopted a resolution on economic development that favors export-led programs. It states:

> A gradual convergence of views on economic policy . . . is emerging. Flexibility, creativity, innovation and openness must be integral parts of our economic systems. . . . Countries have to adapt their national policies to facilitate open exchange . . . and be supportive of investment. . . .

In the 1990s, geographers will watch capital flows carefully, because the pace of development within individual countries is increasingly determined by their ability to attract foreign capital. The world map of wealth could continue to change dramatically.

THE CONTROL OF THE WORLD ECONOMY

Multinational corporations direct an ever-increasing share of the world economy. Most multinationals are still based in the United States, Europe, and Japan, but some poorer nations including Brazil, Mexico, Thailand, and Indonesia have produced their own. The internationalization of stock markets and investment opportunity has dispersed the ownership and control of these corporations, wherever they are headquartered.

In 1990 the world's 600 biggest companies, all multinational, created at least one-fifth of the world's total value added in manufacturing and agriculture (see Figure 13–4). The United Nations estimated that direct employment by multinational corporations in 1989 was 65 million, or 3 percent of the world's labor force.

The nation-states have seen their economic autonomy slip away, and many have come to resent the multinationals. Some critics in the poor countries attack the multinationals as tools of the rich nations. They protest foreign control of their national economies, and they feel further impoverished as multinationals withdraw profits.

Critics exist in the multinationals' home countries too. They argue that whenever a multinational invests abroad, it is "exporting jobs" and also reducing exports of goods. The U.S. experience seems to corroborate this. From 1966 to 1986 the U.S. share of world manufactured exports dropped from 17.1 percent to 11.7 percent. By contrast, U.S. multinational enterprises, exporting both from the United States and from their foreign locations, captured a share of between 17 percent and 19 percent

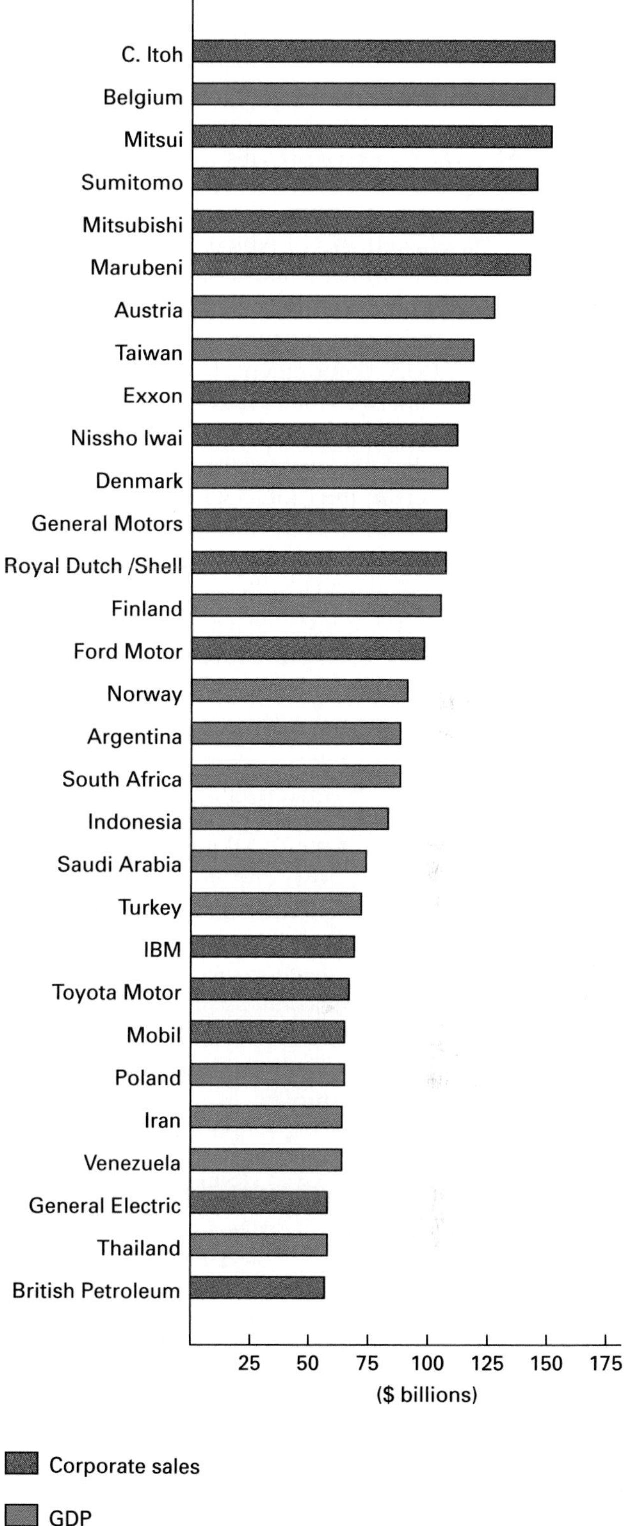

Figure 13–4. Some of the world's largest multinational corporations control economies as large as those of some countries.

of world manufactured exports through those years. By 1986 the sales of products manufactured abroad by foreign subsidiaries of U.S. companies totaled three times the value of U.S. exports. As the dollar fell in the late 1980s, however, exports directly from the United States increased.

Multinational Corporations and National Allegiances

It is becoming difficult to tell which corporations are corporate citizens of which countries (see Table 13–2). Among the largest exporters from Taiwan, for example, are AT&T, RCA and Texas Instruments. These are generally considered U.S. corporations, and yet exports of these firms from Taiwan to the United States account for a good share of the U.S. trade deficit with Taiwan. In 1991, Zenith, the last U.S. company that was manufacturing television sets in the United States moved its operations to Mexico. Meanwhile, foreign-owned corporations continue to manufacture sets in the United States under familiar "American" brand names that these foreign corporations have bought: Magnavox and Sylvania (Dutch), RCA and General Electric (French), and Quasar (Japanese). In 1991 the U.S. government, acting on the behalf of a wholly Japanese-owned typewriter factory in Tennessee, protested the trading practices of a Singapore corporation that is a subsidiary of the U.S. Smith-Corona Corporation, which is 48 percent owned by a British corporation. The United States is taking the position in international negotiations that any company that employs and trains U.S. workers and that adds value in the United States is a U.S. corporation, no matter which flag it flies at its international headquarters, wherever those headquarters are.

In fact multinational corporations do not represent any countries; all corporations represent the profit motive. Directors of corporations define their responsibility as maximizing the return on shareholders' capital, and so they will invest that capital wherever they believe returns will be the highest. As the head of Colgate-Palmolive asserted in 1990: "The United States does not have an automatic call on our resources. There is no mindset that puts this country first." In 1989, 64 percent of the company's sales and 63 percent of its operating profits were generated outside the United States. In 1990 the president of Motorola was asked whether, in the event of a recession, Motorola would close operations abroad first to save U.S. jobs. He replied: "No. We must treat our employees all over the world equally."

Two Case Studies The two following case studies exemplify different principles of international corporate expansion. The manufacture of automobiles in the United States demonstrates foreign direct investment to capture the U.S. market. For foreign manufacturers, investing in the United States substitutes for exporting cars to the United States. The amount of this investment is determined by the relative costs of manufacturing at other international locations and also by the height of any tariff wall the United States erects to protect domestic automobile manufacturing.

McDonald's expands internationally not by exporting hamburgers. It invests in facilities abroad, but it exports techniques of making and selling hamburgers.

The manufacture of automobiles for the U.S. market Automobile manufacturing companies that

TABLE 13–2 ***The 20 Companies with the Largest Share of Their Sales Outside Their Home Country***

Company	Home country	Sales outside home country	Assets outside home country
Nestlé	Switzerland	98.0%	95.0%*
Sandoz	Switzerland	96.0	94.0
SKF	Sweden	96.0	90.0
Hoffmann–La Roche	Switzerland	96.0	60.0
Philips	Netherlands	94.0	85.0*
SmithKline Beecham	Britain	89.0	75.0
ABB	Sweden	85.0*	NA
Electrolux	Sweden	83.0	80.0
Volvo	Sweden	80.0	30.0
ICI	Britain	78.0	50.0
Michelin	France	78.0	NA
Hoechst	West Germany	77.0	NA
Unilever	Britain/Netherlands	75.0*	70.0*
Air Liquide	France	70.0	66.0
Canon	Japan	69.0	32.0
Northern Telecom	Canada	67.1	70.5
Sony	Japan	66.0	NA
Bayer	West Germany	65.4	NA
BASF	West Germany	65.0	NA
Gillette	United States	65.0	63.0

Source: Business Week, May 14, 1990. * *Business Week* estimate.

are headquartered in the United States, Japan, Germany, South Korea, and other countries have invested in manufacturing, assembling, and marketing their products around the world. Furthermore, many of these corporations either own substantial percentages of one another, or else they have established partnerships for the manufacture or marketing of particular products in particular countries.

For example, in 1982 the Japanese Honda corporation opened an automobile assembly plant in Marysville, Ohio (see Figure 13–5). By 1990 that plant assembled more cars than any other plant in the United States. Moreover, the value of local content rose. Originally most of the parts assembled there were imported from Japan, but soon more parts were manufactured there, and the car was increasingly designed in the United States. The plant is not unionized, and it has introduced many management and manufacturing techniques to the United States.

In 1989 the Honda Accord became the first foreign car to become the number-one selling model in the United States. That car was a hybrid built by workers in both Japan and the United States and was even exported around the world from both countries. The two-door version was designed in the United States, manufactured only in the United States, and exported to Japan. The station wagon was designed and is manufactured in the United States for export to Europe. Some European countries have limited their imports of this car as they would a Japanese manufactured product. U.S. Trade Representative Carla Hills has insisted, however, that "A Japanese nameplate car made in our country is an American car."

Most of the Honda Corporation's profits go to Japan, but the company spends great amounts in the United States for salaries, materials, research, and training. The manufacture of popular Japanese-nameplate cars in the United States also lowers the U.S. trade deficit, some 40 percent of which was, in 1989, accounted for by automotive trade. Half of that was with Japan. U.S. production replaces imports. In 1990, U.S. factories of Japanese car manufacturers held about 30 percent of the total U.S. car market, and slightly more than one-half of all Japanese-nameplate cars sold in the United States had been made in the United States.

By 1990 world car manufacturing was so interconnected that no one could tell from a car's nameplate which flag the corporation flew, where the corporation's stockholders lived, or where the car was either designed or manufactured. Chrysler, which was once bailed out by the U.S. government and which used a patriotic theme in its advertising, was using a lower percentage of U.S.-made parts than was General Motors or Ford. Chrysler owned 24 percent of Mitsubishi Motors and, through Mitsubishi, a share of the Korean maker Hyundai. Mitsubishi was making cars under Chrysler's label, and the two companies were jointly operating a factory in Normal, Illinois, producing vehicles under both nameplates. Ford, with one-third of its sales outside the United States, owned all of Jaguar (originally British) and 25 percent of Mazda. Mazda was making cars in the United States for Ford; Ford was making trucks for Mazda; the two companies were trading parts. Each owned a share of Korea's Kia Motors, which produced parts for the Ford Festiva for export to the United States. Ford and Nissan were jointly producing cars in Australia and in the United States. Ford and Volkswagen had merged into a single company in Latin America, which was exporting trucks to the United States.

General Motors owned more than 40 percent of Isuzu, which had a joint venture in the United States with Subaru, which is partly owned by Nissan. General Motors also owned half of Daewoo Motors, Hyundai's major com-

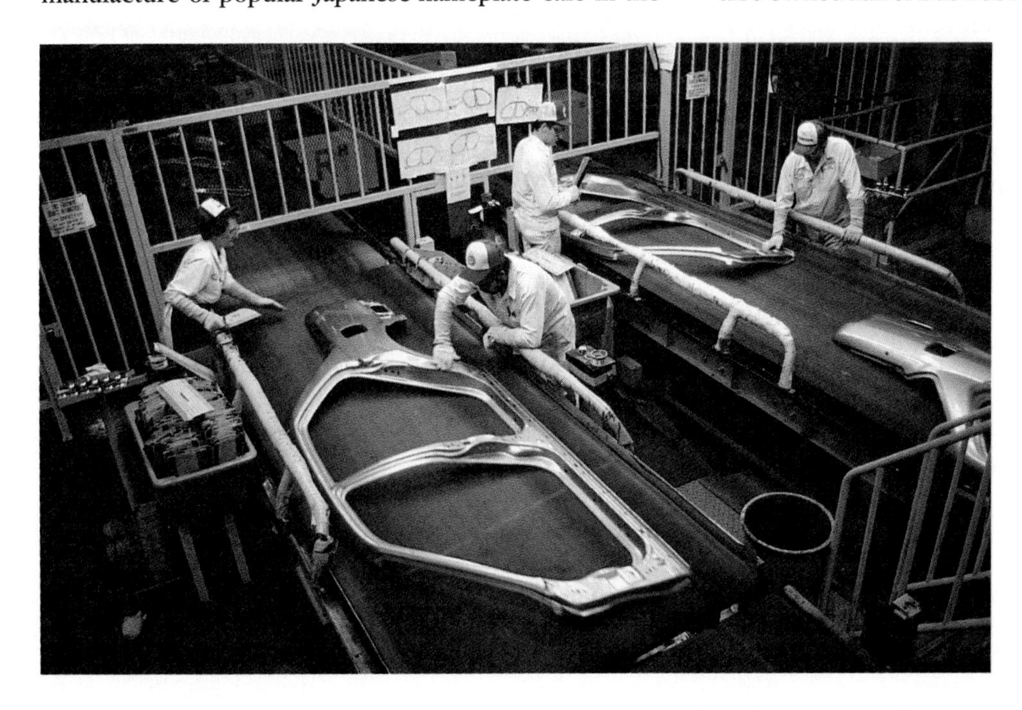

Figure 13–5. The Honda factory in Marysville, Ohio, has brought new management techniques and technologies to the United States. Here steel parts are stamped for assembly. (Courtesy of Andy Snow/SABA.)

petitor in Korea. Daewoo made Nissan cars for Japan and Pontiacs for the United States; it was selling to Isuzu in Japan cars that were designed by GM-Europe. General Motors was making cars together with Toyota in Australia and in the United States. Honda America sold more cars in America than its parent did in Japan, and Honda was exporting cars from the United States.

As long as U.S. residents buy millions of cars each year, thousands of U.S. workers will probably work in auto-manufacturing plants in the United States. We cannot predict, however, what the names of the manufacturing companies will be or who will own them. It is possible that by the year 2000 General Motors, Ford, and Chrysler will not be manufacturing cars in the United States, but that they will have moved all their production to foreign countries, and that all domestic production will be performed by "foreign" firms.

McDonald's McDonald's promises worldwide uniformity of a product that cannot be exported. Therefore, McDonald's must export the technology and skills to manufacture its product. Although 68 percent of the corporation's 1989 profits were still generated in the United States, the chain was represented in 53 countries and was expanding more rapidly overseas. It opened over 300 new shops around the world in 1990 alone.

Manufacturing a standard product in many different countries can be difficult. When McDonald's prepared to open its first hamburger restaurant in Moscow, in January 1990, for example, it was challenged to ensure that customers would be served exactly the same food and drink they would get in any other McDonald's in the world. The company's food specialists worked for 2 years to establish the necessary quality-control infrastructure. The company built a food-processing plant in Moscow to mold its beef patties, chip its french fries, and bake its rolls. The plant included its own dairy. Russian farms were contracted to produce the beef and potatoes. Bull semen was shipped over to guarantee the quality of the cattle. Russet Burbank potatoes were planted. McDonald's staff taught Russian farmers how to extend the cattle's feed cycle and to maximize the potato harvest. Four Russian management trainees attended the company's Hamburger University in Illinois.

The company signed a joint-venture agreement with the Moscow city council to build 20 restaurants. The city would own 51 percent of the total venture, McDonald's 49 percent. McDonald's can use its profits, in rubles, to pay its Russian farmers and staff. Given the black-market exchange rate, customers in Moscow with hard currency can buy the cheapest Big Mac in the world.

By 1992 Moscow's largest McDonald's was selling more sandwiches than any other in the world (30,000 hamburgers per day), and, accompanying the privatization of the Russian economy, McDonald's was gradually subcontracting to Russian suppliers responsibility for providing basic supplies.

McDonald's executives reported that the business skill most difficult to export to Russia was a pleasing attitude. In the communist command economy there had been no system of reward or loss attributable to sales, and so it had made no difference to managers or workers whether the customer had been pleased.

The International Tertiary Sector

When most people hear the words "foreign trade," they think of oil and iron, bananas and coffee, cars and clothes being shipped around the world. It is easy to imagine such "things." The international economy, however, has transcended trade in primary products and manufactured goods. Trade in the tertiary sector, in services, totaled over $600 billion in 1989, and this segment is growing faster than trade in manufactured goods. Services are also taking a growing share of international investment.

In many cases corporations that provide services to manufacturing corporations in their homelands have followed these out into the world arena (see Figure 13–6). This partly explains the internationalization of legal counsel, business consulting, accounting, and advertising (see Figure 13–7). Many professions such as architecture and medical care market their skills around the world. For many nations, welcoming tourists has become a key element of their domestic economies, as has welcoming foreign students to their schools. Financial services, entertainment, and still other services are produced for global markets.

International services demonstrate the same trends as trade in goods. Countries are losing their autonomy to an increasingly global market. Transnational corporations treat the world as "one place" and exploit resources, locate facilities, and market their products accordingly; and the countries struggle to preserve their territorial sovereignty.

The United States is the world's leading exporter of services, and the nation enjoys regular trade surpluses in them. In education, for example, the United States earned a net $4.3 billion in 1990. Other U.S. tertiary exports include real estate development and management, accounting, medical care, business consulting, computer software development, legal services, advertising, security and commodity brokerage, and architectural design (see Figure 13–8). U.S. entertainment productions dominate the world's airwaves and movie screens. Movies, music, television programs, and home videos together account for a significant annual trade surplus.

The Globalization of Finance The FDI that is redrawing the world map of manufacturing is only a fraction of the vast amounts of capital moving around the world. One global capital market has formed just as one global wheat market had formed by 1901. Multinational banking has developed, and telecommunications make it possible

Figure 13–6. Electronic Data Systems Corporation coordinates worldwide computer consulting services from this main operations room in Texas. (Courtesy of Electronic Data Systems.)

to monitor and trade in national currencies, stocks, and bonds listed anywhere in the world in real time. By 1992 over $600 billion was traded daily on the world's foreign exchange markets. This is over 100 times the average daily merchandise trade. Almost $5 trillion worth of stock changed hands on the world's major stock exchanges. The International Organization of Securities Commissions (IOSC) monitors this activity, but no international agency records ownership, even by national source, of these assets.

The internationalization of finance puts financial issues among issues of international trade. Nations are expected to regulate their own stock exchanges, and the home countries of global banks are responsible for supervising all their operations, but countries do not agree on how to do these things. The international headquarters of many giant banks are located in places renowned for lenient banking regulations. The 100-square-mile (260-km^2) British Caribbean colony of the Cayman Islands (population 25,000) is one of the world's greatest financial centers, holding assets of over $400 billion in 1992. It has, however, almost no vaults, tellers, security guards, or even bank buildings. Assets are held electronically in computers. Many banks consist of one-room

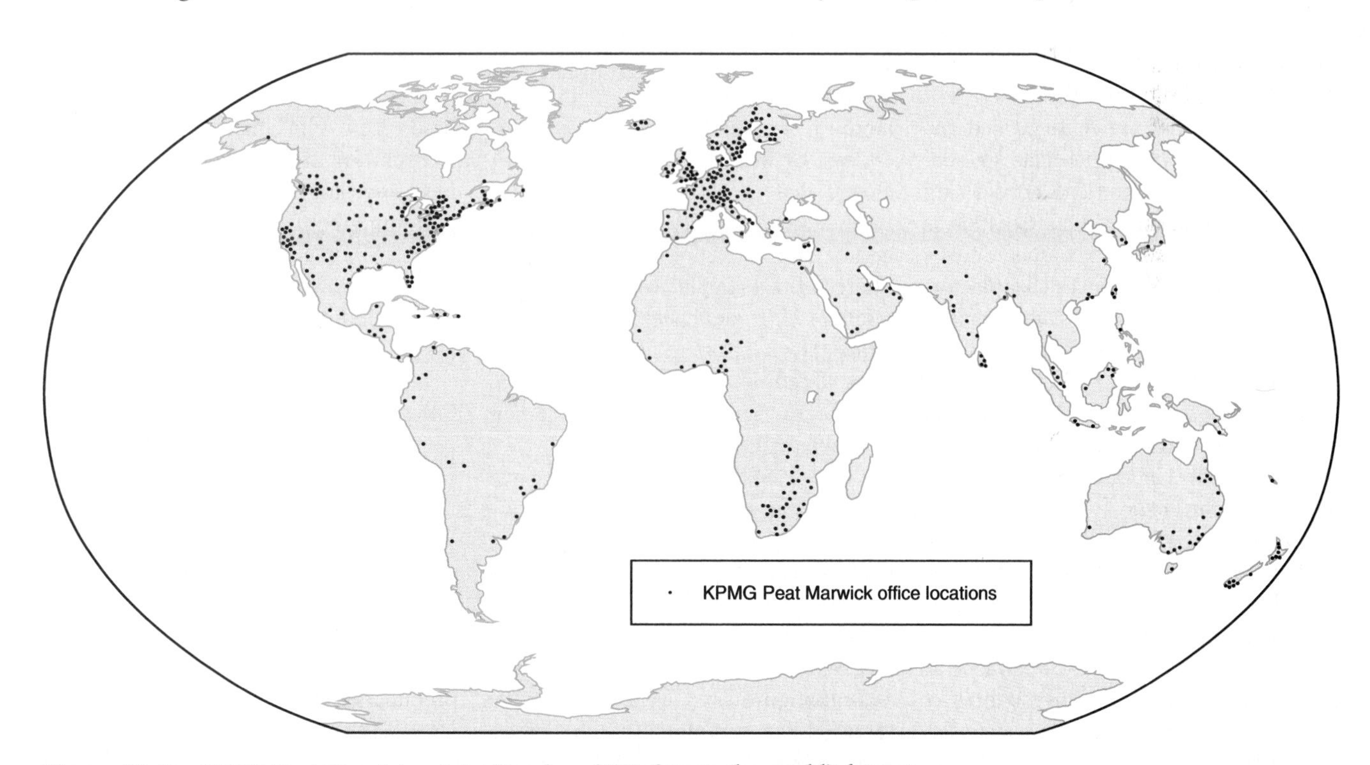

Figure 13–7. KPMG Peat Marwick, a joint Dutch and U.S. firm, is the world's largest accounting firm. It has representative offices in all the economic capitals of the world. (Courtesy of KPMG Peat Marwick.)

Figure 13–8. This stunning glass pyramid is the new entrance to the Louvre art museum in Paris, France. It was designed by the architect I. M. Pei, a U.S. citizen who was born in China but lives in New York. The same year this building opened, Mr. Pei opened a new symphony hall in Dallas and a 70-story skyscraper headquarters for the Bank of China in Hong Kong. Architectural design has become an important service export from the United States. (Courtesy of Pei Cobb Freed & Partners/Diede von Schawen.)

CRITICAL THINKING: *Selling U.S. Culture*

Many Americans are flattered and enriched if U.S. culture products are sold abroad, but many are also considering two critical questions. One question is whether increasing dependence on foreign markets will affect what is produced. This would affect what Americans have an opportunity to see themselves. The second question is whether U.S. citizens should sell the corporations that actually produce the culture. Is the United States selling its cultural birthright?

Studies have found that if a movie studio, for example, is owned by a foreign investor, or if a significant share of the studio's profits come from showing its work abroad, then the treatment of foreign countries and their cultures is notably more sympathetic than when the ownership and intended audience was restricted to the United States. Furthermore, it has been noted that the production of films with sophisticated dialogue has generally given way to comedies with easily accessible visual humor and to simple action pictures. These trends might reflect changing tastes of U.S. audiences, or they might represent an effort to increase foreign sales. In either case, they are changing the nature of the film entertainment available to U.S. audiences.

In 1989, 43 percent of the revenues of the seven largest U.S. movie studios came from abroad. That percentage was rising rapidly, and foreign countries present the greatest opportunity for increasing sales. In 1991 the four largest movie studios in the United States were foreign-owned: MGM/UA, owned by an Italian corporation; 20th Century Fox, Australian; Columbia Pictures, Japanese; and Universal Studios, Japanese. New films made by the Disney Corporation are also financed by Japanese investors. Will Matsushita Corporation, the Japanese owner of Universal Studios, let the studio make a movie about World War II? When the chairman of the board of Matsushita was asked, he said he could not answer.

Questions

Is it bad, good, or a matter of no difference that U.S. film studios grow more sensitive to foreign sensibilities and tastes?

Could this make U.S. moviegoers more cosmopolitan?

Should the United States protect culture producers and forbid their sale to foreign interests?

offices, and confidentiality of transactions is legally protected. International agencies cannot trace the profits of, for example, the international drug trade.

Financial operations depend on cultural codes. In the United States, for example, people are extended credit on the assumption that they will prefer to pay back the debt rather than to escape into legal bankruptcy. If everybody in the United States who could file for bankruptcy did so, the economy would collapse. People in different cultures, however, conduct business differently. Therefore, the internationalization of business can raise problems of cross-cultural misunderstanding or even mistrust. In Muslim areas substantial loans have been made without collateral on the basis of trust among the participants in the transaction or, in the Arab oil countries, in confidence that the rulers, with their immense wealth, guarantee the businesses of their citizens. In several cases rich Arab rulers have in fact paid their citizens' international debts. In Japan, too, and among the Overseas Chinese, business transactions depend on webs of allegiances not easily understood by people of other cultures. Some transactions have caused uproars in the West.

Today Japan is one of the world's principal sources of capital, and so the regulations governing the Tokyo stock market affect the opening and closing of factories throughout the world. The Arab oil states are another source of capital, and so the interest rate on U.S. college tuition loans is affected by Middle Eastern politics. Bank policies in Frankfurt, Germany, and Johannesburg, South Africa, affect the checking account charges of U.S. banks. Financial manipulations in Rio de Janeiro and Bombay affect the security of many U.S. elderly people's pensions, and the interest rates set at the Federal Reserve Bank in Minneapolis affect business taxes in Indonesia. The explanations of the *where* and *why there* of local economic affairs may lie in global finance.

Tourism Tourism is, by some accounts, the world's largest industry. The World Tourist Organization (WTO) estimated that in 1990 international tourist receipts alone totaled over $250 billion, and this did not include travel fares. Tourism is the top earner of foreign exchange for many individual countries. In 1991 tourism became the largest export industry of the United States and earned a $10.6 billion surplus.

Tourism is expected to grow everywhere by at least 5 percent per year, with even higher rates in the poor countries. Tourists flooded Eastern Europe after the opening of the Iron Curtain, and tourist expenditures provided the first substantial injection of capital into the Eastern economies. Poland received 18 million visitors in 1990. Poland does not have accurate accounting of tourist expenditures, but if each tourist spent only $25, that income represented about $12 per Pole.

Citizens of the rich countries account for the largest tourist expenditures. The most popular destinations are also rich countries (see Figure 13–9). When income from tourism is measured as a share of total national exports, however, the importance of tourist income to the poor countries becomes clear (see Table 13–3). Turkey, for instance, quadrupled the value of its exports of goods between 1980 and 1989, but the value of its tourist revenues increased twice as quickly. By 1989 Turkey's revenues from tourism were 22 percent of the value of Turkey's exports of goods. Thailand was the world's fastest growing economy from 1988 to 1990, and despite Thailand's rapid industrialization in those 3 years, tour-

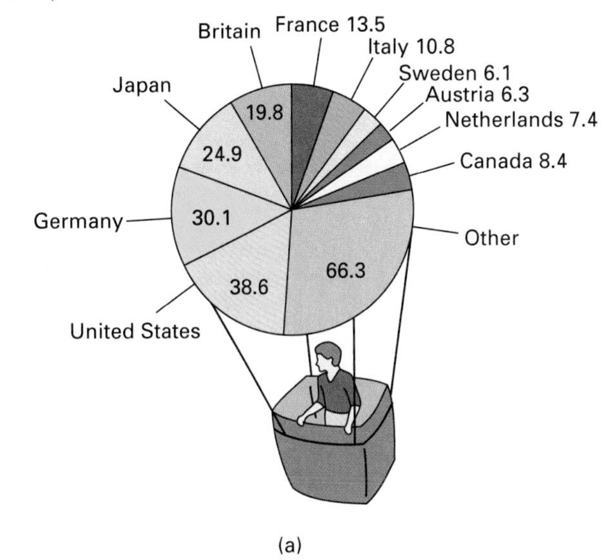

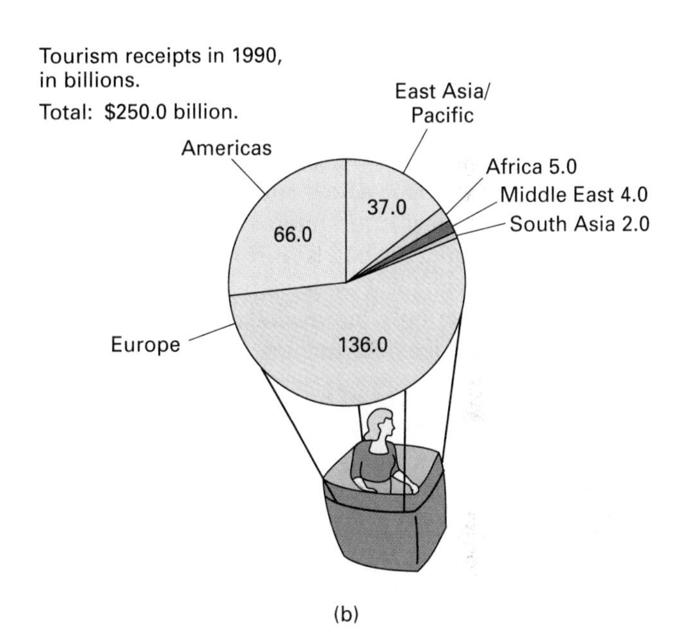

Figure 13–9. Citizens of the rich countries are the world's greatest tourist spenders (a), and the rich countries also attract the greatest amounts of total tourist expenditures (b). (Data from World Tourist Association.)

TABLE 13–3 ***Tourist Income as a Percentage of Exports for Selected Countries, 1990***

Country	Travel and tourism as percent of total exports
Jamaica	36
Spain	25
Greece	22
Portugal	17
Mexico	11
India	9
Philippines	7
Brazil	6
Colombia	6
Egypt	4

Source: American Express Travel Related Services.

ism grew faster. By 1990 it represented 6 percent of Thailand's GDP, and it was the country's leading foreign exchange earner.

Tourism creates many jobs, and it offers extensive examples of the multiplier effects discussed in Chapter 9. One new hotel creates demand for a host of related goods and services: meals, souvenirs, boat or taxi rides, rental cars, and so forth. Tourism may stimulate the local agricultural economy by providing a new market for cash crops and processed food for the hotel market. Tourism is generally a high-value-added and labor-intensive industry that trains local labor. International hotel chains build, own, and operate hotels in some locations, but they generally resign local ownership to local investors. The chains provide only international advertising, reservations, and management training.

Benefits and drawbacks of tourism Tourism does have drawbacks for the host country. Tourism may corrupt local culture, encourage prostitution, and provoke inflation. Too many tourists may overwhelm the natural environment.

Tourism imposes foreign-currency costs on a host country. Some things that tourists need must be imported, and so foreign earnings leak out. As a general rule, however, the more developed and complex the rest of the economy is, the greater the gains from tourism. A modern tourist industry draws on many other industries, and a country as developed as, for instance, Kenya, produces many of these goods locally. A small, poor country does not. As poor countries develop, fewer operating costs leak abroad. As with any other aspect of economic growth, diversity is best.

Today almost all countries are promoting tourism. Even the Chinese are using the Great Wall, built to keep foreigners out, to draw them in.

The "three *A*'s" of tourism The tourist potential of any spot depends on its "three *A*'s": accessibility, accommodations, and attractions. International airline routes are the lifelines of tourist destinations, and destinations must provide sufficient hotel rooms, preferably in a range of prices. A fine convention hall also attracts the conventions and trade shows that have become important economic activities in many cities.

Many tourists find sunshine and a beach sufficient attraction (see Figure 13–10). For these, the map of vacation climates supplemented by a map of the best beaches provides directions to their goals. Recreational opportunities draw millions more to mountains.

Tropical regions or areas spectacularly endowed with wildlife may feature **ecotourism,** which is travel to see distinctive examples of scenery, unusual natural environments, and wildlife. Ecotourism is a rapidly rising share of world tourism. The rich bird life on the islands of Lake Nicaragua, the lions of Kenya, and the gorillas of Rwanda are important tourist destinations. Several countries advertise opportunities for tourists to "hunt" rare animals—but with cameras rather than guns. Belize created a Ministry of Tourism and Environment to safeguard that country's jaguar preserve, barrier reef, and howler monkey sanctuary. Ecotourism is Belize's leading export (see Figure 13–11). Ecotourism today provides the money needed to save many countries' natural environments as well as many species of rare animals.

The degree of cultural diffusion generated by tourism depends partly on the character of the tourists. Some tourists seek a cultural experience completely different from what they experience back home, and they give themselves up to the culture of their destination. Even these tourists, however, change the true character of any place they visit. Other tourists want only a limited experience of anything different from what they know at home. These people are called "home-plus" tourists. They want Egypt, for instance, to have all the conveniences they are accustomed to at home plus the pyramids; Peru to be everything familiar plus Inka ruins. A great deal of their home material culture must be re-created for them in the tourist destinations. Many tourists from the United States, for example, feel most comfortable eating at McDonald's around the world.

Local residents, however, may adopt the foreign material culture that was re-created for the tourists, and in this way they may surrender their own local culture. The local native people everywhere eating at McDonald's exemplify this type of cultural diffusion. Some countries set aside restricted areas where activities are allowed that are only for tourists. Several Islamic countries, for instance, have set aside areas where liquor is allowed.

Many tourists travel just to see someplace different. Wholly fabricated theme parks, such as Disneyworld, provide enjoyment to millions of visitors. Euro Disneyland, 20 miles (32 km) east of Paris, is a self-contained world 20 percent the size of the city of Paris. The Disney Corporation expects 11 million Europeans to visit per year. Nobody, however, wants to turn a whole country into a theme park. Local attractions are most surely unique if they are genuinely part of a place's traditional

Figure 13–10. In the late 1940s the Mexican government chose Acapulco to be the first focus of government-promoted tourism. The city enjoys year-round sunshine and temperatures of around 80° F (27° C), and several sandy beaches stretch for 10 miles (16 km) along its coastline. Luxury hotels are set against a majestic backdrop of mountains and evergreen tropical vegetation. Mexico's foreign exchange earnings from tourism tripled from 1945 to 1950 and are among the world's highest today. Acapulco was an old port city. From 1565 until 1815 galleons sailed annually from here to Manila in the Philippines, exchanging silver for porcelain, spices, and silks. (Courtesy of the Mexican Tourist Office.)

local culture. This ensures that they cannot easily be duplicated elsewhere, and it also encourages the local population, presumably proud of what they have, to be hospitable. Some countries educate their populations about the financial importance of tourists, in the hope that education will inspire courtesy. Jamaica, for example, has a nationwide education program on tourism.

Figure 13–11. The government of Belize depends on income from ecotourism to protect this rare jaguar. (Courtesy of Jeanne White/Photo Researchers.)

The economic importance of tourism is today so great that in many places attributes of local culture that might otherwise die away are preserved almost exclusively for tourists. Traditional local festivals survive as tourist shows. Some countries will even create new aspects of local culture specifically for presentation to tourists. In 1977 the Israeli Minister of Tourism, for example, hired a French chef to invent an Israeli cuisine. The chef devised over 400 recipes that are now featured to tourists in Israel and are even being adopted by Israelis. Events such as this reverse the "normal" process of cultural generation and diffusion outlined in Chapter 5.

Cities attract tourists with their fine cuisine, shopping opportunities, and arts. Cities can import both performing arts festivals and art works. In 1987 Memphis, Tennessee, borrowed from Egypt an exhibition titled "Treasures of Ramses the Great." It attracted visitors who injected $83 million into the city economy, or about $125 per city resident (see Figure 13–12). The city has borrowed internationally and mounted a number of shows. Tourists coming to visit Elvis Presley's home, Graceland, inject an additional estimated $40 million per year into the city's economy.

Overall, the countries best endowed for tourism have both natural and cultural attractions, pleasant climate, good beaches, and reasonably well educated populations. Political stability is a necessity.

The Global Office Many developing countries export clerical work. They have pools of educated young people willing to work for wages lower than those in

Figure 13–12. The city of Memphis, Tennessee, has made tourism one of its key industries. The city has built this massive 32-story pyramid arena and museum to echo its Egyptian namesake. (Courtesy of Memphis Convention and Visitors Bureau.)

the rich countries, but their domestic economies cannot employ them. Large corporations need clerical assistants and computer operators, and telecommunications allows the corporations to tap these pools of labor wherever they are. U.S. companies have been moving clerical operations out of urban centers for decades. Now they can relocate those jobs around the world.

In 1989 Irish computer-science college graduates earned about one-half as much as U.S. graduates. Therefore, New York Life Insurance Company claims are processed in an Irish village connected by computer to New Jersey. The Boeing Corporation, Bechtel, McGraw-Hill, The Travelers Corporation, and other U.S. firms have located office operations in Ireland. The computerization of the catalogues of the New York Public Library, the British Museum Library, and of California's Getty Art Museum was carried out in Manila, the Philippines, by a U.S. firm based in Kansas City. American Airlines' data processing operations are not in Tulsa, where the corporation has its headquarters, but in Barbados and the Dominican Republic.

It is hard to find these jobs in international statistics because if the workers are working for a multinational manufacturing firm, they might show up categorized either as "manufacturing workers" in the manufacturing sector or as "clerks" in the service sector. Nevertheless, the relocation of these jobs can boost national economies and can also spread cultural values and practices to other countries. It is hard to predict where these footloose jobs will be located in the year 2000. Don't be surprised if your auto insurance claim is processed in India or Egypt. These countries are importing computer hardware and diligently training software professionals, and the time-distance and cost-distance to these places is, for electronic transmissions today, negligible.

Governing International Trade

The rules regulating business vary so widely from nation to nation that the terms of trade among countries require political regulation. The United Nations first proposed an International Trade Organization in 1947, and a preliminary **General Agreement on Tariffs and Trade (GATT)** was signed in that year by 23 countries. GATT provided an outline of rules, but the organization to elaborate and enforce the rules was never created. Therefore GATT, which was intended as an interim arrangement, has instead remained as the only instrument laying down internationally accepted trade rules. By 1992, some 108 countries had formally signed the agreement, and most other countries followed without formally signing. Under GATT procedures, trade disputes lead to the convening of panels of neutral specialists at GATT headquarters in Geneva. GATT has no power of enforcement beyond peer pressure, but its decisions are generally accepted. In 1992, for the first time, GATT found even some U.S. state laws—state tax laws regarding rebates to brewers—in violation of international trade laws. The state laws will have to yield to U.S. international obligations.

GATT was created to cover world trade in raw materials and manufactured goods. Since GATT's founding, average tariffs in industrial countries have fallen from 40 percent to 5 percent of the value of goods being traded, and the volume of world trade in goods has grown almost 12-fold. Trade in farm products and in services, however, was outside the GATT structure, and this trade has grown rapidly both in value and as a percentage of all world trade. By 1990, it represented about one-third of world trade. GATT also failed to provide protection for intellectual property, such as patents and copyrights.

U.S. firms claim to lose over $60 billion per year because of foreign copying of their intellectual property such as videos and pharmaceuticals without paying fees. Many nations felt that GATT should be expanded to include these items of trade.

From 1986 through 1990 a round of GATT negotiations (the eighth, known as the "Uruguay Round") met to define terms of international trade in these items. The Uruguay Round foundered in 1990 because too many countries demanded exemptions. Japan and the European countries refused to end farm subsidies; in turn, the poor countries refused to open their markets to service exports from the rich countries. During the 1990s the GATT world trading system will continue to be challenged to reach international agreement on terms of trade.

Other international organizations oversee specific aspects of international commerce, such as the IOSC already mentioned. The London-based International Accounting Standards Committee continues to work to define global corporate accounting procedures. An international uniform code will greatly simplify international business.

International Labor Organization

Since World War II, organized labor unions in different nations have come to profit from national wage standards, conditions of work legislation, social security, and unemployment insurance programs. Economic globalization jeopardizes the gains of national labor unions and of all those who benefit from national social programs. Wages are lower in the poor countries, and so global economic competition could force wage restraint or even concessions from workers in the rich countries. In the discussion of global commodity markets earlier in this chapter, we noted that transportation and communication allow the expansion of market regions within which producers must compete. The same rule applies to wage rates: Workers in the rich countries find themselves competing for jobs with workers in low-wage countries.

Worldwide labor organization has not kept pace with corporate organization. The International Labor Organization (ILO) works to improve international labor conditions. Three main international trade union confederations also exist: The International Confederation of Free Trade Unions, the World Federation of Trade Unions, and the World Confederation of Labor. None of these, however, exercises impressive power or influence.

Some observers of international trade, including many members of the U.S. Congress, have expressed concern about the conditions of labor in developing countries and also the environmental pollution that is accompanying industrialization in many countries. A powerful, centralized world trade organization could guarantee that goods in world trade were produced under internationally acceptable conditions. In the 1930s the United States abolished child labor by forbidding shipment across state lines of goods produced by underaged workers. The rich nations united could similarly forbid shipment in international commerce of goods produced under conditions of inadequate workers' rights or conditions that pollute or damage the environment. The populations of the rich countries, however, enjoy cheap consumer goods imported from the poor countries, and they are unaware of or do not seem to care about the conditions under which the goods are produced. Therefore, such restrictions do not exist.

Nevertheless, discriminatory practices within a country, whether racial, sexual, religious, or any other sort, affect its relations with other countries. Japanese social policies, for instance, have interfered with Japanese relations with the rest of the world. U.S. subsidiaries of Japanese corporations have repeatedly run afoul of U.S. antidiscrimination laws, in cases of both sex and racial discrimination. In 1990 Japan's minister of justice publicly apologized that "Japanese may not have a rich sen-

CRITICAL THINKING: *Multinational Corporate Unions*

International labor solidarity may be promoted, paradoxically, by the multinational corporations. Workers identify themselves as, for instance, Ford workers, no matter which national subsidiary of Ford employs them. A sense of worker solidarity and demands for parity of working conditions and pay arise spontaneously. International labor organization by corporate employer may play a greater role in the future.

Questions

The president of Motorola was quoted earlier in this chapter as insisting that "We must treat our employees all over the world equally." Do you think that Motorola pays its employees the same in California and in Taiwan?

What would the geography of world incomes and welfare be if Motorola and other multinationals treated and paid all of their employees everywhere the same?

What would be the impact on individual nations' national pay scales? Health care? Environmental protection legislation?

Do you think that multinationals would build factories all over the world if they had to treat workers everywhere the same?

sitivity toward racial questions" because the country has little "experience in communicating and intermingling with different nationalities." Multiplying world trade requires international understanding.

THE COLLAPSE OF EMPIRES

In the year 1900 "international affairs" was a matter of agreement among far fewer national governments than exist today. Since 1900, however, several imperial powers have granted independence to colonies they had accumulated since the fifteenth century, and the number of sovereign states has multiplied.

The following discussion will highlight a few of these individual transitions that gave birth to large numbers of new states. We will focus on a few international frameworks of allegiance among states that were previously united in empire and a few areas that remain problems of jurisdiction. The few remnants of empires can be seen in Figure 13–13.

British Empire to Commonwealth

At its peak in 1900 the British Empire covered over one-quarter of the Earth's land surface. This empire has gradually been transformed into a loose association of independent states, the Commonwealth of Nations.

The Commonwealth of Nations has grown steadily as the empire has shrunk. Most of the United Kingdom's former colonies have chosen to join, and today the Commonwealth has 50 members. Most have become republics, but Queen Elizabeth II remains head of state in 16 countries in addition to the United Kingdom, including Canada, Australia, and Jamaica. The British legal system, language, and education system, the Anglican Church, and other cultural traditions linger in each of the countries. The Commonwealth cannot form a common economic or foreign policy, but it offers a framework for consultation and cooperation to achieve common ends where they exist. As the United Kingdom turns toward its European Community partners, however, the future role of the Commonwealth is uncertain.

Britain retains a few other small colonies around the world, and some of them present special problems. Hong Kong is a 409-square-mile (1060-km^2) colony at the mouth of China's Canton River. Since Hong Kong Island was acquired in 1841, Hong Kong has developed great wealth as a transshipment point, a manufacturing center for apparel and consumer electronics, and a financial service center. Hong Kong's 5.7 million people have enjoyed civil liberties.

In 1985 Britain and China agreed that Hong Kong will revert to Chinese rule in 1997. The treaty states that Hong Kong may keep its capitalist system for 50 years, but many people doubt China's intention to honor that agreement. Tens of thousands of people and billions of dollars of capital are flooding out of the colony.

Britain retains the Falkland Islands ("Malvinas" in Spanish) against claims by Argentina, which even invaded the islands in 1982, and Gibraltar against claims to it by

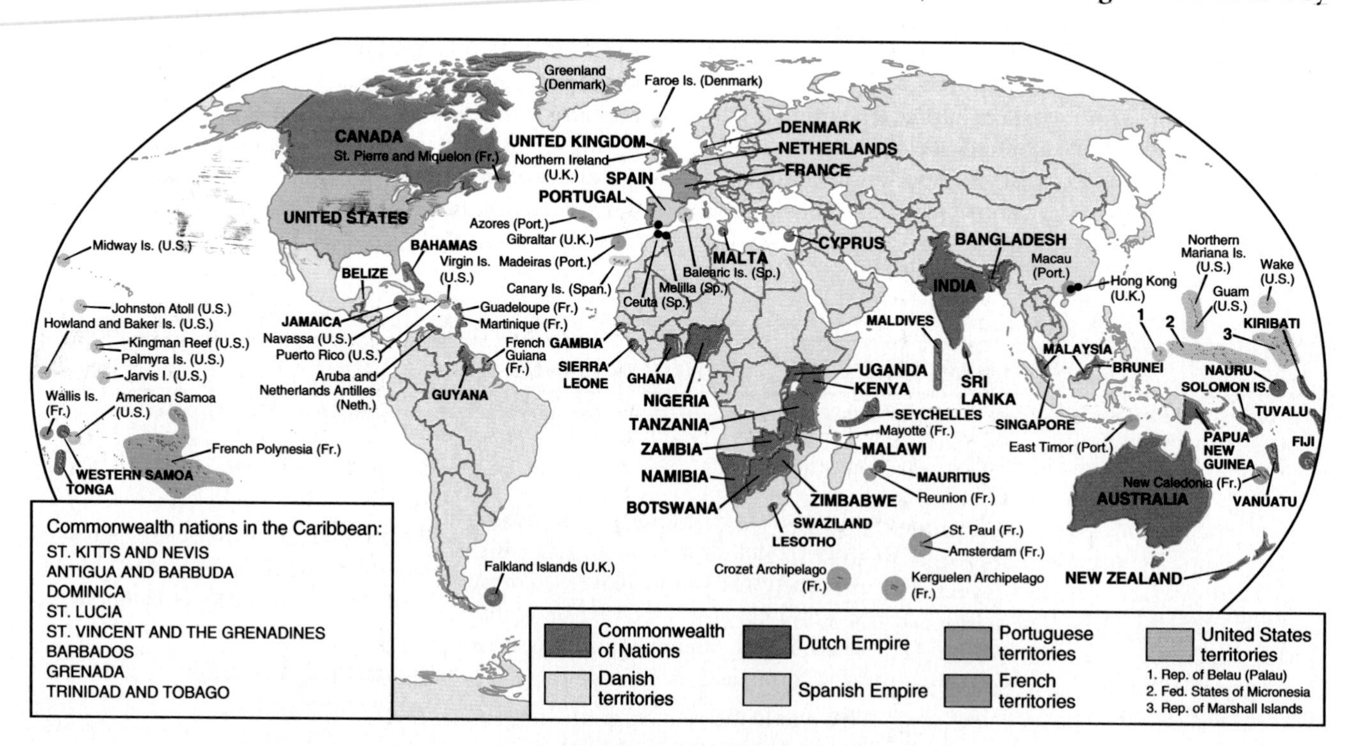

Figure 13–13. This map shows the current members of the Commonwealth of Nations, the international cooperative organization that was born out of the British Empire. Also indicated are the remaining bits of the Spanish, Portuguese, Danish, Dutch, French, and U.S. empires.

Spain. Residents of both colonies have repeatedly voted to continue their current status.

Northern Ireland's status is open to dispute. The United Kingdom gave independence to 26 of Ireland's 32 counties in 1921, but the 6 counties of Northern Ireland (Ulster) elected to remain within the United Kingdom. The Irish Republic regards the 1921 partition as provisional and remains formally committed to unification. Northern Ireland is ruled directly by the government in London, but since 1985 the government of the United Kingdom has recognized the republic's right to be consulted on Ulster affairs. Since the 1960s Ulster has been wracked by violence. The province's Roman Catholics, about 40 percent of the total population, complain of discrimination by the majority Protestants, and paramilitary groups on both sides are guilty of murder and terrorism.

The French Empire

France built up one empire in the seventeenth and eighteenth centuries but lost most of its colonies during an eighteenth-century rivalry with Britain. During the nineteenth century France built up a second empire, beginning with the occupation of Algeria in 1830 and culminating in 1919 with the assumption of rule in Syria, Togoland, and Cameroon.

The French believed that their subject peoples would mature not to independence but to full representation in the government in Paris. The four French colonies of Martinique, Guadaloupe, Réunion, and Guiana were organized as Overseas Departments of France, and each was granted representation in the French National Assembly in 1946.

France granted independence to most of its other colonies in the 1950s, but it offered various forms of continuing alliance. Still today France is the largest donor of aid to African countries and maintains the largest non-African military forces on the continent. Thirteen former colonies use an international currency, the CFA franc, whose exchange rate is stabilized against the French franc. France's future relationship with Africa is in doubt as France, like Britain, is more closely bound in the EC.

The present French Republic encompasses, in addition to mainland France and the four Overseas Departments, two Territorial Collectivities, and four Overseas Territories. France also possesses a number of islands. One of these, New Caledonia, holds 9 percent of world nickel reserves, but the other territories are generally poor and sparsely populated.

The Successor States of the Ottoman Empire

The collapse of the Ottoman Turkish Empire and the birth of a new Turkish nation in the 1920s was discussed in Chapter 5. Most of the Turkish Mideastern Empire was divided between French and British rule, from which new states eventually were granted independence.

Ever since then, the Middle East has been a geopolitical cauldron of competing loyalties. The map of today's states resulted from the interests of the colonial powers rather than the sentiments of the local people (see Figure 13–14). Israel was carved in the region to provide a homeland for the Jewish people, and bitter confrontation over this land—historical Palestine—continues. In many countries, national loyalties are still weak. In much of the Mideast, for example, Arabs bestow their loyalty on units that are much smaller than states (clans, tribes, or families) or on ideas that are much bigger. One such idea is *pan-Arabism*, the notion that some bond exists among all Arabic-speaking peoples that is more legitimate than the modern Arabic states. Even this idea was to some degree a colonial creation. The British exploited it to enlist the Arabs against the Turks in World War I, but it remains the formal creed of the ruling Baath parties in both Iraq and Syria.

The Mideast is still plagued by awkward borders, rulers whose power is based on foreign interests rather than popular support, religious animosities, and political and economic intervention by international oil companies and banks. Several wars have troubled the region since World War II. Israel has been able to defend its independence and also has achieved a genuine economic miracle in making its desert land bloom and in building an economy. To accomplish this, Israel has relied on generous financial assistance from the U.S. government (in 1989 almost $500 per Israeli—22 percent of U.S. total foreign aid assistance) plus billions of additional dollars from U.S. private citizens. Israel seeks security, but that cannot be guaranteed without Arab recognition of Israel's right to exist.

In Lebanon a civil war—inspired partly by religious animosities and partly by the interference of foreign interests—has raged since the early 1980s. Iraq fought against Iran from 1980 to 1988. In 1990 Iraq occupied Kuwait, and in January 1991 the United States and 31 allies acting under UN auspices launched a war to drive Iraq out of Kuwait. That goal was quickly achieved.

Today many Arab regimes face crises of legitimacy. Demands are arising throughout the region for democratization. Western allies who had hoped to build international security structures have to consider that the countries used as building blocks have not completed their own task of nation building. In addition, because of the residue of anti-Westernism, their governments may be weakened by alliances designed to protect them.

The Empire of the United States

The United States organized most of its territories and admitted them to the Union fairly quickly. With the granting of statehood to Alaska and Hawaii in 1959, the United States incorporated even overseas territories.

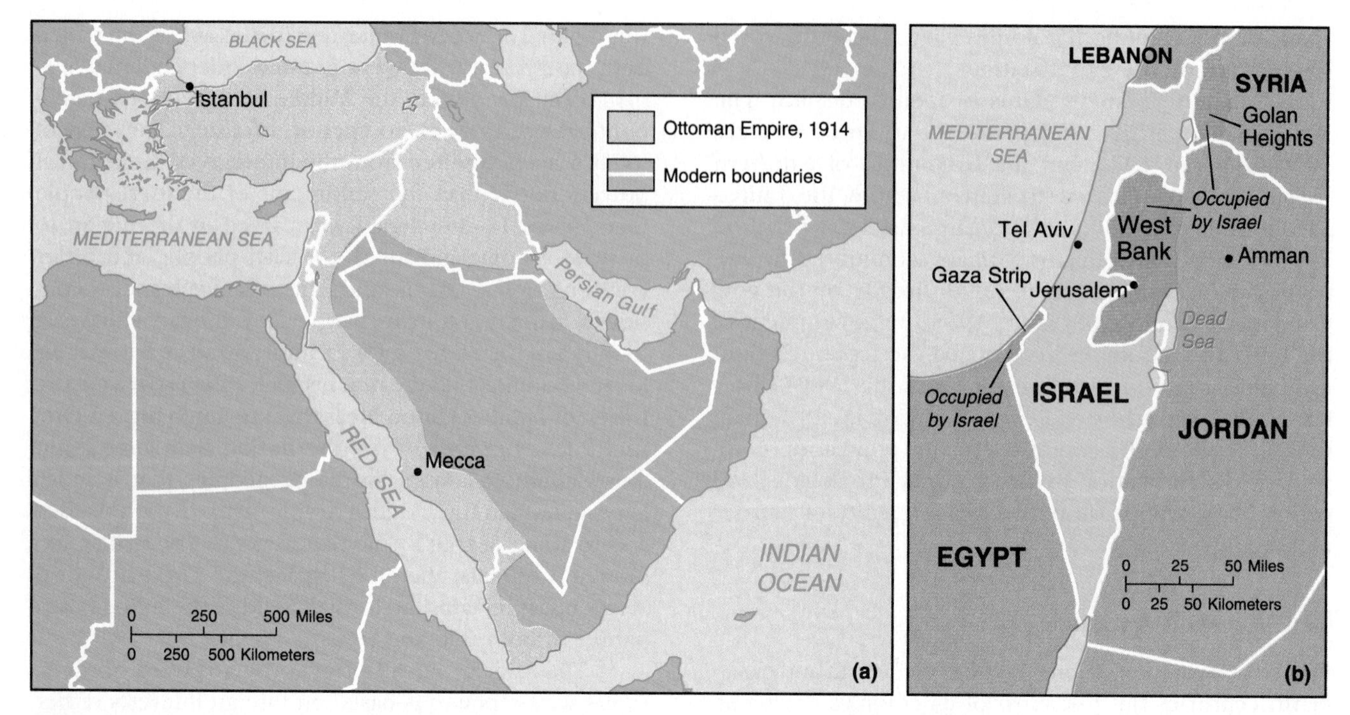

Figure 13–14. (a) In 1914 the Ottoman Turkish Empire still ruled much of the Mideast, but after World War I the British and French subdivided the territory. Several Mideastern borders have been disputed for 75 years. (b) Israel was carved out of the region in 1948, and today it occupies disputed territory seized in war. The United States does not recognize Israel's claim to Jerusalem as its capital, and the U.S. embassy remains in Tel Aviv.

Former Colonies The United States took Cuba from Spain in 1898 but granted full independence in 1934. Today it retains only a naval base in the island's Guantanamo Bay.

The United States also took the Philippines from Spain in 1898. It granted the islands independence in 1946, but it retained two enormous military bases that provided a full 5 percent of that country's GNP. By the late 1980s, however, these bases were becoming targets of anti-American expressions of nationalism. In 1990 a volcanic eruption rendered one base useless, and the United States and the Philippines could not agree on terms to continue the lease on the other. Therefore, the United States is leaving after a presence of almost 100 years.

Panama and Liberia are two independent countries that have been politically tied to the United States so closely for so long that their genuine independence may be questioned. The United States provoked Panama's uprising for independence from Colombia in 1903, and then, by prearrangement, leased the Canal Zone from the new country. A 1979 treaty allows Panama to take responsibility for the canal in 2000, but until then the United States operates it through an independent federal agency supervised by five Americans and four Panamanians. All nine are appointed by the president of the United States.

In 1989 the United States invaded Panama. The ostensible purpose of the invasion was to interdict drug traffic to the United States and to remove from power and bring to the United States for trial a military dictator, Manuel Noriega, as an international drug trafficker. The United States continues to occupy the country.

Some 14 percent of the world's merchant ships are registered in Panama. This would seem to be an astonishingly high percentage for such a small, poor country. Many of these ships, however, are owned by U.S. corporations, which register the ships in Panama to evade U.S. taxes and regulation. The Panamanian flag is known as a *flag of convenience.* Panama's rank among states holding key world patents (see Table 11–3) can also be explained only by the country's quasi-colonial relationship with the United States.

The United States maintains a "special relationship" (an official State Department term) with Liberia, which was founded as a haven for freed U.S. slaves and received independence in 1847. Inequities in wealth and political power fire antagonism between the dominant Americo Liberians (in 1990 only 5 percent of the total population of 2.5 million) and the indigenous Africans, and the indigenous group is itself split among ethnic groups. Civil war split the country in the 1980s, but in 1990 an international peacekeeping force of eight West African states occupied the country.

The United States maintains military facilities in Liberia, and the U.S. dollar is Liberia's official currency. The United States gives Liberia several hundred million dollars in aid each year. The Liberian flag is another flag of convenience. Liberia registers about 6 percent of the world's merchant marine fleet representing 13 percent of total world gross tonnage. The fleets of both Panama and Liberia may actually be owned by corporations from many rich countries, but in both cases U.S. ownership is probably most significant.

U.S. Colonies Today The United States still retains a modest empire of islands. Their total population is under 4 million, most of whom are citizens of the United States. The five largest colonies—Puerto Rico, the Virgin Islands, Guam, Samoa, and the Northern Marianas—have locally elected governors and legislatures. They are subject to U.S. laws, although they have no direct voice in political processes. All use U.S. currency and depend on the United States for their economic well-being.

Puerto Rico The United States took Puerto Rico from Spain in 1898, granted it territorial status in 1917, and in 1952 elevated it to the unique status of a free commonwealth. Puerto Rico enjoys internal autonomy, but Puerto Ricans are also under the jurisdiction of the federal government. They do not vote in presidential elections, and their resident commissioner in Congress can vote only in committee meetings. The United States and Puerto Rico share a common market and monetary system, but Puerto Ricans pay no federal taxes, and certain federal customs and excise duties are paid back into the island's treasury. Sometime in the 1990s Puerto Rico will hold another plebiscite to decide whether Puerto Ricans want full independence, statehood, or a continuation of commonwealth status. In the last plebiscite in 1967 the votes for those three categories were, respectively: 0.06 percent, 38.9 percent, and 60.4 percent.

Most arguments being forwarded in the debate over independence are economic. The densely populated, resource-poor island now enjoys a standard of living far above that of any other Caribbean or Latin American nation, and so it can be considered either as a rich Caribbean country or, conversely, as a poor part of the United States. Statehood would bring Puerto Rico into several federal financial assistance programs from which it is now excluded. That would increase federal welfare expenditures at least tenfold, but Puerto Ricans would have to pay federal income tax.

Statehood would also end federal laws that grant tax reductions to U.S. companies that build facilities on the island. These laws were passed to encourage industrialization, and they have succeeded. Overall, manufacturing accounts for 40 percent of Puerto Rico's economic output and about 18 percent of employment. Agricultural production accounts for only 1.5 percent of output and 4 percent of employment, far below the levels in tropical developing countries.

Puerto Rico's 3.4 million people would make it 28th among the states in population, but its per capita income ($5400 in 1988) was only two-thirds that of Mississippi, currently ranked last. Only 40 percent of Puerto Ricans over 24 have a high school degree, another statistic that would place Puerto Ricans last, behind Kentucky at 53 percent.

Puerto Ricans are Spanish-speaking, and many Puerto Ricans wish to retain basic linguistic and cultural differences. This might make it difficult for Puerto Rico to assimilate fully into U.S. life. Mainlanders have never been asked—nor probably will they be—whether they are willing to accept Puerto Ricans into full Union status. If the Puerto Ricans choose independence, however, terms of continuing financial assistance plus leases on U.S. military bases (a full 13 percent of the territory) will have to be negotiated.

The Monroe Doctrine The United States extends its military power over the entire Western hemisphere by the terms of the Monroe Doctrine of 1823. The Monroe Doctrine declared that the United States considers dangerous to its peace and safety any attempt on the part of European powers to extend their systems of government to any point in the Western Hemisphere. The 1904 Roosevelt Corollary added that "Chronic wrongdoing, or an impotence that results in a general loosening of the ties of civilized society, may . . . require intervention . . . and in the Western Hemisphere the . . . Monroe Doctrine may force the United States . . . to the exercise of an international police power."

The United States has regularly invoked this doctrine and corollary to intervene in and occupy other Western Hemisphere countries. Overall, in the first 9 decades of the twentieth century, the United States invaded and occupied an average of five Latin American countries per decade. U.S. military assistance has also supported many undemocratic Latin American governments.

GLOBAL GOVERNMENT

The Dominican Father Francisco de Vitoria (1486–1546) earned recognition as "The Father of International Law" by arguing that each people has the right to its own ruler, its "natural lord." Relations among nations would have to recognize mutual rights. Subsequent centuries have shown only token respect for this principle.

Discussions of world order are generally based on four principles: (1) state sovereignty, (2) self-determination, (3) democratic governments within states, and (4) the universal protection of human rights. These four principles, however, may be imperfect, and the four are often in conflict with one another. Two or more countries negotiate agreements on specific issues of international concern, but international society as a whole has neither the centralized government, judicial system, and police that characterize a state, nor even a consensus on what constitutes a crime.

As the number of states has multiplied in the twentieth century, the number of actual and possible conflicts among states has grown from several causes. These causes of conflict include aggressive ambitions, as with Iraq in the early 1990s; border disputes or rival claims on territory, as between Chad and Libya; and domestic crises or policies that have effects abroad, causing other states to threaten to intervene, as with Yugoslavia. World flows of refugees are an international problem. The necessity for common action in areas of common concern, such as the oceans or protecting the habitability of the Earth itself, is another pressure for global government.

The first permanent international political structure was the League of Nations. The League was founded by the Allies after World War I, but several nations, including the United States, refused to join. The League provided a forum for discussion of world problems, and it carried out many humanitarian projects. It exercised little real power, however, and was unable to resolve the disputes that eventually led to World War II.

The United Nations

In 1942, 26 states that were allied against the Axis powers joined in a Declaration by the United Nations, pledging to continue their united war effort. A meeting of Allies in San Francisco in 1945 drew up a charter for a United Nations organization to continue in existence after the war, and by the end of 1945 the new organization had 51 members. That number had risen to 166 member states by 1992.

The UN General Assembly serves as a rudimentary legislature of a world government, but it has no power of enforcement (see Figure 13–15). The power that the United Nations exercises even theoretically is vested in the 15-member Security Council. Five members of the council (the United States, China, France, Russia—originally the USSR—and the United Kingdom) hold their seats permanently, and the remaining 10 members are elected for 2-year terms by the General Assembly. This allocation of seats reflects the world allocation of power in 1945, and it may be changed in the 1990s. Germany and Japan, defeated in World War II but two of the richest countries in the world today, may win permanent membership status. The UN Secretariat, with the Secretary General at its head, handles all administrative functions.

The United Nations does not have an international police force, but in a few cases the Security Council has voted for the United Nations to field armies. UN forces fought in Korea from 1950 to 1953, and in 1990 a Security Council resolution gave the United States and its allies sanction to force Iraq to retreat from Kuwait. In a greater number of cases, the United Nations has sent observer troops to patrol world trouble spots. Examples include Yugoslavia in 1992, split by civil war, and the Mediterranean island of Cyprus, which is split by feuding Greeks and Turks.

Figure 13–15. The United Nations found a permanent home on the east side of Manhattan Island, where its headquarters buildings were designed by an international committee of architects, including Le Corbusier. The high slab of the Secretariat dominates the group, with the domed General Assembly to the north. The land legally is not part of the United States of America. (Courtesy of Rafael Macia/ Photo Researchers.)

The International Court of Justice in the Hague is another branch of the United Nations that seems like a government. The litigant states, however, must choose to come before it, and the court has no power of enforcement.

UN Special Agencies The United Nations has more than 40 specialized agencies. Some of these agencies facilitate communications among member states: the Universal Postal Union, the International Civil Aviation Organization, the World Meteorological Organization, and the Intergovernmental Maritime Consultative Organization. In 1992 the UN International Telecommunications Union reached new agreements on the allocation of frequencies in the radio spectrum, thus allowing further international development of cellular telephones, satellite computer transmissions, and digitial radio broadcasting. Other UN agencies encourage international cooperation toward specific humanitarian goals: the Food and Agriculture Organization; the World Tourist Organization; the International Labor Organization; the United Nations Educational, Scientific, and Cultural Organization; the International Atomic Energy Agency; the International Bank for Reconstruction and Development;

and the International Monetary Fund. The UN Conference on Trade and Development (UNCTAD) works to assist the development and trade of the poor countries. In 1992, for example, UNCTAD developed a standardized computer customs collection system, which was soon in use in over 50 countries and facilitated international trade.

Is National Sovereignty Inviolable? The idea of national sovereignty means that international borders are inviolable; states may do whatever they wish to their own people within their own borders. No matter how monstrous any regime might be, no matter how much it persecutes its own people or oppresses minorities, no outside state nor any international agency has the right to interfere. This is true even when internal persecution triggers international flows of refugees. There have been cases of widespread famine tolerated by the starving country's government (Ethiopia in the 1980s) or caused by a civil war (Somalia). In these cases the government's resistance to outside assistance, or chaos in the country, has prevented international rescue efforts from succeeding or even taking place (as in Somalia).

Events surrounding the 1991 UN war against Iraq, however, may represent the closest the United Nations has ever come to breaching the principle of state sovereignty. The United States wanted to replace the Iraqi ruler, Saddam Hussein, but the internationally agreed-on purpose of the war was only to force Iraq to retreat from Kuwait. Iraq was defeated and retreated, but its government remained in power. Then the world stood by and watched as the government of Iraq massacred thousands of Kurds and Shiites who had rebelled during the war.

Speaking before the UN General Assembly on September 23, 1991, U.S. president Bush insisted that sanctions against Iraq remain in force as long as Saddam Hussein "remains in power." Referring to "nationalist passions," the president said that no one can "promise that today's borders will remain fixed." "Despots ignore," he said, "the heartening fact that the rest of the world is embarked on a new age of liberty," and he called upon the United Nations to take up "the important business of promoting values. . . ." This could be interpreted as a challenge for the United Nations to interfere in the internal affairs of member states. Many representatives of despotic regimes in the Great Hall that day must have felt uncomfortable.

JURISDICTION OVER THE EARTH'S OPEN SPACES

Among the most important issues facing international agreement are those concerning the rights of different nations in the Earth's open spaces. These include the Arctic and Antarctica and the world's seas.

The Arctic and Antarctica

Eight countries have territory north of the Arctic Circle: the United States, Canada, Denmark, Finland, Iceland, Norway, Sweden, and Russia. In 1991 these countries signed a pact pledging cooperation in monitoring arctic pollution and protecting the region's plant and animal life. In 1992, however, Russia admitted that for years the Soviet Navy dumped radioactive waste in Arctic waters. International scientists are investigating the results of this action.

Two countries dispute their claims to arctic territory. Canada claims dominium over the uninhabited Arctic Islands north of Canada all the way to the North Pole. The United States has challenged Canada's claims to these lands and has attempted to create a passage for shipping from Alaska's North Slope across to the U.S. East Coast. The issue remains in dispute.

Antarctica has never had any permanent inhabitants, but seven states claim overlapping sovereignty. Australia, New Zealand, Chile, and Argentina, the world's most southerly states, claim sovereignty on the basis of proximity. The claims of Britain, Norway, and France are based on explorations (see Figure 13–16). Neither the United States nor Russia, which support most of the scientific research there, makes any territorial claims, nor

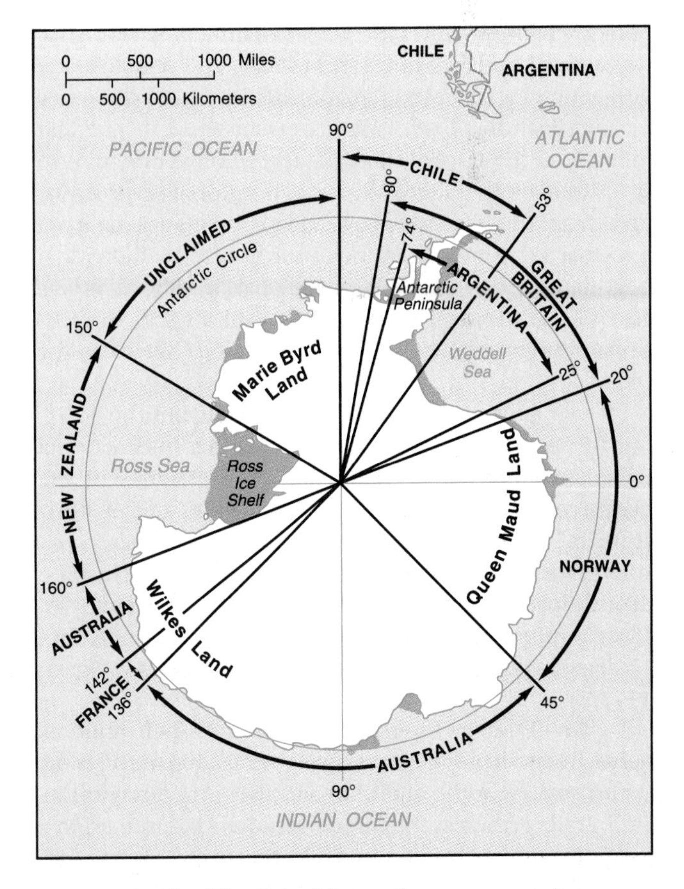

Figure 13–16. The United States does not recognize any of these overlapping claims to Antarctica.

do they recognize other nations' claims. Thirty-nine states have subscribed to an Antarctic Treaty of 1959, which resolves that "Antarctica shall continue forever to be used exclusively for peaceful purposes and shall not become the scene or object of international discord." The treaty prohibits military installations, but it opens the continent to all nations for scientific research and study. The treaty also prohibits signatories from harming the antarctic environment.

Despite the terms of this treaty, however, some nations, led by Malaysia, demand that exploration should be undertaken for antarctic resources and that any resources found should be exploited for the good of all nations. Other nations, led by France and Australia, insist that Antarctica should be set aside as a wilderness preserve.

Territorial Waters

The nation-states have divided up all the land on Earth (except Antarctica), but the question of how far out to sea the territorial claim of a country can reach, is regulated by international agreements. The Dutch jurist Hugo Grotius (1583–1645) was the first person to argue that the world's seas are open space, *mare liberum*, that no political unit can claim. Each coastal state, however, claimed coastal waters to defend itself. A later Dutch jurist, Cornelius van Bynkershoek, accepted this in his book *De Domina Maris* (1702). The limit of sovereignty was set at 3 nautical miles from shore, and this distance was generally accepted for over 200 years. All distances in adjudication of sea rights are measured in nautical miles, an international standard agreed on as one minute of latitude—6076 feet (1852 m). Territorial sea is measured from the low tide mark, and problems arising from irregular coastlines or islands are resolved by individual negotiations. The seaward extension of land borders was also left to bilateral negotiations until a 1958 International Conference on the Law of the Sea set rules for this.

The United States accepted a 3-mile limit in 1793, but in 1945 President Truman broke that international covenant and claimed sole right to the riches of the continental shelf up to 200 nautical miles out. A **continental shelf** is an area of relatively shallow water that surrounds most continents before the continental slope drops more sharply to the deep-sea floor. The 1958 Sea Law Conference agreed on a water depth of 656 feet (200 m) as the definition of a continental shelf (see Figure 13–17).

The Truman Proclamation claimed shelf mineral rights, but it did not claim control over fishing or shipping in the seas over the shelf beyond the 3-mile territorial limit. It nevertheless triggered extended claims to fishing rights by other nations. In 1976 the United States extended its own claims to exclusive fishing rights up to 200 nautical miles out. About 90 percent of the world's marine fish harvest is caught within 200 miles of the coasts (see Table 13–4).

TABLE 13–4 ***The World's Leading Fishing Nations, 1990***

Country	Percent of world marine catch
Japan	14.6
USSR	12.7
United States	6.4
Chile	6.4
Peru	6.0
China	5.9
South Korea	3.6
Thailand	2.7
Norway	2.5
Indonesia	2.4
Denmark	2.2
India	2.1

The UN Convention on the Law of the Sea In 1982 the United Nations proposed a new treaty on the law of the sea, which had been signed by 157 countries by 1992. The 1982 treaty authorizes each coastal state to claim a 200-mile **exclusive economic zone (EEZ),** in which it controls both mining and fishing rights. Therefore, possession of even a small island in the ocean grants a zone of 126,000 square miles of sea around that island (see Figure 13–18). This might help explain why France retains many small island colonies. As Table 13–5 indicates, France has the world's largest EEZ.

Still another clause of the 1982 treaty guarantees ships *innocent passage* through the waters of one state on the way to another. The extension of countries' territorial waters, however, means that countries have claimed many of the world's narrow waterways. They

TABLE 13–5 ***The World's Largest Exclusive Economic Zones***

Country	Area (1000 km²)	Percent of world total
France	12003	10.4
United States	9711	8.4
New Zealand	6663	5.8
Indonesia	5409	4.7
Australia	4496	3.9
CIS states	4490	3.9
Japan	3861	3.3
Kiribati	3550	3.1
Brazil	3168	2.7
Canada	2939	2.5
Mexico	2851	2.5
Papua New Guinea	2367	2.0

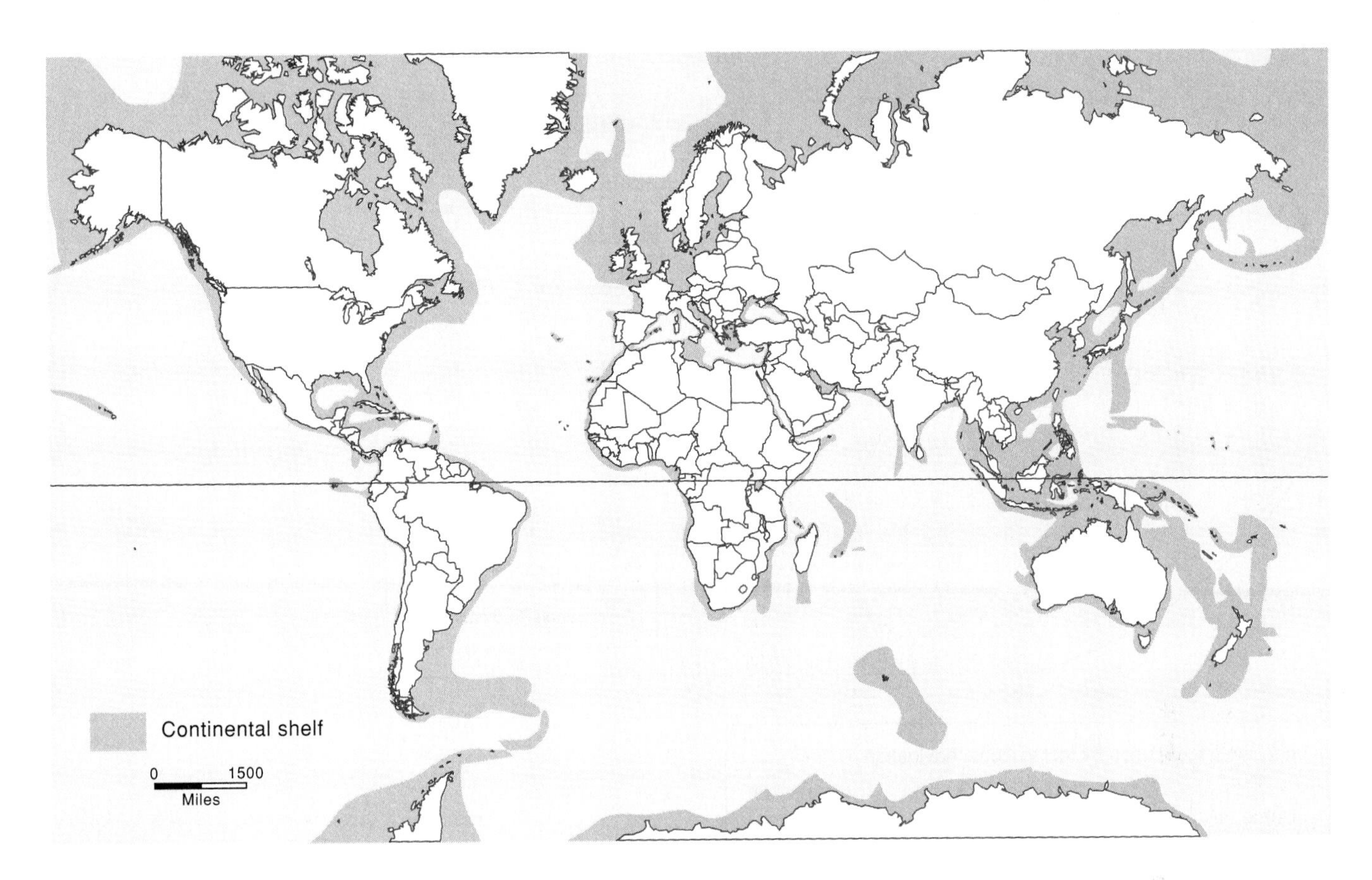

Figure 13–17. The various continental shelves extend out as little as a few hundred yards or as far as over 600 miles.

remain open in times of peace, but they can be closed in time of war (see Figure 13–19).

The United States has refused to sign the 1982 Convention in protest against provisions that call for internationalization of mining seabed resources. It also has boycotted a commission attempting to compromise the disputed provisions. Several of the treaty's other provisions nevertheless serve U.S. interests. For example, one provision allows nations to declare a 12-mile territorial limit. Therefore, in 1983, U.S. president Reagan announced that the United States would regard all but the seabed provisions as law, even though the country had not signed the treaty. In accordance with the treaty, on December 28, 1988, the United States extended its territorial limit to 12 miles. Given the length of the U.S. coastline, this seaward extension enlarged the nation's territory by 185,000 square miles.

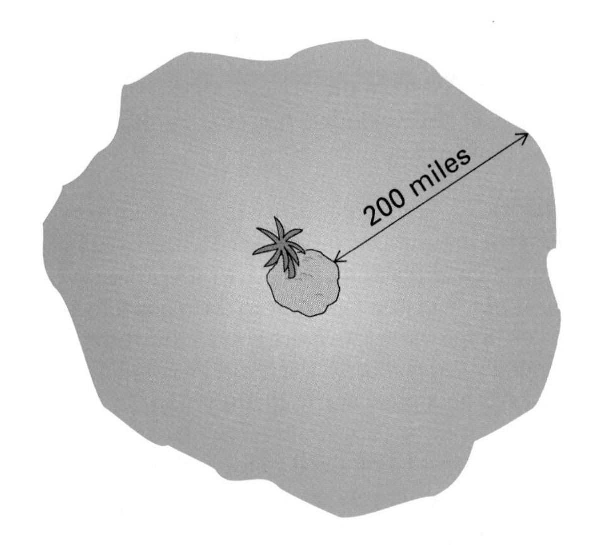

Figure 13–18. Possession of even one small island allows a nation to claim sea resources located within 200 miles in every direction. This amounts to a total area of 126,000 square miles.

Landlocked States Some 40 states are landlocked and without seacoasts. Several other states are not totally landlocked, but their own coasts are unsuitable for port development, and so they have to rely on neighbors' ports. Landlocked states must secure the right to use the high seas (waters beyond territorial waters), the right of innocent passage through the territorial waters of coastal states, port facilities along suitable coasts, and transit facilities from the port to their own territory.

Landlocked or partially landlocked states may gain access to the sea in one of three ways. First, any navigable

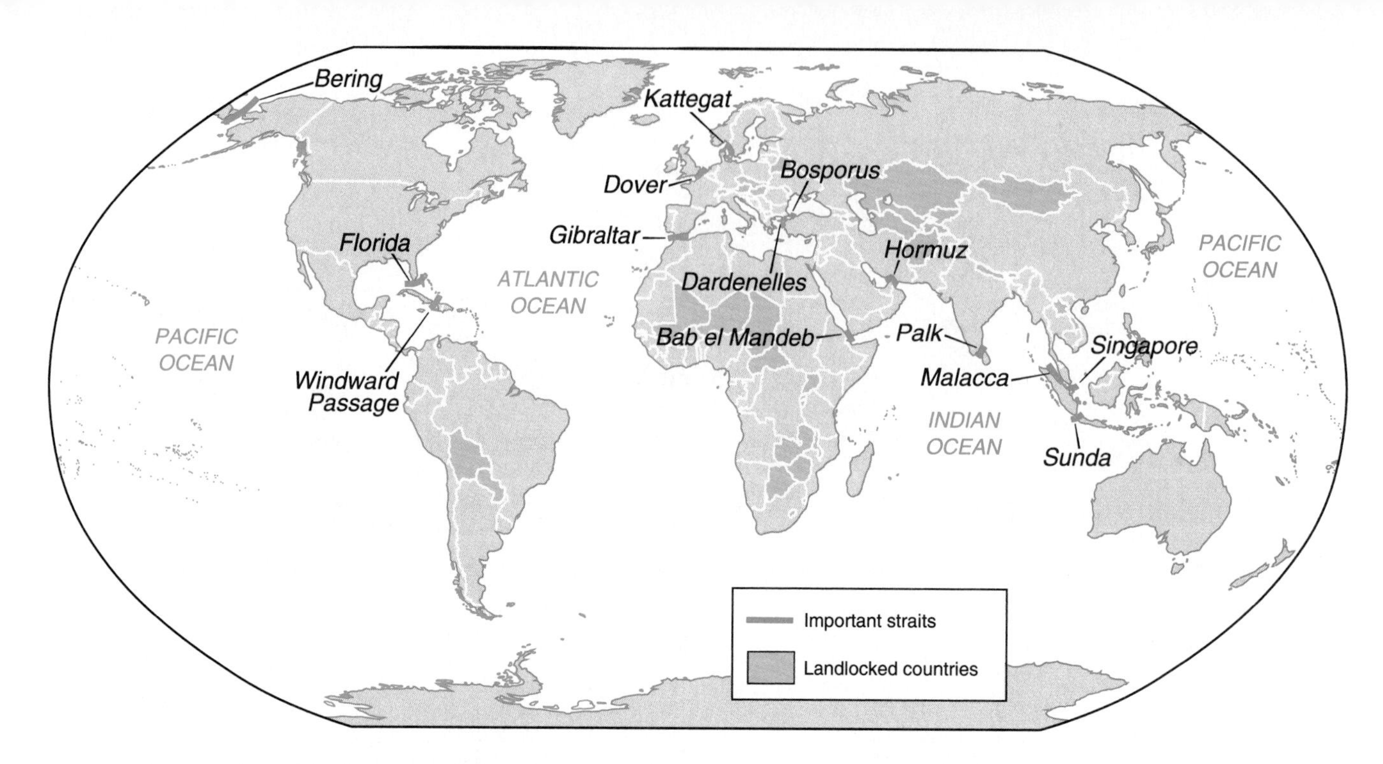

Important straits used for international navigation

Name	Sovereignty	Least width (nautical mi.)
Windward Passage	Cuba, Haiti	46
Florida	U.S.A., Bahamas, Cuba	43
Kattegat	Denmark, Sweden	12
Dover	France, United Kingdom	18
Gibraltar	Spain, Morocco	8
Dardenelles	Turkey	0.6
Bosporus	Turkey	0.45
Bab el Mandeb	Djibouti, Yemen	9
Hormuz	Iran, Oman	26
Malacca	Indonesia, Malaysia	8
Palk	India, Sri Lanka	3
Singapore	Singapore, Indonesia, Malaysia	2
Sunda	Indonesia	4
Bering	U.S.A., Russia	19

Figure 13–19. This map shows the world's landlocked states and some of the world's most important narrow sea lanes. As states have extended their seaward claims to jurisdiction, many of these narrow waterways have been claimed by the adjacent states.

river that reaches to the sea may be declared open to the navigation of all states. Second, a landlocked state may obtain a corridor of land reaching either to the sea or to a navigable river. Third, a landlocked state may be granted port facilities at a specific port plus freedom of transit along a route to that port.

International Rivers Freedom of navigation on rivers that flow through several countries was first proclaimed by France in 1792 and applied to the Scheldt River. The idea was extended to the Rhine in 1816, and the Commission for the Navigation of the Rhine still exists. International commissions regulate navigation on many other rivers, and often these same commissions guard against pollution and regulate the drawing of irrigation waters from the river.

The St. Lawrence Seaway, completed jointly by the United States and Canada in 1959, is an example of international cooperation on an international waterway (see Figure 13–20). The project not only allows seagoing ships to proceed from the Atlantic Ocean into the Great Lakes, but it also included the construction of hydroelectric power plants.

Corridors Several countries have **corridors**—long, thin extensions out to seaports (see Figure 13–21). Some of these, such as Zaire's corridor to the Atlantic Ocean and Colombia's Letitia Corridor to the navigable Amazon River, are important transport routes, but others, such as Namibia's Caprivi Strip to the international Zambezi River, serve no significant traffic function.

Transit Most landlocked states rely on transit rights across another country's territory. Coastal states have signed international conventions promising to assist the movement of goods across their territories from landlocked states without levying discriminatory tolls, taxes, or freight charges. Bolivia, for example, lost access to the sea when it surrendered territory to Chile after a 1904 war, but the Chileans agreed to build a railroad

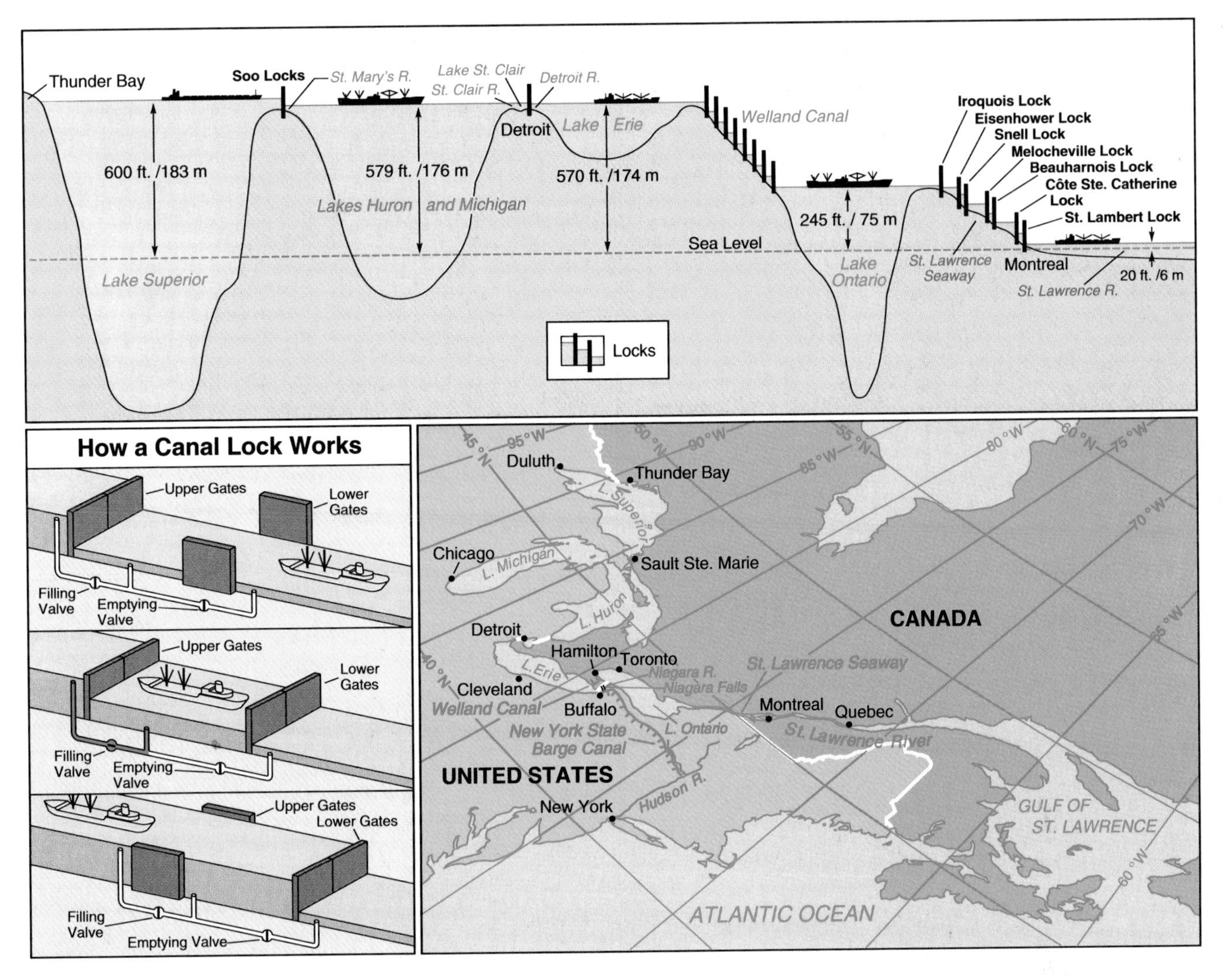

Figure 13–20. The St. Lawrence Seaway allows ocean vessels to reach as far into the continental interior as Duluth, Minnesota.

connecting La Paz to the Chilean port of Arica and to guarantee Bolivia free transit. Argentina grants Bolivia a free zone at the city of Rosario on the Paraná River.

Transit rights can nevertheless be denied during international disputes. India, for example, closed its border with Nepal from March 1989 until June 1990. This caused economic hardship for Nepal, but it will also cause long-term problems for India. When the Nepalese were unable to obtain petroleum, they cut a much higher number of trees in the Himalayan foothills for firewood. This will inevitably trigger floods, destruction, and death downstream in India, and also in neighboring Bangladesh, which was not party to this conflict (see Figure 13–22).

Coastal states can actually profit by granting transit rights. The coastal state's railroads have a captive customer. Coastal states' ports gain extra business and opportunity to serve as break-of-bulk points for processing imported or exported raw materials (see Chapter 9). The fact that a landlocked state misses these opportunities can hinder its economic development. Landlocked states also lose out in the apportionment of the resources of the continental shelves, exclusive economic zones, and fishing opportunities.

High Seas

All nations agree that beyond territorial waters are **high seas,** where all nations should enjoy equal rights. Several problems, however, threaten any area where rights are not specifically assigned. These include depletion of resources (such as overfishing) and pollution.

International conventions regulate each of these potential abuses of the high seas. The 1972 London Dumping Convention, for example, signed by 66 countries, originally banned the dumping of radioactive and other "highly dangerous" wastes at sea, but it has been extended to ban the dumping of all forms of industrial wastes by 1995. A 1982 amendment to the 1946 International Convention for the Regulation of Whaling bans all commercial whaling. In general, however, the en-

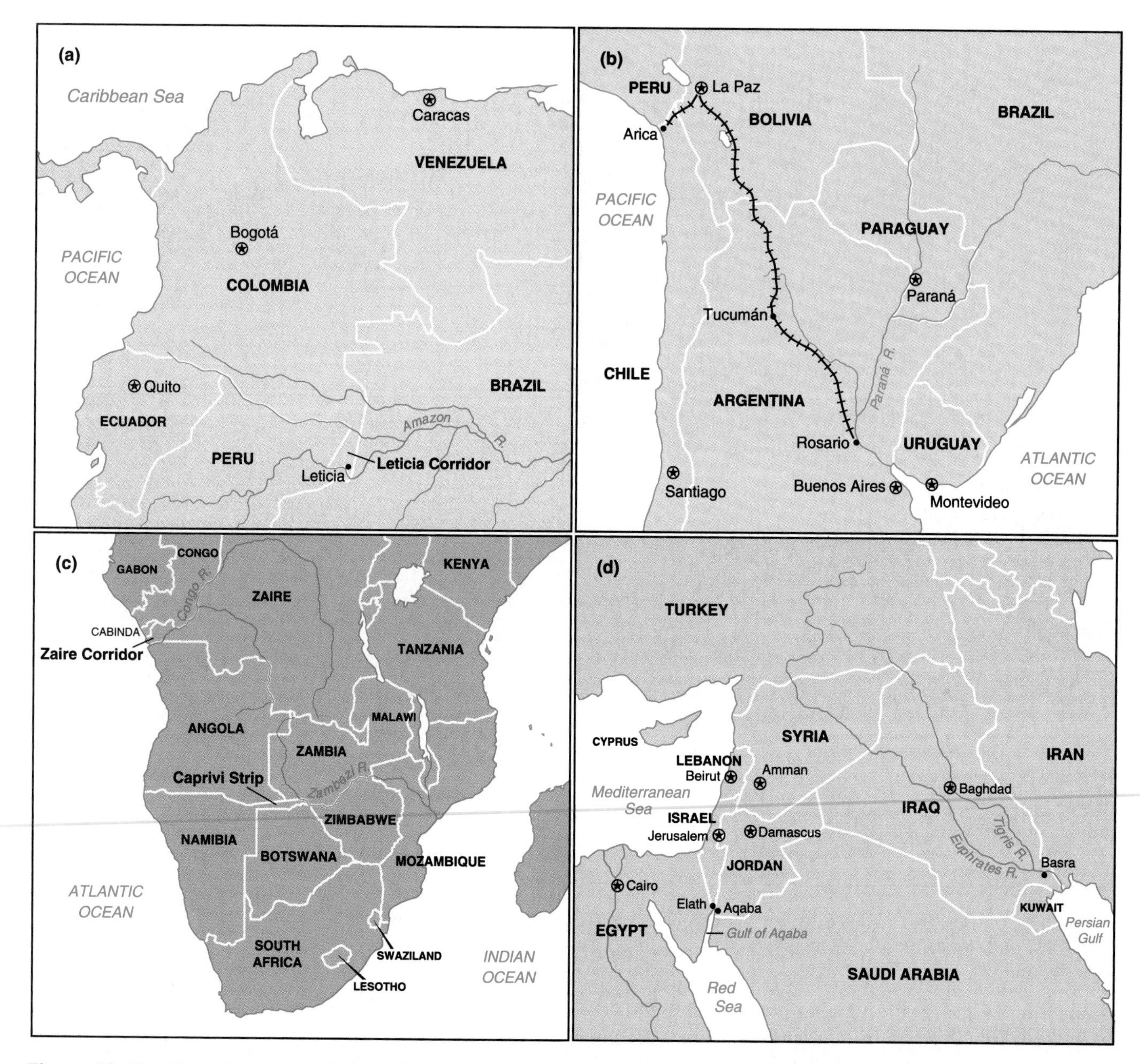

Figure 13–21. Several countries that would otherwise be landlocked have narrow corridors of territory reaching to provide access to the sea or to a navigable river.

forcement of international agreements such as these is uncertain, and their effect is limited.

Airspace

The question of how far up a nation's sovereignty extends is as complex as the question of how far out to sea it extends. Ships enjoy innocent passage through territorial waters, but airplanes have never been granted innocent passage to fly over countries. Airlines must negotiate that right, and states commonly prescribe narrow air corridors, the altitudes at which aircraft must fly, and even the hours of the day the passages are open. In March 1992, however, 25 countries, including the member countries of NATO, 4 former Soviet republics, and several Eastern European countries, signed a new open-skies treaty. This treaty allows these countries to conduct aerial surveillance of one another at short notice. More countries may sign this treaty.

Nobody can agree on the altitude at which national airspace ends. One possible definition of national airspace would limit it to the lowest altitude at which artificial unpowered satellites can be put into orbit at least once around the Earth. That ranges between 70 and 100 miles (113 and 161 km).

No country wants to be the first to militarize space. In 1990, however, U.S. vice president Dan Quayle, chairman of the National Space Council, proclaimed the mil-

Figure 13–22. The inevitable result of the deforestation in Nepal (a) is death and destruction downstream in Bangladesh (b), where the workers are frantically building dikes against the flooding rivers. Bangladesh was not a party to the dispute between Nepal and India that triggered much Nepalese deforestation, and yet its people will suffer from it. (Courtesy of Agency for International Development/(a) Delores Weiss.)

itarization of space almost inevitable: "I think eventually we will probably see agreements to establish self-defense zones in space that will actually divide up space. . . . If others come within our zones, then in fact they would be infringing on our territory."

Countries have so far refrained from extending territorial claims to the moon or planets, but several have insisted that they reserve the right to do so (see Figure 13–23). Our study of geography, therefore, is thus far restricted to the Earth.

Figure 13–23. When U.S. astronaut Neil A. Armstrong first walked on the moon on July 20, 1969, he quickly planted a U.S. flag. The United States denies that the gesture was intended as a claim. (Courtesy of NASA.)

SUMMARY

Many new states have recently been formed, but their independence of action has been undermined by international links. Tensions among economic, cultural, and political organizing activities generate much of today's international news.

International trade has multiplied faster than total world production. At first capital began to be invested internationally, and a world market for primary products, foods, and minerals evolved. Then U.S. industrialists took the lead in economic internationalism. Today multinationals are attempting global sales strategies.

The flow of FDI in the 1980s increased and redistributed the world's productive capacity. The United States is the largest source of FDI and also the largest recipient. The Triad nations are allocating a decreasing share of FDI to the poor countries and instead are increasingly investing in one another. The countries that attract most investment have chosen export-led economic growth. Each Triad member dominates FDI in a cluster of countries. Countries must compete to attract investment.

It is difficult to tell which corporations are corporate citizens of which countries, but the United States argues that any company that employs and trains U.S. workers and adds value in the United States is a U.S. corporation.

Tertiary-sector international trade is growing. The United States is the world's leading exporter of services. One global capital market has formed, putting financial issues among issues of international trade. Citizens of the rich countries account for the largest tourist expenditures, and the most popular destinations are rich countries, but tourist income is particularly important to the poor countries. Tourism has drawbacks for the host country, but almost all countries are promoting it. The tourist potential of any spot depends on its accessibility, accommodations, and attractions. Ecotourism provides the money needed to save many countries' natural environments and rare animals. The degree of cultural diffusion generated by tourism depends partly on the character of the tourists.

GATT defines international trade rules, and it is being expanded to cover trade in agricultural products and in services, and to protect intellectual property. Worldwide labor organization has not kept pace with corporate organization, but labor organization by corporate employer may play a future role.

The British Empire has been transformed into the Commonwealth of Nations, but Britain retains a few other small colonies, as does France. The Middle East is a geopolitical cauldron of competing loyalties. The United States retains a modest empire of islands. Puerto Rico will vote for independence, statehood, or a continuation of commonwealth status.

The number of actual and possible conflicts among states has grown. The UN General Assembly serves as a rudimentary world legislature, but power is vested in the Security Council. The United Nations has many specialized agencies.

The rights of different countries concerning the Earth's open spaces, including the Arctic, Antarctica, and the seas, are negotiated. The 1982 UN Convention on the Law of the Sea authorizes each coastal state to claim a 200-mile exclusive economic zone (EEZ), allows nations to declare a 12-mile territorial limit, and guarantees ships innocent passage. Landlocked states must secure access to the seas and a port facility. Beyond territorial waters are high seas. National airspace might be limited to the lowest altitude at which artificial unpowered satellites can be put into orbit at least once around the Earth.

KEY TERMS

globalization (p. 417)
portfolio holdings (p. 417)
foreign direct investment (FDI) (p. 417)
multinational or transnational corporation (p. 418)
ecotourism (p. 428)
General Agreement on Tariffs and Trade (GATT) (p. 430)
continental shelf (p. 438)
exclusive economic zone (EEZ) (p. 438)
corridor (p. 440)
high seas (p. 441)

QUESTIONS FOR INVESTIGATION AND DISCUSSION

1. Are any products exported from factories in your town? Where do the products go? Who owns the factories? Which stock exchanges trade that stock? If the factories are owned by a large corporation, where else does that corporation have factories making the same product, and which other markets does that other factory serve? If the company that owns that factory threatens to close it and shift production to another factory elsewhere, what steps do you think your local government or workers should take to keep the factory open?

2. Does your region export any agricultural products? If so, what? Where? Where are they processed or packaged?

3. Do any service professionals or service corporations in your town export their services?

4. What percentage of all goods produced in your town is exported? What percentage of your local labor force is dependent on exports? Have these percentages been rising or falling?

5. What percentage of goods in local stores is produced in the United States? Examine in detail a few specific categories of goods: clothing, consumer electronics, or toys, for example.

6. When did the heads of state of the Commonwealth of Nations meet most recently? Which issues did they discuss?

7. Could you recommend a weighted system of voting in the UN General Assembly—by population, wealth, or territory?

8. Investigate a few of the projects of the specialized agencies of the United Nations, such as UNICEF, the FAO, or WHO.

ADDITIONAL READINGS

Ashworth, Gregory, and Brian Goodall, eds. *Marketing Tourism Places.* New York: Routledge, 1990.

Blake, Gerald, ed. *Maritime Boundaries and Ocean Resources.* Totowa, N.J.: Barnes & Noble Books, 1988.

Boo, Elizabeth. *Ecotourism: The Potentials and Pitfalls.* Washington, D.C.: World Wildlife Fund, 1990. 2 vols.

Hamilton, John Maxwell. *Main Street America and the Third World.* Cabin John, Md.: Seven Locks Press, 1988.

Kirdar, Üner, ed. *Change: Threat or Opportunity for Human Progress?* New York: UN Development Programme, 1992. 5 vols.: *Political Change; Ecological Change; Economic Change; Social Change; Market Change.*

Knox, Paul, and John Agnew. *The Geography of the World Economy.* London: Edward Arnold, 1989.

Prescott, John Robert Victor. *Maritime Political Boundaries of the World.* New York: Methuen, 1986.

Ross, Robert J. S., and Kent C. Trachte. *Global Capitalism: The New Leviathan.* Albany: State University of New York Press, 1990.

Taylor, M., and N. Thrift, eds. *The Geography of Multinationals.* London: Croom-Helm, 1982.

United Nations. *The Law of the Sea.* New York, 1983.

UN Conference on Trade and Development. *Handbook of International Trade and Development Statistics.* New York, annually.

Protecting the Global Environment

Political community was defined in Chapter 10 as a willingness to join together and form a government to solve common problems. Chapter 13 listed a number of the problems and challenges facing all humankind that are triggering increasing global political community and cooperation. These include conduct among states; jurisdiction and regulation of use of the Earth's open spaces, the high seas, and Antarctica; and the regulation of the evolving global economy. Protection of the global environment is another challenge to all humankind, and it is probably the one most likely to require international collaboration. Unlike the competitive national goals of building economic or military might, this goal is positive and inclusive, and it cannot be achieved by any one country acting alone. Environmental matters threaten the internal political stability of some countries, and they are a source of friction among countries.

International bilateral cooperation was pioneered when the United States and Canada first allowed a citizen of one country who is affected by environmentally harmful activity in another country to sue in that country to stop or modify that action and to collect damages. Such bilateral cooperation has continued. The United States has acted to reduce emissions from Midwest factories and power plants that have polluted Canada in the form of acid rain—the subject of a focus box in Chapter 2. Air pollution recognizes no international borders. The United States and Canada have also acted jointly to clean up pollution in the Great Lakes, and in 1992 the United States and Mexico signed an agreement to cooperate in cleaning up pollution along that border as well.

Protection of the global environment is another challenge to all humankind, and it is probably the one most likely to require international collaboration.

Global cooperation to protect the environment was first proposed by U.S. president John Kennedy in an address to the UN General Assembly in September 1963. He urged international action "to protect the forest and wild game preserves now in danger of extinction; to improve the marine harvests of food from our oceans; and to prevent the contamination of air and water by industrial as well as nuclear pollution."

The United Nations sponsored a Conference on the Human Environment in Stockholm, Sweden, in 1972, which was attended by more than 100 nations representing over 90 percent of the world's population. The conference issued a 26-point *Declaration on the Human Environment*. That document recognized humankind's "solemn responsibility to protect and improve the environment for present and future generations." The document is in no way binding, but the near-universality of its acceptance marked it as a milestone in global cooperation.

That document recognized humankind's "solemn responsibility to protect and improve the environment for present and future generations."

The next year, in 1973, a Convention on International Trade in Endangered Species of Wild Flora and Fauna united 108 countries in efforts to restrict trade in endangered species or in products made from them. The first world ban on trade in any product achieved under the terms of this convention was a 1990 ban on world trade in ivory (see Figure 1).

The United Nations called a second conference on global environmental matters in Oslo, Norway, in 1988. That conference, titled "Sustainable Development," emphasized that economic growth must go hand-in-hand with preservation of natural resources. The conference encouraged the study of statistics such as the gross sustainable product, discussed in Chapter 11.

A third UN Conference on the Global Environment opened in Rio de Janeiro, Brazil, in June 1992. Brazil was a focal point of world concern about the environment because of the depletion of the rainforest there, as emphasized in the focus box on rainforest depletion in Chapter 2.

In 1987 the world's richest countries assembled in Montreal, Canada, and called for curbing global use of the ozone-destroying chlorofluorocarbons (CFCs) discussed in Chapter 2. The goal would be achieved by

Figure 1. These contraband tusks will be destroyed to discourage the continuing slaughter of rare African elephants. (Courtesy of Camera Pix/Gamma-Liaison.)

banning the importation of products made with them. This conference preceded by 3 years the world ban on ivory trade, and so it marked the first instance of internationally approved sanctions to promote an environmental goal. Meeting again in London in 1990, these countries pledged to halt production by 2000 and to assist the developing countries to do so by 2010. As new scientific evidence underlined the degree of loss of the Earth's protective ozone layer, the rich countries agreed to stop producing CFCs still sooner, by 1995.

The principal international "divide" on many issues of environmental protection is between the rich countries and the poor countries.

The principal international "divide" on many issues of environmental protection is between the rich countries and the poor countries. The rich countries advise the poor countries to develop their economies only in sustainable ways and at sustainable rates. These words of advice, however, seem to contradict both the past experience and even the contemporary behavior of the rich countries. Many of the rich countries polluted their own environments in the process of growing rich, and as their wealth has increased so has their ability to repair environmental damage (see Figure 2). Nevertheless, they are still responsible for much of the Earth-threatening

(a) (b)

Figure 2. These photographs taken (a) ten years ago and (b) recently illustrate the considerable die back in Germany's Black Forest caused by acid rain. (Courtesy of Régis/Bossu/Sygma.)

Figure 3. The smoke from an oil refinery obscures the setting sun in Torrance, California. (Courtesy of Kathleen Campbell/Gamma-Liaison.)

pollution. Their burning of fossil fuels, for example, makes the greatest contribution to global warming gases.

Many of the poor countries are skeptical of the advice from the rich countries. Many of them are in desperate need to increase per capita product, and they might seem willing to do this in any way they can. Perhaps only after having reached some minimum level of per capita product will they be able to consider the sustainability of what they have achieved. Is there a minimum level of per capita welfare below which GSP is simply irrelevant to a struggling population, but above which countries can "afford" to sustain or even repair their environments? Are the rich countries willing to help the poor countries reach this level of per capita development? What is the responsibility of the rich countries to reduce their own pollution before they admonish or advise the poor countries on their policies? (See Figure 3.) If environmental pressures result directly from population increases, is family planning, perhaps, the most important of all potential environmental protection measures? These are among the most important questions facing humankind. Watch for the evolution of a global political community to confront these issues.

ADDITIONAL READINGS

BENEDICK, RICHARD ELLIOT. *Ozone Diplomacy: New Directions in Safeguarding the Planet.* Cambridge: Harvard University Press, 1992.

BORMAN, F. HERBERT, and STEPHEN R. KELLERT, eds. *Ecology, Economics, Ethics: The Broken Circle.* New Haven: Yale University Press, 1992.

GORDON, ANITA, and DAVID SUZUKI. *It's a Matter of Survival.* Cambridge: Harvard University Press, 1992.

MANES, CHRISTOPHER. *Green Rage: Radical Environmentalism and the Unmaking of Civilization.* Boston: Little, Brown, 1992.

MATHEWS, JESSICA T., ed. *Preserving the Global Environment: The Challenge of Shared Leadership.* New York: Norton, 1991.

THE WORLDWATCH INSTITUTE. *State of the World.* New York: W. W. Norton & Co., Inc., annually since 1984.

Glossary

abrasion The chipping and grinding effect of rock fragments as they are swirled or bounced or rolled downstream by moving water.

absolute humidity A direct measure of the water vapor content of air, expressed as the weight of water vapor in a given volume of air, usually as grams of water per cubic meter of air.

absolute location The latitude and longitude of a place; its exact location on the globe.

acculturation The process of adopting some aspect of the culture of another group of people.

adiabatic cooling Cooling by expansion in rising air.

adiabatic warming Warming by contraction in descending air.

aeolian processes Processes related to wind action that are most pronounced, widespread, and effective in dry lands.

African diaspora The migration of black peoples out of Africa, either freely or in slavery. *Diaspora* is the Greek word for "scattering."

agglomeration The bringing together of people and activities in one place for greater convenience.

Agricultural Revolution The tremendous increase in productivity of agriculture that was caused by the application of science and technology. Sometimes called the **scientific revolution in agriculture.**

air mass An extensive body of air that has relatively uniform properties in the horizontal dimension and moves as an entity.

alfisol A widely distributed soil order distinguished by a subsurface clay horizon and a medium-to-generous supply of plant nutrients and water.

alluvium Any stream-deposited sedimentary material.

alphabet A system of writing in which the letters represent sounds.

angiosperms Plants that have seeds encased in some sort of protective body, such as a fruit, a nut, or a seedpod.

animism The belief that all natural objects have souls or spirits.

anticyclone A high-pressure center.

apartheid "Apartness" in Dutch. A policy of legally mandated racial segregation established in South Africa during the 1940s but crumbling in the early 1990s.

aquaculture Herding or domesticating aquatic animals and farming aquatic plants for food.

arboreal Tree-dwelling.

aridisol A soil order occupying dry environments that do not have enough water to remove soluble minerals from the soil; typified by a thin profile that is greatly lacking in organic matter and a sandy texture.

arithmetic density The number of people per unit of area.

ascent Large-scale vertical motions of the atmosphere.

at-large systems of voting Systems of representation in which there are no districts; all members of a government are chosen by the entire electorate.

atmospheric pressure The weight of the air.

autarky Totally self-sufficient economic independence.

average lapse rate The normal expectable rate of temperature decrease in the troposphere, about 3.6°F per 1000 feet (6.5°C per km).

axis (Earth's axis) The diameter line that connects the points of maximum flattening on the Earth's surface.

azimuthal projection A family of maps derived by the perspective extension of the geographic grid from a globe to a plane that is tangent to the globe at some point.

barometer An instrument that measures atmospheric pressure.

basic sector The part of a city's economy that is dedicated to exports of either goods or services.

beach An exposed deposit of loose sediment, normally composed of sand and/or gravel, and occupying the coastal transition zone between land and water.

bedrock Residual rock that has not experienced erosion.

behavioral geography The study of our perceptions of the environment and of how our perceptions influence our behaviors.

biome A large, recognizable assemblage of plants and animals in functional interaction with its environment.

biota The total complex of plant and animal life.

bloc A group of countries that presents a common policy when dealing with other countries.

boreal forest (taiga) An extensive needle-leaf forest in the subarctic regions of North America and Eurasia.

brain drain The migration of well-educated people from less developed to more developed countries.

broadleaf trees Trees that have flat and expansive leaves.

bryophytes Mosses and liverworts.

built environment Roads, buildings, and other human constructions.

buoyancy The capacity to float or rise in a fluid or gas.

capital Money that has been saved or accumulated and is available for investment.

capital-intensive Any activity in which a great deal of capital is invested in machinery relative to the number of workers.

carrying capacity The ability of a place to support people. This is usually an attempt to measure the fertility or productivity of agricultural land.

cartel An organization of producers of a good who collaborate to keep supplies low and prices high.

cartogram A map on which different regions are drawn relative to some value other than their land surface areas. *Area cartograms* are drawn so that the relative size of a place reflects its population, wealth, or some other characteristic.

caste In the Hindu religion, one of the classes of people in a hierarchy of classes.

cavern A subterranean hollow in the Earth, especially a large one.

central business district (CBD) The core of the city, where government offices, retail shops, and office buildings traditionally concentrate: "downtown."

central place theory Deductive geographic theory of Walter Christaller that determines the distribution of cities as central places across an isotropic plain.

centrifugal forces Political, cultural, or economic forces that tend to pull apart the regions or peoples of a state.

centripetal forces Political, cultural, or economic forces that tend to bind the regions and peoples of a state.

chaparral Shrubby vegetation of the mediterranean climatic region of North America.

circulation All transportation and communication of goods, people, ideas, and capital combined.

cirriform cloud A cloud that is thin and wispy, composed of ice crystals rather than water particles, and found at high elevations.

city A concentrated nonagricultural human settlement.

climate An aggregate of day-to-day weather conditions over a long period of time.

climatic climax vegetation A stable association of vegetation in equilibrium with local soil and climate conditions.

cognates Words in two different languages that reveal their common ancestry through similar appearance or sound.

cognitive behavioralism A psychological theory that asserts that people react to their environment as they perceive it. Therefore, in order to understand peoples' actions, you must understand their thoughts.

coincidental Events that occur together or at the same time. There may be a cause-and-effect relationship, but if there is none, the two events are often called "merely coincidental."

cold front The leading edge of a cool air mass actively displacing warm air.

cold war The period after World War II during which a democratic and capitalist "West" and a communist "East" confronted each other. The two blocs never engaged directly in combat, but they competed for economic growth and for influence among the new countries that gained their independence after World War II. In 1990 NATO declared the cold war to be over.

Commercial Revolution The expansion of trade and economic activities that occurred from about 1650 to 1750.

common market A customs union that also has a common system of laws creating similar conditions of economic production within all member countries.

community A group of people held together by traditional networks of rights and responsibilities.

compound A pure chemical substance composed of two or more elements whose composition is constant.

conceptual value added The value added to an item through advertising or any other method of enhancing its status that raises its price without actually transforming the item itself.

condensation Process by which water vapor is converted to liquid water.

condensation nuclei Tiny atmospheric particles of dust, smoke, and salt that serve as collection centers for water molecules.

conformality The property of a map projection that maintains proper shapes of surface features.

conic projection A family of maps in which one or more cones is set tangent to, or intersecting, a portion of the globe and the geographic grid is projected onto the cone(s).

consolidated metropolitan statistical area (CMSA) Two or more contiguous MSAs.

contagious diffusion Same as **contiguous diffusion**.

contiguous diffusion Diffusion that occurs from one place directly to a neighboring place. Also known as **contagious diffusion**.

continental climate A climate typical of the interior of a large landmass, normally characterized by large annual temperature ranges and relatively sparse precipitation.

continental drift Theory that proposes that the present continents were originally connected as one or two large landmasses that have broken up and literally drifted apart over the last several hundred million years.

continental shelf An area of relatively shallow water that surrounds most continents before the continental slope drops more sharply to the deep sea floor. The 1958 Sea Law Conference agreed on a water depth of 656 feet (200 meters) as the definition of a continental shelf.

convective lifting Uplift of parcels of air due to density differences.

convergent lifting The forced rising of air due to crowding in areas of air convergence.

conurbation An area in which a number of cities have grown and merged together.

coral reef A coralline formation that fringes continents and islands in warm-water tropical oceans.

core area The historic homeland of a certain nation that eventually became a nation-state territory for that nation.

Coriolis effect The apparent deflection of free-moving objects to the right in the Northern Hemisphere and to the left in the Southern Hemisphere, in response to the rotation of the Earth.

corridor A thin extension of land from a state to reach out to a seacoast or to a navigable waterway.

Council for Mutual Economic Assistance (COMECON) An economic bloc formed among the Soviet Union and its Eastern European satellites; formally disbanded in 1991.

crude birth rate The annual number of live births per 1000 people.

crude death rate The annual number of deaths per 1000 people.

crustal limit The total amount of a given mineral in the earth's crust.

cultural diffusion The spreading out of aspects of cultures from their hearth areas and their adoption elsewhere.

cultural ecology The study of the ways societies adapt to environments.

cultural imperialism The substitution of one set of cultural traditions for another either by force or by degrading those who fail to acculturate and by rewarding those who do.

cultural landscape A landscape that has been modified by human action.

cultural subnationalism The condition in a country when the entire population is not bound by a shared sense of nationalism but is split by several local primary allegiances.

culture Everything about the way a people live.

culture realm The entire region throughout which a culture prevails.

cumuliform cloud A cloud that is massive and rounded, usually with a flat base and limited horizontal extent, but often billowing upward to great heights.

current A large amount of a gas or liquid moving in a certain direction.

customs union A group of countries that have no customs among themselves and also share a common external tariff.

cyclone Low-pressure center.

cylindrical projection A family of maps derived from the concept of projection onto a paper cylinder that is tangential to, or intersecting with, a globe.

deciduous tree A tree that experiences an annual period in which all leaves die and usually fall from the tree, due either to a cold season or a dry season.

deflation The shifting of loose particles by wind blowing them into the air or rolling them along the ground.

delta A landform at the mouth of a river produced by the sudden dissipation of a stream's velocity and the resulting deposition of the stream's load.

dematerialization The ability of new industrial processes to squeeze more output from each unit of raw material input.

demographic transition A model of a series of stages in the relationship between the crude birth rate, crude death rate, and the rate of natural population increase, which has been observed in countries over time.

demography The study of individual populations in terms of specific characteristics. These characteristics might include the age groups in the population, the ratio between the sexes, income levels, or other characteristics.

denudation The total effect of all actions that lowers the surface of the continents.

dependency ratio The ratio of the combined child population less than 15 years old and adult population over 65 years old to the population of those between 15 and 65 years of age.

deposition Something precipitated or laid down by a natural process.

desert A very arid region.

dew point The critical air temperature at which saturation is reached.

dialect A minor variation within a language.

diastrophism A general term that refers to the deformation of the Earth's crust.

diffusionism The argument that there has been little independent invention of the components of civilization in different parts of the world, but that most attributes of culture have diffused. Diffusionism is generally discredited today.

diminishing returns An economic principle. Diminishing returns exists when, in successively applying equal amounts of one factor of production to the remaining factors, an added application yields a smaller increase in production than the application just preceding.

distance decay The idea that the presence or impact of any cultural attribute diminishes away from its hearth area, just as the volume of a sound diminishes with distance.

division of labor The separation of work into distinct processes and the apportionment of these processes among different individuals in order to increase productive efficiency.

dominium The principle of rule over a defined territory.

doubling time The time it takes a population to double at its existing rate of annual increase.

downdraft A subsiding current of air.

downstream activities An image to indicate the activities that treat or handle goods in some form farther from their original source as raw materials and closer to their final consumption.

dualism The condition in a country where a modern commercial economy overlies a traditional subsistence economy and there is little exchange between the two.

dumping Selling an item for less than it costs to produce.

earthquake Abrupt movement of the Earth's crust.

ecological niche A particular combination of physical, chemical, and biological factors that a species needs in order to thrive.

economic geography The study of how various peoples make their living and what they trade with other peoples.

economies of scale The economic factors that make it possible to lower cost per item manufactured if a manufacturer manufactures a large number of the item.

ecosystem The totality of interactions among organisms and the environment in the area of consideration.

ecotourism Travel to see distinctive examples of scenery, unusual natural environments, or wildlife.

ecumene The areas of densest national population settlement; a "core" area of a country.

edge cities Same as **satellite cities.**

electoral geography The subfield of political geography that studies voting districts and voting patterns.

elliptical projection A family of map projections in which the entire world is displayed in an oval shape.

eluviation The process by which gravitational water picks up fine particles of soil from the upper layers and carries them downward.

emigration The movement of people out of a place.

enculturation The process of teaching youngsters the society's values and traditions, its political and social culture. Also called **socialization**.

endogenous factors The local site factors "generated within." Elements of the specific local environment or of local cultural history and development that affect the present description of a place.

entisol The least developed of all soil orders, with little mineral alteration and no pedogenic horizons. These soils are commonly thin and/or sandy and have limited productivity, although those developed on recent alluvial deposits tend to be quite fertile.

environmental determinism The simplistic belief that human events can be explained entirely as the direct result of the physical environment.

equator The parallel of 0° latitude.

equinox The time of the year when the perpendicular rays of the sun strike the equator, the circle of illumination just touches both poles, and the periods of daylight and darkness are each 12 hours long all over the Earth.

equivalence The property of a map projection that maintains equal areal relationships in all parts of the map.

erosion Detachment and removal of fragmented rock material.

ethnocentrism A tendency to judge foreign cultures by the standards and practices of one's own, and usually to judge them unfavorably.

etymology The study of word origins and history.

European Community (EC) A common market that officially came into existence in 1967. In 1992 its members included Belgium, the Netherlands, Luxembourg, Germany, France, Italy, Denmark, Ireland, the United Kingdom, Greece, Spain, and Portugal.

European Currency Unit (ECU) A monetary unit created by the European Community in 1978. It is a theoretical unit of account and does not exist in banknotes.

European Economic Area (EEA) A free trade area formed between the EC and EFTA.

European Free Trade Association (EFTA) A free trade area formed among Austria, Liechtenstein, Finland, Iceland, Norway, Sweden, and Switzerland.

evaporation Process by which liquid water is converted to gaseous water vapor.

evergreen A tree or shrub that sheds its leaves on a sporadic or successive basis, but at any given time appears to be fully leaved.

exclusive economic zone (EEZ) According to the 1982 United Nations Convention on the Law of the Sea, a zone extending 200 miles out from the coast of each state within which the state may control both mining and fishing rights.

exogenous factors The factors "generated outside" a place that refer to the way a particular place interacts with other places.

export-led method of growth A program that welcomes foreign investment to build factories to manufacture goods for international markets.

external economies Goods or services that can be rented rather than owned.

extratropical anticyclone An extensive migratory high-pressure cell of the midlatitudes that moves generally with the westerlies.

extratropical cyclone Large migratory low-pressure system that occurs within the middle latitudes and moves generally with the westerlies.

extrusive igneous rock Molten rock ejected onto the Earth's surface, solidifying quickly in the open air.

exurbs Towns that make up the outermost ring of expanding metropolitan areas.

fault A fracture or zone of fracture where the rock is forcefully broken with an accompanying displacement; i.e., an actual movement of the crust on one or both sides of the break. The movement can be horizontal or vertical, or a combination of both.

fault line The intersection of a fault zone with the Earth's surface.

fault plane Interface along which movement occurs in faulting.

fault zone Zone of weakness in the crust where faulting may take place.

fauna Animals.

federal government A system of government in which the balance of power between the central government and the local governments lies with the local governments.

felt needs Things that people feel that they must have.

fertility The potential of a soil for plant growth as determined by the availability of nutrients in the soil. See also **productivity**.

fertility rate The number of children born per year per 1000 females in a population.

First World During the cold war, the bloc of allied western nations, for the most part capitalist, democratic, and rich.

fjord A glacial trough that has been partly drowned by the sea.

flood basalt A large-scale outpouring of basaltic lava that may cover an extensive area of the Earth's surface.

floodplain A flattish valley floor covered with stream-deposited sediments and subject to periodic or episodic inundation by overflow from the stream.

flora Plants.

fluvial process Running water—including both overland flow and streamflow.

fog A cloud whose base is at or very near ground level.

food chain Sequential predation in which organisms feed upon one another, with organisms at one level providing food for organisms at the next level, etc. Energy is thus transferred through the ecosystem.

footloose activity Any economic activity that can move or relocate freely.

foreign direct investment (FDI) Investment by foreigners in wholly owned factories that are operated by the foreign owner.

fractional scale Ratio of distance between points on a map and the same points on the Earth's surface; expressed as a ratio or fraction.

free trade area A group of countries that have no internal tariffs but whose members are free to set their own tariffs on trade with the rest of the world.

friction The resistance of a surface to the relative motion of a body moving over it.

friction of distance The cost of moving anything, measured in time, money, or any other unit.

front A zone of discontinuity between unlike air masses.

frontal lifting The forced rise of a mass of air due to the advance of a weather front.

frontier An undeveloped region that may offer potential for settlement; an area into which settlement may be advancing.

fundamentalism Strict adherence to traditional religious beliefs.

futures contracts Obligations to sell a certain amount of a good at a fixed price at a future time.

General Agreement on Tariffs and Trade (GATT) An international instrument laying down internationally accepted trade rules. First signed as an interim agreement by 23 countries in 1947.

gentrification The revitalization of old central-city residential neighborhoods by an influx of white-collar workers.

geographic database An amount of geographic information stored in a geographic information system.

geographic grid A network of lines that intersect at right angles, which enables us to pinpoint the location of any point on the surface.

geographic information systems Collections of data

about the distribution of phenomena stored in computers where they can conveniently be retrieved and manipulated in geographical studies.

geography The study of the interaction of all physical and human phenomena at individual places and of how interactions among places form patterns and organize space.

geomancy The planning or design of cities or settlements in accord with cardinal points on the compass in an effort to obtain harmony with the universe.

geomorphic Long-term processes, both external and internal, that determine the shape and form of the topography.

gerrymandering Drawing voting district lines that include or exclude specific groups of voters so that one group gains an unfair advantage.

ghetto Originally the part of a city where Jews were forced to live. Today, more generally, a residential community of any identifiable group segregated either by force or by choice.

glaciofluvial deposition The action whereby much of the debris that is carried along by glaciers is eventually deposited or redeposited by glacial meltwater.

globalization The expansion of the scale of organization of many activities to cover the whole globe.

government The people who are actually in power at any time.

graphic scale The use of a line marked off in graduated distances as a map scale.

greenbelts Areas of forest or farming preserved around cities.

green revolution The success of botanical science in introducing higher-yielding and hardier strains of crops.

gross domestic product (GDP) The total value of all goods and services produced within a country.

gross national product (GNP) The total income of a country's residents, no matter where it comes from.

gross sustainable product (GSP) The total of a country's GDP minus the value of destroyed or depleted natural resources. This is a new statistic that economists are still working to define.

growth poles Sites that are favored as foci of economic growth.

gymnosperms ("naked seeds") Seed-reproducing plants that carry their seeds in cones.

hail Rounded or irregular pellets or lumps of ice produced in cumulonimbus clouds as a result of active turbulence and vertical air currents. Small ice particles grow by collecting moisture from supercooled cloud droplets.

hard currencies Currencies that are freely traded on international currency exchanges.

hardwood Angiosperm trees that are usually broadleaved and deciduous. Their wood has a relatively complicated structure, but it is not always hard.

hearth area The place where a distinctive culture originates.

hierarchical diffusion Diffusion that occurs downward or upward in a hierarchy of organization. When mapped, it shows up as a network of spots rather than as an ink blot spreading across a map.

high seas The area of international waters beyond the territorial claims of all countries.

hinterland The area surrounding any city to which the city provides services and upon which the city draws for its needs.

historical consciousness A people's sense of their own history; it may affect their behavior.

historical geography The subfield within geography that studies the geography of the past and how geographic distributions have changed through time.

historical materialism A school of history that tries to write a plot for human history based on the idea that human technology has given humankind greater control over the environment and that this progress determines every other aspect of human economies, politics, and even philosophy.

histosol A soil order characterized by organic, rather than mineral, soils, which is invariably saturated with water all or most of the time.

Human Development Index (HDI) An index of human welfare devised by the United Nations Human Development Programme. It compares three statistics among countries: purchasing power (a figure that adjusts income to the local cost of living), life expectancy, and adult literacy.

humidity Water vapor in the air.

hurricane A tropical cyclone affecting North or Central America.

hydrologic cycle A series of storage areas interconnected by various transfer processes, in which there is a ceaseless interchange of moisture in terms of its geographical location and its physical state.

hypothesis A proposition that is tentatively assumed to draw out all its consequences so that it can be tested against all the facts that an investigator can gather.

ice cap A small ice sheet, normally found in the summit area of high mountains.

iconography A set of symbols for a political unit.

igneous rock Rock formed by solidification of molten magma.

illuviation The process by which fine particles of soil from the upper layers are deposited at a lower level.

immigration The movement of people into a place.

import-substitution method of growth A policy of protecting domestic markets and favoring domestic manufacturers until some presumed time at which they can compete with imported competition.

inceptisol An immature order of soils that has relatively faint characteristics; not yet prominent enough to produce diagnostic horizons.

incorporation The process of defining a city territory and establishing a government.

indirect rule Colonial rule through the use of native leaders as intermediaries between the colonial power and the people.

Industrial Revolution The transition from an agricultural and commercial society to an industrial society relying on inanimate power and complex machinery. This occurred first in Europe between about 1750 and 1850.

industrial societies Societies in which the bulk of employment is occupied in the secondary sector.

inertia The force that keeps things stable. Generally, the force that keeps things the way they are, patterns fixed.

infant industries Industries that are getting started in a country. They cannot compete with imports, and so the government may choose to protect them against imported competition.

infant mortality rate The number of infants per 1000 who die before reaching 1 year of age.

informal or underground sector All economic activities that do not appear in official accounts. The people may not have licenses to do what they are doing, or they may be avoiding taxes, or for some other reason their activity escapes official notice.

infrastructure All of a people's assets fixed in place, including railroads, pipelines, highways, airports, and housing.

input-output models Economic models that analyze the flow of money through economies.

insolation Incoming solar radiation.

interculturation Efforts to incorporate some of the practices of one culture into another.

interfluve The higher land above the valley sides that separates adjacent valleys.

internal economies Goods and services that a business enterprise can provide for itself without having to rely on outside suppliers.

international organizations Organizations that coordinate activities among nation-states.

Intertropical Convergence Zone The region where the northeast trades and the southeast trades converge.

intrusive igneous rock Rocks that cool and solidify beneath the Earth's surface.

invertebrates Animals without backbones.

isobar A line joining points of equal atmospheric pressure.

isohyet A line joining points of equal numerical value of precipitation.

isolated city model Deductive image of Johann Heinrich von Thünen in which the existence of a city isolated on an isotropic plain determines the distribution of land uses across that plain.

isotherm A line joining points of equal temperature.

isotropic plain An imaginary, homogeneous flat surface across which population density and resources are uniformly distributed.

jet stream A rapidly moving current of air concentrated along a quasi-horizontal axis in the upper troposphere or in the stratosphere, characterized by strong vertical and lateral wind shears and featuring one or more velocity maxima.

joints Cracks that develop in bedrock due to stress, but in which there is no appreciable movement parallel to the walls of the joint.

labor force participation rate The percentage of the population that is currently employed or looking for a job.

labor-intensive Any activity that employs a high ratio of workers to the amount of capital invested in machinery.

lagoon A body of quiet salt or brackish water in an area between a barrier island or a barrier reef and the mainland.

laissez-faire capitalism A capitalist system of economic organization that minimizes the government's role in the economy.

language A set of words, plus their pronunciation and methods of combining them, that is used and understood to communicate within a group of people.

language family The languages that are related by descent from a common protolanguage.

latitude Distance measured north and south of the equator.

lava Molten magma that is extruded onto the surface of the Earth, where it cools and solidifies.

law A theory that has withstood testing over a long time. Even laws, however, can be revised or rejected if new evidence is discovered.

leaching The process in which gravitational water dissolves soluble materials and carries them downward in solution to be redeposited at lower levels.

legitimacy Rule as a result of acquisition by means that are according to law or custom.

liberation theology A movement within the Roman Catholic Church that recommends political activism to fight poverty. It was named after the book *The Theology of Liberation* by the Peruvian priest Gustavo Gutierrez.

life expectancy The average number of years that a newborn baby within a given population can expect to live.

linear cartograms Cartograms that show either the time or the cost of getting from one point to another.

lingua franca A second language held in common for international discourse.

liquidity The measure of any asset's ability to be converted into other assets or goods.

locational determinants The considerations that are relevant in determining the location of a particular activity.

loess A fine-grained, wind-deposited silt. Loess lacks horizontal stratification, and its most distinctive characteristic is its ability to stand in vertical cliffs.

longitude Distance measured in degrees, minutes, and seconds, east and west from the prime meridian on the Earth's surface.

magma Molten material in the Earth's interior.

map projection A systematic representation of all or part of the three-dimensional Earth surface on a two-dimensional flat surface.

maritime climate A climate typical of an ocean area or the margin of a landmass, normally characterized by small annual temperature ranges and relatively abundant moisture.

market-oriented manufacturing Manufacturing that locates close to the market.

mass wasting The downslope movement of broken rock material by gravity, sometimes lubricated by the presence of water.

material culture All the objects that people use, including clothing, housing, and tools.

material-oriented manufacturing Manufacturing that locates close to the source of the raw material.

mental maps Our ideas and impressions about places. They may or may not be based in real experience of having visited the places, and they may or may not conform to reality.

meritocracy A society in which the most capable people can rise to the top on merit alone.

metals Elements that are usually heavy, reflect light, can be hammered and drawn, and are good conductors of heat and electricity.

metamorphic rock Rock that was originally something else but has been drastically changed by massive forces of heat and/or pressure working on it from within the Earth.

metropolis Greek for "mother city." Name generally given to the world's largest urban areas.

metropolitan statistical area (MSA; originally standard metropolitan statistical area, SMSA) U.S. government term for "an integrated economic and social unit with a recognized large population nucleus." A principal central city and its suburban counties.

millibar An "absolute" measure of pressure, consisting of one-thousandth part of a bar, or 1000 dynes per square centimeter.

minerals Naturally occurring substances obtainable by mining. Organic minerals, including coal, petroleum, and natural gas, contain carbon compounds. Most minerals, however, are inorganic.

model An idealized, simplified representation of a more complex phenomenon, which helps us to understand or analyze that phenomenon.

mollisol A soil order characterized by the presence of a mollic epipedon, which is a mineral surface horizon that is dark, thick, contains abundant humus and base nutrients, and retains a soft character when it dries out.

monotheism Belief in one god.

monsoon A seasonal reversal of winds; a general onshore movement in summer and a general offshore flow in winter, with a very distinctive seasonal precipitation regime.

moraine The largest and generally most conspicuous landform feature produced by glacial deposition, which consists of irregular rolling topography that rises somewhat above the level of the surrounding terrain.

multinational or transnational corporations Enterprises with producing and marketing facilities in several countries.

multiplier effect The fact that an increase in jobs in the basic sector of a city multiplies jobs in the nonbasic sector.

nation A group of people who want to have their own government and rule themselves.

national self-determination U.S. president Woodrow Wilson's term for the belief that each nation should have its own state.

nation-state A state that rules over a territory that contains all the people of a nation.

natural increase (decrease) The difference between the number of births and the number of deaths. It may be expressed as a rate or a percentage of the total population.

natural landscape A landscape without evidence of human activity.

needle-leaf trees Trees adorned with thin slivers of tough, leathery, waxy needles rather than typical leaves.

negative externalities Costs of any activity that are imposed on others beyond the immediate participants in the activity.

nonbasic sector The part of a city's economy dedicated to looking after the needs of the city's own residents.

North Atlantic Treaty Organization (NATO) An international military alliance established in 1949 among Belgium, Denmark, France, Great Britain, Iceland, Italy, Luxembourg, the Netherlands, Norway, Portugal, Canada, and the United States. Greece and Turkey joined in 1951, West Germany in 1955, and Spain in 1982.

nuclear fission A process in which the nueclei of certain heavy atoms decay, giving off energy.

nuclear fusion The process of fusing hydrogen atoms into helium atoms, releasing energy.

occluded front A complex front formed when a cold front overtakes a warm front.

official language The language in which government business is normally conducted, official records are kept, signs are posted, and other official business is transacted.

offshore bar (barrier bar) Long, narrow sand bar built up in shallow offshore waters.

oil shales A geological formation in which great quantities of petroleum are chemically locked.

oligarchy Government by an elite privileged clique.

opportunity cost The amount of return sacrificed by leaving capital invested in one form or activity rather than another.

ores The mixtures of minerals from which metals are extracted.

orographic lifting The forced ascent of air over a topographic barrier.

orthography A system of writing.

outcrop Surface exposure of bedrock.

oxisol The most thoroughly weathered and leached of all soils. This soil order invariably displays a high degree of mineral alteration and profile development.

parallel A circle resulting from an isoline connecting all points of equal latitude.

philological nationalism The idea that "mother tongues" have given birth to nations.

physiological density The density of population per unit of arable land.

photosynthesis The basic process whereby plants produce stored chemical energy from water and carbon dioxide and which is activated by sunlight.

plane of the ecliptic The imaginary plane that passes through the sun and through the Earth at every position in its orbit around the sun.

plane of the equator An imaginary two-dimensional surface that passes through the Earth of 0° of latitude.

plate tectonics A coherent theory of massive crustal rearrangement based on the movement of continent-sized crustal plates.

polar easterlies A global wind system that occupies most of the area between the Polar Highs and about 60° of latitude. The winds move generally from east to west and are typically cold and dry.

polar front The contact between unlike air masses in the subpolar low-pressure zone.

polar front jet stream A rapidly moving current of air in the upper troposphere or stratosphere above the surface position of the polar front.

polar high A high-pressure cell situated over either polar region.

polarity A characteristic of the Earth's axis wherein it always points toward Polaris (the North Star) at every position in the Earth's orbit around the sun.

political community A willingness to join together and form a government to solve problems.

political culture The "unwritten rules" of a regime; the unwritten ways in which, in any culture, written rules are interpreted and actually enforced.

political economy The study of the various methods countries use to organize their economic life.

political geography The branch of geography that deals with the boundaries and subdivisions of political units.

polyglot states Countries that grant equality to two or more official languages.

population geography The study of the distribution of people.

population growth (decrease) For individual countries or regions, the natural population increase or decrease modified by subtracting emigration out of that area or adding immigration into that area.

population projection A prediction of a future population assuming that current trends remain the same or else change in defined ways.

population pyramid A graphic device for illustrating the age and sex composition of a population.

portfolio holdings An investment that represents a minority holding in an enterprise; owners of portfolio holdings do not take an active role in running the enterprise.

positive checks As defined by Thomas Malthus, premature deaths of all types, such as those caused by war, famine, and disease.

positive externalities Benefits of any activity that have effects beyond the immediate participants in that activity.

possibilism The theory that the physical environment itself will neither suggest nor determine what people will attempt, but it may limit what people achieve profitably.

postindustrial society A society in which the bulk of employment is occupied in the tertiary sector or in which the bulk of its output is derived from tertiary-sector activities.

prairie A tall grassland in the midlatitudes.

preindustrial societies An imprecise term used to describe societies in which the bulk of employment is occupied in the primary sector.

pressure gradient The horizontal rate of variation of atmospheric pressure.

preventive checks As defined by Thomas Malthus, human actions designed to limit population growth.

primary sector Activities that extract earth resources, including agriculture, fishing, forestry, and mining.

primary series The natural succession of plant types at any location, which progresses from land without vegetation to a mature climatic climax community. Also known as **prisere**.

primate city One large city that concentrates a high degree of an entire nation's population or of national political, intellectual, or economic life.

prime meridian The meridian passing through the Royal Observatory at Greenwich (England), just east of London, and from which longitude is measured.

prisere Another term for **primary series**.

privatize To sell government assets to private investors, or to assign responsibility for public services to private contractors.

productivity The amount of plants that actually grow in a soil. Productivity is determined by a number of external factors as well as a soil's fertility.

protolanguage The ancestor that is common to any group of several of today's languages.

proxemics The cross-cultural study of the use of space.

pteridophytes Spore-bearing plants such as ferns, horsetails, and clubmosses.

pull factors Factors considered in migration that attract people to new destinations. Pull factors include economic opportunity and the promise of religious and political liberty.

push factors Factors considered in migration that drive people away from wherever they are. Push factors include starvation and political or religious persecution.

race Popular term of dubious real value for categorizing humankind into categories on the basis of secondary physical characteristics such as skin color.

racism A belief in the inherent superiority of one race over another and the linking of human ability, potential, and behavior to racial inheritance.

rain The most common and widespread form of precipitation, consisting of drops of liquid water.

recyclable resources Resources that can be reused, or recycled. Recycling refers both to the reuse of industrial wastes and the reuse of obsolete or discarded goods for their material content.

refugee As defined by a 1951 Geneva Convention, someone who has left his or her home country with "a well-founded fear of being persecuted in his country of origin for reasons of race, religion, nationality, membership of a particular social group, or political opinion." Several countries have expanded the definition to accept persecution due to sexual orientation.

regime Within any state, the general set of rules and regulations for governing.

regional geography Analytical syntheses of individual places, usually beginning with the physical environment and then cataloguing the human activities at that place.

regnum A system of government over a group of people rather than a defined territory.

regolith A layer of broken and partly decomposed rock particles that covers bedrock.

relative humidity An expression of the amount of water vapor in the air in comparison with the total amount that could be there if the air were saturated. This is a ratio that is expressed as a percentage.

relative location The location of a place measured in terms of its accessibility from other places.

remote sensing Study of an object or surface from a distance by using various instruments.

renewable resources Resources such as fish, livestock, and trees, that naturally renew themselves and can, therefore, be harvested at the rate at which they reproduce themselves through time. Supplies will diminish only if they are harvested at a rate above their natural reproduction rate.

replacement rate The total fertility rate that stabilizes a population; it is about 2.1.

reserves The amounts of a mineral that can economically be recovered for use.

resource (1) Generally, anything that can be consumed or put to use by humankind. (2) In reference to mineral resources, the amount of the crustal limit of a given mineral that is currently or potentially extractable.

revolution of rising expectations An increase in the number or quality of goods that people expect to have.

rock Solid material composed of aggregated mineral particles.

Rossby wave A very large north-south undulation of the upper-air westerlies.

sacerdotalism The belief that a church or priests intercede between God and humankind.

sand dune A mound, ridge, or low hill of loose windblown sand.

satellite cities Suburban foci of urban-type activities. Also called **edge cities**.

savanna A low-latitude grassland characterized by tall forms.

scale The numerical representation of the relationship between length measured on a map and the corresponding distance on the ground.

scientific method A set of procedures used in the systematic pursuit of knowledge. The three basic steps in the scientific method are (1) observe the world; (2) explain what you see as simply as possible; and (3) hold on to your hypothesis until it is disproved.

scientific revolution in agriculture See **Agricultural Revolution**.

scrub A collective term for low trees or shrubs.

sea-floor spreading The pulling apart of crustal plates to permit the rise of deep-seated magma to the Earth's surface in midocean areas.

secondary sector Those activities that receive the commodities and raw materials produced by the primary sector and transform them into manufactured goods. Construction is usually also included in this sector.

Second World During the cold war, the bloc of allied communist nations.

sectoral evolution The growth of the secondary and tertiary sectors of a country's economy relative to the primary sector. Later in the process, the tertiary sector grows relative even to the secondary sector.

secularism A lifestyle or policy that purposely ignores or excludes religious considerations.

sedimentary rock Rock formed of sediment that is consolidated by the combination of pressure and cementation.

selva (tropical rainforest) A distinctive assemblage of tropical vegetation that is dominated by a great variety of tall, high-crowned trees.

service sector Same as **tertiary sector**.

shamanism The use of a human medium to communicate with the spirit world.

sinkhole (doline) A small, rounded depression that is formed by the dissolution of surface limestone, typically at joint intersections.

site factors Endogenous factors in the description of a place.

situation factors Exogenous factors in the description of a place.

sleet Small raindrops that freeze during descent.

snow Solid precipitation in the form of ice crystals, small pellets, or flakes, which is formed by the direct conversion of water vapor into ice.

social Darwinism British sociologist Herbert Spencer's extension of Darwin's theory of evolution to the belief that "Nature's law" calls for "the survival of the fittest," even among cultures and whole peoples.

socialization Another term for **enculturation**.

society A group of people who interact as self-interested individuals.

softwoods Gymnosperm trees—nearly all are needle-leaved evergreens—with wood of simple cellular structure but not always soft.

soil An infinitely varying mixture of weathered mineral particles, decaying organic matter, living organisms, gases, and liquid solutions. Soil is that part of the outer skin of the Earth occupied by plant roots.

soil horizon A more or less distinctly recognizable horizontal layer of soil.

soil profile A vertical cross section from the Earth's surface down through the soil layers into the parent material beneath.

sojourners Migrants who intend to stay in their new location only until they can save enough capital to return home to a higher standard of living.

solstice One of those two times of the year in which the sun's perpendicular rays hit the northernmost or southernmost latitudes ($23\frac{1}{2}°$) reached during the Earth's cycle of revolution.

speech community A group of people who regularly speak together or hear one another speak and therefore make the sounds of a language in the same way.

speleothem A feature formed by precipitated deposits of minerals on the wall, floor, or roof of a cave.

spit A linear deposit of marine sediment that is attached to the land at one or both ends.

spodosol A soil order characterized by the occurrence of a spodic subsurface horizon, which is an illuvial layer where organic matter and aluminum accumulate, and which has a dark, sometimes reddish, color.

stability The capacity of air to resist vertical movement.

stalactite A pendant structure hanging downward from a cavern's roof.

stalagmite A projecting structure growing upward from a cavern's floor.

standard language The way any language is spoken and written according to formal rules of diction and grammar.

state An independent political unit that claims exclusive and exhaustive jurisdiction over a defined territory and over all of the people and activities within it; also called a country.

state-directed capitalism A type of national economy in which the government does not own many companies outright, but it plans and regulates the economy.

stationary front The common "boundary" between two air masses in a situation in which neither air mass displaces the other.

steppe A plant association dominated by short grasses and bunchgrasses of the midlatitudes.

storm A disturbance of the normal condition of the atmosphere, normally including strong winds, clouds, and precipitation.

strategic minerals Minerals that are especially important to industrial processes and for which there are no known practicable substitutes.

stratiform cloud A cloud form characterized by clouds that appear as grayish sheets or layers that cover most or all of the sky, rarely being broken into individual cloud units.

subduction Descent of the edge of a crustal plate under the edge of an adjoining plate, presumably involving melting of the subducted material.

subpolar low A zone of low pressure that is situated at about 50° to 60° of latitude in both Northern and Southern hemispheres (also referred to as the **polar front**).

subsidence inversion A temperature inversion that

occurs well above the Earth's surface as a result of air sinking from above.

subtropical high Large semipermanent high-pressure cells centered at about 30° latitude over the oceans, which have average diameters of 2000 miles (3200 km) and are usually elongated east-west.

suburb The areas into which growing cities often spill beyond their legal boundaries. Some suburbs are entirely residential, but others develop their own service centers for the surrounding residential population. Some suburbs may be older cities that were once distinct but have been engulfed by the growth of a larger neighbor, but others are new settlements that incorporate in their own right. What defines an area as suburban is its economic and social integration with a larger population nucleus nearby.

sunset industries In an advanced country, low-technology industries that are being allowed to go broke.

superimposed boundaries The political borders that the colonial powers imposed upon overseas territories, no matter what the level or form of political organization had been among the native peoples.

supranational organizations Organizations that exercise powers over the member nation-states.

syncretic religions Religions that combine elements of two or more faiths.

systematic geography Same as **topical geography**.

taiga (boreal forest) The great northern coniferous forest.

tar sands A geological formation in which great quantities of petroleum are chemically locked.

telecommuting Working at home on a computer terminal attached to a central workplace.

temperature inversion A situation in which temperature increases upward, and the normal condition is inverted.

territoriality The behavior observed in many animals of laying claim to territory and defending it against members of their own species. It is unclear whether humans exhibit biologically determined territoriality.

tertiary sector All activities that provide services or manipulate knowledge. Also known as **service sector**.

theocracy Government by a church organization.

theory A hypothesis that has been given probability by experimental evidence.

Third World Term devised in 1954 to describe the countries that tried to maintain a neutral political position during the cold war between the capitalist democratic countries and the communist countries.

threshold The minimum demand necessary for a product or service to be offered.

threshold principle of states The belief that a nation must have some minimal population and territory to merit self-determination.

thunderstorm A relatively violent convective storm accompanied by thunder and lightning.

till Rock debris that is deposited directly by moving or melting ice, with no meltwater flow or redeposition involved.

topical geography The study of one particular topic of analysis as it varies among all places, as, for instance, the geography of climate, the geography of religion, or urban geography. Also known as **systematic geography**.

toponymy The study of place names.

tornado A localized cyclonic low-pressure cell surrounded by a whirling cylinder of wind spinning so violently that centrifugal force creates partial vacuum within the funnel.

total fertility rate The number of children an average woman in a given society would have over her lifetime.

trade winds The major wind system of the tropics, issuing from the equatorward sides of the Subtropical Highs and diverging toward the west and toward the equator.

tropical cyclone A storm most significantly affecting the tropics and subtropics, which is intense, revolving, rain-drenched, migratory, destructive, and erratic. Such a storm system consists of a prominent low-pressure center that is essentially circular in shape and has a steep pressure gradient outward from the center.

tropical rainforest See **selva.**

troposphere The lowest thermal layer of the atmosphere, in which temperature decreases with height.

tundra A complex mix of very low growing plants, including grasses, forbs, dwarf shrubs, mosses, and lichens, but no trees. Tundra occurs only in the perennially cold climates of high latitudes or high altitudes.

ultisol A soil order similar to alfisols, but more thoroughly weathered and more completely leached of bases.

unitary government A system of government in which the balance of power between a central government and local governments lies at the center.

updraft A rising movement of air.

urban enterprise zones Districts within which manufacturers receive tax subsidies as a way of attracting them to create new jobs.

urban geography The geographic study of cities.

urbanization The process in which an increasing share of a population becomes concentrated in cities.

valley That portion of the total terrain in which a drainage system is clearly established.

value added The difference between the value of a raw material and the value of that material after it has been transported or manufactured into something.

vertebrates Animals that have a backbone that protects their spinal cord—fishes, amphibians, reptiles, birds, and mammals.

vertisol A soil order comprising a specialized type of soil that contains a large quantity of clay and has an exceptional capacity for absorbing water. An alternation of wetting and drying, expansion and contraction, produces a churning effect that mixes the soil constituents, inhibits the development of horizons, and may even cause minor irregularities in the surface of the land.

vulcanism General term that refers to movement of magma from the interior of the Earth to or near the surface.

warm front The leading edge of an advancing warm air mass.

Warsaw Pact An international military alliance formed in 1955 among the USSR, Albania (which withdrew in 1968), Bulgaria, Czechoslovakia, East Germany, Hungary, Poland, and Romania. It formally dissolved in 1991.

water vapor The gaseous state of moisture.

weather The short-term atmospheric conditions for a given time and a specific area.

weathering The physical and chemical disintegration of rock that is exposed to the weather.

westerlies The great wind system of the midlatitudes that flows basically from west to east around the world in the latitudinal zone between about 30° and 60° both north and south of the equator.

wind Air moving horizontally with respect to the Earth's surface.

workers' remittances Money mailed home to their families by people who are working abroad.

zero population growth A stabilization of a population.

Zionism The belief that Jews should have a homeland of their own in Palestine.

zoning Governmental restriction of the use to which various parcels of land may be put.

Index

NOTE: Page numbers in *italics* refer to illustrations.